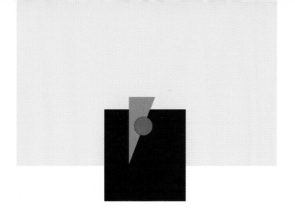

ELEMENTARY STATISTICS

KEY TO EXERCISE SYMBOLS

 biology

 chemistry

 geology

 education

 agriculture

 public opinion

 political science

 business/economics

 physics

 sociology

medicine

 psychology

 environmental studies

 engineering/technical

 everyday life

 sports

 criminal justice

 formula used by computer

 formula used by calculator

 formula for hand calculations

D formula definition

GREEK ALPHABET

alpha	A	(α)	nu	N	ν	
beta	B	(β)	xi	Ξ	ξ	
gamma	Γ	γ	omicron	O	o	
delta	Δ	δ	pi	Π	π	
epsilon	E	(ϵ)	rho	P	(ρ)	
zeta	Z	ζ	sigma	(Σ)	(σ)	
eta	H	η	tau	T	τ	
theta	Θ	θ	upsilon	Υ	υ	
iota	I	ι	phi	Φ	ϕ	
kappa	K	κ	chi	X	(χ)	
lambda	Λ	λ	psi	Ψ	ψ	
mu	M	(μ)	omega	Ω	ω	

Circled letters are used in this textbook.

ELEMENTARY STATISTICS

SEVENTH EDITION

ROBERT JOHNSON

Monroe Community College

An Alexander Kugushev Book

Duxbury Press

An Imprint of Wadsworth Publishing Company

I(T)P ® An International Thomson Publishing Company

Belmont Albany Bonn Boston Cincinnati Detroit London Madrid Melbourne
Mexico City New York Paris San Francisco Singapore Tokyo Toronto Washington

Editorial Assistant Cynthia Mazow
Marketing Manager Joanne Terhaar
Advertising Project Manager Joseph Jodar
Print Buyer Karen Hunt
Permissions Editor Peggy Meehan
Production Greg Hubit Bookworks

Text and Cover Designer Stuart Paterson
Copy Editor Margaret Moore
Compositor Jonathan Peck Typographers
Cover Printer Phoenix Color Corp.
Printer R. R. Donnelley & Sons, Crawfordsville

For more information, contact Duxbury Press at Wadsworth Publishing Company:

Wadsworth Publishing Company
10 Davis Drive
Belmont, California 94002, USA

International Thomson Publishing Europe
Berkshire House 168–173
High Holborn
London, WCIV 7AA, England

Thomas Nelson Australia
102 Dodds Street
South Melbourne 3205
Victoria, Australia

Nelson Canada
1120 Birchmount Road
Scarborough, Ontario
Canada MIK 5G4

International Thomson Editores
Campos Eliseos 385, Piso 7
Col. Polanco
11560 México D.F. México

International Thomson Publishing GmbH
Königswinterer Strasse 418
53227 Bonn, Germany

International Thomson Publishing Asia
221 Henderson Road
#05–10 Henderson Building
Singapore 0315

International Thomson Publishing Japan
Hirakawacho Kyowa Building, 3F
2–2–1 Hirakawacho
Chiyoda-ku, Tokyo 102, Japan

Library of Congress Cataloging-in-Publication Data

Johnson, Robert Russell, [date]
  Elementary statistics / Robert Johnson.—7th ed.
   p. cm.
  Includes index.
   ISBN 0-534-24324-X
   1. Statistics I. Title
QA276.12.J64 1996
519.5—dc20   95-24045

To my mother
and to the memory
of my father

PAST AND PRESENT REVIEWERS FOR *ELEMENTARY STATISTICS*

Nancy Adcox
Mt. San Antonio College

Paul Alper
College of St. Thomas

William D. Bandes
San Diego Mesa College

Barbara Jean Blass
Oakland Community College

Austin Bonis
Rochester Institute of Technology

Louis F. Bush
San Diego City College

Rodney E. Chase
Oakland Community College

David M. Crystal
Rochester Institute of Technology

Joyce Curry and Frank C. Denny
Chabot College

Larry Dorn
Fresno Community College

Shirley Dowdy
West Virginia University

Joan Garfield
University of Minnesota General
College

Carol Hall
New Mexico State University

Hank Harmeling
North Shore Community College

Bryan A. Haworth
California State College at
Bakersfield

John C. Holahan
Xerox Corp.

James E. Holstein
University of Missouri

Robert Hoyt
Southwestern Montana State

Peter Intarapanach
Southern Connecticut
State University

T. Henry Jablonski, Jr.
East Tennessee State University

Sherry Johnson

Meyer M. Kaplan
The William Patterson College of
New Jersey

Michael Karelius
American River College

Anand S. Katiyar
McNeese State University

Gayle S. Kent
Southern Florida College

Raymond Knodel
Bemidji State University

Pat Kuby
Monroe Community College

Robert O. Maier
El Camino College

Mark Anthony McComb
Mississippi College

John Meyer
Muhlenberg College

Jeffrey Mock
Diablo Valley College

David Naccarato
University of New Haven

Harold Nemer
Riverside Community College

Daniel Powers
University of Texas, Austin

Janet M. Rich
Miami-Dade Junior College

Larry J. Ringer
Texas A & M University

John T. Ritschdorff
Marist College

John Rogers
California State Polytechnic Institute
at San Luis Obispo

Thomas Rotolo
University of Arizona

**Barbara F. Ryan and
Thomas A. Ryan**
Pennsylvania State University

Robert J. Salhany
Rhode Island College

Howard Stratton
State University of New York at
Albany

Larry Stephens
University of Nebraska—Omaha

Thomas Sturm
College of St. Thomas

Edward A. Sylvestre
Eastman Kodak Co.

William Tomhave
University of Minnesota

Bruce Trumbo
California State University, Hayward

Richard Uschold
Canisius College

John C. Van Druff
Fort Steilacoom Community College

Philip A. Van Veldhuizen
University of Alaska

John Vincenzi
Saddleback College

Kenneth D. Wantling
Montgomery College

Mary Wheeler
Monroe Community College

Don Williams
Austin College

CONTENTS

PREFACE

PURPOSE AND PREREQUISITES

This book is written for an introductory course for students who need a working knowledge of statistics but do not have a strong mathematical background. Statistical methodology uses many formulas and an occasional algebraic equation. Those students who have not completed intermediate algebra should complete at least one semester of college mathematics as a prerequisite before attempting this course.

MY OBJECTIVES

The primary objective of the seventh edition of *Elementary Statistics* is to present a very readable introduction to statistics that will promote learning and understanding, as well as one that will motivate students by presenting statistics in a context that relates to their personal experiences.

Statistics is a practical discipline that evolves with the changing needs of our society. Today's student is the product of a particular cultural environment and is motivated differently from the student of a few years ago. In this text I present statistics as a useful tool in learning about the world around us. While studying descriptive and inferential concepts, students will become aware of their real-world applications in such fields as the physical and social sciences, business, economics, and engineering.

IMPORTANT ONGOING FEATURES

This seventh edition continues to feature the following elements:

- ▼ A communication style that reflects current student culture;
- ▼ A strong computer flavor, with numerous annotated outputs and computer exercises that can be used as laboratory assignments;
- ▼ Many exercises that are set up for solution in MINITAB, with corresponding instructions;
- ▼ A Chapter 1 that introduces ideas of variability and data collection, as well as basic terms;
- ▼ Case Studies based on situations of interest and using real data;
- ▼ An early descriptive presentation of linear correlation and linear regression in Chapter 3;
- ▼ Brief biographies of four prominent statisticians.

THIS REVISION

Periodically, the original premises behind a textbook need to be reexamined to ensure that they remain valid. In this seventh edition, I have done so, and the result is the most thorough revision *Elementary Statistics* has undergone in the twenty-some years since its initial publication. I hope that the users of the seventh edition will appreciate the following improvements:

▼ A presentation that is more visual, more approachable, and clearer throughout.

▼ The presentation is *more approachable* because it is often less technical, such as in the new discussion of hypothesis testing in Chapter 8.

▼ It is *clearer* because formulas have been expressed in "word algebra," paraphrasing algebraic symbolism in recognition of the student's frequent discomfort with notation. This is done any time a formula of any complexity is introduced, such as in Chapters 2 and 3 with many of the descriptive statistics formulas, when inequality notation with class intervals is presented, and when probability statements are introduced.

▼ It is more *visual* as well. There are new graphic displays that compare similar statistics (Figure 2-27), that organize procedures for finding percentiles (Figure 2-27), and that organize and compare the various statistical tests (Figures 9-1, 9-10, 10-1). There are simulations that visually demonstrate—for example, level of confidence, alpha and *p*-value (Illustrations 8-3, 8-14, 8-20).

▼ A rewritten Chapter 8 provides a more natural flow of ideas from sampling distribution, to estimation, to *p*-values, to hypothesis testing. It also contains more drill exercises, broken down into smaller steps.

▼ The exercises have been expanded and improved, from simple drills to critical thinking questions. The simplest exercises help students acquire confidence in computations and single-step questions; some appear in the margin of the text for the students to do as they read the text material. The end-of-section and end-of-chapter exercises have been expanded to include many real-life data situations and critical thinking questions all aimed at getting the student involved in real-world statistical applications.

▼ More real data are used and all significant data sets are included on a data disk.

▼ There is an increased number of exercises that demonstrate statistical theory through computer simulation. Several of these could be used as special assignments or as lab exercises. Karl Pearson tossed a coin 24,000 times, observing heads/tails. It must have taken days for him to have realized the power of the law of large numbers; Exercise 9.67 allows the student to repeat his experiment in seconds—even repeat it several times in a few minutes. There are simulation exercises designed to help the student understand the importance of assumptions to a statistical test (Exercises 9.42 and 9.112).

▼ Two specially designed tables help students determine *p*-values so that their initial experience is not overwhelming. Tables 5 and 7 allow *p*-values to be read directly for tests involving the *z* or the *t* statistic.

▼ The mean and standard deviation of frequency distributions are now in a separate section.

▼ The focus on interpreting computer output has been increased.

▼ The MINITAB, Version 10.2, instructions are presented with both session commands and menu commands being displayed.

▼ The Pareto diagram, a popular graph in industry, is presented in Chapter 2.

▼ New physical page layout makes use of the margins with the addition of margin exercises and margin "conversation."

▼ A Chapter Case Study illustrates an application of the material to be studied in the chapter and comes full circle when the student completes the corresponding exercise at the end of the chapter.

TO THE INSTRUCTOR

A primary objective of this book is to offer a truly readable presentation of elementary statistics. The chapters are designed to interest and involve students and to guide them step by step, in a logical manner, through the material. Each chapter includes the following main features:

▼ A *Chapter Outline* shows the student what to look for.

▼ A *Chapter Case Study* illustrates an application of the material to be studied in that chapter and leaves the student with unanswered questions that the student will be able to answer when the chapter has been completed. The unanswered questions form an end-of-chapter exercise that brings the student full circle with the concepts involved.

▼ *Chapter Objectives* motivate the student to study the specific topics to be learned in the chapter.

▼ Completely *worked-out examples* present solutions to illustrate concepts and also to demonstrate the applications of statistics in real-world situations.

▼ The in-section *Case Studies* are brief versions of actual published articles focusing on the use of statistics. They are accompanied by margin exercises that the student is encouraged to complete when reading the Case Study.

▼ *End-of-section Exercises* give students the opportunity to practice concepts that are presented in the section.

▼ An *In Retrospect* section summarizes the chapter's material and relates it to the chapter's objectives.

▼ *Chapter Exercises* give students the opportunity to integrate conceptual and computational skills and to choose the appropriate method from those covered in the chapter.

▼ A *Vocabulary List* helps students review key terms.

▼ Three *Chapter Quizzes* provide students with self-evaluation of their mastery of the material.

▼ I have included at the end of each part a *Working With Your Own Data* section for student exploration. These sections provide a personalized learning experience by directing students to collect their own data and apply techniques they have been studying. In my experience, students retain much more by applying methods just learned in regular homework assignments to data they already understand and are familiar with.

This book reflects the computer's growing role in the teaching of statistics. A data disk containing all the data sets from the text is available.

▼ Some exercises are designed to encourage students to take advantage of the "friendly" power of the computer.

▼ MINITAB commands are introduced in the text, and students are encouraged to use them as much as possible. Several exercises list the commands needed for that exercise.

The first three chapters are introductory by nature. Chapter 3 is a descriptive (first-look) presentation of bivariate data. This material is presented at this point in the book because students often ask about the relationship between two sets of data (such as heights and weights).

In the chapters on probability (4 and 5), the concepts of permutations and combinations are deliberately avoided in Chapter 4. The binomial coefficient is introduced in the probability chapters in connection with the binomial probability distribution. Information about counting techniques can be found in Appendix A.

The instructor has several options in the selection of topics to be studied in a given course. Chapters 1 through 9 are considered to be the basic core of a course (some sections of Chapters 2, 3, 4, and 6 may be omitted without affecting continuity). Following the completion of Chapter 9, any combination of Chapters 10 through 14 may be studied. However, there are two restrictions: Chapter 3 must be studied prior to Chapter 13, and Chapter 10 must precede Chapter 12.

The suggestions of instructors using the previous editions have been invaluable in helping me improve the text for the present revision. Should you or any of your students have comments or suggestions, I would be most grateful to receive them. Please address such communications to me at Monroe Community College, Rochester, New York 14623.

SUPPLEMENTS

The Statistical Tutor is a student manual that:

a. Contains the *complete solutions* to all margin exercises and the odd-numbered exercises (the same exercises whose answers are in the back of the book).

b. Contains many helpful hints and suggestions to guide you through the learning process. It includes many summaries and overviews.

c. Contains several review lessons to help you refresh on materials studied previously in other courses.

The Instructor's Manual is also intended to be uncommonly helpful. It contains:

a. Everything that is in the *Statistical Tutor*.
b. The complete solutions to all even-numbered exercises.
c. Many helpful teaching suggestions that an instructor might incorporate. The notes specifically intended for the instructor are set in a type font different from that for the student material.

The **Test Bank** contains a combination of true-false, multiple-choice, short answer, matching, and computational test questions for each chapter in the text.

Transparency Masters highlighting important illustrations and pertinent examples are available for this edition.

The **MINITAB Student Supplement** is provided for those interested in teaching or learning the course interactively with the computer. This supplement is a text-specific introduction to the MINITAB statistical analysis system and is keyed to text discussion and examples.

ACKNOWLEDGMENTS

I owe a debt to many other books. Many of the ideas, principles, examples, and developments that appear in this text stem from thoughts provoked by these sources.

It is a pleasure to acknowledge the aid and encouragement I have received throughout the development of this text from my students and colleagues at Monroe Community College. A special thanks to those who read and offered suggestions about this and the previous editions, whose names are listed on page vi, before the Contents.

In addition, I would especially like to thank Patricia Kuby for her invaluable assistance throughout the development of this seventh edition and in particular for developing the *Statistical Tutor* and *Instructor's Manual*; Larry Stephens for his assistance with the exercises; August J. Zarcone for developing the *Test Bank*; Greg Hubit and Margaret Moore for the quality work that they have put into the production of this edition; and Alex Kugushev for his leadership.

Thanks also to the many authors and publishers who so generously extended reproduction permission for the case studies and tables used in the text. These acknowledgments are specified individually throughout the text.

The last and certainly the most significant of all—thank you to my loving wife, Barbara, for her assistance and for her incredible amount of patience through these last few months.

Robert Johnson

Your Guide to Getting the Most Out of Elementary Statistics, Seventh Edition

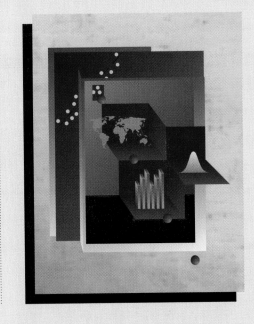

Before you begin reading this textbook, here is a lesson you might enjoy: Statistics is not math. Statistics is simply a scientific tool for describing the world we live in. Sometimes it requires math, other times it doesn't.

Perhaps you've also heard that statistics is just formalized common sense. This is true, but there are more specific applications, too—particularly in business and the physical and social sciences. This textbook has statistics examples from these disciplines as well as many other fields and professions.

Elementary Statistics, Seventh Edition will introduce you to two very important types of statistics that people use every day—known as descriptive and inferential statistics. These two types of statistics allow researchers to make meaningful comparisons between data and ultimately enable them to draw reasonable conclusions. This is the power of statistics, and you are about to learn it.

To get the most out of this book, you should familiarize yourself with its many learning features. Some of them are graphical, some are based on computer simulation, and others are anecdotal. On the following pages, you will find examples of several features this book contains and suggestions on how to make the best use of them. Take a moment to look them over.

GETTING STARTED—READING, CHAPTER BY CHAPTER

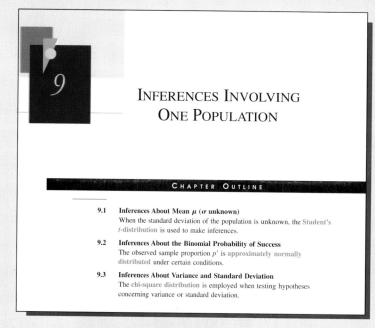

9

INFERENCES INVOLVING ONE POPULATION

CHAPTER OUTLINE

9.1 Inferences About Mean μ (σ unknown)
When the standard deviation of the population is unknown, the Student's *t*-distribution is used to make inferences.

9.2 Inferences About the Binomial Probability of Success
The observed sample proportion p' is approximately normally distributed under certain conditions.

9.3 Inferences About Variance and Standard Deviation
The chi-square distribution is employed when testing hypotheses concerning variance or standard deviation.

Every chapter in the text opens with two important study tools.

CHAPTER OUTLINES appear at the beginning of each chapter to give you a schematic overview of what you are about to learn. These outlines will give you a head start on learning important terms and concepts that appear in the chapter.

CHAPTER OBJECTIVES prepare you for new material by describing why the material you are about to read is important. Use *Chapter Objectives* to plan how you will review and study each chapter.

● CHAPTER OBJECTIVES

Imagine that you took an exam at the last meeting of your favorite class. Today your instructor returns your exam paper and it has a grade of 78. If you are like most students in this situation, as soon as you see your grade you want to know how your grade compares to those of the rest of the class and you immediately ask: "What was the average exam grade?" Your instructor replies, "The class average was 68." Since 78 is 10 points above the average, you ask: "How close to the top is my grade?" Your instructor replies that the grades ranged from 42 to 87 points. The accompanying figure summarizes the information we have so far.

```
                              Yours
      |------------------------|----|---|
      42                       68   78  87
      Low                   Average    High
```

A third question that is sometimes asked is "How are the grades distributed?" Your instructor replies that half the class had grades between 65 and 75. With this information you conclude that your grade is fairly good.

The preceding illustration demonstrates the basic process of statistics and how students make use of statistics on a regular basis: There is a set of data (set of exam scores), the data is described with a few descriptive statistics (average grade and high and low grades), and based on this information students are able to draw conclusions about their relative success.

A large part of this chapter is devoted to learning how to present and describe sets of data. There are four basic types of descriptive statistics: (1) measures of central tendency, (2) measures of dispersion (spread), (3) measures of position, and (4) types of distribution. The measures of central tendency are used to describe or locate the "middle" of the data values. The measures of dispersion measure the spread or variability within the data values. When very large sets of data are involved, the measures of position become useful. For example, when college board exams are taken, thousands of scores result. Your concern is the position of your score with respect to all the others. Averages, ranges, and so on, are not enough. A measure of position will tell you that your score is better than a certain percentage of the other scores. For example, your score is better than 85% of all the scores. The fourth concept, the type of distribution, describes the data by telling you whether the values are evenly distributed or clustered (bunched) around a certain value. Another very important part of this chapter will explain how to interpret the findings so that we know what the data is telling us about the sampled population.

In this chapter, we deal with single-variable data, which is data collected for one numerical variable at a time. In Chapter 3 we will learn to work with two variables at a time and learn about their relationship.

STATISTICS AT WORK IN THE WORLD—CASE STUDIES

This text contains two types of Case Studies designed to teach you important concepts and to demonstrate how statistics work in the real world.

At the start of every chapter you will find a **Chapter Case Study**. It covers many of the concepts you will study in the chapter and gets you to start thinking—through specific questions—about the statistical concepts found in the chapter. At the end of the chapter, after you've done your reading and practiced the techniques presented, you can revisit the *Chapter Case Study* questions, finding that you should now be able to answer them from a statistical standpoint.

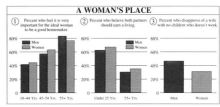

Chapter Case Study

THE AMERICAN GENDER EVOLUTION: GETTING WHAT WE WANT

THE *NEW WOMAN* SURVEY

The results reported here are based on a telephone survey of 1,201 adults supervised by the polling firm of Yankelovich Clancy Shulman and conducted during March 1990. It has a sampling error of ±3 percent and all statistics exclude "don't know" answers.

THE CHANGING FACE OF MARRIAGE

Marriage is as important as ever, but in the past 20 years, the roles of husband and wife—and their expectations of marriage—have changed dramatically. Today, couples often fall into one of two categories—Traditional and Egalitarian—which reflect opposing ideals of marriage. Other couples, struggling to integrate old and new ideals, fall somewhere in between, in marriages that could be called Transitional.

In the Traditional marriage, the wife generally derives her identity through her family and her role as wife and mother. The husband defines himself through his work and his role as breadwinner.

Many older Americans—those over age 55—have Traditional marriages. They are most likely to say, for instance, that the ideal woman is a good homemaker (see Chart 1).

In an Egalitarian marriage, the couple views the union as an equal partnership. Both husband and wife gain a sense of identity through work and family; they share the burden of breadwinning as well as the pleasure of nurturing children.

Egalitarian marriages are most likely to be found among younger Americans—those under age 45. . . .

. . . Most American women and men under age 45 believe that *both* partners should be responsible for earning a living (see Chart 2). . . .

It's inevitable that over the next decade more couples will be hoping to find an Egalitarian marriage. This year, 65 percent of American wives under 65 are working outside the home, according to the U.S. Bureau of Labor Statistics. As more wives remain in the work force, fewer Americans will expect men to be the sole breadwinner. . . .

Perhaps that is why some men go even further and insist that wives are obligated to work. Almost half of American men today actually disapprove of a wife with no children who doesn't work; so do about one third of women (see Chart 3). . . .

A WOMAN'S PLACE

① Percent who feel it is very important for the ideal woman to be a good homemaker
② Percent who believe both partners should earn a living
③ Percent who disapprove of a wife with no children who doesn't work.

Source: "The American Gender Evolution: Getting What We Want," New Woman (November 1990). Reprinted by permission.

Today's newspapers and magazines often report the findings of survey polls about various aspects of life. When an article reports 62% or a graph shows 62%, did you ever wonder what is this 62%: Is it a parameter or a statistic? Somewhere in the article you might find bits of information like: "telephone survey of 1201 adults" and "has a sampling error of ±3 percent." What does this information tell us? How does this information relate to the statistical inferences that we have been learning about? How do terms like point estimate, level of confidence, maximum error of estimate, and confidence interval estimation relate to the values reported in the article as it applies to this situation involving percentage of people? Is percentage of people a population parameter, and if so, is it related to any of the parameters that we have studied? We will return to these questions at the end of the chapter.

Case Study **Tricky Graph Number 2**

2-5

The histogram shows the number of drivers fatally injured in automobile accidents in New York State in 1968.

EXERCISE 2.142

If the vertical scale of the fatal-accident histogram were to be changed to "number of fatalities per 10,000 drivers," the graph would not be a histogram. Explain why.

Notice the "normal-appearing" distribution of ages of fatally injured drivers. The reason for grouping the ages in this manner is completely unclear. Certainly the age of fatally injured automobile drivers is not expected to be normally distributed. Notice too that the class width varies considerably. At first glance, this histogram seems to indicate that 30- to 39-year-olds have the "most fatal accidents." This may or may not be the case. Perhaps this age group has more drivers than any of the other groups. The group with the most drivers might be expected to have the highest number of driver fatalities. A vertical scale that showed the number of driver fatalities per thousand drivers might be more representative. An age grouping that created classifications with the same number of drivers in each class might also be more representative. It all depends on the intended purpose.

Throughout the text, you will also find case studies that deal with statistical concepts as they are being presented. Margin exercises accompany these case studies and are an excellent way to practice solving problems.

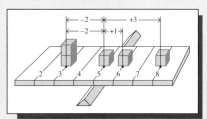

You will also find graphical displays in this text of **charts, graphs, and tables**. Graphics are very important in statistics. They are the pictures that allow you to see vast amounts of data in an easy-to-understand format.

UNDERSTANDING STATISTICS

Coefficient of linear
correlation

variables are not related linearly, and therefore there is no linear correlation.

The **coefficient of linear correlation** r is the numerical measure of the strength of the linear relationship between two variables. The coefficient reflects the consistency of the effect that a change in one variable has on the other. The value of the linear correlation coefficient helps us to answer the question "Is there a linear correlation between the two variables under consideration?" The linear correlation coefficient r always has a value between -1 and $+1$. A value of $+1$ signifies a **perfect positive correlation**, and a value of -1 shows a **perfect negative correlation**. If as x increases there is a general increase in the value of y, then r will be positive in value. For example, a positive value of r would be expected for age and height

Perfect positive and
negative correlation

All the important **definitions** and key terms in this textbook are highlighted for your quick reference. In addition, a **Vocabulary List** near the end of every chapter lets you review key terms for exams.

VOCABULARY LIST

Be able to define each term. Pay special attention to the key terms, which are printed in red. In addition, describe in your own words, and give an example of, each term. Your examples should not be ones given in class or in the textbook. Page numbers indicate the first appearance of a term.

attribute data (p. 17)
census (p. 24)
data (p. 15)
descriptive statistics (p. 6)
experiment (p. 16)
finite population (p. 15)
inferential statistics (p. 6)
infinite population (p. 15)
judgment sample (p. 24)
numerical data (p. 17)
parameter (p. 16)
population (p. 14)
probability (p. 28)

probability sample (p. 25)
qualitative variable (p. 17)
quantitative variable (p. 17)
random (p. 25)
random sample (p. 25)
representative sampling frame (p. 24)
sample (p. 15)
sampling frame (p. 24)
statistic (p. 16)
statistics (p. 6)
variable (p. 15)
variability (p. 22)

The best way to learn statistics is to practice—and once you think you understand a concept, practice it some more for good measure.

This text contains hundreds of **exercises** with many applications. You will find special **icons** next to some exercises. These indicate the type of application—business, sports, criminal justice, etc.—the exercise covers.

You will also find some exercises in the book's margins; they have been placed there for you to practice on key concepts right away—should you decide to solve them before reading the entire chapter. These exercises are an excellent way to reinforce statistical ideas and give you the opportunity to decide what is the best way for you to approach your homework.

EXERCISES

2.129 According to the empirical rule, practically all the data should lie between $(\bar{x} - 3s)$ and $(\bar{x} + 3s)$. The range also accounts for all the data.
   a. What relationship should hold (approximately) between the standard deviation and the range?
   b. How can you use the results of (a) to estimate the standard deviation in situations when the range is known?

2.130 The average clean-up time for a crew of a medium-size firm is 84.0 hours and the standard deviation is 6.8 hours. Assuming that the empirical rule is appropriate,
   a. what proportion of the time will it take the clean-up crew 97.6 or more hours to clean the plant?
   b. the total clean-up time will fall within what interval 95% of the time?

2.131 Chebyshev's theorem can be stated in a form equivalent to that given on p. 104. For example, to say that "at least 75% of the data fall within two standard deviations of the mean" is equivalent to stating that "at most, 25% will be more than two standard deviations away from the mean."
   a. At most, what percentage of a distribution will be three or more standard deviations from the mean?
   b. At most, what percentage of a distribution will be four or more standard deviations from the mean?

2.132 An article titled "Effects of Modifying Instruction in a College Classroom" (*Psychological Reports*, 1986) describes an introductory psychology course in which 232 students received instruction consisting of lectures accompanied by unit testing to mastery and by student proctors. At the end of the course, students assigned ratings from 1 (significantly agree) to 5 (significantly disagree) to several statements. One such statement was "Material was presented at an appropriate level and pace." The mean and standard deviation for the responses to this statement were 1.83 and 0.44. According to Chebyshev's theorem, at least what percentage of the ratings was a 1 or a 2?

2.133 In an article titled "Development of the Breast-Feeding Attrition Prediction Tool" (*Nursing Research*, March/April 1994, Vol. 43, No. 2), the results of administering the Breast-Feeding Attrition Prediction Tool (BAPT) to 72 women with prior successful experience breast-feeding are reported. The mean score for the Positive Breast-feeding Sentiment (PBS) score for this sample is reported to equal 356.3 with a standard deviation of 65.9.
   a. According to Chebyshev's theorem, at least what percent of the PBS scores are between 224.5 and 488.1?
   b. If it is known that the PBS scores are normally distributed, what percent of the PBS scores are between 224.5 and 488.1?

2.134 An article titled "Computer-Enhanced Algebra Resources: The Effects on Achievement and Attitudes" (*International Journal of Math Education in Science and Technology*, 1980, Vol. 11, No. 4) compared algebra courses that used computer-assisted instruction with courses that do not. The scores that the computer-assisted instruction group made on an achievement test consisting of 50 problems had these summary statistics:

$$n = 57, \quad \bar{x} = 23.14, \quad s = 7.02$$

   a. Find the limits within which at least 75% of the scores fell.

Doing Statistics on the Computer and on Your Own

as *units squared*. In our example of pounds, this would be *pounds squared*. As you can see, the unit has very little meaning.

MINITAB (Release 10) commands to find the standard deviation of the data listed in C1.

Session commands	*Menu commands*
Enter: `STDEv C1`	Choose: `Calc > Column Stats`
	Select: `Standard deviation`
	Enter: `Input variable, C1`

Additional statistics may be found using the following commands:

Session commands		*Menu Commands*
`COUNt` - Number of data in column		Choose: `Calc > Column Stats`
`SUM`	Sum of the data in column	`Sum`
`MINImum`	Smallest value in column	`Minimum`
`MAXImum`	Largest value in column	`Maximum`
`RANGe`	Range of values in column	`Range`
`SSQ`	Sum of squared x-values, Σx^2	`Sum of squares`

Most statistical calculations are now done by computers. In this book you will find numerous references to *MINITAB*—a statistics software package that lets you focus on learning concepts while it handles the calculation and graphing of data.

Working with Your Own Data

Each semester, new students enter your college environment. You may have wondered, "What will the student body be like this semester?" As a beginning statistics student, you have just finished studying three chapters of basic descriptive statistical techniques. Let's use some of these techniques to describe some characteristics of your college's student body.

A | Single Variable Data

1. Define the population to be studied.
2. Choose a variable to define. (You may define your own variable, or you may use one of the variables in the accompanying table if you are not able to collect your own data. Ask your instructor for guidance.)
3. Collect 35 pieces of data for your variable.
4. Construct a stem-and-leaf display of your data. Be sure to label it.
5. Calculate the value of the measure of central tendency that you believe best answers the question "What is the average value of your variable?" Explain why you chose this measure.
6. Calculate the sample mean for your data (unless you used the mean in Question 5).
7. Calculate the sample standard deviation for your data.
8. Find the value of the 85th percentile, P_{85}.
9. Construct a graphic display (other than a stem-and-leaf) that you believe "best" displays your data. Explain why the graph best presents your data.

B | Bivariate Data

1. Define the population to be studied.
2. Choose and define two quantitative variables that will produce bivariate data. (You may define your own variables, or you may use two of the variables in the accompanying table if you are not able to collect your own data. Ask your instructor for guidance.)
3. Collect 15 ordered pairs of data.
4. Construct a scatter diagram of your data. (Be sure to label it completely.)
5. Using a table to assist with the organization, calculate the extensions x^2, xy, and y^2, and the summations of x, y, x^2, xy, and y^2.
6. Calculate the linear correlation coefficient r.
7. Calculate the equation of the line of best fit.
8. Draw the line of best fit on your scatter diagram.

At the end of each of this text's four parts, you will find a **Working With Your Own Data** section designed to encourage you to explore statistics on your own. These sections give you the opportunity to collect your own data and to apply the techniques you've learned in the text.

An **In Retrospect** section summarizes the concepts you have learned in the chapter, pointing out their relationship to material covered previously.

Chapter Exercises are based on all the techinques you studied in the chapter. They provide you with the opportunity to identify which procedure to use to produce desired results.

Vocabulary Lists help you decide how much of the material you truly understand. Here's a test: Try to define all the vocabulary in a chapter to a friend. If you can do it, fine. If not, you know you need more practice.

Finally, end-of-chapter **Quizzes**— ranked A, B, and C—are self-checks; they will tell how ready you really are for class exams. Quiz A is conceptual. Quiz B is for methods. And Quiz C encourages you to apply your knowledge. Correct responses for quizzes are in the back of the text.

Also, a **Statistical Tutor** is available for this text. The Tutor will help you practice important concepts, and it reinforces your instructor's lessons and the chapter presentations in the text. Ask your instructor or college bookstore how to order a copy.

IN RETROSPECT

You have been introduced to some of the more common techniques of descriptive statistics. There are far too many specific types of statistics used in nearly every specialized field of study for us to review here. We have outlined the uses of only the most universal statistics. Specifically, you have seen several basic graphic techniques (circle and bar graphs, Pareto diagrams, dotplots, stem-and-leaf displays, histograms, ogives, and box-and-whiskers) that are used to present sample data in picture form. You have also been introduced to some of the more common measures of central tendency (mean, median, mode, midrange, and midquartile), measures of dispersion (range, variance, and standard deviation), and measures of position (quartiles, percentiles, and z-score).

You should now be aware that an average can be any one of five different statistics, and you should understand the distinctions among the different types of averages. The article "'Average' Means Different Things" in Case Study 2-2 (p. 73) discusses four of the averages studied in this chapter. You might reread it now and find that it has more meaning and is of more interest. It will be time well spent!

You should also have a feeling for, and an understanding of, the concept of a standard deviation. You were introduced to Chebyshev's theorem and the empirical rule for this purpose.

The exercises in this chapter (as in others) are extremely important; they will help you nail down the concepts studied before you go on to learn how to use these ideas in later chapters. A good understanding of the descriptive techniques presented in this chapter is fundamental to your success in the later chapters.

CHAPTER EXERCISES

2.145 Samples A and B are shown in the following table. Notice that the two samples are the same except that the 8 in A has been replaced by a 9 in B.

| A: | 2 | 4 | 5 | 5 | 7 | 8 |
| B: | 2 | 4 | 5 | 5 | 7 | 9 |

What effect does changing the 8 to a 9 have on each of the following statistics?
a. mean   b. median   c. mode   d. midrange
e. range   f. variance   g. standard deviation

2.146 Samples C and D are shown in the following table. Notice that the two samples are alike except for two values.

| C: | 20 | 60 | 60 | 70 | 90 |
| D: | 20 | 30 | 70 | 90 | 90 |

VOCABULARY LIST

Be able to define each term. Pay special attention to the key terms, which are printed in red. In addition, describe in your own words, and give an example of, each term. Your examples should not be ones given in class or in the textbook.

The bracketed numbers indicate the chapter(s) in which the term appeared previously, but you should define the terms again to show increased understanding of their meaning. Page numbers indicate the first appearance of the term in Chapter 9.

calculated value [8] (p. 441)
chi-square (p. 474)
chi-square distribution (p. 471)
conclusion [8] (p. 443)
confidence interval [8] (p. 439)
critical region [8] (p. 445)
critical value [8] (p. 443)
decision [8] (p. 443)
degrees of freedom (p. 434)
hypothesis test [8] (p. 441)
inference [8] (p. 432)
level of confidence [8] (p. 439)
level of significance [8] (p. 441)
maximum error of estimate [8] (p. 440)
observed binomial probability (p') (p. 454)
one-tailed test [8] (p. 442)

parameter [1, 8] (p. 437)
p-value [8] (p. 442)
proportion [6] (p. 454)
random variable [5, 6] (p. 454)
response variable [1] (p. 454)
sample size [8] (p. 434)
sample statistic [1, 2] (p. 455)
σ known [8] (p. 432)
σ unknown (p. 432)
standard error [7, 8] (p. 432)
standard normal, z [2, 6, 8] (p. 459)
Student's t (p. 432)
test statistic [8] (p. 441)
two-tailed test [8] (p. 442)

QUIZ A

Answer "True" if the statement is always true. If the statement is not always true, replace the words shown in bold with words that make the statement always true.

9.1 The Student's t-distribution is an approximately normal distribution but is more **dispersed** than the normal distribution.

9.2 The **chi-square** distribution is used for inferences about the mean when σ is unknown.

9.3 The **Student's** t-distribution is used for all inferences about a population's variance.

9.4 If the test statistic falls in the critical region, the null hypothesis has **been proved true**.

9.5 When the test statistic is t and the number of degrees of freedom gets very large, the critical value of t is very close to that of z.

DESCRIPTIVE STATISTICS

A typical objective in statistics is to describe "the population" based on information obtained by observing relatively few individual elements. We must learn how to sort out the generalizations contained within the clues provided by the sample data and "paint" a picture of the population. We study the sample, but it's the population that is of primary interest to us.

When one embarks on a statistical solution to a problem, a certain sequence of events must develop: (1) The situation being investigated is carefully and fully defined, (2) a sample is collected from the population following an established and appropriate procedure, (3) the sample data are converted into usable information (this usable information, either numerical or pictorial,

is called sample statistics*), and (4) the theories of statistical inference are applied to the sample information in order to draw conclusions about the sampled population (these conclusions or answers are called* inferences*). This sequence of events is illustrated by the Statistical Process diagram on the following page.*

The first part of this textbook, Chapters 1–3, concentrates on the first three of the four events identified above. The second part, Chapters 4–7, deals with probability theory, the theory on which statistical inferences rely. The third and fourth parts, Chapters 8–10 and 11–14, survey the various types of inferences that can be made from sample information.

The Statistical Process

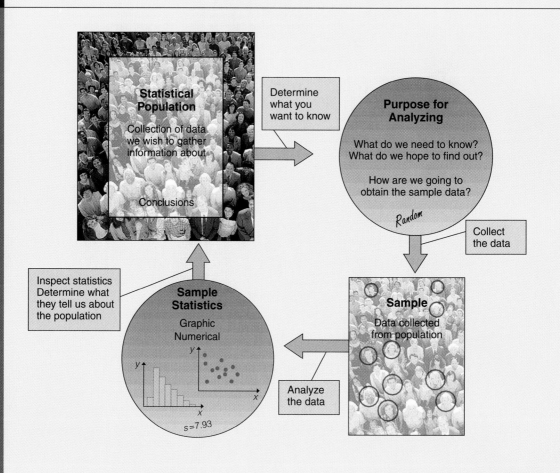

SIR FRANCIS GALTON

SIR FRANCIS GALTON, English anthropologist and a pioneer of human intelligence studies, was born on February 16, 1822, in a village near Birmingham, England, to Samuel Tertiles and Anne (Violetta) Galton.

Galton's family included men and women of exceptional ability, one of whom was his cousin Charles Darwin. Although his family life was happy, and he was grateful to his parents for all they had done for him, he felt little use for the conventional religious and classical education he was given.

As a teen, Galton toured a number of medical institutions in Europe and began his medical training in hospitals in Birmingham and London. He continued his medical studies until the death of his father. Having been left a sizable fortune, he decided to discontinue medical training and pursue his love of traveling. Several years of these travels led him to the exploration of primitive parts of southwestern Africa, where he gained valuable information that earned him recognition and a fellowship in the Royal Geographical Society. In 1853, at the age of only 31, Galton was awarded a gold medal from the Society in recognition of his hard work and many achievements. It was also in 1853 that Galton married Louisa Butler and they settled in London; their marriage remained childless.

Although Galton made important contributions to many fields of knowledge, he was best known for his work in eugenics; he spent most of the latter part of his life researching and promoting his belief that inheritance played a major role in the intelligence of man. Galton, a pioneer in the development of some of the refined statistical techniques that we use today, used the laws of probability to support his theory. His application of research techniques—curves of normal distribution, correlation coefficients, and percentile grading—to a large population revealed important facts about the intellectual and physical charactertistics that are passed from one generation to the next and the ways in which children differ from their parents. He also discovered, with the use of the graph, that characteristics of two different generations could be plotted against one another, revealing important information.

Sir Francis Galton died in England on January 17, 1911, leaving behind a collection of 9 books, approximately 200 articles and lectures, and a valuable legacy of statistical techniques and knowledge.

STATISTICS

Chapter Case Study

DESCRIBING OURSELVES STATISTICALLY

We all think of ourselves as being very different from everyone else. To a certain extent this is true, but no matter how different you are, it is likely that you can relate to one or more of the statistics presented in the following article from USA Today.

FACTS PAINT USA'S PORTRAIT

People in the USA like to walk, sip soft drinks and eat fatty foods.

Those and thousands of other measures of U.S. life are in the new *Statistical Abstract of the United States*.

Among the facts in the 1,011 pages and 1,410 tables in the 114th edition:

- Walking was the No. 1 exercise in 1992, followed by swimming and bicycling.
- Soft drinks are the preferred beverage, with 44.1 million gallons consumed in 1992. Alcohol, coffee, and milk are next popular.
- The average American in 1990 ate 165 grams of fat a day, above the 141 grams of the 1950s.

- 98.3% of households have TV; 61% have cable.
- Washington, D.C., workers had the highest average pay in 1992, $37,971. Lowest: South Dakota, $18,016.
- The average bank robbery in 1992 got $3,325.
- A fourth of the 9.9 million women due child support in 1989 didn't get it.
- New Hampshire had the most women in its legislature last year, 142; Louisiana had the fewest, seven.

Source: Margaret Usdansky, *USA Today*, 10-23-94.

The above and thousands of other measures are used to describe life in the United States. What is the statistical population? Are these measures parameters or statistics? What is the difference? What variable is related to the fact that the average bank robbery got $3325? What kind of variable was used when 98.3% of households were found to have TVs?

At the end of the chapter we will return to the above questions.

● CHAPTER OBJECTIVES

●

The purpose of this introductory chapter is to (1) create an initial image of the field of statistics, an image that will grow and develop; (2) introduce several basic vocabulary words used in studying statistics, such as *population*, *variable*, and *statistic*; and (3) present some initial ideas and concerns about the processes used to obtain sample data.

1.1 | WHAT IS STATISTICS?

Statistics is the universal language of the sciences. As potential users of statistics, we need to master both the "science" and the "art" of using statistical methodology correctly. Careful use of statistical methods will enable us to obtain accurate information from data. These methods include (1) carefully defining the situation, (2) gathering data, (3) accurately summarizing the data, and (4) deriving and communicating meaningful conclusions.

Statistics

Statistics involves information, numbers to summarize this information, and their interpretation. The word ***statistics*** has different meanings to people of varied backgrounds and interests. To some people it is a field of "hocus-pocus" whereby a person in the know overwhelms the rest of us. To others it is a way of collecting and displaying large amounts of information. And to still another group it is a way of "making decisions in the face of uncertainty." In the proper perspective, each of these points of view is correct.

Descriptive and inferential statistics

The field of statistics can be roughly subdivided into two areas: descriptive statistics and inferential statistics. **Descriptive statistics** is what most people think of when they hear the word *statistics*. It includes the collection, presentation, and description of sample data. The term **inferential statistics** refers to the technique of interpreting the values resulting from the descriptive techniques and making decisions and drawing conclusions about the sampled population.

Statistics is more than just numbers—it is data, what is done to data, what is learned from the data, and the resulting conclusions. Let's use the following definition:

STATISTICS

The science of collecting, describing, and interpreting data.

Before we begin our detailed study, let's look at a few illustrations of how and when statistics can be applied.

Case Study

Measuring Physical Discomfort

1-1

Sitting in an uncomfortable seat for long periods of time is no fun. Slipping into a seat on a jet airliner when you are larger than average can be outright painful.

By Del Jones, *USA Today*, 5-27-94

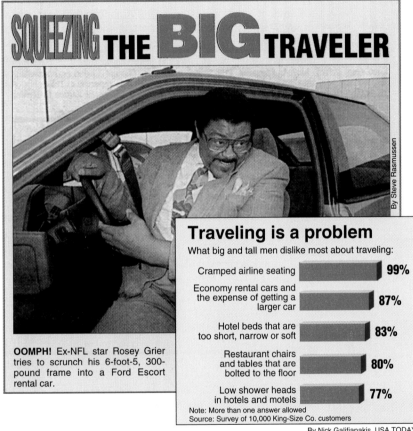

By Steve Rasmussen

OOMPH! Ex-NFL star Rosey Grier tries to scrunch his 6-foot-5, 300-pound frame into a Ford Escort rental car.

Traveling is a problem

What big and tall men dislike most about traveling:

Cramped airline seating	**99%**
Economy rental cars and the expense of getting a larger car	**87%**
Hotel beds that are too short, narrow or soft	**83%**
Restaurant chairs and tables that are bolted to the floor	**80%**
Low shower heads in hotels and motels	**77%**

Note: More than one answer allowed
Source: Survey of 10,000 King-Size Co. customers

By Nick Galifianakis, USA TODAY

EXERCISE 1.1

Refer to "Traveling is a problem."

a. Who was surveyed?

b. How many were surveyed?

c. Explain the meaning of "Cramped airline seating: 99%."

d. Why are so many percentages reported?

The world is small for big and tall business travelers. Ask Rosey Grier, a 6-foot-5, 300-pound former NFL defensive lineman, who never met a running back he couldn't tackle. But he can't win against airline seats built for 5-foot-9, 170-pounders.

Grier's not alone. 13 million other men are at least 6-foot-2 or 225 pounds.

Airline seats have never been roomy. "They were originally designed to fit a very well-known athlete: (jockey) Willie Shoemaker," jokes Ed Perkins, editor of Consumer Reports Travel Letter, which measures the distance between seats every two years. Over the past 20 years, airlines have gradually squeezed about 10% more seats into their jets. All the while, they have inched the rows on all jets closer together. Rows are 4 inches closer together than they were 20 years ago.

Case Study

1-2

Explaining Our Sexual Behavior

Sex has been said to be "everybody's favorite topic."

SEX IN THE SNORING '90s

A new survey of American men shocks the nation with what it didn't find. They're the sorts of questions any average American male asks: How many times a night should I be having sex? In what grade should I start? . . . But there is nowhere to turn for answers. If you wanted to know something the government considered important, like the size of the sardine catch, you could look it up in a minute. However, when you don't pay taxes on it, the government has no reason to keep track of it. Now, in what the authors call the first scientifically valid survey of its kind, 3,321 American men in their 20s and 30s were questioned about their sexual practices, and for the first time in the history of sex research, the results are not astounding.

THE REPORT

In the most complete study since the Kinsey report, a new survey charts American sexual activity. The results are—yawn—striking.

1

is the median number of times men said they had sex per week.

7.3

is the median number of sexual partners, over lifetime, per man.
(Guttmacher, 1993)

INFIDELITY

How faithful are the women of America? It depends on whom you ask. Two studies produced very different answers.

26%

Married women who have had extramarital affairs.
(Janus, 1993)

39%

Married women who have cheated on their husbands.
(Cosmopolitan Reader's Survey, 1993)

EXERCISE 1.2

Refer to "Sex in the Snoring '90s."

a. What group was surveyed?

b. Why do seemingly similar surveys produce such different results?

One thing the Battelle study did reveal was the huge philosophical gap between two schools of sex research. One group conducts personal interviews with subjects chosen on the basis of census data to represent a scientific sample of the population, which is supposed to provide better statistical validity. The other uses anonymous mail-in questionnaires, which seem to elicit juicier responses. The former generally publish papers in scientific journals, while the latter typically write articles in women's magazines and best-selling books.

Janus, who distributed his survey through graduate students and in piles left in doctors' offices, asserts that people are more likely to be truthful on an anonymous questionnaire than in face-to-face interviews. The Batelle group's interviews were all conducted in the respondents' homes, by female interviewers who knocked on the door with no previous introduction. Thirty percent of the men contacted in this way refused to participate. Janus suggests that the missing information is probably in that 30 percent. On the other hand, the people motivated to pick up, fill out and mail in a questionnaire on sex represent a self-selected group—selected, in part, by being interested in sex. Janus' data also show that his subjects were considerably richer, less Protestant and more Jewish than the nation as a whole.

Source: Jerry Adler, *Newsweek*, April 26, 1993.

Case Study Describing Our Ever-Changing Political Opinions

1-3

Political mood: Growing cynicism

The electorate nationally is feeling politically alienated and in an anti-incumbent mood, a poll released Wednesday shows. But interest in politics is climbing. The mood of the electorate:

Government is not really run to benefit all people

Year	%
1988	44%
1990	45%
1992	54%
1994	57%

Federal government controls too much of daily life

Year	%
1988	61%
1990	62%
1992	64%
1994	69%

New people are needed in Washington, even if less effective

Year	%
1988	51%
1990	47%
1992	56%
1994	60%

It is my duty as citizen to always vote

Year	%
1988	88%
1990	85%
1992	91%
1994	93%

Elected officials don't care what people think

Year	%
1988	51%
1990	53%
1992	62%
1994	66%

Pretty interested in local politics

Year	%
1988	72%
1990	70%
1992	73%
1994	76%

Source: Times Mirror Center for the People & Press nationwide surveys of 3,800 adults, conducted in July. Margin of error: ± 3 percentage points.

By Nick Galifianakis, USA TODAY

EXERCISE 1.3

Explain how this information shows a "growing cynicism."

The political mood of the American electorate is continuously changing. There seems to be an endless stream of polls being reported showing the latest change in the president's popularity and the current opinion on all aspects of the political scene.

The uses of statistics are unlimited. It is much harder to name a field in which statistics is not used than it is to name one in which statistics plays an integral part. The following are a few examples of how and where statistics are used:

▼ In education descriptive statistics are frequently used to describe test results.
▼ In science the data resulting from experiments must be collected and analyzed.
▼ In government many kinds of statistical data are collected all the time. In fact, the U.S. government is probably the world's greatest collector of statistical data.

A very important part of the statistical process is that of studying the statistical results and formulating appropriate conclusions. These conclusions must then be communicated to others. Nothing is gained from research unless the findings are shared with others.

Case Study

Two of the Three R's

The article "Survey Says Computers, Math Valued More" is about the importance of writing (the very skill we use when reporting results to others) in high school curricula, and it clearly states one conclusion being drawn as a result of this survey.

SURVEY SAYS COMPUTERS, MATH VALUED MORE

Learning to write well is less important than math and computers, students and their parents think, and that helps lead to students' poor reading and writing skills.

That's one of the main findings of the third annual *AFT/Chrysler Report on Kids, Parents and Reading*, a survey conducted among 500 teacher-members of the American Federation of Teachers.

Teachers place writing with math and reading as the three most important skills; parents rank writing fourth, after math, reading, and computers, and students rank writing fifth, below science.

EXERCISE 1.4

Refer to "Who values writing?"

a. What appears to be the purpose of this survey?

b. Identify one conclusion that is expressed.

c. What evidence supports that conclusion?

Who values writing?

Teachers assign writing a much higher place among the skills that young people need than do parents and students themselves. "The ability to write develops the ability to think, to make distinctions," a high school English teacher said in responding to the survey.

Most important skills for young people to learn

	Teachers			Parents	Youth
	Most Important	Second Most Important	Total Important	Total Important	Total Important
	%	%	%	%	%
Reading	56	13	69	62	34
Math	11	40	51	54	65
Writing	13	22	35	20	21
Computers	5	12	17	40	34
Science	2	3	5	10	24
History	1	1	2	4	13

From research conducted by Peter D. Hart Research Associates, Inc., Washington, D.C., in September 1994.

Statistics are being reported everywhere—newspapers, magazines, television. We read and hear about all kinds of new research results, especially health-related findings.

Case Study

1-5

Tricky Business

"One ounce of statistics technique requires one pound of common sense for proper application."

Harvard Health Letter, Special Supplement, October 1994.

Because we are mortal and . . . live in an imperfect world, risk will always be with us. . . . As we go about our lives, we weigh the relative risks and benefits of our actions all the time. Most often we act on imperfect and incomplete information. We do the best we can. Fortunately, even if we make the "wrong" decision, it is likely to turn out to be all right. Something else, completely unexpected, will undoubtedly get us in the end. . . .

Commercial flight is one of very few areas where the degree of risk has been calculated and reduced about as far as is practical, and any further significant reductions would be prohibitively expensive. Once we walk off the airplane, however, our risks vary dramatically and are much more difficult to fathom, so we constantly make decisions based on more or less educated guesses.

Common Sense

So, . . . how can responsible individuals with no special expertise make intelligent decisions about all the information and misinformation that bombards us?

The same humble horse sense that keeps most of us from sticking our hand into the fire is an invaluable tool for sorting out what we read and hear. It's important to remember that news, by its very definition, is something new and unusual. No wonder that newspaper reports so often depart from what common sense tells us or what most experts believe. After all, the hundredth study showing a relationship between cholesterol and heart disease is hardly news, but the one study that fails to make such a connection is likely to become a headline. Clearly it would be silly for people to drastically change their lives on the basis of one newspaper article or a lone scientific study.

That doesn't mean we should throw out everything we read or hear. Once medical experts have reached consensus on a particular health issue, their message is amplified by the popular press and codified in guidelines issued by government agencies and national organizations. Today, for instance, there is widespread agreement that having a high blood cholesterol level is a risk factor for heart disease. . . . The health and science pages are filled with articles on how excess cholesterol should be managed. Even advertisements and labels on food products are broadcasting the cholesterol message.

EXERCISE   1.5

Find a recent newspaper article that illustrates an "apples are bad" type of report.

Statistics can be fun—and entertaining, too.

STATISTICAL
SNAPSHOT

And Then There's "Entertainment Statistics"

The 1990s success story for the news media business has been the evolution of "entertainment news." The tabloids have reached new heights of popularity, but the even bigger success story belongs to television's entertainment news stories: newsmagazine shows and talk shows. There are so many of them that they're barely countable. You, of course, realize that the success entertainment news has enjoyed is largely due to "entertainment statistics." (By the way, you will not be tested on these statistics, guaranteed!)

On world news:

© 1995 (Bill Plympton from the Cartoon Bank)

EXERCISES

1.6 Determine which of the following statements is descriptive in nature and which is inferential. Refer to "Traveling is a problem" in Case Study 1-1 (p. 7).
   a. 99% of all big and tall travelers dislike cramped airline seating the most.
   b. 99% of the 10,000 King-Size Co. customers disliked cramped airline seating the most.

1.7 Determine which of the following statements is descriptive in nature and which is inferential. Refer to "Political mood: Growing Cynicism" in Case Study 1-3 (p. 9).
   a. In 1988, 44% of all Americans thought that government was not really run to benefit all people.
   b. In 1994, 60% of the 3800 polled thought that new people are needed in Washington, even if they are less effective.

1.8 Refer to the USA Snapshot® "Reacting to crime" (p. 13).
   a. What group of people were polled?
   b. How many people were polled?
   c. What information was obtained from each person?
   d. Explain the meaning of "55% Carry less cash."
   e. How many people answered "Carry less cash"?
   f. Why do the reported values (55%, 29%, 28%, . . .) add up to more than 100%?

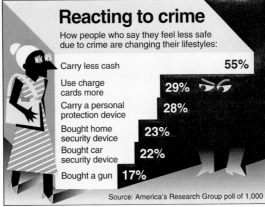

USA SNAPSHOTS®

A look at statistics that shape the nation

Reacting to crime

How people who say they feel less safe due to crime are changing their lifestyles:

Carry less cash	**55%**
Use charge cards more	**29%**
Carry a personal protection device	**28%**
Bought home security device	**23%**
Bought car security device	**22%**
Bought a gun	**17%**

Source: America's Research Group poll of 1,000

By Sam Ward, USA TODAY

 1.9 Percentages do not always tell the whole story.

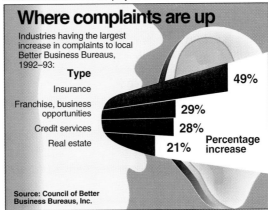

USA SNAPSHOTS®

A look at statistics that shape your finances

Where complaints are up

Industries having the largest increase in complaints to local Better Business Bureaus, 1992–93:

Type	
Insurance	**49%**
Franchise, business opportunities	**29%**
Credit services	**28%**
Real estate	**21%**

Percentage increase

Source: Council of Better Business Bureaus, Inc.

By John Riley and Elys A. McLean, USA TODAY

a. What impression do you get from the statistics (percentages) reported in "Where complaints are up"?

b. Suppose that in 1992 there were 39 and in 1993 there were 58 complaints. What is the percentage of increase?

c. Suppose that in 1992 there were 490 and in 1993 there were 593 complaints. What is the percentage of increase?

d. Compare the increase in complaints listed in (b) and (c).

e. Do you have the same impression about the information presented in the bar graph after seeing the number of complaints in (b) and (c)? Explain.

1.10 The importance of eating healthful foods and having a healthful lifestyle is a topic in today's newspapers and magazines.

Pass a bagel, hold the coffee

By Nancy Hellmich
USA Today

The USA's breakfast of champions now includes more bananas and bagels — and less coffee.

It studied food consumption reports of 2,000 households during a 14-day period:

• Coffee is still the most popular drink, poured at 40% of all a.m. meals — down from 60% in 1984 largely because it's less popular among those under 35.

• Bananas are now the No. 1 fruit, replacing grapefruit — the favorite in 1984.

• Bagels are the fastest growing food on the menu. People ate an average of 7 bagels per person in 1993, up from 2.6 in 1984.

• Fried eggs are down to 11 per person in 1993, from 23 in 1984. Bacon has also taken a beating, down to 13 servings per person in 1993 from 21 in 1984.

• Doughnuts, sweet rolls and Danish are still popular, with a consumption of 12 per person in '93, from 11 in '84.

a. Who was surveyed?
b. What information was collected from each household?
c. What conclusion do you draw from the five survey summary statements with regard to healthful foods?
d. What conclusion do you draw from the five survey summary statements with regard to convenience?

1.11 If you were an independent researcher, how would you test the accuracy of the statement made in Case Study 1-1 that "airline seats [were] built for 5-foot-9, 170-pounders"? Include who you would survey and what information you would collect.

1.12 How would you obtain an answer to the question "Is the American electorate in an anti-incumbent mood?" Include who you would survey and what information you would collect.

1.2 | INTRODUCTION TO BASIC TERMS

In order to begin our study of statistics we need to first define a few basic terms.

POPULATION

A collection, or set, of individuals or objects or events whose properties are to be analyzed.

The population is the complete collection of individuals or objects that are of interest to the sample collector. The concept of a population is the most fundamental idea in statistics. The population of concern must be carefully defined and is considered fully defined only when its membership list of elements is specified. The set of "all students who have ever attended a U.S. college" is an example of a well-defined population.

Typically, we think of a population as a collection of people. However, in statstics the population could be a collection of animals, or manufactured objects, or whatever. For example, the set of all redwood trees in California could be a population.

There are two kinds of populations, finite and infinite. When the membership of a population can be (or could be) physically listed, the population is said to be *finite*. When the membership is unlimited, the population is *infinite*. The books in your college library are a finite population; the OPAC (Online Public Access Catalog, the computerized card catalog) lists the exact membership. All the registered voters in the United States form a very large finite population; if necessary, a composite of all voter lists from all voting precincts across the United States could be compiled. On the other hand, the population of all people who might use aspirin and the population of all 40-watt light bulbs to be produced by Sylvania are infinite.

Finite
Infinite

SAMPLE

A subset of a population.

A sample consists of the individuals, objects, or measurements selected by the sample collector from the population.

VARIABLE

A characteristic of interest about each individual element of a population or sample.

A student's age at entrance into college, the color of the student's hair, the student's height, and the student's weight are four variables.

DATA (SINGULAR)

The value of the variable associated with one element of a population or sample. This value may be a number, a word, or a symbol.

For example, Bill Jones entered college at age "23," his hair is "brown," he is "71 inches" tall, and he weighs "183 pounds." These four pieces of data are the values for the four variables as applied to Bill Jones.

DATA (PLURAL)

The set of values collected for the variable from each of the elements belonging to the sample.

The set of 25 heights collected from 25 students is an example of a set of data.

EXPERIMENT

A planned activity whose results yield a set of data.

This includes both the activities for selecting the elements and obtaining the data values.

PARAMETER

A numerical value summarizing all the data of an entire population.

The "average" age at time of admission of all students who have ever attended our college or the "proportion" of students who were over 21 years of age when they entered college are examples of two different population parameters. A parameter is a value that describes the entire population. Often a Greek letter is used to symbolize the name of a parameter. These symbols will be assigned as we study individual parameters.

For every parameter there is a corresponding sample statistic. The statistic describes the sample the same way the parameter describes the population.

STATISTIC

A numerical value summarizing the sample data.

The "average" height found by using the set of 25 heights is an example of a sample statistic. A statistic is a value that describes a sample. Most sample statistics are found with the aid of formulas and are typically assigned symbolic names using letters of the English alphabet (for example, $\bar{x}$, s, and r).

▼| ILLUSTRATION 1-1

Parameter or statistic
Variable
Population

A statistics student is interested in finding out something about the average dollar value of cars owned by the faculty members of our college. Each of the eight terms just described can be identified in this situation.

1. The *population* is the collection of all cars owned by all faculty members at our college.
2. A *sample* is any subset of that population. For example, the cars owned by members of the mathematics department would be a sample.
3. The *variable* is the "dollar value" of each individual car.
4. One *data* would be the dollar value of a particular car. Mr. Jones's car, for example, is valued at $9400.
5. The *data* would be the set of values that correspond to the sample obtained (9400; 8700; 15,950; . . .).
6. The *experiment* would be the methods used to select the cars forming the sample and determining the value of each car in the sample. It could be carried out by questioning each member of the mathematics department, or in other ways.
7. The *parameter* about which we are seeking information is the "average" value of all cars in the population.
8. The *statistic* that will be found is the "average" value of the cars in the sample.

Yes, from one statement we get all of this.

▲ |

There are basically two kinds of variables: (1) variables that obtain *qualitative* information and (2) variables that obtain *quantitative* information.

QUALITATIVE, OR ATTRIBUTE, VARIABLE

A variable that categorizes or describes an element of a population. Arithmetic operations, such as addition and averaging, are not meaningful for data resulting from a qualitative variable.

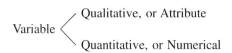

$$\text{Variable} \left\langle \begin{array}{l} \text{Qualitative, or Attribute} \\ \text{Quantitative, or Numerical} \end{array} \right.$$

EXERCISE   1.13

Name two attribute variables about its customers that a newly opened department store might find informative to study.

A sample of four hair-salon customers were surveyed for their "hair color" and "hometown." Both variables are examples of qualitative (attribute) variables, as they both describe some characteristic of the person. The data collected were: {blonde, brown, black, brown} and {Brighton, Columbus, Albany, Jacksonville}. [It would be meaningless to find the sample average by adding and dividing by 4. For example, (blonde + brown + black + brown)/4 is undefined.]

QUANTITATIVE, OR NUMERICAL, VARIABLE

A variable that quantifies an element of a population. Arithmetic operations, such as addition and averaging, are meaningful for data resulting from a **quantitative** variable.

Case Study   Eating Together: Still Important

1-6

The January 6, 1991, issue of the Democrat and Chronicle *presented the results of a* New York Times/CBS News *poll regarding today's attitudes about family togetherness. The graphic shows the results of four variables.*

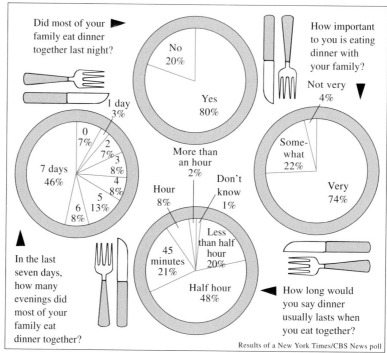

Did most of your family eat dinner together last night?

No 20%

Yes 80%

1 day 3%

How important to you is eating dinner with your family?

Not very 4%

Some-what 22%

Very 74%

0 7%
2 7%
3 8%
4 8%
5 13%
6 8%
7 days 46%

More than an hour 2%

Don't know 1%

Hour 8%

In the last seven days, how many evenings did most of your family eat dinner together?

45 minutes 21%

Less than half hour 20%

Half hour 48%

How long would you say dinner usually lasts when you eat together?

Results of a New York Times/CBS News poll

Joan Bossert Democrat and Chronicle

Source: Joan Bossert, "Eating Together: Still Important," *Democrat and Chronicle,* January 6, 1991, copyright 1991 by *Democrat and Chronicle.* Reprinted by permission.

EXERCISE   1.14

a. Name the four variables.
b. What kind of variable is each?

EXERCISE   1.15

Name two numerical variables about its customers that a newly opened department store might find informative to study.

"Total cost" of textbooks purchased by each student for this semester's classes is an example of a quantitative (numerical) variable. A sample resulted in the following data: $238.87, $94.57, $139.24. [To find the "average cost," simply add the three numbers and divide by three: $(238.87 + 94.57 + 139.24)/3 = \$157.56.$]

Don't let the appearance of the data fool you in regard to their type. Qualitative variables are not always easy to recognize; sometimes they appear as numbers. The above sample of hair colors could be coded: 1 = black, 2 = blonde, 3 = brown. The sample data would then appear as {2, 3, 1, 3}, but they are still attribute data. Calculating the "average hair color" $[(2 + 3 + 1 + 3)/4 = 9/4 = 2.25]$ is still meaningless. The hometowns could have been identified using ZIP codes. The average ZIP code doesn't make sense either; therefore, ZIP code numbers would be qualitative, too.

Case Study   **Pounding Those Drives**

1.7

The September 16, 1994, issue of USA Today *presented a graphic on the statistics about how far two professional golfers drive a golf ball. Two variables are reported: the distance and the distance per pound each can drive a golf ball. It appears that the purpose of the distance per pound statistic is to standardize the information according to size of person to make it fair to compare.*

EXERCISE   1.16

a. What other measures of size are there that might play a role in drive distance?

b. If Daly is 5′11″ and Robbins is 5′6″, compare them using "yards per inch."

USA SNAPSHOTS®

A look at statistics that shape the sports world

Pounding those drives

Golfer John Daly leads all PGA Tour players in driving distance, but for her size, LPGA leader Kelly Robbins gets the most yardage:

Avg. driving yards
252.6
Yards per pound
1.804
Kelly Robbins (140 pounds)

Avg. driving yards
290.2
Yards per pound
1.658
John Daly (175 pounds)

Source: USA TODAY research   By John Riley and Julie Stacey, USA TODAY

Let's look at another example. Suppose that after surveying a parking lot, I summarized the sample data by reporting 5 red, 8 blue, 6 green, and 2 yellow cars. You must look at each individual source to determine the kind of information being collected. One specific car was red; "red" is the data from that one car. And red is an attribute. Thus, this collection (5 red, 8 blue, and so on) is a summary of the attribute data.

Another example of information that is deceiving is an identification number. Flight #249 and Room 168 appear to be numerical data. The 249 does not describe any property of the flight—late or on time, quality of snack served, number of passengers, or anything else about the flight. The flight number only identifies a specific flight, which might be the source of data for variables like "number of passengers on board." Driver's license numbers, Social Security numbers, and bank account numbers are all identification numbers and do not themselves serve as variables.

Remember to inspect the individual variable and an individual data, and you should have little trouble distinguishing between qualitative and quantitative variables.

EXERCISES

1.17 A drug manufacturer is interested in the proportion of persons who have hypertension (elevated blood pressure) whose condition can be controlled by a new drug the company has developed. A study involving 5000 individuals with hypertension is conducted, and it is found that 80% of the individuals are able to control their hypertension with the drug. Assuming that the 5000 individuals are representative of the group who have hypertension, answer the following questions:
   a. What is the population?
   b. What is the sample?
   c. Identify the parameter of interest.
   d. Identify the statistic and give its value.
   e. Do we know the value of the parameter?

1.18 The admissions office wants to estimate the cost of textbooks for students at our college. Let the variable x be the total cost of all textbooks purchased by a student this semester. The plan is to randomly identify 100 students and obtain their total textbook costs. The average cost for the 100 students will be used to estimate the average cost for all students.
   a. Describe the parameter the admissions office wishes to estimate.
   b. Describe the population.
   c. Describe the variable involved.
   d. Describe the sample.
   e. Describe the statistic and how you would use the 100 data collected to calculate the statistic.

1.19 A quality-control technician selects assembled parts from an assembly line and records the following information concerning each part:

   A: defective or nondefective
   B: the employee number of the individual who assembled the part
   C: the weight of the part

   a. What is the population?
   b. Is the population finite or infinite?
   c. What is the sample?
   d. Classify the responses for each of the three variables as either attribute or quantitative data.

1.20 Select ten students currently enrolled at your college and collect data for these three variables:

   X: number of courses enrolled in
   Y: total cost of textbooks and supplies for courses
   Z: method of payment used for textbooks and supplies

   a. What is the population?
   b. Is the population finite or infinite?
   c. What is the sample?
   d. Classify the responses for each of the three variables as either attribute or quantitative data.

1.21 *Parade* magazine (Nov. 13, 1994) reported the results of a survey of 372 households conducted for *Parade* by Mark Clements Research, Inc. Of those surveyed, 63% said they were familiar with the "food pyramid" developed by the U.S. Department of Agriculture.

a. Do the above results concerning the food pyramid refer to attribute or numerical data?

b. Describe the parameter of interest.

c. Describe the statistic and give its value.

 1.22 An article in *Time* magazine (Aug. 8, 1994) reports the results of a study of 6000 people published in the *Journal of the American Medical Association*. According to the study, 23% of Americans age 65 and over are using medications that are notorious for triggering insomnia, fainting spells, or even amnesia among the elderly.

a. What is the population?

b. What is the sample?

c. What is the characteristic of interest about each individual element of the population?

d. Identify the parameter of interest.

e. Identify the statistic and give its value.

1.23 Identify each of the following as examples of (1) attribute (qualitative) or (2) numerical (quantitative) variables:

a. the breaking strength of a given type of string

b. the hair color of children auditioning for the musical *Annie*

c. the number of stop signs in towns of less than 500 people

d. whether or not a faucet is defective

e. the number of questions answered correctly on a standardized test

f. the length of time required to answer a telephone call at a certain real estate office

1.24 Identify each of the following as examples of (1) attribute (qualitative) or (2) numerical (quantitative) variables:

a. a poll of registered voters as to which candidate they support

b. the length of time required for a wound to heal when using a new medicine

c. the number of telephone calls arriving at a switchboard per ten-minute period

d. the distance first-year college women can kick a football

e. the number of pages per job coming off a computer printer

f. the kind of tree used as a Christmas tree

1.25 Suppose a 12-year-old asked you to explain the difference between a sample and a population.

a. What information should your answer include?

b. What reasons would you give him or her for why one would take a sample instead of surveying every member of the population?

1.26 Suppose a 12-year-old asked you to explain the difference between a statistic and a parameter.

a. What information should your answer include?

b. What reasons would you give him or her for why one would report the value of a statistic instead of a parameter?

1.3 | MEASURABILITY AND VARIABILITY

Within a set of data, we always expect variation. If little or no variation is found, we would guess that the measuring device is not calibrated with a small enough unit. For example, we take a carton of a favorite candy bar (24) and weigh each bar individually. We observe that each of the 24 candy bars weighs 7/8 ounce, to the nearest 1/8 ounce. Does this mean that the bars are all identical in weight? Not really. Suppose that we were to weigh them on an analytical balance that weighs to the nearest ten-thousandth of an ounce. Now their weights would be variable.

Variability

It does not matter what the response variable is; there will be **variability** in the data if the tool of measurement is precise enough. One of the primary objectives in statistical analysis is that of measuring variability. For example, in the study of quality control, measuring variability is an absolute essential. Controlling (or reducing) the variability in a manufacturing process is a field all its own, statistical process control.

EXERCISES

 1.27 Suppose we measure the weights (in pounds) of the individuals in each of the following groups:

Group 1: cheerleaders for National Football League teams
Group 2: players for National Football League teams

For which group would you expect the data to have more variability? Explain why.

 1.28 Suppose you were trying to decide which of two machines to purchase. Furthermore, suppose the length to which the machines cut a particular product part was important. If both machines produced parts that had the same length on the average, what other consideration regarding the lengths would be important? Why?

1.29 Teachers use examinations to measure a student's knowledge about their subject. Explain how "a lack of variability in the resulting scores might indicate that the exam was not a very effective measuring device."

 1.30 A coin-operated coffee vending machine dispenses, on the average, 6 oz of coffee per cup. Can this statement be true of a vending machine that occasionally dispenses only enough to barely fill the cup half full (say, 4 oz)? Explain.

1.4 | DATA COLLECTION

One of the first problems a statistician faces is that of obtaining data. Data doesn't just happen, it must be collected. It is important to obtain good data since the inferences ultimately made will be based on the statistics obtained from the data. These inferences can only be as good as the data.

The collection of data for statistical analysis is an involved process and includes the following steps:

1. Defining the objectives of the survey or experiment.
    Examples: comparing the effectiveness of a new drug to the effectiveness of the standard drug; estimating the average household income in our county.
2. Defining the variable and the population of interest.
    Examples: length of recovery time for patients suffering from a particular disease; total income for households in our county.
3. Defining the data-collection and data-measuring schemes.
    This includes sampling procedures, sample size, and the data-measuring device (questionnaire, telephone, and so on).
4. Determining the appropriate descriptive or inferential data-analysis techniques.

Case Study

1-8

What Every Woman Should Know About Stockbrokers

A study by Money *magazine suggests that "stockbrokers are more courteous to women clients than men," but they "take men more seriously."*

EXERCISE 1.31

Refer to "What Every Woman Should Know About Stockbrokers."

a. What is the population of interest?

b. What information was collected?

Money magazine hired one hundred testers—50 men and 50 women—sent them into 50 randomly chosen brokerage offices in Los Angeles and Chicago. Among their findings:

▼ Sixty percent of the men were asked about their tolerance of risk in investing, against only 48% of the women.

▼ Brokers tended to explain prospective investments more fully to the men than to women. For instance, those who recommended government bond funds told 82% of their male clients how they work. Only 50% of the females got the explanation.

▼ On the other hand, brokers were less likely to allow interruptions when the prospective client was a woman.

Source: Money, June 1993.

Often an analyst is stuck with data already collected, possibly even data collected for other purposes, which makes it impossible to determine whether or not the data are "good." Collecting the data yourself using approved techniques is much preferred. Although this text will be concerned chiefly with various data-analysis techniques, you should be aware of the concerns of data collection.

The following illustration describes the population and the variable of interest for a specific investigation:

▼ ILLUSTRATION 1-2

The admissions office at our college wishes to estimate the current "average" cost of textbooks per semester, per student. The population of interest is the "currently enrolled student body," and the variable is the "total amount spent for textbooks" by each student this semester. ▲

Experiment

The two methods used to collect data are experiments and surveys. In an **experiment**, the investigator controls or modifies the environment and observes the effect on the variable under study. We often read about laboratory results obtained by using white rats to test different doses of a new medication and its effect on blood pressure. The experimental treatments were designed specifically to obtain the data needed to study the effect on the variable. In a **survey**, data are obtained by sampling some of the population of interest. The investigator, however, does not modify the environment.

Survey

Census

If every element in the population can be listed, or enumerated, and observed, then a **census** is compiled. A census is a 100% survey. A census for the population in Illustration 1-2 (p. 23) could be obtained by contacting each student on the registrar's computer printout of all registered students. However, censuses are seldom used because they are often very difficult and time-consuming to compile, and therefore very expensive. Imagine the task of compiling a census of every person who is a potential client at brokerage firms in Case Study 1-8 (p. 23). Instead of a census, a sample survey is usually conducted.

Sampling frame

When selecting a sample for a survey, it is necessary to construct a **sampling frame**.

SAMPLING FRAME

A list of the elements belonging to the population from which the sample will be drawn.

Ideally, the sampling frame should be identical to the population with every element of the population listed once and only once. In Illustration 1-2, the registrar's computer list will serve as a sampling frame for the admissions office. In this case, the census becomes the sampling frame. In other situations, a census may not be so easy to obtain, because a complete list is not available. Lists of registered voters or the telephone directory are sometimes used as sampling frames of the general public. One of these might be used in Case Study 1-6 (p. 18). Depending on the nature of the information sought, the list of registered voters or the telephone directory may or may not serve as a good sampling frame. Since only the elements in the frame have a chance to be selected as part of the sample, **it is important that the sampling frame be representative of the population**.

Sample design

Once a representative sampling frame has been established, we proceed with selecting the sample elements from the sampling frame. This selection process is called the **sample design**. There are many different types of sample designs; however, they all fit into two categories: *judgment samples* and *probability samples*.

JUDGMENT SAMPLES

Samples that are selected on the basis of being "typical."

When a judgment sample is drawn, the person selecting the sample chooses items that he or she thinks are representative of the population. The validity of the results from a judgment sample reflects the soundness of the collector's judgment.

PROBABILITY SAMPLES

Samples in which the elements to be selected are drawn on the basis of probability. Each element in a population has a certain probability of being selected as part of the sample.

One of the most common methods used to collect data is the *random sample*.

RANDOM SAMPLE

A sample selected in such a way that every element in the population has an equal probability of being chosen. Equivalently, all samples of size *n* have an equal chance of being selected. Random samples are obtained either by sampling with replacement from a finite population or by sampling without replacement from an infinite population.

Inherent in the concept of randomness is the idea that the next result (or occurrence) is not predictable. When a random sample is drawn, every effort must be made to ensure that each element has an equal probability of being selected and that the next result does not become predictable. The proper procedure for selecting a random sample is to use a random-number generator or a table of random numbers. Mistakes are frequently made because the term *random* (equal chance) is confused with *haphazard* (without pattern).

To select a simple random sample, first assign a number to each element in the sampling frame. This is usually done sequentially using the same number of digits for each element. Then go to a table of random numbers and select as many numbers with that number of digits as are needed for the sample size desired. Each numbered element in the sampling frame that corresponds to a selected random number is chosen for the sample.

▼ | ILLUSTRATION 1 - 3

Let's return to Illustration 1-2 (p. 23). Mr. Car, who works in the admissions office, has obtained a computer list of this semester's full-time enrollment. There are 4265 student names on the list. He numbered the students 0001, 0002, 0003, and so on, up to 4265; then, using four-digit random numbers, he identified a sample: 1288, 2177, 1952, 2463, 1644, 1004, and so on were selected. (See the *Statistical Tutor* for a discussion of the use of the random-number table.)

▲ |

NOTE In this text, all the statistical methods assume that random sampling has been used to collect the data.

Case Study

1-9

Kinder, Gentler U.S. Turns Sour

Public opinion polls appear in our newspapers nearly every day. Sometimes the results are newsworthy, and sometimes the results are not, but they are made into news anyway. "Kinder, Gentler U.S. Turns Sour" reports on a Times Mirror survey, and there is every reason to believe these results are newsworthy. However, next to these results are instructions for a local follow-up survey. This local follow-up survey may result in interesting numbers, but due to the method of collection, the results will be unreliable.

EXERCISE 1.32

Refer to "Kinder, Gentler U.S. Turns Sour."

a. What is the population of interest?

b. What does the statistic 69% mean?

c. Why is a "phone in" survey not a reliable survey technique?

Kinder, gentler U.S. turns sour

Americans angry, unsure financially, new poll indicates

THE NEW YORK TIMES

WASHINGTON—As they turn inward and worry about their own financial difficulties, American voters have become less **NATION** compassionate about the problems of the poor and minorities, a study of public opinion shows.

The finding was part of a wide-ranging study, conducted by the Times Mirror Center for The People & The Press and made public yesterday, that found an electorate that is "angry, self-absorbed and politically" unanchored.

The poll also found a striking decline in public support for social welfare programs. Fifty-seven percent said it was the responsibility of government to take care of people who cannot take care of themselves, down from 69 percent in 1992 and 71 percent in 1987.

For this survey, 3,800 adults nationwide were interviewed by telephone from July 12 to July 25.

What do you think?

One of the quesions in a new poll by the Times Mirror Center for The People & The Press asked, "Do you think the government should take care of people who cannot take care of themselves?" What do you think? To answer:

YES, call:
(716) 777-3055

NO, call:
(716) 777-3056

You may call from a Touch-Tone, rotary or TDD phone. Calls will be taken until 9 p.m. This is only a sampling of reader opinion and is not a scientific survey. Results will be in tomorrow's *Democrat and Chronicle* and *Times-Union*.

EXERCISES

1.34   a. What is a sampling frame?

b. What did Mr. Car use for a sampling frame in Illustration 1-3?

1.35   Consider a simple population consisting of only the numbers 1, 2, and 3 (an unlimited number of each). There are nine different samples of size two that could be drawn from this population: (1, 1), (1, 2), (1, 3), (2, 1), (2, 2), (2, 3), (3, 1), (3, 2), (3, 3).

a. Explain why the above list of samples represents all possible random samples of size two that can be randomly selected from the population of {1, 2, 3}.

b. If the population consists of the numbers 1, 2, 3, and 4, list all the samples of size two that could possibly be selected.

c. If the population consists of the numbers 1, 2, and 3, list all the samples of size three that could possibly be selected.

d. If the population consists of the numbers 1, 2, 3, and 4, list all the samples of size three that could possibly be selected.

Case Study

1-10

People Have to Tell the Truth or Statistics Don't Mean Anything

Statistical results are only as good as the sample information collected is accurate, as so well described by David Hoff in "People Have to Tell the Truth or Statistics Don't Mean Anything."

I read a story in a local newspaper the other day from The Associated Press about a poll of married men and women. . . . Nine of 10 people surveyed claimed they had never been unfaithful, and four of five said they would marry the same person again.

The next day I read another story, this one in *The Wall Street Journal*, about the heavyweight of greeting cards, Hallmark. The story said Hallmark was getting into divorce announcements in a big way in 1990 because, according to company literature, "about 50 percent of all marriages end in divorce—and people need a way to communicate about it."

A week later I read a story in *Parade* magazine that stated "two out of three married men today reportedly commit adultery." The story also said that in a 1986 survey of married women age 25 to 50, "a whopping 41 percent of these wives said they were cheating or had cheated on their husbands in the past."

So, I ask, what can you make of this? One story says nine out of 10 people claim they have never been unfaithful, another says 41 percent of the women and 66 percent of the men are cheating. One story says four out of five spouses would marry the same mate again, and the other says divorce is so prevalent, it's a business opportunity. Whom do you believe?

I'll go with Hallmark.

A glaring problem with polls is that to be accurate or close to accurate, those surveyed must tell the truth.

In the poll of married people, three out of four respondents said their spouse was their best friend. That's 75 percent. But the divorce rate is 50 percent. Either 25 percent are lying, or they divorce their best friend. Also, 64 percent said their marriage was "very happy," and 71 percent said "they're trying very hard to improve their marriage." If Hallmark believed those numbers, it wouldn't be jumping into the divorce card business.

So credibility in polls—at least polls about our love life—is as stable as a teen-age romance, but that doesn't mean we'll stop reporting them.

They are not without some flicker of reality, and they do make fun reading.

Source: David Hoff, "People Have to Tell the Truth or Statistics Don't Mean Anything," *The New York Times.*

EXERCISE 1.33

Describe two polling techniques that could easily result in very inaccurate statistics.

 1.36 A wholesale food distributor in a large metropolitan area would like to test the demand for a new food product. He distributes food through five large supermarket chains. The food distributor selects a sample of stores located in areas where he believes the shoppers are receptive to trying new products. What type of sampling does this represent?

1.37 A random sample could be very difficult to obtain. Why?

1.38 Why is the random sample so important in statistics?

1.39 The Design, Sample, and Method section of an article titled "Making Behavior Changes After a Myocardial Infarction" (*Western Journal of Nursing Research*, Aug. 1993) discusses the selection of 16 informants who constituted 8 family dyads. The article states that "to initiate contact with informants, names of persons who met the criteria were obtained from the medical records of a cardiac rehabilitation center in central Texas. Potential informants were then contacted by telephone to obtain preliminary consent. Confidentiality and anonymity of inform-ants were ensured by coding the data to identify informants and link dyads."

   a. Is this a judgment sample or a probability sample?

   b. Is it appropriate to perform statistical inference using this sample? Justify your answer.

1.40 An article titled "Surface Sampling in Gravel Streams" (*Journal of Hydraulic Engineering*, April 1993) discusses grid sampling and areal sampling. Grid sampling involves the removal by hand of stones found at specific points. These points are established on the gravel surface through the use of a wire mesh or by using predetermined distances on a survey tape. The material collected by grid sampling is usually analyzed as a frequency distribution by number. An areal sample is collected by removing all the particles found in a predetermined area of channel bed. The material recovered is most often analyzed as a frequency distribution by weight. Would you categorize these sample designs as judgment samples or probability samples?

1.41 The telephone book might not be a representative sampling frame. Explain why.

1.42 The election board's voter registration list is not a census of the adult population. Explain why.

1.5 | COMPARISON OF PROBABILITY AND STATISTICS

Probability and statistics are two separate but related fields of mathematics. It has been said that "probability is the vehicle of statistics." That is, if it were not for the laws of probability, the theory of statistics would not be possible.

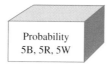

EXERCISE 1.43

Which of the following illustrates probability? statistics? A—How likely is heads to occur when a coin is tossed? B—The weights of 35 babies are studied in an attempt to estimate gain in the first month after birth.

Let's illustrate the relationship and the difference between these two branches of mathematics by looking at two boxes. We know the probability box contains five blue, five red, and five white poker chips. Probability tries to answer questions such as "If one chip is drawn from this box, what is the chance that it is blue?" On the other hand, in the statistics box we don't know what the combination of chips is. We draw a sample and, based on the findings in the sample, make conjectures about what we believe to be in the box. Note the difference: Probability asks about the chance that something specific (a sample) will happen when you know the possibilities (that is, you know the population). Statistics, on the other hand, asks you to draw a sample, describe the sample (descriptive statistics), and then make inferences about the population based on the information found in the sample (inferential statistics).

▼

EXERCISES

1.44   Classify each of the following as a probability or a statistics problem.
   a. determining whether a new drug shortens the recovery time from a certain illness.
   b. determining the chance that heads will result when a coin is flipped
   c. determining the amount of waiting time required to check out at a certain grocery store
   d. determining the chance that you will be dealt a "blackjack"

1.45   Classify each of the following as a probability or a statistics problem.
   a. determining how long it takes to handle a typical telephone inquiry at a real estate office
   b. determining the length of life for the 100-watt light bulbs a company produces
   c. determining the chance that a blue ball will be drawn from a bowl that contains 15 balls, of which 5 are blue
   d. determining the shearing strength of the rivets that your company just purchased for building airplanes

1.6 | STATISTICS AND THE COMPUTER

In recent years, the computer has had a tremendous effect on almost every aspect of life. The field of statistics is no exception. As you will see, the field of statistics uses many techniques that are repetitive in nature: calculations of numerical statistics, procedures for constructing graphic displays of data, and procedures that are followed to formulate statistical inferences. The computer is good at performing these sometimes long and tedious operations. If that computer has one of the standard statistical packages on line, it will make the analysis easy to perform.

Your local computer center can provide you with a list of what is available. Some of the more readily available packaged programs are: MINITAB, SYSTAT, STATA, SAS (Statistical Analysis System), Statgraphics, and SPSS (Statistical Package for the Social Sciences). The examples of computer output you will see throughout this textbook will all be MINITAB generated.

There is a great temptation to use the computer to analyze any and all sets of data and then treat the results as though the statistics are correct. Remember the old adage "Garbage in, garbage out!" **Responsible use of statistical methodology is very important. The burden is on the user to ensure that the appropriate methods are correctly applied and that accurate conclusions are drawn and communicated to others.**

Throughout this textbook, as statistical procedures are studied, you will find the information needed to have a computer complete the same procedures using the MINITAB software. Specific details on use of computers available to you needs to be obtained from your instructor or from your local computer lab person. Specific details about the use of MINITAB are available by using the Help system in the MINITAB software.

An explanation of the typographical conventions that will be used in this textbook is given below.

Session Commands

▼ **Boldface type**: indicates letters, numbers, or symbols you must enter using the keyboard.

Most session commands are simple, easy-to-remember 'words' like PLOT, LET, or SORT. The session commands require the use of the first four (sometimes three) letters followed by arguments. The letters of the command name that must be used appear in boldface **CAPS**; letters in lowercase are optional. Arguments can be columns (C), constants (K), names, etc. MINITAB is not case sensitive, so you may type in either upper- or lowercase letters. You type in the commands at the **MTB >** prompt and subcommands at the **SUB >** prompt. When the next line is to be a subcommand, a command line must end with a 'semicolon.' The last subcommand must end with a 'period.'

Menu Commands

▼ Choose: tells you to make a menu selection by using either a keyboard entry or a mouse "point and click" entry.
For example: Choose: **Stat > SPC > Pareto Chart** instructs you to open the Stat Menu, choose SPC, and then choose Pareto Chart.

For Both Session and Menu Commands

▼ Type: instructs you to enter using the keyboard. It is important to type this information exactly as shown, including punctuation and spaces.
▼ Press: indicates you should press the specified key.
▼ Hold: indicates you are to hold down the specified key while you press another key; then release both keys.

EXERCISES

1.46 How have computers increased the usefulness of statistics to professionals such as researchers, government workers who analyze data, statistical consultants, and so on?

1.47 How might computers help you in statistics?

IN RETROSPECT

You should now have a general feeling of what statistics is about—an image that will grow and change as you work your way through this book. You know what a sample and a population are, the distinction between qualitative (attribute) and quantitative (numerical) variables. You even know the difference between statistics and probability (although we will not study probability in detail until Chapter 4). You should also have an appreciation for and a partial understanding of how important random samples are in statistics.

Throughout the chapter you have seen numerous articles that represent various aspects of statistics. The USA SNAPSHOTS® picture a variety of information about ourselves as we describe ourselves and other aspects of the world around us. Statistics can be entertaining—for example, the statistical snapshot "And Then There's 'Entertainment Statistics.'" Statistics are everywhere: on the news ("stock market was up . . ."), in commercials ("four out of five doctors say . . ."). The examples are endless. Look around and find some examples of statistics in your daily life (see Exercises 1.61 and 1.62, p. 33).

CHAPTER EXERCISES

 1.48 "Describing Ourselves Statistically," the chapter case study on page 5, contains several statistics that describe the life of the average American.
- a. In general, what is the statistical population?
- b. Do the measures reported (98.3% have TV, 165 grams of fat, $37,971) appear to be parameters or statistics? Why? What is the difference?
- c. What variable was used to collect the data related to the "average bank robbery got $3325"?
- d. What kind of variable was used to collect data when "98.3% of households" was found?
- e. Does your household have a cable hook-up for your TV? How would your answer have been used to calculate the 61% reported?

1.49 We want to describe the so-called typical student at your college. Describe a variable that measures some characteristic of a student and results in
- a. attribute data   b. numerical data

 1.50 A candidate for a political office claims that he will win the election. A poll is conducted and 35 of 150 voters indicate that they will vote for the candidate, 100 voters indicate that they will vote for his opponent, and 15 voters are undecided.
- a. What is the population parameter of interest?
- b. What is the value of the sample statistic that might be used to estimate the population parameter?
- c. Would you tend to believe the candidate based on the results of the poll?

 1.51 A researcher studying consumer buying habits asks every 20th person entering Publix Supermarket how many times per week he or she goes grocery shopping. She then records the answer as T.
- a. Is $T = 3$ an example of (1) a sample, (2) a variable, (3) a statistic, (4) a parameter, or (5) a piece of data?

Suppose the researcher questions 427 shoppers during the survey.
- b. Give an example of a question that can be answered using the tools of descriptive statistics.
- c. Give an example of a question that can be answered using the tools of inferential statistics.

1.52   A researcher studying the attitudes of parents of preschool children interviews a random sample of 50 mothers, each having one preschool child. He asks each mother, "How many times did you compliment your child yesterday?" He records the answer as *C*.

    a. Is *C* (1) a piece of data, (2) a statistic, (3) a parameter, (4) a variable, or (5) a sample?

    b. Give an example of a question that can be answered using the tools of descriptive statistics.

    c. Give an example of a question that can be answered using the tools of inferential statistics.

1.53   The August 29/September 5, 1994 issue of *U.S. News & World Report* references a study by health economists at the University of Southern California that indicated that Alzheimer's disease cost the nation $82.7 billion a year in medical expenses and lost productivity. Patients' earning loss was $22 billion, the value of time of unpaid caregivers was $35 billion, and the cost of paid care was $24 billion.

    a. What is the population?

    b. What is the response variable?

    c. What is the parameter?

    d. What is the statistic?

1.54   A USA Snapshot® from *USA Today* (Nov. 1, 1994) described the greatest sources of stress in starting a company. According to the snapshot, the CEOs of *Inc.* magazine's 500 fastest-growing private companies gave the following responses: 50% said company finances, 23% said the need to succeed, 10% said time commitments, 9% said personal relationships, and 8% were classified as "other." Would the data from this article be classified as qualitative or quantitative?

1.55   An article titled "Want a Job in Food?" found in *Parade* magazine (Nov. 13, 1994) references a recent study at the University of California involving 2000 young men. The study found that in 2000 young men who did not go to college, of those who took restaurant jobs (typically as fast-food counter workers), one in two reached a higher-level blue-collar job and one in four reached a managerial position within four years.

    a. What is the population?

    b. What is the sample?

    c. Is this a judgment sample or a probability sample?

1.56   The June 1994 issue of *Good Housekeeping* reported on a new study on rape. The study found that women who screamed, bit, kicked, or ran were more likely to avoid rape than women who tried pleading, crying, or offered no resistance, and they were no more apt to be injured. The authors, however, cautioned that the study could not be interpreted as proof that all women should forcefully resist. The study involved 150 Omaha, Nebraska, police reports of rape or attempted rape.

    a. Are the data in this study attribute or numerical?

    b. Is this a judgment or a probability sample?

1.57   The USA Snapshot® "Room service, please" (p. 33) presents two statistics, 11% and 31%.

    a. What population(s) was being surveyed?

    b. Can you conclude that more women than men use room service? Explain.

    c. Name three possible reasons that might explain the large difference in the two percentages.

    d. What conclusion(s) can be justified from the evidence presented?

USA SNAPSHOTS®
A look at statistics that shape your finances

Room service, please
Business travelers who
use room service:

31%

Women

11%

Men

Note: Based
on survey of
60 women
and 57 men

Source: Novotel New York   By Patti Stang and Suzy Parker, USA TODAY

USA SNAPSHOTS®
A look at statistics that shape our lives

Weighing in
What percentage of U.S.
adults fall within their
recommended weight range?

A. **73%**

B. **41%**

C. **19%**

Source: Princeton
Survey Research
Associates/
Prevention Index

Answer **C**

By Dierdre Schwiesow and Bob Laird, USA TODAY

1.58 The USA Snapshot® "Weighing in" asks the reader to estimate the percentage of adults whose weight falls within the recommended range.
  a. What variable is being studied? Describe it.
  b. Is the variable qualitative or quantitative? Explain the difference.
  c. How was the value of the statistic determined?

1.59 Describe, in your own words, and give an example of each of the following terms. Your examples should not be ones given in class or in the textbook.
  a. variable   b. data   c. sample
  d. population   e. statistic   f. parameter

1.60 Describe, in your own words, and give an example of the following terms. Your examples should not be ones given in class or in the textbook.
  a. random sample   b. probability sample   c. judgment sample

1.61 Find an article or an advertisement in a newspaper or a magazine that exemplifies the use of statistics.
  a. Identify and describe one statistic reported in the article.
  b. Identify and describe the variable related to the statistic in (a).
  c. Identify and describe the sample related to the statistic in (a).
  d. Identify and describe the population from which the sample in (c) was taken.

1.62 a. Find an article in a newspaper or a magazine that exemplifies the use of statistics in a way that might be considered "entertainment" or "recreational." Describe why you think this article fits one of these categories.
  b. Find an article in a newspaper or a magazine that exemplifies the use of statistics and is presenting an unusual finding as the result of a study. Describe why these results are "newsworthy."

VOCABULARY LIST

Be able to define each term. Pay special attention to the key terms, which are printed in red. In addition, describe in your own words, and give an example of, each term. Your examples should not be ones given in class or in the textbook. Page numbers indicate the first appearance of a term.

attribute data (p. 17)
census (p. 24)
data (p. 15)
descriptive statistics (p. 6)
experiment (p. 16)
finite population (p. 15)
inferential statistics (p. 6)
infinite population (p. 15)
judgment sample (p. 24)
numerical data (p. 17)
parameter (p. 16)
population (p. 14)
probability (p. 28)

probability sample (p. 25)
qualitative variable (p. 17)
quantitative variable (p. 17)
random (p. 25)
random sample (p. 25)
representative sampling frame (p. 24)
sample (p. 15)
sampling frame (p. 24)
statistic (p. 16)
statistics (p. 6)
variable (p. 15)
variability (p. 22)

QUIZ A

Answer "True" if the statement is always true. If the statement is not always true, replace the words shown in bold with words that make the statement always true.

1.1   **Inferential** statistics is the study and description of data that result from an experiment.

1.2   **Descriptive statistics** is the study of a sample that enables us to make projections or estimates about the population from which the sample was drawn.

1.3   A **population** is typically a very large collection of individuals or objects about which we desire information.

1.4   A statistic is the calculated measure of some characteristic of a **population**.

1.5   A parameter is the measure of some characteristic of a **sample**.

1.6   As a result of surveying 50 freshmen, it was found that 16 had participated in interscholastic sports, 23 had served as officers of classes and clubs, and 18 had been in school plays during their high school years. This is an example of **numerical data**.

1.7   The "number of rotten apples per shipping crate" is an example of a **qualitative** variable.

1.8   The "thickness of a sheet of sheet metal" used in a manufacturing process is an illustration of a **quantitative** variable.

1.9   A **representative** sample is a sample obtained in such a way that all individuals had an equal chance to be selected.

1.10   The basic objective of **statistics** is that of obtaining a sample, inspecting this sample, and then making inferences about the unknown characteristics of the population from which the sample was drawn.

Quiz B

The owners of Corner Convenience Stores are concerned about the quality of service their customers receive. In order to study the service received, they collected samples for each of several variables.

1.1   Classify each of the following variables as being (A) qualitative, (B) quantitative, or (C) not a variable:
　　　　a. method of payment for purchases (cash, credit card, check)
　　　　b. ZIP code for the customer's home mailing address
　　　　c. amount of sales tax on purchase
　　　　d. number of items purchased
　　　　e. customer's driver's-license number

1.2   The mean checkout time for all customers at Corner Convenience Stores is to be estimated by using the mean checkout time required by 75 randomly selected customers. Match the items in column 2 with the statistical terms in column 1.

	1	2
___	data (one)	(a) the 75 customers
___	data (set)	(b) the mean time for all customers
___	experiment	(c) two minutes, one customer's checkout time
___	parameter	(d) the mean time for the 75 customers
___	population	(e) all customers at Corner Convenience
___	sample	(f) the checkout time for one customer
___	statistic	(g) the 75 times
___	variable	(h) the process used to select the 75 customers and measure their times

Quiz C

Write a brief paragraph in response to each question.

1.1   The *population* and the *sample* are both sets of objects. Describe the relationship between them and give an example.

1.2   The *variable* and the *data* for a specific situation are closely related. Explain this relationship and give an example.

1.3   The *data*, the *statistic*, and the *parameter* are all values used to describe a statistical situation. How does one distinguish among these three terms? Give an example.

1.4   What conditions are required in order for a sample to be a random sample?

2

DESCRIPTIVE ANALYSIS
AND PRESENTATION OF
SINGLE-VARIABLE DATA

MIDDLE CLASS NOT EASILY DEFINED

Have you ever wondered about the amount of annual income the people around you make? You say you know of someone who must make close to $100,000; this person owns several cars, a boat, a very nice house in a wealthier neighborhood and describes himself as "middle class." You also know someone who maybe makes $20,000 annually; this person says, "I don't have an abundance of money, but I'm a self-provider." Many would describe both as middle class even though they're much different. Such descriptions of the USA's 68.5 million families appeared in USA Today, *December 14, 1994.*

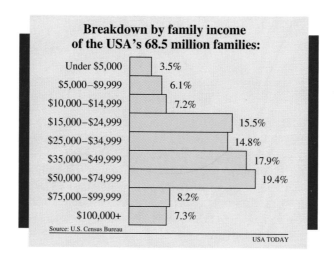

Breakdown by family income of the USA's 68.5 million families:

Income	Percent
Under $5,000	3.5%
$5,000–$9,999	6.1%
$10,000–$14,999	7.2%
$15,000–$24,999	15.5%
$25,000–$34,999	14.8%
$35,000–$49,999	17.9%
$50,000–$74,999	19.4%
$75,000–$99,999	8.2%
$100,000+	7.3%

Source: U.S. Census Bureau

USA TODAY

Below is a sample of 50 family annual incomes, to the nearest thousand dollars, for 1994.

23	29	8	51	87	58	41	37	17	12	34	65	46
16	25	39	53	30	19	3	52	70	101	66	73	26
15	12	42	105	98	22	14	6	16	70	76	36	35
112	18	31	48	4	35	9	45	80	10	52		

How would you summarize this data so that the information contained within the data can best be understood?

We will learn various methods of describing data in this chapter and will return to this case study at the end of the chapter. (See Exercise 2.165, p. 121.)

● CHAPTER OBJECTIVES

Imagine that you took an exam at the last meeting of your favorite class. Today your instructor returns your exam paper and it has a grade of 78. If you are like most students in this situation, as soon as you see your grade you want to know how your grade compares to those of the rest of the class and you immediately ask: "What was the average exam grade?" Your instructor replies, "The class average was 68." Since 78 is 10 points above the average, you ask: "How close to the top is my grade?" Your instructor replies that the grades ranged from 42 to 87 points. The accompanying figure summarizes the information we have so far.

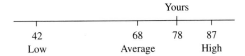

A third question that is sometimes asked is "How are the grades distributed?" Your instructor replies that half the class had grades between 65 and 75. With this information you conclude that your grade is fairly good.

The preceding illustration demonstrates the basic process of statistics and how students make use of statistics on a regular basis: There is a set of data (set of exam scores), the data is described with a few descriptive statistics (average grade and high and low grades), and based on this information students are able to draw conclusions about their relative success.

A large part of this chapter is devoted to learning how to present and describe sets of data. There are four basic types of descriptive statistics: (1) measures of central tendency, (2) measures of dispersion (spread), (3) measures of position, and (4) types of distribution. The measures of central tendency are used to describe or locate the "middle" of the data values. The measures of dispersion measure the spread or variability within the data values. When very large sets of data are involved, the measures of position become useful. For example, when college board exams are taken, thousands of scores result. Your concern is the position of your score with respect to all the others. Averages, ranges, and so on, are not enough. A measure of position will tell you that your score is better than a certain percentage of the other scores. For example, your score is better than 85% of all the scores. The fourth concept, the type of distribution, describes the data by telling you whether the values are evenly distributed or clustered (bunched) around a certain value. Another very important part of this chapter will explain how to interpret the findings so that we know what the data is telling us about the sampled population.

In this chapter, we deal with single-variable data, which is data collected for one numerical variable at a time. In Chapter 3 we will learn to work with two variables at a time and learn about their relationship.

GRAPHIC PRESENTATION OF DATA

2.1 | GRAPHS, PARETO DIAGRAMS, AND STEM-AND-LEAF DISPLAYS

Once the sample data has been collected, we must "get acquainted" with it. One of the most helpful ways to become acquainted is to use an initial exploratory data-analysis technique that will result in a pictorial representation of the data. The resulting displays visually reveal patterns of behavior of the variable being studied. There are several graphic (pictorial) ways to describe data. The method used is determined by the type of data and the idea to be presented.

NOTE   There is no single correct answer when constructing a graphic display. The analyst's judgment and the circumstances surrounding the problem play a major role in the development of the graphic.

Qualitative Data

CIRCLE GRAPHS AND BAR GRAPHS

Graphs that are used to summarize attribute data. **Circle graphs** (pie diagrams) show the amount of data that belong to each category as a proportional part of a circle. **Bar graphs** show the amount of data that belong to each category as proportionally sized rectangular areas.

▼ | ILLUSTRATION   2 - 1

Table 2-1 lists the number of cases of each type of operation performed at General Hospital last year.

TABLE 2-1

Operations Performed at General Hospital Last Year

Type of Operation	Number of Cases
Thoracic	20
Bones and joints	45
Eye, ear, nose, and throat	58
General	98
Abdominal	115
Urologic	74
Proctologic	65
Neurosurgery	23
Total	498

FIGURE 2-1

Circle Graph

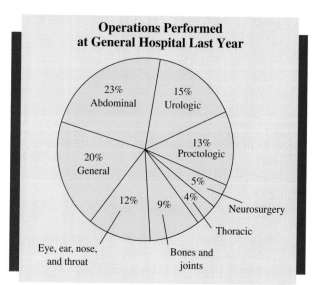

These data are displayed on a circle graph in Figure 2-1 with each type of operation represented by a relative proportion of the circle and reported in a percentage. Figure 2-2 displays the same "type of operation" data but in the form of a bar graph.

FIGURE 2-2

Bar Graph

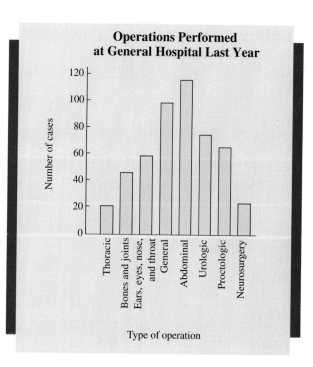

When the bar graph is presented in the form of a **Pareto diagram**, it presents additional and very helpful information.

EXERCISE   2.3

In your opinion, does the circle graph (Exercise 2.1) or the bar graph (Exercise 2.2) result in a better representation of the information? Explain.

PARETO DIAGRAM

A bar graph with the bars arranged from the most numerous category to the least numerous category. It includes a line graph displaying the cumulative percentages and counts for the bars.

▼ ILLUSTRATION   2 - 2

The FBI reported (*USA Today*, 6-29-94) the number of hate crimes by category for 1993. The Pareto diagram in Figure 2-3 shows the 6746 categorized hate crimes, their percentages, and cumulative percentages.

FIGURE 2-3

Pareto Diagram

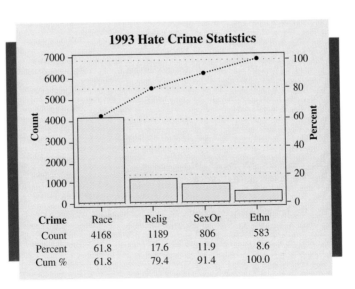

EXERCISE   2.4

A shirt inspector at a clothing factory categorized the last 500 defects as: 67—missing button, 153—bad seam, 258—improperly sized, 22—fabric flaw. Construct a Pareto diagram for this information.

1993 Hate Crime Statistics

Crime	Race	Relig	SexOr	Ethn
Count	4168	1189	806	583
Percent	61.8	17.6	11.9	8.6
Cum %	61.8	79.4	91.4	100.0

▲ |

The Pareto diagram is popular in quality-control applications. A Pareto diagram showing types of defects will show the types of defects that have the greatest effect on the defective rate in order of effect. It will be easy to see which defects should be targeted in order to most effectively lower the defective rate.

MINITAB (Release 10) commands to construct a Pareto diagram with the categories listed in C1 and their corresponding counts or frequencies listed in C2.

Session commands   *Menu commands*

```
Enter:                        Choose: Stat > SPC > Pareto Chart
  %PARETO C1;                 Select: Chart defect table
  COUNts C2;                  Enter:  Label in C1
  TITLe 'your title'.                 Frequency in C2
                                      Your Title
```

Quantitative Data

One major reason for constructing a graph of quantitative data is to display its distribution.

DISTRIBUTION

The pattern of variability displayed by the data of a variable. The distribution displays the frequency of each value of the variable.

DOTPLOT DISPLAY

Displays the data of a sample by representing each piece of data with a dot positioned along a scale. This scale can be either horizontal or vertical. The frequency of the values is represented along the other scale.

▼| ILLUSTRATION 2-3

A sample of 19 exam grades was randomly selected from a large class:

76	74	82	96	66	76	78	72	52	68
86	84	62	76	78	92	82	74	88	

Figure 2-4 is a dotplot of the 19 exam scores.

FIGURE 2-4

Dotplot

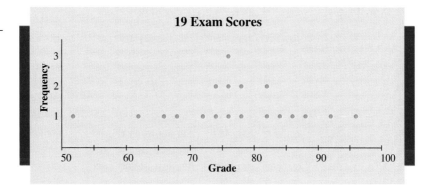

▲| Notice how the data are "bunched" near the center and more "spread out" near the extremes.

EXERCISE 2.5

The number of points scored during each game by a high school basketball team last season was:

56 54 61 71 46 61
55 68 60 66 54
61 52 36 64 51.
Construct a dotplot of these data.

The dotplot display is a convenient technique to use as you first begin to analyze the data. It results in a picture of the data as well as sorts the data into numerical order. (To *sort* data is to list the data in rank order according to numerical value.)

MINITAB (Release 10) commands to construct a dotplot with the data listed in C1.

Session commands   *Menu commands*

Enter: DOTPlot C1   Choose: Graph > Char. Graph > Dotplots
                           Enter: Variable in C1

In recent years a technique known as the **stem-and-leaf display** has become very popular for summarizing numerical data. It is a combination of a graphic technique and a sorting technique. These displays are simple to create and use, and are well suited to computer applications.

STEM-AND-LEAF DISPLAY

Pictures the data of a sample using the actual digits that make up the data values. Each numerical data is divided into two parts: The leading digit(s) becomes the *stem*, and the trailing digit(s) becomes the *leaf*. The stems are located along the main axis, and a leaf for each piece of data is located so as to display the distribution of the data.

▼| ILLUSTRATION   2-4

Let's construct a stem-and-leaf display for the 19 exam scores given in Illustration 2-3.

76	74	82	96	66	76	78	72	52	68
86	84	62	76	78	92	82	74	88	

At a quick glance we see that there are scores in the 50s, 60s, 70s, 80s, and 90s. Let's use the first digit of each score as the stem and the second digit as the leaf. Typically, the display is constructed in a vertical position. Draw a vertical line and place the stems, in order, to the left of the line.

```
5 |
6 |
7 |
8 |
9 |
```

Next we place each leaf on its stem. This is done by placing the trailing digit on the right side of the vertical line opposite its corresponding leading digit. Our first data value is 76; 7 is the stem and 6 is the leaf. Thus, we place a 6 opposite the stem 7.

```
7 | 6
```

The next data value is 74, so a leaf of 4 is placed on the 7 stem next to the 6.

```
7 | 6 4
```

The next data value is 82, so a leaf of 2 is placed on the 8 stem.

```
7 | 6 4
8 | 2
```

We continue until each of the other 16 leaves is placed on the display. Figure 2-5 shows the resulting stem-and-leaf display.

EXERCISE   2.6

What stem value and what leaf value will be used to represent the next piece of data, 96? 66?

FIGURE 2-5

Stem-and-Leaf

19 Exam Scores

```
5 | 2
6 | 6 8 2
7 | 6 4 6 8 2 6 8 4
8 | 2 6 4 2 8
9 | 6 2
```

From Figure 2-5, we see that the grades are centered around the 70s. In this case, all scores with the same tens digit were placed on the same branch, but this may not always be desired. Suppose we reconstruct the display; this time instead of grouping ten possible values on each stem, let's group the values so that only five possible values could fall on each stem. Do you notice a difference in the appearance of Figure 2-6?

FIGURE 2-6

Stem-and-Leaf

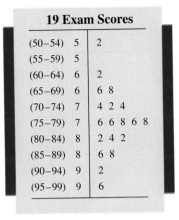

19 Exam Scores		
(50–54)	5	2
(55–59)	5	
(60–64)	6	2
(65–69)	6	6 8
(70–74)	7	4 2 4
(75–79)	7	6 6 8 6 8
(80–84)	8	2 4 2
(85–89)	8	6 8
(90–94)	9	2
(95–99)	9	6

▲ | The general shape is approximately symmetrical about the high 70s. Our information is a little more refined, but basically we see the same distribution.

It is fairly typical of many variables to display a distribution that is concentrated (mounded) about a central value and then in some manner be dispersed in both directions. Often a graphic display reveals something that the anaylst may or may not have anticipated. Illustration 2-5 demonstrates what generally occurs when two populations are sampled together.

▼ | ILLUSTRATION 2-5

A random sample of 50 college students was selected. Their weights were obtained from their medical records. The resulting data are listed in Table 2-2.

TABLE 2-2

Weights of 50 College Students

EXERCISE 2.7

Construct a stem-and-leaf display of the number of points scored during each basketball game last season: 56 54 61 71 46 61 55 68 60 66 54 61 52 36 64 51.

Student	1	2	3	4	5	6	7	8	9	10
Male/Female	F	M	F	M	M	F	F	M	M	F
Weight	98	150	108	158	162	112	118	167	170	120
Student	11	12	13	14	15	16	17	18	19	20
Male/Female	M	M	M	F	F	M	F	M	M	F
Weight	177	186	191	128	135	195	137	205	190	120
Student	21	22	23	24	25	26	27	28	29	30
Male/Female	M	M	F	M	F	F	M	M	M	M
Weight	188	176	118	168	115	115	162	157	154	148
Student	31	32	33	34	35	36	37	38	39	40
Male/Female	F	M	M	F	M	F	M	F	M	M
Weight	101	143	145	108	155	110	154	116	161	165
Student	41	42	43	44	45	46	47	48	49	50
Male/Female	F	M	F	M	M	F	F	M	M	M
Weight	142	184	120	170	195	132	129	215	176	183

This data set is on your Data Disk.

Notice that the weights range from 98 to 215 pounds. Let's group the weights on stems of ten units using the hundreds and the tens digits as stems and the units digit as the leaf. See Figure 2-7. The leaves have been arranged in numerical order.

FIGURE 2-7

Stem-and-Leaf

**Weights of
50 College Students (lb)
Stem-and-Leaf of WEIGHT
N = 50   Leaf Unit = 1.0**

9	8
10	1 8 8
11	0 2 5 5 6 8 8
12	0 0 0 8 9
13	2 5 7
14	2 3 5 8
15	0 4 4 5 7 8
16	1 2 2 5 7 8
17	0 0 6 6 7
18	3 4 6 8
19	0 1 5 5
20	5
21	5

MINITAB (Release 10) commands to construct a stem-and-leaf diagram with the data listed in C1.

Session commands   *Menu commands*

Enter: STEM C1;   Choose: Graph > Char.Graph > Stem
      INCRement 10.   Enter: Variable C1
                           Increment 10

EXERCISE 2.8

What do you think "leaf unit = 1.0" means in Figure 2-7?

Close inspection of Figure 2-7 suggests that two overlapping distributions may be involved. That is exactly what we have: a distribution of female weights and a distribution of male weights. Figure 2-8 shows a "back-to-back" stem-and-leaf display of this set of data and makes it obvious that two distinct distributions are involved.

FIGURE 2-8

"Back-to-Back"
Stem-and-Leaf

Female		Male
Weights of 50 College Students (lb)		
Female		Male
8	09	
1 8 8	10	
0 2 5 5 6 8 8	11	
0 0 0 8 9	12	
2 5 7	13	
2	14	3 5 8
	15	0 4 4 5 7 8
	16	1 2 2 5 7 8
	17	0 0 6 6 7
	18	3 4 6 8
	19	0 1 5 5
	20	5
	21	5

Figure 2-9, a "side-by-side" dotplot of the same 50 weight data, shows the same distinction between the two subsets.

FIGURE 2-9

Dotplots with
Common Scale

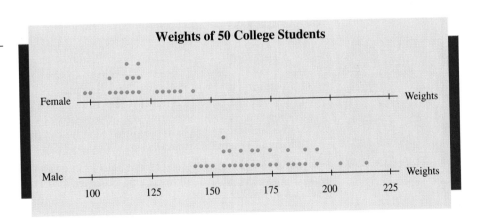

Based on the information shown in Figures 2-8 and 2-9, and on what we know about people's weight, it seems reasonable to conclude that female college students weigh less than male college students. Situations involving more than one set of data are discussed further in Chapter 3.

MINITAB (Release 10) commands to construct a multiple dotplot, one above the other, with the data listed in C1 and their corresponding categories listed in C2.

Session commands   *Menu commands*

Enter: DOTPlotC1;   Choose: **Graph > Char. Graph > Dotplots**
       SAME scale;   Enter: Variable in C1
       BY C2.   By variable C2
                                  Select: Same

EXERCISES

2.9   A computer-anxiety questionnaire was given to 200 students in a course in which computers are used. One of the questions was "I enjoy using computers." The responses to this particular question were

Response	Number
Strongly agree	50
Agree	75
Slightly agree	25
Slightly disagree	15
Disagree	15
Strongly disagree	20

  a. Construct a bar graph that shows these responses.
  b. Calculate the percentage for each response, and construct a circle graph for these responses.
  c. Compare the two graphs you constructed in (a) and (b); which one seems to be the most informative? Explain why.

2.10 A sample of student-owned General Motors automobiles was identified and the make of each noted. The resulting sample follows (Ch = Chevrolet, P = Pontiac, O = Oldsmobile, B = Buick, Ca = Cadillac):

Ch	B	Ch	P	Ch	O	B	Ch	Ca	Ch
B	Ca	P	O	P	P	Ch	P	O	O
Ch	B	Ch	B	Ch	P	O	Ca	P	Ch
O	Ch	Ch	B	P	Ch	Ca	O	Ch	B
B	O	Ch	Ch	O	Ch	Ch	B	Ch	B

  a. Find the number of cars of each make in the sample.
  b. What percentage of these cars were Chevrolets? Pontiacs? Oldsmobiles? Buicks? Cadillacs?
  c. Draw a bar graph showing the percentages found in (b).

2.11 The June 13, 1994, issue of *Time* magazine gives the results of a poll of 600 adult Americans concerning their interest in soccer. The results are as follows:

How interested are you in soccer?	
Very interested	9%
Somewhat interested	23%
Not very interested	20%
Not interested at all	46%
Not sure	2%

  a. Change the percents to frequencies.
  b. Using the percentages, construct a circle graph for the above poll.
  c. Using the frequencies, construct a bar graph for the above poll.

2.12 According to a USA Snapshot® in *USA Today*, 10-7-94, 54% of the parents found the neighborhood they live in very safe for kids. Thirty-eight percent found their neighborhood somewhat safe, 5% found their neighborhood not too safe, and 2% found their neighborhood not safe at all. Since the percentages add up to 99%, the remaining 1% was likely "Do not know" responses. Assuming that 1000 individuals were surveyed,
  a. Construct a circle graph for this study.
  b. How many parents said their neighborhood was not safe?
  c. State and explain any conclusions you draw from this information.

2.13 The USA Snapshot® "How to say I love you" reports the results of a David Michaelson & Associates survey for Ethel M Chocolates, on the best way to show affection.

Best way to show affection	Give gift	Hold hands	Hugging/kissing	Smiling	Other
Percent who said	10%	10%	51%	20%	9%

Draw a Pareto diagram picturing this information.

2.14 Medical-record practitioners are continually plagued with the problem of incomplete and delinquent medical files. Programs with positive incentives have been introduced to combat this problem. "The effects of positive incentive programs on physician chart completion" (*Topics in Health Record Management*, Sept. 1990) reported on the difficulties of implementing these programs in Florida. Draw a Pareto diagram of this information.

If you use MINITAB, use %PARETO

Barriers	Number	Percentage
Lack of administrative support	10	43.5
Time	5	21.7
Money	4	17.4
Physician apathy	3	13.1
Rewards problem behavior	1	4.3
Total	23	100.0

2.15   The January 10, 1991 USA Snapshot® "What's in U.S. landfills" reports the percentages of each type of waste in our landfills: food—4%, glass—2%, metal—14%, paper—38%, plastic—18%, yard waste—11%, other—13%.

  a. Construct a Pareto diagram displaying this information.

  b. Because of the size of the "other" category, the Pareto diagram may not be the best graph to use. Explain why, and describe what additional information is needed to make the Pareto diagram more appropriate.

2.16   The final-inspection defect report for assembly-line A12 is reported on a Pareto diagram.

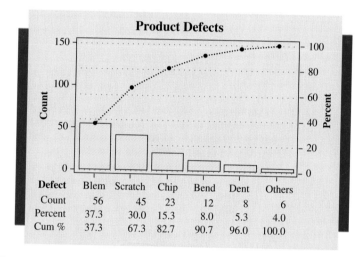

  a. What is the total defect count in the report?

  b. Verify the 30.0% listed for "scratch."

  c. Explain how the "cum % for bend" value of 90.7% was obtained and what it means.

  d. Management has given the production line the goal of reducing their defects by 50%. What two defects would you suggest they give special attention to in working toward this goal? Explain.

2.17   A city police officer, using radar, checked the speed of cars as they were traveling down a city street:

Many of these data sets are on Data Disk.

27	23	22	38	43	24
25	23	22	52	31	30
29	28	27	25	29	28
26	33	25	27	25	
21	23	24	18	23	

Construct a dotplot of these data.

2.18   As baseball players, Hank Aaron, Babe Ruth, and Roger Maris were well known for their ability to hit homeruns. Listed below is the number of homeruns each player hit in each Major League season in which he played in 75 or more games.

If you use Minitab, use DOTPlot and SAME. Reminder: Data is on Data Disk.

Aaron	13	27	26	44	30	39	40	34	45	44	24	32
	44	39	29	44	38	47	34	40	20	12	10	
Ruth	11	29	54	59	35	41	46	25	47	60		
	54	46	49	46	41	34	22					
Maris	14	28	16	39	61	33	23	26	13	9	5	

a. Construct a dotplot of these three sets of data, using the same axis.
b. Make a case for each of the following statements: "Aaron was the homerun king!" "Ruth was the homerun king!" "Maris was the homerun king!" In what way do the dotplots support each statement?

2.19 A computer was used to construct the following dotplot.

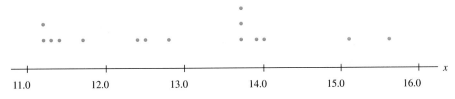

a. How many data are shown?
b. List the values of the five smallest data.
c. What is the value of the largest data?
d. What value occurred the greatest number of times? How many times did it occur?

2.20 Delco Products, a division of General Motors, produces commutators designed to be 18.810 mm in overall length. (A commutator is a device used in the electrical system of an automobile.) The following sample of 35 commutators was taken while monitoring the manufacturing process:

The Overall Length of Commutators

18.802	18.810	18.780	18.757	18.824	18.827	18.825
18.809	18.794	18.787	18.844	18.824	18.829	18.817
18.785	18.747	18.802	18.826	18.810	18.802	18.780
18.830	18.874	18.836	18.758	18.813	18.844	18.861
18.824	18.835	18.794	18.853	18.823	18.863	18.808

Source: With permission of Delco Products Division, GMC.

Use a computer to construct a dotplot of this data.

2.21 On the first day of class last semester, 30 students were asked for their one-way travel times from home to college (to the nearest five minutes). The resulting data were as follows:

20	20	30	25	20	25
35	25	15	25	25	40
25	30	15	20	45	25
15	20	20	20	20	25
5	20	20	10	5	20

Construct a stem-and-leaf display.

2.22 The following amounts are the fees charged by Quik Delivery for the 40 small packages it delivered last Thursday afternoon.

4.03	3.56	3.10	6.04	5.62	3.16	2.93	3.82
4.30	3.86	4.57	3.59	4.57	6.16	2.88	5.03
5.46	3.87	6.81	4.91	3.62	3.62	3.80	3.70
4.15	2.07	3.77	5.77	7.86	4.63	4.81	2.86
5.02	5.24	4.02	5.44	4.65	3.89	4.00	2.99

Construct a stem-and-leaf display.

2.23 MINITAB was used to construct a stem-and-leaf display of these data:

```
With data in column C1:

MTB>    STEM C1;
SUBC>   INCREMENT = .1.
Stem-and-Leaf of C1      N = 16
Leaf Unit = 0.010
   1    59  7
   4    60  148
  (5)   61  02669
   7    62  0247
   3    63  58
   1    64  3
```

a. What is the meaning of INCREMENT = .1?
b. What is the meaning of Leaf Unit = 0.010?
c. How many data are shown on this stem-and-leaf?
d. List the first four data values.

2.24 A term often used in solar energy research is *heating-degree-days*. This concept is related to the difference between an indoor temperature of 65°F and the average outside temperature for a given day. If the average outside temperature is 5°F, this would give 60 heating-degree-days. The annual heating-degree-day normals for several Nebraska locations are shown on the following stem-and-leaf display constructed by using MINITAB (see next page).
a. What is the meaning of INCREMENT = 100?
b. What is the meaning of Leaf Unit = 10?
c. List the first four data values.
d. List all the data values that occurred more than once.

```
With data in column C1:

MTB>    STEM C1;
SUBC>   INCREMENT = 100.
Stem-and-leaf of C1      N = 25
Leaf Unit = 10
     2    60   78
     7    61   03699
     9    62   69
    11    63   26
   (3)    64   233
    11    65   48
     9    66   8
     8    67   249
     5    68   18
     3    69   145
```

2.2 | FREQUENCY DISTRIBUTIONS, HISTOGRAMS, AND OGIVES

Lists of large sets of data do not present much of a picture. Sometimes we want to condense the data into a more manageable form. This can be accomplished with the aid of a **frequency distribution**.

TABLE 2-3

Ungrouped Frequency Distribution

x	f
0	1
1	3
2	8
3	5
4	3

FREQUENCY DISTRIBUTION

A listing, often expressed in chart form, that pairs each value of a variable with its frequency.

To demonstrate the concept of a frequency distribution, let's use this set of data:

3	2	2	3	2
4	4	1	2	2
4	3	2	0	2
2	1	3	3	1

EXERCISE 2.25

Form an ungrouped frequency distribution of the resulting data:
1, 2, 1, 0, 4, 2, 1, 1, 0, 1, 2, 4

If we let x represent the variable, we can use a frequency distribution to represent this set of data by listing the x values with their frequencies. For example, the value 1 occurs in the sample three times; therefore, the frequency for $x = 1$ is 3. The complete set of data is represented by the frequency distribution shown in Table 2-3 above.

The frequency f is the number of times the value x occurs in the sample. Table 2-3 is an **ungrouped frequency distribution**—"ungrouped" because each value of x in the distribution stands alone. When a large set of data has many different x-values instead of a few repeated values, as in the previous example, we can group the values into a set of classes and construct a **grouped frequency distribution**. The stem-and-leaf display in Figure 2-5 (p. 44) shows, in picture form, a grouped frequency distribution. Each stem represents a class. The number of leaves on each stem is the same as the frequency for that same class. The data represented in Figure 2-5 are listed as a frequency distribution in Table 2-4.

TABLE 2-4

Grouped Frequency
Distribution

		Class	Frequency
50 or more to less than 60	⟶	$50 \leq x < 60$	1
60 or more to less than 70	⟶	$60 \leq x < 70$	3
70 or more to less than 80	⟶	$70 \leq x < 80$	8
80 or more to less than 90	⟶	$80 \leq x < 90$	5
90 or more to less than 100	⟶	$90 \leq x < 100$	2
			19

EXERCISE 2.26

Referring to
Table 2-4:

a. Explain what $f = 8$ represents.
b. Find the Σf. Explain what Σf represents.
c. How many data are there?

The stem-and-leaf process can be used to construct a frequency distribution; however, the stem representation is not compatible with all class widths. For example, class widths of 3, 4, or 7 are awkward to use. Thus, sometimes it is advantageous to have a separate procedure for constructing a grouped frequency distribution.

▼   **ILLUSTRATION 2-6**

To illustrate this grouping (or classifying) procedure, let's use a sample of 50 final exam scores taken from last semester's elementary statistics class. Table 2-5 lists the 50 scores.

TABLE 2-5

Statistics Exam
Scores

60	47	82	95	88	72	67	66	68	98
90	77	86	58	64	95	74	72	88	74
77	39	90	63	68	97	70	64	70	70
58	78	89	44	55	85	82	83	72	77
72	86	50	94	92	80	91	75	76	78

The basic guidelines to follow in constructing a grouped frequency distribution are:

1. Each class should be of the same width.
2. Classes should be set up so that they do not overlap and so that each piece of data belongs to exactly one class.
3. For the exercises given in this textbook, 5 to 12 classes are most desirable since all samples contain fewer than 125 data. (The square root of n is a reasonable guideline for *number of classes* with samples of fewer than 150 data.)
4. Use a system that takes advantage of a number pattern, to guarantee accuracy. (This will be demonstrated in the following example.)
5. When it is convenient, an even class width is often advantageous.

Procedure

1. Identify the high and the low scores ($H = 98$, $L = 39$), and find the range. Range $= H - L = 98 - 39 = 59$.
2. Select a number of classes ($m = 7$) and a class width ($c = 10$) so that the product ($mc = 70$) is a bit larger than the range (range $= 59$).
3. Pick a starting point. This starting point should be a little smaller than the lowest score L. Suppose that we start at 35; counting from there by 10s (the class width), we get 35, 45, 55, 65, . . . , 95, 105. These are called the **class boundaries**. Observations that fall on class boundaries are placed into the class interval to the right, except for the interval to the farthest right, where it is placed in the interval to the left.

Class boundaries

Our classes for Illustration 2-6 are

35 or more to less than 45 → $35 \leq x < 45$ $75 \leq x < 85$
45 or more to less than 55 → $45 \leq x < 55$ $85 \leq x < 95$
55 or more to less than 65 → $55 \leq x < 65$ *95 or more to and including 105* → $95 \leq x \leq 105$
65 or more to less than 75 → $65 \leq x < 75$

NOTES

1. At a glance you can check the number pattern to determine whether the arithmetic used to form the classes was correct (35, 45, 55, . . . , 105).
2. The class width is the difference between the upper and lower class boundaries.
3. Many combinations of class widths, number of classes, and starting points are possible when classifying data. *There is no one best choice.* Try a few different combinations, and use good judgment to decide on the one to use.

Class width

Case Study   Home Sweet Home

2-1

EXERCISE   2.27

a. Express the category "5–7 years" shown in "Home sweet home" using the interval notation $a \leq x < b$. (Remember, a child is age 7 for many days, right up to the day of her eighth birthday.)

b. Express the information on the circle graph as a grouped frequency distribution.

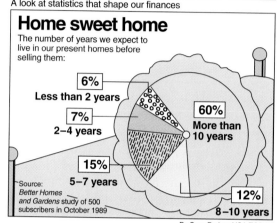

USA SNAPSHOTS®
A look at statistics that shape our finances

Home sweet home
The number of years we expect to live in our present homes before selling them:

6%
Less than 2 years

7%
2–4 years

15%
5–7 years

60%
More than 10 years

12%
8–10 years

Source: *Better Homes and Gardens* study of 500 subscribers in October 1989

By Suzy Parker, USA TODAY

Source: Copyright 1990, USA TODAY. Reprinted with permission.

A March 1990 issue of USA Today *presented this graphic, which shows a relative frequency distribution in the form of a circle graph. Each section of the circle represents a class interval of the values of the variable, the number of years owners expect to live in their homes, and the percentages are the relative frequencies for each interval. This same information could be expressed as a grouped frequency distribution.*

Once the classes are set up, we need to sort the data into classes. The method used to sort will depend on the current format of the data: If the data are ranked, the frequencies can be counted; if the data are not ranked, we will tally the data to find the frequency numbers. When classifying data, it helps to use a standard chart (see Table 2-6).

TABLE 2-6

Standard Chart for Frequency Distribution

Class Number	Class Tallies	Boundaries	Frequency
1	||	$35 \leq x < 45$	2
2	||	$45 \leq x < 55$	2
3	卌 ||	$55 \leq x < 65$	7
4	卌 卌 |||	$65 \leq x < 75$	13
5	卌 卌 |	$75 \leq x < 85$	11
6	卌 卌 |	$85 \leq x < 95$	11
7	||||	$95 \leq x \leq 105$	4
			50

NOTES

1. If the data have been ranked (list form, dotplot, or stem-and-leaf), tallying is unnecessary; just count the data belonging to each class.
2. If the data are not ranked, be careful as you tally.
3. The frequency f for each class is the number of pieces of data that belong in that class.
4. The sum of the frequencies should be exactly equal to the number of pieces of data n ($n = \Sigma f$). This summation serves as a good check.

NOTE   See the *Statistical Tutor* for information about the Σ (read "sum") notation.

Class mark

Each class needs a single numerical value to represent all the data values that fall into that class. The **class mark** (sometimes called *class midpoint*) is the numerical value that is exactly in the middle of each class and is found by adding the class boundaries and dividing by 2. Table 2-7 shows an additional column for the class mark, x.

TABLE 2-7

Frequency Distribution with Class Marks

Class Number	Class Boundaries	Frequency f	Class Mark x
1	$35 \leq x < 45$	2	40
2	$45 \leq x < 55$	2	50
3	$55 \leq x < 65$	7	60
4	$65 \leq x < 75$	13	70
5	$75 \leq x < 85$	11	80
6	$85 \leq x < 95$	11	90
7	$95 \leq x \leq 105$	4	100
		50	

EXERCISE 2.28

a. 65 belongs to which class?
b. Explain the meaning of "$65 \leq x < 75$."
c. Explain what "class width" is, and describe four ways that it shows up.

▲

As a check of your arithmetic, successive class marks should be a class width apart, which is 10 in this illustration (40, 50, 60, . . . , 100 is a recognizable pattern).

NOTE   Now you can see why it is helpful to have an even class width. An odd class width would have resulted in a class mark with an extra digit. (For example, the class 45–54 is 9 wide and the class mark is 49.5.)

Note that when we classify data into classes, we lose some information. Only when we have all the raw data do we know the exact values that were actually observed in each class. For example, we put a 47 and a 50 into class number 2, with class boundaries of 45 and 55. Once they are placed in the class, their values are lost to us and we use the class mark, 50, as their representative value.

HISTOGRAM

A bar graph representing a frequency distribution of a quantitative variable. A histogram is made up of the following components:

1. A title, which identifies the population or sample of concern.
2. A vertical scale, which identifies the frequencies in the various classes.
3. A horizontal scale, which identifies the variable *x*. Values for the class boundaries or class marks may be labeled along the *x*-axis. Use whichever method of labeling the axis best presents the variable.

The frequency distribution from Table 2-7 appears in histogram form in Figure 2-10.

FIGURE 2-10

Frequency Histogram

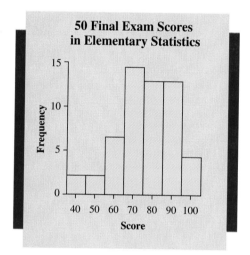

50 Final Exam Scores in Elementary Statistics

Relative frequency

EXERCISE 2.29

Draw a frequency histogram of the annual salaries for resort-club managers. Label class boundaries.

Ann. Sal. ($1000)				
15–25	25–35	35–45	45–55	55–65
No. Mgrs.				
12	37	26	19	6

(*Note:* Be sure to identify both scales so that the histogram tells the complete story.)

Sometimes the relative frequency of a value is important. The **relative frequency** is a proportional measure of the frequency of an occurrence. It is found by dividing the class frequency by the total number of observations. Relative frequency can be expressed as a common fraction, in decimal form, or as a percentage. For example, in Illustration 2-6 the frequency associated with the third class (55–65) is 7. The relative frequency for the third class is 7/50, or 0.14, or 14%. Relative frequencies are often useful in a presentation because nearly everybody understands fractional parts when expressed as percents. Relative frequencies are particularly useful when comparing the frequency distributions of two different size sets of data. Figure 2-11 is a relative frequency histogram of the sample of the 50 final exam scores from Table 2-7.

FIGURE 2-11

Relative Frequency
Histogram

EXERCISE   2.30

Explain the similarit-
ies and the differ-
ences between
Figures 2-10 and
2-11.

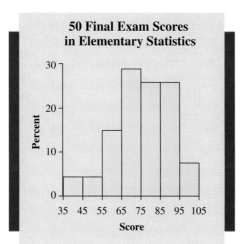

A stem-and-leaf display contains all the information in a histogram. Figure 2-5 shows the stem-and-leaf display constructed in Illustration 2-4. Figure 2-5 has been rotated 90° and labels have been added to form the histogram shown in Figure 2-12.

FIGURE 2-12

Modified Stem-and-
Leaf

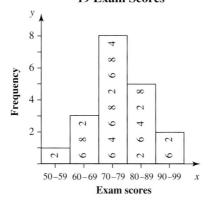

MINITAB (Release 10) commands to construct a histogram with the data listed in C1.

Session commands

Enter: HISTogram C1;

       FREQuency or PERCent;

       MIDPoint A : B/C;
       or

Menu commands

Choose: **Graph > Histogram**

Enter: Graph variable, **C1**

Choose: Annotation > Title,
Enter: **Your title**

(*continued*)

MINITAB *(continued)*

CUTPoint A:B/C;

TITLe 'your title'.

Choose: Options
  Type of histogram,
    Choose: **Frequency** or **Percent**
  Type of interval,
    Choose: **Midpoint** or **Cutpoint**
  Definition of intervals,
    Choose: **Automatic**
         or **Number of intervals**,
           enter: **N**
         or **Midpt/cutpt**,
           enter: **A:B/C**

Choose one of the three.

Notes: 1. *Midpoint* is the class mark, and *cutpoints* are the class boundaries.
       2. *Percent* is relative frequency.
       3. *A* = smallest class mark or boundary, *B* = largest class mark or boundary, *C* = class width you want to specify.
       4. *Automatic* means MINITAB will make all the choices; *N = number of intervals*, the number of classes you want used.

MINITAB (Release 10) commands to construct a histogram of a frequency distribution with class marks in C1 and frequencies in C2. By adding the midpoint of the next smaller class at the smaller end of the frequency distribution and by adding the midpoint of the next larger at the larger end, both with frequencies of zero, PLOT will draw the histogram showing the end classes of full width.

Session commands

Enter: **Plot C2*C1;**
       AREA;
       STEP 0;
       TYPE 0.

Menu commands

Choose: **Graph > Plot**
Select: Data Display: **AREA**
Edit Attributes:
  Choose: Fill Type: **None**
  Connection Function: **Step**

Histograms are valuable tools. For example, the histogram of the sample should have a distribution shape that is very similar to that of the population from which the sample was drawn. If the reader of a histogram is at all familiar with the variable involved, he or she will usually be able to interpret several important facts. Figure 2-13 presents histograms with descriptive labels resulting from their geometric shape.

Briefly, the terms used to describe histograms are as follows:

▼ **Symmetrical**: Both sides of this distribution are identical.
▼ **Uniform (rectangular):** Every value appears with equal frequency.
▼ **Skewed:** One tail is stretched out longer than the other. The direction of skewness is on the side of the longer tail.

FIGURE 2-13   Shapes of Histograms

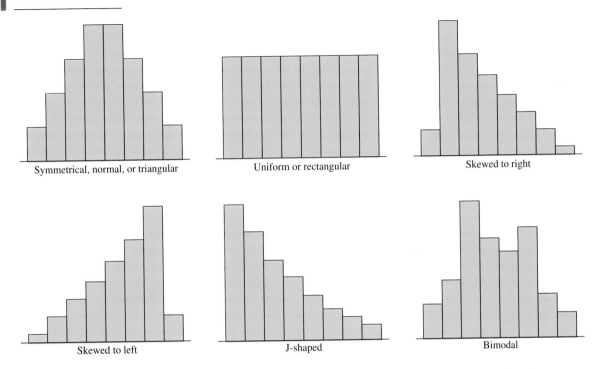

Symmetrical, normal, or triangular

Uniform or rectangular

Skewed to right

Skewed to left

J-shaped

Bimodal

▼ **J-shaped:** There is no tail on the side of the class with the highest frequency.
▼ **Bimodal:** The two most populous classes are separated by one or more classes. This situation often implies that two populations are being sampled.
▼ **Normal:** A symmetrical distribution is mounded up about the mean and becomes sparse at the extremes. (Additional properties are discussed later.)

NOTES

1. The **mode** is the value of the piece of data that occurs with the greatest frequency. (Mode will be discussed in Section 2.3, p. 71.)
2. The **modal class** is the class with the highest frequency.
3. A **bimodal distribution** has two high-frequency classes separated by classes with lower frequencies.

Another way to express a frequency distribution is to use a **cumulative frequency distribution**.

CUMULATIVE FREQUENCY DISTRIBUTION

A frequency distribution that pairs cumulative frequencies with values of the variable.

Cumulative frequency

The **cumulative frequency** for any given class is the sum of the frequency for that class and the frequencies of all classes of smaller values. Table 2-8 shows the cumulative frequency distribution from Table 2-7.

TABLE 2-8

Using Frequency Distribution to Form a Cumulative Frequency Distribution

Class Number	Class Boundaries	Frequency f	Cumulative Frequency	
1	$35 \leq x < 45$	2	2	(2)
2	$45 \leq x < 55$	2	4	(2 + 2)
3	$55 \leq x < 65$	7	11	(4 + 7)
4	$65 \leq x < 75$	13	24	(11 + 13)
5	$75 \leq x < 85$	11	35	(24 + 11)
6	$85 \leq x < 95$	11	46	(35 + 11)
7	$95 \leq x \leq 105$	4	50	(46 + 4)
		50		

Cumulative relative frequency distribution

The same information can be presented by using a **cumulative relative frequency distribution** (see Table 2-9). This combines the cumulative frequency and the relative frequency ideas.

TABLE 2-9

Cumulative Relative Frequency Distribution

Class Number	Class Boundaries	Cumulative Relative Frequency	*Cumulative frequencies are for interval 35 up to upper boundary of that class*
1	$35 \leq x < 45$	2/50, or 0.04	← *from 35 up to less than 45*
2	$45 \leq x < 55$	4/50, or 0.08	← *from 35 up to less than 55*
3	$55 \leq x < 65$	11/50, or 0.22	← *from 35 up to less than 65*
4	$65 \leq x < 75$	24/50, or 0.48	
5	$75 \leq x < 85$	35/50, or 0.70	⋮
6	$85 \leq x < 95$	46/50, or 0.92	
7	$95 \leq x \leq 105$	50/50, or 1.00	← *from 35 up to and including 105*

EXERCISE 2.32

Express this frequency distribution as a cumulative frequency distribution:

Ann. Salary ($1000)					
15–25	25–35	35–45	45–55	55–65	
No. Managers					
12	37	26	19	6	

OGIVE (pronounced o'jīv)

A line graph of a cumulative frequency or cumulative relative frequency distribution. An ogive has the following components:

1. A title, which identifies the population or sample.
2. A vertical scale, which identifies either the cumulative frequencies or the cumulative relative frequencies. (Figure 2-14 shows an ogive with cumulative relative frequencies.)
3. A horizontal scale, which identifies the upper class boundaries. Until the upper boundary of a class has been reached, you cannot be sure

you have accumulated all the data in that class. Therefore, the horizontal scale for an ogive is always based on the upper class boundaries.

NOTE   Every ogive starts on the left with a relative frequency of zero at the lower class boundary of the first class and ends on the right with a relative frequency of 100% at the upper class boundary of the last class.

FIGURE 2-14

Ogive

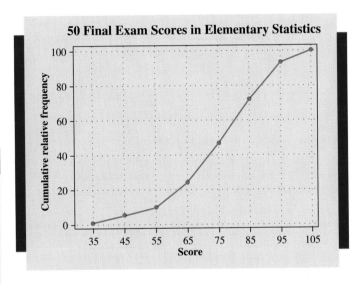

50 Final Exam Scores in Elementary Statistics

EXERCISE   2.33

Express this frequency distribution as a cumulative relative frequency distribution:

Ann. Salary ($1000)				
15–25	25–35	35–45	45–55	55–65

No. Managers				
12	37	26	19	6

REMEMBER   All graphic representations of sets of data need to be completely self-explanatory. That includes a descriptive meaningful title and proper identification of the vertical and horizontal scales.

EXERCISE   2.34

Construct an ogive for the cumulative relative frequency distribution found in answering Exercise 2.33.

MINITAB (Release 10) commands to construct an ogive with the class boundaries listed in C1 and the cumulative percentages listed in C2.

Session commands

Enter: PLOT C2 * C1;
       CONNect;
       TITLe 'your title'.

Menu commands

Choose: Graph > Plot
Graph variable,
  Enter: y, C2 and x, C1
Data display, Choose: Connect
Annotation > Title,
  Enter: your title

NOTE
1. Use percentages; that is, use 25% for the relative frequency of 0.25.
2. Enter 0 (zero) for the percent paired with the first class boundary.

EXERCISES

2.35 In the May 1990 issue of the journal *Social Work*, the following ungrouped frequency distribution was reported:

Number of Children Living at Home	Mexican- American Women
0	23
1	22
2	17
3	7
4	1

Source: Copyright 1990, National Association of Social Workers, Inc. *Social Work*.

Construct a histogram of this distribution.

2.36 An article in the *Therapeutic Recreation Journal* reported the following distribution for the variable "number of persistent disagreements between 66 patients and their therapeutic recreation specialist":

Number of items	Frequency		Number of items	Frequency
0	2		6	8
1	2		7	11
2	4		8	7
3	10		9	3
4	7		10	1
5	9		11	2

Source: Reprinted with permission of the National Recreation and Park Association, Alexandria, VA, from Pauline Petryshen and Diane Essex-Sorlie, "Persistent Disagreement Between Therapeutic Recreation Specialists and Patients in Psychiatric Hospitals," *Therapeutic Recreation Journal*, Vol. XXIV, Third Quarter, 1990.

Draw a histogram of this ungrouped frequency distribution.

2.37 The ages of 50 dancers who responded to a call to audition for a musical comedy are:

Many of these data sets are on Data Disk.

21	19	22	19	18	20	23	19	19	20
19	20	21	22	21	20	22	20	21	20
21	19	21	21	19	19	20	19	19	19
20	20	19	21	21	22	19	19	21	19
18	21	19	18	22	21	24	20	24	17

a. Prepare an ungrouped frequency distribution of these ages.
b. Prepare an ungrouped relative frequency distribution of the same data.
c. Prepare a cumulative relative frequency distribution of the same data.
d. Prepare a relative frequency histogram of these data.
e. Prepare an ogive of these data.

2.38 The opening-round scores for the Ladies' Professional Golf Association tournament at Locust Hill Country Club were posted as follows:

69	73	72	74	77	80	75	74	72	83	68	73
75	78	76	74	73	68	71	72	75	79	74	75
74	74	68	79	75	76	75	77	74	74	75	75
72	73	73	72	72	71	71	70	82	77	76	73
72	72	72	75	75	74	74	74	76	76	74	73
74	73	72	72	74	71	72	73	72	72	74	74
67	69	71	70	72	74	76	75	75	74	73	74
74	78	77	81	73	73	74	68	71	74	78	70
68	71	72	72	75	74	76	77	74	74	73	73
70	68	69	71	77	78	68	72	73	78	77	79
79	77	75	75	74	73	73	72	71	68	70	71
78	78	76	74	75	72	72	72	75	74	76	77
78	78										

If you use MINITAB, try Table C1.

a. Form an ungrouped frequency distribution of these scores.
b. Draw a histogram of the first-round golf scores. Use the frequency distribution from (a).

2.39 The KSW computer-science aptitude test was given to 50 students. The following frequency distribution resulted from their scores.

KSW Test Score	Frequency
0–4	4
4–8	8
8–12	8
12–16	20
16–20	6
20–24	3
24–28	1

a. What are the class boundaries for the class of largest frequency?
b. Give all class marks associated with this frequency distribution.
c. What is the class width?
d. Give the relative frequencies for the classes.
e. Draw a relative frequency histogram of the test scores.

2.40 The USA Snapshot® "Nuns an aging order" reports that the median age of the USA's 94,022 Roman Catholic nuns is 65 years and that the percentage of nuns in the country by age group is:

Under 50	51–70	Over 70	Refused to give age
16%	42%	37%	5%

This information is based on a survey of 1049 Roman Catholic nuns.
Suppose the survey had resulted in the frequency distribution shown on page 66 (52 ages unknown).

MINITAB's PLOT command will draw this histogram. Be sure to add class midpoints of 15 and 95 with frequencies of zero.

Age	20–30	30–40	40–50	50–60	60–70	70–80	80–90
Freq.	34	58	76	187	254	241	147

a. Draw and completely label a frequency histogram.
b. Draw and completely label a relative frequency histogram of the same distribution.
c. Carefully examine the two histograms, (a) and (b), and explain why one of them might be easier to understand. (Retain these solutions for use in answering Exercise 2.103, p. 92.)

2.41   The speeds of 55 cars were measured by a radar device on a city street:

27	23	22	38	43	24	35	26	28	18	20
25	23	22	52	31	30	41	45	29	27	43
29	28	27	25	29	28	24	37	28	29	18
26	33	25	27	25	34	32	36	22	32	33
21	23	24	18	48	23	16	38	26	21	23

If you use MINITAB, use HIST and CUTP 12:54/6.

a. Classify these data into a grouped frequency distribution by using class boundaries 12-18-24 ··· 54. (Retain this solution for use in answering Exercise 2.101, p. 92.)
b. Find the class width.
c. For the class 24–30, find (1) the class mark, (2) the lower class boundary, and (3) the upper class boundary.
d. Construct a frequency histogram of these data.

2.42   The hemoglobin A_{1c} test, a blood test given to diabetics during their periodic checkups, indicates the level of control of blood sugar during the past two to three months. The following data were obtained for 40 different diabetics at a university clinic that treats diabetic patients.

6.5	5.0	5.6	7.6	4.8	8.0	7.5	7.9	8.0	9.2
6.4	6.0	5.6	6.0	5.7	9.2	8.1	8.0	6.5	6.6
5.0	8.0	6.5	6.1	6.4	6.6	7.2	5.9	4.0	5.7
7.9	6.0	5.6	6.0	6.2	7.7	6.7	7.7	8.2	9.0

a. Classify these A_{1c} values into a grouped frequency distribution using the classes 3.7–4.7, 4.7–5.7, and so on.
b. What are the class marks for these classes?
c. Construct a frequency histogram of these data.

2.43   All of the third-graders at Roth Elementary School were given a physical-fitness strength test. These data resulted:

12	22	6	9	2	9	5	9	3	5	16	1	22
18	6	12	21	23	9	10	24	21	17	11	18	19
17	5	14	16	19	19	18	3	4	21	16	20	15
14	17	4	5	22	12	15	18	20	8	10	13	20
6	9	2	17	15	9	4	15	14	19	3	24	

a. Construct a dotplot
b. Prepare a grouped frequency distribution using classes 1–4, 4–7, and so on, and draw a histogram of the distribution. (Retain for use in answering Exercise 2.65, p. 76.)
c. Prepare a grouped frequency distribution using classes 0–3, 3–6, 6–9, and so on, and draw a histogram of the distribution.
d. Prepare a grouped frequency distribution using classes −2.5–2.5, 2.5–7.5, 7.5–12.5, and so on, and draw a histogram of the distribution.

e. Prepare a grouped frequency distribution using classes of your choice, and draw a histogram of the distribution.

f. Describe the shape of the histogram found in (b), (c), (d), and (e) separately. Relate the distribution seen in the histogram to the distribution seen in the dotplot.

g. Discuss how the number of classes used and the choice of class boundaries used affect the appearance of the resulting histogram.

2.44 The Children's Defense Fund report listed the "percentage of child support collected in 1992 in each of 50 states and DC," according to *USA Today*, 6-17-94, as follows:

23.8	17.4	8.7	23.5	14.1	14.0	19.6	27.1	11.3	16.9	16.8
32.5	29.2	9.2	14.0	22.7	25.1	16.7	15.6	21.7	24.9	20.2
17.6	33.6	9.3	19.9	24.8	18.8	19.5	28.6	20.1	17.2	17.7
19.2	21.0	20.4	14.9	17.7	30.6	8.6	24.1	28.2	11.9	12.7
22.7	40.3	22.9	33.7	19.5	31.3	22.3				

a. Prepare a grouped frequency distribution.

b. Prepare a histogram of the frequency distribution found in (a).

2.45 The following 40 amounts are the fees that Fast Delivery charged for delivering small freight items last Thursday afternoon.

If you use MINITAB, use subcommand PERCent.

4.03	3.56	3.10	6.04	5.62	3.16	2.93	3.82
4.30	3.86	4.57	3.59	4.57	5.16	2.88	5.02
5.46	3.87	6.81	4.91	3.62	3.62	3.80	3.70
4.15	4.07	3.77	5.77	7.86	4.63	4.81	2.86
5.02	5.24	4.02	5.44	4.65	3.89	4.00	2.99

a. Classify these data into a grouped frequency distribution.

b. Construct a relative frequency histogram of these data.

2.46 An August 1994 issue of *USA Today* reported the median price of used homes sold in each of 133 metro areas across the USA between April and June 1994 as released by the National Association of Realtors. (Prices have been rounded to the nearest $1000.)

365 is for Honolulu, the only noncontinental area represented.

87	110	109	66	214	80	94	112	124	95	119
76	67	194	68	100	97	181	89	82	79	81
75	92	80	107	77	146	97	89	103	88	97
74	96	61	86	71	115	82	84	74	96	75
105	77	82	85	88	76	85	92	88	134	365
82	92	82	76	86	88	133	75	110	87	74
188	81	115	77	87	102	175	110	101	72	138
81	96	161	191	142	77	175	106	58	67	75
92	78	69	118	91	118	117	112	132	113	93
129	87	84	126	59	96	79	177	256	95	156
71	80	70	94	79	110	73	87	84	119	99
75	76	64	130	90	73	159	52	115	75	130
63										

a. Classify these data into a grouped frequency distribution.

b. Express the distribution as a relative frequency distribution.

c. Construct a relative frequency histogram.

(Retain these solutions for use in answering Exercise 2.106, p. 94.)

2.47   a. Prepare a cumulative relative frequency distribution for the variable "number of persistent disagreements between patients and their specialists" in Exercise 2.36.
b. Construct an ogive of the distribution.

2.48   a. Prepare a cumulative relative frequency distribution for the variable "KSW test score"
in Exercise 2.39.
b. Construct an ogive of the distribution.

NUMERICAL DESCRIPTIVE STATISTICS

2.3 | MEASURES OF CENTRAL TENDENCY

Measures of central tendency are numerical values that locate, in some sense, the
middle of a set of data. The term *average* is often associated with all measures of
central tendency.

MEAN

The average with which you are probably most familiar. It is represented
by $\bar{x}$ (read "x bar" or "sample mean"). The mean is found by adding all the
values of the variable x (this sum of x values is symbolized Σx) and dividing
by the number of these values, n. We express this is formula form as

sample mean: $x\text{-}bar = \dfrac{sum\ of\ x}{number}$

$$\bar{x} = \frac{\Sigma x}{n} \qquad\qquad \textbf{(2-1)}$$

NOTES

1. See the *Statistical Tutor* for information about the Σ notation.
2. The population mean, μ (lowercase mu, Greek alphabet), is the mean of all
x values for the entire population.

▼| ILLUSTRATION   2-7

A set of data consists of the five values 6, 3, 8, 6, and 4. Find the mean.

SOLUTION Using formula (2-1), we find

$$\bar{x} = \frac{\Sigma x}{n} = \frac{6 + 3 + 8 + 6 + 4}{5} = \frac{27}{5} = \textbf{5.4}$$

▲| Therefore, the mean of this sample is 5.4.

A physical representation of the mean can be constructed by thinking of a number line balanced on a fulcrum. A weight is placed on a number on the line corresponding to each number in the sample of Illustration 2.7. In Figure 2-15 there is one weight each on the 3, 8, and 4 and two weights on the 6, since there were two 6s in the sample. The mean is the value that balances the weights on the number line—in this case, 5.4.

FIGURE 2-15

Physical
Representation of
Mean

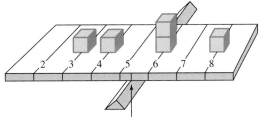

$\bar{x} = 5.4$ (the center of gravity, or balance point)

EXERCISE   2.49

The number of children, x, belonging to each of eight families registering for swimming was 1, 2, 1, 3, 2, 1, 5, 3. Find the mean, $\bar{x}$.

MINITAB (Release 10) commands to find the mean of the data listed in C1.

Session commands

Enter: MEAN C1

Menu commands

Choose: `Calc > Column Statistics`
Select: `Mean`
Enter: `Input variable, C1`

MEDIAN

The value of the data that occupies the middle position when the data are ranked in order according to size. It is represented by $\tilde{x}$ (read "x tilde" or "sample median").

NOTE   The population median, M (uppercase mu, Greek alphabet), is the data value in the middle position of the entire ranked population.

Procedure for Finding the Median
Step 1: Rank the data.
Step 2: Determine the depth of the median.

The *depth* (number of positions from either end), or position, of the median is determined by the formula

depth of median: *depth of median* $= \dfrac{number + 1}{2}$

$$d(\tilde{x}) = \dfrac{n + 1}{2} \qquad \text{(2-2)}$$

The median's depth (or position) is found by adding the position numbers of the smallest (1) and largest (n) data values and dividing by 2. (n is the same number as the number of pieces of data.)

Step 3: Determine the value of the median.
Count over the ranked data, locating the data in the $d(x)$th position. The median will be the same regardless of which end of the ranked data (high or low) you count from. In fact, counting from both ends will serve as an excellent check.

The following two illustrations will demonstrate this procedure as it applies to both odd-numbered and even-numbered sets of data.

▼ | ILLUSTRATION   2 - 8

Find the median for the set of data {6, 3, 8, 5, 3}.

SOLUTION

STEP 1   The data, ranked in order of size, are 3, 3, 5, 6, and 8.

STEP 2   Depth of the median: $d(\tilde{x}) = \dfrac{n + 1}{2} = \dfrac{5 + 1}{2} = \mathbf{3}$

STEP 3   That is, the median is the third number from either end in the ranked data, or $\tilde{x} = \mathbf{5}$. Notice that the median essentially separates the ranked set of data into two subsets of equal size (see Figure 2-16).

FIGURE 2-16
_____

Median of
{3, 3, 5, 6, 8}   ▲ |

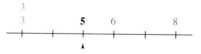

$\tilde{x} = \mathbf{5}$ (the middle value; 2 data are smaller, 2 are larger)

EXERCISE 2.50

Find the median height of a basketball team: 73, 76, 72, 70, and 74 inches.

NOTE   The value of $d(\tilde{x})$ is the depth of the median, NOT the value of the median, $\tilde{x}$.
     As in Illustration 2-8, when n is odd, the depth of the median, $d(\tilde{x})$, will always be an integer. However, when n is even, the depth of the median, $d(\tilde{x})$, will always be a half-number, as shown in Illustration 2-9.

▼ | ILLUSTRATION 2-9

Find the median for the sample 9, 6, 7, 9, 10, 8.

SOLUTION

STEP 1 The data, ranked in order of size, are 6, 7, 8, 9, 9, and 10.

STEP 2 Depth of the median: $d(\tilde{x}) = \dfrac{n + 1}{2} = \dfrac{6 + 1}{2} = 3.5$

STEP 3 That is, the median is halfway between the third and fourth pieces of data. To find the number halfway between any two values, add the two values together and divide by 2. In this case, add the third value (8) and the fourth value (9), then

FIGURE 2-17

Median of {6, 7, 8, 9, 9, 10}

$\tilde{x} = \mathbf{8.5}$ (value in middle; 3 data smaller, 3 larger)

divide by 2. The median is $\tilde{x} = \dfrac{8 + 9}{2} = \mathbf{8.5}$, a number halfway between the "middle"

two numbers (see Figure 2-17). Notice that the median again separates the ranked set of data into two subsets of the same size.

▲ |

EXERCISE 2.51

Find the median rate paid at Jim's Burgers if the workers' hourly rates are 4.25, 4.15, 4.90, 4.25, 4.60, 4.50, 4.60, 4.75.

MINITAB (Release 10) commands to find the median of the data listed in C1.

Session commands

Enter: MEDIan C1

Menu commands

Choose: Calc > Column Statistics
Select: Median
Enter: Input variable, C1

MODE

The mode is the value of x that occurs most frequently.

In the set of data from Illustration 2-8—3, 3, 5, 6, 8—the mode is 3 (see Figure 2-18).

FIGURE 2-18
_____

Mode of
{3, 3, 5, 6, 8}

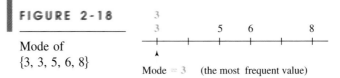

Mode = 3   (the most frequent value)

In the sample 6, 7, 8, 9, 9, 10, the mode is 9. In this sample, only the 9 occurs more than once; in the data from Illustration 2-8, only the 3 occurs more than once. If two or more values in a sample are tied for the highest frequency (number of occurrences), we say there is **no mode**. For example, in the sample 3, 3, 4, 5, 5, 7, both the 3 and the 5 appear an equal number of times. There is no one value that appears most often; thus, this sample has no mode.

No mode

EXERCISE 2.52

The number of cars per apartment, owned by a sample of tenants in a large complex is 1, 2, 1, 2, 2, 2, 1, 2, 3, 2. What is the mode?

EXERCISE 2.53

USA Today, August 1994, reported on "How much airlines spend, per passenger, on inflight meals: lowest, $0.13 and highest, $10.62. Find the midrange.

MIDRANGE

The number exactly midway between a lowest value data *L* and a highest value data *H*. It is found by averaging the low and the high values.

$$midrange = \frac{low\ value\ +\ high\ value}{2}$$

$$midrange = \frac{L + H}{2} \qquad (2\text{-}3)$$

For the set of data from Illustration 2-8—3, 3, 5, 6, 8—$L = 3$ and $H = 8$ (see Figure 2-19). Therefore, the midrange is

$$midrange = \frac{L + H}{2} = \frac{3 + 8}{2} = \textbf{5.5}$$

FIGURE 2-19
_____

Midrange of
{3, 3, 5, 6, 8}

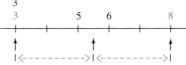

Midrange = **5.5**   (midway between the extremes)

EXERCISE 2.54

Find the mean, median, mode, and midrange for the sample data 9, 6, 7, 9, 10, 8.

The four measures of central tendency represent four different methods of describing the middle. These four values may be the same, but more likely they will result in different values. For the sample data from Illustration 2-9, the mean $\bar{x}$ is 8.2, the median $\tilde{x}$ is 8.5, the mode is 9, and the midrange is 8. Their relationship to each other and to the data is shown in Figure 2-20.

FIGURE 2-20
_____

Measures of Central Tendency for {6, 7, 8, 9, 9, 10}

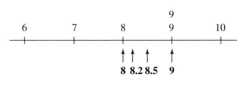

Case Study

2-2

"Average" Means Different Things

When it comes to convenience, few things can match that wonderful mathematical device called averaging.

How handy it is! With an average you can take a fistful of figures on any subject—temperatures, incomes, velocities, populations, light-years, hair-breadths, anything at all that can be measured— and compute one figure that will represent the whole fistful.

But there is one thing to remember. There are several kinds of measures ordinarily known as averages. And each gives a different picture of the figures it is called on to represent.

Take an example. Here are the annual incomes of ten families:

$54,000	$39,000	$37,500	$36,750	$35,250
$31,500	$31,500	$31,500	$31,500	$25,500

What would this group's "typical" income be? Averaging would provide the answer, so let's compute the typical income by the simpler and most frequently used kinds of averaging.

The arithmetic mean. When anyone cites an average without specifying which kind, you can probably assume that he has the arithmetic mean in mind. It is the most common form of average, obtained by adding items in the series, then dividing by the number of items. In our example, the sum of the ten incomes divided by 10 is $35,400. The mean is representative of the series in the sense that the sum of the amounts by which the higher figures exceed the mean is exactly the same as the sum of the amounts by which the lower figures fall short of the mean.

The median. As you may have observed, six families earn less than the mean, four earn more. You might very well wish to represent this varied group by the income of the family that is right smack dab in the middle of the whole bunch. To do this, you need to find the median. It would be easy if there were 11 families in the group. The family sixth from highest (or sixth from lowest) would be in the middle and have the median income. But with ten families there is no middle family. So you add the two central incomes ($31,500 and $35,250 in this case) and divide by 2. The median works out to $33,375.

(continued)

ROUND-OFF RULE

When rounding off an answer, let's agree to keep one more decimal place in our answer than was present in our original information. To avoid round-off buildup, round off only the final answer, not the intermediate steps. That is, avoid use of a rounded value to do further calculations. In our previous examples, the data were composed of whole numbers; therefore, those answers that have decimal values should be rounded to the nearest tenth. See the *Statistical Tutor* for specific instructions on how to perform the rounding off.

The midrange. Another number that might be used to represent the group is the midrange, computed by calculating the figure that lies halfway between the highest and lowest incomes. To find this figure, add the highest and lowest incomes ($54,000 and $25,500), divide by 2, and you have the amount that lies halfway between the extremes, $39,750.

The mode. So, three kinds of averages, and not one family actually has an income matching any of them. Say you want to represent the group by stating the income that occurs most frequently. That kind of representativeness is called a mode. In this example $31,500 would be the modal income. More familiies earn that income than any other.

Four different averages, each valid, correct, and informative in its way. But how they differ!

arithmetic mean	$35,400
median	$33,375
midrange	$39,750
mode	$31,500

And they would differ still more if just one family in the group were a millionaire—or one were jobless!

So there are three lessons to take away from today's class in averages. First, when you see or hear an average, find out which average it is. Then you'll know what kind of picture you are being given. Second, think about the figures being averaged so you can judge whether the average used is appropriate. And third, don't assume that a literal mathematical quantification is intended every time somebody says "average." It isn't. All of us often say "the average person" with no thought of implying a mean, median, or mode. All we intend to convey is the idea of other people who are in many ways a great deal like the rest of us.

Source: Reprinted by permission from CHANGING TIMES magazine (March 1980 issue). Copyright by The Kiplinger Washington Editors.

EXERCISES

2.55 Consider the sample 2, 4, 7, 8, 9. Find the following:
a. the mean $\bar{x}$ b. the median $\tilde{x}$
c. the mode   d. the midrange

2.56 Consider the sample 6, 8, 7, 5, 3, 7. Find the following:
a. the mean $\bar{x}$ b. the median $\tilde{x}$
c. the mode   d. the midrange

2.57 Fifteen randomly selected college students were asked to state the number of hours they slept last night. The resulting data are 5, 6, 6, 8, 7, 7, 9, 5, 4, 8, 11, 6, 7, 8, 7. Find the following:
a. the mean $\bar{x}$ b. the median $\tilde{x}$
c. the mode   d. the midrange

2.58 An article titled "Financing Your Kids' College Education" (*Farming*, Sept./Oct. 1994) listed the following in-state tuition and fees per school year for 14 land-grant universities: 1554, 2291, 2084, 4443, 2884, 2478, 3087, 3708, 2510, 2055, 3000, 2052, 2550, 2013.

    a. Find the mean in-state tuition and fees per school year.
    b. Find the median in-state tuition and fees per school year.
    c. Find the midrange in-state tuition and fees per school year.
    d. Find the mode, if one exists, per school year.

2.59   *USA Today* reported the following statistics about the changes in median home prices for metro areas during the first quarter of 1994.

    Biggest increase: Biloxi-Gulfport, Miss. 19.1%   Biggest decrease: Detroit, Mich. −8.4%

Based on this information, find the average change in the median metro home price during the first quarter of 1994.

2.60   *USA Today*, 10-28-94, reported on the average annual pay received by all workers covered by state and federal unemployment insurance for the 50 states. Connecticut had the highest with $33,169, South Dakota had the lowest with $18,613.
    a. Estimate the national average and the midrange for the states.
    b. The national average was reported to be $26,362. What can you conclude about the distribution of the state averages based on the relationship between the midrange and the national average?

2.61   Recruits for a police academy were required to undergo a test that measures their exercise capacity. The exercise capacity (measured in minutes) was obtained for each of 20 recruits:

Many of these data sets are on Data Disk.

| 25 | 27 | 30 | 33 | 30 | 32 | 30 | 34 | 30 | 27 |
| 26 | 25 | 29 | 31 | 31 | 32 | 34 | 32 | 33 | 30 |

    a. Find the mean, median, mode, and midrange.
    b. Construct a dotplot of these data, and locate the mean, median, mode, and midrange on the graph.
    c. Describe the relationship between the four "averages" (similarity) and what properties the data show that cause the four averages to be so similar.
(Retain these solutions for use in answering Exercise 2.79, p. 84.)

2.62   *Atlantic Monthly* (Nov. 1990) contains an article titled "The Case for More School Days." The number of days in the standard school year is given for several different countries as follows:

Country	n (days)/yr	Country	n (days)/year
Japan	243	New Zealand	190
West Germany	226–240	Nigeria	190
South Korea	220	British Columbia	185
Israel	216	France	185
Luxembourg	216	Ontario	185
Soviet Union	211	Ireland	184
Netherlands	200	New Brunswick	182
Scotland	200	Quebec	180
Thailand	200	Spain	180
Hong Kong	195	Sweden	180
England/Wales	192	United States	180
Hungary	192	French Belgium	175
Swaziland	191	Flemish Belgium	160
Finland	190		

(continued)

a. Find the mean and median number of days per year of school for the countries listed. (Use the midpoint of the 226–240 interval for West Germany when computing your answers.)

b. Construct a stem-and-leaf display of these data.

c. Describe the relationship between the mean and the median and what properties of the data cause the mean to be larger than the median.

(Retain these solutions for use in answering Exercise 2.80, p. 84.)

2.63 You are responsible for planning the parking needed for a new 256-unit apartment complex and you're told to base the needs on the statistic "average number of vehicles per household is 1.9."

a. Which average (mean, median, mode, midrange) will be helpful to you? Explain.

b. Explain why "1.9" cannot be the median, the mode, or the midrange for the variable "number of vehicles."

c. If the owner wants parking that will accommodate 90% of all the tenants who own vehicles, how many spaces must you plan for?

2.64 Starting with the data values 70 and 100, add three data values to your sample so that the sample has: (Justify your answer in each case.)

a. a mean of 100

b. a median of 70

c. a mode of 87

d. a midrange of 70

e. a mean of 100 and a median of 70

f. a mean of 100 and a mode of 87

g. a mean of 100 and a midrange of 70

h. a mean of 100, a median of 70, and a mode of 87

2.65 All of the third-graders at Roth Elementary School were given a physical-fitness strength test. The following data resulted:

12	22	6	9	2	9	5	9	3	5	16	1	22
18	6	12	21	23	9	10	24	21	17	11	18	19
17	5	14	16	19	19	18	3	4	21	16	20	15
14	17	4	5	22	12	15	18	20	8	10	13	20
6	9	2	17	15	9	4	15	14	19	3	24	

a. Construct a dotplot.

b. Find the mode.

c. Prepare a grouped frequency distribution using classes 1–4, 4–7, and so on, and draw a histogram of the distribution. (See Exercise 2.43, p. 66.)

d. Describe the distribution; specifically, is the distribution bimodal (about what values)?

e. Compare your answers in (a) and (c), and comment on the relationship between the mode and the modal values that occurred in these data.

f. Could a discrepancy like this occur when using an ungrouped frequency distribution? Explain.

g. Explain why, in general, the mode of a set of data does not necessarily tell us the same information as the modal values do.

2.66 Case Study 2-2 uses a sample of ten annual incomes to discuss the four averages.

a. Calculate the mean, median, mode, and midrange for the ten incomes. Compare your results with those found in the article.

b. What is there about the distribution of these ten data that causes the four averages to be so different?

2.4 | MEASURES OF DISPERSION

Measures of
dispersion

Having located the "middle" with the measures of central tendency, our search for information from data sets now turns to the measures of dispersion (spread). The **measures of dispersion** include the *range*, *variance*, and *standard deviation*. These numerical values describe the amount of spread, or variability, that is found among the data: Closely grouped data have relatively small values, and more widely spread-out data have larger values. The closest grouping occurs when the data has no dispersion (all data are the same value) for which the measure of dispersion will be zero. There is no limit to how widely spread out the data can be; therefore, measures of dispersion can be very large.

EXERCISE 2.67

USA Today, 1-10-95, reported the cost per square foot of U.S. city office space to be lowest in Houston ($13.15) and highest in Washington, D.C. ($32.43). Find the range.

RANGE

The difference in value between the highest-valued (H) and the lowest-valued (L) pieces of data:

$$range = high\ value - low\ value$$

$$range = H - L \tag{2-4}$$

The sample 3, 3, 5, 6, 8 has a range of

$$H - L = 8 - 3 = \mathbf{5}$$

The range of 5 tells us that these data all fall within a 5-unit interval (see Figure 2-21).

FIGURE 2-21

Range of {3, 3, 5, 6, 8}

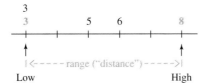

The other measures of dispersion to be studied in this chapter are measures of dispersion about the mean. To develop a measure of dispersion about the mean, let's first answer the question "How far is each x from the mean?"

DEVIATION FROM THE MEAN

A deviation from the mean, $x - \bar{x}$, is the difference between the value of x and the mean $\bar{x}$.

Each individual value x deviates from the mean by an amount equal to $(x - \bar{x})$. This deviation $(x - \bar{x})$ is zero when x is equal to the mean $\bar{x}$. The deviation $(x - \bar{x})$ is positive if x is larger than $\bar{x}$ and negative if x is smaller than $\bar{x}$.

Consider the sample 6, 3, 8, 5, 3. Using formula (2-1), $\bar{x} = \dfrac{\sum x}{n}$, we find that the mean is 5. Each deviation, $(x - \bar{x})$, is then found by subtracting 5 from each x-value.

Data	x	6	3	8	5	3
Deviation	$x - \bar{x}$	1	−2	3	0	−2

Figure 2-22 shows the four nonzero deviations from the mean.

FIGURE 2-22

Deviations from the Mean

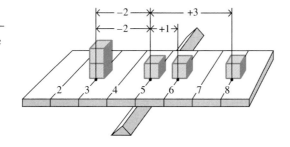

To describe the "average" value of these deviations we might use the mean deviation, the sum of the deviations divided by n, $\dfrac{\sum(x - \bar{x})}{n}$. However, since the sum of the deviations, $\sum(x - \bar{x})$, is exactly zero, the mean deviation will also be zero. As a matter of fact, it will always be zero. Since the mean deviation will always be zero, it will not be a useful statistic.

The sum of the deviations $\sum(x - \bar{x})$, is always zero, because of the neutralizing effect between the deviations of x-values smaller than the mean (which are negative) and those x-values larger than the mean (which are positive). This neutralizing effect can be removed if we do something to make all the deviations positive. This can be accomplished in two ways. First, by using the absolute value of the deviation, $|x - \bar{x}|$, we can treat each deviation for its "size" or distance only. For our illustration we obtain the following absolute deviations.

EXERCISE 2.68

a. The data value $x = 45$ has a deviation value of 12. Explain the meaning of this.

b. The data value $x = 84$ has a deviation value of −20. Explain the meaning of this.

Data	x	6	3	8	5	3
Absolute Value of Deviation:	$|x - \bar{x}|$	1	2	3	0	2

MEAN ABSOLUTE DEVIATION

EXERCISE 2.69

The mean of the absolute values of the deviations from the mean.

$$mean\ absolute\ deviation = \frac{sum\ of\ (absolute\ values\ of\ deviations)}{number}$$

The summation $\Sigma(x - \bar{x})$ is always zero. Why? Think back to the definition of the mean (p. 68) and see if you can justify this statement.

$$mean\ absolute\ deviation = \frac{\sum |x - \bar{x}|}{n} \qquad (2\text{-}5)$$

For our example, the sum of the absolute deviations is 8 $(1 + 2 + 3 + 0 + 2)$ and

$$mean\ absolute\ deviation = \frac{\sum |x - \bar{x}|}{n} = \frac{8}{5} = 1.6$$

Although this particular measure of spread is not used too frequently, it is a measure of dispersion. It tells us the mean "distance" the data is from the mean.

A second way to eliminate the positive–negative neutralizing effect is to square each of the deviations; squared deviations will all be nonnegative (positive or zero) values. The squared deviations are used to find the *variance*.

SAMPLE VARIANCE

The sample variance, s^2, is the mean of the squared deviations, calculated using $n - 1$ as the divisor.

$$sample\ variance:\quad s\text{-}squared = \frac{sum\ of\ (deviations)^2}{number - 1}$$

$$s^2 = \frac{\sum (x - \bar{x})^2}{n - 1} \qquad (2\text{-}6)$$

where n is the sample size, that is, the number of data in the sample.

The variance of our sample 6, 3, 8, 5, 3 is found in Table 2-10 using formula (2-6).

TABLE 2-10

Calculating Variance Using Formula (2-6)

Step 1. Find Σx.	Step 2. Find $\bar{x}$.	Step 3. Find each $x - \bar{x}$.	Step 4. Find $\Sigma(x - \bar{x})^2$.	Step 5. Sample variance.
6		$6 - 5 = 1$	$(1)^2 = 1$	
3	$\bar{x} = \dfrac{\sum x}{n}$	$3 - 5 = -2$	$(-2)^2 = 4$	$s^2 = \dfrac{\sum (x - \bar{x})^2}{n - 1} = \dfrac{18}{4} = 4.5$
8		$8 - 5 = 3$	$(3)^2 = 9$	
5	$\bar{x} = \dfrac{25}{5}$	$5 - 5 = 0$	$(0)^2 = 0$	
3		$3 - 5 = -2$	$(-2)^2 = 4$	
$\Sigma x = 25$	$\bar{x} = 5$	$\Sigma(x - \bar{x}) = 0$ ck	$\Sigma(x - \bar{x})^2 = 18$	

EXERCISE 2.70

Use formula (2-6) to find the variance for the sample {1, 3, 5, 6, 10}.

NOTES

1. The sum of all the x's is used to find $\bar{x}$.
2. The sum of the deviations, $\Sigma(x - \bar{x})$, is always zero, provided the exact value of $\bar{x}$ is used. Use this as a check in your calculations, as we did in Table 2-10.
3. If a rounded value of $\bar{x}$ is used, then the $\Sigma(x - \bar{x})$ will not always be exactly zero. It will, however, be reasonably close to zero.

The set of data in Exercise 2.70 is more dispersed than the set in Table 2-10, and therefore its variance is larger. A comparison of these two samples is shown in Figure 2-23.

FIGURE 2-23

Comparison of Data

STANDARD DEVIATION

The standard deviation of a sample, s, is the positive square root of the variance:

sample standard deviation: s = *square root of sample variance*

$$s = \sqrt{s^2} \tag{2-7}$$

For the samples shown in Figure 2-23, the standard deviations are $\sqrt{4.5}$, or **2.1**, and $\sqrt{11.5}$, or **3.4**.

NOTE The numerator for sample variance, $\Sigma(x - \bar{x})^2$, is often called the "sum of squares for x" and symbolized by SS(x). Thus, formula (2-6) can be expressed

sample variance: $s^2 = \dfrac{SS(x)}{n - 1}$,   where SS($x$) = $\Sigma(x - \bar{x})^2$ (2-8)

The formulas for variance can be modified into other forms for easier use in various situations. For example, suppose that we have the sample 6, 3, 8, 5, 2. The variance for this sample is computed in Table 2-11.

The arithmetic for this example has become more complicated because the mean contains nonzero digits to the right of the decimal point. However, the "sum of squares for x," the numerator of formula (2-6), can be rewritten:

sum of squares: $SS(x) = \Sigma x^2 - \dfrac{\left(\Sigma x\right)^2}{n}$ (2-9)

TABLE 2-11

Calculating Variance Using Formula (2-6)

Step 1. Find Σx.	Step 2. Find $\bar{x}$.	Step 3. Find each $x - \bar{x}$.	Step 4. Find $\Sigma(x - \bar{x})^2$.	Step 5. Sample variance.
6	$\bar{x} = \dfrac{\Sigma x}{n}$	$6 - 4.8 = 1.2$	$(1.2)^2 = 1.44$	$s^2 = \dfrac{\Sigma(x - \bar{x})^2}{n - 1}$
3		$3 - 4.8 = -1.8$	$(-1.8)^2 = 3.24$	
8	$\bar{x} = \dfrac{24}{5}$	$8 - 4.8 = 3.2$	$(3.2)^2 = 10.24$	$= \dfrac{22.8}{4} = 5.7$
5		$5 - 4.8 = 0.2$	$(0.2)^2 = 0.04$	
2	$\bar{x} = 4.8$	$2 - 4.8 = -2.8$	$(-2.8)^2 = 7.84$	
$\Sigma x = 24$		$\Sigma(x - \bar{x}) = 0$ ck	$\Sigma(x - \bar{x})^2 = 22.80$	

Combining formulas (2-8) and (2-9) yields the shortcut formula:

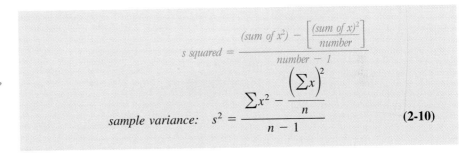

$$s \text{ squared} = \frac{(\text{sum of } x^2) - \left[\dfrac{(\text{sum of } x)^2}{\text{number}}\right]}{\text{number} - 1}$$

$$\text{sample variance:} \quad s^2 = \frac{\Sigma x^2 - \dfrac{\left(\Sigma x\right)^2}{n}}{n - 1} \tag{2-10}$$

Formulas (2-9) and (2-10) are called "shortcut" because they bypass the calculation of $\bar{x}$.

The computations for $SS(x)$, s^2, and s using formulas (2-9), (2-10), and (2-7) are performed as shown in Table 2-12.

TABLE 2-12

Calculating Standard Deviation Using the Shortcut Method

Step 1. Find Σx.	Step 2. Find Σx^2.	Step 3. Sum of squares.	Step 4. Variance.	Step 5. Standard deviation.
6	$6^2 = 36$	$SS(x) = \Sigma x^2 - \dfrac{\left(\Sigma x\right)^2}{n}$	$s^2 = \dfrac{\Sigma x^2 - \dfrac{\left(\Sigma x\right)^2}{n}}{n - 1}$	$s = \sqrt{s^2} = \sqrt{5.7} = 2.4$
3	$3^2 = 9$			
8	$8^2 = 64$	$= 138 - \dfrac{(24)^2}{5}$		
5	$5^2 = 25$		$= \dfrac{22.8}{4} = 5.7$	
2	$2^2 = 4$	$= 138 - 115.2$		
$\Sigma x = 24$	$\Sigma x^2 = 138$	$= 22.8$		

EXERCISE 2.71

Use formula (2-10) to find the variance of the sample {1, 3, 5, 6, 10}. Compare the results to Exercise 2.70.

The unit of measure for the standard deviation is the same as the unit of measure for the data. For example, if our data are in pounds, then the standard deviation s will also be in pounds. The unit of measure for variance might then be thought of as *units squared*. In our example of pounds, this would be *pounds squared*. As you can see, the unit has very little meaning.

MINITAB (Release 10) commands to find the standard deviation of the data listed in C1.

Session commands   *Menu commands*

Enter: STDEv C1   Choose: Calc > Column Stats
                                      Select: Standard deviation
                                      Enter: Input variable, C1

Additional statistics may be found using the following commands:

Session commands   *Menu Commands*

COUNt - Number of data in   Choose: Calc > Column Stats
        column

SUM   Sum of the data in column   Sum
MINImum   Smallest value in column   Minimum
MAXImum   Largest value in column   Maximum
RANGe   Range of values in column   Range
SSQ   Sum of squared x-values, Σx^2 Sum of squares

MULTIPLE FORMULAS

The reason we have multiple formulas is for convenience—that is, convenience relative to the situation. The following statements will help you decide which formula to use.

1. When you are working on a computer and using statistical software, you will generally store all the data values first. The computer is very capable with repeated operations and can "revisit" the stored data as often as necessary to complete a procedure. The computations for sample variance will be done using formula (2-6) following the process shown in Table 2-10. (See Exercise 2.85, p. 85.)
2. When you are working on a calculator with built-in statistical functions, the calculator must perform all necessary operations on each piece of data as the data are entered (most handheld calculators do not have the ability to store data). Then after all data has been entered, the computations will be completed using the appropriate summations. The computations for sample variance will be completed using formula (2-10), following the procedure shown in Table 2-12. (See Exercise 2.86, p. 85.)

3. If you are doing the computations either by hand or with the aid of a calculator, but not using statistical functions, the most convenient formula to use will depend on how many data there are and how convenient the numerical values are to work with.

When a formula has multiple forms, look for one of these icons:

 will be used to identify the formula most likely to be used by a computer.

 will be used to identify the formula most likely to be used by a calculator.

 will be used to identify the formula most likely to be convenient for hand calculations.

 will be used to identify the "definition" formula.

EXERCISES

2.72 Consider the sample 2, 4, 7, 8, 9. Find the following:
    a. range   b. variance s^2, using formula (2-6)   c. standard deviation s

2.73 Consider the sample 6, 8, 7, 5, 3, 7. Find the following:
    a. range   b. variance s^2, using formula (2-6)   c. standard deviation s

2.74 Given the sample 7, 6, 10, 7, 5, 9, 3, 7, 5, 13, find the following:
    a. variance s^2, using formula (2-6)   b. variance s^2, using formula (2-10)
    c. standard deviation s

2.75 Fifteen randomly selected college students were asked to state the number of hours they slept last night. The resulting data are 5, 6, 6, 8, 7, 7, 9, 5, 4, 8, 11, 6, 7, 8, 7. Find the following:
    a. variance s^2, using formula (2-6)   b. variance s^2, using formula (2-10)
    c. standard deviation s

2.76 An article titled "Financing Your Kids' College Education" (*Farming*, Sept./Oct. 1994) listed the following in-state tuition and fees per school year for 14 land-grant universities: 1554, 2291, 2084, 4443, 2884, 2478, 3087, 3708, 2510, 2055, 3000, 2052, 2550, 2013. Find the following:
    a. variance s^2 b. standard deviation s

2.77 Adding (or subtracting) the same number from each value in a set of data does not affect the measures of variability for that set of data.
    a. Find the variance of the following set of annual heating-degree-day data:

6017, 6173, 6275, 6350, 6001, 6300

    b. Find the variance of the following set of data [obtained by subtracting 6000 from each value in (a)]:

17, 173, 275, 350, 1, 300

2.78   An article titled "Railroads on the Fast Track" (*USA Today*, 10-7-94) gave the miles of track operated by the ten largest railroad companies. The table of results is given below.

Company	Miles of Track Operated
CSX Transportation	32,844
Burlington Northern	32,791
Union Pacific	30,315
Norfolk Southern	25,855
Conrail	24,183
Southern Pacific	18,768
Santa Fe Pacific	15,367
Chicago and North Western	8,620
Soo Line	7,265
Illinois Central	4,764

a. Find the mean and the standard deviation for the miles of track operated for these ten largest railroad companies.
b. Construct a graph that displays the spread of the data.
c. Discuss the relationship among the spread of the data, the mean, and the standard deviation.

2.79   Recruits for a police academy were required to undergo a test that measures their exercise capacity. The exercise capacity (measured in minutes) was obtained for each of 20 recruits:

25	27	30	33	30	32	30	34	30	27
26	25	29	31	31	32	34	32	33	30

a. Find the range.
b. Find the variance.
c. Find the standard deviation.
d. Using the dotplot drawn in answering Exercise 2.61 (p. 75), (1) draw a line representing the range, and (2) draw a line starting at the mean whose length represents the value of the standard deviation.
e. Describe how the distribution of data, the range, and the standard deviation are related.

2.80   a. Find the range and the standard deviation for the number of days per year of school, using the data in Exercuse 2.62 (p. 75).
b. Draw lines on the stem-and-leaf diagram drawn in answering Exercise 2.62 that represent the range and the standard deviation. Remember, the standard deviation is a measure for the mean.
c. Describe the relationship among the distribution of the data, the range, and the standard deviation.

2.81   Consider the following two sets of data:

Set 1:	46	55	50	47	52
Set 2:	30	55	65	47	53

Both sets have the same mean, which is 50. Compare these measures for both sets: $\Sigma(x - \bar{x})$, $\Sigma|x - \bar{x}|$, SS(x), and range. Comment on the meaning of these comparisons.

2.82 Comment on the following statement: "The mean loss for customers at First State Bank (which was not insured) was \$150. The standard deviation of the losses was −\$125."

2.83 Start with $x = 100$ and add 4 x-values, making a sample of five data such that
　　a. $s = 0$.　　b. $0 < s < 1$.　　c. $5 < s < 10$.　　d. $20 < s < 30$.

2.84 Each of two samples has a standard deviation of 5. If the two sets of data are made into one set of ten data, will the new sample have a standard deviation that is (a) less than, (b) about the same as, or (c) larger than the original standard deviation of 5? Make up two sets of five data each whose standard deviation is 5 and justify your answer. Include the calculations.

2.85 With the following set of data in column 1 on the data worksheet:

　　　　7　　6　　10　　7　　5　　9　　3　　7　　5　　13

enter these MINITAB commands on the session window:

```
LET K1 = SUM(C1)/COUNt(C1)
LET C2 = C1 − K1
LET C3 = C2**2
LET K2 = SUM(C3)/(COUNt(C1) − 1)
LET K3 = SQRT(K2)
PRINt C1-C3
PRINt K1-K3
STDEV C1
```

　　a. Explain what each command instructed the computer to do and what resulted.
　　b. Explain how each command relates to formula (2-6) and the process shown in Table 2-10.
　　c. Explain why this formula is well suited when using a computer.

2.86 With the following set of data in column 1 on the data worksheet:

　　　　7　　6　　10　　7　　5　　9　　3　　7　　5　　13

enter these MINITAB commands on the session window:

```
LET C2 = C1**2
LET K1 = COUNt(C1)
LET K2 = SUM(C1)
LET K3 = SUM(C2)
LET K4 = (K3 − (K2**2/K1))/(K1 − 1)
LET K5 = SQRT(K4)
PRINT C1 C2
PRINT K1-K5
STDEV C1
```

　　a. Explain what each command instructed the computer to do and what resulted.
　　b. Explain how each command relates to formula (2-10) and the process shown in Table 2-12.
　　c. Explain why this formula is best suited when using a calculator.

2.5 | MEAN AND STANDARD DEVIATION OF FREQUENCY DISTRIBUTION

When the sample data are in the form of a frequency distribution, we will need to make a slight adaptation to formulas (2-1) and (2-10) in order to find the mean, the variance, and the standard deviation.

▼| ILLUSTRATION   2 - 10

Find the mean, the variance, and the standard deviation for the sample data represented by the frequency distribution in Table 2-13.

TABLE 2-13

Ungrouped Frequency Distribution

x	f
1	5
2	9
3	8
4	6
	$\Sigma f = 28$

REMEMBER   This frequency distribution represents a sample of 28 values: five 1's, nine 2's, eight 3's, and six 4's.

In order to calculate the sample mean $\bar{x}$ and sample variance s^2 using formulas (2-1) and (2-10), we need the sum of the 28 x-values, Σx, and the sum of the 28 x-squared values, Σx^2.

The summations, Σx and Σx^2, could be found as follows:

$$\Sigma x = \underbrace{1 + 1 + \cdots + 1}_{5 \text{ of them}} + \underbrace{2 + 2 + \cdots + 2}_{9 \text{ of them}} + \underbrace{3 + 3 + \cdots + 3}_{8 \text{ of them}} + \underbrace{4 + 4 + \cdots + 4}_{6 \text{ of them}}$$

$$\Sigma x = (5)(1) + (9)(2) + (8)(3) + (6)(4)$$

$$\Sigma x = 5 + 18 + 24 + 24 = \mathbf{71}$$

$$\Sigma x^2 = \underbrace{1^2 + \cdots + 1^2}_{5 \text{ of them}} + \underbrace{2^2 + \cdots + 2^2}_{9 \text{ of them}} + \underbrace{3^2 + \cdots + 3^2}_{8 \text{ of them}} + \underbrace{4^2 + \cdots + 4^2}_{6 \text{ of them}}$$

$$\Sigma x^2 = (5)(1) + (9)(4) + (8)(9) + (6)(16)$$

$$\Sigma x^2 = 5 + 36 + 72 + 96 = \mathbf{209}$$

However, we will use the frequency distribution to determine these summations by expanding it to become an *extensions table*. The extensions xf and x^2f are formed by multiplying row by row; then three column totals are found. The objective of the extensions table is to obtain these three column totals. (See Table 2-14.)

TABLE 2-14

Ungrouped Frequency Distribution: Extensions xf and x^2f

x	f	xf	x^2f
1	5	5	5
2	9	18	36
3	8	24	72
4	6	24	96
	$\Sigma f = 28$	$\Sigma xf = 71$	$\Sigma x^2f = 209$

number of data   *sum of x, using frequencies*   *sum of x^2, using frequencies*

EXERCISE 2.87

A survey asking the "number of telephones" per household, x, was conducted; results are shown here as a frequency distribution.

x	f
0	1
1	3
2	8
3	5
4	3

a. Complete the extensions table.
b. Find the three summations, Σf, Σxf, Σx^2f, for the frequency distribution.
c. Describe what each of the following represents: $x = 4$, $f = 8$, Σf, Σxf.

NOTES

1. The extensions in the xf column are the subtotals of the like values.
2. The extensions in the x^2f column are the subtotals of the like squared values.
3. The three column totals, Σf, Σxf, and Σx^2f, are the same values as were previously known as n, Σx, and Σx^2, respectively. That is, $\Sigma f = n$, the number of pieces of data; $\Sigma xf = \Sigma x$, the sum of the data; and $\Sigma x^2f = \Sigma x^2$, the sum of the squared data.
4. Think of the f in the summation expressions Σxf and Σx^2f as an indication that the sums were obtained with the use of a frequency distribution.
5. The sum of the x-column is NOT a meaningful number. The x-column lists each possible value of x once, not accounting for the repeated values.

To find the mean of a frequency distribution, formula (2-1) on page 68 is modified to indicate the use of the frequency distribution.

$$x\ bar = \frac{sum\ of\ x,\ using\ frequencies}{number}$$

mean of frequency distribution: $\bar{x} = \dfrac{\Sigma xf}{\Sigma f}$ **(2-11)**

The mean value of x for the frequency distribution in Table 2-14 is found by using formula (2-11).

mean: $\bar{x} = \dfrac{\Sigma xf}{\Sigma f} = \dfrac{71}{28} = 2.536 = $ **2.5**

To find the variance of the frequency distribution, formula (2-10) is modified to indicate the use of the frequency distribution.

EXERCISE 2.88

Explain why (a) the "sum of the x-column" has no relationship to the "sum of the data" and (b) the "Σxf" represents the "sum of the data" represented by the frequency distribution in Exercise 2.87.

$$s\text{ squared} = \frac{(\text{sum of }x^2\text{, using frequencies}) - \left[\dfrac{(\text{sum of }x\text{, using frequencies})^2}{\text{number}}\right]}{\text{number} - 1}$$

variance of frequency distribution:

$$s^2 = \frac{\Sigma x^2 f - \dfrac{\left(\Sigma xf\right)^2}{\Sigma f}}{\Sigma f - 1} \tag{2-12}$$

The variance of x for the frequency distribution in Table 2-14 is found by using the formula (2-12).

$$\text{variance:}\quad s^2 = \frac{\Sigma x^2 f - \dfrac{\left(\Sigma xf\right)^2}{\Sigma f}}{\Sigma f - 1} = \frac{209 - \dfrac{(71)^2}{28}}{28 - 1} = \frac{28.964}{27} = 1.073 = \mathbf{1.1}$$

The standard deviation of x for the frequency distribution in Table 2-14 is found by using formula (2-7), the positive square root of variance.

$$\text{standard deviation: } s = \sqrt{s^2} = \sqrt{1.073} = 1.036 = \mathbf{1.0}$$

▼ **ILLUSTRATION   2 - 11**

Find the mean, variance, and standard deviation of the sample of 50 exam scores using the grouped frequency distribution found in Table 2-7 (p. 57).

SOLUTION We will use an extensions table to find the three summations in the same manner we did in Illustration 2-10. The class marks will be used as the representative values for the classes.

TABLE 2-15

Frequency Distribution of 50 Exam Scores:

Class Number	Class Marks, x	f	xf	$x^2 f$
1	40	2	80	3200
2	50	2	100	5000
3	60	7	420	25200
4	70	13	910	63700
5	80	11	880	70400
6	90	11	990	89100
7	100	4	400	40000
		$\Sigma f = 50$	$\Sigma xf = 3780$	$\Sigma x^2 f = 296600$

EXERCISE 2.89

Find the mean of the data shown in the frequency distribution in Exercise 2.87.

The mean value of x for the frequency distribution in Table 2-15 is found by using formula (2-11).

EXERCISE   2.90

Find the variance for the data shown in the frequency distribution in Exercise 2.87.

mean: $\bar{x} = \dfrac{\sum xf}{\sum f} = \dfrac{3780}{50} = \mathbf{75.6}$

The variance of x for the frequency distribution in Table 2-15 is found by using formula (2-12).

variance: $s^2 = \dfrac{\sum x^2 f - \dfrac{\left(\sum xf\right)^2}{\sum f}}{\sum f - 1} = \dfrac{296{,}600 - \dfrac{3780^2}{50}}{50 - 1} = \dfrac{10{,}832}{49} = 221.0612 = \mathbf{221.1}$

The standard deviation of x for the frequency distribution in Table 2-15 is found by using formula (2-7).

standard deviation: $s = \sqrt{s^2} = \sqrt{221.0612} = 14.868 = \mathbf{14.9}$

▲ |

EXERCISE   2.91

Find the standard deviation for the data shown in the frequency distribution in Exercise 2.87.

EXERCISE   2.92

Find the mean, variance, and standard deviation of the data shown in the following frequency distribution.

Class	f
2–6	2
6–10	10
10–14	12
14–18	9
18–22	7

MINITAB (Release 10) commands to find the mean, variance, and standard deviation of a frequency distribution where the class marks are listed in C1 and the frequencies in C2.

Session commands

To obtain the extensions table:
```
    LET C3 = C1*C2
    LET C4 = C1*C3
    LET K1 = SUM(C2)
    LET K2 = SUM(C3)
    LET K3 = SUM(C4)
    PRINt C1-C4
    PRINt K1-K3
```

To find mean:
```
    LET K4 = K2/K1
    PRINt K4
```

To find variance:
```
    LET K5 = (K3-(K2**2/K1))/(K1-1)
    PRINt K5
```

To find standard deviation:
```
    LET K6 = SQRT(K5)
    PRINt K6
```

Helpful Hint: When using a series of session commands, instead of entering them one at a time, go to an editable area on the session window (any place above the MTB > prompt), and type all of the commands, one command to a line. When you have them all typed correctly, highlight the entire set of commands, COPY them, and PASTE them at the active MTB > prompt. Press the ENTER key and the computer will execute the entire set of the commands in sequence.

EXERCISES

2.93   A survey of medical doctors asked the number of children each had fathered. The results are summarized by this ungrouped frequency distribution:

Number of Children	0	1	2	3	4	6
Number of Doctors	15	12	26	14	4	2

Calculate the sample mean, variance, and standard deviation for the number of children the doctors had fathered.

2.94   The weight gains (in grams) for chicks fed on a high-protein diet were as follows:

Weight Gain	Frequency
12.5	2
12.7	6
13.0	22
13.1	29
13.2	12
13.8	4

a. Find the mean.
b. Find the variance.
c. Find the standard deviation.

2.95   The October 15, 1993, issue of *Library Journal* (Vol. 118, No. 17) gives the following table for the salaries of minority placements by type of library.

Library Type	Number	Average Salary
Academic	46	$27,825
Public	34	24,657
School	23	30,336
Special	16	29,406
Other	4	25,200

a. Find the total of all salaries for the above 123 individuals.
b. Find the mean salary for the above 123 individuals.
c. What is the modal library type? Explain.
d. Find the standard deviation for the above 123 salaries.

2.96   The November 4–6, 1994, issue of *USA Today* gave the following age distirbution for individuals aged 30–64 in 1994. Its source is the U.S. Census Bureau.

Age	1994 Population (in millions)
30–49	80.5
50–54	13.2
55–64	21.0

Estimate the mean of this distribution by replacing each age group by its midpoint.

2.97 Find the variance and the standard deviation for the following grouped frequency distribution.

Class Limits	f
3–6	2
6–9	10
9–12	12
12–15	9
15–18	7

2.98 Find the mean and the variance for the following grouped frequency distribution.

Class Limits	f
2–6	7
6–10	15
10–14	22
14–18	14
18–22	2

2.99 A sheet-metal firm employs several troubleshooters to make emergency repairs of furnaces. Typically, the troubleshooters take many short trips. For the purpose of estimating travel expenses for the coming year, the firm took a sample of 20 travel-expense vouchers related to troubleshooting. The following information resulted.

Dollar Amount on Voucher	Number of Vouchers
$0.00–10.00	2
10.00–20.00	8
20.00–30.00	7
30.00–40.00	2
40.00–50.00	1
Total in sample	20

Calculate the mean and the standard deviation for these travel-expense dollar amounts.

2.100 The following distribution of commuting distances was obtained for a sample of Mutual of Nebraska employees.

Distance (miles)	Frequency
1.0–3.0	2
3.0–5.0	6
5.0–7.0	12
7.0–9.0	50
9.0–11.0	35
11.0–13.0	15
13.0–15.0	5

Find the mean and standard deviation for the commuting distances.

2.101 Find the mean and standard deviation for the set of speeds for the 55 cars given in Exercise 2.41. Use the frequency distribution found in answering 2.41 on page 66.

2.102 A quality-control technician selected 25 one-pound boxes from a production process and found the following distribution of weights (in ounces).

Weight	Frequency
15.95–15.98	2
15.98–16.01	4
16.01–16.04	15
16.04–16.07	3
16.07–16.10	1

Find the mean and standard deviation weight for this distribution.

2.103 The USA Snapshot® "Nuns an aging order" reports that the median age of the USA's 94,022 Roman Catholic nuns is 65 years and that the percentage of nuns in the country by age group is:

Under 50	51–70	Over 70	Refused to give age
16%	42%	37%	5%

This information is based on a survey of 1049 Roman Catholic nuns.

Suppose the survey had resulted in the frequency distribution on the next page (52 ages unknown). (See histogram drawn as answer to Exercise 2.40, p. 65.)

    a. Find the mean, median, mode, and midrange for this distribution of ages.

    b. Find the variance and standard deviation.

Age	20–30	30–40	40–50	50–60	60–70	70–80	80–90
Freq.	34	58	76	187	254	241	147

2.104   The amount of money adults say they will spend on gifts during this holiday was described in "What 'Santa' Will Spend," *USA Today*, 11-23-94.

	Nothing	$1–$300	$301–$600	$601–$1000	Over $1000	Didn't Know
Percent who said	1%	24%	30%	20%	14%	11%

Average: $734

Let the following distribution represent that part of the sample who did know:

x, amount	0	150	450	800	1500
Frequency	1	24	30	20	14

a. Find the mean of the frequency distribution.
b. Do you believe the average reported could have been the mean? Explain.
c. Find the median of the frequency distribution.
d. Do you believe the average reported could have been the median? Explain.
e. Find the mode of the frequency distribution.
f. Do you believe the average reported could have been the mode? Explain.
g. Could the average reported have been the midrange? If so, what was the largest amount of money reported?

2.105   The ages of several thousand level 1 employees is being described by using the following random sample.

Ages	48	33	47	49	33	44	51	40	45	50	32	33	44
	32	47	55	33	44	42	28	37	46	35	40	40	41
	47	38	35	40	39	38	35	39	43				

Chris has been assigned the task of calculating the mean and the standard deviation, but he is not sure if he should use a grouped frequency distribution or not.
a. Find the mean and the standard deviation without using a grouped frequency distribution.
b. Find the mean and the standard deviation using the frequency distribution. (Form a grouped frequency distribution using 28–32, 32–36, . . . as the classes.)
c. Compare the two sets of answers. Use of the frequency distribution caused an error. What percent is this error?

(continued)

d. Look at the seven data that fell into class 40–44. Find the Σx and Σx^2 for these seven data. What values were substituted for these two summations when the frequency distribution was used?

e. What caused answer (b) to be different from answer (a)? Explain this difference.

f. Why are both sets of answers estimates? And why is the error introduced by the frequency distribution therefore not a big deal?

g. Which method would you advise Chris to use?

2.106 Use a computer to calculate the mean, the variance, and the standard deviation for the frequency distribution of prices of previously owned homes found in Exercise 2.46, page 67.

2.6 | MEASURES OF POSITION

Measures of position are used to describe the position a specific data value possesses in relation to the rest of the data. **Quartiles** and **percentiles** are two of the most popular measures of position.

QUARTILES

Values of the variable that divide the ranked data into quarters; each set of data has three quartiles. The *first quartile*, Q_1, is a number such that at most 25% of the data are smaller in value than Q_1 and at most 75% are larger. The *second quartile* is the median. The *third quartile*, Q_3, is a number such that at most 75% of the data are smaller in value than Q_3 and at most 25% are larger (see Figure 2-24).

FIGURE 2-24

Quartiles

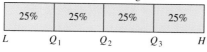

The procedure for determining the value of the quartiles is the same as that for percentiles and is shown in the following description of percentiles.

PERCENTILES

Values of the variable that divide a set of ranked data into 100 equal subsets; each set of data has 99 percentiles (see Figure 2-25). The kth percentile, P_k, is a value such that at most $k\%$ of the data are smaller in value than P_k and at most $(100 - k)\%$ of the data are larger (see Figure 2-26).

FIGURE 2-25

Percentiles

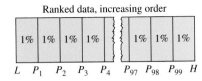

Percentiles

FIGURE 2-25

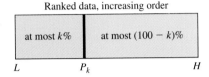

FIGURE 2-26

kth Percentile

NOTES

1. The 1st quartile and the 25th percentile are the same; that is, $Q_1 = P_{25}$. Also, $Q_3 = P_{75}$.
2. The median, the 2nd quartile, and the 50th percentile are all the same: $\tilde{x} = Q_2 = P_{50}$. Therefore, when asked to find P_{50} or Q_2, use the procedure for finding the median.

The procedure for determining the value of any kth percentile (or quartile) involves four basic steps as outlined on the diagram in Figure 2-27. Illustration 2-12 demonstrates the procedure.

FIGURE 2-27

Finding P_k
Procedure

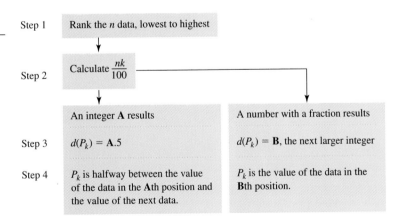

▼| ILLUSTRATION   2 - 12

Using the sample of 50 elementary statistics final exam scores listed in Table 2-16, find the first quartile Q_1, the 58th percentile P_{58}, and the third quartile Q_3.

TABLE 2-16

Raw Scores for Elementary Statistics Exam

60	72	90	95	77	97	58	85	72	80
47	67	77	74	39	70	78	82	86	91
82	66	86	72	90	64	89	83	50	75
95	68	58	88	63	70	44	72	94	76
88	98	64	74	68	70	55	77	92	78

EXERCISE 2.107

Using the concept of depth, describe the position of 91 in the set of 50 exam scores in two different ways.

SOLUTION

STEP 1   Rank the data: A ranked list may be formulated (see Table 2-17), or a graphic display showing the ranked data may be used. The dotplot or the stem-and-leaf are handy for this purpose. The stem-and-leaf is especially helpful since it gives depth numbers counted from both extremes when it is computer generated (see Figure 2-28).

TABLE 2-17 Ranked Data: Statistics Exam Scores

39	64	72	78	89
44	66	72	80	90
47	[67]	74	82	90
50	68	74	82	91
55	68	75	83	92
58	70	76	85	94
58	70	77	86	95
60	70	77	[86]	95
63	72	[77]	88	97
64	72	[78]	88	98

FIGURE 2-28

Final Exam Scores

Stem-and-leaf of Score $N = 50$
Leaf Unit = 1.0

1	3	9
2	4	4
3	4	7
4	5	0
7	5	5 8 8
11	6	0 3 4 4
15	6	6 [7] 8 8
24	7	0 0 0 2 2 2 2 4 4
(7)	7	5 6 7 7 [7] 8 8
19	8	0 2 2 3
15	8	5 6 [6] 8 8 9
9	9	0 0 1 2 4
4	9	5 5 7 8

13th position from L

29th and 30th position from L

13th position from H

Find Q_1:

STEP 2   Find $\dfrac{nk}{100}$: $\dfrac{nk}{100} = \dfrac{(50)(25)}{100} = $ **12.5** ($n = 50$ and $k = 25$, since $Q_1 = P_{25}$.)

STEP 3   Find the depth of Q_1: $d(Q_1) = $ **13** (Since 12.5 contains a fraction, **B** is next larger integer, 13.)

STEP 4 Find Q_1: Q_1 is the 13th value, counting from L (see Table 2-17 or Figure 2-28). $Q_1 = \mathbf{67}$

Find P_{58}:

EXERCISE 2.108

Find P_{20} and P_{35} for the exam scores in Table 2-17.

STEP 2 Find $\frac{nk}{100}$: $\frac{nk}{100} = \frac{(50)(58)}{100} = \mathbf{29}$ ($n = 50$ and $k = 58$, since P_{58}.)

STEP 3 Find the depth of P_{58}: $d(P_{58}) = \mathbf{29.5}$ [Since **A** = 29 (an integer), add .5, use 29.5.]

STEP 4 Find P_{58}: P_{58} is the value halfway between the values of the 29th and the 30th pieces of data counting from L (see Table 2-17 or Figure 2-28).

$$P_{58} = \frac{77 + 78}{2} = 77.5$$

Optional technique: When k is greater than 50, subtract k from 100 and use $(100 - k)$ in place of k in step 2. The depth is then counted from the largest valued data H. Find Q_3 using the optional technique:

EXERCISE 2.109

Find P_{80} and P_{95} for the exam scores in Table 2-17.

STEP 2 Find $\frac{nk}{100}$: $\frac{nk}{100} = \frac{(50)(25)}{100} = \mathbf{12.5}$ ($n = 50$ and $k = 75$, since $Q_3 = P_{75}$, and $k > 50$; use $100 - k = 100 - 75 = 25$.)

STEP 3 Find the depth of Q_3 from H: $d(Q_3) = \mathbf{13}$

STEP 4 Find Q_3: Q_3 is the 13th value, counting from H (see Table 2-17 or Figure 2-28). $Q_3 = \mathbf{86}$ ▲

An additional measure of central tendency, the **midquartile**, can now be defined.

Midquartile

MIDQUARTILE

The numerical value midway between the first quartile and the third quartile.

$$\text{midquartile} = \frac{Q_1 + Q_3}{2} \quad \textbf{(2-13)}$$

▼ ILLUSTRATION 2-13

Find the midquartile for the set of 50 exam scores given in Illustration 2-12.

SOLUTION $Q_1 = 67$ and $Q_3 = 86$, as found in Illustration 2-12. Thus,

$$\text{midquartile} = \frac{Q_1 + Q_3}{2} = \frac{67 + 86}{2} = \mathbf{76.5}$$

▲

The median, the midrange, and the midquartile are not necessarily the same value. They are each middle values, but by different definitions of the middle. Figure 2-29 summarizes the relationship of these three statistics as applied to the 50 exam scores from Illustration 2-12.

FIGURE 2-29

Final Exam Scores

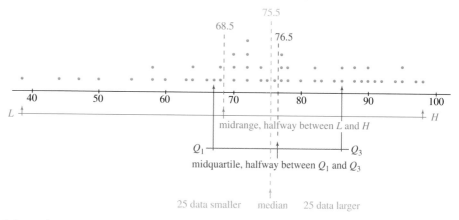

EXERCISE   2.110

What property does the distribution need for the median, the mid-range, and the midquartile to all be the same value?

A 5-number summary is very effective in describing a set of data. It is easy information to obtain and is very informative to the reader.

5-NUMBER SUMMARY

The 5-number summary is composed of:

1. L, the smallest value in the data set,
2. Q_1, the first quartile (also called P_{25}, 25th percentile),
3. $\tilde{x}$, the median,
4. Q_3, the third quartile (also called P_{75}, 75th percentile),
5. H, the largest value in the data set.

The 5-number summary for the set of 50 exam scores in Illustration 2-12 is

39	67	75.5	86	98
L	Q_1	$\tilde{x}$	Q_3	H

Interquartile range

Notice that these five numerical values divide the set of data into four subsets, with one-quarter of the data in each subset. From the 5-number summary we can observe how much the data is spread out in each of the quarters. An additional measure of dispersion can now be defined. The **interquartile range** is the difference between the first and third quartiles. It is the range of the middle 50% of the data.

The 5-number summary is even more informative when it is displayed on a diagram drawn to scale. One of the computer-generated graphic displays that accomplishes this is known as the **box-and-whisker display**.

Box-and-whisker
display

BOX-AND-WHISKER DISPLAY

A graphic representation of the 5-number summary. The five numerical values (smallest, first quartile, median, third quartile, and largest) are located on a scale, either vertical or horizontal. The box is used to depict the middle half of the data that lies between the two quartiles. The whiskers are line segments used to depict the other half of the data: One line segment represents the quarter of the data that is smaller in value than the first quartile, and a second line segment represents the quarter of the data that is larger in value than the third quartile.

Figure 2-30 pictures a box-and-whiskers display of the 50 exam scores.

FIGURE 2-30
─────────────
Box-and-Whiskers

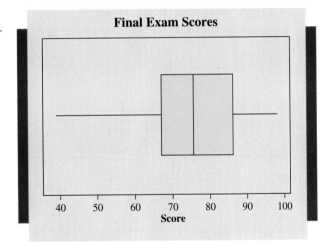

EXERCISE 2.111

Draw a box-and-whisker display for the set of data whose 5-number summary is 42–62–72–82–97.

MINITAB (Release 10) commands to find the 5-number summary values (plus a few other things) for the data in C1.

Session commands

Enter: DESCribe C1

MINITAB (Release 10) commands to construct a box-and-whisker diagram with the data listed in C1.

Session commands	Menu commands

Enter: BOXPlot C1;
      TITLe 'your title';
      TRANspose.
	Choose: Graph > Boxplot
	Enter variable: C1
	Options:
	   Annotations > Title
	   Enter: your title
	Options: Select Transpose

z-score

The position of a specific value can be measured in terms of the mean and standard deviation using the **z-score**, also called the *standard score*.

---

z-SCORE

The position a particular value of x has relative to the mean measured in standard deviations. The z- score is found by the formula

$$z = \frac{\text{value} - \text{mean}}{\text{st. dev.}} = \frac{x - \bar{x}}{s} \qquad \textbf{(2-14)}$$

▼| ILLUSTRATION   2 - 14

Find the standard scores for (a) 92 and (b) 72 with respect to a sample of exam grades that have a mean score of 75.9 and a standard deviation of 11.1.

SOLUTION

a. $x = 92, \bar{x} = 75.9; s = 11.1$. Thus, $z = \dfrac{x - \bar{x}}{s} = \dfrac{92 - 75.9}{11.1} = \dfrac{16.1}{11.1} = \textbf{1.45}$.

b. $x = 72, \bar{x} = 75.9; s = 11.1$. Thus, $z = \dfrac{x - \bar{x}}{s} = \dfrac{72 - 75.9}{11.1} = \dfrac{-3.9}{11.1} = \textbf{-0.35}$.

This means that the score 92 is approximately one-and-one-half standard deviations above the mean, while the score 72 is approximately one-third of a standard deviation below the mean.

▲|

NOTES

1. Typically, the calculated value of z is rounded to the nearest hundredth.
2. z-scores typically range in value from approximately -3.00 to $+3.00$.

EXERCISE 2.112

Find the *z*-score for
test scores of 92
and 63 on a test
with a mean of 72
and a standard
deviation of 12.

Since the *z*-score is a measure of relative position with respect to the mean, it can be used to help make a comparison of two raw scores that come from separate populations. For example, suppose that you want to compare a grade you received on a test with a friend's grade on a comparable exam in her course. You received a raw score of 45 points; she obtained 72 points. Is her grade better? We need more information before we can draw such a conclusion. Suppose that the mean on the exam you took was 38 and the mean on her exam was 65. Your grades are both 7 points above the mean, so we still can't draw a definite conclusion. However, the standard deviation on the exam you took was 7 points, and it was 14 points on your friend's exam. This means that your score is one (1) standard deviation above the mean ($z = 1.0$), whereas your friend's grade is only one-half of a standard deviation above the mean ($z = 0.5$). Since your score has the "better" relative position, we would conclude that your score is slightly better than your friend's score. (Again, this is speaking from a relative point of view.)

Additional MINITAB (Release 10) session commands that can be very helpful in analyzing data are as follows:

Session commands	*Description*
Enter: SORT C1 C2	Sorts data in C1 into ascending order and stores it in C2.
TABLe C1	Forms ungrouped frequency distribution of integer data.
PRINt C1	Prints data in C1 horizontally on session window.
PRINt C1 C2	Prints data in C1 and C2 vertically on session window.
PRINt C1 - C5	Prints data in all columns vertically on session window.

MINITAB (Release 10) can be used to generate random samples of K data from a specified theoretical population and the data will be put into C1.

Session commands	*Menu commands*
Enter: RANDom K observations into C1;	Choose: Calc > Random Data
Identify type of data using one of these subcommands:	Select: Type of data wanted
Normal μ σ.	Enter: Number of rows(data) K
Uniform L to H.	To be stored in C1
EXPOnential μ.	Population parameters
GAMMa A B.	(μ, σ, L, H, A, or B)

MINITAB (Release 10) can be used to select random samples of K data from a specified column of existing data. Selection will be without replacement unless you request replacement.

(continued)

MINITAB *(continued)*

Session commands	*Menu commands*
If sampling without replacement: Enter: SAMPle K of C1 into C2	Choose: Calc > Rand Data > Samp Columns Enter: Number of rows(data): K From: C1
If sampling with replacement: Enter: SAMPle K of C1 into C2; REPLace.	To be stored in: C2 Select: Replacement, if wanted

EXERCISES

2.113 The following data are the yields, in pounds, of hops.

3.9	3.4	5.1	2.7	4.4
7.0	5.6	2.6	4.8	5.6
7.0	4.8	5.0	6.8	4.8
3.7	5.8	3.6	4.0	5.6

a. Find the first and the third quartiles of the yield.
b. Find the midquartile.
c. Find the following percentiles: (1) P_{15}, (2) P_{33}, (3) P_{90}.

2.114 A research study of manual dexterity involved determining the time required to complete a task. The time required for each of 40 disabled individuals is as follows (data are ranked):

7.1	7.2	7.2	7.6	7.6	7.9	8.1	8.1	8.1	8.3
8.3	8.4	8.4	8.9	9.0	9.0	9.1	9.1	9.1	9.1
9.4	9.6	9.9	10.1	10.1	10.1	10.2	10.3	10.5	10.7
11.0	11.1	11.2	11.2	11.2	12.0	13.6	14.7	14.9	15.5

Find: a. Q_1 b. Q_2 c. Q_3 d. P_{95} e. the 5-number summary
f. Draw the box-and-whisker display.

2.115 Consider the following set of ignition times that were recorded for a synthetic fabric.

30.1	30.1	30.2	30.5	31.0	31.1	31.2	31.3	31.3	31.4
31.5	31.6	31.6	32.0	32.4	32.5	33.0	33.0	33.0	33.5
34.0	34.5	34.5	35.0	35.0	35.6	36.0	36.5	36.9	37.0
37.5	37.5	37.6	38.0	39.5					

Find: a. the median   b. the midrange   c. the midquartile   d. the 5-number summary
e. Draw the box-and-whisker display.

2.116 An article titled "Can Primary Care Physicians' Questions Be Answered Using the Medical Journal Literature?" (*Bulletin of the Medical Library Association*, Vol. 82, No. 2, April 1994) gave the following information concerning total search and selection time (in minutes) for finding relevant medical-journal literature for physician-submitted questions: mean = 43.2, 10th percentile = 16.0, and 90th percentile = 74.4.

a. Is the distribution of times symmetrical? Explain.

Suppose the total search and selection times for 15 physician-submitted questions were as follows: 35, 55, 25, 75, 80, 50, 45, 65, 70, 40, 30, 45, 30, 55, 70.

b. Find the mean of this sample.

c. Find the 10th percentile and the 90th percentile for this sample of times.

d. Are the sample data symmetrically distributed? Explain your answer using the answers found in (b) and (c) and at least one graph of the data.

2.117 A sample has a mean of 50 and a standard deviation of 4.0. Find the z-score for each value of x.

a. $x = 54$ b. $x = 50$ c. $x = 59$ d. $x = 45$

2.118 An exam produced grades with a mean score of 74.2 and a standard deviation of 11.5. Find the z-score for each of the following test scores, x:

a. $x = 54$ b. $x = 68$ c. $x = 79$ d. $x = 93$

2.119 A nationally administered test has a mean of 500 and a standard deviation of 100. If your standard score on this test was 1.8, what was your test score?

2.120 A sample has a mean of 120 and a standard deviation of 20.0. Find the value of x that corresponds to each of these standard scores:

a. $z = 0.0$ b. $z = 1.2$ c. $z = -1.4$ d. $z = 2.05$

2.121 a. What does it mean to say that $x = 152$ has a standard score of $+1.5$?

b. What does it mean to say that a particular value of x has a z-score of -2.1?

c. In general, the standard score is a measure of what?

2.122 In a study involving mastery learning (*Research in Higher Education*, Vol. 20, No. 4, 1984), 34 students took a pretest. The mean score was 11.04, and the standard deviation was 2.36. Find the z-score for scores of 9 and 15 on the 20-question pretest.

2.123 Which x-value has the higher value relative to the set of data from which it comes?

A: $x = 85$, where mean $= 72$ and standard deviation $= 8$

B: $x = 93$, where mean $= 87$ and standard deviation $= 5$

2.124 Which x-value has the lower relative position with respect to the set of data from which it comes?

A: $x = 28.1$, where $\bar{x} = 25.7$ and $s = 1.8$

B: $x = 39.2$, where $\bar{x} = 34.1$ and $s = 4.3$

2.7 | INTERPRETING AND UNDERSTANDING STANDARD DEVIATION

Standard deviation is a measure of fluctuation (dispersion) in the data. It has been defined as a value calculated with the use of formulas. But you may wonder what it really is. It is a kind of yardstick by which we can compare the variability of one set of data with another. This particular "measure" can be understood further by examining two statements: **Chebyshev's theorem** and the **empirical rule**.

EXERCISE   2.125

Instructions for an essay assignment include the statement "The length is to be within 25 words of 200." What values of *x*, number of words, satisfy these instructions?

CHEBYSHEV'S THEOREM

The proportion of any distribution that lies within k standard deviations of the mean is at least $1 - \frac{1}{k^2}$, where k is any positive number larger than 1. This theorem applies to all distributions of data.

This theorem says that within two standard deviations of the mean ($k = 2$) you will always find at least 75% (that is, 75% or more) of the data.

$$1 - \frac{1}{k^2} = 1 - \frac{1}{2^2} = 1 - \frac{1}{4} = \frac{3}{4} = 0.75, \textbf{ at least 75\%}$$

Figure 2-31 shows a mounded distribution that illustrates at least 75%.

FIGURE 2-31

Chebyshev's Theorem with $k = 2$

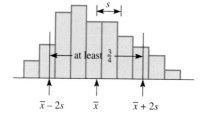

If we consider the interval enclosed by three standard deviations on either side of the mean ($k = 3$), the theorem says that we will always find at least 89% (that is, 89% or more) of the data.

$$1 - \frac{1}{k^2} = 1 - \frac{1}{3^2} = 1 - \frac{1}{9} = \frac{8}{9} = 0.89, \textbf{ at least 89\%}$$

Figure 2-32 shows a mounded distribution that illustrates at least 89%.

FIGURE 2-32

Chebyshev's Theorem with $k = 3$

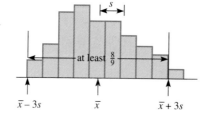

EMPIRICAL RULE

If a variable is normally distributed, then: within one standard deviation of the mean there will be approximately 68% of the data; within two standard

EXERCISE   2.126

According to Chebyshev's theorem, what proportion of a distribution will be within $k = 4$ standard deviations of the mean?

deviations of the mean there will be approximately 95% of the data; and within three standard deviations of the mean there will be approximately 99.7% of the data. [This rule applies specifically to a normal (bell-shaped) distribution, but it is frequently applied as an interpretive guide to any mounded distribution.]

Figure 2-33 shows the intervals of one, two, and three standard deviations about the mean of an approximately normal distribution. Usually these proportions do not occur exactly in a sample, but your observed values will be close when a large sample is drawn from a normally distributed population.

FIGURE   2-33

Empirical Rule

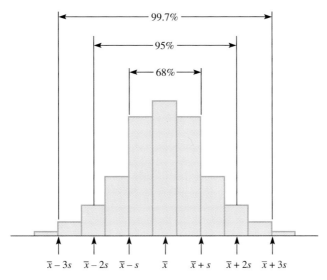

EXERCISE   2.127

a. What proportion of a normal distribution is greater than the mean?
b. What proportion is within one standard deviation of the mean?
c. What proportion is greater than a value that is one standard deviation below the mean?

The empirical rule can be used to determine whether or not a set of data is approximately normally distributed. Let's demonstrate this application by working with the distribution of final exam scores that we have been using throughout this chapter. The mean, $\bar{x}$, was found to be 75.6, and the standard deviation, s, was 14.9. The interval from one standard deviation below the mean, $\bar{x} - s$, to one standard deviation above the mean, $\bar{x} + s$, is $75.6 - 14.9 = 60.7$ to $75.6 + 14.9 = 90.5$. This interval (60.7 to 90.5) includes 61, 62, 63, . . . , 89, 90. Upon inspection of the ranked data (Table 2-17, p. 96), we see that 35 of the 50 data pieces, or 70%, lie within one standard deviation of the mean. Further, $\bar{x} - 2s = 75.6 - (2)(14.9) = 75.6 - 29.8 = 45.8$, to $\bar{x} + 2s = 75.6 + 29.8 = 105.4$, and the interval from 45.8 to 105.4, two standard deviations about the mean, includes 48 of the 50 data pieces, or 96%. All 50 data, or 100%, are included within three standard deviations of the mean (from 30.9 to 120.3). This information can be placed in a table for comparison with the values given by the empirical rule (see Table 2-18).

TABLE 2-18

Observed
Percentages
Interval Versus
the Empirical

Interval	Empirical Rule Percentage	Percentage Found
$\bar{x} - s$ to $\bar{x} + s$	≈ 68	70
$\bar{x} - 2s$ to $\bar{x} + 2s$	≈ 95	96
$\bar{x} - 3s$ to $\bar{x} + 3s$	≈ 99.7	100

The percentages found are reasonably close to those of the empirical rule. By combining this evidence with the shape of the histogram, we can safely say that the final exam data are approximately normally distributed.

If a distribution is approximately normal, it will be nearly symmetrical and the mean will divide the distribution in half (the mean and the median are the same in a symmetrical distribution). This allows us to refine the empirical rule. Figure 2-34 shows this refinement.

FIGURE 2-34

Refinement of
Empirical Rule

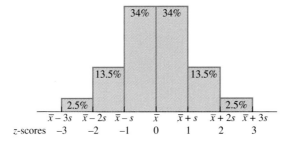

There is also a graphic way to test for normality. This is accomplished by drawing a relative frequency ogive of the grouped data on *probability paper* (which can be purchased at your college bookstore). On this paper the vertical scale is measured in percentages and is placed on the right side of the graph paper. All the directions and guidelines given on pages 62–63 for drawing an ogive must be followed. An ogive of the statistics final exam scores is drawn and labeled on a piece of probability paper in Figure 2-35. (Dashed lines are for later references to this drawing.)

Probability paper

To **test for normality**, first draw a straight line from the lower-left corner to the upper-right corner of the graph connecting the "next-to-end" points of the ogive. Then, if the ogive lies close to this straight line, the distribution is said to be approximately normal. The ogive for an exactly normal distribution will trace the straight line. The dashed line (I) in Figure 2-35 is the straight-line test for normality. The ogive suggests that the distribution of exam scores is approximately normal. (*Warning:* This graphic technique is very sensitive to the scale used along the horizontal axis.)

Test for normality

The ogive is a handy tool to use for finding percentiles of a frequency distribution. If the cumulative frequency distribution is drawn on probability paper, as in Figure 2-35, it is a simple matter to determine any kth percentile. Locate the value k on the vertical scale along the right-hand side, and follow the horizontal line for k until

FIGURE 2-35

Ogive on
Probability Paper

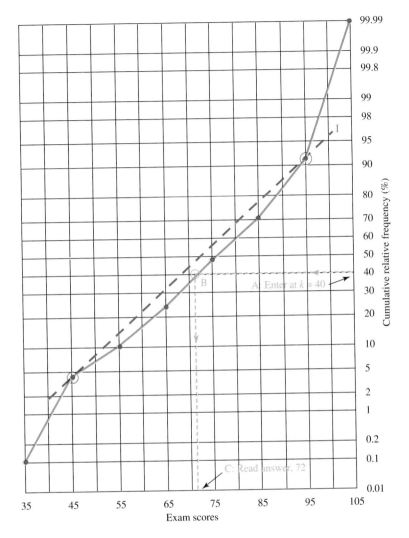

it intersects the line of the ogive. Then follow the vertical line that passes through this point of intersection to the bottom of the graph. Read the value of x from the horizontal scale. This value of x is the value of the kth percentile. For example, let's find the value of P_{40} for the data shown in Figure 2-35. First locate the value of k (40) along the right-hand side of the graph (point A). Follow the horizontal line at k (40) until it intersects the graph of the ogive (point B). Then determine the value of x that corresponds to this point of intersection by reading it from the scale on the x-axis (point C): $P_{40} = 72$. This method can be used to find percentiles, quartiles, or the median. (*Note:* Because of the grouping into classes, the results may differ slightly from the answers obtained when using ranked data.)

EXERCISE 2.128

Using the ogive in Figure 2-35, estimate the value of P_{90}.

EXERCISES

2.129 According to the empirical rule, practically all the data should lie between $(\bar{x} - 3s)$ and $(\bar{x} + 3s)$. The range also accounts for all the data.
  a. What relationship should hold (approximately) between the standard deviation and the range?
  b. How can you use the results of (a) to estimate the standard deviation in situations when the range is known?

2.130 The average clean-up time for a crew of a medium-size firm is 84.0 hours and the standard deviation is 6.8 hours. Assuming that the empirical rule is appropriate,
  a. what proportion of the time will it take the clean-up crew 97.6 or more hours to clean the plant?
  b. the total clean-up time will fall within what interval 95% of the time?

2.131 Chebyshev's theorem can be stated in a form equivalent to that given on p. 104. For example, to say that "at least 75% of the data fall within two standard deviations of the mean" is equivalent to stating that "at most, 25% will be more than two standard deviations away from the mean."
  a. At most, what percentage of a distribution will be three or more standard deviations from the mean?
  b. At most, what percentage of a distribution will be four or more standard deviations from the mean?

2.132 An article titled "Effects of Modifying Instruction in a College Classroom" (*Psychological Reports*, 1986) describes an introductory psychology course in which 232 students received instruction consisting of lectures accompanied by unit testing to mastery and by student proctors. At the end of the course, students assigned ratings from 1 (significantly agree) to 5 (significantly disagree) to several statements. One such statement was "Material was presented at an appropriate level and pace." The mean and standard deviation for the responses to this statement were 1.83 and 0.44. According to Chebyshev's theorem, at least what percentage of the ratings was a 1 or a 2?

2.133 In an article titled "Development of the Breast-Feeding Attrition Prediction Tool" (*Nursing Research*, March/April 1994, Vol. 43, No. 2), the results of administering the Breast-Feeding Attrition Prediction Tool (BAPT) to 72 women with prior successful experience breast-feeding are reported. The mean score for the Positive Breast-feeding Sentiment (PBS) score for this sample is reported to equal 356.3 with a standard deviation of 65.9.
  a. According to Chebyshev's theorem, at least what percent of the PBS scores are between 224.5 and 488.1?
  b. If it is known that the PBS scores are normally distributed, what percent of the PBS scores are between 224.5 and 488.1?

2.134 An article titled "Computer-Enhanced Algebra Resources: The Effects on Achievement and Attitudes" (*International Journal of Math Education in Science and Technology*, 1980, Vol. 11, No. 4) compared algebra courses that used computer-assisted instruction with courses that do not. The scores that the computer-assisted instruction group made on an achievement test consisting of 50 problems had these summary statistics:

$$n = 57, \qquad \bar{x} = 23.14, \qquad s = 7.02$$

  a. Find the limits within which at least 75% of the scores fell.

b. If the scores are normally distributed, what percentage of the scores will be below 30.16?

2.135 Sixty college freshmen were asked to give the number of children in their families (number of their brothers and sisters plus 1). The data collected follow:

1	6	3	5	5	3	4	1	2	7	3	2
3	4	5	3	1	3	2	1	4	4	2	2
3	9	4	3	3	5	3	5	7	3	1	1
3	5	2	6	4	3	3	3	3	3	2	3
4	3	5	7	3	2	1	2	3	2	4	3

a. Construct an ungrouped frequency distribution of these data.
b. Use the ungrouped frequency distribution found in (a) to find the mean and standard deviation of the data.
c. Find the values of $\bar{x} - s$ and $\bar{x} + s$.
d. How many of the 60 pieces of data have values between $\bar{x} - s$ and $\bar{x} + s$? What percentage of the sample is this?
e. Find the values of $\bar{x} - 2s$ and $\bar{x} + 2s$.
f. How many of the 60 pieces of data have values between $\bar{x} - 2s$ and $\bar{x} + 2s$? What percentage of the sample is this?
g. Find the values of $\bar{x} - 3s$ and $\bar{x} + 3s$.
h. What percentage of the sample has values between $\bar{x} - 3s$ and $\bar{x} + 3s$?
i. Compare the answers found in (f) and (h) to the results predicted by Chebyshev's theorem.
j. Compare the answers found in (d), (f), and (h) to the results predicted by the empirical rule. Does the result suggest an approximately normal distribution?

2.136 On the first day of class last semester, 50 students were asked for the one-way distance from home to college (to the nearest mile). The resulting data follow:

6	5	3	24	15	15	6	2	1	3
5	10	9	21	8	10	9	14	16	16
10	21	20	15	9	4	12	27	10	10
3	9	17	6	11	10	12	5	7	11
5	8	22	20	13	1	8	13	4	18

a. Construct a grouped frequency distribution of the data by using 1–4 as the first class.
b. On probability paper, draw and label carefully the ogive depicting the distribution of one-way distances.
c. Does this distribution appear to have an approximately normal distribution? Explain.
d. Estimate the values of P_{10}, Q_1, $\bar{x}$, Q_3, and P_{90} from the ogive drawn in (b).
e. Calculate the mean and standard deviation.
f. Determine the values of $\bar{x} \pm 2s$, and determine the percentage of data within two standard deviations of the mean using the ogive.

2.137 The Empirical Rule states that the one, two, and three standard deviation intervals about the mean will contain 68%, 95%, and 99.7% respectively.

a. Use this set of MINITAB commands to randomly generate a sample of 100 data from a normal distribution with mean 50 and standard deviation 10. The histogram will be constructed using class boundaries which are multiples of the standard deviation 10 above and be-

To enter several commands repeatedly, see Helpful Hint, p. 89.

```
RANDom 100 C1;
  NORMal 50 10.
HISTogram C1;
  CUTPoint 10:90/10.
MEAN C1
STDev C1
```

(continued)

low the mean 50. Inspect the histogram to determine the percentage of the data that fell within each of the one, two, and three standard deviation intervals. How closely do the three percentages compare to the percentages claimed in the empirical rule?

b. Repeat part (a). Did you get results similar to those in part (a)? Explain.

c. Consider repeating (a) several more times. Are the results similar each time? If so, in what way?

d. What do you conclude about the truth of the empirical rule?

2.138   Chebyshev's theorem states that "at least $1 - \dfrac{1}{k^2}$" of the data of a distribution will lie within k standard deviations of the mean.

```
RANDom 100 C1;
 EXPO 50.
HISTogram C1;
 CUTPoint 0:300/50.
MEAN C1
STDev C1
```

a. Use this set of MINITAB commands to randomly generate a sample of 100 data from a non-normal distribution with mean 50 and standard deviation 50. The histogram will be constructed using class boundaries which are multiples of the standard deviation 50 above and below the mean 50. Inspect the histogram to determine the percentage of the data that fell within each of the one, two, three, and four standard deviation intervals. How closely do these percentages compare to the percentages claimed in Chebyshev's theorem and the empirical rule?

b. Repeat part (a). Did you get results similar to those in part (b)? Explain.

c. Consider repeating (a) several more times. Are the results similar each time? If so, in what way?

d. What do you conclude about the truth of Chebyshev's theorem and the empirical rule?

2.8 | THE ART OF STATISTICAL DECEPTION

"There are three kinds of lies—lies, damned lies, and statistics." These remarkable words spoken by Disraeli (19th-century British prime minister) represent the cynical view of statistics held by many people. Most people are on the consumer end of statistics and therefore have to "swallow" them.

Good Arithmetic, Bad Statistics

Let's explore an outright statistical lie. Suppose that a small business firm employs eight people who earn between $300 and $350 per week. The owner of the business pays himself $1250 per week. He reports to the general public that the average wage paid to the employees of his firm is $430 per week. That may be an example of good arithmetic, but it is also an example of bad statistics. It is a misrepresentation of the situation, since only one employee, the owner, receives more than the mean salary. The public will think that most of the employees earn about $430 per week.

EXERCISE 2.139

Is it possible for eight employees to earn between $300 and $350, and one earn $1250 per week, and for the mean to be $430? Verify your answer.

Graphic representations can be tricky and misleading. The frequency scale (which is usually the vertical axis) should start at zero in order to present a total picture. Usually, graphs that do not start at zero are used to save space. Nevertheless, this can be deceptive. Graphs in which the frequency scale starts at zero tend to emphasize the side of the numbers involved, whereas graphs that are chopped off may tend to emphasize the variation in the number without regard to the actual size of the number.

Case Study

2-3

Good Information, Wrong Graph

Good information can become an outright lie when it is presented with the wrong graph. Such is the case with this circle graph. Circle graphs represent the "whole" as well as the parts.

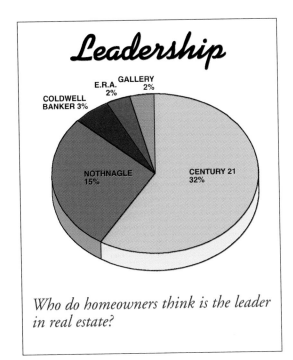

Leadership

Who do homeowners think is the leader in real estate?

EXERCISE 2.140

What's wrong with this graph? (*Hint:* Is 32% more than half?) Describe two ways to correct this situation.

Insufficient Information

A real estate agent describes a mountainside piece of land as a perfect place for a nice summer home and tells you that the average high temperature during the months of July and August is 77 degrees. If the standard deviation is 5 degrees, this is not too bad. Assuming that these daily highs are approximately normally distributed, the temperature will range from three standard deviations below to three standard deviations above the mean temperature. The range would be from 62 to 92 degrees, with only a few days above 87 and a few days below 67. However, what the agent didn't tell you is that the standard deviation of temperature highs is 10 degrees, which means the daily high temperatures (if normally distributed) range from 47 degrees to 107 degrees. These extremes are not too pleasant!

What it all comes down to is that statistics, like all languages, can be and is abused. In the hands of the careless, the unknowledgeable, or the unscrupulous, statistical information can be as false as "damned lies."

Darrell Huff's "How to Lie With Statistics" is an easy, fun to read, clever book that every student of statistics should read.

Case Study   Tricky Graph Number 1

2-4

The graph in Figure 1 was presented in an advertisement. Without careful study, a viewer might conclude that there is a great difference in the percentage of Chevys, Fords, Toyotas, and Nissan trucks still on the road. Figure 2 presents the same information. Do you get the same impression? The two graphs do not show different information; they merely create different impressions. The variability among the four percentages is the same in both figures.

FIGURE 1   A Graph That Doesn't Start at Zero

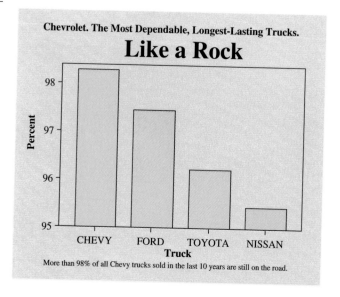

FIGURE 2   The Same Graph, But Starting at Zero

EXERCISE 2.141

Explain how Figure 1 visually emphasizes variability and de-emphasizes the size of the numbers, whereas Figure 2 emphasizes equally large numbers and deemphasizes the variability that exists.

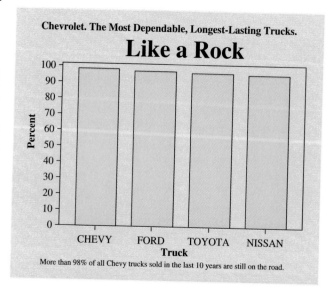

Case Study

Tricky Graph Number 2

The histogram shows the number of drivers fatally injured in automobile accidents in New York State in 1968.

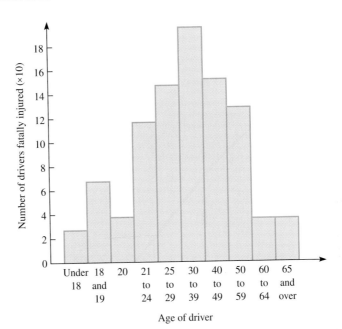

Age of driver

EXERCISE 2.142

If the vertical scale of the fatal-accident histogram were to be changed to "number of fatalities per 10,000 drivers," the graph would not be a histogram. Explain why.

Notice the "normal-appearing" distribution of ages of fatally injured drivers. The reason for grouping the ages in this manner is completely unclear. Certainly the age of fatally injured automobile drivers is not expected to be normally distributed. Notice too that the class width varies considerably. At first glance, this histogram seems to indicate that 30- to 39-year-olds have the "most fatal accidents." This may or may not be the case. Perhaps this age group has more drivers than any of the other groups. The group with the most drivers might be expected to have the highest number of driver fatalities. A vertical scale that showed the number of driver fatalities per thousand drivers might be more representative. An age grouping that created classifications with the same number of drivers in each class might also be more representative. It all depends on the intended purpose.

Case Study

Tricky Graph Number 3

There are situations in which a change in the class width might be appropriate for the graphic representation. However, one must be extremely careful when calculating the values of the numerical measures. The annual salaries paid to the employees of a large firm is a situation for which a change in class width might be appropriate. A large firm has employees with salaries ranging from $8000 to $175,000. Figure 1 shows all salaries sorted into ten classes of equal width. Because of the skewed nature of the data, 90% of the employees are in the first two classes. To present a more accurate picture, one might consider a classification system of 8000—10,000—12,000—15,000—18,000—21,000—25,000—30,000—36,000—45,000—65,000—90,000—125,000 and up. This system should allow for more accurate description of the lower salaries and will also allow the higher salaries to be lumped together. There are, generally speaking, usually only a few very high salaries. Figure 2 shows the histogram using this second classification system; it is not an easy histogram to comprehend.

FIGURE 1

Annual
Salaries and
Wages
Grouped
Using
Uniform Class
Widths

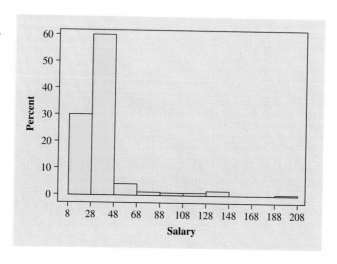

FIGURE 2

Annual
Salaries and
Wages
Grouped
Using
Unequal Class
Widths

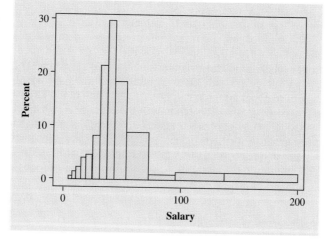

EXERCISES

2.143   The histogram presented as the chapter case study, page 37, qualifies as a "tricky graph."
    a. How does this graph violate the guidelines for drawing histograms?
    b. Why do you think the graph was drawn as it was?

2.144   Find an article or an advertisement containing a graph that in some way misrepresents the information of statistics. Describe how this graph misrepresents the facts.

IN RETROSPECT

You have been introduced to some of the more common techniques of descriptive statistics. There are far too many specific types of statistics used in nearly every specialized field of study for us to review here. We have outlined the uses of only the most universal statistics. Specifically, you have seen several basic graphic techniques (circle and bar graphs, Pareto diagrams, dotplots, stem-and-leaf displays, histograms, ogives, and box-and-whiskers) that are used to present sample data in picture form. You have also been introduced to some of the more common measures of central tendency (mean, median, mode, midrange, and midquartile), measures of dispersion (range, variance, and standard deviation), and measures of position (quartiles, percentiles, and z-score).

You should now be aware that an average can be any one of five different statistics, and you should understand the distinctions among the different types of averages. The article "'Average' Means Different Things" in Case Study 2-2 (p. 73) discusses four of the averages studied in this chapter. You might reread it now and find that it has more meaning and is of more interest. It will be time well spent!

You should also have a feeling for, and an understanding of, the concept of a standard deviation. You were introduced to Chebyshev's theorem and the empirical rule for this purpose.

The exercises in this chapter (as in others) are extremely important; they will help you nail down the concepts studied before you go on to learn how to use these ideas in later chapters. A good understanding of the descriptive techniques presented in this chapter is fundamental to your success in the later chapters.

CHAPTER EXERCISES

2.145   Samples A and B are shown in the following table. Notice that the two samples are the same except that the 8 in A has been replaced by a 9 in B.

     A: 2   4   5   5   7   8
     B: 2   4   5   5   7   9

What effect does changing the 8 to a 9 have on each of the following statistics?
   a. mean   b. median   c. mode   d. midrange
   e. range   f. variance   g. standard deviation

2.146   Samples C and D are shown in the following table. Notice that the two samples are alike except for two values.

     C: 20   60   60   70   90
     D: 20   30   70   90   90

What effect does changing the two 60's to 30 and 90 have on each of the following statistics?
   a. mean   b. median   c. mode   d. midrange
   e. range   f. variance   g. standard deviation

2.147   The addition of a new accelerator is claimed to decrease the drying time of latex paint by more than 4%. Several test samples were conducted with the following percentage decreases in drying time:

5.2	6.4	3.8	6.3	4.1	2.8	3.2	4.7

a. Find the sample mean.
b. Find the sample standard deviation.
c. Do you think these percentages average 4 or more? Explain.

(Retain these solutions for use in answering Exercise 9.26.)

2.148   Gasoline pumped from a supplier's pipeline is supposed to have an octane rating of 87.5. On 13 consecutive days, a sample was taken and analyzed with the following results:

Many of these data sets are on Data Disk.

88.6	86.4	87.2	88.4	87.2	87.6	86.8
86.1	87.4	87.3	86.4	86.6	87.1	

a. Find the sample mean.
b. Find the sample standard deviation.
c. Do you think these readings seem to average 87.5? Explain.

(Retain these solutions for use in answering Exercise 9.38.)

2.149   "Light Fast Food Can Be Heavy on Percentage of Fat Calories" (*USA Today*, 5-16-91) listed the percentage of calories derived from fat in commonly eaten fast food chicken items as follows:

43	32	55	40	43	26	50	60	34	55	53

Find the following sample statistics for the variable, percentage of calories from fat.
a. mean   b. standard deviation

2.150   The percentage earned on average shareholders' equity (the rate of return on common stock) for the Pacific Lighting Corporation for the last several years has been

8.6	9.8	10.9	11.8	13.7	13.9	14.6

Determine:
a. the median rate of return for these years
b. the mean rate of return for these years
c. the mean absolute deviation for these data

2.151   The number of pigs born per litter last year at Circle J Pig Farm are given in the following list.

11	8	13	14	11	8	14	7	11	13	10	8
7	4	9	11	10	12	6	12	5	10	9	10
12	10	3	6	9	10	10	13	12	9	12	7
9	5	12	7	11	9	4	8	13	12	11	
13	11	12	8	6	13	11	6	12	7	5	

If you use MINITAB, try TABLE.

a. Construct an ungrouped frequency distribution of these data.
b. How many litters of pigs were there last year?
c. What is the meaning and value of Σf?

2.152 The following set of data gives the ages of 118 known offenders who committed an auto theft last year in Garden City, Michigan.

11	14	15	15	16	16	17	18	19	21	25	36
12	14	15	15	16	16	17	18	19	21	25	39
13	14	15	15	16	17	17	18	20	22	26	43
13	14	15	15	16	17	17	18	20	22	26	46
13	14	15	16	16	17	17	18	20	22	27	50
13	14	15	16	16	17	17	19	20	23	27	54
13	14	15	16	16	17	18	19	20	23	29	59
13	15	15	16	16	17	18	19	20	23	30	67
14	15	15	16	16	17	18	19	21	24	31	
14	15	15	16	16	17	18	19	21	24	34	

a. Find the mean.   b. Find the median.   c. Find the mode.
d. Find Q_1 and Q_3.   e. Find P_{10} and P_{95}.

2.153 The stopping distance on a wet surface was determined for 25 cars each traveling at 30 miles per hour. The data (in feet) are shown on the following stem-and-leaf display:

```
 6 | 3 7 6 3 9
 7 | 4 2 0 1 1 2 0 5
 8 | 5 4 5 5 6
 9 | 4 1 0 0 5
10 | 5 4
```

Find the mean and the standard deviation of these stopping distances.

2.154 Compute the mean and the standard deviation for the following set of data. Then find the percentage of the data that is within two standard deviations of the mean.

```
1 | .4 .7 .1
2 | .4 .5
3 | .5 .0 .4 .1
4 | .4
5 | .5 .8 .7
6 | .8 .8 .2 .8 .6
7 | .5
8 |
9 | .4
```

2.155 The December 14, 1994, USA Snapshot® "A home in the city" reported an average 49% of residents of U.S. cities—excluding suburbs—own their homes. Cities with highest and lowest percentages of homeowners are: Cheyenne, Wyoming (65%), Boise, Idaho (64%), New York City (31%), and Newark, New Jersey (28%).
a. Define the terms *population* and *variable* with regard to this information.
b. Are the four numbers reported (65%, 64%, 31%, 28%) data or statistics? Explain.
c. Is the average, 49%, a data, a statistic, or a parameter value? Explain why.
d. Is the average value the midrange?

2.156 Ask one of your instructors for a list of exam grades (15 to 25 grades) from a class.
a. Find five measures of central tendency.
b. Find the three measures of dispersion.

(continued)

c. Construct a stem-and-leaf display. Does this diagram suggest that the grades are normally distributed?

d. Find the following measures of location: (1) Q_1 and Q_3, (2) P_{15} and P_{60}, (3) the standard
score z for the highest grade.

2.157   The distribution of credit hours, per student, taken this semester at a certain college was as
follows:

If you use MINITAB, see sessions commands on page 89.

Credit Hours	Frequency
3	75
6	150
8	30
9	50
12	70
14	300
15	400
16	1050
17	750
18	515
19	120
20	60

Find: a. mean   b. median   c. mode   d. midrange

2.158   A survey of 32 workers at building 815 of Eastman Kodak Company was taken last May.
Each worker was asked: "How many hours of television did you watch yesterday?" The results
were as follows:

0	0	$\frac{1}{2}$	1	2	0	3	$2\frac{1}{2}$
0	0	1	$1\frac{1}{2}$	5	$2\frac{1}{2}$	0	2
$2\frac{1}{2}$	1	0	2	0	$2\frac{1}{2}$	4	0
6	$2\frac{1}{2}$	0	$\frac{1}{2}$	1	$1\frac{1}{2}$	0	2

a. Construct a stem-and-leaf display.   b. Find the mean.

c. Find the median.   d. Find the mode.

e. Find the midrange.

f. Which one of the measures of central tendency would best represent the average viewer
if you were trying to portray the typical television viewer? Explain.

g. Which measure of central tendency would best describe the amount of television
watched? Explain.

h. Find the range.   i. Find the variance.

j. Find the standard deviation.

2.159   Given the following frequency distribution of ages, x, in years, of cars found in a parking lot,

x	1	2	3	4	5	6	7	8	9	10	11
f	16	20	18	12	9	7	6	3	5	3	1

a. draw a histogram of the data.

b. find the five measures of central tendency.

c. find Q_1 and Q_3.   d. find P_{15} and P_{12}.
e. find the three measures of dispersion (range, s^2, s).

2.160 An article in *Therapeutic Recreation Journal* reports a distribution for the variable, number of persistent disagreements. Sixty-six patients and their therapeutic recreation specialist each answered a checklist of problems with yes or no. Disagreement occurs when the specialist and the patient did not respond identically to an item on the checklist. It becomes a persistent disagreement if the item remains in disagreement after a second interview.
a. Draw a dotplot of this sample data.
b. Find the median number of persistent disagreements.
c. Find the mean number of persistent disagreements.
d. Find the standard deviation for the number of persistent disagreements.
e. Draw a vertical line on the dotplot at the mean.
f. Draw a horizontal line segment on the dotplot whose length represents the standard deviation (start at the mean).

Number of items	Frequency
0	2
1	2
2	4
3	10
4	7
5	9
6	8
7	11
8	7
9	3
10	1
11	2

Source: Data reprinted with permission of the National Recreation and Park Association, Alexandria, VA, from Pauline Petryshen and Diane Essex-Sorlie, "Persistent Disagreement Between Therapeutic Recreation Specialists and Patients in Psychiatric Hospitals," *Therapeutic Recreation Journal*, Vol. XXIV, Third Quarter, 1990.

2.161 *USA Today*, 10-25-94, reported in the Snapshot® "Mystery of the remote" that 44% of the families surveyed never misplaced the family television remote control, while 38% misplaced it one to five times weekly and 17% misplaced it more than five times weekly. One percent of the families surveyed didn't know. Suppose you took a survey that resulted in the following data. Let x be the number of times per week that the family's television remote control gets misplaced. The survey resulted in the following data.

x	0	1	2	3	4	5	6	7	8	9
f	220	92	38	21	24	30	34	20	16	5

a. Construct a histogram.
b. Find the mean, median, mode, midrange.
c. Find the variance and standard deviation.
d. Find Q_1, Q_3, and P_{90}.
e. Find the midquartile.
f. Find the five-number summary and draw a box-and-whisker display.

 2.162   In the May 1990 issue of the journal *Social Work*, Marlow reports the following results:

Number of Children Living at Home	Mexican-American Women	Anglo-American Women
0	23	38
1	22	9
2	17	15
3	7	9
4	1	1

Copyright 1990, National Association of Social Workers, Inc. *Social Work.*

a. Construct a histogram for each of the preceding distributions. Draw them on the same axis, using two different colors, so that you can compare their distributions.
b. Calculate the mean and standard deviation for the Mexican-American data.
c. Calculate the mean and standard deviation for the Anglo-American data.
d. Do these two distributions seem to be different? Cite specific reasons for your answer.

 2.163   The length of life of 220 incandescent 60-watt lamps was obtained and yielded the frequency distribution shown in the following table.

Class Limit	Frequency
500–600	3
600–700	7
700–800	14
800–900	28
900–1000	64
1000–1100	57
1100–1200	23
1200–1300	13
1300–1400	7
1400–1500	4

a. Construct a histogram of these data using a vertical scale for relative frequencies.
b. Find the mean length of life.   c. Find the standard deviation.

2.164 The following table shows the age distribution of heads of families.

Age of Head of Family (years)	Number
20–25	23
25–30	38
30–35	51
35–40	55
40–45	53
45–50	50
50–55	48
55–60	39
60–65	31
65–70	26
70–75	20
75–80	16
	450

a. Find the mean age of the heads of families.
b. Find the standard deviation.

2.165 The sample of 1994 family annual incomes that were listed in the chapter case study, page 37, can be described using both graphic and numeric techniques that we have studied in this chapter.

a. Construct at least two graphic displays that present these data and their distribution.
b. Which graph drawn in (a) best presents the data, in your opinion? Explain why you made this choice.
c. Calculate at least three different measures of central tendency for this sample.
d. Which measure of central tendency, found in (c), best presents the concept of "average" annual family income? Explain why you made this choice.
e. Calculate at least three different measures of dispersion for the sample.
f. Which measure, found in (e), best presents the amount of dispersion in the data? Explain your choice.
g. Find the boundaries of the middle 50% of the sample.
h. Make a case for defining "middle class" to be the middle 50% and describe this group.

2.166 The lengths (in millimeters) of 100 brown trout in pond 2-B at Happy Acres Fish Hatchery on June 15 of last year were as follows:

15.0	15.3	14.4	10.4	10.2	11.5	15.4	11.7	15.0	10.9
13.6	10.5	13.8	15.0	13.8	14.5	13.7	13.9	12.5	15.2
10.7	13.1	10.6	12.1	14.9	14.1	12.7	14.0	10.1	14.1
10.3	15.2	15.0	12.9	10.7	10.3	10.8	15.3	14.9	14.8
14.9	11.8	10.4	11.0	11.4	14.3	15.1	11.5	10.2	10.1
14.7	15.1	12.8	14.8	15.0	10.4	13.5	14.5	14.9	13.9
10.1	14.8	13.7	10.9	10.6	12.4	14.5	10.5	15.1	15.8
12.0	15.5	10.8	14.4	15.4	14.8	11.4	15.1	10.3	15.4
15.0	14.0	15.0	15.1	13.7	14.7	10.7	14.5	13.9	11.7
15.1	10.9	11.3	10.5	15.3	14.0	14.6	12.6	15.3	10.4

(continued)

a. Find the mean.
b. Find the median.
c. Find the mode.
d. Find the midrange.
e. Find the range.
f. Find Q_1 and Q_3.
g. Find the midquartile
h. Find P_{35} and P_{64}.
i. Construct a grouped frequency distribution that uses 10.0–10.5 as the first class.
j. Construct a histogram of the frequency distribution.
k. Construct a cumulative relative frequency distribution.
l. Construct an ogive of the cumulative relative frequency distribution.
m. Find the mean of the frequency distribution.
n. Find the standard deviation of the frequency distribution.

 2.167 "Grandparents are grand" appeared as a USA Snapshot® in *USA Today*, 12-9-94, and reported on "how much grandparents say they spend annually on gifts and entertainment for each of their grandchildren."

Dollar Interval	$0–$100	$101–$200	$201–$500	$501 or more
Percent	54.4%	16.5%	15.7%	4.7%

Suppose that this information was obtained from a sample of 1000 grandparents and that the 8.7% who did not answer spent nothing. Use values of $0, $50, $150, $350, $750 as class marks, and estimate the sample mean and the standard deviation for the variable x, amount spent.

 2.168 "Trade show viewers," a USA Snapshot® that appeared in *USA Today*, 6-13-94, reported that a record 85 million people will attend trade shows in the USA during 1994.

Time Spent Viewing Show	4 or fewer hr	4–8 hr	More than 8 hr	No answer
Percent of Sample	0.21	0.38	0.39	0.02

Suppose the above sample data resulted from a survey of 200 visitors.
a. Form a frequency distribution representing the 98% of the sample that resulted in data.
b. Use the distribution in (a) to estimate the mean amount of time spent viewing.
c. Use the distribution in (a) to estimate the standard deviation for time spent viewing.

 2.169 The 1994 Indianapolis 500 lineup of 33 race cars had the following qualifying speeds (*USA Today*, 5-27-94):

221.0	221.1	221.2	221.3	221.3	221.4	221.4	221.5	222.1	222.3	222.4
222.5	222.5	222.7	222.7	222.9	223.0	223.1	223.2	223.3	223.4	223.5
223.7	223.7	223.8	224.0	224.1	224.2	226.2	226.3	227.3	227.6	228.0

a. Prepare a grouped frequency distribution for the qualifying speeds.
b. Draw a frequency histogram for the distribution in (a).
c. Use the grouped frequency distribution in (a) to calculate the sample mean and standard deviation.

2.170   Earnings per share for 40 firms in the radio and transmitting equipment industry follow:

4.62	0.25	1.07	5.56	0.10	1.34	2.50	1.62
1.29	2.11	2.14	1.36	7.25	5.39	3.46	1.93
6.04	0.84	1.91	2.05	3.20	−0.19	7.05	2.75
9.56	3.72	5.10	3.58	4.90	2.27	1.80	0.44
4.22	2.08	0.91	3.15	3.71	1.12	0.50	1.93

  a. Prepare a frequency distribution and a frequency histogram for these data.
  b. Which class of your frequency distribution contains the median?

2.171   For a normal (or bell-shaped) distribution, find the percentile rank that corresponds to
  a. $z = 2$ b. $z = -1$

2.172   For a normal (or bell-shaped) distribution, find the z-score that corresponds to the kth percentile:
  a. $k = 20$ b. $k = 95$

2.173   Bill and Rob are good friends, although they attend different high schools in their city. The city school system uses a battery of fitness tests to test all high school students. After completing the fitness tests, Bill and Rob are comparing their scores to see who did better in each event. They need help.

	Sit-ups	Pull-ups	Shuttle Run	50-Yard Dash	Softball Throw
Bill	$z = -1$	$z = -1.3$	$z = 0.0$	$z = 1.0$	$z = 0.5$
Rob	61	17	9.6	6.0	179 ft
Mean	70	8	9.8	6.6	173 ft
Standard Deviation	12	6	0.6	0.3	16 ft

Bill received his test result in z-scores, whereas Rob was given raw scores. Since both boys understand raw scores, convert Bill's z-scores to raw scores in order to make an accurate comparison.

2.174   Twins Jean and Joan Wong are in fifth grade (different sections), and the class has been given a series of ability tests.

	Results	
Skill	**Jean: z-Score**	**Joan: Percentile**
Fitness	2.0	99
Posture	1.0	69
Agility	1.0	88
Flexibility	−1.0	35
Strength	0.0	50

If the scores for these ability tests are approximately normally distributed, which girl has the higher relative score on each of the skills listed? Explain your answers.

2.175   Chebyshev's theorem guarantees what proportion of a distribution will be included between the following?

      a. $\bar{x} - 2s$ and $\bar{x} + 2s$ b. $\bar{x} - 3s$ and $\bar{x} + 3s$

2.176   The empirical rule indicates that we can expect to find what proportion of the sample to be included between the following?

      a. $\bar{x} - s$ and $\bar{x} + s$ b. $\bar{x} - 2s$ and $\bar{x} + 2s$ c. $\bar{x} - 3s$ and $\bar{x} + 3s$

2.177   Why is it that the z-score for a value belonging to a normal distribution usually lies between -3 and 3?

2.178   The mean mileage per tire is 30,000 miles and the standard deviation is 2500 miles for a certain tire.

      a. If we assume that the mileage is normally distributed, approximately what percentage of all such tires will give between 22,500 and 37,500 miles?

      b. If we assume nothing about the shape of the distribution, approximately what percentage of all such tires will give between 22,500 and 37,500 miles?

2.179   Manufacturing specifications are often based on the results of samples taken from satisfactory pilot runs. The following data resulted from just such a situation in which eight pilot batches were completed and sampled. The resulting particle sizes, in angstroms (where $1\text{Å} = 10^{-8}$ cm) were

      3923   3807   3786   3710   4010   4230   4226   4133

      a. Find the sample mean.

      b. Find the sample standard deviation.

      c. Assuming that particle size has an approximately normal distribution, determine the manufacturing specs that bounds 95% of the particle sizes (that is, find the 95% interval, $\bar{x} + 2s$).

2.180   Delco Products, a division of General Motors, produces a bracket that is used as part of a power doorlock assembly. The length of this bracket is constantly being monitored. A sample of 30 power door brackets resulted in the following lengths (in millimeters).

11.86	11.88	11.88	11.91	11.88	11.88
11.88	11.88	11.88	11.86	11.88	11.88
11.88	11.88	11.86	11.83	11.86	11.86
11.88	11.88	11.88	11.83	11.86	11.86
11.86	11.88	11.88	11.86	11.88	11.83

Source: With permission of Delco Products Division, GMC.

      a. Without doing any calculations, what would you estimate for a sample mean?

      b. Construct an ungrouped frequency distribution.

      c. Draw a histogram of this frequency distribution.

      d. Use the frequency distribution and calculate the sample mean and standard deviation.

      e. Determine the limits of the $\bar{x}$ and $3s$ interval and mark this interval on the histogram.

      f. The product specification limits are 11.7–12.3. Does the sample indicate that production is within these requirements? Justify your answer.

2.181 The dollar value of each state's export business for 1993, in billions of dollars, was reported in *USA Today*, 6-27-94 as follows:

2.5	0.8	5.8	1.1	68.1	6.2	10.2	3.5	14.7	6.1	0.2
1.2	20.3	8.4	1.9	3.1	3.3	3.2	1.1	2.7	11.6	25.3
10.0	0.8	4.7	0.2	1.7	0.5	1.1	14.5	0.4	40.7	8.0
0.3	17.5	2.3	6.2	13.2	0.9	3.2	0.2	6.1	35.6	2.0
2.3	8.2	27.7	0.8	5.8	0.9					

a. Prepare a graphic display of these data that you believe pictures the distribution fairly.
b. Find the mean and the median.
c. Find the standard deviation.
d. If you wanted to indicate the total amount of exports for all states, would you report the mean or the median as the average value? Explain.
e. If you wanted to indicate the typical amount exported by one state, would you report the mean or the median as the average value? Explain.

2.182 The *Democrat and Chronicle*, 1-14-95, published the median prices for single-family homes in the towns surrounding Rochester, New York, for January 1994 and January 1995.

1995	93450	86850	89900	71000	94900	62900	67900	86500
	250000	80450	87000	90000	137500	181000	195000	94200
	85000	72000						
1994	105500	98000	85000	83000	93000	69900	85000	104700
	77900	70000	145000	154900	152500	153200	162000	

a. Construct the graphic display for both sets of data so that you can compare 1994 and 1995 median prices. (Use a graph of your choice, but use the same scale for both.)
b. Based on the small amount of information shown in your graphs, how do you think 1995 home prices compare to those of a year earlier? Explain.

2.183 The dotplot below shows the number of attempted passes thrown by the quarterbacks for 22 of the NFL teams that played on one particular Sunday afternoon.

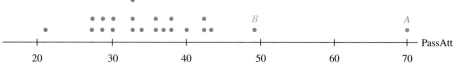

a. Describe the distribution, including how points A and B seem to relate to the others.
b. If you remove point A, and maybe B, would you say the remaining data have an approximately normal distribution? Explain.
c. Based on the information about distributions that Chebyshev's theorem and the empirical rule give us, how typical an event do you feel point A represents? Explain.

2.184 Lee Trevino played in 17 golfing events during the first six months of 1994, and his earnings for each of those events was as follows:

29500	72000	120000	50400	8250	47667	90000	10200	115000
22500	22500	112500	48600	105000	13650	50600	157500	

    a. Calculate several numerical statistics, and draw at least one graph of these data.
    b. Use the information found in (a) and prepare a short descriptive report that best describes Lee Trevino's earnings for the first half of 1994.

2.185 Starting with the data values of 70 and 85, add three data values to your sample so that the sample has: (justify your answer in each case)
    a. a standard deviation of 5.
    b. a standard deviation of 10.
    c. a standard deviation of 15.
    d. Compare your three samples and the variety of values needed to obtain each of the required standard deviations.

2.186 Make up a set of 18 data (think of them as exam scores) so that the sample meets each of these sets of criteria:
    a. Mean is 75, standard deviation is 10.
    b. Mean is 75, maximum is 98, minimum is 40, standard deviation is 10.
    c. Mean is 75, maximum is 98, minimum is 40, standard deviation is 15.
    d. How are the data in the sample for (b) different from those in (c)?

2.187 The Sagarin college basketball ratings for the difficulty of all 302 Division I basketball teams' schedules are listed below as they appeared in *USA Today*, 12-29-94.

73.28	74.84	81.55	71.67	75.72	80.61	72.53	70.12	79.31	75.83	76.51	67.33
80.24	69.28	74.55	67.29	79.29	69.45	73.65	69.98	72.77	74.14	73.85	70.19
72.42	65.60	77.88	70.66	80.17	66.13	72.76	65.13	61.80	81.40	75.22	72.44
68.98	71.26	76.46	68.40	71.24	75.13	73.56	76.52	70.82	68.56	74.08	68.16
68.62	62.38	71.75	62.66	68.43	79.65	61.70	67.67	69.51	75.73	75.15	65.69
75.94	74.16	69.19	72.68	68.15	71.81	73.68	70.57	69.84	65.10	64.56	61.87
78.27	67.98	57.87	74.17	82.64	69.64	75.30	72.76	77.47	75.11	73.62	78.69
74.52	79.86	75.54	73.25	72.10	78.93	75.38	71.17	71.54	70.79	78.08	72.26
72.26	67.08	75.51	70.87	72.27	72.17	77.09	65.90	73.72	66.02	73.88	71.05
60.97	78.36	72.81	68.16	76.17	69.59	73.08	69.02	77.11	74.12	74.43	66.42
79.04	75.83	67.36	70.16	71.09	76.48	72.39	72.61	66.61	87.48	68.93	75.79
68.48	62.74	76.24	68.65	76.00	74.23	77.21	68.19	78.05	66.06	73.70	67.84
65.48	77.76	82.19	64.88	73.81	60.43	73.44	73.77	70.04	79.65	72.55	69.77
71.85	74.28	74.34	71.02	69.82	74.69	70.16	72.67	72.34	63.34	70.00	76.29
77.98	81.48	75.49	73.24	79.61	64.24	79.10	76.34	67.20	72.02	67.11	80.95
69.72	86.78	79.24	74.53	68.43	69.37	71.08	78.40	75.62	73.49	73.02	74.02
65.80	72.52	67.14	74.30	65.20	68.64	73.04	73.15	64.39	77.42	73.75	63.23
82.62	68.28	70.45	64.71	65.40	68.09	72.99	67.67	71.31	72.75	65.10	67.84
62.87	71.29	65.76	66.55	67.62	80.37	67.93	71.63	72.51	68.83	76.45	74.29
71.44	75.62	77.00	63.28	68.31	77.90	79.92	77.90	78.76	70.80	69.51	73.41
78.41	66.85	85.80	67.52	73.63	77.97	76.22	78.22	71.04	67.95	68.89	72.77
70.96	74.41	68.35	69.08	75.21	80.29	71.08	68.07	59.40	68.40	68.38	69.83
72.32	69.97	67.97	72.27	70.68	70.49	66.59	68.99	72.00	61.96	82.29	63.29
75.10	68.76	62.47	65.82	71.33	68.59	72.21	73.07	68.55	73.23	74.09	75.43
77.12	74.12	73.02	65.32	68.79	75.83	70.85	52.73	80.21	76.84	81.74	86.85
77.92	75.07										

    a. Construct the dotplot and a histogram of this population. (Use class boundaries 51–54– 57– . . . –90.) Since this is a very large set of data, let's think of it as a population. Very often in statistics we do not know the population, and we rely on samples to tell us about the population.

b. If you use MINITAB, use command SAMPle 20 C1 C2

b. Select a random sample of 20 data from the above population, and construct a dotplot and a histogram of the sample [use the same classes as used in part (a)].

c. Select three more random samples of size 20 and repeat (b).

d. Find the mean, median, minimum, and maximum value for the population and each of the four samples.

e. Compare the distributions of each sample to the population answering the main question "Is a sample a reasonable estimate of the population?" (How close is "close enough"? If you were allowed to "move" or "change" two or three of the data in each sample, could you make the histograms of each sample look very much like the population?)

2.188

a. Use a computer to generate a random sample of 500 values of a normally distributed variable x with a mean of 100 and a standard deviation of 20. Construct a histogram of the 500 values.

Let's consider the 500 x-values found in (a) as a population.

b. Randomly select a sample of 30 values from the population found in (a). Construct a histogram of the sample using the same class intervals as used in (a).

c. Repeat part (b) three times.

d. Calculate several values (mean, median, maximum, minimum, standard deviation, etc.) describing the population and each of the four samples. (use DESCribe.)

e. Do you think a sample of 30 data adequately represent a population? [Compare each of the four samples found in (b) and (c) to the population.]

IF you use MINITAB, see Helpful Hint on page 89.

```
a. MINITAB commands to
   generate a "population."

   RANDom 500 C1;
    NORMal 100 20.
   HISTogram C1;
    CUTPoints 20:180/10.
```

```
b. MINITAB commands to
   randomly select a
   sample 30 data from C1
   and draw histogram

   SAMPle 30 C1 C2
   HISTogram C2;
    CUTPoints 20:180/10.
```

2.189   Repeat Exercise 2.188 using a different sample size. You might try a few different sample sizes; $n = 10$, $n = 15$, $n = 20$, $n = 40$, $n = 50$, $n = 75$. What effect does increasing the sample size have on the effectiveness of the sample to depict the population? Explain.

2.190   Repeat Exercise 2.188 using populations with different shaped distributions.

a. Use a uniform or rectangular distribution. (Replace the subcommands used in Exercise 2.188; in place of NORMal use: **UNIForm 50 150.** and in place of CUTPoints, on two occasions, use: **CUTPoints 50:150/10.**)

b. Use a skewed distribution. (Replace the subcommands used in Exercise 2.186; in place of NORMal use: **GAMMa 25 4.** and in place of CUTPoints, on two occasions, use: **CUTPoints 50:180/10.**)

c. Use a J-shaped distribution. (Replace the subcommands used in Exercise 2.188; in place of NORMal use: **EXPO 50.** and in place of CUTPoints, on two occasions, use: **CUTPoints 0:250/10.**)

d. Does the shape of the distribution of the population have an effect on how well a sample of size 30 represents the population? Explain.

e. What effect do you think changing the sample size would have on the effectiveness of the sample to depict the population? Try a few different sample sizes. Do the results agree with your expectations? Explain.

VOCABULARY LIST

Be able to define each term. Pay special attention to the key terms, which are printed in red. In addition, describe in your own words, and give an example of, each term. Your examples should not be ones given in class or in the textbook. Page numbers indicate the first appearance of the term.

bar graph (p. 39)
bell-shaped distribution (p. 105)
bimodal frequency distribution (p. 61)
box-and-whisker plot (p. 99)
Chebyshev's theorem (p. 104)
circle graph (p. 39)
class (p. 54)
class boundary (p. 55)
class mark (p. 57)
class width (p. 55)
cumulative frequency distribution (p. 61)
cumulative relative frequency
    distribution (p. 62)
depth (p. 70)
deviation from the mean (p. 77)
distribution (p. 42)
dotplot (p. 42)
empirical rule (p. 104)
5-number summary (p. 98)
frequency (p. 53)
frequency distribution (p. 53)
frequency histogram (p. 58)
grouped frequency distribution (p. 54)
histogram (p. 58)
interquartile range (p. 98)
mean (p. 68)
mean absolute deviation (p. 79)
measure of central tendency (p. 68)
measure of dispersion (p. 77)
measure of position (p. 94)
median (p. 69)
midquartile (p. 97)

midrange (p. 72)
modal class (p. 61)
mode (p. 71)
normal distribution (p. 61)
ogive (p. 62)
Pareto diagrams (p. 41)
percentile (p. 94)
probability paper (p. 106)
qualitative data (p. 39)
quantitative data (p. 42)
quartile (p. 94)
range (p. 77)
rectangular distribution (p. 61)
relative frequency (p. 58)
relative frequency distribution
    (p. 58)
relative frequency histogram
    (p. 59)
single-variable data (p. 38)
skewed distribution (p. 60)
standard deviation (p. 80)
standard score (p. 100)
stem-and-leaf display (p. 43)
summation (p. 57)
tally (p. 56)
test for normality (p. 106)
ungrouped frequency distribution
    (p. 54)
variance (p. 79)
x bar ($\bar{x}$) (p. 68)
z-score (p. 100)

QUIZ A

Answer "True" if the statement is always true. If the statement is not always true, replace the words in bold with the words that make the statement always true.

2.1 The **mean** of a sample always divides the data into two equal halves—half larger and half smaller in value than itself.

2.2 A measure of **central tendency** is a quantitative value that describes how widely the data are dispersed about a central value.

2.3 The sum of the squares of the deviations from the mean, $\Sigma(x - \bar{x})^2$, will **sometimes** be negative.

2.4 For any distribution, the sum of the deviations from the mean equals **zero**.

2.5 The standard deviation for the set of values 2, 2, 2, 2, and 2 is **2**.

2.6 On a test John scored at the 50th percentile and Jorge scored at the 25th percentile; therefore, John's test score was **twice** Jorge's test score.

2.7 The frequency of a class is the number of pieces of data whose values fall within the **boundaries** of that class.

2.8 **Frequency distributions** are used in statistics to present large quantities of repeating values in a concise form.

2.9 The unit of measure for the standard score is always in **standard deviations**.

2.10 For a bell-shaped distribution, the range will be approximately equal to **six standard deviations**.

QUIZ B

The results of a consumer study completed at Corner Convenience Store were reported in the accompanying histogram. Find the answer to each of the following.

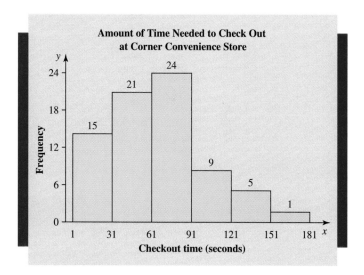

2.1
    a. What is the class width?
    b. What is the class mark for the class 31–61?
    c. What is the upper boundary for the class 61–91?
    d. What is the frequency of the class 1–31?
    e. What is the frequency of the class containing the largest observed value of x?
    f. What is the lower boundary of the class with the largest frequency?
    g. How many pieces of data are shown in this histogram?
    h. What is the value of the mode?
    i. What is the value of the midrange?
    j. Estimate the value of the 90th percentile, P_{90}.

2.2 A sample of the purchases for several Corner Convenience Store customers resulted in the following sample data:

x = number of items purchased per customer

x	f
1	6
2	10
3	9
4	8
5	7

  a. What does the "2" represent?
  b. What does the "9" represent?
  c. How many customers were used to form this sample?
  d. How many items were purchased by the customers in this sample?
  e. What is the largest number of items purchased by one customer?

Find each of the following (show formulas and work):
  f. mode   g. median   h. midrange
  i. mean   j. variance   k. standard deviation

2.3 Given the set of data 4, 8, 9, 8, 6, 5, 7, 5, 8, find each of the following sample statistics.
  a. mean   f. P_{40}
  b. median   g. variance
  c. mode   h. standard deviation
  d. midrange   i. range
  e. first quartile

2.4 a. Find the standard score for the value x = 452 relative to its sample, where the sample mean is 500 and the standard deviation is 32.
  b. Find the value of x that corresponds to the standard score of 1.2, where the mean is 135 and the standard deviation is 15.

QUIZ C

Answer all questions.

2.1 The Corner Convenience Store kept track of the number of paying customers it had during the noon hour each day for 100 days. The following are the resulting statistics rounded to the nearest integer:

  mean = 95   standard deviation = 12
  median = 97   first quartile = 85
  mode = 98   third quartile = 107
  midrange = 93   range = 56

  a. The Corner Convenience Store served what number of paying customers during noon hour more often than any other number? Explain how you determined your answer.

    b. On how many days was there between 85 and 107 paying customers during the noon hour? Explain how you determined your answer.

    c. What was the greatest number of paying customers during any one noon hour? Explain how you determined your answer.

    d. For how many of the 100 days was the number of paying customers within three standard deviations of the mean ($\bar{x} \pm 3s$)? Explain how you determined your answer.

2.2 Mr. VanCott started his own machine shop several years ago. His business grew and has become very successful in recent years. Currently he employs 14 people, including himself, and pays the following annual salaries:

Owner, President	$80,000	Worker	$25,000
Business Manager	50,000	Worker	25,000
Production Manager	40,000	Worker	25,000
Shop Foreman	35,000	Worker	20,000
Worker	30,000	Worker	20,000
Worker	30,000	Worker	20,000
Worker	28,000	Worker	20,000

    a. Calculate the four "averages": mean, median, mode, midrange.

    b. Draw a dotplot of these salaries and locate each of the four averages on it.

    c. Suppose you were the feature writer assigned to write this week's feature story on Mr. VanCott's machine shop, one of a series on local small businesses that are prospering. You plan to interview Mr. VanCott, his business manager, the shop foreman, and one of the newer workers. Which statisical average do you think each will give as their answer when asked, "What is the average annual salary paid to the employees here at VanCott's?" Explain why each person interviewed has a different perspective and why this viewpoint will cause each to cite a different statistical average.

    d. What is there about the distribution of these annual salaries that causes the four "average values" to be so different from each other?

2.3 Create a set of data containing three or more values:

    a. where the mean is 12 and the standard deviation is zero.

    b. where the mean is 20 and the range is 10.

    c. where the mean, median, and mode are all equal.

    d. where the mean, median, and mode are all different.

    e. where the mean, median, and mode are all different and the median is the largest and the mode is the smallest of the three.

    f. where the mean, median, and mode are all different and the mean is the largest and the median is the smallest.

2.4 A set of test papers was machine scored. Later it was discovered that two points should be added to each score. Student A thought the mean score should also be increased by two points. Student B added that the standard deviation should also be increased by 2 points. Who is right? Justify your answer.

2.5 Student A stated that "both the standard deviation and the variance preserved the same unit of measurement as the observed data." Student B disagreed, arguing that "the unit of measurement for variance was a meaningless unit of measurement." Who is right? Justify your answer.

DESCRIPTIVE ANALYSIS AND PRESENTATION OF BIVARIATE DATA

CHAPTER OUTLINE

3.1 Bivariate Data
Two variables are paired together for analysis.

3.2 Linear Correlation
Does an increase in the value of one variable indicate a change in the value of the other?

3.3 Linear Regression
The line of best fit is a mathematical expression for the relationship between two variables.

CHART YOUR UPPER LIMIT

*H*ave you, or someone you know, ever wondered if you should lose a few pounds? The experts at the American Health Foundation offer the following chart as a guide to the upper limit you should let your weight reach.

NIP WEIGHT GAIN IN THE BUD FOR BETTER HEALTH

The first time your weight edges up five to seven pounds over what's healthy, you'd better get a grip or you could be in a lifelong battle.

And if you're already seriously overweight, losing eight to 16 pounds will benefit you.

Those are among the conclusions of a panel of 18 medical and obesity experts, convened by the American Health Foundation, a non-profit group.

The experts reviewed information on obesity, concluding that people tend to set unattainable goals. So they defined upper limits on "healthy" weights for people of different heights, which they consider attainable, maintainable goals. Risks begin escalating when you go above these.

If you are below these weights, it doesn't mean you should go up. There is some data that even lower weights may convey other health benefits.

Chart your upper limit

A panel of obesity experts has come up with recommendations for upper limits of healthy weights for adults. The weights represent a statistical target that may require adjustment based on age, gender and health.

Height	Weight
4-foot-11	124 pounds
5 feet	128 pounds
5-foot-1	132 pounds
5-foot-2	136 pounds
5-foot-3	141 pounds
5-foot-4	145 pounds
5-foot-5	150 pounds
5-foot-6	155 pounds
5-foot-7	159 pounds
5-foot-8	164 pounds
5-foot-9	169 pounds
5-foot-10	174 pounds
5-foot-11	179 pounds
6 feet	184 pounds
6-foot-1	189 pounds
6-foot-2	194 pounds
6-foot-3	200 pounds
6-foot-4	205 pounds

Source: American Health Foundation

Study the chart, and you will see that as a person's height increases, the recommended weight limit seems to increase in a steady pattern; mostly at a rate of either four or five pounds per inch. This seems to suggest that there is some mathematical relationship between a person's height and weight. What kind of relationship is this? How can this relationship be expressed (other than in chart form)? (See Exercise 3.67, p. 177.)

Source: Nanci Hellmich, *USA Today*, 10-4-94.

● CHAPTER OBJECTIVES

In the field of statistics there are many problems that require a combined analysis of two variables. In business, in education, and in many other fields, we often want to answer such questions as "Are these two variables related? If so, how are they related? Are these variables correlated?" The relationships being discussed are not cause-and-effect relationships, but rather are the mathematical relationships that allow us to predict the behavior of one variable based on knowledge about another variable.

Let's look at a few specific illustrations.

▼ ILLUSTRATION 3-1

Are female voter opinions on the U.S. president's current position regarding tax increases related to political party affiliation?

▼ ILLUSTRATION 3-2

As a person grows taller, he or she usually gains weight. Someone might ask, "Is there a relationship between height and weight?"

▼ ILLUSTRATION 3-3

Research doctors test new drugs (old ones, too) by prescribing different dosages and observing the responses of their patients. One question we could ask here is "Does the drug dosage prescribed determine the amount of recovery time the patient needs?"

▼ ILLUSTRATION 3-4

A high school guidance counselor would like to predict the academic success that students graduating from her school will have in college. In cases like this, the predicted value (grade-point average at college) depends on many traits of the students: (1) how well they did in high school, (2) their intelligence, (3) their desire to ▲ succeed at college, and so on.

These questions all require the analysis of bivariate data to obtain the answers. In this chapter we take a first look at the techniques of tabling and graphing bivariate data and the descriptive aspects of correlation and regression analysis. The objectives of this chapter are (1) to be able to represent bivariate data in tabular and graphic form, (2) to gain an understanding of the distinction between the basic purposes of correlation analysis and regression analysis, and (3) to become familiar with the ideas of descriptive presentation. With these objectives in mind, we will restrict our discussion to the simplest and most basic form of correlation and regression analysis—the bivariate linear case.

3.1 | BIVARIATE DATA

BIVARIATE DATA

Bivarate data consist of the values of two different response variables that are obtained from the same population element.

Each of the two variables may be either qualitative or quantitative in nature. As a result, three combinations of variable types can form bivariate data:

1. Both variables are qualitative (attribute).
2. One variable is qualitative (attribute) and the other is quantitative (numerical).
3. Both variables are quantitative (both numerical).

In this section we study tabular and graphic methods for displaying each of these combinations of bivariate data.

Two Qualitative Variables

When bivariate data result from two qualitative (attribute or categorical) variables, the data are often arranged on a *cross-tabulation* or *contingency table*. Let's look at an illustration.

▼| ILLUSTRATION   3 - 5

Thirty students from our college were randomly identified and classified according to two variables: (1) gender (M/F) and (2) major (liberal arts, business administration, technology), as shown in Table 3-1. These 30 bivariate data can be summarized on a 2 × 3 cross-tabulation table (the two rows represent the two gender categories of male and female, and the three columns represent the three major categories of liberal arts, business administration, and technology) by determining how many students fit categorically into each cell. Adams is male (M) and liberal arts (LA) and is classified in the cell in the first row, first column. See the red tally mark in Table 3-2. The other 29 students are classified (tallied, shown in black) in a similar fashion.

TABLE 3-1

Gender and Major of 30 College Students

Name	Gender	Major	Name	Gender	Major
Adams	M	LA	Kee	M	BA
Argento	F	BA	Kleeberg	M	LA
Baker	M	LA	Light	M	BA
Bennett	F	LA	Linton	F	LA
Brock	M	BA	Lopez	M	T
Brand	M	T	McGowan	M	BA
Chun	F	LA	Mowers	F	BA
Crain	M	T	Ornt	M	T
Cross	F	BA	Palmer	F	LA
Ellis	F	BA	Pullen	M	T
Feeney	M	T	Rattan	M	BA
Flanigan	M	LA	Sherman	F	LA
Hodge	F	LA	Small	F	T
Holmes	M	T	Tate	M	BA
Jopson	F	T	Yamamoto	M	LA

TABLE 3-2

Cross-Tabulation of Gender and Major (tallied)

		Major		
		Liberal Arts	**Business Ad.**	**Technology**
Gender	**Male**	‖‖‖ (5)	‖‖‖‖ (6)	‖‖‖‖‖ (7)
	Female	‖‖‖‖ (6)	‖‖‖‖ (4)	‖‖ (2)

The resulting 2 × 3 cross-tabulation (contingency table), Table 3-3, shows the frequency for each cross category of the two variables along with the row and column totals, called *marginal totals* (or *marginals*). The total of the marginal totals is the *grand total* and is equal to *n*, the sample size.

TABLE 3-3

Cross-Tabulation of Gender and Major (frequencies)

		Major			
		Liberal Arts	**Business Ad.**	**Technology**	**Row Totals**
Gender	**Male**	5	6	7	18
	Female	6	4	2	12
	Column totals	11	10	9	30

Contingency tables often show percentages (relative frequencies). These percentages can be based on the entire sample or on the subsample (row or column) classifications.

Percentages Based on the Grand Total (Entire Sample)   The contingency table shown in Table 3-3 can easily be converted to percentages of the grand total by dividing each frequency by the grand total and multiplying by 100. For example, 6 becomes 20% $\left[\left(\dfrac{6}{30}\right) \times 100 = 20\right]$. See Table 3-4.

TABLE 3-4

Cross-Tabulation of Gender and Major (relative frequencies; % of grand total)

		Major			Row Totals
		Liberal Arts	**Business Ad.**	**Technology**	
Gender	**Male**	17%	20%	23%	60%
	Female	20%	13%	7%	40%
	Column totals	37%	33%	30%	100%

With the table expressed in percentages of the grand total, we can easily see that 60% of the sample were male, 40% were female, 30% were technology majors, and so on. These same statistics (numerical values describing sample results) can be shown in a bar graph (see Figure 3-1).

FIGURE 3-1

Bar Graph

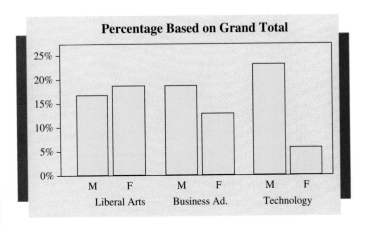

EXERCISE 3.1

In a national survey 500 men and 500 women—married, but not to each other—were asked to rate their spouse as a love partner.

	Very Good	Good	Other
Husband	230	173	97
Wife	240	177	83

Express the table as percentages of the total.

Table 3-4 and Figure 3-1 show the distribution of male liberal arts students, female liberal arts students, male business administration students, and so on relative to the entire sample.

Percentages Based on Row Totals   The entries in the same contingency table, Table 3-3, can be expressed as percentages of the row totals (or gender) by dividing each row entry by that row's total and multiplying by 100. Table 3-5 is based on row totals.

From Table 3-5 we see that 28% of the male students were majoring in liberal arts while 50% of the female students were majoring in liberal arts. These same statistics are shown in the bar graph in Figure 3.2.

Table 3-5 and Figure 3-2 show the distribution of the three majors for male and female students separately.

TABLE 3-5

Cross-Tabulation of
Gender and Major
(% of row totals)

		Major			
		Liberal Arts	**Business Ad.**	**Technology**	**Row Totals**
Gender	**Male**	28%	33%	39%	100%
	Female	50%	33%	17%	100%
	Column totals	37%	33%	30%	100%

FIGURE 3-2

Bar Graph

EXERCISE   3.2

Express the table in Exercise 3.1 as percentages of the row totals. Why might one prefer the table to be expressed this way?

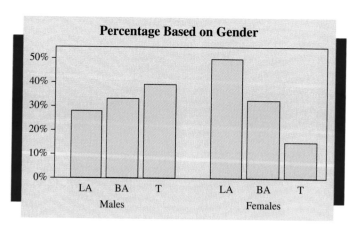

Percentages Based on Column Totals The entries in the contingency table in Table 3-3 can also be expressed as percentages of the column totals (or major) by dividing each column entry by that column's total and multiplying by 100. Table 3-6 is based on column totals.

TABLE 3-6 Cross-Tabulation of Gender and Major (% of column totals)

		Major			
		Liberal Arts	**Business Ad.**	**Technology**	**Row Totals**
Gender	**Male**	45%	60%	78%	60%
	Female	55%	40%	22%	40%
	Column totals	100%	100%	100%	100%

EXERCISE 3.3

Express the table in Exercise 3.1 as percentages of the column totals. Why might one prefer the table to be expressed this way?

From Table 3-6 we see that 45% of the liberal arts students were male while 55% of the liberal arts students were female. These same statistics are shown in the bar graph in Figure 3-3.

FIGURE 3-3

Bar Graph

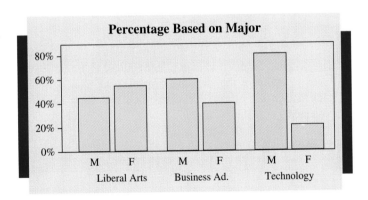

Table 3-6 and Figure 3-3 show the distribution of male and female students for each major separately.

MINITAB (Release 10) commands to construct a cross-tabulation table with the row variable categories listed in C1 and the column variable categories listed in C2.

Session commands

Enter: TABle C1 C2;
       COUNts;
       ROWPercents;
       COLPercents;
       TOTPercents.

Menu commands

Choose: Stat > Tables > CrossTab
Enter: Variables: C1 C2
Select: Counts
       Row Percents
       Column Percents
       Total Percents

Suggestion: The four subcommands (COUNt, ROWPercent, COLPercent, TOTPercent) can be used together; however, the resulting table will be much easier to read if you use one subcommand at a time. Always use a period to end the last subcommand.

One Qualitative and One Quantitative Variable

When bivariate data result from one qualitative and one quantitative variable, the quantitative values are viewed as separate samples, each set identified by levels of the qualitative variable. Each sample is described using the techniques from Chapter 2, and the results are displayed for side-by-side comparison.

▼ ILLUSTRATION   3 - 6

The distance required to stop a 3000-pound automobile on wet pavement was measured to compare the stopping capability of three different tread designs. Tires of each design were tested repeatedly on the same automobile on a controlled wet pavement.

TABLE 3-7

Stopping Distances for Three Tread Designs	Design A (*n* = 6)		Design B (*n* = 6)		Design C (*n* = 6)	
	37	36	33	35	40	39
	34	40	34	42	41	41
	38	32	38	34	40	43

The design of the tread is a qualitative variable with three levels of response, and the stopping distance is a quantitative variable. The distribution of the stopping distances for tread design A is to be compared with the distribution of stopping distances for each of the other tread designs. This comparison may be accomplished with both numerical and graphical techniques. Some of the available options are shown in Figure 3-4, Table 3-8, and Table 3-9.

FIGURE 3-4

Dotplot and Box-and-Whisker

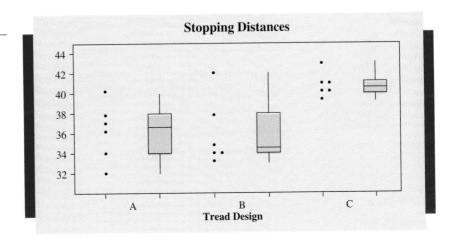

TABLE 3-8

5-Number Summary for Each Design

	Design A	Design B	Design C
High	40	42	43
Q_3	38	38	41
Median	36.5	34.5	40.5
Q_1	34	34	40
Low	32	33	39

TABLE 3-9

Mean and Standard Deviation for Each Design ▲ |

	Design A	Design B	Design C
Mean	36.2	36.0	40.7
St. Dev.	2.9	3.4	1.4

EXERCISE 3.4

The January unemployment rate for U.S. cities:
Eastern: 5.5, 6.6, 5.3, 5.4, 5.8
Western: 5.4, 6.0, 6.4, 5.6, 6.9.
Display as dotplots using same scale; compare means and medians.

Much of the information described above can be demonstrated using many other statistical techniques such as stem-and-leaf displays, histograms, and other numerical statistics.

Case Study

3-1

Visual Preference as a Test of Infant Word Comprehension

Table 3, a cross-tabulation table, displays the frequency of number of words comprehended at 14 and 20 months of age. The marginal totals present two separate frequency distributions for the variables: "14-month score" along the right side and "20-month score" along the bottom.

Most infants begin to understand words late in the first year and produce them early in the second year. Many theorists believe that early production vocabulary is a relatively small subset of comprehension vocabulary. . . . That is, a child who produces only 3 or 4 recognizable words may comprehend the meaning of 10 times that number.

The three experimenters presented explore the usefulness of a visual preference technique for assessing word comprehension in infants. . . . In Experiment 2, word comprehension was assessed longitudinally, at 14 and 20 months, to see if individual differences at 14 months predicted word comprehension scores at 20 months.

EXERCISE   3.5

Explain how the frequency numbers in Table 3 suggest that "children with low comprehension scores at 14 months were most likely to be among those with low comprehension scores at 20 months."

METHOD

Subjects

The original sample consisted of 100 infants assessed within 2 weeks of their 14th month. Ninety-one children returned to the laboratory for a second assessment at 20 months. . . .

Table 3 displays the cross-tabulation of comprehension scores at 14 and 20 months. Word comprehension scores at 14 months ranged between 0 and 5, with an average of 2.31 words comprehended. At 20 months, scores ranged between 1 and 7, with an average of 3.17 words comprehended. . . . Table 3 suggests the nature of this relation: children with low comprehension scores at 14 months were more likely to be among those with low comprehension scores at 20 months.

▎**TABLE   3**

Distribution of Word Comprehension Scores at 14 and 20 Months

Words Comprehended at 14 Months	Words Comprehended at 20 Months							Total
	1	2	3	4	5	6	7	
0	1	0	1	2	0	0	0	4
1	4	8	8	1	1	1	0	23
2	5	5	5	3	3	3	1	25
3	2	1	6	2	3	0	2	16
4	1	3	6	2	2	0	0	14
5	0	1	1	0	1	1	1	5
Total	13	18	27	10	10	5	4	87

Source: J. Steven Reznick, *Applied Psycholinguistics* 11 (1990), 145–166. Reprinted by permission.

MINITAB (Release 10) commands to compare the values in C1 as
separated into sets identified by categories listed in C2.

Session commands   *Menu commands*

Enter: BOXPlot C1;   Choose: Graph > Boxplot
       BY C2;   Enter: Y: C1   X: C2

       DOTPlot C1;   Choose: GRAPH > Char. Graph > Dotplot
       BY C2;   Enter: Variable: C1
       SAME.   By Variable C2

                            Select: Same scale

       DESCribe C1;
       BY C2.

Two Quantitative Variables

Ordered pairs
Input and output
variables

When the bivariate data is the result of two quantitative variables, it is customary to
express the data mathematically as **ordered pairs** (x, y), where x is the **input variable**
(sometimes called the independent variable) and y is the **output variable** (sometimes
called the *dependent variable*). The data are said to be ordered because one value,
x, is always written first. They are said to be paired because for each x value there
is a corresponding y value from the same source. For example, if x is height and y
is weight, a height and corresponding weight are recorded for each person. The input
variable x is measured or controlled in order to predict the output variable y. For
example, in Illustration 3-3 the researcher can control the amount of drug prescribed.
Therefore, the amount of drug would be referred to as x. In the case of height and
weight, either variable could be treated as input, the other as output, depending on the
question being asked. However, different results will be obtained from the regression
analysis, depending on the choice made.

In problems that deal with two quantitative variables, we will present our sample
data pictorially on a *scatter diagram*.

EXERCISE   3.6

Which variable,
height or weight,
would you use as
the input variable?
Explain why.

SCATTER DIAGRAM

A plot of all the ordered pairs of bivariate data on a coordinate axis system.
The input variable x is plotted on the horizontal axis, and the output variable y is plotted on the vertical axis.

EXERCISE   3.7

Draw a coordinate
axis and plot the
points (0, 6), (3, 5),
(3, 2), (5, 0).

NOTE   When constructing a scatter diagram, it is convenient to construct scales so
that the range of the y-values along the vertical axis is equal to or slightly shorter
than the range of the x-values along the horizontal axis. This creates a "window of
data" that is approximately square.

▼ | ILLUSTRATION   3-7

In Mr. Chamberlain's physical fitness course, several fitness scores were taken. The following sample is the number of push-ups and sit-ups done by ten randomly selected students:

(27, 30), (22, 26), (15, 25), (35, 42), (30, 38),
(52, 40), (35, 32), (55, 54), (40, 50), (40, 43)

Table 3-10 shows these sample data, and Figure 3-5 shows a scatter diagram of these data.

TABLE 3-10

Data for Push-ups and Sit-ups

	Student									
	1	**2**	**3**	**4**	**5**	**6**	**7**	**8**	**9**	**10**
Push-ups (x)	27	22	15	35	30	52	35	55	40	40
Sit-ups (y)	30	26	25	42	38	40	32	54	50	43

FIGURE 3-5

Scatter Diagram

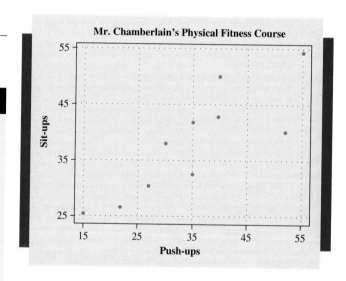

EXERCISE 3.8

Does studying for an exam pay off?

a. Draw a scatter diagram of the number of hours studied, x, compared to the exam grade, y, received.

x	2	5	1	4	2
y	80	80	70	90	60

b. Explain what you can conclude based on the pattern of data shown on the scatter diagram drawn in (a).

MINITAB (Release 10) commands to construct a scatter diagram with the x-variable listed in C1 and the y-variable listed in C2.

Session commands

```
Enter: PLOT C2 * C1;
       TITLe 'your title'.
```

Menu commands

```
Choose: Graph > Plot
Enter:  Variable: Y: C2  X: C1
Annotation > Title: your title
```

EXERCISES

3.9 Vita Health Food Store conducted a market survey. The results are in the accompanying table.

If you are using MINITAB, try TABLE. See page 140. Many of these sets are on Data Disk.

  a. Form a cross-tabulation of the variables, gender of customer, and made a purchase (yes/no). Express the results in frequencies, showing marginal totals.
  b. Express the contingency table found in (a) in percentages based on the grand total.
  c. Express the contingency table found in (a) in percentages based on the marginal total for gender.
  d. Draw a bar graph showing the results from (c).

Vita Health Food Store Market Research Survey Findings

Person	Purchase	Gender	Age	Person	Purchase	Gender	Age
1	no	M	43	31	no	F	19
2	yes	M	58	32	yes	F	59
3	no	F	34	33	no	F	32
4	no	M	66	34	yes	M	46
5	yes	M	46	35	no	F	53
6	yes	F	52	36	no	M	40
7	no	F	18	37	yes	F	50
8	no	F	50	38	no	F	52
9	yes	F	39	39	yes	M	68
10	no	M	46	40	no	F	35
11	yes	M	62	41	yes	F	32
12	no	M	40	42	no	F	37
13	yes	M	47	43	no	F	39
14	no	M	64	44	no	M	28
15	no	M	21	45	no	M	68
16	no	F	42	46	yes	F	64
17	no	M	19	47	no	M	19
18	yes	F	69	48	yes	M	57
19	no	M	59	49	no	F	45
20	yes	M	54	50	no	F	22
21	no	F	48	51	yes	F	63
22	yes	F	49	52	no	M	35
23	yes	F	42	53	no	F	29
24	no	F	49	54	no	M	51
25	yes	F	61	55	yes	F	43
26	yes	M	56	56	no	M	30
27	no	F	44	57	yes	M	70
28	yes	F	60	58	no	F	21
29	yes	F	48	59	no	F	47
30	no	M	50	60	no	M	59

3.10 Using Vita's results on the table above,

a. Form a cross-tabulation of the variables, age of customer (use categories: under 35, 35–50, over 50) and made a purchase (yes/no). Express the results in frequencies, showing marginal totals.

b. Express the contingency table found in (a) in percentages based on the grand total.

c. Express the contingency table found in (a) in percentages based on the marginal total for age.

d. Draw a bar graph showing the results from (c).

3.11 A statewide survey was conducted to investigate the relationship between viewers' preferences for ABC, CBS, NBC, or PBS for news information and their political party affiliation. The results are shown in tabular form:

	ABC	CBS	NBC	PBS
Democrat	200	200	250	150
Republican	450	350	500	200
Other	150	400	100	50

a. How many viewers were surveyed?

b. Why is this bivariate data? What type of variable is each one?

c. How many preferred to watch CBS?

d. What percentage of the survey was Republican?

e. What percentage of the Democrats preferred ABC?

3.12 Consider the accompanying contingency table, which presents the results of an advertising survey about the use of credit by Martan Oil Company customers:

	Number of Purchases at Gasoline Station Last Year					
Preferred Method of Payment	**0–4**	**5–9**	**10–14**	**15–19**	**20 and Over**	**Sum**
Cash	150	100	25	0	0	275
Oil company card	50	35	115	80	70	350
National or bank credit card	50	60	65	45	5	225
Sum	250	195	205	125	75	850

a. How many customers were surveyed?

b. Why is this bivariate data? What type of variable is each one?

c. How many customers preferred to use an oil-company credit card?

d. How many customers made 20 or more purchases last year?

e. How many customers preferred to use an oil-company credit card and made only between five and nine purchases last year?

f. What does the 80 in the fourth cell in the second row mean?

3.13 What effect does the minimum amount have on the interest rate being offered on six-month CDs? The *Pittsburgh Post-Gazette*, 12-25-94, listed the following advertised rate of return, *y*, for a minimum deposit of $500, $1000, or $2500, *x* in $100.

x	25	25	10	25	25	10	25	25	10	25	25	10	25	25
y	5.13	5.00	5.45	5.10	5.50	5.11	5.04	4.91	5.00	5.30	5.10	5.17	5.06	5.39

x	25	10	10	10	10	10	10	5	5	5	5	5	5
y	5.00	5.32	5.10	5.32	5.28	5.06	5.13	5.06	5.12	4.00	5.26	5.16	5.06

x	5	5	5	10	25	25	10	5	10	5	5	10	25
y	4.55	5.00	5.12	5.10	5.11	5.22	5.06	5.20	5.32	4.94	5.25	5.28	4.75

If you are using MINITAB, try the commands on page 143.

a. Prepare a dotplot of the three sets of data using a common scale.
b. Prepare a 5-number summary and a boxplot of the three sets of data. Use the same scale for the boxplots.
c. Describe any differences you see between the three sets of data.

3.14 Can a woman's height be predicted using her mother's height? The heights of some mother–daughter pairs are listed; *x* is the mother's height and *y* is the daughter's height.

x	63	63	67	65	61	63	61	64	62	63
y	63	65	65	65	64	64	63	62	63	64

x	64	63	64	64	63	67	61	65	64	65	66
y	64	64	65	65	62	66	62	63	66	66	65

a. Draw two dotplots using the same scale showing the two sets of data side by side.
b. What can you conclude from seeing the two sets of heights shown as separate sets this way? Explain.
c. Draw a scatter diagram of these data as ordered pairs.
d. What can you conclude from seeing the data presented as ordered pairs? Explain.

3.15 The following data compare the height, weight, and age of the players on the two teams that played in the 1994 World Cup finals, from Italy and Brazil:
   a. Compare each of the three variables—height, weight, and age—using either a dotplot or a histogram (use the same scale).
   b. Based on what you see in graphs in (a), can you detect a substantial difference between the two teams in regard to these three variables? Explain.
   c. Explain why the data, as used in (a), was not bivariate data.

Player	Ht-Ital	Wt-Ital	Age-Ital	Ht-Braz	Wt-Braz	Age-Braz
1	73	192	27	71	176	28
2	72	168	27	69	152	29
3	67	144	25	67	163	31
4	70	163	28	74	183	29
5	73	170	26	71	176	26
6	69	155	34	71	167	30
7	73	165	27	70	145	30
8	71	161	30	70	164	30
9	70	163	34	67	158	27
10	69	159	27	74	191	29
11	69	159	22	66	154	28
12	74	170	28	74	194	29
13	73	159	22	71	164	28
14	73	174	27	69	163	24
15	69	157	24	73	169	24
16	68	150	30	66	156	24
17	69	161	31	67	147	27
18	72	173	25	67	158	25
19	70	163	33	70	158	28
20	67	150	26	67	165	17
21	66	143	28	70	174	25
22	71	176	25	72	169	35

3.16

If you are using MINITAB, try PLOT.

a. Draw a scatter diagram showing height, x, and weight, y, for the Italian World Cup soccer team using the data in Exercise 3.15.

b. Draw a scatter diagram showing height, x, and weight, y, for the Brazilian World Cup soccer team using the data in Exercise 3.15.

c. Explain why the data, as used in (a) and (b), are bivariate data.

3.17 The accompanying data show the number of hours x studied for an exam and the grade y received in the exam. (y is measured in 10s; that is, $y = 8$ means that the grade, rounded to the nearest 10 points, is 80.) Draw the scatter diagram. (Retain this solution for use in answering Exercise 3.31, p. 158.)

x	2	3	3	4	4	5	5	6	6	6	7	7	7	8	8
y	5	5	7	5	7	7	8	6	9	8	7	9	10	8	9

3.18 An experimental psychologist asserts that the older a child is the fewer irrelevant answers he or she will give during a controlled experiment. To investigate this claim, the following data were collected. Draw a scatter diagram. (Retain this solution for use in answering Exercise 3.32, p. 158.)

Age (x)	2	4	5	6	6	7	9	9	10	12
Number of Irrelevant Answers (y)	12	13	9	7	12	8	6	9	7	5

3.19 In a study involving children's fear related to being hospitalized, the age and the score each child made on the Child Medical Fear Scale (CMFS) were:

Age (x)	8	9	9	10	11	9	9	9	11	11
CMFS score (y)	31	25	40	27	35	29	25	34	27	36

Construct a scatter diagram of these data.

3.20 A sample of 15 upperclass students who commute to classes was selected at registration. They were asked to estimate the distance (x) and the time (y) required to commute each day to class (see the following table). Construct a scatter diagram depicting these data.

Distance, x (nearest mile)	Time, y (nearest 5 minutes)	Distance, x (nearest mile)	Time, y (nearest 5 minutes)
18	20	2	5
8	15	15	25
20	25	16	30
5	20	9	20
5	15	21	30
11	25	5	10
9	20	15	20
10	25		

3.21 Walter Payton was one of the NFL's greatest running backs. Below are listed the number of carries and the total yards gained in each of his 13 years with the Chicago Bears.

Number Carries	196	311	339	333	369	317	339	148	314	381	324	321	146
Total Yards	679	1390	1852	1359	1610	1460	1222	596	1421	1684	1551	1333	586

a. Construct a scatter diagram depicting these data.
b. How would you describe this scatter diagram? What do you see that is unusual about the scatter diagram?
c. What circumstances might explain this "two group" appearance of the data points? Explain.

3.22 Tony Dorsett was also a great NFL running back. Below are listed the number of carries and the total yards gained in each of his 12 years in the NFL.

Number Carries	208	290	250	278	342	177	289	302	305	184	130	181
Total Yards	1007	1325	1107	1185	1646	745	1321	1189	1307	748	456	703

a. Construct a scatter diagram depicting these data.
b. How would you describe this scatter diagram?
c. What circumstances might explain this "straight line" appearance of the data points? Explain.

3.2 | LINEAR CORRELATION

Linear correlation analysis

Independent and dependent variables

No correlation

The primary purpose of **linear correlation analysis** is to measure the strength of a linear relationship between two variables. Let's examine some scatter diagrams demonstrating different relationships between input, or **independent variables**, x, and output, or **dependent variables**, y. If as x increases there is no definite shift in the values of y, we say there is **no correlation**, or no relationship between x and y. If as x increases there is a shift in the values of y, there is a *correlation*. The correlation is *positive* when y tends to increase and *negative* when y tends to decrease. If the ordered pairs (x, y) tend to follow a straight-line path, there is a **linear correlation**. The preciseness of the shift in y as x increases determines the strength of the linear correlation. The scatter diagrams in Figure 3-6 demonstrate these ideas.

Linear correlation

Perfect linear correlation occurs when all the points fall exactly along a straight line, as shown in Figure 3-7. This can be either positive or negative, depending on whether y increases or decreases as x increases. If the data form a straight horizontal or vertical line, there is no correlation, since one variable has no effect on the other, as shown in Figure 3-8 on page 152.

Scatter diagrams do not always appear in one of the forms shown in Figures 3-6, 3-7, and 3-8. Sometimes they suggest relationships other than linear, as in Figure 3-9 on page 152. There appears to be a definite pattern; however, the two variables are not related linearly, and therefore there is no linear correlation.

Coefficient of linear correlation

The **coefficient of linear correlation** r is the numerical measure of the strength of the linear relationship between two variables. The coefficient reflects the consistency of the effect that a change in one variable has on the other. The value of the linear correlation coefficient helps us to answer the question "Is there a linear correlation between the two variables under consideration?" The linear correlation coefficient r always has a value between -1 and $+1$. A value of $+1$ signifies a **perfect positive correlation**, and a value of -1 shows a **perfect negative correlation**. If as x increases there is a general increase in the value of y, then r will be positive in value. For example, a positive value of r would be expected for age and height

Perfect positive and negative correlation

FIGURE 3-6

Scatter Diagrams and Correlation

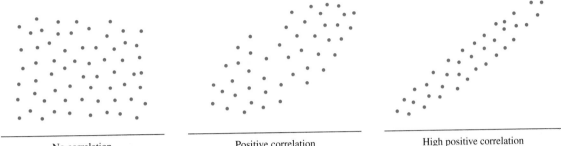

No correlation

Positive correlation

High positive correlation

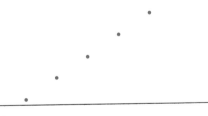

Negative correlation

High negative correlation

FIGURE 3-7

(a) Perfect Positive
Correlation;
(b) Perfect
Negative
Correlation

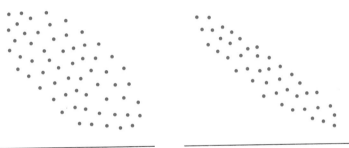

(a)

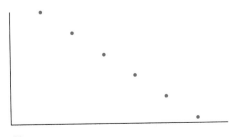

(b)

FIGURE 3-8

(a) Horizontal—No
Correlation;
(b) Vertical—No
Correlation

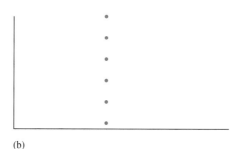

(a)

(b)

FIGURE 3-9

No Linear
Correlation

of children, because as children grow older, they grow taller. Also, consider the age x and resale value y of an automobile. As the car ages, its resale value decreases. Since as x increases, y decreases, the relationship results in a negative value for r.

Pearson's product
moment formula

The value of r is defined by **Pearson's product moment formula**:

$$r = \frac{\sum (x - \bar{x})(y - \bar{y})}{(n - 1)s_x s_y} \qquad (3\text{-}1)$$

NOTES

1. s_x and s_y are the standard deviations of the x and y variables.
2. The development of this formula is discussed in Chapter 13.

To calculate r, we will use an alternative formula, formula (3-2), that is equivalent to formula (3-1). As preliminary calculations, we will separately calculate three sums of squares and then substitute them into formula (3-2) to obtain r.

$$r = \frac{sum\ of\ squares\ for\ xy}{\sqrt{(sum\ of\ squares\ for\ x)(sum\ of\ squares\ for\ y)}}$$

$$r = \frac{SS(xy)}{\sqrt{SS(x)SS(y)}} \qquad \text{(3-2)}$$

$$sum\ of\ squares\ for\ x = sum\ of\ x^2 - \frac{(sum\ of\ x)^2}{n}$$

where

$$SS(x) = \sum x^2 - \frac{\left(\sum x\right)^2}{n} \qquad \text{(2-9)}$$

$$sum\ of\ squares\ for\ y = sum\ of\ y^2 - \frac{(sum\ of\ y)^2}{n}$$

> SS(x) is the numerator of variance, page 80.

$$SS(y) = \sum y^2 - \frac{\left(\sum y\right)^2}{n} \qquad \text{(3-3)}$$

$$sum\ of\ squares\ for\ xy = sum\ of\ xy - \frac{(sum\ of\ x)(sum\ of\ y)}{n}$$

$$SS(xy) = \sum xy - \frac{\sum x \sum y}{n} \qquad \text{(3-4)}$$

▼ ILLUSTRATION 3 - 8

Find the linear correlation coefficient for the push-up/sit-up data in Illustration 3-7, page 144.

SOLUTION First, we need to construct an extensions table (Table 3-11) listing all the pairs of values (x, y) to aid in finding the extensions x^2, xy, and y^2 and the five column totals.

TABLE 3-11

Extensions Table for Finding Five Summations

Student	Push-ups (x)	x^2	Sit-ups (y)	y^2	xy
1	27	729	30	900	810
2	22	484	26	676	572
3	15	225	25	625	375
4	35	1,225	42	1,764	1,470
5	30	900	38	1,444	1,140
6	52	2,704	40	1,600	2,080
7	35	1,225	32	1,024	1,120
8	55	3,025	54	2.916	2,970
9	40	1,600	50	2,500	2,000
10	40	1,600	43	1,849	1,720
	$\Sigma x = 351$	$\Sigma x^2 = 13,717$	$\Sigma y = 380$	$\Sigma y^2 = 15,298$	$\Sigma xy = 14,257$
	↑ sum of x	↑ sum of x^2	↑ sum of y	↑ sum of y^2	↑ sum of xy

Second, to complete the preliminary calculations, substitute the five summations (the five column totals) from the extensions table into formulas (2-9), (3-3), and (3-4), and calculate the three sums of squares.

Save these prelim-inary calculations for later use in regression.

$$SS(x) = \sum x^2 - \frac{\left(\sum x\right)^2}{n} = 13,717 - \frac{(351)^2}{10} = 1396.9$$

$$SS(y) = \sum y^2 - \frac{\left(\sum y\right)^2}{n} = 15,298 - \frac{(380)^2}{10} = 858.0$$

$$SS(xy) = \sum xy - \frac{\sum x \sum y}{n} = 14,257 - \frac{(351)(380)}{10} = 919.0$$

Third, substitute the three sums of squares into formula (3-2) and obtain the value of the correlation coefficient.

$$r = \frac{SS(xy)}{\sqrt{SS(x)SS(y)}} = \frac{919.0}{\sqrt{(1396.9)(858.0)}} = 0.8394 = \mathbf{0.84}$$

▲ |

NOTE   Typically, r is rounded to the nearest hundredth.

EXERCISE 3.23

Does studying for an exam pay off? The number of hours studied, x, is com-pared to the exam grade, y:

x	2	5	1	4	2
y	80	80	70	90	60

a. Complete the pre-liminary calcula-tions: extensions, five sums, and SS(x), SS(y), SS(xy).

b. Find r.

The value of the calculated linear correlation coefficient helps us answer the question "Is there a linear correlation between the two variables under consideration?" When the calculated value of r is close to zero, we conclude that there is little or no linear correlation. As the calculated value of r changes from 0.0 toward either $+1.0$ or -1.0, it indicates an increasingly stronger linear correlation between the two variables. From a graphical viewpoint, when we calculate r, we are measuring how well a straight line describes the scatterplot of ordered pairs. As the value of r changes from 0.0 toward $+1.0$ or -1.0, the data points creating a pattern move closer to a straight line. Further discussion involving the interpretation of the calculated value of r is found in Chapter 13.

MINITAB (Release 10) commands to calculate the correlation coefficient with the *x*-variable listed in C1 and the *y*-variable listed in C2.

Session commands   *Menu commands*

Enter: CORRelation C1 C2   Choose: Stat > BasicStat. > Correl
                            Enter: Variables: C1 C2

Use the following sets of *session commands*, with *x* listed in C1 and *y* listed in C2, to complete the preliminary calculations: the extensions table and the three sums of squares.

The extensions table

```
LET C3 = C1*C1
LET C4 = C1*C2
LET C5 = C2*C2
LET K1 = SUM(C1)
LET K2 = SUM(C2)
LET K3 = SUM(C3)
LET K4 = SUM(C4)
LET K5 = SUM(C5)
PRINt C1-C5
PRINt K1-K5
```

The three sums of squares

```
SS(x);
   LET K6 = K3 − (K1**2/COUNT(C1))
   PRINt K6
SS(y);
   LET K7 = K5 − (K2**2/COUNT(C1))
   PRINt K7
SS(xy);
   LET K8 = K4 − (K1*K2/COUNT(C1))
   PRINt K8
```

For a convenient way to handle sets of commands, see Helpful Hints on page 89.

Case Study   Points and Fouls in Basketball

3-2

Albert Shulte discusses the relationship between the number of personal fouls committed and total points scored during a season by the members of a junior-varsity basketball team. It appears, from Figure 1, that the players score about three points for every personal foul committed. The data for the second year, Table 2, seem to indicate approximately the same relationship.

FIGURE 1

First Year

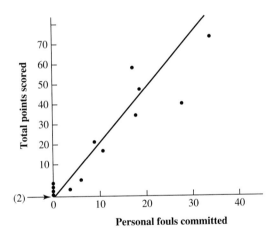

TABLE 2

Second Year

Player	Total Points	Personal Fouls	Player	Total Points	Personal Fouls
Brummett	2	1	McPartlin	35	5
Cooper	75	24	Pointer	46	20
Felice	0	1	Schuback	0	1
Hook	59	18	Wilson	2	3
Hurd	9	9	Zuelch	57	22
Kampsen	7	3			

EXERCISE 3.24

a. Draw a scatter diagram of the data in Table 2.

b. Calculate the correlation coefficient for the data in Table 2.

c. Does this mean that an increase in number of fouls will cause the player to score more points? What does a strong correlation coefficient mean? Explain in your own words.

Sometimes you may think that two different sets of numbers are related in some way, although not perfectly. How can you decide if they are related? If they are related, is there any reason why they should be, or is it purely accidental? Does a change in one of the variables cause a change in the other?

The data come from the records of a junior-varsity basketball team. . . for two separate years. . . . A quick look at Table 2 makes one feel that a strong relationship exists between the number of points . . . and the number of personal fouls. . . . In Figure 1, the data for the first year have been plotted. . . .

Now let's think a bit more about the relationship that seems to exist. . . . The table and the figure both indicate that the more fouls [a player] commits, the more points he scores. . . . Surely no coach would coach a player to make lots of fouls in the hope that he would therefore score more points. The crucial word in the previous sentence is "therefore." Is the fact that a player commits more fouls the reason that he scores more points? Of course not. But maybe both of these . . . are the results of some other act. Perhaps . . . more game time.

A graph such as that in Figure 1 can show that a high degree of correlation exists, but it does not tell us why.

Source: Albert P. Shulte, "Points and Fouls in Basketball," in *Exploring Data*, from *Statistics by Example*. Edited and prepared by the Joint Committee on the Curriculum in Statistics and Probability of the American Statistical Association and the National Council of Teachers of Mathematics. Copyright 1973 by Addison-Wesley Publishing Co., Inc. Reprinted by permission.

Estimating the Linear Correlation Coefficient Visually

With a formula as complex as formula (3-2), it would be very convenient to be able to inspect the scatter diagram of the data and estimate the calculated value of r. This would serve as a check on calculations. The following **method for estimating r** is quick and generally yields a reasonable estimate when the "window of data" is approximately square.

Procedure

1. Lay two pencils on your scatter diagram. Keeping them parallel, move them to a position so that they are as close together as possible yet have all the points on the scatter diagram between them. (See Figure 3-10.)

2. Visualize a rectangular region that is bounded by the two pencils and that ends just beyond the points on the scatter diagram. (See the shaded portion of Figure 3-10.)

FIGURE 3-10

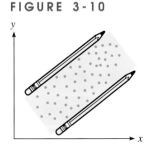

3. Estimate how many times longer the rectangle is than it is wide. An easy way to do this is to mentally mark off squares in the rectangle. (See Figure 3.11.) Call this number of multiples k.

EXERCISE 3.25

Estimate the correlation coefficient for each of the following:

4. The value of r may be estimated as $\pm\left(1 - \dfrac{1}{k}\right)$.

5. The sign assigned to r is determined by the general position of the length of the rectangular region. If it lies in an increasing position (see Figure 3-12), r will be positive; if it lies in a decreasing position, r will be negative. *If the rectangle is in either a horizontal or a vertical position, then r will be zero, regardless of the length–width ratio.*

NOTE This estimation technique does not replace the calculation of r. It is very sensitive to the "spread" of the diagram. However, if the "window of data" is approximately square, this approximation will be helpful when used as a mental estimate or check.

Let's use this method to estimate the value of the linear correlation coefficient for the relationship between the number of push-ups and sit-ups. As shown in Figure 3-13, we find that the rectangle is approximately 3.5 times longer than it is wide; that is, $k \approx 3.5$, and the rectangle lies in an increasing position. Therefore, our estimate for r is

$$r \approx +\left(1 - \frac{1}{3.5}\right) \approx +0.7$$

FIGURE 3-11

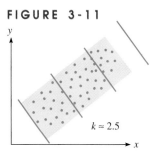

$k \approx 2.5$

FIGURE 3-12

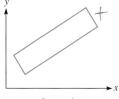

Increasing

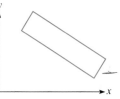

Decreasing

FIGURE 3-13

Push-ups Versus Sit-ups for Ten Students

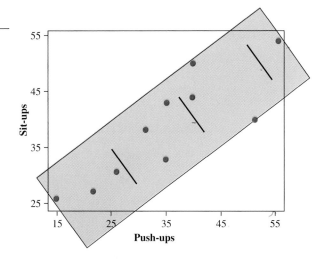

EXERCISES

3.26 How would you interpret the findings of a correlation study that reported a linear correlation coefficient of -1.34?

3.27 How would you interpret the findings of a correlation study that reported a linear correlation coefficient of $+0.3$?

3.28 Explain why it makes sense for a set of data to have a correlation coefficient of zero when the data show a very definite pattern as in Figure 3-9.

3.29 In an article titled "Self-Reported Fears of Hospitalized School-Age Children" (*Journal of Pediatric Nursing*, Vol. 9, No. 2, 1994), the authors report a correlation of 0.10 between the ages of children and the score they made on the Child Medical Fear Scale (CMFS). Suppose the ages and CMFS scores for ten children were as follows (same data as Exercise 3.19):

If you use MINITAB, try the set of commands given on page 155.

CMFS Score	31	25	40	27	35	29	25	34	27	36
Age	8	9	9	10	11	9	9	9	11	11

Find a. $SS(x)$ b. $SS(y)$ c. $SS(xy)$ d. the value of r for these data

3.30 An article titled "A Profile of Mood in Ambulatory Nursing Home Residents" (*Archives of Psychiatric Nursing*, Vol. 8, No. 5, 1994) discusses the administration of the Profile of Mood States (POMS) test to 54 nursing-home residents 60 years of age or older. The POMS test has six subscales of mood. The article reports the correlation between the Anger–Hostility subscale and the Depression–Dejection subscale as 0.5536. Suppose the POMS test was administered to eight nursing-home residents with these subscale scores:

Anger–Hostility Score	7	14	15	10	12	17	15	20
Depression–Dejection Score	12	18	14	16	13	19	16	17

Find a. $SS(x)$ b. $SS(y)$ c. $SS(xy)$ d. the value of r for these data

3.31 a. Use the scatter diagram you drew in answering Exercise 3.17 (p. 148) to estimate r for the sample data relating the number of hours studied and the exam grade.
b. Calculate r.

Have you tried CORRelation?

3.32 a. Use the scatter diagram you drew in answering Exercise 3.18 (p. 148) to estimate r for the sample data relating the number of irrelevant anwers and the child's age.
b. Calculate r.

3.33 Consider the following data, which give the weight (in thousands of pounds) x and gasoline mileage (miles per gallon) y for ten different automobiles.

x	2.5	3.0	4.0	3.5	2.7	4.5	3.8	2.9	5.0	2.2
y	40	43	30	35	42	19	32	39	15	14

Find a. SS(x)   b. SS(y)   c. SS(xy)   d. Pearson's product moment r

3.34 A marketing firm wished to determine whether or not the number of television commercials broadcast were linearly correlated to the sales of its product. The data, obtained from each of several cities, are shown in the following table.

City	A	B	C	D	E	F	G	H	I	J
No. TV Commercials (x)	12	6	9	15	11	15	8	16	12	6
Sales Units (y)	7	5	10	14	12	9	6	11	11	8

a. Draw a scatter diagram.   b. Estimate r.   c. Calculate r.

3.35 An article titled "Leader Power, Commitment Satisfaction, and Propensity to Leave a Job Among U.S. Accountants" (*Journal of Social Psychology*, Vol. 133, No. 5, Oct. 1993) reported a linear correlation coefficient of $-.61$ between satisfaction with work scores and propensity to leave a job scores. Suppose similar assessments of work satisfaction, x, and propensity to leave a job, y, gave the following scores.

x	12	24	17	28	24	36	20
y	44	36	25	23	32	17	24

a. Find the linear correlation between x and y.
b. What does the value of this correlation coefficient seem to be telling us? Explain.

3.36 An article titled "College Recreation Facility Survey" (*Athletic Business*, April 1994) reported the following results from 358 four-year colleges and universities in the United States and Canada.

Enrollment	Number of Schools	Total Square Feet Devoted to Recreation per School
0–1,249	58	47,864
1,250–2,499	53	71,828
2,500–4,999	53	89,716
5,000–9,999	62	101,016
10,000–17,999	68	127,952
18,000 or over	64	200,896

a. Using the midpoint of the first five enrollment classes and 25,000 in place of the class 18,000 or over for the x-values and the total square feet devoted to recreation per school for the y-values, find the linear correlation coefficient between x and y.

b. What does the value of this correlation seem to be telling us? Explain.

3.3 | LINEAR REGRESSSION

Although the correlation coefficient measures the strength of a linear relationship, it does not tell us about the mathematical relationship between the two variables. In Section 3.2, the correlation coefficient for the push-up/sit-up data was found to be 0.84 (see p. 154). This implies that there is a linear relationship between the number of push-ups and the number of sit-ups a student does. The correlation coefficient does not help us predict the number of sit-ups a person can do based on knowing he or she can do 28 push-ups. **Regression analysis** finds the equation of the line that best describes the relationship between the two variables. One use of this equation is to make predictions. There are many situations in which we make use of these predictions regularly: for example, predicting the success a student will have in college based on high school results and predicting the distance required to stop a car based on its speed. Generally, the exact value of y is not predictable and we are usually satisfied if the predictions are reasonably close.

Regression analysis

The relationship between these two variables will be an algebraic expression describing the mathematical relationship between x and y. Here are some examples of various possible relationships, called **models** or **prediction equations**:

Models
Prediction equations

Linear:	$\hat{y} = b_0 + b_1 x$
Quadratic:	$\hat{y} = a + bx + cx^2$
Exponential:	$\hat{y} = a(b^x)$
Logarithmic:	$\hat{y} = a \log_b x$

Figures 3-14, 3-15, and 3-16 show patterns of bivariate data that appear to have a relationship, whereas in Figure 3-17, the variables do not seem to be related.

If a straight-line model seems appropriate, the best-fitting straight line is found by using the **method of least squares**. Suppose that $\hat{y} = b_0 + b_1 x$ is the equation of a straight line, where $\hat{y}$ (read "y hat") represents the **predicted value** of y that corresponds to a particular value of x. The **least squares criterion** requires that we find the constants b_0 and b_1 such that the sum $\Sigma(y - \hat{y})^2$ is as small as possible. (See Figure 3-18, p. 161.)

Method of least squares
Predicted value
Least squares criterion

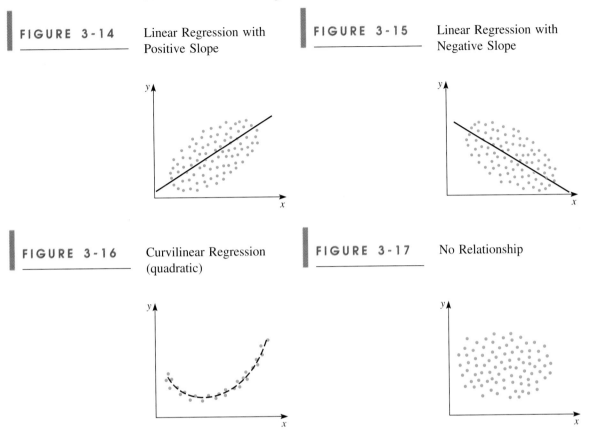

FIGURE 3-14   Linear Regression with Positive Slope

FIGURE 3-15   Linear Regression with Negative Slope

FIGURE 3-16   Curvilinear Regression (quadratic)

FIGURE 3-17   No Relationship

Figure 3-18 shows the distance of an observed value of y from a predicted value of $\hat{y}$. The length of this distance represents the value $(y - \hat{y})$ (shown as the red line segment in Figure 3-18). Note that $(y - \hat{y})$ is positive when the point (x, y) is above the line and negative when (x, y) is below the line.

FIGURE 3-18

Observed and Predicted Values of y

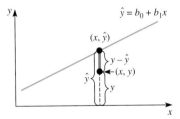

Figure 3-19 (p. 162) shows a scatter diagram with what might appear to be the line of best fit, along with the ten individual $(y - \hat{y})$'s. (Positive ones are shown in red, negative ones in green.) The sum of the squares of these differences is minimized (made as small as possible) if the line is indeed the line of best fit.

Figure 3-20 shows the same data points as Figure 3-19 with the ten individual $(y - \hat{y})$'s associated with a line that is definitely not the line of best fit. (The value of $\Sigma(y - \hat{y})^2$ is 149, much larger than 23 from Figure 3.19.) Every different line

FIGURE 3-19   The Line of Best Fit

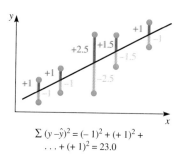

$$\Sigma (y - \hat{y})^2 = (-1)^2 + (+1)^2 + \\ \ldots + (+1)^2 = 23.0$$

FIGURE 3-20   Not the Line of Best Fit

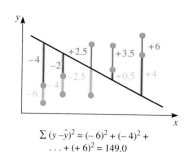

$$\Sigma (y - \hat{y})^2 = (-6)^2 + (-4)^2 + \\ \ldots + (+6)^2 = 149.0$$

Line of best fit
Slope
y-Intercept

drawn through this set of ten points will cause a different value for $\Sigma (y - \hat{y})^2$. Our job is to find the one line that will result in $\Sigma (y - \hat{y})^2$ being the least possible value.

The **equation of the line of best fit** is determined by its **slope** (b_1) and its **y-intercept** (b_0). (See the *Statistical Tutor* for a review of the concepts of slope and intercept of a straight line.) The values of the constants—slope and y-intercept—that satisfy the least squares criterion are found by using these formulas:

$$\text{slope:} \quad b_1 = \frac{\Sigma (x - \bar{x})(y - \bar{y})}{\Sigma (x - \bar{x})^2} \tag{3-5}$$

$$y\text{-}intercept = \frac{(sum\ of\ y) - [(slope)(sum\ of\ x)]}{number}$$

$$y\text{-intercept:} \quad b_0 = \frac{\Sigma y - \left(b_1 \cdot \Sigma x\right)}{n} \tag{3-6}$$

(The derivation of these formulas is beyond the scope of this text.)

We will use a mathematical equivalent of formula (3-5) to find slope b_1 that uses the sums of squares found in the preliminary calculations for correlation:

$$\text{slope:} \quad b_1 = \frac{SS(xy)}{SS(x)} \tag{3-7}$$

Notice that the numerator of formula (3-7) is the SS(xy) formula (3-4) and the denominator is formula (2-9) from the correlation coefficient calculations. Thus, if you have previously calculated the linear correlation coefficient using the procedure outlined on pages 153 to 154, you can easily find the slope of the line of best fit. If you did not previously calculate r, set up a table similar to Table 3-11 (p. 153) and complete the necessary preliminary calculations.

Now let's consider the data in Illustration 3-7 (p. 144) and the question of predicting a student's sit-ups based on the number of push-ups. We want to find the line of best fit, $\hat{y} = b_0 + b_1 x$. The preliminary calculations have already been completed in Table 3-11 and on page 154. To calculate the slope, b_1, using formula (3-7), recall $SS(xy) = 919.0$ and $SS(x) = 1396.9$.

$$\text{slope:} \quad b_1 = \frac{SS(xy)}{SS(x)} = \frac{919.0}{1396.9} = 0.6579 = \textbf{0.66}$$

To calculate the y-intercept, b_0, using formula (3-6), recall $\Sigma x = 351$ and $\Sigma y = 380$ from the extensions table.

$$y\text{-intercept:} \quad b_0 = \frac{\Sigma y - \left(b_1 \cdot \Sigma x\right)}{n} = \frac{380 - (0.6579)(351)}{10}$$

$$= \frac{380 - 230.9229}{10} = 14.9077 = \textbf{14.9}$$

Thus the equation of the line of best fit is

$$\hat{y} = \textbf{14.9} + \textbf{0.66}x$$

NOTES

1. Remember to keep at least three extra decimal places while doing the calculations to ensure an accurate answer.
2. When rounding off the calculated values of b_0 and b_1, always keep at least two significant digits in the final answer.

Now that we know the equation for the line of best fit, let's draw the line on the scatter diagram so that we can visualize the relationship between the line and the data. Two points will be needed in order to draw the line on the diagram. Select two convenient x-values, one near each extreme of the domain ($x = 10$ and $x = 60$ are good choices for this illustration), and find their corresponding y-values.

For $x = 10$: $\hat{y} = 14.9 + 0.66x = 14.9 + 0.66(10) = 21.5$; (10, 21.5)

For $x = 60$: $\hat{y} = 14.9 + 0.66x = 14.9 + 0.66(60) = 54.5$; (60, 54.5)

These two points (10, 21.5) and (60, 54.5) are then located on the scatter diagram (use a blue + to distinguish them from data points) and the line of best fit is drawn (shown in red in Figure 3-21).

There are some additional facts about the least squares method that we need to discuss.

1. The slope b_1 represents the predicted change in y per unit increase in x. In our example, $b_1 = 0.66$; thus, if a student had done an additional ten push-ups (x), we would predict that he or she would have done approximately an additional seven (0.66×10) sit-ups (y).
2. The y-intercept is the value of y where the line of best fit intersects the y-axis. (The y-intercept is easily seen on the scatter diagram, shown in a green + in Figure 3-21, when the vertical scale is located above $x = 0$.) However,

---

EXERCISE 3.37

The formulas for finding the slope and the y-intercept of the line of best fit use both summations, Σ's, and sums of squares, SS( )'s. It is important to know the difference. In reference to Illustration 3-8:

a. Find three pairs of values:
Σx^2, SS(x);
Σy^2, SS(y); and
Σxy, SS(xy).
b. Explain the difference between the numbers for each pair of numbers.

EXERCISE 3.38

The values of x used to find points for graphing the line $\hat{y} = 14.9 + 0.66x$ are arbitrary. Suppose you choose to use $x = 20$ and $x = 50$.

a. What are the corresponding $\hat{y}$-values?
b. Locate these two points on Figure 3-21. Are these points on the line of best fit? Explain why or why not.

FIGURE 3-21

Line of Best Fit for Push-ups Versus Sit-ups

EXERCISE 3.39

Does studying for an exam pay off? The number of hours studied, *x*, is compared to the exam grade, *y*:

x	2	5	1	4	2
y	80	80	70	90	60

a. Find the equation for the line of best fit.

b. Draw the line of best fit on the scatter diagram of the data drawn in Exercise 3.8 (p. 144).

c. Based on what you see in answers (a) and (b), does studying for an exam pay off? Explain.

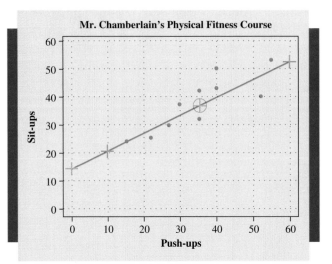

Mr. Chamberlain's Physical Fitness Course

in interpreting b_0 you first must consider whether $x = 0$ is a realistic *x*-value before you conclude that you would predict $\hat{y} = b_0$ if $x = 0$. To predict that if a student did no push-ups he or she would still do approximately 15 sit-ups ($b_0 = 14.9$) is probably incorrect. Second, the *x*-value of zero is outside the domain of the data on which the regression line is based. In predicting *y* based on an *x*-value, check to be sure that the *x*-value is within the domain of the *x*-values observed.

3. The line of best fit will always pass through the point $(\bar{x}, \bar{y})$. When drawing the line of best fit on your scatter diagram, use this point as a check. For our illustration, $\bar{x} = \Sigma x/n = 351/10 = 35.1$, $\bar{y} = \Sigma y/n = 380/10 = 38.0$; therefore, $(\bar{x}, \bar{y}) = (35.1, 38.0)$, as shown in green ⊕ in Figure 3-21.

▼ Illustration 3-9

In a random sample of eight college women, each was asked for her height (to the nearest inch) and her weight (to the nearest five pounds). The data obtained are shown in Table 3-12. Find an equation to predict the weight of a college woman based on her height (the equation of the line of best fit), and draw it on the scatter diagram in Figure 3-22.

TABLE 3-12

Data for College Women's Heights and Weights

	1	2	3	4	5	6	7	8
Height (*x*)	65	65	62	67	69	65	61	67
Weight (*y*)	105	125	110	120	140	135	95	130

SOLUTION Before we start the process of finding the equation for the line of best fit, it is often helpful to draw the scatter diagram, which will give you visual insight about the relationship between the two variables. The scatter diagram for the data on the height and weight of college women, shown in Figure 3-22, indicates that the linear model is appropriate.

FIGURE 3-22

Scatter Diagram

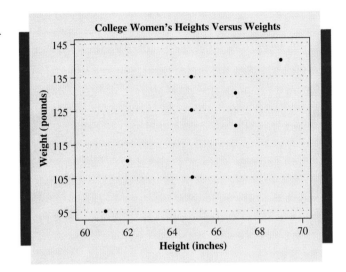

To find the equation for the line of best fit, we first need to complete the preliminary calculations, as shown in Table 3-13.

TABLE 3-13

Preliminary
Calculations Needed
to Find b_1 and b_0

Student	Height (x)	x^2	Weight (y)	xy
1	65	4,225	105	6,825
2	65	4,225	125	8,125
3	62	3,844	110	6,820
4	67	4,489	120	8,040
5	69	4,761	140	9,660
6	65	4,225	135	8,775
7	61	3,721	95	5,795
8	67	4,489	130	8,710
	$\Sigma x = 521$	$\Sigma x^2 = 33,979$	$\Sigma y = 960$	$\Sigma xy = 62,750$

The other preliminary calculations including finding SS(x), formula (2-9), and SS(xy), formula (3-4).

$$SS(x) = \sum x^2 - \frac{\left(\sum x\right)^2}{n} = 33,979 - \left[\frac{521^2}{8}\right] = 48.875$$

$$SS(xy) = \sum xy - \frac{\sum x \sum y}{n} = 62,750 - \left[\frac{(521)(960)}{8}\right] = 230.0$$

Second, we need to find the slope and the y-intercept using formulas (3-7) and (3-6).

$$\text{slope:}\quad b_1 = \frac{SS(xy)}{SS(x)} = \frac{230.0}{48.875} = 4.706 = \mathbf{4.71}$$

$$y\text{-intercept:}\quad b_0 = \frac{\sum y - \left(b_1 \cdot \sum x\right)}{n} = \frac{960 - (4.706)(521)}{8} = -186.478 = \mathbf{-186.5}$$

Thus, the equation of the line of best fit is

$$\hat{y} = \mathbf{-186.5 + 4.71x}$$

To draw the line of best fit on the scatter diagram, we need to locate two points. Substitute two values for x, for example, 60 and 70, into the equation for the line of best fit and obtain two corresponding values for $\hat{y}$:

$$\hat{y} = -186.5 + 4.71x = -186.5 + (4.71)(60) = -186.5 + 282.6 = 96.1 = 96 \text{ and}$$
$$\hat{y} = -186.5 + 4.71x = -186.5 + (4.71)(70) = -186.5 + 329.7 = 143.2 = 143$$

The values (60, 96) and (70, 143) represent two points (designated by a + and shown in red in Figure 3-23) that enable us to draw the line of best fit.

FIGURE 3-23

Scatter Diagram
with Line of Best
Fit

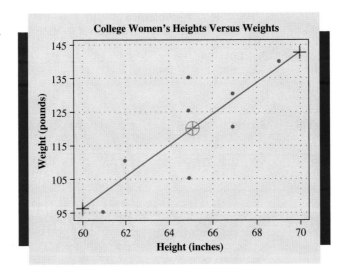

NOTE   In Figure 3-23, $(\bar{x}, \bar{y}) = (65.1, 120)$ is also on the line of best fit. It is the green cross in the circle. (Use $\bar{x}, \bar{y}$ as a check on your work.)

Making Predictions

One of the main purposes for obtaining a regression equation is for making predictions. Once a linear relationship has been established and the value of the input

EXERCISE   3.40

If all students who
can do 40 push-ups
are asked to do as
many sit-ups as
possible:

a. How many sit-
   ups do you ex-
   pect each can
   do?
b. Will they all be
   able to do the
   same number?
c. Explain the mean-
   ing of answer (a).

variable x is known, we can predict a value of y ($\hat{y}$). For example, in the physical fitness illustration, the equation was found to be $\hat{y} = 14.9 + 0.66x$. If student A can do 25 push-ups, how many sit-ups do you therefore predict that A will be able to do? The predicted value is

$$\hat{y} = 14.9 + 0.66x = 14.9 + (0.66)(25) = 14.9 + 16.5 = 31.4 = \mathbf{31}$$

(You should not expect this predicted value to exactly occur; rather, it is the average number of sit-ups that you would expect from all students who could do 25 push-ups.)

When making predictions on the line of best fit, observe the following restrictions:

1. The equation should be used to make predictions only about the population from which the sample was drawn. For example, using the relationship between the height and the weight of college women to predict the weight of professional athletes given their height would be questionable.
2. The equation should be used only to cover the sample domain of the input variable. For example, for Illustration 3-9 the prediction that a college woman of height zero weighs -186.5 pounds is nonsense. Do not use a height outside the sample domain of 61 to 69 inches to predict weight. On occasion you might wish to use the line of best fit to estimate values outside the domain interval of the sample. This can be done, but you should do it with caution and only for values close to the domain interval.
3. If the sample was taken in 1994, do not expect the results to have been valid in 1929 or to hold in 2004. The women of today may be different from the women of 1929 and the women of 2004.

MINITAB (Release 10) commands to find the equation of the line of best fit for ordered pairs with x-values listed in C1 and y-values listed in C2.

Session commands   *Menu commands*

Enter: **REGRess C2 1 C1**   Choose: **Stat > Regress > Regress**

If you plan to draw a   Enter
scatter diagram, you will   variables: Response (y): **C2**
need to use subcommand   Predictors (x): **C1**
FITS to find the points   Select: **FITS**, if wanted.
for the line of best fit.

Enter: **REGRess C2 1 C1;**
        FITS C3.

MINITAB (Release 10) commands to draw the scatter diagram with the line of best fit superimposed on the data points; x-values are listed in C1, y-values in C2, and y-hat values in C3. (Note: The above: REGRess, with FITS, must be used first to create C3.)

(continued)

MINITAB *(continued)*

Session commands

Enter: PLOT C2*C1;
       LINE C1 C3;
       TITLe 'your title'.

Menu commands

Choose:Graph > Plot
Graph Variables: Enter: (y) C2
                       (x) C1
Select: Annotation:
    Title: **your title**
    Line: Points: Enter: C1 C3
            Type: Solid

Case Study

3-3

EXERCISE 3.41

a. Sketch a scatter diagram for family income vs. price of car, year vs. price of car, year vs. family income (*x* vs. *y*).

b. Are the relationships linear? Explain.

c. Does the average price of a new car seem to be approximately one-half the annual median family income? Explain.

Thirty Years of Car Buying

The following table lists the average price of a new car and the median annual family income for each of several years. The average price of a new car seems to be approximately one-half the annual median family income for the same year. Do you think there is a linear relationship?

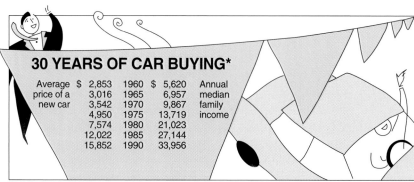

30 YEARS OF CAR BUYING*

Average price of a new car		Annual median family income
$ 2,853	1960	$ 5,620
3,016	1965	6,957
3,542	1970	9,867
4,950	1975	13,719
7,574	1980	21,023
12,022	1985	27,144
15,852	1990	33,956

*Statistics from the U.S. Department of Commerce, supplied by the Motor Vehicle Manufacturers Association.

Source: New Woman, November 1990. Statistics from the U.S. Department of Commerce, supplied by the Motor Vehicles Manufacturers Association of the United States. Graphics by Laura Wallace.

Estimating the Line of Best Fit

As with the approximation of *r*, estimating the line of best fit and its equation should be used only as a mental estimate or check. On the scatter diagram of the data, draw the straight line that appears to be the line of best fit. [*Hint:* If you draw a line parallel to and halfway between your pencils (whose location was described in Section 3.2, p. 156, for estimating r), you will have a reasonable estimate for the line of best fit.] Then use this line to approximate the equation of the line of best fit.

Figure 3-24 shows the pencils and the resulting estimated line for Illustration 3-9. This line can now be used to approximate the equation.

FIGURE 3-24

Estimating the Line of Best Fit for the College Women Data

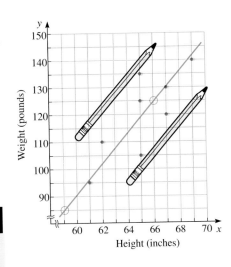

EXERCISE 3.42

The graph's y-intercept is -250, not approximately 80, as might be read from Figure 3-24. Why?

EXERCISE 3.43

The choice of the two points, (x_1, y_1) and (x_2, y_2), is somewhat arbitrary. When different points are selected, slightly different values for b_0 and b_1 will result, but they should be approximately the same. Use points (61, 95) and (67, 130) and find the slope and the y-intercept.

First, locate two points, (x_1, y_1) and (x_2, y_2), along the line and determine their coordinates. Two such points, circled in Figure 3-24, have the coordinates (59, 85) and (66, 125). These two pairs of coordinates can now be used in the following formula to obtain an estimate for the slope b_1.

$$\text{estimate the slope, } b_1: \quad b_1 \approx \frac{y_2 - y_1}{x_2 - x_1} = \frac{125 - 85}{66 - 59} = \frac{40}{7} = 5.7$$

Using this result, the coordinates of one of the points, and the following formula, we can determine an estimation for the y-intercept b_0:

$$\text{estimate the } y\text{-intercept, } b_0: \quad b_0 \approx y - b_1 \cdot x = 85 - (5.7)(59)$$
$$= 85 - 336.3 = -251.3, \text{ or } -250$$

We now estimate the equation for the line of best fit to be $\hat{y} = -250 + 5.7x$. This should serve as a crude estimate.

EXERCISES

3.44   Draw a scatter diagram for these data:

x	2	12	4	6	9	4	11	3	10	11	3	1	13	12	14	7	2	8
y	4	8	10	9	10	8	8	5	10	9	8	3	9	8	8	11	6	9

Would you be justified in using the techniques of linear regression on these data to find the line of best fit? Explain.

3.45   Ajay used linear regression to help him understand his monthly telephone bill. The line of best fit was $\hat{y} = 23.65 + 1.28x$; x is the number of long-distance calls made during a month and y is the total telephone cost for a month. In terms of number of long distance calls and cost:

   a. Explain the meaning of the y-intercept, 23.65.
   b. Explain the meaning of the slope, 1.28.

3.46   In Illustration 3-9, the slope of the line of best fit was found to be +4.71. Explain what a slope of 4.71 indicates.

3.47   A study was conducted to investigate the relationship between the cost y (in tens of thousands of dollars) per unit of equipment manufactured and the number of units produced per run x. The equation of the line of best fit was found to be $\hat{y} = 7.31 - 0.01x$, and x was observed for values between 10 and 200. If a production run is scheduled to produce 50 units, what would you predict the cost per unit to be?

3.48   A study was conducted to investigate the relationship between resale price y (in hundreds of dollars) and the age x (in years) of midsize luxury American-made automobiles. The equation of the line of best fit was $\hat{y} = 185.7 - 21.52x$.

   a. Find the resale value of such a car when it is three years old.
   b. Find the resale value of such a car when it is six years old.
   c. What is the average annual decrease in the resale price of these cars?

3.49   An article titled "Microbioluminometry Assay (MBA): Determination of Erythromycin Activity in Plasma or Serum" (*Journal of Pharmaceutical Sciences*, Dec. 1989) compares the MBA assay with another method, agar diffusion plate assay, for determining the erythromycin activity in plasma or serum. The MBA technique is economial because it requires less sample and reagents.

   If X represents the agar diffusion plate assay in micrograms per milliliter and Y represents the MBA assay in micrograms per milliliter, then the results may be expressed as

$$Y = 0.054128 + 0.92012X \quad r = 0.9525 \quad n = 206$$

   If the agar diffusion plate assay determined the level of erythromycin to be 1.200, predict what the MBA assay would be.

3.50   In the article "Beyond Prediction: The Challenge of Minority Achievement in Higher Education" (Lunneborg and Lunneborg, *Journal of Multicultural Counseling and Development*), the relationship between high school GPA and first-year university GPA was investigated for the following groups: Asian Americans, blacks, chicanos, Native Americans, and whites. For the 43 Native Americans, the correlation was found to be 0.26 and the equation of the line of best fit was found to be university GPA = 1.85 + 0.30 × high school GPA. Both GPAs were 4.0 scales.

Use the equation of the line of best fit to predict the mean first-year university GPA for all Native Americans who had a high school GPA equal to 3.0.

3.51   An article in *USA Today*, 10-26-94, titled "Rivals Target 'Access Fees'" reported the 1993 total revenue (in billions) and the 1993 access fees (in billions) for seven local phone firms. Access fees are fees that local phone firms charge long-distance carriers. The total revenue and access fees for the seven firms are as follows:

Have you tried the MINITAB commands yet?

Local Phone Firm	Total Revenue	Access Fees
PacTel	9.2	2.3
US West	10.3	2.9
Southwestern Bell	10.7	2.7
Ameritech	11.7	2.7
Bell Atlantic	13.0	3.1
NYNEX	13.4	3.4
BellSouth	15.9	3.9

Let x = total revenue and y = access fees, and find the equation of the line of best fit.

3.52   People not only live longer today but also live longer independently. The May/June 1989 issue of *Public Health Reports* published an article titled "A Multistate Analysis of Active Life Expectancy." Two of the variables studied were a person's current age and the expected number of years remaining.

Age x	65	67	69	71	73	75	77	79	81	83
Years Remaining y	16.5	15.1	13.7	12.4	11.2	10.1	9.0	8.4	7.1	6.4

a. Draw a scatter diagram.

b. Calculate the equation of best fit.

c. Draw the line of best fit on the scatter diagram.

d. What is the expected years remaining for a person who is 70 years old? Find the answer in two different ways: Use the equation from (b) and use the line on the scatter diagram from (c).

e. Are you surprised that the data all lie so close to the line of best fit? Explain why the ordered pairs follow the line of best fit so closely.

3.53   A record of maintenance costs is kept for each of several cash registers throughout a department store chain. A sample of 14 registers gave the following data:

Age x (years)	Maintenance Cost y (dollars)	Age x (years)	Maintenance Cost y (dollars)
6	142	2	99
7	231	1	114
1	73	9	191
3	90	3	160
6	176	8	155
4	132	9	231
5	167	8	202

a. Draw a scatter diagram that shows these data.
b. Calculate the equation of the line of best fit.
c. A particular cash register is eight years old. How much maintenance (cost) do you predict it will require this year?
d. Interpret your answer to (c).

3.54   The following data are the ages and the asking prices for 19 used foreign compact cars:

Age x (years)	Price y ($\times$ \$100)	Age x (years)	Price y ($\times$ \$100)
3	68	6	42
5	52	8	22
3	63	5	50
6	24	6	36
4	60	5	46
4	60	7	36
6	28	4	48
7	36	7	20
2	68	5	36
2	64		

a. Draw a scatter diagram.
b. Calculate the equation of the line of best fit.
c. Graph the line of best fit on the scatter diagram.
d. Predict the average asking price for all such foreign cars that are five years old. Obtain this answer in two ways: Use the equation from (b) and use the line drawn in (c).

3.55 An article titled "Unions and Ethnic Diversity: The Israeli Case of East European Immigrants" (*Journal of Applied Behavioral Science*, Vol. 30, No. 2, June 1994) gives the linear correlation coefficient between the age and the income of the subjects in the study as 0.25. Suppose the following table gives the income (in thousands of dollars) and the age for ten individuals.

Income	25	34	46	41	30	44	27	24	47	31
Age	23	27	35	25	25	46	35	30	37	42

    a. Find the linear correlation between age and income.
    b. Find the equation of the line of best fit, where x = age and y = income.

3.56 An article titled "Women, Work and Well-Being: The Importance of Work Conditions (*Health and Social Behavior*, Vol. 35, No. 3, Sept. 1994) studied 202 full-time homemakers and 197 employed wives. The linear correlation coefficient between family income and education was reported to equal 0.43 for the participants in the study. A similar study involving eight individuals gave the following results. (x represents the years of education, and y represents the family income in thousands of dollars).

x	12	13	10	14	11	14	16	16
y	34	45	36	47	43	35	50	42

    a. Find the linear correlation between x and y.
    b. Find the equation of the line of best fit.

IN RETROSPECT

To sum up what we have just learned, there is a distinct difference between the purpose of regression analysis and the purpose of correlation. In regression analysis, we seek a relationship between the variables. The equation that represents this relationship may be the answer desired, or it may be the means to the prediction that is desired. In correlation analysis, on the other hand, we simply ask: Is there a linear correlation between the two variables?

The chapter case study and the articles that form the other case studies show a variety of applications for the techniques of correlation and regression. These articles are worth reading again. When bivariate data appear to fall along a straight line on the scatter diagram, they suggest a linear relationship. But this is not proof of cause and effect. Clearly, if a basketball player commits too many fouls, he will not be scoring more points. He will be in foul trouble and "riding the pine" with no chance to score. It also seems reasonable that the more game time he has, the more points he will score and the more fouls he will commit. Thus, a positive correlation and a positive regression relationship will exist between these two variables.

The bivariate linear methods we have studied thus far have been presented for the purpose of a first, descriptive look. More details must, by necessity, wait until additional developmental work has been completed. After completing this chapter, you should have a basic understanding of bivariate data, how they are different from just two sets of data, how to present them, what correlation and regression analysis are, and how each is used. The evaluation and the interpretation of the meaning and the meaningfulness of these results are studied in Chapter 13. After a few more exercises to practice the methods of this chapter, we will study some basic probability.

CHAPTER EXERCISES

3.57   Gun laws, reasons for them, and demonstrations against them have all made the news frequently in recent years. How did people become so interested in gun-related sports? Is their interest related to their environment as a youth? The cross-tabulation table below summarizes the results when gun enthusiasts were asked the question "Who introduced you to gun-related sports?"

	Parents	**Relatives**	**Friends**	**Combination**	**Others**
Rural	75	46	24	37	2
Nonrural	30	32	44	57	7

    a. Find the marginal totals.
    b. Express the table as percentages of the grand total.
    c. Express the table as percentages of who influenced youth.
    d. Express the table as percentages of each type of environment.
    e. Draw a bar graph based on the type of environment as a youth.

3.58   "Fear of the dentist" (or the dentist's chair) is an emotion felt by many people of all age groups. A survey of 100 individuals in each of five age groups was conducted about this fear, and the results were as follows:

	Elementary	**Jr. High**	**Sr. High**	**College**	**Adult**
No. Who Fear	37	28	25	27	21
No. Who Do Not Fear	63	72	75	73	79

    a. Find the marginal totals.
    b. Express the table as percentages of the grand total.
    c. Express the table as percentages of each group's marginal totals.
    d. Express the table as percentages of those who fear and those who do not fear.
    e. Draw a bar graph based on age groups.

3.59 When was the last time you saw your doctor? That was the question asked for the survey summarized below:

		Time Since Last Consultation with Your Physician		
		Less Than 6 Months	6 Months to Less Than 1 Year	1 Year or More
Age	Under 28 years	413	192	295
	28–40	574	208	218
	Over 40	653	288	259

    a. Find the marginal totals.

    b. Express the table as percentages of the grand total.

    c. Express the table as percentages of each age group's marginal totals.

    d. Express the table as percentages of each time period.

    e. Draw a bar graph based on the grand total.

 3.60 Part of quality control is keeping track of what is occurring. The contingency table below shows the number of rejected castings that occurred last month.

	Causes for Rejection of Casting		
	1st Shift	2nd Shift	3rd Shift
Sand	87	110	72
Shift	16	17	4
Drop	12	17	16
Corebreak	18	16	33
Broken	17	12	20
Other	8	18	22

    a. Find the marginal totals.

    b. Express the table as percentages of the grand total.

    c. Express the table as percentages of each shift's marginal totals.

    d. Express the table as percentages of each type of rejection.

    e. Draw a bar graph based on the shifts.

3.61 Determine whether each of the following questions requires correlation analysis or regression analysis to obtain an answer.

    a. Is there a correlation between the grades a student attained in high school and the grades he or she attained in college?

    b. What is the relationship between the weight of a package and the cost of mailing it first class?

    c. Is there a linear relationship between a person's height and shoe size?

    d. What is the relationship between the number of worker-hours and the number of units of production completed?

    e. Is the score obtained on a certain aptitude test linearly related to a person's ability to perform a certain job?

3.62 An automobile owner records the number of gallons of gasoline, x, required to fill the gasoline tank and the number of miles traveled, y, between fill-ups.

    a. If she does a correlation analysis on the data, what would be her purpose and what would be the nature of her results?

    b. If she does a regression analysis on the data, what would be her purpose and what would be the nature of her results?

3.63 The following data were generated using the equation $y = 2x + 1$.

x	0	1	2	3	4
y	1	3	5	7	9

A scatterplot of these data results in five points that fall perfectly on a straight line. Find the correlation coefficient and the equation of the line of best fit.

3.64 Consider this set of bivariate data:

x	1	1	3	3
y	1	3	1	3

    a. Draw a scatter diagram.

    b. Calculate the correlation coefficient.

    c. Calculate the line of best fit.

3.65 Start with the point (5, 5) and add at least four ordered pairs, (x, y), to make a set of ordered pairs that display the following properties. Show that your sample satisfies the requirements.

    a. The correlation of x and y is 0.0.

    b. The correlation of x and y is $+1.0$.

    c. The correlation of x and y is -1.0.

    d. The correlation of x and y is between -0.2 and 0.0.

    e. The correlation of x and y is between $+0.5$ and $+0.7$.

3.66 Start with the point (5, 5) and add at least four ordered pairs, (x, y), to make a set of ordered pairs that display the following properties. Show that your sample satisfies the requirements.

    a. The correlation of x and y is between $+0.9$ and $+1.0$, and the slope of the line of best fit is $+0.5$.

    b. The correlation of x and y is between $+0.5$ and $+0.7$, and the slope of the line of best fit is $+0.5$.

    c. The correlation of x and y is between -0.7 and -0.9, and the slope of the line of best fit is -0.5.

    d. The correlation of x and y is between $+0.5$ and $+0.7$, and the slope of the line of best fit is -1.0.

3.67 A panel of experts came up with the chart of upper limits on weights for better health, and it appeared in the chapter case study on page 133.
   a. Using the techniques studied in this chapter—scatter diagram, correlation coefficient, line of best fit—describe the information on the chart.
   b. What does the value of the correlation coefficient tell you? Explain.
   c. What does the value of the slope for the line of best fit tell you? Explain.
   d. If I am 5-foot-10 in height, explain what the given chart tells me and how I would use the statistics found in answering part (a) to tell me the same thing.

 3.68 "Fast-Food Fat Counts Full of Surprises," in *USA Today*, 10-20-94, compared some of the popular fast-food items in calories and fat.

Calories (x)	270	420	210	450	130	310	290	450	446	640	233
Fat (y)	9	20	10	22	6	25	7	20	20	38	11

x	552	360	838	199	360	345	552
y	55	6	20	12	36	28	22

   a. Draw a scatter diagram of these data.
   b. Calculate the linear coefficient, r.
   c. Find the equation of the line of best fit.
   d. Explain the meaning of the above answers.

 3.69 An article titled "Measuring the Monday Blues: Validation of a Job Satisfaction Scale for the Human Services" (*Social Work Research*, Vol. 18, No. 1, March 1994) measured the job satisfaction of subjects with a 14-question survey. The article reported a linear correlation coefficient between the job satisfaction score and the salary of the participants in the study of .13. The following data represent the job satisfaction scores, y, and the salaries, x, for a sample of similar individuals.

x	23	31	33	18	24	40	29	37
y	7	12	6	5	12	14	10	7

   a. Find the linear correlation coefficient between x and y.
   b. Find the equation of the line of best fit.

 3.70 A biological study of a minnow called the blacknose dace was conducted. The length, y, in millimeters and the age, x, to the nearest year were recorded.
   a. Draw a scatter diagram of these data.
   b. Calculate the correlation coefficient.
   c. Find the equation of the line of best fit.
   d. Explain the meaning of the above answers.

x	0	3	2	2	1	3	2	4	1	1
y	25	80	45	40	36	75	50	95	30	15

 3.71   A survey was conducted at a large metropolitan university campus. Twenty-four students were interviewed. Two of the questions asked were: "How many hours per week are you employed?" and "How many credit hours are you currently registered for?"

Hours Employed (x)	20	40	35	15	40	20	20	0	20	40
Credit hours (y)	6	3	6	9	6	6	3	15	6	9

x	10	20	30	40	15	0	0	0	10	40	0	0	30	25
y	9	3	6	6	3	12	15	18	6	6	21	12	6	9

a. Draw a scatter diagram of these data.
b. Is there a correlation between x and y?
c. What conclusions do you believe are supported by answers to (a) and (b)? Explain.

 3.72   "Meaning Construction in School Literacy Tasks: A Study of Bilingual Students," from the *American Educational Research Journal* (Fall 1990) reported the following data for a sample of 12 fifth-graders of Mexican heritage and all from bilingual homes.

	Percentile Scores		
Number of Years in U.S.	Reading	Language	English Envisionment*
11	60	60	4.5
5	22	24	3.5
4	52	50	5.0
6	1	17	2.5
12	20	63	3.5
5	26	31	3.5
11	30	63	2.5
9	20	25	3.0
10	20	53	3.5
11	17	20	2.0
11	17	29	2.0
11	13	15	2.5

*Ability to unfold the meaning of what they are reading.

a. Construct a scatter diagram for reading, x, and language, y.
b. Construct a scatter diagram for reading, x, and English envisionment, y.
c. Construct a scatter diagram for years, x, and English envisionment, y.
d. Calculate the linear correlation coefficient for reading, x, and language, y.
e. Calculate the linear correlation coefficient for reading, x, and English envisionment, y.
f. Calculate the linear correlation coefficient for years, x, and English envisionment, y.

3.73 a. Verify, algebraically, that formula (3-2) for calculating r is equivalent to the definition formula (3-1).
b. Verify, algebraically, that formula (3-7) is equivalent to formula (3-5).

3.74 The following equation gives a relationship that exists between b_1 and r:

$$r = b_1 \sqrt{\frac{SS(x)}{SS(y)}}$$

a. Verify this equation for these data:

x	4	3	2	3	0
y	11	8	6	7	4

b. Verify this equation using formulas (3-2) and (3-7).

VOCABULARY LIST

Be able to define each term. Pay special attention to key terms, which are printed in red. In addition, describe in your own words, and give an example of, each term. Your examples should not be the ones given in class or in the textbook. Page numbers indicate the first appearance of the term.

bivariate data (p. 135)
coefficient of linear correlation (p. 150)
contingency table (p. 136)
correlation (p. 150)
correlation analysis (p. 150)
dependent variable (p. 143)
independent variable (p. 143)
input variable (p. 143)
least squares criterion (p. 160)
linear correlation (p. 150)
line of best fit (p. 162)
method of least squares (p. 160)
negative correlation (p. 151)

ordered pair (p. 143)
output variable (p. 143)
Pearson's product moment r (p. 152)
positive correlation (p. 151)
predicted value (p. 160)
prediction equation (p. 160)
regression (p. 160)
regression analysis (p. 160)
scatter diagram (p. 143)
slope, b_1 (p. 162)
y-intercept, b_0 (p. 162)

QUIZ A

Answer "True" if the statement is always true. If the statement is not always true, replace the words shown in bold with words that make the statement always true.

3.1 **Correlation** analysis is a method of obtaining the equation that represents the relationship between two variables.

3.2 The linear correlation coefficient is used to determine the **equation that represents** the relationship between two variables.

3.3 A correlation coefficient of **zero** means that the two variables are perfectly correlated.

3.4 Whenever the slope of the regression line is zero, the **correlation coefficient** will also be zero.

3.5 When r is positive, b_1 will always be **negative**.

3.6 The **slope** of the regression line represents the amount of change expected to take place in y when x increases by one unit.

3.7 When the calculated value of r is positive, the calculated value of b_1 will be **negative**.

3.8 Correlation coefficients range between **0 and +1**.

3.9 The value being predicted is called the **input variable**.

3.10 The line of best fit is used to predict the **average value of y** that can be expected to occur at a given value of x.

QUIZ B

3.1

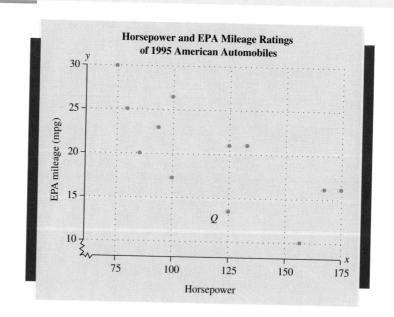

a. Match the items described in column 2 with the terms in column 1.

Column 1	Column 2
_____ Population	A. The horsepower rating for an automobile
_____ Sample	B. All 1995 American-made automobiles
_____ Input Variable	C. The EPA mileage rating for an automobile
_____ Output Variable	D. The 1995 automobiles whose ratings are shown on the scatter diagram

b. Find the sample size.
c. What is the smallest value reported for the output variable?
d. What is the largest value reported for the input variable?
e. Does the scatter diagram suggest a positive (P), negative (N), or zero (Z) linear correlation coefficient?
f. What are the coordinates of point Q?
g. Will the slope for the line of best fit be positive (P), negative (N), or zero (Z)?
h. Will the intercept for the line of best fit be positive (P), negative (N), or zero (Z)?

3.2 A research group reports the correlation coefficient for two variables to be 2.3. What can you conclude from this information?

3.3 For the bivariate data, the extensions, and the totals shown on the table, find the following:

x	y	x^2	xy	y^2
2	6	4	12	36
3	5	9	15	25
3	7	9	21	49
4	7	16	28	49
5	7	25	35	49
5	9	25	45	81
6	8	36	48	64
28	49	124	204	353

a. $SS(x)$
b. $SS(y)$
c. $SS(xy)$
d. the linear correlation coefficient, r
e. the slope, b_1
f. the y-intercept, b_0
g. the equation of the line of best fit

QUIZ C

3.1   A test was administered to measure the mathematics ability of the people in a certain town. Some of the townspeople were totally surprised to find out that their test results and their shoe sizes correlated strongly. Explain why a strong positive correlation should not have been such a surprise.

3.2   Student A collected a set of bivariate data and calculated r, the linear correlation coefficient. The resulting value was -1.78. Student A proclaimed that this value indicated that there was no correlation between the two variables since the value of r was not between -1.0 and $+1.0$. Student B argued that -1.78 was impossible and that only values of r near zero implied no correlation. Who is correct? Justify your answer.

3.3   The linear correlation coefficient, r, is a numerical value that ranges from -1.0 to $+1.0$. Write a sentence or two describing the meaning of r for each of these values:
        a. -0.93 b. $+0.89$ c. -0.03 d. $+0.08$ e. -2.3

3.4   Make up a set of three or more ordered pairs such that
        a. $r = 0.0$ b. $r = +1.0$ c. $r = -1.0$ d. $b_1 = 0.0$

WORKING WITH YOUR OWN DATA

Each semester, new students enter your college environment. You may have wondered, "What will the student body be like this semester?" As a beginning statistics student, you have just finished studying three chapters of basic descriptive statistical techniques. Let's use some of these techniques to describe some characteristics of your college's student body.

A | SINGLE VARIABLE DATA

1. Define the population to be studied.
2. Choose a variable to define. (You may define your own variable, or you may use one of the variables in the accompanying table if you are not able to collect your own data. Ask your instructor for guidance.)
3. Collect 35 pieces of data for your variable.
4. Construct a stem-and-leaf display of your data. Be sure to label it.
5. Calculate the value of the measure of central tendency that you believe best answers the question "What is the average value of your variable?" Explain why you chose this measure.
6. Calculate the sample mean for your data (unless you used the mean in Question 5).
7. Calculate the sample standard deviation for your data.
8. Find the value of the 85th percentile, P_{85}.
9. Construct a graphic display (other than a stem-and-leaf) that you believe "best" displays your data. Explain why the graph best presents your data.

B | BIVARIATE DATA

1. Define the population to be studied.
2. Choose and define two quantitative variables that will produce bivariate data. (You may define your own variables, or you may use two of the variables in the accompanying table if you are not able to collect your own data. Ask your instructor for guidance.)
3. Collect 15 ordered pairs of data.
4. Construct a scatter diagram of your data. (Be sure to label it completely.)
5. Using a table to assist with the organization, calculate the extensions x^2, xy, and y^2, and the summations of x, y, x^2, xy, and y^2.
6. Calculate the linear correlation coefficient r.
7. Calculate the equation of the line of best fit.
8. Draw the line of best fit on your scatter diagram.

The following table of data was collected on the first day of class last semester. You may use it as a source for your data if you are not able to collect your own.

The MINITAB command SAMPle will select your random sample.

▼ Variable A = student's sex (male/female)
▼ Variable B = student's age at last birthday
▼ Variable C = number of completed credit hours toward degree
▼ Variable D = "Do you have a job (full/part time)?" (yes/no)
▼ Variable E = number of hours worked last week, if D = yes
▼ Variable F = wages (before taxes) earned last week, if D = yes

Student	A	B	C	D	E	F	Student	A	B	C	D	E	F
1	M	21	16	No			38	F	42	34	Yes	40	244
2	M	18	0	Yes	10	34	39	M	25	60	Yes	60	503
3	F	23	18	Yes	46	206	40	M	39	32	Yes	40	500
4	M	17	0	No			41	M	29	13	Yes	39	375
5	M	17	0	Yes	40	157	42	M	19	18	Yes	51	201
6	M	40	17	No			43	M	25	0	Yes	48	500
7	M	20	16	Yes	40	300	44	F	18	0	No		
8	M	18	0	No			45	M	32	68	Yes	44	473
9	F	18	0	Yes	20	70	46	F	21	0	No		
10	M	29	9	Yes	8	32	47	F	26	0	Yes	40	320
11	M	20	22	Yes	38	146	48	M	24	11	Yes	45	330
12	M	34	0	Yes	40	340	49	F	19	0	Yes	40	220
13	M	19	31	Yes	29	105	50	M	19	0	Yes	10	33
14	M	18	0	No			51	F	35	59	Yes	25	88
15	M	20	0	Yes	48	350	52	F	24	6	Yes	40	300
16	F	27	3	Yes	40	130	53	F	20	33	Yes	40	170
17	M	19	10	Yes	40	202	54	F	26	0	Yes	52	300
18	F	18	16	Yes	40	140	55	F	17	0	Yes	27	100
19	M	19	4	Yes	6	22	56	M	25	18	Yes	41	355
20	F	29	9	No			57	M	24	0	No		
21	F	21	0	Yes	20	80	58	M	21	0	Yes	30	150
22	F	39	6	No			59	M	30	12	Yes	48	555
23	M	23	34	Yes	42	415	60	F	19	0	Yes	38	169
24	F	31	0	Yes	48	325	61	M	32	45	Yes	40	385
25	F	22	7	Yes	40	195	62	M	26	90	Yes	40	340
26	F	27	75	Yes	20	130	63	M	20	64	Yes	10	45
27	F	19	0	No			64	M	24	0	Yes	30	150
28	M	22	20	Yes	40	470	65	M	20	14	No		
29	F	60	0	Yes	40	390	66	M	21	70	Yes	40	340
30	M	25	14	No			67	F	20	13	Yes	40	206
31	F	24	45	No			68	F	33	3	Yes	32	246
32	M	34	4	No			69	F	25	68	Yes	40	330
33	M	29	48	No			70	F	29	48	Yes	40	525
34	M	22	80	Yes	40	336	71	F	40	0	Yes	40	400
35	M	21	12	Yes	26	143	72	F	36	3	Yes	40	300
36	F	18	0	No			73	F	35	0	Yes	40	280
37	M	18	0	Yes	13	65	74	F	28	0	Yes	40	350

Student	A	B	C	D	E	F	Student	A	B	C	D	E	F
75	M	40	64	Yes	40	390	88	F	27	9	Yes	40	260
76	F	31	0	Yes	40	200	89	F	26	3	Yes	40	240
77	F	32	0	Yes	40	270	90	F	23	9	Yes	40	330
78	F	37	0	Yes	24	150	91	M	41	3	Yes	23	253
79	F	35	0	Yes	40	350	92	M	39	0	Yes	40	110
80	M	21	72	Yes	45	470	93	M	21	0	Yes	40	246
81	F	27	0	Yes	40	550	94	F	32	0	Yes	40	350
82	F	42	47	Yes	37	300	95	F	48	58	Yes	40	714
83	F	41	21	Yes	40	250	96	F	26	0	Yes	32	200
84	M	36	0	Yes	40	400	97	F	27	0	Yes	40	350
85	M	25	16	Yes	40	480	98	F	52	56	Yes	40	390
86	F	18	0	Yes	45	189	99	F	34	27	Yes	8	77
87	M	22	0	Yes	40	385	100	F	49	3	Yes	24	260

## PART TWO

PROBABILITY

*B*efore continuing our study of statistics, we must make a slight detour and study some basic probability. Probability is often called the "vehicle" of statistics; that is, the probability associated with chance occurrences is the underlying theory for statistics. Recall that in Chapter 1 we described probability as the science of making statements about what will occur when samples are drawn from known populations. Statistics was described as the science of selecting a sample and making inferences about the unknown population from which it is drawn. To make these inferences, we need to study sample results in situations in which the population is known so that we will be able to understand the behavior of chance occurrences.

In Part Two we study the basic theory of probability (Chapter 4), probability distributions of discrete variables (Chapter 5), and probability distributions for continuous random variables (Chapter 6). Following this brief study of probability, we will study the techniques of inferential statistics in Part Three.

KARL PEARSON

KARL PEARSON, known as one of the fathers of modern statistics, was born March 27, 1857, in London, the second son of prominent attorney William Pearson and his wife, Fanny Smith. Karl was tutored at home until, at age nine, he entered University College School in London. In 1875, following a year of illness that required him to be privately tutored, he was awarded a scholarship to King's College, Cambridge. There, in May 1879, he earned his B.A. (with honors) in mathematics; he then went on to earn an M.A. in law in 1882.

After receiving his law degree, he moved to Heidelberg, Germany, where he became proficient in literature, philosophy, physics, and metaphysics, as well as in German history and folklore.

Pearson returned to University College, where, in 1884, he was appointed Goldsmid professor of applied mathematics and mechanics. In addition, Pearson lectured in geometry at Gresham College, London, from 1891 to 1894. Later, in 1911, he relinquished the Goldsmid chair to become the first Galton Professor of Eugenics.

In 1896 Pearson was elected to the Royal Society and was awarded the Society's Darwin Medal in 1898. In 1900 Pearson invented the *chi-square* (denoted by χ^2); it is the oldest inference procedure still used in its original form and is often used in today's economics and business applications. Around that time, he also discovered *random phenomena* (or probability), by tossing a coin 24,000 times to determine the frequency of its landing "heads up" as opposed to "tails up." Result: 12,012 heads, a relative frequency of 0.5005.

It was during Pearson's association with Sir Francis Galton that he developed the *linear correlation coefficient* (sometimes referred to as the *Pearson product moment correlation coefficient*, in his honor). He was editor of, and a major contributor to, the statistical journal BIOMETRIKA, which he co-founded with fellow statisticians Galton and Weldon. Pearson returned to London in 1933 where he died on April 27, 1936.

Pearson's only son, Egon S. (second oldest of three children), born in 1895 to Karl and his wife, Maria Sharpe, also became a well-known statistician.

PROBABILITY

Chapter Case Study

LOTTERY RESULTS

*S*tate-run lotteries have become BIG, in fact, VERY BIG business. USA Today, on January 23, 1995, listed "state lottery numbers for Friday, Saturday, and Sunday, Jan. 20–22." The list included 208 different games from 35 different states and 6 different games that were interstate. Four states had only one game listed, while 17 states listed between 7 and 13 different games. There appears to be a LOTTO fever going around!! The listing below appeared in the Pittsburgh Post-Gazette, December 24, 1994, and reported an analysis of Pennsylvania's "Big 4" lottery. This game has been played 3345 times since November 22, 1980, and no one has ever, at least as of December 24, 1994, picked all four numbers and won the BIG $ grand prize.

Lottery Results for Friday, Dec. 23, 1994

PENNSYLVANIA LOTTERY

Big 4

9-6-3-3

Number of winners: 559
Money paid out: $147,900

Last time numbers hit straight:
Never

Big 4 Analysis

(Times that each number has been picked in the first, second, third, or fourth positions, and total times drawn since the game began Nov. 22, 1980.)

No.	First	Second	Third	Fourth	Total
0	343	312	352	328	1335
1	326	330	351	357	1364
2	347	323	315	344	1329
3	320	327	350	351	1348
4	304	345	331	318	1298
5	321	348	322	343	1334
6	339	306	329	316	1290
7	348	346	351	311	1356
8	337	367	329	350	1383
9	360	341	315	327	1343

What is the likeliness of picking the correct four single-digit numbers? Are the numbers equally likely to be winners? Does this game appear to be random? Does it seem reasonable that there has been no straight winner in 14 years? (See Exercise 4.136, p. 245.)

CHAPTER OBJECTIVES

You may already be familiar with some ideas of probability, because probability is part of our everyday culture. We constantly hear people making probability-oriented statements such as

- ▼ "Our team will probably win the game tonight."
- ▼ "There is a 40% chance of rain this afternoon."
- ▼ "I will most likely have a date for the winter weekend."
- ▼ "If I park in the faculty parking area, I will probably get a ticket."
- ▼ "I have a 50-50 chance of passing today's chemistry exam."

Everyone has made or heard these kinds of statements. What exactly do they mean? Do they, in fact, mean what they say? Some statements may be based on scientific information and others on subjective prejudice. Whatever the case may be, they are probabilistic inferences—not fact but conjectures.

In this chapter we study the basic concept of probability and the rules that apply to the probability of both simple and compound events.

CONCEPTS OF PROBABILITY

4.1 | THE NATURE OF PROBABILITY

Let's consider an experiment in which we toss two coins simultaneously and record the number of heads that occur. The only possible outcomes are 0H (zero heads), 1H (one head), and 2H (two heads). Let's toss the two coins 10 times and record our findings.

2H,   1H,   1H,   2H,   1H,   0H,   1H,   1H,   1H,   2H

Summary:

Outcome	Frequency
2H	3
1H	6
0H	1

Suppose that we repeat this experiment 19 times. Table 4-1 shows the totals for 20 sets of 10 tosses. (Trial 1 shows the totals from our first experiment.)

TABLE 4-1 Experimental Results of Tossing Two Coins

Outcome	Trial																				Total
	1	2	3	4	5	6	7	8	9	10	11	12	13	14	15	16	17	18	19	20	
2H	3	3	5	1	4	2	4	3	1	1	2	5	6	3	1	4	1	0	3	1	53
1H	6	5	5	5	5	7	5	5	5	5	8	4	3	7	5	1	5	4	5	9	104
0H	1	2	0	4	1	1	1	2	4	4	0	1	1	0	4	5	4	6	2	0	43

The total of 200 tosses of the pair of coins resulted in 2H on 53 occasions, 1H on 104 occasions, and 0H on 43 occasions. We can express these results in terms of relative frequencies and show the results using a histogram, as in Figure 4-1.

FIGURE 4-1

Relative Frequency Histogram for Coin-Tossing Experiment

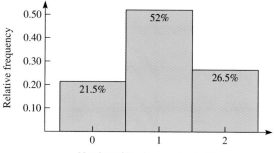

Number of heads on each set of tosses

What conclusions can be reached? If we look at the individual sets of 10 tosses, we notice a large variation in the number of times each of the events (2H, 1H, and 0H) occurred. In both the 0H and the 2H categories, there were as many as 6 occurrences and as few as 0 occurrences in a set of 10 tosses. In the 1H category there were as few as 1 occurrence and as many as 9 occurrences.

If we were to continue this experiment for several hundred more tosses, what would you expect to happen in terms of the relative frequencies of these three events? It looks as if we have approximately a 1 : 2 : 1 ratio in the totals of Table 4-1. We might therefore expect to find the relative frequency for 0H to be approximately $\frac{1}{4}$, or 25%; the relative frequency for 1H to be approximately $\frac{1}{2}$, or 50%; and the relative frequency for 2H to be approximately $\frac{1}{4}$, or 25%. These relative frequencies accurately reflect the concept of probability.

Many probability experiments can be simulated by using random numbers tables or by having a computer randomly generate number values representing the various experimental outcomes. The key to either method is to maintain the probabilities. For example, the preceding tossing of two coins experiment could be simulated by letting odd integers represent heads (H) and even integers, tails (T). [$P(H) = \frac{1}{2}$ and $P(\text{odd}) = \frac{1}{2}$, $P(T) = \frac{1}{2}$ and $P(\text{even}) = \frac{1}{2}$; therefore, we have assigned random digits to the events so as to maintain the probabilities of heads and tails.] Since we are tossing two coins, we will need a two-digit random integer, the first digit for the first coin and the second digit for the second coin, for each toss of the two coins. Thus, the random integer 45 would represent T, H, or 1H. This would work with both the random-number table and the computer. However, if the computer is used, by letting 1 = H and 0 = T on each toss of a coin, and by randomly generating two columns, one for each coin, the computer will do all the tally work, too. By adding the two columns together, we will find the total to be the number of heads seen in each toss of the two coins. See the commands that follow.

MINITAB (Release 10) commands to generate random integers 0 and 1 in C1 and C2, add C1 and C2, to obtain the number of heads on the two coins, and then TALLy the findings to give us the results of the experiment.

Session commands   *Menu commands*

Enter: RANDom 200 C1 C2;   Choose: Calc > Random Data > Integer
      INTEger 0 1.   Enter: Generate: 200
      LET C3 = C1 + C2   Store in: C1 C2
      TALLy C3;   Minimum value: 0
      COUNts;   Maximum value: 1
      PERCents.   Use session command:
                      LET C3 = C1 + C2
                Choose: Stat > Table > Tally
                Enter: Variable: C3
                Select: Count and Percent

EXERCISES

4.1 Toss a single coin 10 times and record H (head) or T (tail) after each toss. Using your results, find the relative frequency of
    a. heads   b. tails

4.2 Roll a single die 20 times and record a 1, 2, 3, 4, 5, or 6 after each roll. Using your results, find the relative frequency of
    a. 1   b. 2   c. 3   d. 4   e. 5   f. 6

4.3 Place three coins in a cup, shake and dump them out, and observe the number of heads showing. Record 0H, 1H, 2H, or 3H after each trial. Repeat the process 25 times. Using your results, find the relative frequency of
    a. 0H   b. 1H   c. 2H   d. 3H

4.4 Place a pair of dice in a cup, shake and dump them out. Observe the sum of dots. Record 2, 3, 4, . . . , 12. Repeat the process 25 times. Using your results, find the relative frequency for each of the values: 2, 3, 4, 5, . . . , 12.

4.5 Use either the random numbers table (Appendix B) or a computer to simulate:
    a. The rolling of a die 50 times; express your results as relative frequencies.
    b. The tossing of a coin 100 times; express your results as relative frequencies.

4.6   Use either the random numbers table (Appendix B) or a computer to simulate the random selection of 100 single-digit numbers, 0 through 9.
   a. List your 100 digits.
   b. Prepare a relative frequency distribution of the 100 digits.
   c. Prepare a relative frequency histogram of the distribution in (b).

4.2 | PROBABILITY OF EVENTS

We are now ready to define what is meant by probability. Specifically, we talk about "the probability that a certain event will occur."

PROBABILITY THAT AN EVENT WILL OCCUR

The relative frequency with which that event can be expected to occur.

The probability of an event may be obtained in three different ways: (1) empirically, (2) theoretically, and (3) subjectively. The first method was illustrated in the experiment in Section 4-1 and might be called **experimental**, or **empirical probability**. This is the *observed relative frequency with which an event occurs*. In our coin-tossing illustration, we observed exactly one head (1H) on 104 of the 200 tosses of the pair of coins. The observed empirical probability for the occurrence of 1H was 104/200, or 0.52.

Experimental, or empirical, probability

When the value assigned to the probability of an event results from experimental data, we will identify the probability of the event with the symbol $P'(\ )$.

NOTE: The *prime notation* is used to denote empirical probabilities.

The value assigned to the probability of event A as a result of experimentation can be found by means of the formula

$$P'(A) = \frac{n(A)}{n} \tag{4-1}$$

where $n(A)$ is the number of times that event A is observed and n is the number of times the experiment is attempted.

EXERCISE 4.7

If you roll a die 40 times and 9 of the rolls result in a "5," what empirical probability was observed for the event "5"?

Consider the rolling of a die. Define event A as the occurrence of a 1. In a single roll of a die, there are six possible outcomes. Assuming that the die is symmetrical, each number should have an equal likelihood of occurring. Intuitively, the probability of A, or the expected relative frequency of a 1, is $\frac{1}{6}$. (Later we will formalize this calculation.)

EXERCISE 4.8

Explain why an empirical probability, an observed proportion, and a relative frequency are actually three different names for the same thing.

What does this mean? Does it mean that once in every six rolls a 1 will occur? No, it does not. Saying that the probability of a 1, $P(1)$, is $\frac{1}{6}$ means that in the long run the proportion of times that a 1 occurs is approximately $\frac{1}{6}$. How close to $\frac{1}{6}$ can we expect the observed relative frequency to be?

Table 4-2 (p. 194) shows the number of 1's observed in each set of six rolls of a die (column 1), an observed relative frequency for each set of six rolls (column 2), and a cumulative relative frequency (column 3). Each trial is a set of six rolls. Figure 4-2a shows the fluctuation of the observed probability for event A on each of the 20 trials (column 2, Table 4-2). Figure 4-2b shows the fluctuation of the cumulative relative frequency (column 3, Table 4-2). Notice that the observed relative frequency on each trial of six rolls of a die tends to fluctuate about $\frac{1}{6}$. Notice also that the observed values on the cumulative graph seem to become more stable; in fact, they become relatively close to the expected $\frac{1}{6}$, or $0.166\overline{6} = 0.167$.

TABLE 4-2

Experimental Results of Rolling a Die Six Times in Each Trial

Trial	(1) Number of 1's Observed	(2) Relative Frequency	(3) Cumulative Relative Frequency
1	1	1/6	1/6 = 0.17
2	2	2/6	3/12 = 0.25
3	0	0/6	3/18 = 0.17
4	1	1/6	4/24 = 0.17
5	0	0/6	4/30 = 0.13
6	1	1/6	5/36 = 0.14
7	2	2/6	7/42 = 0.17
8	2	2/6	9/48 = 0.19
9	0	0/6	9/54 = 0.17
10	0	0/6	9/60 = 0.15
11	1	1/6	10/66 = 0.15
12	0	0/6	10/72 = 0.14
13	2	2/6	12/78 = 0.15
14	1	1/6	13/84 = 0.15
15	1	1/6	14/90 = 0.16
16	3	3/6	17/96 = 0.18
17	0	0/6	17/102 = 0.17
18	1	1/6	18/108 = 0.17
19	0	0/6	18/114 = 0.16
20	1	1/6	19/120 = 0.16

FIGURE 4-2

Fluctuations Found in the Die-Tossing Experiment

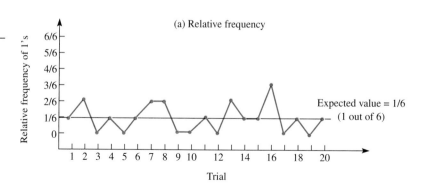

FIGURE 4-2

Continued

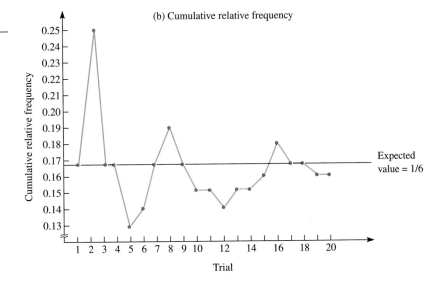

(b) Cumulative relative frequency

Cumulative relative frequency

Expected value = 1/6

Trial

A cumulative graph such as Figure 4-2b demonstrates the idea of long-term average. When only a few rolls were observed (as on each trial, Figure 4-2a), the probability $P'(A)$ fluctated between 0 and $\frac{1}{2}$. As the experiment was repeated, however, the cumulative graph suggests a stabilizing effect on the observed cumulative probability. This stabilizing effect, or **long-term average** value, is often referred to as the *law of large numbers*.

Long-term average

LAW OF LARGE NUMBERS

If the number of times an experiment is repeated is increased, the ratio of the number of successful occurrences to the number of trials will tend to approach the theoretical probability of the outcome for an individual trial.

The law of large numbers is telling us that the larger the number of experimental trials n, the closer the empirical probability $P'(A)$ is expected to be to the true probability $P(A)$. There are many applications of this concept. The die-tossing experiment is an example in which we can easily compare the results to what we expect to happen; this gives us a chance to verify the claim of the law of large numbers. Illustration 4-1 is an illustration in which we live with the results obtained from large sets of data when the theoretical expectation is unknown.

▼ ILLUSTRATION   4-1

The key to establishing proper life insurance rates is using the probability that the insureds will live one, two, or three years, and so forth, from the time they purchase policies. These probabilities are derived from actual life and death statistics and hence are empirical probabilities. They are published by the government and are extremely important to the life insurance industry.

EXERCISES

4.9 Explain what is meant by the statement "When a single die is rolled, the probability of a 1 is $\frac{1}{6}$."

4.10 Explain what is meant by the statement "When one coin is tossed one time, there is a 50-50 chance of getting a tails."

4.11 According to *U.S. News & World Report* (12-3-90), 36 million Americans were victims of serious crimes in 1989, including nearly 19,000 who were murdered. If you assume that there were 200 million Americans in 1989 and that one of them is selected at random, what is the probability that the one selected was
    a. a victim of a serious crime?
    b. a murder victim?

4.12 The April 1994 issue of *Athletic Business* contains the results of a survey sent to 358 recreation directors at four-year colleges and universities in the United States and Canada. A summary of the enrollments at the schools follows.

Enrollment	Number of Schools	Percent
0–1249	58	16.2
1250–2499	53	14.8
2500–4999	53	14.8
5000–9999	62	17.3
10,000–17,999	68	19.0
18,000 or over	64	17.9
Total	358	100.0

One of the 358 responding schools is selected at random. Find the probabilities of the following events.
    a. The school had an enrollment of less than 2500.
    b. The school had an enrollment of 10,000 or more.
    c. The school had an enrollment of between 2500 and 9999 inclusive.

4.13 Take two dice (one white and one some color) and roll them 50 times, recording the results as ordered pairs [(white, black), for example; (3, 5) represents 3 on the white die and 5 on the color die]. (You could simulate these 50 rolls using a random numbers table or a computer.) Then calculate the following observed probabilities:
    a. P'(white die is an odd number)
    b. P'(sum is 6)
    c. P'(both dice show odd number)
    d. P'(number on color die is larger than number on white die)

4.14 Use a random numbers table or a computer to simulate the rolling of a pair of dice 100 times.
   a. List the results of each roll as an ordered pair and the sum.
   b. Prepare an ungrouped frequency distribution and a histogram of the results.
   c. Describe how these results compare to what you expect to occur when two dice are rolled.

MINITAB commands
RANDom 100 C1 C2;
   INTEger 1 6.
LET C3 = C1 + C2
PRINt C1 - C3
TALLy C3
HISTogram C3;
   MIDPoints 2:12.

4.15 Using a coin, perform the die experiment discussed on pages 193–195. Toss a coin 10 times, observing the number of heads (or put 10 coins in a cup, shake, and dump them into a box, and use each toss for a block of 10), and record the results. Repeat until you have 200 tosses. Chart and graph the data as individual sets of 10 and as cumulative relative frequencies. Do your data tend to support the claim that $P(\text{head}) = \frac{1}{2}$? Explain.

4.16 Let's estimate the probability that a thumbtack lands "point up," ⬇ (as opposed to "point down," (🖈)) when tossed and lands on a hard surface. Using a thumbtack, perform the die experiment discussed on pages 193–195. Toss the thumbtack 10 times, observing the number of "point up" (or put 10 identical thumbtacks in a cup, shake, and dump them into a box, and use each toss for a block of 10), and record the results. Repeat until you have 200 tosses. Chart and graph the data as individual sets of 10 and as cumulative relative frequencies. What do you believe the $P(⬇)$ to be? Explain.

4.3 | Simple Sample Spaces

Let's return to an earlier question: What values might we expect to be assigned to the three events (0H, 1H, 2H) associated with our coin-tossing experiment? As we inspect these three events, we see that they do not tend to happen with the same relative frequency. Why? Suppose that the experiment of tossing two pennies and observing the number of heads had actually been carried out using a penny and a nickel—two distinct coins. Would this have changed our results? No, it would have had no effect on the experiment. However, it does show that there are more than three possible outcomes.

When a penny is tossed, it may land as heads or tails. When a nickel is tossed, it may also land as heads or tails. If we toss them simultaneously, we see that there are actually four different possible outcomes. These four outcomes match up with the previous events as follows:

1. heads on penny and heads on nickel—2H
2. heads on penny and tails on nickel—1H
3. tails on penny and heads on nickel—1H
4. tails on penny and tails on nickel—0H

Ordered pair

In this experiment with the penny and the nickel, let's use an **ordered pair** notation. The first listing will correspond to the penny and the second will correspond to the nickel. Thus, (H, T) represents the event that a head occurs on the penny and a tail occurs on the nickel. Our listing of events for the tossing of a penny and a nickel looks like this:

$$(H, H), (H, T), (T, H), (T, T)$$

What we have accomplished here is a listing of what is known as the *sample space* for this experiment.

EXPERIMENT

Any process that yields a result or an observation.

OUTCOME

A particular result of an experiment.

SAMPLE SPACE

Sample points

The set of all possible outcomes of an experiment. The sample space is typically called S and may take any number of forms: a list, a tree diagram, a lattice grid system, and so on. The individual outcomes in a sample space are called **sample points**. $n(S)$ is the number of sample points in sample space S.

EVENT

Any subset of the sample space. If A is an event, then $n(A)$ is the number of sample points that belong to event A.

Regardless of the form in which they are presented, the outcomes in a sample space can never overlap. Also, all possible outcomes must be represented. These characteristics are called *mutually exclusive* and *all inclusive*, respectively. A more detailed explanation of these characteristics will be presented later; for the moment, however, an intuitive grasp of their meaning is sufficient.

Now let's look at some illustrations of probability experiments and their associated sample spaces.

▼ | EXPERIMENT   4 - 1

A single coin is tossed once and the outcome—a head (H) or a tail (T)—is recorded.

Sample space: $S = \{H, T\}$ and $n(S) = 2$

▼ | EXPERIMENT   4 - 2

Two coins, one penny and one nickel, are tossed simultaneously and the outcome for each coin is recorded using ordered pair notation: (penny, nickel). The sample space is shown here in two different ways:

Tree diagram

TREE DIAGRAM REPRESENTATION*   **Listing**

Penny   Nickel   Outcomes $S = \{(H, H), (H, T), (T, H), (T, T)\}$

	H	H, H
H	T	H, T
T	H	T, H
	T	T, T

(four branches, each branch   and $n(S) = 4$
shows a possible outcome)

EXERCISE   4.17

You are to select one single-digit number randomly. List the sample space.

Notice that both representations show the same four possible outcomes. For example, the top branch on the tree diagram shows heads on both coins, as does the first ordered pair in the listing.

▼ | EXPERIMENT   4 - 3

A die is rolled one time and the number of spots on the top face observed. The sample space is

$$S = \{1, 2, 3, 4, 5, 6\} \quad \text{and} \quad n(S) = 6$$

▼ | EXPERIMENT   4 - 4

A box contains three poker chips (one red, one blue, one white), and two are drawn *with replacement*. (This means that one chip is selected, its color is observed, and then the chip is replaced in the box.) The chips are scrambled before a second chip is selected and its color observed. The sample space is shown in two different ways:

*See the *Statistical Tutor* for information about tree diagrams.

EXERCISE 4.18

A penny is tossed
and a die rolled. List
the sample spaces
as a tree diagram.

TREE DIAGRAM REPRESENTATION

First drawing	Second drawing	Outcomes
R	R	R, R
	B	R, B
	W	R, W
B	R	B, R
	B	B, B
	W	B, W
W	R	W, R
	B	W, B
	W	W, W and $n(S) = 9$

▼ EXPERIMENT 4-5

EXERCISE 4.19

Draw a tree diagram
representing the pos-
sible arrangements
of boys and girls
from oldest to youn-
gest for a family of:

a. two children
b. three children

A box contains one each of a red, a blue, and a white poker chip. Two chips are drawn simultaneously or one at a time without replacement (meaning one chip is selected, and then a second is selected without replacing the first). The sample space is shown in two ways:

TREE DIAGRAM REPRESENTATION

First drawing	Second drawing	Outcomes
R	B	R, B
	W	R, W
B	R	B, R
	W	B, W
W	R	W, R
	B	W, B and $n(S) = 6$

This experiment is the same as Experiment 4-4 except that the first chip is not replaced before the second selection is made.

▼ EXPERIMENT 4-6

Three coins are tossed, or one coin is tossed three times, with head (H) or tail (T) observed on each coin. The sample space is shown as follows:

TREE DIAGRAM REPRESENTATION

EXERCISE 4.20

A box contains one each of $1, $5, $10, and $20 bills.

a. One is selected at random; list the sample space.
b. Two bills are drawn at random (without replacement); list the sample space as a tree diagram.
c. Two bills are drawn at random (without replacement); list the sample space as a chart.

	First toss	Second toss	Third toss	Outcomes
			H	H, H, H
		H		
			T	H, H, T
	H			
			H	H, T, H
		T		
			T	H, T, T
			H	T, H, H
		H		
			T	T, H, T
	T			
			H	T, T, H
		T		
			T	T, T, T and $n(S) = 8$

▼ E X P E R I M E N T 4 - 7

Two dice (one white, one black) are each rolled one time and the number of dots showing on each die is observed. The sample space is shown by a chart representation:

CHART REPRESENTATION

and $n(S) = 36$

▼ EXPERIMENT   4 - 8

Two dice are rolled and the sum of their dots is observed. The sample space is

$$S = \{2, 3, 4, 5, 6, 7, 8, 9, 10, 11, 12] \text{ and } n(S) = 11$$

▲ (or the 36-point sample space listed in Experiment 4-7).

You will notice that two different sample spaces are suggested for Experiment 4-8. Both of these sets satisfy the definition of a sample space and thus either could be used. We will learn later why the 36-point sample space is more useful than the other.

▼ EXPERIMENT   4 - 9

A weather forecaster predicts that there will be a measurable amount of precipitation or no precipitation on a given day. The sample space is

$$S = \{\text{precipitation, no precipitation}\} \text{ and } n(S) = 2$$

▼ EXPERIMENT   4 - 10

The 6024 students at a nearby college have been cross-tabulated according to gender and their college status:

	Full-time Student	Part-time Student
Female	2136	548
Male	2458	882

and $n(S) = 6024$

One student is to be picked at random from the student body.

▼ EXPERIMENT   4 - 11

A lucky customer will get to randomly select one key from a barrel containing a key to each car on Used Car Charlie's lot. The accompanying Venn diagram summarizes Charlie's inventory.

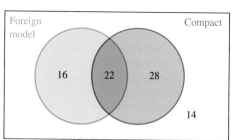

and $n(S) = 80$

▲

Special attention should always be given to the sample space. Like the statistical population, the sample space must be well defined. Once the sample space is defined, you will find the remaining work much easier.

EXERCISES

4.21   The face cards are removed from a regular deck and then 1 card is selected from this set of 12 face cards. List the sample space for this experiment.

4.22   An experiment consists of drawing one marble from a box that contains a mixture of red, yellow, and green marbles.
   a. List the sample space.
   b. Can we be sure that each outcome in the sample space is equally likely?
   c. If two marbles are drawn from the box, list the sample space.

4.23   In the July 23, 1990, issue of *Time*, it was reported that 600,000 new cases of skin malignancies would be diagnosed in the United States in 1990. Most of the malignancies would be either malignant melanoma, basal-cell carcinoma, or squamous-cell carcinoma. The remaining malignancies were classified as "other." If two malignancy cases are randomly selected from the records of a dermatologist, give the sample space for the possible combinations of selected records.

4.24   a. A balanced coin is tossed twice. List a sample space showing the possible outcomes.
   b. A biased coin (it favors heads in a ratio of 3 to 1) is tossed twice. List a sample space showing the possible outcomes.

4.25   An experiment consists of two trials. The first is tossing a penny and observing heads or tails; the second is rolling a die and observing a 1, 2, 3, 4, 5, or 6. Construct the sample space.

4.26   A computer generates (in random fashion) pairs of integers. The first integer is between 1 and 5, inclusive, and the second is between 1 and 4, inclusive. Represent the sample space on a coordinate axis system where x is the first number and y is the second number.

4.27   A coin is tossed and a head or a tail observed. If a head results, the coin is tossed a second time. If a tail results on the first toss, a die is rolled.
   a. Construct the sample space for this experiment.
   b. What is the probability that a die is rolled in the second stage of this experiment?

4.28   A box stored in a warehouse contains 100 identical parts of which 10 are defective and 90 are nondefective. Three parts are selected without replacement. Construct a tree diagram representing the sample space.

4.29   Use a computer (or random numbers table) to simulate 200 trials of the experiment described in Exercise 4.25, the tossing of a penny and the rolling of a die. Let 1 = H, 2 = T for the penny, and 1, 2, 3, 4, 5, 6 for the die. Report your results using a cross-tabulated table showing the frequency of each outcome.
   a. Find the relative frequency for heads.
   b. Find the relative frequency for 3.
   c. Find the relative frequency for (H, 3).

```
MINITAB commands
RANDom 200 C1;
   INTEger 1 2.
RANDom 200 C2;
   INTEger 1 6.
TABLe C1 C2;
   COUNts;
   TOTPercent.
```

4.30 Use a computer (or random numbers table) to simulate the experiment described in Exercise 4.26; x is an integer 1 to 5, and y is an integer 1 to 4. Generate a list of 100 random x-values and 100 y-values. Make a list of the resulting 100 ordered pairs of integers.

  a. Find the relative frequency for $x = 2$.
  b. Find the relative frequency for $y = 3$.
  c. Find the relative frequency for the ordered pair (2, 3).

If you use MINITAB, try RANDom, INTEger, TABLe, COUNts, TOTPercent.

4.4 | RULES OF PROBABILITY

Let's return now to the concept of probability and relate it to the sample space. Recall that the probability of an event was defined as the relative frequency with which the event could be expected to occur.

In the sample space associated with Experiment 4-1, the tossing of one coin, we find two possible outcomes: heads (H) and tails (T). We have an "intuitive feeling" that these two events will occur with approximately the same frequency. The coin is a symmetrical object and therefore would not be expected to favor either of the two outcomes. We would expect heads to occur $\frac{1}{2}$ of the time. Thus, the probability that a head will occur on a single toss of a coin is thought to be $\frac{1}{2}$.

Equally likely events

This description is the basis for the second technique for assigning the probability of an event. In a sample space containing **sample points that are equally likely to occur**, the probability $P(A)$ of an event A is the ratio of the number $n(A)$ of points that satisfy the definition of event A to the number $n(S)$ of sample points in the entire sample space. That is,

$$P(A) = \frac{n(A)}{n(S)} \tag{4-2}$$

Theoretical probability

This formula gives a **theoretical probability** value of event A's occurrence.

If we apply formula (4-2) to the equally likely sample space for the tossing of two coins (Experiment 4-2), we find the theoretical probabilities $P(0H)$, $P(1H)$, and $P(2H)$ discussed previously.

$$P(0H) = \frac{n(0H)}{n(S)} = \frac{1}{4}, \quad P(1H) = \frac{n(1H)}{n(S)} = \frac{2}{4} = \frac{1}{2}, \quad P(2H) = \frac{n(2H)}{n(S)} = \frac{1}{4}$$

The use of formula (4-2) requires the existence of a sample space in which each outcome is equally likely. Thus, when dealing with experiments that have more than one possible sample space, it is helpful to construct a sample space in which the sample points are equally likely.

Consider Experiment 4-8, where two dice were rolled. If you list the sample space as the 11 sums, the sample points are not equally likely. If you use the 36-point sample space, all the sample points are equally likely as in Experiment 4-7.

For example, the sum of 2 represents {(1, 1)}; the sum of 3 represents {(2, 1), (1, 2)}; and the sum of 4 represents {(1, 3), (3, 1), (2, 2)}. Thus, we can use formula (4-2) and the 36-point sample space to obtain the probabilities for the 11 sums.

$$P(2) = \frac{n(2)}{n(S)} = \frac{1}{36}, \quad P(3) = \frac{n(3)}{n(S)} = \frac{2}{36}, \quad P(4) = \frac{n(4)}{n(S)} = \frac{3}{36}$$

EXERCISE 4.31

Find the probabilities $P(5)$, $P(6)$, $P(7)$, $P(8)$, $P(9)$, $P(10)$, $P(11)$, $P(12)$ for the sum of two dice.

and so forth.

In many cases the assumption of equally likely events does not make sense. The sample points in Experiment 4-8 are not equally likely, and there is no reason to believe that the sample points in Experiment 4-9 are equally likely.

What do we do when the sample space elements are not equally likely or not a combination of equally likely events? We could use empirical probabilities. But what do we do when no experiment has been done or can be performed?

Let's look again at Experiment 4-9. The weather forecaster often assigns a probability to the event "precipitation." For example, "there is a 20% chance of rain today," or "there is a 70% chance of snow tomorrow." In such cases the only method available for assigning probabilities is personal judgment. These probability assign-

Subjective probability

ments are called **subjective probabilities**. The accuracy of subjective probabilities depends on the individual's ability to correctly assess the situation.

Often, personal judgment of the probability of the possible outcomes of an experiment is expressed by comparing the likelihood among the various outcomes. For example, the weather forecaster's personal assessment might be that "it is five times more likely to rain (R) tomorrow than not rain (NR)"; $P(R) = 5 \cdot P(NR)$. If this is the case, what values should be assigned to $P(R)$ and $P(NR)$? To answer this question, we need to review some of the ideas about probability that we've already discussed.

1. Probability represents a relative frequency.
2. $P(A)$ is the ratio of the number of times an event can be expected to occur divided by the number of trials.
3. The numerator of the probability ratio must be a positive number or zero.
4. The denominator of the probability ratio must be a positive number (greater than zero).
5. The number of times an event can be expected to occur in n trials is always less than or equal to the total number of trials.

Thus, it is reasonable to conclude that a probability is always a numerical value between zero and one.

Property 1 $0 \le P(A) \le 1$

NOTES
1. The probability is zero if the event cannot occur.
2. The probability is one if the event occurs every time.

Property 2 $\displaystyle\sum_{\text{all outcomes}} P(A) = 1$

Property 2 states that if we add the probabilities of each sample point in the sample space, the total probability must equal one. This makes sense because when we sum all the probabilities, we are asking, "What is the probability the experiment will yield an outcome?" and this will happen every time.

EXERCISE 4.32

If four times as many students pass a statistics course as fail, and one statistics student is selected at random, what is the probability that the student will pass statistics?

Now we are ready to assign probabilities to $P(R)$ and $P(NR)$. The events R and NR cover the sample space, and the weather forecaster's personal judgment was

$$P(R) = 5 \cdot P(NR)$$

From Property 2, we know that

$$P(R) + P(NR) = 1$$

By substituting $5 \cdot P(NR)$ for $P(R)$, we get

$$5 \cdot P(NR) + P(NR) = 1$$
$$6 \cdot P(NR) = 1$$
$$P(NR) = \tfrac{1}{6}$$

$$P(R) = 5 \cdot P(NR) = 5\left(\frac{1}{6}\right) = \frac{5}{6}$$

Odds

Odds

The statement "It is five times more likely to rain tomorrow (R) than not rain (NR)" is often expressed as "The odds are 5 to 1 in favor of rain tomorrow" (also written $5:1$). **Odds** is simply another way of expressing probabilities. The relationships among *odds for an event*, *odds against an event*, and the *probability of an event* are expressed in the following rules.

If the *odds in favor of an event A* are *a to b*, then

　1. The odds against event A are *b to a*.
　2. The probability of event A is

$$P(A) = \frac{a}{a + b}$$

　3. The probability that event A will not occur is

$$P(A \text{ does not occur}) = \frac{b}{a + b}$$

EXERCISE 4.33

The odds for the 49ers winning next year's Super Bowl are 2 to 7.

a. What is the probability the 49ers will win next year's Super Bowl?

b. What are the odds against the 49ers winning next year's Super Bowl?

To illustrate these rules, consider the statement "The odds favoring rain tomorrow are 5 to 1." Using the preceding notation, $a = 5$ and $b = 1$. Therefore, the probability of rain tomorrow is $\frac{5}{5+1}$, or $\frac{5}{6}$. The odds against rain tomorrow are 1 to 5 (or $1:5$), and the probability that there is no rain tomorrow is $\frac{1}{5+1}$, or $\frac{1}{6}$.

COMPLEMENT OF AN EVENT

The set of all sample points in the sample space that do not belong to event A. The complement of event A is denoted by $\overline{A}$ (read "A complement").

Case Study **Trying to Beat the Odds**

4-1

Many young men aspire to become professional athletes. Only a few make it to the big time as indicated in the following graph. For every 2400 college senior basketball players, only 64 make a professional team; that translates to a probability of only 0.027 (64/2400).

EXERCISE 4.34

Referring to "Trying to Beat the Odds," find:

a. The probability that a high school senior basketball player makes a pro team.
b. The odds that a player who makes a college basketball team plays as a senior.
c. The odds against a college senior basketball player making a pro team.

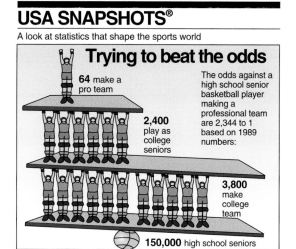

USA SNAPSHOTS®

A look at statistics that shape the sports world

Trying to beat the odds

64 make a pro team

The odds against a high school senior basketball player making a professional team are 2,344 to 1 based on 1989 numbers:

2,400 play as college seniors

3,800 make college team

150,000 high school seniors

Source: NCAA   By Julie Stacey, USA TODAY

Source: Copyright 1990, USA TODAY. Reprinted with permission.

For example, the complement of the event "success" is "failure"; the complement of "heads" is "tails" for the tossing of one coin; the complement of "at least one head" on 10 tosses of a coin is "no heads."

By combining the information in the definition of complement with Property 2 (p. 205), we can say that

$$P(A) + P(\overline{A}) = 1.0 \quad \text{for any event A}$$

It then follows that

EXERCISE 4.35

Find the probability of drawing a non-face card from a well-shuffled deck of 52 playing cards.

$$P(\overline{A}) = 1 - P(A) \tag{4-3}$$

NOTE Every event A has a complementary event $\overline{A}$. Complementary probabilities are very useful when the question asks for the probability of "at least one." Generally this represents a combination of several events, but the complementary event "none" is a single outcome. It is easier to solve for the complementary event and get the answer by using formula (4-3).

▼ | ILLUSTRATION 4-2

Two dice are rolled. What is the probability that the sum is at least 3 (that is, 3 or larger)?

SOLUTION Rather than finding the probability for each of the sums 3 and larger, it will be much simpler to find the probability that the sum is 2 (less than 3) and then use formula (4-3), letting "at least 3" be $\overline{A}$.

$$P(A) = \frac{1}{36} \quad \text{(as found on page 204)}$$

▲ |

$$P(\overline{A}) = 1 - P(A) = 1 - \frac{1}{36} = \frac{35}{36} \quad \text{[using formula (4-3)]}$$

EXERCISES

▼

4.36 A box contains marbles of five different colors: red, green, blue, yellow, and purple. There is an equal number of each color. Assign probabilities to each element in the sample space.

4.37 Suppose that a box of marbles contains an equal number of red marbles and yellow marbles but twice as many green marbles as red marbles. Draw one marble from the box and observe its color. Assign probabilities to the elements in the sample space.

4.38 One card is to be drawn from a well-shuffled deck of 52 playing cards. Find the probability the card drawn is:
   a. a heart.
   b. a five.
   c. a red card.
   d. a red number card.

4.39 A single die is rolled. Find the probability the number on top is:
   a. a 3.
   b. an odd number.
   c. a number less than 5.
   d. a number no more than 3.

4.40 A transportation engineer in charge of a new traffic-control system expresses the subjective probability that the system functions correctly 99 times as often as it malfunctions.
   a. Based on this belief, what is the probability that the system functions properly?
   b. Based on this belief, what is the probability that the system malfunctions?

4.41 Events A, B, and C are defined on sample space S. Their corresponding sets of sample points do not intersect and their union is S. Further, event B is twice as likely to occur as event A, and event C is twice as likely to occur as event B. Determine the probability of each of these three events.

4.42 Three coins are tossed and the number of heads observed is recorded. Find the probability for each of the possible results, 0H, 1H, 2H, and 3H.

4.43 Let *x* be the success rating of a new television show. The accompanying table lists the subjective probabilities assigned to each *x* for a particular new show as assigned by three different media people. Which of these sets of probabilities are inappropriate because they violate a basic rule of probability? Explain.

	Judge		
Success Rating (*x*)	**A**	**B**	**C**
Highly successful	0.5	0.6	0.3
Successful	0.4	0.5	0.3
Not successful	0.3	−0.1	0.3

4.44 Two dice are rolled (Experiment 4-7). Find the probabilities in parts (b) through (e). Use the sample space given on page 201.
   a. Why is the set {2, 3, 4, . . . , 12} not a useful sample space?
   b. *P*(white die is an odd number)
   c. *P*(sum is 6)
   d. *P*(both dice show odd numbers)
   e. *P*(number on black die is larger than number on white die)
   f. Explain why these answers and the answers found in Exercise 4.13 are not exactly the same.

4.45 A group of files in a medical clinic classifies the patients by gender and by type of diabetes (I or II). The groupings may be shown as follows. The table gives the number in each classification.

	Type of Diabetes	
Gender	**I**	**II**
Male	25	20
Female	35	20

If one file is selected at random, find the probability that
   a. the selected individual is female.
   b. the selected individual is a Type II.

4.46 A lottery is conducted and 500 tickets are sold. The stubs to the tickets are well shuffled, and the winner is chosen by randomly selecting one stub. If you bought 25 tickets, what is the probability that you will win?

4.47 Danny Sheridan's odds against winning the NBA 1994–95 championship were reported in *USA Today* (11-2-94). Some of the odds he gave are: Phoenix: 4 : 1; New York: 5 : 1; Houston: 6 : 1; Indiana: 9 : 1; Atlanta: 20 : 1; Boston: 1000 : 1; Dallas: 750,000 : 1.
   a. What are the odds favoring Phoenix winning?
   b. What probability does Sheridan give Dallas of winning?
   c. What is the probability that Atlanta will not win?

4.48   In the USA Shapshot® "Who is killed by firearms?" *USA Today* (1-17-95) reported: per 100,000 population: total—14.8, male—26.0, female–4.1.
    a. Express the three rates as probabilities.
    b. Explain why "male rate" + "female rate" ≠ "total rate."

4.49   According to an article in *Glamour* (April 1991), one out of every nine people diagnosed with AIDS during 1991 will be female. Based on this information, find the probability that an individual diagnosed with AIDS in 1991 will be male.

4.50   According to *Science News* (Nov. 1990), sleep apnea affects 2 million individuals in the United States. The sleep disorder interrupts breathing and can awaken its sufferers as often as five times an hour. Many people do not recognize the condition even though it causes loud snoring. Assuming there are 200 million people in the United States, what is the probability that an individual chosen at random will not be affected by sleep apnea?

CALCULATING PROBABILITIES
OF COMPOUND EVENTS

Compound events   **Compound events** are formed by combining several simple events. We will study the following three compound events in the remainder of this chapter:

1. the probability that either event A or event B will occur, *P*(A or B)
2. the probability that both events A and B will occur, *P*(A and B)
3. the probability that event A will occur given that event B has occurred, *P*(A|B)

NOTE   In determining which compound probability we are seeking, it is not enough to look for the words *either/or, and,* or *given.* We must carefully examine the question asked to determine what combination of events is called for.

4.5 | MUTUALLY EXCLUSIVE EVENTS AND THE ADDITION RULE

Mutually Exclusive Events

MUTUALLY EXCLUSIVE EVENTS

Events defined in such a way that the occurrence of one event precludes the occurrence of any of the other events. (In short, if one of them happens, the others cannot happen.)

The following illustrations give examples of events to help you understand the concept of mutually exclusive.

▼| ILLUSTRATION   4-3

EXERCISE   4.51

One student is selected from the student body of your college. The student is: M—male, F—female, S—registered for statistics.

a. Are events M and F mutually exclusive? Explain.

b. Are events M and S mutually exclusive? Explain.

c. Are events F and S mutually exclusive? Explain.

d. Are events M and F complementary? Explain.

e. Are events M and S complementary? Explain.

f. Are complementary events also mutually exclusive events? Explain.

g. Are mutually exclusive events also complementary events? Explain.

A group of 200 college students is known to consist of 140 full-time (80 female and 60 male) students and 60 part-time (40 female and 20 male) students. From this group one student is to be selected at random.

	200 College Students		
	Full-time	**Part-time**	**Total**
Female	80	40	120
Male	60	20	80
Total	140	60	200

Two events related to this selection are defined. Event A is "the student selected is full-time" and event B is "the student selected is a part-time male." Since no student is both "full-time" and "part-time male," the two events A and B are mutually exclusive events.

A third event, event C, is defined to be "the student selected is female." Now let's consider the two events A and C. Since there are 80 students that are "full-time" and "female," the two events A and C are not mutually exclusive events. This "intersection" of A and C can be seen on the accompanying Venn diagram or in the preceding table.

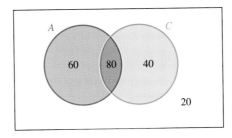

▼| ILLUSTRATION   4-4

Consider an experiment in which two dice are rolled. Three events are defined:

   A: The sum of the numbers on the two dice is 7.
   B: The sum of the numbers on the two dice is 10.
   C: Each of the two dice shows the same number.

Let's determine whether these three events are mutually exclusive.

   We can show three events are mutually exclusive by showing that each pair of events is mutually exclusive. Are events A and B mutually exclusive? Yes, they are, since the sum on the two dice cannot be both 7 and 10 at the same time. If a sum of 7 occurs, it is impossible for the sum to be 10.

Figure 4-3 presents the sample space for this experiment. This is the same sample space shown in Experiment 4-7, except that ordered pairs are used in place of the pictures. The ovals, diamonds, and rectangles show the ordered pairs that are in events A, B, and C, respectively. We can see that events A and B do not **intersect**. Therefore, they are mutually exclusive. Point (5, 5) satisfies both events B and C. Therefore, B and C are not mutually exclusive. Two dice can each show a 5, which satisfies C; and the total satisfies B. Since we found one pair of events that are not mutually exclusive, events A, B, and C are not mutually exclusive.

FIGURE 4-3

Sample Space for the Roll of Two Dice

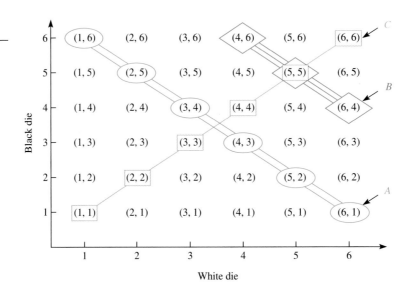

Addition rule

Addition Rule

Let us now consider the compound probability $P(A \text{ or } B)$, where A and B are mutually exclusive events.

▼ ILLUSTRATION 4-5

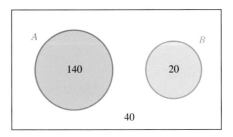

In Illustration 4-3 we considered an experiment in which one student was to be selected at random from a group of 200. Event A was "student selected is full-time,"

and event B was "student selected is part-time and male." The probability of event A, $P(A)$, is $\frac{140}{200}$, or 0.7, and the probability of event B, $P(B)$, is $\frac{20}{200}$, or 0.1. Let's find the probability of A or B, $P(A \text{ or } B)$. It seems reasonable to add the two probabilities 0.7 and 0.1 to obtain an answer of $P(A \text{ or } B) = 0.8$. This is further justified by looking at the sample space. We see there is a total of 160 that either are full-time or are male part-time students $\left(\frac{160}{200} = 0.8\right)$. Recall that these two events, A and B, are mutually exclusive.

▲

When events are *not mutually exclusive*, we cannot find the probability that one or the other occurs by simply adding the individual probabilities. Why not? Let's look at an illustration and see what happens when events are not mutually exclusive.

▼   ILLUSTRATION   4 - 6

Using the sample space and the events defined in Illustration 4-3, find the probability that the student selected is "full-time" or "female," $P(A \text{ or } C)$.

SOLUTION If we look at the sample space, we see that $P(A) = \frac{140}{200}$, or 0.7, and that $P(C) = \frac{120}{200}$, or 0.6. If we add the two numbers 0.7 and 0.6, we will get 1.3, a number larger than 1. We also know, from basic properties of probability, that probability numbers are never larger than 1. So what happened? If we take another look at the sample space, we will see that 80 of the 200 students have been counted twice if we add $\frac{140}{200}$ and $\frac{120}{200}$. Only 180 students are "full-time" or "female." Thus, the probability of A or C is

$$P(A \text{ or } C) = \frac{180}{200} = 0.9$$

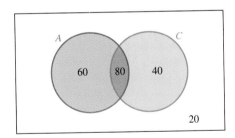

▲

We can add probabilities to find the probability of an "or" compound event, but we must make an adjustment in situations such as the previous example.

GENERAL ADDITION RULE

Let A and B be two events defined in a sample space S.

$$P(A \text{ or } B) = P(A) + P(B) - P(A \text{ and } B) \qquad \textbf{(4-4a)}$$

SPECIAL ADDITION RULE

Let A and B be two events defined in a sample space. If A and B are *mutually exclusive* events, then

$$P(A \text{ or } B) = P(A) + P(B) \tag{4-4b}$$

This can be expanded to consider *more than two* mutually exclusive events:

$$P(A \text{ or } B \text{ or } C \text{ or } \dots \text{ or } E) = P(A) + P(B) + P(C) + \dots + P(E) \tag{4-4c}$$

The key to this formula is the property "mutually exclusive." If two events are mutually exclusive, there is no double counting of sample points. If events are not mutually exclusive, then when probabilities are added, the double counting will occur. Let's look at some examples.

In Figure 4-4, events A and B are not mutually exclusive. The probability of the event "A or B," $P(A \text{ or } B)$, is represented by the union (total) of the shaded regions. The probability of the event "A and B," $P(A \text{ and } B)$, is represented by the area contained in region II. The probability $P(A)$ is represented by the area of circle A. That is, $P(A) = P(\text{region I}) + P(\text{region II})$. Furthermore, $P(B) = P(\text{region II}) + P(\text{region III})$. And $P(A \text{ or } B)$ is the sum of the probabilities associated with the three regions:

$$P(A \text{ or } B) = P(\text{I}) + P(\text{II}) + P(\text{III})$$

FIGURE 4-4

Nonmutually
Exclusive Events

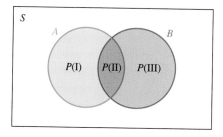

However, if $P(A)$ is added to $P(B)$, we have

$$P(A) + P(B) = [P(\text{I}) + P(\text{II})] + [P(\text{II}) + P(\text{III})]$$
$$= P(\text{I}) + 2P(\text{II}) + P(\text{III})$$

This is the double count previously mentioned. However, if we subtract one measure of region II from this total, we will be left with the correct value.

The addition formula (4-4b) is a special case of the more general rule stated in formula (4-4a). If A and B are mutually exclusive events, $P(A \text{ and } B) = 0$. (They cannot both happen at the same time.) Thus, the last term in formula (4-4a) is zero when events are mutually exclusive.

▼ | I L L U S T R A T I O N   4 - 7

One white die and one black die are rolled. Find the probability that the white die shows a number smaller than 3 or the sum of the dice is greater than 9.

S O L U T I O N 1 A = white die shows a 1 or a 2; B = sum of both dice is 10, 11, or 12.

$$P(A) = \frac{12}{36} = \frac{1}{3} \quad \text{and} \quad P(B) = \frac{6}{36} = \frac{1}{6}$$

$$P(A \text{ or } B) = P(A) + P(B) - P(A \text{ and } B)$$

$$= \frac{1}{3} + \frac{1}{6} - 0 = \frac{1}{2}$$

[$P(A$ and $B) = 0$, since the events do not intersect.]

S O L U T I O N 2

$$P(A \text{ or } B) = \frac{n(A \text{ or } B)}{n(S)} = \frac{18}{36} = \frac{1}{2}$$

(Look at the sample space, Figure 4-3, p. 212, and count.)

▼ | I L L U S T R A T I O N   4 - 8

EXERCISE 4.52

A pair of dice is rolled. Event T is defined as the occurrence of a "total of 10 or 11," and event D is the occurrence of "doubles." Find the probability $P(T$ or $D)$.

A pair of dice are rolled. Define events as: A—sum of 7, C—doubles, E—sum of 8.

a. Which pairs of events, A and C, A and E, C and E, are mutually exclusive? Explain.

b. Find the probabilities $P(A$ or $C)$, $P(A$ or $E)$, and $P(C$ or $E)$.

S O L U T I O N Look at the sample space of 36 ordered pairs for the rolling of two dice in Figure 4-3. Event T occurs if any one of 5 ordered pairs occurs: (4, 6), (5, 5), (6, 4), (5, 6), (6, 5). Therefore, $P(T) = \frac{5}{36}$. Event D occurs if any one of 6 ordered pairs occurs: (1, 1), (2, 2), (3, 3), (4, 4), (5, 5), (6, 6). Therefore, $P(D) = \frac{6}{36}$. Notice, however, that these two events are not mutually exclusive. The two events "share" the point (5, 5). Thus, the probability $P(T$ and $D) = \frac{1}{36}$. As a result, the probability $P(T$ or $D)$ will be found using formula (4-4a).

$$P(T \text{ or } D) = P(T) + P(D) - P(T \text{ and } D)$$

$$= \frac{5}{36} + \frac{6}{36} - \frac{1}{36} = \frac{10}{36} = \frac{5}{18}$$

▲ | (Look at the sample space in Figure 4.3 and verify $P(T$ or $D) = \frac{5}{18}$.)

EXERCISES

4.53   Determine whether or not each of the following pairs of events is mutually exclusive.
   a. Five coins are tossed: "one head is observed," "at least one head is observed."
   b. A salesperson calls on a client and makes a sale: "the sale exceeds $100," "the sale exceeds $1000."
   c. One student is selected at random from a student body: the person selected is "male," the person selected is "over 21 years of age."
   d. Two dice are rolled: the total showing is "less than 7," the total showing is "more than 9."

4.54   Determine whether each of the following sets of events is mutually exclusive.
   a. Five coins are tossed: "no more than one head is observed," "two heads are observed," "three or more heads are observed."
   b. A salesperson calls on a client and makes a sale: the amount of the sale is "less than $100," is "between $100 and $1000," is "more than $500."
   c. One student is selected at random from the student body: the person selected is "female," is "male," or is "over 21."
   d. Two dice are rolled: the number of dots showing on each die are "both odd," are "both even," are "total seven," or are "total eleven."

4.55   Explain why $P(A \text{ and } B) = 0$ when events A and B are mutually exclusive.

4.56   Explain why $P(A \text{ occurring, when } B \text{ has occurred}) = 0$ when events A and B are mutually exclusive.

4.57   If $P(A) = 0.3$ and $P(B) = 0.4$, and if A and B are mutually exclusive events, find the following:
   a. $P(\overline{A})$ b. $P(\overline{B})$ c. $P(A \text{ or } B)$ d. $P(A \text{ and } B)$

4.58   If $P(A) = 0.4$, $P(B) = 0.5$, and $P(A \text{ and } B) = 0.1$, find $P(A \text{ or } B)$.

4.59   One student is selected at random from a student body. Suppose the probability that this student is female is 0.5 and the probability that this student is working a part-time job is 0.6. Are the two events "female" and "working" mutually exclusive events? Explain.

4.60   The following table summarizes the teaching experience and educational background of the teachers in a public school.

	Teaching Experience	
Educational Background	**Less Than 5 Years**	**5 Years or More**
Less than a master's degree	75	40
Master's degree or more	55	30

Let A be the event that a teacher, selected at random, "has less than a master's degree" and let B represent "the teacher has less than five years experience." Find:
   a. $P(A \text{ and } B)$ b. $P(A \text{ or } B)$ c. $P(\overline{B})$

4.61 An article titled "The Mystery" appeared in the September 19, 1994, issue of *Newsweek*. The article gave the death risk per flight for United States airlines as 1 in 7 million, for international airlines in industrialized countries as 1 in 3.5 million, and for international airlines in developing countries as 1 in 500,000. The statistic "death risk" was defined as the probability that someone who randomly selected one of the airline's flights over the 20-year study period would be killed in flight. Find the following probabilities for this 20-year study period.
  a. being killed in a United States airline flight
  b. not being killed in a United States airline flight
  c. being killed in a United States airline flight or in a developing-countries airline flight.

4.62 An article titled "Pain and Pain-Related Side Effects in an ICU and on a Surgical Unit: Nurses' Management" (*American Journal of Critical Care*, January 1994, Vol. 3, No. 1) gave the following table summarizing the study participants. (*Note:* ICU is an acronym for intensive care unit.)

Gender	ICU	Surgical Unit
Female	9	6
Male	11	18

One of these participants is randomly selected. Answer the following questions.
  a. Are the events "being a female" and "being in the ICU" mutually exclusive?
  b. Are the events "being in the ICU" and "being in the surgical unit" mutually exclusive?
  c. Find *P*(ICU or female).
  d. Find *P*(ICU or male).

4.63 A parts store sells both new and used parts. Sixty percent of the parts in stock are used. Sixty-one percent are used or defective. If 5% of the store's parts are defective, what percentage are both used and defective?

4.64 Union officials report that 60% of the workers at a large factory belong to the union, 90% make over $12 per hour, and 40% belong to the union and make over $12 per hour. Do you believe these percentages? Explain.

4.6 | INDEPENDENCE, THE MULTIPLICATION RULE, AND CONDITIONAL PROBABILITY

Consider this example. The event that a 2 shows on a white die is A, and the event that a 2 shows on a black die is B. If both die are rolled once, what is the probability that two 2's occur?

$$P(A) = \frac{1}{6} \quad \text{and} \quad P(B) = \frac{1}{6}$$

$$P(A \text{ and } B) = \frac{n(A \text{ and } B)}{n(S)} = \frac{1}{36}$$

Notice that by multiplying the probabilities of the simple events the correct answer is found for *P*(A and B). Multiplication does not always work, however. For example, P(sum of 7 and double) when two dice are rolled is zero (as seen in Figure 4-3). However, if $P(7)$ is multiplied by P(double), we obtain $\left(\frac{1}{6}\right)\left(\frac{1}{6}\right) = \frac{1}{36}$.

Multiplication does not work for P(sum of 10 and double), either. By definition and by inspection of the sample space, we know that P(10 and double) $= \frac{1}{36}$ (the point (5, 5) is the only element). However, if we multiply $P(10)$ by P(double), we obtain $\left(\frac{3}{36}\right)\left(\frac{6}{36}\right) = \frac{1}{72}$. The probability of this event cannot be both values.

The property that is required for multiplying probabilities is **independence**. Multiplication worked in the one foregoing example because the events were independent. In the other two examples, the events were not independent and multiplication gave us incorrect answers.

NOTE   There are several situations that result in the compound event "and." Some of the more common ones are: (1) A followed by B, (2) A and B occurred simultaneously, (3) the intersection of A and B, (4) both A and B, and (5) A but not B (equivalent to A and not B).

Independence and Conditional Probabilities

INDEPENDENT EVENTS

Two events A and B are independent events if the occurrence (or nonoccurrence) of one does not affect the probability assigned to the occurrence of the other.

Sometimes independence is easy to determine, for example, if the two events being considered have to do with unrelated trials, such as the tossing of a penny and a nickel. The results on the penny in no way affect the probability of heads or tails on the nickel. Similarly, the results on the nickel have no effect on the probability of heads or tails on the penny. Therefore, the results on the penny and the results on the nickel are *independent*. However, if *events* are defined as combinations of outcomes from separate trials, the independence of the events may or may not be so easy to determine. The separate results of each trial (dice in the next illustration) may be independent, but the compound events defined using both trials (both dice) may or may not be independent.

Lack of independence, called **dependence**, is demonstrated by the following illustration. Reconsider the experiment of rolling two dice and observing the two events "sum of 10" and "double." As stated previously, $P(10) = \frac{3}{36} = \frac{1}{12}$ and P(double) $= \frac{6}{36} = \frac{1}{6}$. Does the occurrence of 10 affect the probability of a double? Think of it this way. A sum of 10 has occurred; it must be one of the following: {(4, 6), (5, 5), (6, 4)}. One of these three possibilities is a double. Therefore, we must conclude that the P(double, *knowing* 10 has occurred), written P(double|10), is $\frac{1}{3}$. Since $\frac{1}{3}$ does not equal the original probability of a double, $\frac{1}{6}$, we can conclude that the event "10" has an effect on the probability of a double. Therefore, "double" and "10" are dependent events.

Whether or not events are independent often becomes clear by examining the events in question. Rolling one die does not affect the outcome of a second roll. However, in many cases, independence is not self-evident, and the question of independence itself may be of special interest. Consider the events "having a checking account at a bank" and "having a loan account at the same bank." Having a checking account at a bank may increase the probability that the same person has a loan account. This has practical implications. For example, it would make sense to advertise loan programs to checking-account clients if they are more likely to apply for loans than are people who are not customers of the bank.

One approach to the problem is to *assume* independence or dependence. The correctness of the probability analysis depends on the truth of the assumption. In practice, we often assume independence and compare *calculated* probabilities with *actual frequencies* of outcomes in order to infer whether the assumption of independence is warranted.

CONDITIONAL PROBABILITY

The symbol $P(A|B)$ represents the probability that A will occur given that B has occurred. This is called a **conditional probability**.

The previous definition of independent events can now be written in a more formal manner.

INDEPENDENT EVENTS

Two events A and B are independent events if

$$P(A|B) = P(A) \text{ or if } P(B|A) = P(B) \qquad \textbf{(4-5)}$$

Let's consider conditional probability. Take, for example, the experiment in which a single die is rolled: $S = \{1, 2, 3, 4, 5, 6\}$. Two events that can be defined for this experiment are B = "an even number occurs" and A = "a 4 occurs." Then $P(A) = \frac{1}{6}$. Event A is satisfied by exactly one of the six equally likely sample points in S. The conditional probability of A given B, $P(A|B)$, is found in a similar manner, but the list of possible events is no longer the sample space. Think of it this way. A die is rolled out of your sight, and you are told the condition, the number showing is even, that is, event B has occurred. Knowing this condition, you are asked to assign a probability to the event that the even number is a 4. There are only three possibilities in the current (or reduced) sample space, $\{2, 4, 6\}$. Each of the three outcomes is equally likely; thus, $P(A|B) = \frac{1}{3}$.

We can write this as

$$P(A|B) = \frac{n(A \cap B)}{n(B)} \qquad \textbf{(4-6a)}$$

or equivalently,

$$P(A|B) = \frac{P(A \text{ and } B)}{P(B)} \qquad \textbf{(4-6b)}$$

Thus, for our example,

$$P(A|B) = \frac{\frac{1}{6}}{\frac{1}{2}} = \frac{1}{3}$$

▼| ILLUSTRATION 4-9

In a sample of 150 residents, each person was asked if he or she favored the concept of having a single countywide police agency. The county is composed of one large city and many suburban townships. The residence (city or outside the city) and the responses of the residents are summarized in Table 4-3. If one of these residents was to be selected at random, what is the probability that the person will (a) favor the concept? (b) favor the concept if the person selected is a city resident? (c) favor the concept if the person selected is a resident from outside the city? (d) Are the events F (favor the concept) and C (reside in city) independent?

TABLE 4-3

Sample Results for
Illustration 4-9

Residence	Opinion		
	Favor (F)	Oppose ($\overline{\text{F}}$)	Total
In city (C)	80	40	120
Outside of city ($\overline{\text{C}}$)	20	10	30
Total	100	50	150

EXERCISE 4.65

300 viewers were asked if they were satisfied with coverage of a recent disaster.

Gender	Female	Male
Satisfied	80	55
Not satisfied	120	45

One viewer is to be randomly selected from those surveyed.

a. Find P(Satisfied).
b. Find P(S|female).
c. Find P(S|male).
d. Is event S independent of gender? Explain.

SOLUTION

(a) $P(F)$ is the proportion of the total sample that favor the concept. Therefore,

$$P(F) = \frac{n(F)}{n(S)} = \frac{100}{150} = \frac{2}{3}$$

(b) $P(F|C)$ is the probability that the person selected favors the concept given that he or she lives in the city. The sample space is reduced to the 120 city residents in the sample. Of these, 80 favored the concept; therefore,

$$P(F|C) = \frac{n(F \text{ and } C)}{n(C)} = \frac{80}{120} = \frac{2}{3}$$

(c) $P(F|\overline{C})$ is the probability that the person selected favors the concept, knowing that the person lives outside the city. The sample space is reduced to the 30 noncity residents; therefore,

$$P(F|\overline{C}) = \frac{n(F \text{ and } \overline{C})}{n(\overline{C})} = \frac{20}{30} = \frac{2}{3}$$

(d) All three probabilities have the same value, $\frac{2}{3}$. Therefore, we can say that the events F (favor) and C (reside in city) are independent. The location of residence did not affect $P(F)$.

▲ |

Multiplication Rule

GENERAL MULTIPLICATION RULE

Let A and B be two events defined in sample space S. Then

$$P(A \text{ and } B) = P(A) \cdot P(B|A) \qquad \textbf{(4-7a)}$$

or

$$P(A \text{ and } B) = P(B) \cdot P(A|B) \qquad \textbf{(4-7b)}$$

EXERCISE 4.66

A and B are independent events, and $P(A) = 0.7$ and $P(B) = 0.4$. Find $P(A \text{ and } B)$.

If events A and B are independent, then the general multiplication rule [formula (4-7)] reduces to the *special multiplication rule*, formula (4-8).

SPECIAL MULTIPLICATION RULE

EXERCISE 4.67

R and H are events with $P(R) = 0.6$ and $P(H|R) = 0.25$. Find $P(R \text{ and } H)$.

Let A and B be two events defined in sample space S. If A and B are *independent events*, then

$$P(A \text{ and } B) = P(A) \cdot P(B) \qquad \textbf{(4-8a)}$$

This formula can be expanded. If A, B, C, . . . , G are independent events, then

$$P(A \text{ and } B \text{ and } C \text{ and } \ldots \text{ and } G) = P(A) \cdot P(B) \cdot P(C) \cdots P(G)$$
$$\textbf{(4-8b)}$$

▼ | ILLUSTRATION   4 - 10

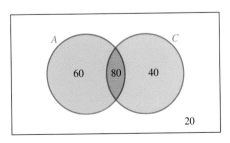

One student is selected at random from a group of 200 known to consist of 140 full-time (80 female and 60 male) students and 60 part-time (40 female and 20 male)

students. (See Illustration 4-3.) Event A is "the student selected is full-time," and event C is "the student selected is female."

  (a) Are events A and C independent? (b) Find the probability $P(A \text{ and } C)$ using the multiplication rule.

SOLUTION 1

  (a) First find the probabilities $P(A)$, $P(C)$, and $P(A|C)$.

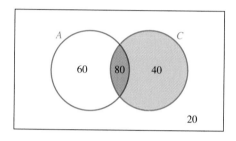

$$P(A) = \frac{n(A)}{n(S)} = \frac{140}{200} = 0.7$$

$$P(C) = \frac{n(C)}{n(S)} = \frac{120}{200} = 0.6$$

$$P(A|C) = \frac{n(A \text{ and } C)}{n(S)} = \frac{80}{120} = 0.67$$

  A and C are dependent events since $P(A) \neq P(A|C)$.

  (b) $P(A \text{ and } C) = P(C) \cdot P(A|C)$

$$= \frac{120}{200} \cdot \frac{80}{120} = \frac{80}{200} = \mathbf{0.4}$$

EXERCISE 4.68

a. Find $P(A \text{ and } C)$ in Illustration 4-10 using the sample space and formula (4-2).
b. Does your answer in (a) agree with the solution in Illustration 4-10? Explain.

SOLUTION 2

  (a) First find the probabilities $P(A)$, $P(C)$, and $P(C|A)$.

$$P(A) = \frac{n(A)}{n(S)} = \frac{140}{200} = 0.7$$

$$P(C) = \frac{n(C)}{n(S)} = \frac{120}{200} = 0.6$$

$$P(C|A) = \frac{n(C \text{ and } A)}{n(A)} = \frac{80}{140} = 0.57$$

A and C are dependent events since $P(C) \neq P(C|A)$

(b) $P(A \text{ and } C) = P(A) \cdot P(C|A)$

$$= \frac{140}{200} \cdot \frac{80}{140} = \frac{80}{200} = \textbf{0.4}$$

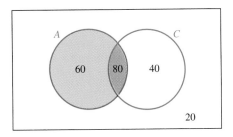

▼ ┃ ILLUSTRATION 4 - 11

One white and one black die are rolled. Find the probability that the sum of their numbers is 7 and that the number on the black die is larger than the number on the white die.

SOLUTION A = "sum is 7"; B = "black number larger than white number." The "and" requires the use of the multiplication rule. However, we do not yet know whether events A and B are independent. (Refer to Figure 4-3 for the sample space of this experiment.) We see that $P(A) = \frac{6}{36} = \frac{1}{6}$. Also, $P(A|B)$ is obtained from the reduced sample space, which includes 15 points above the gray diagonal line. Of the 15 equally likely points, 3 of them—(1, 6), (2, 5), and (3, 4)—satisfy event A. Therefore, $P(A|B) = \dfrac{n(A \text{ and } B)}{n(B)} = \frac{3}{15} = \frac{1}{5}$. Since this is a different value than $P(A)$, the events are dependent. So we must use formula (4-7b) to obtain $P(A \text{ and } B)$.

▲ ┃
$$P(A \text{ and } B) = P(B) \cdot P(A|B) = \frac{15}{36} \cdot \frac{3}{15} = \frac{3}{36} = \frac{1}{12}$$

NOTES
1. Independence and mutually exclusive are two very different concepts.
   a. Mutually exclusive says the two events cannot occur together; that is, they have no intersection.
   b. Independence says each event does not affect the other event's probability.
2. $P(A \text{ and } B) = P(A) \cdot P(B)$ when A and B are independent.
   a. Since $P(A)$ and $P(B)$ are not zero, $P(A \text{ and } B)$ is nonzero.
   b. Thus, independent events have an intersection.
3. Events cannot be both mutually exclusive and independent. Therefore,
   a. if two events are independent, then they are not mutually exclusive.
   b. if two events are mutually exclusive, then they are not independent.

Case Study   New York PICK 10

4-2

New York State, like many other states and cities, runs several lottery games for the purpose of raising money to finance education. The PICK 10 game is currently one of several New York State games that a person can play. The game is explained and the chances of winning each of the various prizes is stated on the ticket. Did you know that you could win a lottery by being totally wrong? The rules say that if you pick ten numbers and zero of them match the numbers drawn, you win $4 on a $1 ticket with a 1 in 22 chance. Can you verify the probability of $\frac{1}{22}$? To have zero matches means that all ten numbers picked did not match the ten numbers picked by the lottery.

$$P(\text{zero matches}) = P(A \text{ and } B \text{ and } C \text{ and } D \text{ and } \ldots \text{ and } J),$$
$$(\text{all nonmatches})$$

$$= \left(\frac{60}{80}\right)\left(\frac{59}{79}\right)\left(\frac{58}{78}\right)\left(\frac{57}{77}\right)\left(\frac{56}{76}\right) \cdots \left(\frac{51}{71}\right)$$

$$= 0.04579$$

$$= \frac{1}{21.839}, \text{ or approximately } \frac{1}{22}$$

EXERCISE   4.69

Case Study 4-2 describes rules for playing New York State's Pick 10 lottery. Many states have similar lottery games.

a. If someone plays Pick 10, that person has a 1 in 8,911,711 chance of winning a half-million-dollar prize. Verify this probability.

b. Verify that the chance of winning the $6000 prize is 1 in 163,381.

c. Verify that the overall chance of winning something is 1 in 17.

HOW TO PLAY PICK 10
(a) Each play card has 4 games, you may play 1, 2, 3 or all 4 games. Each game costs $1.00 to play.
(b) Select 10 numbers in each game you want to play.
(c) Select the number of days you want to play these numbers.
(d) Use only blue or black ballpoint pen or pencil for marking. Red ink will not be accepted.
(e) Present your completed play card to your Pick 10 Agent for processing.
(f) If you can't think of 10 numbers, just ask for Quick Pick. The computer will choose your numbers for you.

HOW TO WIN
(a) Each night, the Lottery randomly selects 20 numbers from a field of 80; these 20 are the winning numbers.
(b) You may win if all, some or none of your selections match any of the winning numbers.

PRIZE LEVELS AND CHANCES OF WINNING

WINNING NUMBERS MATCHED PER GAME	PRIZE	CHANCES OF WINNING
10	$500,000	1:8,911,711
9	$ 6,000	1:163,381
8	$ 300	1:7,384
7	$ 40	1:621
6	$ 10	1:87
0	$ 4	1:22

Overall chances of winning: 1 in 17

Source: New York State Pick 10 Play Card courtesy of NYS Division of the Lottery.

EXERCISES

4.70   a. Describe in your own words what it means for two events to be mutually exclusive.
b. Describe in your own words what it means for two events to be independent.
c. Explain how mutually exclusive and independence are two very different properties.

4.71   Determine whether or not each of the following pairs of events is independent:
a. rolling a pair of dice and observing a "1" on the first die and a "1" on the second die
b. drawing a "spade" from a regular deck of playing cards and then drawing another "spade" from the same deck without replacing the first card
c. same as (b) except the first card is returned to the deck before the second drawing
d. owning a red automobile and having blonde hair
e. owning a red automobile and having a flat tire today
f. studying for an exam and passing the exam

4.72   Determine whether or not the following pairs of events are independent:
a. rolling a pair of dice and observing a "2" on one of the dice and having a "total of 10"
b. drawing one card from a regular deck of playing cards and having a "red" card and having an "ace"
c. raining today and passing today's exam
d. raining today and playing golf today
e. completing today's homework assignment and being on time for class.

4.73   If $P(A) = 0.3$ and $P(B) = 0.4$ and A and B are independent events, what is the probability of each of the following?
a. $P(A \text{ and } B)$ b. $P(B|A)$ c. $P(A|B)$

4.74   Suppose that $P(A) = 0.3$, $P(B) = 0.4$, and $P(A \text{ and } B) = 0.12$.
a. What is $P(A|B)$?   b. What is $P(B|A)$?   c. Are A and B independent?

4.75   Suppose that $P(A) = 0.3$, $P(B) = 0.4$, and $P(A \text{ and } B) = 0.20$.
a. What is $P(A|B)$?   b. What is $P(B|A)$?   c. Are A and B independent?

4.76   Suppose that A and B are events and that the following probabilities are known: $P(A) = 0.3$, $P(B) = 0.4$, and $P(A|B) = 0.2$. Find $P(A \text{ or } B)$.

4.77   A single card is drawn from a standard deck. Let A be the event that "the card is a face card" (a jack, a queen, or a king), B be the occurrence of a "red card," and C represent "the card is a heart." Check to determine whether the following pairs of events are independent or dependent:
a. A and B   b. A and C   c. B and C

4.78   A box contains four red and three blue poker chips. What is the probability when three are selected randomly that all three will be red if we select each chip
a. with replacement?   b. without replacement?

4.79   The December 1994 issue of *The American Spectator* quotes a poll by the Times-Mirror Center for the People and the Press as finding that 71% of Americans believe that the press "gets in the way of society solving its problems."
a. If two Americans are randomly selected, find the probability that both will believe that the press "gets in the way of society solving its problems."

(continued)

b. If two Americans are selected, find the probability that neither of the two will believe that the press "gets in the way of society solving its problems."

c. If three are selected, what is the probability that all three believe the press gets in the way?

4.80 The August 1, 1994, issue of *The New Republic* gives the results of a U.S. Justice Department study which states that among white spousal murder victims, 62% are female. If the records of three victims are randomly selected from a large data base of such murder victims, what is the probability that all three victims are male?

4.81 An article involving smoking cessation intervention in *Heart & Lung* (March/April 1994, Vol. 23, No. 2) divided 80 subjects into a two-way classification as follows:

| | Diagnosis | | |
Group	Cardiovascular	Oncology	General Surgery
Experimental	10	14	13
Usual Care	12	16	15

Suppose one of these 80 subjects is selected at random. Find the probabilities of the following events.

a. The subject is not in the experimental group.

b. The subject is in the experimental group and has an oncology diagnosis.

c. The subject is in the experimental group or has a cardiovascular diagnosis.

4.82 A study concerning coping strategies of abstainers from alcohol appeared in *Image, the Journal of Nursing Scholarship* (Vol. 25, No. 1, Spring 1993). The study involved 23 subjects who were classified according to sex as well as marital status as shown in the following table.

Marital Status	Men	Women
Currently married	10	3
Divorced/separated	3	6
Never married	1	0

One of the subjects is selected at random. Find:

a. the probability that the subject is currently married given that the individual is a man.

b. the probability that the subject is a woman given that the individual is divorced/separated.

c. the probability that the subject is a man given that the individual has never married.

4.83 The owners of a two-person business make their decisions independently of each other, then compare their decisions. If they agree, the decision is made; if they do not agree, then further consideration is necessary before a decision is reached. If they each have a history of making the right decision 60% of the time, what is the probability that together they

a. make the right decision on the first try?

b. make the wrong decision on the first try?

c. delay the decision for further study?

4.84 An article in the *Sunday World-Herald* (Omaha, Nebraska, 11-27-94) quoted a poll by the Times Mirror Center for People and the Press as finding that 70% of those surveyed felt that they had learned enough during the campaign to make an informed choice in their vote on November 8, 1994. Suppose that the 70% finding is true for all Americans. If two Americans were selected randomly,

　　a. what is the probability that both would feel they had learned enough during the campaign to make an informed choice in their vote on November 8, 1994?

　　b. what is the probability that neither would feel he or she had learned enough to make an informed choice?

　　c. what is the probability that one of the two would feel he or she had learned enough to make an informed choice?

4.85 Consider the set of integers 1, 2, 3, 4, and 5.

　　a. One integer is selected at random. What is the probability that it is odd?

　　b. Two integers are selected at random (one at a time with replacement so that each of the five is available for a second selection).

Find the probability that (1) neither is odd; (2) exactly one of them is odd; (3) both are odd.

4.86 A box contains 25 parts, of which 3 are defective and 22 are nondefective. If 2 parts are selected without replacement, find the following probabilities:

　　a. P(both are defective)　　　b. P(exactly one is defective)

　　c. P(neither is defective)

4.87 "Lying Is Just a Way of Life" is a report in *USA Today* (4-29-91). A new survey finds that 91% in the United States say they lie routinely, and 36% of those confess to dark, important lies. Based on these figures, if one person is picked at random, what is the probability that he or she routinely tells dark, important lies?

4.88 You are a contestant on a game show and given a choice of three doors. Behind one of them is a car and behind the other two are goats. You pick door number 3, and the host, knowing what's behind each door, has door number 2 opened, revealing a goat. He then asks you if you would like to keep door number 3 or switch to door number 1. Should you switch? What is the probability that door number 3 has the car? What is the probability that door number 1 has the car?

4.7 | COMBINING THE RULES OF PROBABILITY

Many probability problems can be represented by tree diagrams. In these instances, the addition and multiplication rules can be applied quite readily. To illustrate the use of tree diagrams in solving probability problems, let's use Experiment 4-5. Two poker chips are drawn from a box containing one each of red, blue, and white chips. The tree diagram representing this experiment (Figure 4-4) shows a first drawing and then a second drawing. One chip was drawn on each drawing and not replaced.

After the tree has been drawn and labeled, we need to assign probabilities to each branch of the tree. If we assume that it is equally likely that any chip would be drawn at each stage, we can assign a probability to each branch segment of the tree, as shown in Figure 4-5. Notice that a set of branches that initiate from a single point has a total probability of 1. In this diagram there are four such sets of branch segments. The tree diagram shows six different outcomes. Reading down: branch (1) shows (R, B), branch (2) shows (R, W), and so on. (*Note:* Each outcome for the experiment is represented by a branch that begins at the common starting point and ends at the terminal points at the right.)

FIGURE 4-4

All Possible Combinations That Can Be Drawn

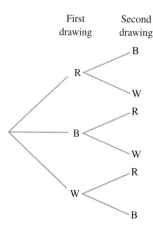

FIGURE 4-5

Probabilities of All Possible Combinations

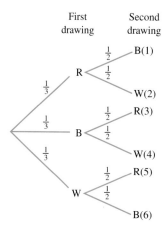

The probability associated with outcome (R, B), P(R on first drawing and B on second drawing), is found by multiplying P(R on first drawing) by P(B on second drawing | R on first drawing). These are the two probabilities $\frac{1}{3}$ and $\frac{1}{2}$ shown on the two branch segments of branch (1) in Figure 4-5. The $\frac{1}{2}$ is the conditional probability asked for by the multiplication rule. Thus, we will multiply along the branches.

Some events will be made up of more than one outcome from our experiment. For example, suppose that we had asked for the probability that one red chip and one blue chip are drawn. You will find two outcomes that satisfy this event, branch (1) or branch (3). With "or" we will use the addition rule (4-4b). Since the branches of a tree diagram represent mutually exclusive events, we have

$$P(\text{one R and one B}) = P[(R_1 \text{ and } B_2) \text{ or } (B_1 \text{ and } R_2)]$$

$$= \left(\frac{1}{3}\right)\left(\frac{1}{2}\right) + \left(\frac{1}{3}\right)\left(\frac{1}{2}\right) = \frac{1}{6} + \frac{1}{6} = \frac{1}{3}$$

FIGURE 4-6

Multiply Along the
Branches

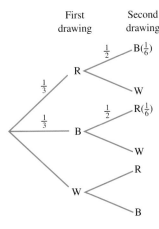

First drawing   Second drawing

NOTES

1. Multiply along the branches $\left(\text{on Figure 4-6; } \frac{1}{3} \cdot \frac{1}{2}\right)$.
2. Add across the branches $\left(\text{on Figure 4-6; } \frac{1}{6} + \frac{1}{6}\right)$.

Now let's consider an example that places all the rules in perspective.

▼ | ILLUSTRATION 4 - 12

A firm plans to test a new product in one randomly selected market area. The market areas can be categorized on the basis of location and population density. The number of markets in each category is presented in Table 4-4.

TABLE 4-4

Number of Markets
by Location and by
Population Density

Location	Population Density		Total
	Urban (U)	**Rural (R)**	
East (E)	25	50	75
West (W)	20	30	50
Total	45	80	125

What is the probability that the test market selected is in the East, $P(E)$? In the West, $P(W)$? What is the probability that the test market is in an urban area, $P(U)$? In a rural area, $P(R)$? What is the probability that the market is a western rural area, $P(W \text{ and } R)$? What is the probability it is an eastern or urban area, $P(E \text{ or } U)$? What is the probability that if it is in the East, it is an urban area, $P(U|E)$? Are "location" and "population density" independent? (What do we mean by independence or dependence in this situation?)

SOLUTION The first four probabilities, $P(E)$, $P(W)$, $P(U)$, and $P(R)$, represent "or" questions. For example, $P(E)$ means that the area is an eastern urban area or an eastern rural area. Since in this and the other three cases, the two components are mutually exclusive (an area can't be both urban and rural), the desired probabilities can be found by simply adding. In each case the probabilities are added across all the rows or columns of the table. Thus, the totals are found in the total column or row.

$$P(\text{E}) = \frac{75}{125} \quad \text{(total for East divided by total number of markets)}$$

$$P(\text{W}) = \frac{50}{125} \quad \text{(total for West divided by total number of markets)}$$

$$P(\text{U}) = \frac{45}{125} \quad \text{(total for urban divided by total number of markets)}$$

$$P(\text{R}) = \frac{80}{125} \quad \text{(total for rural divided by total number of markets)}$$

Now we solve for $P(\text{W and R})$. There are 30 western rural markets and a total of 125 markets. Thus,

$$P(\text{W and R}) = \frac{n(\text{W and R})}{n(S)} = \frac{30}{125}$$

Note that $P(\text{W}) \cdot P(\text{R})$ does *not* give the right answer $\left[\left(\frac{50}{125}\right)\left(\frac{80}{125}\right) = \frac{32}{125}\right]$. Therefore, "location" and "population density" are dependent events.

$P(\text{E or U})$ can be solved in several different ways. The most direct way is to simply examine the table and count the number of markets that satisfy the condition that they are in the East or they are urban. We find 95, [25 + 50 + 20], Thus,

$$P(\text{E or U}) = \frac{n(\text{E or U})}{n(S)} = \frac{95}{125}$$

Note that the first 25 markets were both in the East and urban; thus, E and U are not mutually exclusive events.

Another way to solve for $P(\text{E or U})$ is to use the addition formula:

$$P(\text{E or U}) = P(\text{E}) + P(\text{U}) - P(\text{E and U})$$

which yields

$$\frac{75}{125} + \frac{45}{125} - \frac{25}{125} = \frac{95}{125}$$

A third way to solve the problem is to recognize that the complement of (E or U) is (W and R). Thus, $P(\text{E or U}) = 1 - P(\text{W and R})$. Using the previous calculation, we get $1 - \frac{30}{125} = \frac{95}{125}$.

Finally, we solve for $P(\text{U}|\text{E})$. Looking at Table 4-4, we see that there are 75 markets in the East. Of the 75 eastern markets, 25 are urban. Thus,

$$P(\text{U}|\text{E}) = \frac{n(\text{U and E})}{n(\text{E})} = \frac{25}{75}$$

The conditional probability formula could also be used:

$$P(U|E) = \frac{P(U \text{ and } E)}{P(E)}$$

$$= \frac{\frac{25}{125}}{\frac{75}{125}} = \frac{\mathbf{25}}{\mathbf{75}}$$

"Location" and "population density" are not independent events. They are dependent. This means that the probability of these events is affected by the occurrence of each other.

▲

Although each rule for computing compound probabilities has been discussed separately, you should not think they are only used separately. In many cases they are combined to solve problems. Consider the following two illustrations.

▼   I L L U S T R A T I O N   4 - 13

A production process produces an item. On the averge, 20% of all items produced are defective. Each item is inspected before being shipped. The inspector misclassifies an item 10% of the time; that is,

$$P(\text{classified good}|\text{defective item}) = P(\text{classified defective}|\text{good item})$$
$$= 0.10$$

What proportion of the items will be "classified good"?

S O L U T I O N What do we mean by the event "classified good"?

$$G = \text{item good}$$
$$D = \text{item defective}$$
$$CG = \text{item classified good by inspector}$$
$$CD = \text{item classified defective by inspector}$$

CG consists of two possibilities: "the item is good and is correctly classified good" or "the item is defective and misclassified good." Thus,

$$P(CG) = P[(CG \text{ and } G) \text{ or } (CG \text{ and } D)]$$

Since the two possibilities are mutually exclusive, we can start by using the addition rule, formula (4-4b).

$$P(CG) = P(CG \text{ and } G) + P(CG \text{ and } D)$$

The condition of an item and its classification by the inspector are not independent. The multiplication rule for dependent events must be used. Therefore,

$$P(CG) = [P(G) \cdot P(CG|G)] + [P(D) \cdot P(CG|D)]$$

Substituting the known probabilities, we get

$$P(CG) = [(0.8)(0.9)] + [(0.2)(0.1)]$$
$$= 0.72 + 0.02$$
$$= \mathbf{0.74}$$

FIGURE 4-7

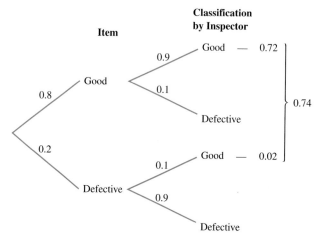

That is, 74% of the items are classified good.

▼ | ILLUSTRATION   4 - 14

Reconsider Illustration 4-13. Suppose that only items that pass inspection are shipped. Items not classified good are scrapped. What is the quality of the shipped items? That is, what percentage of the items shipped are good, $P(G|CG)$?

SOLUTION Using the conditional probability formula (4-6b),

$$P(G|CG) = \frac{P(G \text{ and } CG)}{P(CG)}$$

$$P(G|CG) = \frac{P(G) \cdot P(CG|G)}{P(CG)} \qquad \text{(See Illustration 4-13.)}$$

$$= \frac{(0.8)(0.9)}{0.74} = 0.9729 = \mathbf{0.973}$$

▲ | In other words, 97.3% of all items shipped will be good. Inspection increases the quality of items shipped from 80% good to 97.3% good.

The Reverend Thomas Bayes (1702–1761), an English Presbyterian minister and mathematician, developed an expanded form for conditional probabilities. This

Bayes's rule

expanded rule, called *Bayes's rule*, allows us to revise (or adjust) the probabilities assigned to events in accordance with new information.

BAYES'S RULE

$$P(A_i | B) = \frac{P(A_i) \cdot P(B | A_i)}{\Sigma[P(A_i) \cdot P(B | A_i)]} \qquad (4\text{-}9)$$

where $A_1, \ldots, A_n$ is an all-inclusive set of possible outcomes given B.

Let's take another look at Illustration 4-14 to explain Bayes's rule.

▼ | ILLUSTRATION 4 - 15

Consider the situation described in Illustration 4-13 and suppose that only items classified as good after inspection are shipped. What percentage of the items shipped are good, $P(G | CG)$? What percentage of those shipped are defective, $P(D | CG)$?

SOLUTION First, let's match up the events of the problem to the Bayes's rule notation. The events Bayes identifies as A_i are A_1 = G (good) and A_2 = D (defective), an all-inclusive set of events. The given or conditional event B is CG, classified good.

A tabular approach will be used to help organize the solution. To construct the table, start by listing all possible outcomes A_i that can occur given event B (that is, "shipped") in the first column. See Table 4-5. In the second column, list the initial probabilities of the A_i outcomes. In the third column we list the conditional probability that B happens for each A_i, $P(B | A_i)$. [For our illustration, $P(B | A_1) = P(CG | G)$ and $P(B | A_2) = P(CG | D)$.] These first three columns represent the information obtained from the problem.

TABLE 4-5

Tabular Presentation of Given Information

(1) A_i, **Possible Outcomes**	(2) $P(A_i)$	(3) $P(B | A_i)$
A_1, item good	0.8	0.9
A_2, item defective	0.2	0.1
Total	1.0 (ck)	

To solve for the conditional probabilities $P(A_i | B)$, the first calculation is to multiply each number of column (2) by the number in the same row of column (3). This product is placed in column (4) of the table (see Table 4-6). The column is labeled $P(A_i$ and B). The values calculated represent the probability that both A_i and B will occur. Thus, 72% of the items produced will be good and classified good; 2% of the items will be defective and classified good.

TABLE 4-6

Tabular Solution of Bayes's Rule

(1) A_i, Possible Outcomes	(2) $P(A_i)$	(3) $P(B\|A_i)$	(4) $P(A_i \text{ and } B)$ $= P(A_i)P(B\|A_i)$	(5) $P(A_i\|B)$
A_1, item good	0.8	0.9	0.72	$0.72/0.74 = $ **0.973** $= P(G\|\text{shipped})$
A_2, item defective	0.2	0.1	0.02	$0.02/0.74 = $ **0.027** $= P(D\|\text{shipped})$
Total	1.0 (ck)		$0.74 = P(B)$	1.000 (ck)

EXERCISE 4.89

Describe the similarities between the methods used in Illustrations 4-13 and 4-14 and the methods used in Illustration 4-15.

The second step is to add column (4). The sum represents $P(B)$. Thus, 74% of the items produced will be classified good.

Finally, the answers we are looking for, the conditional probabilities $P(A_i|B)$, are obtained by dividing each number in column (4) by the total of column (4). The results are placed in column (5) and are the answers. Thus, 0.973 is the proportion of items classified good that are good. And 0.027 is the proportion of items classified good that are actually defective.

▲ | **NOTE**   The total of columns (2) and (5) must equal 1. The total of column (3) need not equal 1.

Bayes's rule is of special interest because it gives us a mechanism to revise initial probability estimates when new information is learned, as we see in the next illustration.

▼ | 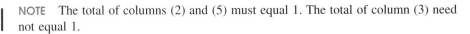 I L L U S T R A T I O N   4 - 16

Prior

Revised, or posterior

Consider the situation in which we feel that the probability that a stock is a good buy is 0.4. That is, our **prior** (before new information) probabilities are $P(\text{good buy}) = 0.4$ and $P(\text{bad buy}) = 0.6$. Now we find out that an investment service that has a record of being right 80% of the time recommends the stock. What should be our **revised**, or **posterior** (after new information), probability that the stock is a good buy, that is, $P(\text{good buy}|\text{investment service recommends it})$, or $P(A_i|B)$?

S O L U T I O N   Using the Bayesian tabular analysis, we find that the revised, or posterior, probability is 0.727 that the stock is a good buy and 0.273 that it is a bad buy (see Table 4-7 on page 236).

Case Study

Ask Marilyn

Marilyn presents a very simple and easy to understand solution to the game show problem "Should you switch?" Her seemingly simple answer stirred up lots of controversy. The question is: "You're on a game show and given a choice of three doors. Behind one is a car; behind the others are goats. You pick Door No. 1, and the host, who knows what's behind them, opens No. 3, which is a goat. He then asks if you want to pick No. 2. Should you switch?" Marilyn's answer was:

The benefits of switching are readily proved by playing through the six games that exhaust all the possibilities. For the first three games, you choose No. 1 and switch each time; for the second three games, you choose No. 1 and "stay" each time, and the host always opens a loser. Here are the results (each row is a game):

DOOR 1	DOOR 2	DOOR 3
AUTO	GOAT	GOAT
Switch and you lose.		
GOAT	AUTO	GOAT
Switch and you win.		
GOAT	GOAT	AUTO
Switch and you win.		
AUTO	GOAT	GOAT
Stay and you win.		
GOAT	AUTO	GOAT
Stay and you lose.		
GOAT	GOAT	AUTO
Stay and you lose.		

When you switch, you win two out of three times and lose one time in three; but when you don't switch, you only win one in three times and lose two in three. Try it yourself.

Alternatively, you actually can play the game with another person acting as the host with three playing cards—two jokers for the goats and an ace for the auto.

Source: Reprinted with permission from *Parade*, copyright © 1990.

Marilyn's answer relies on the following assumptions: (1) after you choose a door, the host will always reveal one of the goats and give you a chance to switch doors, and (2) the host will not take advantage of his knowledge about the location of the car and "lead" you into the choice of his preference. These assumptions have a substantial effect on the solution. By knowing where the car is, the host is "almost" in control. He has many options: (1) he could immediately open the door of your first choice (car or goat, his choice); (2) if you have chosen a goat or the car, he can make or not make dollar offers to buy the door back in ways that "talk" you into whatever will make an exciting show. Remember, it is a game show produced for entertainment. Perhaps now you see why this seemingly simple probability question is not so simple.

The last paragraph of the article suggests a way to simulate the show. Try it with a friend; alternate being host and contestant, alternate rules for the show. If you get real serious, keep a record of the result and the "rules" being used; then you can determine your own probability answers.

EXERCISE 4.90

"Monty Hall's three doors" have been fun to discuss ever since 1963 when the show started. Using the assumption that "after you choose a door, the host will always reveal one of the goats and give you a chance to switch doors;"

a. Construct a tree diagram showing the possibilities when playing the "switch" game plan.

b. Construct a tree diagram showing the possibilities when playing by the "stay" game plan.

c. Use the tree diagrams to verify Marilyn's probabilities, under our assumption.

d. In your own words, explain why these probabilities are not correct.

TABLE 4-7

Tabular Analysis for Illustration 4-16

A_i	$P(A_i)$	$P(B \mid A_i)$	$P(A_i)P(B \mid A_i)$	$P(A_i \mid B)$
A_1, good buy	0.4	0.8	0.32	$\dfrac{0.32}{0.44} = 0.727$
A_2, bad buy	0.6	0.2	0.12	$\dfrac{0.12}{0.44} = 0.273$
Total	1.0 (ck)		$0.44 = P(B)$	1.000 (ck)

▲ |

In analyzing the use of Bayes's rule to revise prior probability estimates in light of new information, we note the following relationship: The stronger the prior probability, the less the effect of the new information on changing the probability. Also, the more conclusive the new information, the greater the impact on the revised probability.

EXERCISES

4.91 If $P(A) = 0.4$ and $P(B) = 0.5$, and if A and B are independent events, find $P(A \text{ or } B)$.

4.92 $P(G) = 0.5$, $P(H) = 0.4$, and $P(G \text{ and } H) = 0.1$ (see diagram).

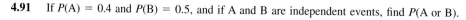

  a. Find $P(G \mid H)$. b. Find $P(H \mid G)$. c. Find $P(\overline{H})$.
  d. Find $P(G \text{ or } H)$. e. Find $P(G \text{ or } \overline{H})$.
  f. Are events G and H mutually exclusive? Explain.
  g. Are events G and H independent? Explain.

4.93 $P(R) = 0.5$, $P(S) = 0.3$, and events R and S are independent.
  a. Find $P(R \text{ and } S)$ b. Find $P(R \text{ or } S)$. c. Find $P(\overline{S})$.
  d. Find $P(R \mid S)$. e. Find $P(\overline{S} \mid R)$.
  f. Are events R and S mutually exclusive? Explain.

4.94 $P(M) = 0.3$, $P(N) = 0.4$, and events M and N are mutually exclusive.
  a. Find $P(M \text{ and } N)$. b. Find $P(M \text{ or } N)$. c. Find $P(M \text{ or } \overline{N})$.
  d. Find $P(M \mid N)$. e. Find $P(M \mid \overline{N})$.
  f. Are events M and N independent? Explain.

4.95 Two flower seeds are randomly selected from a package that contains five seeds for red flowers and three seeds for white flowers.

Draw a tree diagram.

  a. What is the probability that both seeds will result in red flowers?
  b. What is the probability that one of each color is selected?
  c. What is the probability that both seeds are for white flowers?

4.96 The probability that a certain door is locked is 0.6. The key to the door is one of five unidentified keys hanging on a key rack. Two keys are randomly selected before approaching the door. What is the probability that the door may be opened without returning for another key?

4.97 A, B, and C each in turn toss a balanced coin. The first one to throw a head wins.

Draw a tree diagram.

  a. What are their respective chances of winning if each tosses only one time?
  b. What are their respective chances of winning if they continue, given a maximum of two tosses each?

4.98 A coin is flipped three times.
  a. Draw a tree diagram that represents all possible outcomes.
  b. Identify all branches that represent the event "exactly one head occurred."
  c. Find the probability of "exactly one head occurred."

4.99 Box 1 contains two red balls and three green balls, and Box 2 contains four red balls and one green ball. One ball is randomly selected from Box 1 and placed in Box 2. Then one ball is randomly selected from Box 2. What is the probability that the ball selected from Box 2 is green?

4.100 A company that manufactures shoes has three factories. Factory 1 produces 25% of the company's shoes, Factory 2 produces 60%, and Factory 3 produces 15%. One percent of the shoes produced by Factory 1 are mislabeled, 0.5% of those produced by Factory 2 are mislabeled, and 2% of those produced by Factory 3 are mislabeled. If you purchase one pair of shoes manufactured by this company, what is the probability that the shoes are mislabeled?

4.101 An article titled "A Puzzling Plague" found in the January 14, 1991, issue of *Time*, stated that one out of every ten American women will get breast cancer. It also states that of those who do, one out of four will die of it. Use these probabilities to find the probability that a randomly selected American woman will
  a. never get breast cancer.
  b. get breast cancer and not die of it.
  c. get breast cancer and die from it.

4.102 An article in the March 25, 1991, issue of *US News & World Report* discussed the use of clotbusters to save the lives of Americans who have heart attacks. The following projections are given.

Treatment	Percent Surviving
A: with current clotbuster use	66
B: with optimum clotbuster use	70
C: with optimum clotbuster use plus prompt medical care	85

Suppose these projections are correct and that in a study 50% received treatment A, 30% received treatment B, and 20% received treatment C. What is the probability that a given individual in this study survived?

4.103 Use the table at the top of page 238 on worker satisfaction in the Russell Microprocessor Company.
  a. Find the probability that an unskilled worker is satisfied with the work.
  b. Find the probability that a skilled woman employee is satisfied with the work.
  c. Is satisfaction for women employees independent of their being skilled or unskilled?

	Male		Female		
	Skilled	**Unskilled**	**Skilled**	**Unskilled**	**Total**
Satisfied	350	150	25	100	625
Unsatisfied	150	100	75	50	375
Total	500	250	100	150	1000

4.104   Given the following:

$$P(A_1) = 0.2 \qquad P(A_2) = 0.4 \qquad P(A_3) = 0.3 \qquad P(A_4) = 0.1$$
$$P(B|A_1) = 0.5 \qquad P(B|A_2) = 0.4 \qquad P(B|A_3) = 0.2 \qquad P(B|A_4) = 0.1$$

Find $P(A_1|B)$, $P(A_2|B)$, $P(A_3|B)$, and $P(A_4|B)$.

4.105   In an article titled "Why Quitting Means Gaining" (*Time*, March 25, 1991), it was reported that giving up cigarette smoking often results in gaining weight. In examining a group of quitters, the following data were found.

	Weight Gain			
	Major	**Significant**	**Moderate**	**Slight**
Men	9%	14%	22%	55%
Women*	12	11	26	50

*Due to rounding, numbers for women do not total 100%.

Suppose that 60% of the group were men and 40% were women. If a participant were randomly selected and found to have experienced
   a. a major weight gain, find the probability that it was a male.
   b. a slight weight gain, find the probability that it was a woman.

4.106   Given the information in the accompanying table, compute $P(A_1|UF)$ and $P(A_2|UF)$ by filling in the rest of the table. (UF = unfavorable survey results.)

	$P(A_i)$	$P[(UF)|A_i]$	$P(A_i \text{ and } UF)$	$P(A_i|UF)$
A_1, profitable	0.6	0.4		
A_2, not profitable	0.4	0.7		

IN RETROSPECT

You have now studied the basic concepts of probability. These fundamentals need to be understood to allow us to continue our study of statistics. Probability is the vehicle of statistics, and we have begun to see how probabilistic events occur. We have explored theoretical and experimental probabilities for the same event. Does the experimental probability turn out to have the same value as the theoretical? Not exactly, but we have seen that over the long run it does have approximately the same value.

You must, of course, know and understand the basic definition of probability as well as understand the properties of mutual exclusiveness and independence as they apply to the concepts presented in this chapter.

From reading this chapter you should be able to relate the "and" and the "or" of compound events to the multiplication and addition rules. You should also be able to calculate conditional probabilities.

In the next three chapters we will look at distributions associated with probabilistic events. This will prepare us for the statistics that will follow. We must be able to predict the variability that the sample will show with respect to the population before we will be successful at "inferential statistics," in which we describe the population based on the sample statistics available.

CHAPTER EXERCISES

4.107 A fair coin is tossed 100 million times. Let B be the total number of heads observed. Identify each of the following statements as true or false. Explain your answer (no computations required).

  a. It is very probable that B is very close (within a few thousand) to 50 million.

  b. It is very probable that $\dfrac{B}{100 \text{ million}}$ is very close to $\frac{1}{2}$; maybe between 0.49 and 0.51.

  c. Since H's and T's fall with complete irregularity, one cannot say anything about what happens in 100 million tosses. It is just as likely that B is equal to one value as to any other value; for example, $P(B = 2) = P(B = 50 \text{ million})$.

  d. H's and T's fall about equally often. Therefore, if on the first ten tosses we get all H's, it is more probable that the eleventh toss will yield a T than an H.

4.108 Probabilities for events A, B, and C are distributed as shown on the following figure. Find

  a. $P(A \text{ and } B)$.   b. $P(A \text{ or } C)$.   c. $P(A|C)$.

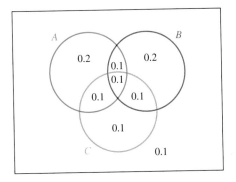

4.109 Suppose a certain ophthalmic trait is associated with eye color. Three hundred randomly selected individuals are studied with results as follows:

| | Eye Color | | | |
Trait	Blue	Brown	Other	Totals
Yes	70	30	20	120
No	20	110	50	180
Totals	90	140	70	300

a. What is the probability that a person selected at random has blue eyes?
b. What is the probability that a person selected at random has the trait?
c. Are events A (has blue eyes) and B (has the trait) independent? Justify your answer.
d. How are the two events A (has blue eyes) and C (has brown eyes) related (independent, mutually exclusive, complementary, or all inclusive)? Explain why or why not each term applies.

4.110 The employees at a large university were classified according to age as well as to whether they belonged to administration, faculty, or staff.

| | Age Group | | | |
	20–30	31–40	41–50	51 or Over
Administration	2	24	16	17
Faculty	1	40	36	28
Staff	16	20	14	2

For a randomly selected employee, find the probability that he or she
a. was in the administration or was 51 or over.
b. was not a faculty member.
c. was a faculty member given that the individual was 41 or over.

4.111 Events R and S are defined on the sample space. If $P(R) = 0.2$ and $P(S) = 0.5$, explain why each of the following statements is either true or false.
a. If R and S are mutually exclusive, then $P(R \text{ or } S) = 0.10$.
b. If R and S are independent, then $P(R \text{ or } S) = 0.6$.
c. If R and S are mutually exclusive, then $P(R \text{ and } S) = 0.7$.
d. If R and S are mutually exclusive, then $P(R \text{ or } S) = 0.6$.

4.112 Show that if event A is a subset of event B, then $P(A \text{ or } B) = P(B)$.

4.113 Let's assume that there are three traffic lights between your house and a friend's house. As you arrive at each light, it may be red (R) or green (G).
a. List the sample space showing all possible sequences of red and green lights that could occur on a trip from your house to your friend's. (RGG represents red at the first light and green at the other two.)
Assuming that each element of the sample space is equally likely to occur,
b. What is the probability that on your next trip to your friend's house you will have to stop for exactly one red light?
c. What is the probability that you will have to stop for at least one red light?

4.114 Assuming that a woman is equally likely to bear a boy as a girl, use a tree diagram to compute the probability that a four-child family consists of one boy and three girls.

4.115 Suppose that when a job candidate comes to interview for a job at RJB Enterprises, the probability that he or she will want the job (A) after the interview is 0.68. Also, the probability that RJB wants the candidate (B) is 0.36. The probability $P(A|B)$ is 0.88.
    a. Find $P(A \text{ and } B)$.   b. Find $P(B|A)$.
    c. Are events A and B independent? Explain.
    d. Are events A and B mutually exclusive? Explain.
    e. What would it mean to say A and B are mutually exclusive events in this exercise?

4.116 A traffic analysis at a busy traffic circle in Washington, D.C., showed that 0.8 of the autos using the circle entered from Connecticut Avenue. Of those entering the traffic circle from Connecticut Avenue, 0.7 continued on Connecticut Avenue at the opposite side of the circle. What is the probability that a randomly selected auto observed in the traffic circle entered from Connecticut and will continue on Connecticut?

4.117 A fair die is rolled five times. What is the probability that a 6 occurred for the first time on the fifth roll?

4.118 In the game of craps, you win on the first roll of a pair of dice if the sum of 7 or 11 occurs. You lose on the first roll if the sum 2, 3, or 12 occurs.
    a. What is the probability that you win on the first roll?
    b. What is the probability that you lose on the first roll?

4.119 According to the National Cancer Data Base report for Hodgkin's disease (*CA—A Cancer Journal for Clinicians*, Jan./Feb. 1991), the highest percentage of patients (31%) were 20 to 29 years of age, and they had a three-year observed survival rate of 91%. What is the probability that an individual who has been diagnosed with Hodgkin's disease is between 20 and 29 years of age *and* will survive for three years?

4.120 The probability that thunderstorms are in the vicinity of a particular midwestern airport on an August day is 0.70. When thunderstorms are in the vicinity, the probability that an airplane lands on time is 0.80. Find the probability that thunderstorms are in the vicinity and the plane lands on time.

4.121 Tires salvaged from a train wreck are on sale at the Getrich Tire Company. Of the 15 tires offered in the sale, 5 tires have suffered internal damage and the remaining 10 are damage free. If you were to randomly select and purchase 2 of these tires,
    a. what is the probability that the tires you purchase are both damage free?
    b. what is the probability that exactly 1 of the tires you purchase is damage free?
    c. what is the probability that at least 1 of the tires you purchase is damage free?

4.122 According to automobile accident statistics, one out of every six accidents results in an insurance claim of $100 or less in property damage. Three cars insured by an insurance company are involved in different accidents. Consider these two events:

    A: The majority of claims exceed $100.
    B: Exactly two claims are $100 or less.

    a. List the sample points for this experiment.
    b. Are the sample points equally likely?
    c. Find $P(A)$ and $P(B)$.
    d. Are A and B independent? Justify your answer.

4.123 One thousand persons screened for a certain disease are given a clinical exam. As a result of the exam, the sample of 1000 persons is distributed according to height and disease status.

Height	Disease Status				
	None	**Mild**	**Moderate**	**Severe**	**Totals**
Tall	122	78	139	61	400
Medium	74	51	90	35	250
Short	104	71	121	54	350
Totals	300	200	350	150	1000

Use this information to estimate the probability of being medium or short in height and of having moderate or severe disease status.

4.124 The following table shows the sentiments of 2500 wage-earning employees at the Spruce Company on a proposal to emphasize fringe benefits rather than wage increases during their impending contract discussions.

Employee	Opinion			
	Favor	**Neutral**	**Opposed**	**Total**
Male	800	200	500	1500
Female	400	100	500	1000
Total	1200	300	1000	2500

a. Calculate the probability that an employee selected at random from this group will be opposed.

b. Calculate the probability that an employee selected at random from this group will be female.

c. Calculate the probability that an employee selected at random from this group will be opposed, given that the person is male.

d. Are the events "opposed" and "female" independent? Explain.

4.125 A shipment of grapefruit arrived containing the following proportions of types: 10% pink seedless, 20% white seedless, 30% pink with seeds, 40% white with seeds. A grapefuit is selected at random from the shipment. Find the probability that

a. it is seedless.   b. it is white.

c. it is pink and seedless.   d. it is pink or seedless.

e. it is pink, given that it is seedless.   f. it is seedless, given that it is pink.

4.126 Salespersons Adams and Jones call on three and four customers, respectively, on a given day. Adams could make 0, 1, 2, or 3 sales, whereas Jones could make 0, 1, 2, 3, or 4 sales. The sample space listing the number of possible sales for each person on a given day is given in the following table:

			Jones		
Adams	**0**	**1**	**2**	**3**	**4**
0	0, 0	1, 0	2, 0	3, 0	4, 0
1	0, 1	1, 1	2, 1	3, 1	4, 1
2	0, 2	1, 2	2, 2	3, 2	4, 2
3	0, 3	1, 3	2, 3	3, 3	4, 3

(3, 1 stands for 3 sales by Jones and 1 sale by Adams.) Assume that each sample point is equally likely. Let's define the events:

$$A = \text{at least one of the salespersons made no sales}$$
$$B = \text{together they made exactly three sales}$$
$$C = \text{each made the same number of sales}$$
$$D = \text{Adams made exactly one sale}$$

Find the following probabilities by *counting* sample points:

a. $P(A)$ b. $P(B)$ c. $P(C)$

d. $P(D)$ e. $P(A \text{ and } B)$ f. $P(B \text{ and } C)$

g. $P(A \text{ or } B)$ h. $P(B \text{ or } C)$ i. $P(A|B)$

j. $P(B|D)$ k. $P(C|B)$ l. $P(B|A)$

m. $P(C|A)$ n. $P(A \text{ or } B \text{ or } C)$

Are the following pairs of events mutually exclusive? Explain.

o. A and B   p. B and C   q. B and D

Are the following pairs of events independent? Explain.

r. A and B   s. B and C   t. B and D

 4.127 A testing organization wishes to rate a particular brand of television. Six TVs are selected at random from stock. If nothing is found wrong with any of the six, the brand is judged satisfactory.

  a. What is the probability that the brand will be rated satisfactory if 10% of the TVs actually are defective?

  b. What is the probability that the brand will be rated satisfactory if 20% of the TVs actually are defective?

  c. What is the probability that the brand will be rated satisfactory if 40% of the TVs actually are defective?

4.128 Coin A is loaded in such a way that $P(\text{heads})$ is 0.6. Coin B is a balanced coin. Both coins are tossed. Find the following:

  a. the sample space that represents this experiment; assign a probability measure to each outcome

  b. $P(\text{both show heads})$

  c. $P(\text{exactly one head shows})$

  d. $P(\text{neither coin shows a head})$

  e. $P(\text{both show heads}|\text{coin A shows a head})$

  f. $P(\text{both show heads}|\text{coin B shows a head})$

  g. $P(\text{heads on coin A}|\text{exactly one head shows})$

4.129   Professor French forgets to set his alarm with a probability of 0.3. If he sets the alarm it rings with a probability of 0.8. If the alarm rings, it will wake him on time to make his first class with a probability of 0.9. If the alarm does not ring, he wakes in time for his first class with a probability of 0.2. What is the probability that Professor French wakes in time to make his first class tomorrow?

4.130   A two-page typed report contains an error on one of the pages. Two proofreaders review the copy. Each has an 80% chance of catching the error. What is the probability that the error will be identified if

   a. each reads a different page?

   b. they each read both pages?

   c. the first proofreader randomly selects a page to read, then the second proofreader randomly selects a page unaware of which page the first selected?

4.131   For a particular population, 30% are in the age group 0–20 years, 40% are in the age group 21–40 years, and the remainder are over 40 years old. Given that an individual from this population is in the 0–20 age group, the probability that the individual has an abnormal result on a glucose tolerance test is 0.05. The conditional probability for an abnormal glucose tolerance test is 0.04 for the 21–40 age group and 0.10 for the over-40 age group. For a randomly selected individual from this population, what is the probability that he or she has an abnormal glucose tolerance test?

4.132   Solve this exercise by using Bayes's rule in tabular form. The treasurer's initial opinion is that there is a 30% chance that an investment will exceed expectations, a 50% chance that it will equal expectations, and a 20% chance that it will return less than expected. A private investment consulting service reviews the investment and reports that it should equal expectations. In the past the consultants were correct 60% of the time, underestimated the return 10% of the time, and overestimated the return 30% of the time. What should be the treasurer's revised probabilities?

4.133   Ninety percent of the insulators produced by Superior Insulator Company are satisfactory. The firm hires an inspector. The inspector inspects all the insulators and correctly classifies an item 90% of the time; that is, $P(\text{classify good}|\text{good}) = P(\text{classify defective}|\text{defective}) = 0.9$. Items classified good are shipped and those classified defective are scrapped.

   a. What percentage of items shipped can be expected to be good?

   b. What percentage of items scrapped can be expected to be good?

4.134   The firm in Exercise 4.133 hires a second inspector, who has the same accuracy record. The second inspector inspects all insulators independently of the first inspector. What percentage of items shipped and what percentage of items scrapped can be expected to be good if items are shipped only if

   a. both inspectors independently say they are good?

   b. at least one inspector says they are good?

4.135   In sports, championships are often decided by two teams playing each other in a championship series. Often the fans of the losing team claim they were unlucky and their team is actually the better team. Suppose team A is the better team, and the probability it will beat team B in any given game is 0.6. What is the probability that the better team A will lose the series if it is

   a. a one-game series?

   b. a best out of three series?

   c. a best out of seven series?

d. Suppose the probability that A would beat B in any given game were actually 0.7. Recompute (a) through (c).

e. Suppose the probability that A would beat B in a given game were actually 0.9. Recompute (a) through (c).

f. What is the relationship between the "best" team winning and the number of games played? The best team winning and the probabilities that each will win?

4.136 The Pennsylvania Lottery game Big 4 has been played for 14 years. The chapter case study on page 189 lists the number of times that each single-digit number has been the winning number for each of the four positions. The frequencies for each number in each position range from 304 to 367.

a. Do the frequencies of each number as a winner in the first position appear to indicate that the numbers occur randomly as first position winners? What statistical evidence can you find to justify your answer? Present a convincing case.

b. Do the frequencies of each number as a winner in the second, third, and fourth positions appear to indicate that the numbers occur randomly as position winners? What evidence can you find to justify your answer?

c. Each single-digit number has appeared as a winning number a different number of times ranging from 1290 to 1383. Do you think these numbers vary sufficiently to make a case that the digits do not occur with equal probability? Present evidence to support your answer.

Don't forget about the descriptive techniques learned in earlier chapters.

VOCABULARY LIST

Be able to define each term. Pay special attention to the key terms, which are printed in red. In addition, describe in your own words, and give an example of, each term. Your examples should not be ones given in class or in the textbook.

The bracketed numbers indicate the chapter in which the first term appeared, but you should define the terms again to show increased understanding of their meaning. Page numbers indicate the first appearance of the term in Chapter 4

addition rule (p. 213)
all-inclusive events (p. 198)
Bayes's rule (p. 233)
complementary event (p. 206)
compound event (p. 210)
conditional probability (p. 219)
dependent events (p. 218)
empirical probability (p. 193)
equally likely events (p. 204)
event (p. 198)
experiment [1] (p. 198)
experimental probability (p. 193)
general multiplication rule (p. 221)
independence (p. 218)
independent events (p. 218)
intersection (p. 212)
law of large numbers (p. 195)
long-term average (p. 195)

multiplication rule (p. 221)
mutually exclusive events (p. 210)
odds (p. 206)
ordered pair (p. 197)
outcome (p. 198)
prior probability (p. 234)
posterior probability (p. 234)
probability of an event (p. 193)
relative frequency [2] (p. 193)
revised probability (p. 234)
sample point (p. 198)
sample space (p. 198)
special multiplication rule (p. 221)
subjective probability (p. 205)
theoretical probability (p. 204)
tree diagram (p. 199)
Venn diagram (p. 202)

QUIZ A

Answer "True" if the statement is always true. If the statement is not always true, replace the words shown in bold with the words that make the statement always true.

4.1 The probability of an event is a **whole number**.

4.2 The concepts of probability and relative frequency as related to an event are very **similar**.

4.3 The **sample space** is the theoretical population for probability problems.

4.4 The sample points of a sample space are **equally likely** events.

4.5 The value found for experimental probability will **always be** exactly equal to the theoretical probability assigned to the same event.

4.6 The probabilities of complementary events always **are equal**.

4.7 If two events are mutually exclusive, they are also **independent**.

4.8 If events A and B are **mutually exclusive**, the sum of their probabilities must be exactly one.

4.9 If the sets of sample points belonging to two different events do not intersect, the events are **independent**.

4.10 A compound event formed by use of the word *and* requires the use of the **addition rule**.

QUIZ B

4.1 A computer is programmed to generate the eight single-digit integers 1, 2, 3, 4, 5, 6, 7, and 8 with equal frequency. Consider the experiment—"the next integer generated." Define:

$$\text{Event A} = \text{"Odd number"} = \{1, 3, 5, 7\}$$
$$\text{Event B} = \text{"Number more than 4"} = \{5, 6, 7, 8\}$$
$$\text{Event C} = \text{"1 or 2"} = \{1, 2\}$$

Find:

a. $P(A)$ b. $P(B)$ c. $P(C)$ d. $P(\overline{C})$
e. $P(A \text{ and } B)$ f. $P(A \text{ or } B)$ g. $P(B \text{ and } C)$ h. $P(B \text{ or } C)$
i. $P(A \text{ and } C)$ j. $P(A \text{ or } C)$ k. $P(A|B)$ l. $P(B|C)$
m. $P(A|C)$

n. Are events A and B mutually exclusive? Explain.
o. Are events B and C mutually exclusive? Explain.
p. Are events A and C mutually exclusive? Explain.
q. Are events A and B independent? Explain.
r. Are events B and C independent? Explain.
s. Are events A and C independent? Explain.

4.2 Given that events A and B are mutually exclusive and $P(A) = 0.4$ and $P(B) = 0.3$, find
a. $P(A \text{ and } B)$ b. $P(A \text{ or } B)$
c. $P(A|B)$ d. Are A and B independent? Explain.

4.3 Given that events C and D are independent and $P(C) = 0.2$ and $P(D) = 0.7$:
a. Find $P(C \text{ and } D)$.   b. Find $P(C \text{ or } D)$.
c. Find $P(C|D)$.   d. Are C and D mutually exclusive? Explain.

4.4 Given events E and F with probabilities $P(E) = 0.5$, $P(F) = 0.4$, and $P(E \text{ and } F) = 0.2$:
   a. Find $P(E \text{ or } F)$.   b. Find $P(E|F)$.
   c. Are E and F mutually exclusive? Explain.
   d. Are E and F independent? Explain.

4.5 Given events G and H with probabilities $P(G) = 0.3$, $P(H) = 0.2$, and $P(G \text{ and } H) = 0.1$:
   a. Find $P(G \text{ or } H)$.   b. Find $P(G|H)$.
   c. Are G and H mutually exclusive? Explain.
   d. Are G and H independent? Explain.

4.6 Janice wants to beocme a police officer. She must pass a physical exam and then a written exam. Records show the probability of passing the physical exam is 0.85 and that once the physical is passed the probability of passing the written exam is 0.60. What is the probability that Janice passes both exams?

QUIZ C

4.1 Explain briefly how you would decide which of the following two events is the more unusual:

   A: a 90-degree day in Vermont, or
   B: a 100-degree day in Florida

4.2 Student A says that "independence" and "mutually exclusive" are basically the same thing; namely, both mean "neither event has anything to do with the other one." Student B argues that although Student A's statement has some truth in it, Student A has missed the point of these two properties. Student B is correct. Carefully explain why.

4.3 Using complete sentences, describe in your own words
   a. mutually exclusive events.   b. independent events.
   c. the probability of an event.   d. a conditional probability.

PROBABILITY DISTRIBUTIONS (DISCRETE VARIABLES)

Chapter Case Study

TELEVISION'S HARD SELL

D*id you ever see one of those infomercials on television and wonder how often people bought the product or service being presented? The USA Snapshot® that appeared in* USA Today *on October 21, 1994, "Television's hard sell," describes "How many times buyers see an infomercial before they purchase its product or service."*

USA SNAPSHOTS®

A look at statistics that shape the nation

Infomercials 1D

Television's hard sell
How many times buyers see an infomercial before they purchase its product or service:

Once	27%
Twice	31%
Three times	18%
Four times	9%
Five or more	15%

Source: National Infomercial Marketing Association

By Cindy Hall and Web Bryant, USA TODAY

a. What is the probability that a buyer watched only once before buying?

b. What is the probability that a viewer watching for the first time will buy?

c. What percentage of the buyers watched the infomercial three or more times before purchasing?

d. Is this a binomial probability experiment?

d. Let x be the number of times a buyer watched before making a purchase. Is this a probability distribution?

f. Find the mean and the standard deviation of x.

(See Exercise 5.97, p. 282.)

● CHAPTER OBJECTIVES

Chapter 2 dealt with frequency distributions of data sets, and Chapter 4 dealt with the fundamentals of probability. Now we are ready to combine these ideas to form probability distributions, which are much like relative frequency distributions. The basic difference between probability and relative frequency distributions is that probability distributions are theoretical probabilities (populations) whereas relative frequency distributions are empirical probabilities (samples).

In this chapter we investigate discrete probability distributions and study measures of central tendency and dispersion for such distributions. Special emphasis is given to the binomial random variable and its probability distribution, since it is the most important discrete random variable encountered in most fields of application.

5.1 | RANDOM VARIABLES

If each outcome of a probability experiment is assigned a numerical value, then as we observe the results of the experiment we are observing a random variable. This numerical value is the *random variable*.

EXERCISE 5.1

You are to survey the students at your college with regard to the number of courses each is enrolled in for this semester. Identify the random variable of interest, and list its possible values.

RANDOM VARIABLE

A variable that assumes a unique numerical value for each of the outcomes in the sample space of a probability experiment.

In other words, a random variable is used to denote the outcomes of a probability experiment. It can take on any numerical value that belongs to the set of all possible outcomes of the experiment. (It is called "random" because the value it assumes is the result of a chance, or random, event.) Each event in a probability experiment must also be defined in such a way that only one value of the random variable is assigned to it, and every event must have a value assigned to it.

The following illustrations demonstrate several random variables.

▼ | ILLUSTRATION 5-1

We toss five coins and observe the "number of heads" visible. The random variable x is the number of heads observed and may take on integer values from 0 to 5.

▼ | ILLUSTRATION   5 - 2

Let the "number of phone calls received" per day by a company be a random variable. Integer values ranging from zero to some very large number are possible values.

▼ | ILLUSTRATION   5 - 3

Let the "length of the cord" on an electric appliance be a random variable. The random variable will be a numerical value between 12 and 72 inches for most appliances.

▼ | ILLUSTRATION   5 - 4

Let the "qualifying speed" for race cars trying to qualify for the Indianapolis 500 be a random variable. Depending on how fast the driver can go, the speeds will be approximately 220 and faster and be measured in miles per hour (to the nearest
▲ | thousandth of a mile).

EXERCISE   5.2

You are to survey the students at your college with regard to the weight of books and supplies they are carrying as they attend class today. Identify the random variable of interest, and list its possible values.

Numerical random variables can be subdivided into two classifications: *discrete* random variables and *continuous* random variables.

DISCRETE RANDOM VARIABLE

A quantitative random variable that can assume a countable number of values. Intuitively, the discrete random variable can assume the values corresponding to isolated points along a line interval. That is, there is a gap between any two values.

CONTINUOUS RANDOM VARIABLE

A quantitative random variable that can assume an uncountable number of values. Intuitively, the continuous random variable can assume any value along a line interval, including every possible value between any two values.

EXERCISE   5.3

The variables in Exercises 5.1 and 5.2 are either discrete or continuous. Which are they and why?

In many cases, the two variables can be distinguished by deciding whether the variables are related to a count or a measurement. The variables in Illustrations 5-1 and 5-2 are both discrete; the values of the variables will be found by counting the number of heads seen or by counting the number of telephone calls that occur. (When counting, fractional values cannot occur; thus, there are gaps between the values that can occur.) The variables in Illustrations 5-3 and 5-4 are continuous random variables; the values of the variables will be found by measuring the length of the cord and the speed. (When measuring, any fractional value can occur; thus, every value along the number line is possible.)

When trying to determine whether a variable is discrete or continuous, remember to look at the variable and think about the values that might occur. Do not look at values that have occurred; they can be very misleading, because you will be tempted to think about the wrong aspect.

Remember that a discrete random variable "has a gap between values." For example, in Illustrations 5-1 and 5-2, no numerical value between 3 and 4 can occur. (3.7 heads is not possible. 3.52 telephone calls is not possible. Only integer values are possible; thus, there are gaps between possible values.)

Remember that a continuous random variable is related to a continuous number line and all the numbers represented by it. When all real numbers, rational or irrational, are possible, then the variable is continuous. Think about the possible lengths of appliance cord that could occur in Illustration 5-3; the length could be any number of inches and fractional part thereof, as long as the ruler we are measuring with has increments accurate enough to measure the exact length. The number of possible values is unlimited.

Consider the variable "judge's score" at a figure-skating competition. If we look at some scores that have previously occurred, 9.9, 9.5, 8.8, 10.0, and we see the presence of decimals, we might think that all fractions are possible and conclude the variable is continuous. This is not true, however. A score of 9.134 is not possible; thus, there are gaps between the possible values and the variable is discrete.

EXERCISE 5.4

a. Explain why the variable "score" for the home team at a basketball game is discrete.

b. Explain why the variable "number of minutes to commute to work" for the workers at a local industry is continuous.

EXERCISES

5.5 A social worker is involved in a study about family structure. She obtains information regarding the number of children per family for a certain community from the census data. Identify the random variable of interest, determine whether it is discrete or continuous, and list its possible values.

5.6 An experiment involves the testing of a new on/off switch. The switch is flipped on and off until it breaks, and the flip on which the break occurs is noted. Identify the random variable of interest, determine whether it is discrete or continuous, and list its possible values.

5.7 An archer shoots arrows at a bull's-eye of a target and measures the distance from the center of the target to the arrow. Identify the random variable of interest, determine whether it is discrete or continuous, and list its possible values.

5.8 "How women define holiday shopping," a USA Snapshot® (12-9-94) reported that 50% said "a pleasure," 22% said "a chore," 19% said "no big deal," and 8% said "a nightmare." The percentages do not sum to 100% due to round-off error.

a. What is the variable involved, and what are the possible values?

b. Why is this variable not a random variable?

5.2 | PROBABILITY DISTRIBUTIONS OF A DISCRETE RANDOM VARIABLE

Recall the coin-tossing experiment we used at the beginning of Section 4-1. Two coins were tossed and no heads, one head, or two heads were observed. If we define the random variable x to be the number of heads observed when two coins are tossed, x can take on the values 0, 1, or 2. The probability of each of these three events is the same as we calculated in Chapter 4 (p. 204):

$$P(x = 0) = P(0H) = \tfrac{1}{4}$$
$$P(x = 1) = P(1H) = \tfrac{1}{2}$$
$$P(x = 2) = P(2H) = \tfrac{1}{4}$$

These probabilities can be listed in any number of ways. One of the most convenient is a table format known as a *probability distribution* (see Table 5-1). Can you see why the name "probability distribution" is used?

TABLE 5-1

Probability Distribution: Tossing Two Coins	x	$P(x)$
	0	0.25
	1	0.50
	2	0.25

PROBABILITY DISTRIBUTION

A distribution of the probabilities associated with each of the values of a random variable. The probability distribution is a theoretical distribution; it is used to represent populations.

The values of the random variable are mutually exclusive events.

In an experiment in which a single die is rolled and the number of dots on the top surface is observed, the random variable is the number observed. The probability distribution for this random variable is shown in Table 5-2.

TABLE 5-2

Probability Distribution: Rolling a Die	x	$P(x)$
	1	$\frac{1}{6}$
	2	$\frac{1}{6}$
	3	$\frac{1}{6}$
	4	$\frac{1}{6}$
	5	$\frac{1}{6}$
	6	$\frac{1}{6}$

EXERCISE 5.9

Express the tossing of one coin as a probability distribution of x, the number of heads occurring (that is, H = 1 and T = 0).

Sometimes it is convenient to write a rule that algebraically expresses the probability of an event in terms of the value of the random variable. This expression is typically written in formula form and is called a *probability function*.

PROBABILITY FUNCTION

A rule that assigns probabilities to the values of the random variables.

A probability function can be as simple as a list pairing the values of a random variable with their probabilities. Tables 5-1 and 5-2 show two such listings. However, a probability function is most often expressed in formula form.

Consider a die that has been modified so that it has one face with one dot, two faces with two dots, and three faces with three dots. Let x be the number of dots observed when this die is rolled. The probability distribution for this experiment is presented in Table 5-3.

TABLE 5-3

Probability Distribution: Rolling a Modified Die

x	P(x)
1	$\frac{1}{6}$
2	$\frac{2}{6}$
3	$\frac{3}{6}$

Each of the probabilities can be represented by the value of x divided by 6. That is, each P(x) is equal to the value of x divided by 6, where x = 1, 2, or 3. Thus,

$$P(x) = \frac{x}{6} \quad \text{for} \quad x = 1, 2, \text{ or } 3$$

is the formula expression for the probability function of this experiment.

The probability function for the experiment of rolling one ordinary die is

$$P(x) = \tfrac{1}{6} \quad \text{for} \quad x = 1, 2, 3, 4, 5, \text{ or } 6$$

Constant function

EXERCISE 5.10

Express $P(x) = \frac{1}{6}$, for x = 1, 2, 3, 4, 5, or 6, in distribution form.

This particular function is called a **constant function** because the value of P(x) does not change as x changes.

Every probability function must display the two basic properties of probability. These two properties are (1) the probability assigned to each value of the random variable must be between 0 and 1, inclusive, that is,

Property 1: $0 \le \text{each } P(x) \le 1$

and (2) the sum of the probabilities assigned to all the values of the random variable must equal 1, that is,

Property 2: $\displaystyle\sum_{\text{all } x} P(x) = 1$

▼ | ILLUSTRATION   5 - 5

Is $P(x) = \dfrac{x}{10}$, for $x = 1, 2, 3,$ or 4, a probability function?

The values of the random variable are all inclusive.

SOLUTION   To answer this question we need only test the function in terms of the two basic properties. The probability distribution is shown in Table 5-4.

TABLE 5-4

Probability Distribution for
$P(x) = \dfrac{x}{10}$, for
$x = 1, 2, 3,$ or 4

x	$P(x)$
1	$\frac{1}{10} = 0.1$ ✓
2	$\frac{2}{10} = 0.2$ ✓
3	$\frac{3}{10} = 0.3$ ✓
4	$\frac{4}{10} = 0.4$ ✓
	$\frac{10}{10} = 1.0$ (ck)

Property 1 is satisfied, since 0.1, 0.2, 0.3, and 0.4 are all numerical values between 0 and 1. (See the ✓ showing each value was checked.) Property 2 is also satisfied, since the sum of all four probabilities is exactly 1. (See the (ck) showing the sum was checked.) Since both properties are satisfied, we can conclude that $P(x) = \dfrac{x}{10}$, for $x = 1, 2, 3,$ or 4 is a probability function.
   What about $P(x = 5)$ (or any value other than $x = 1, 2, 3,$ or 4) for the function $P(x) = \dfrac{x}{10}$, for $x = 1, 2, 3,$ or 4? $P(x = 5)$ is considered to be zero. That is, the probability function provides a probability of zero for all values of x other than the values specified.

▲ |

Probability distributions can be presented graphically. Regardless of the specific graphic representation used, the values of the random variable are plotted on the horizontal scale, and the probability associated with each value of the random

FIGURE 5-1

Line Representation: Probability Distribution for
$P(x) = \dfrac{x}{10}$, for
$x = 1, 2, 3,$ or 4
(see next page)

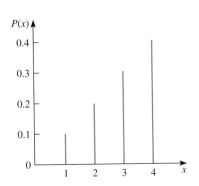

variable is plotted on the vertical scale. The probability distribution of a discrete random variable could be presented by a set of line segments drawn at the values of x and whose lengths represent the probability of each x. Figure 5-1 (p. 255) shows the probability distribution of $P(x) = \dfrac{x}{10}$, for $x = 1, 2, 3,$ or 4.

Probability histogram

However, a regular **histogram** is more frequently used to present probability distributions. Figure 5-2 shows the probability distribution of Figure 5-1 in histogram form.

FIGURE 5-2

Histogram: Probability Distribution for $P(x) = \dfrac{x}{10}$, for $x = 1, 2, 3,$ or 4

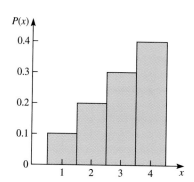

The histogram of a probability distribution uses the physical area of each bar to represent its assigned probability. The bar for $x = 2$ is 1 unit wide (from 1.5 to 2.5) and is 0.2 unit high. Therefore, its area (length × width) is (1)(0.2) = 0.2, the probability assigned to $x = 2$. The areas of the other bars can be determined in similar fashion. This area representation will be an important concept in Chapter 6 when we begin to work with continuous random variables.

EXERCISE 5.11

a. Construct a histogram of the probability distribution $P(x) = \frac{1}{6}$, for $x = 1, 2, 3, 4, 5,$ or 6.

b. Describe the shape of the histogram in (a).

MINITAB (Release 10) commands to generate random data in C3 from a discrete probability distribution with the possible values of the random variable in C1 and their corresponding probabilities listed in C2.

Session commands   *Menu commands*

Enter: RANDom 25 obs into C3;   Choose: Calc>Rand Data>Discr
       DISCrete C1 C2.   Enter: Generate: 25 (number
                                           wanted)
                                           Store in: C3
                                           Values (of x) in: C1
                                           Probabilities in: C2

Case Study   Who Needs the Ambulance?

5-1

Robert Giordana used a relative frequency histogram to help him explain how his ambulance services are used. The histogram is of a discrete variable (number of trips per day) and shows, in percentages, the relative frequency of days with various numbers of service trips.

The Austin City Ambulance Company last week appealed to the City Council for additional municipal funding. Mr. Robert Giordana, the company's business manager, stated that while people see ambulances on the streets occasionally, they rarely have any real concept of the frequency with which an ambulance is called upon for assistance.

EXERCISE 5.12

a. Express the histogram in Case Study 5-1 as a probability distribution.

b. Explain how the distribution implies the conclusion "that there was only one day every three weeks when no trips were made."

In surveying the company's records, the Council found that one ambulance responds to between one and six calls for help on a typical day. The records for a recent six-month period showed that an ambulance made three trips on 25% of the days and four trips on 21% of the days. It was further revealed that there was only one day in every three weeks when no trips were made. But on one day in the same three weeks, they made seven or more trips.

Mr. Giordana reminded the Council that the company was working hard for the community all year long.

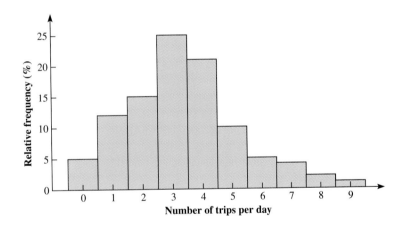

EXERCISES

5.13   Census data are often used to obtain probability distributions for various random variables. Census data for families with a combined income of $50,000 or more in a particular state show that 20% have no children, 30% have one child, 40% have two children, and 10% have three children. From this information, construct the probability distribution for x, where x represents the number of children per family for this income group.

5.14   Test the following function to determine whether it is a probability function. If it is not, try to make it into a probability function. $R(x) = 0.2$ for $x = 0, 1, 2, 3,$ or 4.
   a. List the distribution of probabilities.
   b. Sketch a histogram.

5.15   Test the following function to determine if it is a probability function.

$$P(x) = \frac{x^2 + 5}{50}, \text{ for } x = 1, 2, 3, \text{ or } 4$$

    a. List the probability distribution.
    b. Sketch a histogram.

5.16   Test the following function to determine whether it is a probability function. If it is not, try to make it into a probability function.

$$S(x) = \frac{6 - |x - 7|}{36} \text{ for } x = 2, 3, 4, 5, 6, 7, \ldots, 11, \text{ or } 12$$

    a. List the distribution of probabilities and sketch a histogram.
    b. Do you recognize $S(x)$? If so, identify it.

 5.17   "Kids who smoke," a USA Snapshot®, (4-25-94), reports the percentage of children in each age group who smoke.

Age, x	Percent Who Smoke
12	1.7
13	4.9
14	8.9
15	16.3
16	25.2
17	37.2

Is this a probability distribution? Explain why or why not.

 5.18   "Who isn't covered," a USA Snapshot®, (6-10-94), reported the percentage of people in each age group not covered by health insurance as found by the U.S. Census Bureau (1992 figures).

Age	Percent Not Covered
Under 18	12.4
18–24	28.9
25–34	20.9
35–44	15.5
45–54	14.0
55–64	12.9
over 65	1.2

If age is the variable x, explain why or why not this information is a probability distribution.

5.19   a. Use a computer (or random numbers table) and generate a random sample of 25 observations drawn from the discrete probability distribution

x	1	2	3	4	5
$P(x)$	0.2	0.3	0.3	0.1	0.1

MINITAB commands to use

```
a. RANDom 25 C3;
     DISCrete C1 C2.
b. TABLe C3;
     TOTPercent.
c. HISTogram C3;
     PERCent;
     MIDPoint 1:5.
```

Compare the resulting data to your expectations.

   b. Form a relative frequency distribution of the random data.

   c. Construct a probability histogram of the given distribution and a relative frequency histogram of the observed data using class marks of 1, 2, 3, 4, and 5.

   d. Compare the observed data with the theoretical distribution. Describe your conclusions.

   e. Repeat parts (a) through (d) several times with $n = 25$. Describe the variability you observe between samples.

   f. Repeat parts (a) through (d) several times with $n = 250$. Describe the variability you see between samples of this much larger size.

5.20   a. Use a computer (or random numbers table) and generate a random sample of 100 observations drawn from the discrete probability population $P(x) = (5 - x)/10$, for $x = 1, 2, 3,$ or 4. List the resulting sample.

Use MINITAB commands in Exercise 5.19; change the arguments.

   b. Form a relative frequency distribution of the random data.

   c. Form a probability distribution of the expected probability distribution. Compare the resulting data to your expectations.

   d. Construct a probability histogram of the given distribution and a relative frequency histogram of the observed data using class marks of 1, 2, 3, and 4.

   e. Compare the observed data with the theoretical distribution. Describe your conclusions.

   f. Repeat parts (a) through (d) several times with $n = 100$. Describe the variability you observe between samples.

5.3 | MEAN AND VARIANCE OF A DISCRETE PROBABILITY DISTRIBUTION

Recall that in Chapter 2 we calculated several numerical sample statistics (mean, variance, standard deviation, and others) to describe empirical sets of data. Probability distributions represent theoretical populations, the counterpart to samples. We use the population parameters—mean, variance, and standard deviation—to describe these probability distributions the same as we use sample statistics to describe samples.

NOTES

Sample statistic

1. $\bar{x}$ is the mean of the sample.
2. s^2 and s are the variance and standard deviation of the sample.
3. $\bar{x}, s^2,$ and s are called **sample statistics**.
4. μ (lowercase Greek letter "mu") is the mean of the population.
5. σ^2 ("sigma squared") is the variance of the population.

(continued)

Population parameter

6. σ (lowercase Greek letter "sigma") is the standard deviation of the population.
7. μ, σ^2, and σ as called **population parameters**. (A parameter is a constant. μ, σ^2, and σ are typically unknown values.)

The *mean of the probability distribution of a discrete random variable*, or the *mean of a discrete random variable*, is found in a manner somewhat similar to that used to find the mean of a frequency distribution.

MEAN OF A DISCRETE RANDOM VARIABLE

The mean, μ, of a discrete random variable x is found by multiplying each possible value of x by its own probability and then adding all the products together.

mean of x: mu = sum of (x multiplied by its own probability)

$$\text{mean of } x: \quad \mu = \Sigma[xP(x)] \tag{5-1}$$

The *variance* of a discrete random variable is defined in much the same way as the variance of sample data, the "mean of the squared deviations from the mean."

VARIANCE OF A DISCRETE RANDOM VARIABLE

Variance, σ^2, of a discrete random variable x is found by multiplying each possible value of the squared deviation from the mean, $(x - \mu)^2$, by its own probability and then adding all the products together.

Variance: sigma squared = sum of (squared deviation times probability)

D

$$\text{variance:} \quad \sigma^2 = \Sigma[(x - \mu)^2 P(x)] \tag{5-2}$$

Formula (5-2) is often not convenient to use; it can be reworked into the following form(s):

Variance: sigma squared = sum of (x² times probability) − [sum of (x times probability)]²

$$\sigma^2 = \Sigma[x^2 P(x)] - \{\Sigma[xP(x)]\}^2 \tag{5-3a}$$

EXERCISE 5.21

or

$$\sigma^2 = \Sigma[x^2 P(x)] - \mu^2 \tag{5-3b}$$

Verify that formulas (5-3a) and (5-3b) are equivalent to formula (5-2).

STANDARD DEVIATION OF A DISCRETE RANDOM VARIABLE

The positive square root of variance.

$$\text{standard deviation:} \quad \sigma = \sqrt{\sigma^2} \tag{5-4}$$

▼ | ILLUSTRATION   5 - 6

Find the mean, variance, and standard deviation of the probability function

$$P(x) = \frac{x}{10}, \text{ for } x = 1, 2, 3, \text{ or } 4$$

SOLUTION We will find the mean using formula (5-1), the variance using formula (5-3a), and the standard deviation using formula (5-4). The most convenient way to organize the products and find the totals needed is to expand the probability distribution into an extensions table (see Table 5-5).

TABLE 5-5

Extensions Table:
Probability
Distribution,

$P(x) = \dfrac{x}{10}$, for

$x = 1, 2, 3, \text{ or } 4$

x	$P(x)$	$xP(x)$	x^2	$x^2P(x)$
1	$\frac{1}{10}$	$\frac{1}{10}$	1	$\frac{1}{10}$
2	$\frac{2}{10}$	$\frac{4}{10}$	4	$\frac{8}{10}$
3	$\frac{3}{10}$	$\frac{9}{10}$	9	$\frac{27}{10}$
4	$\frac{4}{10}$	$\frac{16}{10}$	16	$\frac{64}{10}$
Totals	$\Sigma P(x) = \frac{10}{10} = 1.0$ (ck)	$\Sigma[xP(x)] = \frac{30}{10}$		$\Sigma[x^2P(x)] = \frac{100}{10}$

EXERCISE 5.22

a. Form the probability distribution table for $P(x) = \dfrac{x}{6}$, for $x = 1, 2, 3$.
b. Find the extensions $xP(x)$ and $x^2P(x)$ for each x.
c. Find $\Sigma[xP(x)]$ and $\Sigma[x^2P(x)]$.

EXERCISE 5.23

Find the mean for $P(x) = \dfrac{x}{6}$, for $x = 1, 2, 3$ (use the results of Exercise 5.22).

EXERCISE 5.24

Find the variance for $P(x) = \dfrac{x}{6}$, for $x = 1, 2, \text{ or } 3$ (use the results of Exercise 5.22).

Find the mean of x: The $xP(x)$ column contains the value of each x multiplied by its corresponding probability, and the sum at the bottom is the value needed by formula (5-1).

$$\mu = \Sigma[xP(x)] = \frac{30}{10} = \mathbf{3.0}$$

Find the variance of x: The totals at the bottom of the $xP(x)$ and $x^2P(x)$ columns are substituted into formula (5-3a).

$$\sigma^2 = \Sigma[x^2P(x)] - \{\Sigma[xP(x)]\}^2$$

$$\sigma^2 = \frac{100}{10} - \left(\frac{30}{10}\right)^2 = 10 - (3)^2 = 10 - 9 = \mathbf{1.0}$$

Find the standard deviation of x: Use formula (5-4).

$$\sigma = \sqrt{\sigma^2} = \sqrt{1.0} = \mathbf{1.0}$$ ▲ |

NOTES
1. The purpose of the extensions table is to organize the process of finding the three column totals: $\Sigma[P(x)]$, $\Sigma[xP(x)]$, and $\Sigma[x^2P(x)]$.
2. The other columns, x and x^2, *should not* be totaled.
3. $\Sigma[P(x)]$ will always be 1.0; use this only as a check.
4. $\Sigma[xP(x)]$ and $\Sigma[x^2P(x)]$ are used to find the mean and variance of x.

▼| ILLUSTRATION   5-7

A coin is tossed three times. Let the "number of heads" occurring in those three tosses be the random variable x. Find the mean, variance, and standard deviation of x.

SOLUTION There are eight possible outcomes to this experiment: One results in $x = 0$, three in $x = 1$, three in $x = 2$, or one in $x = 3$. Therefore, the probabilities for this random variable are $\frac{1}{8}$, $\frac{3}{8}$, $\frac{3}{8}$, or $\frac{1}{8}$. (See Experiment 4-6, page 200.) The probability distribution associated with this experiment is shown in Figure 5-3 and in Table 5-6. The necessary extensions and summations for the calculation of its mean, variance, and standard deviation are also shown in Table 5-6.

FIGURE 5-3

Probability
Distribution:
Three Coins

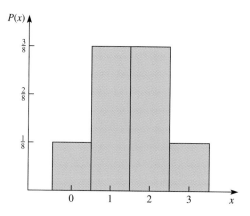

TABLE 5-6

Extensions Table
of Probability
Distribution of
Three Coins

x	$P(x)$	$xP(x)$	x^2	$x^2P(x)$
0	$\frac{1}{8}$	$\frac{0}{8}$	0	$\frac{0}{8}$
1	$\frac{3}{8}$	$\frac{3}{8}$	1	$\frac{3}{8}$
2	$\frac{3}{8}$	$\frac{6}{8}$	4	$\frac{12}{8}$
3	$\frac{1}{8}$	$\frac{3}{8}$	9	$\frac{9}{8}$

Totals $\Sigma[P(x)] = \frac{8}{8} = 1.0$ ⓒⓚ $\Sigma[xP(x)] = \frac{12}{8} = \mathbf{1.5}$ $\Sigma[x^2P(x)] = \frac{24}{8} = 3.0$

The mean is found using formula (5-1):

$$\mu = \Sigma[xP(x)] = \mathbf{1.5}$$

This result, 1.5, is the mean number of heads expected per experiment of three coins.

The variance is found using formula (5-3a):

$$\sigma^2 = \Sigma[x^2 P(x)] - \{\Sigma[xP(x)]\}^2$$
$$\sigma^2 = 3.0 - (1.5)^2 = 3.0 - 2.25 = \mathbf{0.75}$$

The standard deviation is found using formula (5-4):

$$\sigma = \sqrt{\sigma^2} = \sqrt{0.75} = 0.866 = \mathbf{0.87}$$

That is, 0.87 is the standard deviation expected among the number of heads observed per experiment of three coins.

EXERCISES

5.27   Given the probability function

$$P(x) = \frac{5 - x}{10} \quad \text{for} \quad x = 1, 2, 3, \text{ or } 4$$

find the mean and standard deviation.

5.28   Given the probability function

$$R(x) = 0.2 \quad \text{for} \quad x = 0, 1, 2, 3, \text{ or } 4$$

find the mean and standard deviation.

5.29   a. Draw a histogram of the probability distribution for the single-digit random numbers (0, 1, 2, . . . , 9).
   b. Calculate the mean and standard deviation associated with the population of single-digit random numbers.
   c. Represent (1) the location of the mean on the histogram with a vertical line and (2) the magnitude of the standard deviation with a line segment.
   d. How much of this probability distribution is within two standard deviations of the mean?

5.30   The number of calls x to arrive at a switchboard during any 1-minute period is a random variable and has the following probability distribution.

x	0	1	2	3	4
$P(x)$	0.1	0.2	0.4	0.2	0.1

   a. Find the mean and standard deviation of x.
   b. Draw the histogram of $P(x)$, and show the location of μ with a vertical line and the value of σ with a horizontal line segment drawn on the histogram.

5.31   The random variable A has the following probability distribution.

A	1	2	3	4	5
$P(A)$	0.6	0.1	0.1	0.1	0.1

  a. Find the mean standard deviation of A.
  b. How much of the probability distribution is within two standard deviations of the mean?
  c. What is the probability that A is between $\mu - 2\sigma$ and $\mu + 2\sigma$?

5.32   The random variable $\bar{x}$ has the following probability distribution.

$\bar{x}$	1	2	3	4	5
$P(\bar{x})$	0.6	0.1	0.1	0.1	0.1

  a. Find the mean and standard deviation of $\bar{x}$.
  b. What is the probability that $\bar{x}$ is between $\mu - \sigma$ and $\mu + \sigma$?

5.33   A USA Snapshot® (11-1-94) titled "How many telephones we have" reported that 1% have none, 11% have one, 31% have two, and 57% have three or more. Let x equal the number of phones per home, and replace the category "three or more" with exactly "three."
  a. Find the mean and standard deviation for the random variable x.
  b. Explain the effect of replacing the category "three or more" with "three" had on the distribution of x, the mean, and the standard deviation.

5.34   a. Use the probability distribution shown below and describe in your own words how the mean of the variable x is found.

x	1	2	3	4
$P(x)$	0.1	0.2	0.3	0.4

  b. Find the deviation from the mean for each x-value.
  c. Find the value of each "squared deviation from the mean."
  d. Recalling your answer to (a), find the mean of the variable "squared deviation."
  e. Your answer in (d) should be the same as the answer found in Illustration 5-6, page 261. "Variance" was the name given to the "mean of the squared deviations." Explain how formula (5-2) expresses the variance as a mean.

5.4 | THE BINOMIAL PROBABILITY DISTRIBUTION

Consider the following probability experiment. Your instructor gives the class a surprise four-question multiple-choice quiz. You have not studied the material being quizzed and therefore decide to answer the four questions by randomly guessing the answers without reading the questions or the answers.

That's right, guess!

ANSWER PAGE TO QUIZ

Directions: Circle the best answer to each question.

1.	A	B	C
2.	A	B	C
3.	A	B	C
4.	A	B	C

Circle your answers before continuing.

Before we look at the correct answers to the quiz and find out how you did, let's think about some of the things that might happen if a quiz is answered this way.

1. How many of the four questions are you likely to have answered correctly?
2. How likely are you to have more than half of the answers correct?
3. What is the probability that you selected the correct answers to all four questions?
4. What is the probability that you selected wrong answers for all four questions?
5. If an entire class answers the quiz by guessing, what do you think the class "average" number of correct answers will be?

To find the answers to all of these questions, let's start with a tree diagram picturing the sample space showing all 16 possible ways to answer the four-question quiz. Each of the four questions was answered with the correct answer (C) or was answered with a wrong answer (W). See Figure 5-4 on page 266.

Let's convert the information on the tree diagram into a probability distribution. Let x be the "number of correct answers" on one person's quiz when the quiz was taken by randomly guessing. The random variable x may take on any one of the values 0, 1, 2, 3, 4 for each quiz. Figure 5-4 shows 16 branches representing five different values of x. Notice that the event $x = 4$, "four correct answers," is represented by the top branch of the tree diagram, and that event $x = 0$, "zero correct answers," is shown on the bottom branch. The other events, "one correct answer," "two correct answers," and "three correct," are each represented by several branches of the tree. We find that the event $x = 1$ occurs on four different branches, event $x = 2$ occurs on six branches, and $x = 3$ occurs on four branches.

Since each individual question has only one correct answer among the three possible answers, the probability of selecting the correct answer to an individual question is $\frac{1}{3}$. The probability that a wrong answer is selected on each question is $\frac{2}{3}$. The probabilities of each value of x can be found by calculating the

FIGURE 5-4

Tree Diagram:
Possible Answers to
a Four-Question
Quiz

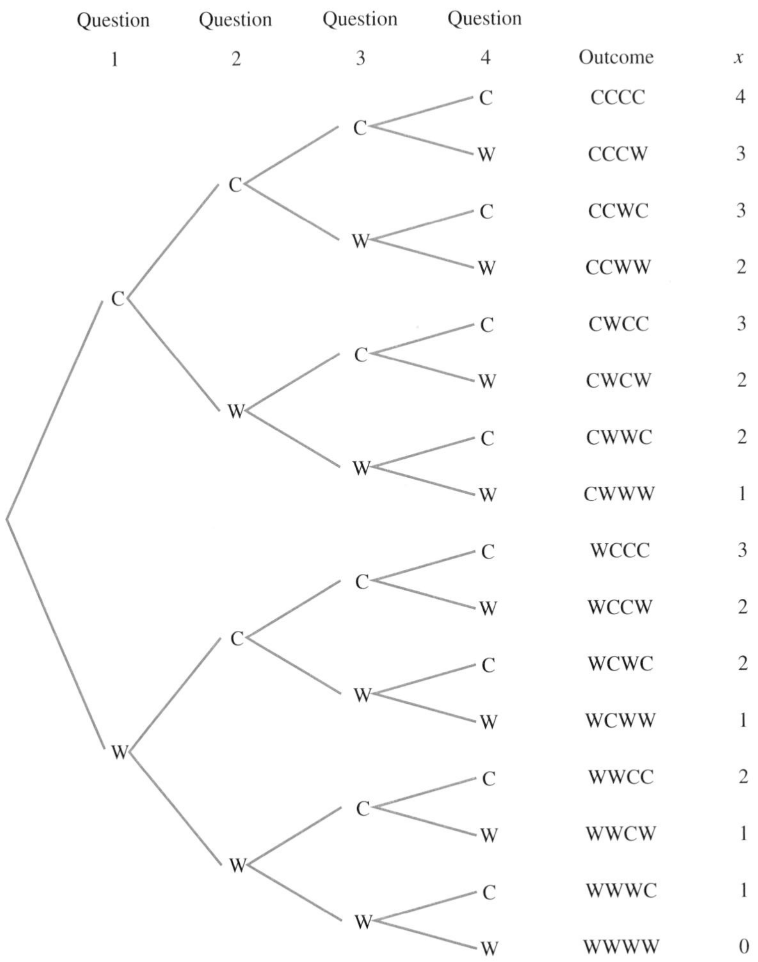

Question 1	Question 2	Question 3	Question 4	Outcome	x
			C	CCCC	4
		C	W	CCCW	3
			C	CCWC	3
		W	W	CCWW	2
C			C	CWCC	3
		C	W	CWCW	2
	W		C	CWWC	2
		W	W	CWWW	1
			C	WCCC	3
		C	W	WCCW	2
	C		C	WCWC	2
		W	W	WCWW	1
W			C	WWCC	2
		C	W	WWCW	1
	W		C	WWWC	1
		W	W	WWWW	0

WWWW represents Wrong$_1$ and Wrong$_2$ and Wrong$_3$ and Wrong$_4$; therefore, its probability is found using the multiplication rule, formula (4-8b).

probabilities of all branches and then combining the probabilities for branches of like
x-values. The calculations follow, and the resulting probability distribution appears
in Table 5-7.

$P(x = 0)$ is the probability that zero questions are answered correctly and four
are answered wrong (one branch, WWWW):

EXERCISE 5.35

Explain why the four
questions represent
independent trials.

$$P(x = 0) = \frac{2}{3} \times \frac{2}{3} \times \frac{2}{3} \times \frac{2}{3} = \left(\frac{2}{3}\right)^4 = \frac{16}{81} \approx \mathbf{0.198}$$

Note   The answering of each individual question is a separate and independent
event, thereby allowing us to use formula (4-8b) and multiply the probabilities.

$P(x = 1)$ is the probability that exactly one question is answered correctly and
the other three are answered wrong (there are four branches on Figure 5-4 where this
occurs—CWWW, WCWW, WWCW, WWWC—and each has the same probability):

EXERCISE 5.36

Explain why the num-
ber 4 is multiplied
into the $P(x = 1)$.

$$P(x = 1) = (4) \times \frac{1}{3} \times \frac{2}{3} \times \frac{2}{3} \times \frac{2}{3} = (4)\left(\frac{1}{3}\right)^1\left(\frac{2}{3}\right)^3 = \frac{32}{81} \approx \mathbf{0.395}$$

$P(x = 2)$ is the probability that exactly two questions are answered correctly and the other two are answered wrong (there are six branches on Figure 5-4 where this occurs—CCWW, CWCW, CWWC, WCCW, WCWC, WWCC—and each has the same probability):

$$P(x = 2) = (6) \times \frac{1}{3} \times \frac{1}{3} \times \frac{2}{3} \times \frac{2}{3} = (6)\left(\frac{1}{3}\right)^2\left(\frac{2}{3}\right)^2 = \frac{24}{81} \approx \mathbf{0.296}$$

$P(x = 3)$ is the probability that exactly three questions are answered correctly and the other one is answered wrong (there are four branches on Figure 5-4 where this occurs—CCCW, CCWC, CWCC, WCCC—and each has the same probability):

$$P(x = 3) = (4) \times \frac{1}{3} \times \frac{1}{3} \times \frac{1}{3} \times \frac{2}{3} = (4)\left(\frac{1}{3}\right)^3\left(\frac{2}{3}\right)^1 = \frac{8}{81} \approx \mathbf{0.099}$$

$P(x = 4)$ is the probability that all four questions are answered correctly (there is only one branch where all four are correct—CCCC):

$$P(x = 4) = \frac{1}{3} \times \frac{1}{3} \times \frac{1}{3} \times \frac{1}{3} = \left(\frac{1}{3}\right)^4 = \frac{1}{81} \approx \mathbf{0.012}$$

*We can add probabil-
ities of discrete ran-
dom variables since
each x-value is mutu-
ally exclusive of the
other [addition rule,
formula (4-4b, c)].*

TABLE 5-7

Probability Distribution for the Four-Question Quiz	x	$P(x)$
	0	0.198
	1	0.395
	2	0.296
	3	0.099
	4	0.012
		1.000 (ck)

Now we can answer the five questions, on page 265, that were asked about the four-question quiz.

Answer 1: The most likely occurrence would be to get one answer; it has a probability of 0.395. Zero, one, or two correct answers are expected to result approximately 89% of the time ($0.198 + 0.395 + 0.296 = 0.889$).

Answer 2: To have more than half correct is represented by $x = 3$, or 4; their total probability is 0.111. (You will pass this quiz only 11% of the time by random guessing.)

EXERCISE 5.37

Where did $\frac{1}{3}$ and 4
come from? Why
multiply them to find
an expected
average?

Answer 3: $P(\text{all 4 correct}) = P(x = 4) = 0.012$ (All correct only 1% of the time.)

Answer 4: $P(\text{all 4 wrong}) = P(x = 0) = 0.198$. (That's almost 20% of the time.)

Answer 5: The class average would be expected to be $\frac{1}{3}$ of 4, or 1.33, correct answers.

The correct answers to the quiz are B, C, B, A. How many correct answers did you have? Which branch of the tree in Figure 5-4 represents your quiz results? You might ask several people to answer the quiz for you by guessing the answers. Then construct an observed relative frequency distribution and compare it to the distribution shown in Table 5-7.

Many experiments are composed of repeated trials whose outcomes can be classified in one of two categories, **success** or **failure**. Examples of such experiments include the previous experiments of tossing coins and other more practical experiments, such as determining whether a product did its prescribed job or did not do it, whether quiz-question answers are right or wrong. There are experiments in which the trials have many outcomes that, under the right conditions, may fit this general description of being classified in one of two categories. For example, when we roll a single die, we usually consider six possible outcomes. However, if we are interested only in knowing whether a "one" shows or not, there are really only two outcomes: the "one" shows or "something else" shows. The experiments just described are called *binomial probability experiments*.

Success or failure

EXERCISE 5.38

Ask 20 or more people to take the quiz, independently; grade their quizzes; form a frequency distribution of the variable "number of correct answers on each quiz."

BINOMIAL PROBABILITY EXPERIMENT

An experiment that is made up of repeated trials that possess the following properties:

1. There are n repeated independent trials.
2. Each **trial** has two possible outcomes (success, failure).
3. $P(\text{success}) = p$, $P(\text{failure}) = q$, and $p + q = 1$.
4. The **binomial random variable** x is the count of the number of successful trials that occur; x may take on any integer value from zero to n.

NOTES

1. Properties 1 and 2 are the two basic properties of any binomial experiment.
2. Property 3 concerns the algebraic notation for each trial.
3. Property 4 concerns the algebraic notation for the complete experiment.
4. It is of utmost importance that both x and p be associated with "success."

The four-question quiz qualifies as a binomial experiment made up of four trials when all four of the answers are obtained by random guessing.

Property 1: A trial is the answering of one question, repeated $n = 4$ times. The trials are independent since the probability of a correct answer on any one question is not affected by the answers on other questions.

Property 2: Two outcomes on each trial: success = C, correct answer, failure = W, wrong answer.

Property 3: $p = P(\text{correct}) = \frac{1}{3}$ and $q = P(\text{wrong}) = \frac{2}{3}$, on each trial. [$p + q = 1$; ✓]

Property 4: x = number of correct answers for the quiz (total experiment) and can be any integer value from 0 to 4.

NOTE   Independent trials mean that the result of one trial does not affect the probability of success of any other trial in the experiment. In other words, the probability of "success" remains constant throughout the entire experiment.

▼| ILLUSTRATION   5 - 8

Consider the experiment of rolling a die 12 times and observing a "one" or "something else." When all 12 rolls have been completed, the number of "ones" is reported. The random variable x would be the number of times that a "one" is observed in the $n = 12$ trials. Since "one" is the outcome of concern, it is considered "success"; therefore, $p = P(\text{one}) = \frac{1}{6}$ and $q = P(\text{not one}) = \frac{5}{6}$. This experiment is binomial.

▼| ILLUSTRATION   5 - 9

If you were an inspector on a production line in a plant where television sets were manufactured, you would be concerned with identifying the number of defective television sets. You probably would define "success" as the occurrence of a defective television. This is not what we normally think of as success, but if we count "defective" sets in a binomial experiment, we must define "success" as a "defective." The random variable x indicates the number of defective sets found per lot of n sets, while $p = P(\text{television is defective})$ and $q = P(\text{television is good})$.

▲|

EXERCISE   5.39

Identify the properties that make flipping a coin 50 times and keeping track of heads a binomial experiment.

The key to working with any probability experiment is its probability distribution. All binomial probability experiments have the same properties, and therefore the same organization scheme could be used to represent all of them. The binomial probability function will allow us to find the probabilities for each possible value of x.

BINOMIAL PROBABILITY FUNCTION

For a binomial experiment, let p represent the probability of a "success" and q represent the probability of a "failure" on a single trial; then $P(x)$, the probability that there will be exactly x successes in n trials, is

$$P(x) = \binom{n}{x}(p^x)(q^{n-x}), \text{ for } x = 0, 1, 2, \ldots, \text{ or } n \qquad \textbf{(5-5)}$$

When you look at the probability function, you notice that it is the product of three basic factors:

1. the number of ways that exactly x successes can occur in n trials, $\binom{n}{x}$
2. the probability of exactly x successes, p^x
3. the probability that failure will occur on the remaining $(n - x)$ trials, q^{n-x}

The number of ways that exactly x successes can occur in a set of n trials is represented by the symbol

$$\binom{n}{x}$$

Binomial coefficient

which must always be a positive integer. This term is called the **binomial coefficient** and is found by using the formula

$$\binom{n}{x} = \frac{n!}{x!(n - x)!} \tag{5-6}$$

EXERCISE   5.40

Find the value of
a. 4!
b. $\binom{4}{3}$

NOTE $n!$ is an abbreviation for the product of the sequence of integers starting with n and ending with 1. For example, $3! = 3 \cdot 2 \cdot 1 = 6$, and $5! = 5 \cdot 4 \cdot 3 \cdot 2 \cdot 1 = 120$. There is one special case, 0!, which is defined to be 1. For further information about factorial notation, see the *Statistical Tutor*.

Also, see the *Statistical Tutor* for general information on the binomial coefficient.

The values for $n!$ and $\binom{n}{x}$ can readily be found using most scientific calculators.

EXERCISE   5.41

Use the probability function for three coins and verify the probabilities for $x = 0$, 2, and 3.

Let's reconsider Illustration 5-7 (p. 262); a coin is tossed three times and we observe the number of heads that occur in the three tosses. This is a binomial experiment because it displays all the properties of a binomial experiment:

1. There are $n = 3$ repeated independent trials (each is a separate toss, and the outcome of any one trial has no effect on the probability of another).
2. Each trial (each toss of the coin) has two outcomes: success = heads (what we are counting) and failure = tails.
3. The probability of success is $p = P(\text{H}) = 0.5$, and the probability of failure is $q = P(\text{T}) = 0.5$, and $p + q = 0.5 + 0.5 = 1$.
4. The random variable x is the number of heads that occur in the three trials. x will assume exactly one of the values 0, 1, 2, or 3 when the experiment is complete.

In Table 5-6 (p. 262), $P(1) = \frac{3}{8}$. Here, $P(1) = 0.375$. $\frac{3}{8} = 0.375$.

The binomial probability function for the tossing of three coins is

$$P(x) = \binom{n}{x}(p)^x(q)^{n-x} = \binom{3}{x}(.5)^x(.5)^{3-x}, \text{ for } x = 0, 1, 2, \text{ or } 3$$

Let's find the probability of $x = 1$ using the preceding binomial probability function.

$$P(x = 1) = \binom{3}{1}(0.5)^1(0.5)^2 = 3(0.5)(0.25) = \mathbf{0.375}$$

Compare this to the value found in Illustration 5-7 (p. 262).

▼| ILLUSTRATION   5 - 10

Consider an experiment that calls for drawing five cards, one at a time with replacement, from a well-shuffled deck of playing cards. The drawn card is identified as a spade or not a spade, returned to the deck, the deck reshuffled, and so on. The random variable x is the number of spades observed in the set of five drawings. Is this a binomial experiment? Let's identify the various properties.

1. There are five repeated drawings; $n = 5$. These individual trials are independent, since the drawn card is returned to the deck and the deck reshuffled before the next drawing.
2. Each drawing is a trial, and each drawing has two outcomes, "spade" or "not spade."
3. $p = P(\text{spade}) = \frac{13}{52}$ and $q = P(\text{not spade}) = \frac{39}{52}$. ($p + q = 1$)
4. x is the number of spades recorded in the five trials, with possible values 0, 1, 2, . . . , 5.

The binomial probability function is

$$P(x) = \binom{5}{x}\left(\frac{13}{52}\right)^x\left(\frac{39}{52}\right)^{5-x} = \binom{5}{x}\left(\frac{1}{4}\right)^x\left(\frac{3}{4}\right)^{5-x}, \text{ for } x = 0, 1, \ldots, \text{ or } 5$$

EXERCISE 5.42

a. Calculate $P(4)$ and $P(5)$ for Illustration 5-10.

b. Verify that the six probabilities $P(0)$, $P(1)$, $P(2)$, . . . , $P(5)$ form a probability distribution.

$$P(0) = \binom{5}{0}\left(\frac{1}{4}\right)^0\left(\frac{3}{4}\right)^5 = (1)(1)(0.2373) = \mathbf{0.2373}$$

$$P(1) = \binom{5}{1}\left(\frac{1}{4}\right)^1\left(\frac{3}{4}\right)^4 = (5)(0.25)(0.3164) = \mathbf{0.3955}$$

$$P(2) = \binom{5}{2}\left(\frac{1}{4}\right)^2\left(\frac{3}{4}\right)^3 = (10)(0.0625)(0.421875) = \mathbf{0.2637}$$

$$P(3) = \binom{5}{3}\left(\frac{1}{4}\right)^3\left(\frac{3}{4}\right)^2 = (10)(0.015625)(0.5625) = \mathbf{0.0879}$$

The two remaining probabilities are left for you (Exercise 5.42).   ▲|

The preceding distribution of probabilities indicates that the single most likely value of x is 1, the event of observing exactly one spade in a hand of five cards. What is the least likely number of spades that would be observed?

▼| ILLUSTRATION   5 - 11

The manager of Steve's Food Market guarantees that none of his cartons of eggs containing a dozen eggs will contain more than one bad egg. If a carton contains more than one bad egg, he will replace the whole dozen and allow the customer to keep the original eggs. If the probability that an individual egg is bad is 0.05, what is the probability that the manager will have to replace a given carton of eggs?

SOLUTION Assuming that this is a binomial experiment, let x be the number of bad eggs found in a carton of a dozen eggs, $p = P(\text{bad}) = 0.05$, and let the inspection of each egg be a trial resulting in finding a "bad" or "not bad" egg. To find the probability that the manager will have to make good on his guarantee, we need the probability function associated with this experiment:

$$P(x) = \binom{12}{x}(0.05)^x(0.95)^{12-x} \quad \text{for} \quad x = 0, 1, 2, \ldots, \text{ or } 12$$

The probability that the manager will replace a dozen eggs is the probability that $x = 2, 3, 4, \ldots$, or 12. Recall that $\Sigma P(x) = 1$, that is,

$$P(0) + P(1) + P(2) + \cdots + P(12) = 1$$
$$P(\text{replacement}) = P(2) + P(3) + \cdots + P(12) = 1 - [(P(0) + P(1)]$$

Finding $P(x = 0)$ and $P(x = 1)$ and subtracting them from 1 is easier than finding each of the other probabilities:

$$P(x) = \binom{12}{x}(0.05)^x(0.95)^{12-x}$$

$$P(0) = \binom{12}{0}(0.05)^0(0.95)^{12} = \mathbf{0.540}$$

$$P(1) = \binom{12}{1}(0.05)^1(0.95)^{11} = \mathbf{0.341}$$

NOTE   The value of many binomial probabilities, for values of $n \leq 15$ and common values of p, are found in Table 2 of Appendix B. In this example, we have $n = 12$ and $p = 0.05$, and we want the probabilities for $x = 0$ and 1. We need to locate the section of Table 2 where $n = 12$, find the column marked $p = 0.05$, and read the numbers opposite $x = 0$ and 1. We find .540 and .341, as shown in Table 5-8. (Look up these values in Table 2.)

TABLE 5-8   Abbreviated Portion of Table 2 in Appendix B, Binomial Probabilities

n	x	0.01	0.05	0.10	0.20	0.30	0.40	p 0.50	0.60	0.70	0.80	0.90	0.95	0.99	x
	⋮	⋮													⋮
→ 12	0	.886	.540	.282	.069	.014	.002	0+	0+	0+	0+	0+	0+	0+	0
→	1	.107	.341	.377	.206	.071	.017	.003	0+	0+	0+	0+	0+	0+	1
	2	.006	.099	.230	.283	.168	.064	.016	.002	0+	0+	0+	0+	0+	2
	3	0+	.017	.085	.236	.240	.142	.054	.012	.001	0+	0+	0+	0+	3
	4	0+	.002	.021	.133	.231	.213	.121	.042	.008	.001	0+	0+	0+	4
	⋮	⋮													⋮

Now's let's return to our illustration:

$$P(\text{replacement}) = 1 - (0.540 + 0.341) = \mathbf{0.119}$$

If $p = 0.05$ is correct, the manager of Steve's Food Market will be busy replacing cartons of eggs. If he replaces 11.9% of all the cartons of eggs he sells, he certainly

EXERCISE 5.43

What would be the manager's "risk" if he bought eggs with $P(\text{bad}) = 0.01$ using the "more than one" guarantee?

will be giving away a substantial proportion of his eggs. This suggests that he should adjust his guarantee. For example, if he were to replace a carton of eggs only when four or more were found bad, he would expect to replace only 3 out of 1000 cartons $[1.0 - (0.540 + 0.341 + 0.099 + 0.017)]$, or 0.3% of the cartons sold. Notice that he will be able to control his "risk" (probability of replacement) if he adjusts the value of the random variable stated in his guarantee.

▲

NOTE   A convenient notation to identify the binomial probability distribution for a binomial experiment with $n = 12$ and $p = 0.30$ is **B(12, 0.30)**.

EXERCISE 5.44

Use Table 5-8 to find the probability that $x = 3$ in a binomial experiment where $n = 12$ and $p = 0.30$: $P(x = 3 \mid B(12, 0.30))$.

MINITAB (Release 10) commands to determine binomial probabilities or cumulative binomial probabilities for x-values listed in C1. Probabilities will be listed in C2 for B(n,p).

Session commands	*Menu commands*
Enter: PDF C1 C2; BINOmial n p.	Choose: Calc > Prob. Dist. > Binomial Select: Probability Enter: Number of trials: n Probability: p
or Enter: CDF C1 C2; BINOmial n p.	Choose: Calc > Prob. Dist. > Binomial Select: Cumulative Probability Enter: Number of trials: n Probability: p

EXERCISES

5.45   Evaluate each of the following.

    a. 4!   b. 7!   c. 0!   d. $\dfrac{6!}{2!}$

    e. $\dfrac{5!}{3!2!}$ f. $\dfrac{6!}{4!(6-4)!}$ g. $(0.3)^4$ h. $\binom{7}{3}$

    i. $\binom{5}{2}$ j. $\binom{3}{0}$ k. $\binom{4}{1}(0.2)^1(0.8)^3$ l. $\binom{5}{0}(0.3)^0(0.7)^5$

5.46   Show that each of the following is true for any values of n and k. Use two specific sets of values for n and k to show that each is true.

    a. $\binom{n}{0} = 1$ and $\binom{n}{n} = 1$ b. $\binom{n}{1} = n$ and $\binom{n}{n-1} = n$ c. $\binom{n}{k} = \binom{n}{n-k}$

 5.47   A carton containing 100 T-shirts is inspected. Each T-shirt is rated "first quality" or "irregular." After all 100 T-shirts have been inspected, the number of irregulars is reported as a random variable. Explain why x is a binomial random variable.

5.48   A die is rolled 20 times and the number of "fives" that occurred is reported as being the random variable. Explain why x is a binomial random variable.

5.49   Four cards are selected, one at a time, from a standard deck of 52 cards. Let x represent the number of aces drawn in the set of 4 cards.
   a. If this experiment is completed without replacement, explain why x is not a binomial random variable.
   b. If this experiment is completed with replacement, explain why x is a binomial random variable.

5.50   The employees at a General Motors assembly plant are polled as they leave work. Each is asked, "What brand of automobile are you riding home in?" The random variable to be reported is the number of each brand mentioned. Is x a binomial random variable? Justify your answer.

5.51   Consider a binomial experiment made up of three trials with outcomes of success S and failure F, where $P(S) = p$ and $P(F) = q$.
   a. Complete the accompanying tree diagram. Label all branches completely.

   b. In column (b) of the tree diagram, express the probability of each outcome represented by the branches as a product of powers of p and q.
   c. Let x be the random variable, the number of successes observed. In column (c), identify the value of x for each branch of the tree diagram.
   d. Notice that all the products in column (b) are made up of three factors and that the value of the random variable is the same as the exponent for the number p.
   e. Write the equation for the binomial probability function for this situation.

5.52   Draw a tree diagram picturing a binomial experiment of four trials.

5.53   If x is a binomial random variable, calculate the probability of x for each case.
   a. $n = 4$, $x = 1$, $p = 0.3$ b. $n = 3$, $x = 2$, $p = 0.8$
   c. $n = 2$, $x = 0$, $p = \frac{1}{4}$ d. $n = 5$, $x = 2$, $p \frac{1}{3}$
   e. $n = 4$, $x = 2$, $p = 0.5$ f. $n = 3$, $x = 3$, $p = \frac{1}{6}$

5.54   If x is a binomial random variable, use Table 2 in Appendix B to determine the probability of x for each case.
   a. $n = 10$, $x = 8$, $p = 0.3$ b. $n = 8$, $x = 7$, $p = 0.95$
   c. $n = 15$, $x = 3$, $p = 0.05$ d. $n = 12$, $x = 12$, $p = 0.99$
   e. $n = 9$, $x = 0$, $p = 0.5$ f. $n = 6$, $x = 1$, $p = 0.01$
   g. Explain the meaning of the symbol $0+$ that appears in Table 2.

5.55   Test the following function to determine whether or not it is a binomial probability function. List the distribution of probabilities and sketch a histogram.

$$T(x) = \binom{5}{x}\left(\frac{1}{2}\right)^{x}\left(\frac{1}{2}\right)^{5-x} \quad \text{for} \quad x = 0, 1, 2, 3, 4, \text{ or } 5$$

5.56   Let x be a random variable with the following probability distribution.

x	0	1	2	3
$P(x)$	0.4	0.3	0.2	0.1

Does x have a binomial distribution? Justify your answer.

 5.57   Ninety percent of the trees planted by a landscaping firm survive. What is the probability that eight or more of the ten trees they just planted will survive? (Find the answer by using a table.)

 5.58   In California, 30% of the people have a certain blood type. What is the probability that exactly 5 out of a randomly selected group of 14 Californians will have that blood type? (Find the answer by using a table.)

 5.59   On the average, 1 out of every 10 boards purchased by a cabinet manufacturer is unusable for building cabinets. What is the probability that 8, 9, or 10 of a set of 11 such boards are usable? (Find the answer by using a table.)

 5.60   A local polling organization maintains that 90% of the eligible voters have never heard of John Anderson, who was a presidential candidate in 1980. If this is so, what is the probability that in a randomly selected sample of 12 eligible voters 2 or fewer have heard of John Anderson?

 5.61   In the biathlon event of the Olympic Games, a participant skis cross-country and on four intermittent occasions stops at a rifle range and shoots a set of five shots. If the center of the target is hit, no penalty points are assessed. If a particular man has a history of hitting the center of the target with 90% of his shots, what is the probability that he will hit the center of the target with

   a. all five of his next set of five shots?
   b. at least four of his next set of five shots? (Assume independence.)

 5.62   A basketball player has a history of making 80% of the foul shots taken during games. What is the probability that he will miss three of the next five foul shots he takes?

 5.63   A machine produces parts of which 0.5% are defective. If a random sample of ten parts produced by this machine contains two or more defectives, the machine is shut down for repairs. Find the probability that the machine will be shut down for repairs based on this sampling plan.

 5.64   The survival rate during a risky operation for patients with no other hope of survival is 80%. What is the probability that exactly four of the next five patients survive this operation?

 5.65   According to an article in the February 1991 issue of *Reader's Digest*, Americans face a 1 in 20 chance of acquiring an infection while hospitalized. If the records of 15 randomly selected hospitalized patients are examined, find the probability that

   a. none of the 15 acquired an infection while hospitalized.
   b. one or more of the 15 acquired an infection while hospitalized.

5.66 An article in the *Omaha World-Herald* (12-1-94) stated that only about 60% of the individuals needing a bone marrow transplant find a suitable donor when they turn to registries of unrelated donors. In a group of 10 individuals needing a bone marrow transplant,
  a. what is the probability that *all 10* will find a suitable donor among the registries of unrelated donors?
  b. what is the probability that *exactly 8* will find a suitable donor among the registries of unrelated donors?
  c. what is the probability that *at least 8* will find a suitable donor among the registries of unrelated donors?
  d. what is the probability that *no more than 5* will find a suitable donor among the registries of unrelated donors?

5.67 If boys and girls are equally likely to be born, what is the probability that in a randomly selected family of six children, there will be boys? (Find the answer by using a formula.)

5.68 One-fourth of a certain breed of rabbits are born with long hair. What is the probability that in a litter of six rabbits, exactly three will have long hair? (Find the answer by using a formula.)

5.69 Suppose that you take a five-question multiple-choice quiz by guessing. Each question has exactly one correct answer of the four alternatives given. What is the probability that you guess more than one-half of the answers correctly? (Find the answer by using a formula.)

5.70 As a quality-control inspector for toy trucks, you have observed that wooden wheels bored off-center occur about 3% of the time. If six wooden wheels are used on each toy truck produced, what is the probability that a randomly selected set of wheels has no off-center wheels?

5.71 According to the USA Snapshot® "Knowing drug addicts," 45% of Americans know somebody who became addicted to a drug other than alcohol. Assuming this to be true, what is the probability that
  a. exactly 3 of a random sample of 5 know someone who became addicted? Calculate the value.
  b. exactly 7 of a random sample of 15 know someone who became addicted? Estimate using Table 2 in Appendix B.
  c. at least 7 of a random sample of 15 know someone who became addicted? Estimate using Table 2.
  d. no more than 7 of a random sample of 15 know someone who became addicted? Estimate using Table 2.

5.72 According to *Financial Executive* (July/August 1993) disability causes 48% of all mortgage foreclosures. Given that 20 mortgage foreclosures are audited by a large lending institution,
  a. find the probability that 5 or fewer of the foreclosures are due to a disability.
  b. find the probability that at least 3 are due to a disability.

5.73 Use a computer to find the probabilities for all possible *x*-values for a binomial experiment where $n = 30$ and $p = 0.35$.

```
MINITAB commands
SET C1
0:30
END
PDF C1 C2;
  BINOmial 30 .35.
```

5.74 An article titled "Sex Surveys: Does Anyone Tell the Truth?" appeared in the July 1993 issue of *American Demographics*. The article quotes a 1991 study by Batelle Memorial Institute of Human Affairs Research Centers that states 2.3% of men aged 20 to 39 are gay. Assuming this percent to be correct, in a random sample of 50 men aged 20 to 39, what is the probability of finding 2 or fewer men who are gay?

5.75 An article titled "Mom, I Want to Live with My Boyfriend" appeared in the February 1994 issue of *Reader's Digest*. The article quoted a Columbia University study that found that only 19% of the men who lived with their girlfriends eventually walked down the aisle with them. Suppose 25 men are interviewed who have lived with a girlfriend in the past. What is the probability that 5 or fewer of them married the girlfriend?

5.76 Use a computer to find the probabilities for all possible x-values for a binomial experiment where $n = 45$ and $p = 0.125$. Explain why there are so many 1.000 listed. Explain what is represented by each number listed.

```
MINITAB commands
SET C1
0:45
END
CDF C1 C2;
   BINOmial 45 .125.
```

5.77 State a very practical reason why the defective item in an industrial situation would be defined to be the "success" in a binomial experiment.

5.78 If the binomial $(q + p)$ is squared, the result is $(q + p)^2 = q^2 + 2qp + p^2$. For the binomial experiment with $n = 2$, the probability of no successes in two trials is q^2 (the first term in the expansion), the probability of one success in two trials is $2qp$ (the second term in the expansion), and the probability of two successes in two trials is p^2 (the third term). Find $(q + p)^3$ and compare its terms to the binomial probabilities for $n = 3$ trials.

5.5 | MEAN AND STANDARD DEVIATION OF THE BINOMIAL DISTRIBUTION

Mean and standard deviation of binomial distribution

The **mean** and **standard deviation** of a theoretical binomial probability distribution can be found by using these two formulas:

$$\mu = np \tag{5-7}$$

$$\sigma = \sqrt{npq} \tag{5-8}$$

The formula for the mean seems appropriate, the number of trials multiplied by the probability of "success." (Recall that the mean number of correct answers on the binomial quiz (Answer 5, p. 267) was expected to be $\frac{1}{3}$ of 4, $4\left(\frac{1}{3}\right)$, or np.) The formula for the standard deviation is not as easily understood. Thus, at this point it is appropriate to look at an example, which demonstrates that formulas (5-7) and (5-8) yield the same results as formulas (5-1) and (5-3a and 5-4).

Returning to Illustration 5-7 (p. 262), x is the number of heads seen when tossing three coins, $n = 3$ and $p = \frac{1}{2} = 0.5$. Using formula (5-7), we find the mean of x to be

$$\mu = np = (3)(0.5) = \mathbf{1.5}$$

Using formula (5-8), we find the standard deviation of x to be

$$\sigma = \sqrt{npq} = \sqrt{(3)(0.5)(0.5)} = \sqrt{0.75} = 0.866 = \mathbf{0.87}$$

Look back at the solution for Illustration 5-7 (p. 262). Note that the results are the same, regardless of the formula you use. However, formulas (5-7) and (5-8) are much easier to use when x is a binomial random variable.

▼| ILLUSTRATION   5 - 12

EXERCISE 5.79

Find the mean and standard deviation for the binomial random variable x with $n = 30$ and $p = 0.6$.

Find the mean and standard deviation of the binomial distribution when $n = 20$ and $p = \frac{1}{5}$ (or 0.2, in decimal form). Recall that the "binomial distribution where $n = 20$ and $p = 0.2$" has a probability function

$$P(x) = \binom{20}{x}(0.2)^x(0.8)^{20-x} \quad \text{for} \quad x = 0, 1, 2, \ldots, \text{or } 20$$

and a corresponding distribution with 21 x-values and 21 probabilities as shown in the following distribution chart, Table 5-9, and on the histogram in Figure 5-5.

TABLE 5-9
Binomial $n = 20$, $p = 0.2$

x	$P(x)$
0	0.012
1	0.058
2	0.137
3	0.205
4	0.218
5	0.175
6	0.109
7	0.055
8	0.022
9	0.007
10	0.002
11	0+
12	0+
13	0+
⋮	⋮
20	0+

FIGURE 5-5
Histogram

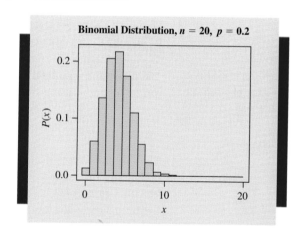

EXERCISE 5.80

Consider the binomial distribution where $n = 11$ and $p = 0.05$.

a. Find the mean and standard deviation using formulas (5-7) and (5-8).
b. Using Table 2 in Appendix B, list the probability distribution and draw a histogram.
c. Locate μ and σ on the histogram.

Let's find the mean and the standard deviation of this distribution of x using formulas (5-7) and (5-8).

$$\mu = np = (20)\left(\frac{1}{5}\right) = \mathbf{4.0}$$

$$\sigma = \sqrt{npq} = \sqrt{(20)\left(\frac{1}{5}\right)\left(\frac{4}{5}\right)} = \sqrt{\frac{80}{25}} = \sqrt{3.2} = \mathbf{1.79}$$

▲|

Figure 5-6 shows the mean (vertical blue line) and the size of the standard deviation (horizontal red line segment) relative to the probability distribution of the variable x.

FIGURE 5-6

Histogram

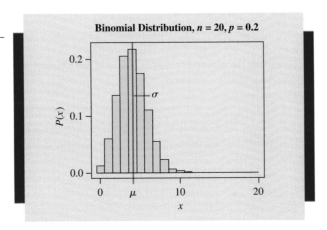

EXERCISES

5.81   Consider the binomial distribution where $n = 11$ and $p = 0.05$ (see Exercise 5.80).
   a. Use the distribution [Exercise 5.80(b) or Table 2] and find the mean and standard deviation using formulas (5-1) and (5-3a and 5-4).
   b. Compare the results of (a) to the answers found in Exercise 5.80(a).

5.82   Given the binomial probability function

$$T(x) = \binom{5}{x} \cdot \left(\frac{1}{2}\right)^x \cdot \left(\frac{1}{2}\right)^{5-x} \quad \text{for} \quad x = 0, 1, 2, 3, 4, 5$$

   a. calculate the mean and standard deviation of the random variable by using formulas (5-1) and (5-3a and 5-4).
   b. calculate the mean and standard deviation using formulas (5-7) and (5-8).
   c. compare the results of (a) and (b).

5.83   Find the mean and standard deviation of x for each of the following binomial random variables:
   a. the number of aces seen in 100 draws from a well-shuffled bridge deck (with replacement).
   b. the number of cars found to have unsafe tires among the 400 cars stopped at a roadblock for inspection. Assume that 6% of all cars have one or more unsafe tires.
   c. the number of melon seeds that germinate when a package of 50 seeds is planted. The package states that the probability of germination is 0.88.

5.84   Find the mean and standard deviation for each of the following binomial random variables:
   a. the number of tails seen in 50 tosses of a quarter.
   b. the number of defective televisions in a shipment of 125. The manufacturer claimed that 98% of the sets were operative.
   c. the number of operative televisions in a shipment of 125. The manufacturer claimed that 98% of the sets were operative.
   d. How are questions (b) and (c) related?

5.85 A binomial random variable has a mean equal to 200 and a standard deviation of 10. Find the values of *n* and *p*.

5.86 The probability of success on a single trial of a binomial experiment is known to be $\frac{1}{4}$. The random variable *x*, number of successes, has a mean value of 80. Find the number of trials involved in this experiment and the standard deviation of *x*.

5.87 Seventy-five percent of the foreign-made autos sold in the United States in 1984 are now falling apart.
    a. Determine the probability distribution of *x*, the number of these autos that are falling apart in a random sample of five cars.
    b. Draw a histogram of the distribution.
    c. Calculate the mean and standard deviation of this distribution.

5.88 A binomial random variable *x* is based on 15 trials with the probability of success equal to 0.3. Find the probability that this variable will take on a value more than two standard deviations from the mean.

5.89 An article titled "The Scourge of Illegitimacy" (*Reader's Digest*, March 1994) states that 30% of all live births are of children born to unmarried mothers. Assuming this percentage to be correct, find the mean and standard deviation of the number of live births to unmarried mothers for all samples of size 500 randomly selected from the birth certificates for babies born in 1994.

5.90 Imprints Galore buys T-shirts (to be imprinted with item of customer's choice) from a manufacturer with the guarantee that the shirts have been inspected and that no more than 1% are imperfect in any way. The shirts arrive in boxes of 12. Let *x* be the number of imperfect shirts found in any one box.
    a. List the probability distribution and draw the histogram of *x*.
    b. What is the probability that any one box has no imperfect shirts?
    c. What is the probability that any one box has no more than one imperfect shirt?
    d. Find the mean and standard deviation of *x*.
    e. What proportion of the distribution is between $\mu - \sigma$ and $\mu + \sigma$?
    f. What proportion of the distribution is between $\mu - 2\sigma$ and $\mu + 2\sigma$?
    g. How does this information relate to the empirical rule and Chebyshev's theorem? Explain.
    h. Use a computer to simulate Imprints Galore buying 200 boxes of shirts and observing *x*, the number of imperfect shirts per box of 12. Describe how the information from the simulation compares to what was expected [answers (a) through (g) describe the expected results].
    i. Repeat (h) several times. Describe how these results compare to those of (a)–(g) and of (h).

See Helpful Hint on page 89.

Note: The binomial variable *x* cannot take on the value −1. The use of −1 (the next would-be class midpoint to left of 0) allows MINITAB to draw the histogram of a probability distribution. Without −1, PLOT will draw only half of the bar representing *x* = 0.

```
MINITAB commands

a. SET C1
   -1:12              See Note.
   END
   PDF C1 C2;
     BINOmial 12 .01.
   PLOT C2*C1;
     AREA;
     STEP 0;
     TYPE 0.
c. CDF C1 C3;
     BINOmial 12 .01.
h. RANDom 200 C4;
     BINOmial 12 .01.
   TALLy C4
   MEAN C4
   STDEv C4
   HISTogram C4;
     PERCent;
     MIDPoint 0:12.
```

IN RETROSPECT

In this chapter we combined concepts of probability with some of the ideas presented in Chapter 2. We now are able to deal with distributions of probability values and to find means, standard deviations, and so on.

In Chapter 4 we explored the concepts of mutually exclusive events and independent events. The addition and multiplication rules were used on several occasions in this chapter, but very little was said about mutual exclusiveness or independence. Recall that every time we add probabilities, as we did in each of the probability distributions, we need to know that the associated events are mutually exclusive. If you look back over the chapter, you will notice that the random variable actually requires events to be mutually exclusive; therefore, no real emphasis was placed on this concept. The same basic comment can be made in reference to the multiplication of probabilities and the concept of independent events. Throughout this chapter, probabilities were multiplied and occasionally independence was mentioned. Independence, of course, is necessary in order to be able to multiply probabilities.

If we were to take a close look at some of the sets of data in Chapter 2, we would see that several of these problems could be reorganized to form probability distributions. For example: (1) Let x be the number of credit hours for which a student is registered in this semester, with the percentage of the student body being reported for each value of x. (2) Let x be the number of correct passageways through which an experimental laboratory animal passes before taking a wrong one. (3) Let x be the number of trips per day for the ambulance service (Case Study 5-1). The list of examples is endless.

We are now ready to extend these concepts to continuous random variables, which we will do in Chapter 6.

CHAPTER EXERCISES

5.91 What are the two basic properties of every probability distribution?

5.92 a. Explain the difference and the relationship between a *probability distribution* and a *probability function*.

b. Explain the difference and the relationship between a *probability distribution* and a *frequency distribution*, and explain how they relate to a *population* and a *sample*.

5.93 Verify whether or not the following is a probability function. State your conclusion and explain.

$$f(x) = \frac{\frac{3}{4}}{x!(3-x)!} \quad \text{for} \quad x = 0, 1, 2, 3$$

5.94   Determine which of the following are probability functions.
   a. $f(x) = 0.25$ for $x = 9, 10, 11, 12$
   b. $f(x) = (3 - x)/2$ for $x = 1, 2, 3, 4$
   c. $f(x) = (x^2 + x + 1)/25$ for $x = 0, 1, 2, 3$

5.95   The number of ships to arrive at a harbor on any given day is a random variable represented by x. The probability distribution for x is

x	10	11	12	13	14
$P(x)$	0.4	0.2	0.2	0.1	0.1

Find the probability that on a given day
   a. exactly 14 ships arrive.   b. at least 12 ships arrive.
   c. at most 11 ships arrive.

5.96   The number of data-entry mistakes per hour made by an individual entering data at a CRT terminal is a random variable represented by x and has the following probability distribution.

x	0	1	2	3
$P(x)$	0.40	0.30	0.25	0.05

   a. Find the mean number of mistakes for one-hour sessions.
   b. Find the probability of at least one mistake during a particular one-hour session.

5.97   Did you ever wonder "How many times buyers see an infomercial before they purchase its product or service?" The USA Snapshot® "Television's hard sell" (10-21-94) answers that question (see the chapter case study, page 249). According to the National Infomercial Marketing Association:

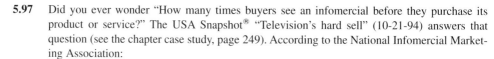

Times Watched Before Buy	1	2	3	4	5 or more
Percentage of Buyers	0.27	0.31	0.18	0.09	0.15

   a. What is the probability that a buyer watched only once before buying?
   b. What is the probability that a viewer watching for the first time will buy?
   c. What percentage of the buyers watched the infomercial three or more times before purchasing?
   d. Is this a binomial probability experiment?
   e. Let x be the number of times a buyer watched before making a purchase. Is this a probability distribution?
   f. Assign $x = 5$ for "5 or more," and find the mean and standard deviation of x.

5.98 When the average U.S. teen-age girl buys new jeans, finding the "Perfect fit ain't easy," according to a USA Snapshot® (8-19-94), which reported on "how many they try on before finding a pair that fits."

Number of Tries	3 or less	4 to 7	8 or more
Percent Who Said	0.61	0.32	0.07

Let $x = 1.5$, 5.5, and 9.5 be the class marks for the three classes.
    a. Find the mean and standard deviation of x.
    b. Because of the "grouped" nature of the probability distribution, the values calculated for the mean and standard deviation are estimates. How different do you believe the true mean and standard deviation are from the values calculated? Explain.

5.99 A doctor knows from experience that 10% of the patients to whom he gives a certain drug will have undesirable side effects. Find the probabilities that among the ten patients to whom he gives the drug,
    a. at most two will have undesirable side effects.
    b. at least two will have undesirable side effects.

5.100 In a recent survey of women, 90% admitted that they had never looked at a copy of *Vogue* magazine. Assuming that this is accurate information, what is the probability that a random sample of three women will show that fewer than two have looked at the magazine?

5.101 Of the Heisman Trophy winners in college football, it is considered that only 0.35 have had successful professional football careers.
    a. What is the probability that none of the next five winners will have successful pro careers?
    b. What is the probability that more than three of the next five winners will have successful pro careers?

5.102 Seventy percent of those seeking a driver's license admitted that they would not report someone if he or she copied some answers during the written exam. You have just entered the room and see ten people waiting to take the written exam. What is the probability that, if the incident happened, five of the ten would not report what they saw?

5.103 The engines on an airliner operate independently. The probability that an individual engine operates for a given trip is 0.95. A plane will be able to complete a trip successfully if at least one-half of its engines operate for the entire trip. Determine whether a four-engine or a two-engine plane has the higher probability of a successful trip.

5.104 A USA Snapshot® titled "Stress does not love company" (11-3-94) answered the question "How people say they prefer to spend stressful times." Forty-eight percent responded "alone," 29% responded "with family," 18% responded "with friends," and 5% responded "other/don't know." Ten individuals are randomly selected and asked the question "How do you prefer to spend stressful times?"
    a. What is the probability that two or fewer will respond by saying "alone"?
    b. Explain why this question can be answered using binomial probabilities.

5.105 The town council has nine members. A proposal must have at least two-thirds of the votes to be accepted. A proposal to establish a new industry in this town has been tabled. If we know that two members of the town council are opposed and that the others randomly vote "in favor" and "against," what is the probability that the proposal will be accepted?

5.106 The "Health Update" section of *Better Homes and Gardens* (July 1990) reported that patients who take long half-life tranquilizers are 70% more likely to suffer falls resulting in hip fractures than those taking similar drugs with a short half-life. It was also reported that in a Massachusetts study, 30% of nursing-home patients who used tranquilizers used the long half-life ones. Suppose that in a survey of 15 nursing-home patients in New York who used tranquilizers, it was found that 10 of the 15 used long half-life tranquilizers.

    a. If the 30% figure for Massachusetts also holds in New York, find the probability of finding 10 or more in a random sample of 15 who use long half-life tranquilizers.

    b. What might you infer from your answer in (a)?

5.107 A box contains ten items of which three are defective and seven are nondefective. Two items are selected without replacement, and x is the number of defectives in the sample of two. Explain why x is not a binomial random variable.

5.108 A large shipment of radios is accepted upon delivery if an inspection of ten randomly selected radios yields no more than one defective radio.

    a. Find the probability that this shipment is accepted if 5% of the total shipment is defective.

    b. Find the probability that this shipment is not accepted if 20% of this shipment is defective.

    c. The binomial probability distribution is often used in situations similar to this one, namely, large populations sampled without replacement. Explain why the binomial yields a good estimate.

5.109 A discrete random variable has a standard deviation equal to 10 and a mean equal to 50. Find $\Sigma\, x^2 P(x)$.

5.110 A binomial random variable is based on $n = 20$ and $p = 0.4$. Find $\Sigma\, x^2 P(x)$.

5.111 For years, the manager of a certain company had sole responsibility for making decisions with regard to company policy. This manager has a history of making the correct decision with a probability of p. Recently company policy has changed, and now all decisions are to be made by majority rule of a three-person committee.

    a. Each member makes a decision independently, and each has a probability of p of making the correct decision. What is the probability that the committee's majority decision will be correct?

    b. If $p = 0.1$, what is the probability that the committee makes the correct decision?

    c. If $p = 0.8$, what is the probability that the committee makes the correct decision?

    d. For what value of p is the committee more likely to make the correct decision by majority rule than the former manager?

    e. For what values (there are three) of p is the probability of a correct decision the same for the manager and for the committee? Justify your answer.

5.112 Suppose one member of the committee in Exercise 5.111 always makes her decision by rolling a die. If the die roll results in an even number, she votes for the proposal, and if an odd number occurs, she votes against it. The other two members still decide independently and have a probability of p of making the correct decision.

    a. What is the probability that the committee's majority decision will be correct?

    b. If $p = 0.1$, what is the probability that the committee makes the correct decision?

    c. If $p = 0.8$, what is the probability that the committee makes the correct decision?

    d. For what values of p is the committee more likely to make the correct decision by majority rule than the former manager?

    e. For what values of p is the probability of a correct decision the same for the manager and for the committee? Justify your answer.

    f. Why is the answer to (e) different than the answer to Exercise 5.111(e)?

5.113 A business firm is considering two investments. It will choose the one that promises the greater payoff. Which of the investments should it accept? (Let the mean profit measure the payoff.)

Invest in Tool Shop		Invest in Book Store	
Profit	**Probability**	**Profit**	**Probability**
$100,000	0.10	$400,000	0.20
50,000	0.30	90,000	0.10
20,000	0.30	−20,000	0.40
−80,000	0.30	−250,000	0.30
Total	1.00	*Total*	1.00

5.114 Bill has completed a ten-question multiple-choice test on which he answered seven questions correctly. Each question had one correct answer to be chosen from five alternatives. Bill says that he answered the test by randomly guessing the answers without reading the questions or answers.

    a. Define the random variable x to be the number of correct answers on this test, and construct the probability distribution if the answers were obtained by random guessing.

    b. What is the probability that Bill guessed seven of the ten answers correctly?

    c. What is the probability that anybody can guess six or more answers correctly?

    d. Do you believe that Bill actually randomly guessed as he claims? Explain.

5.115 A random variable that can assume any one of the integer values 1, 2, ... , n with equal probabilities of $1/n$ is said to have a uniform distribution. The probability function is written $P(x) = 1/n$, for $x = 1, 2, 3, \ldots, n$. Show that $\mu = (n + 1)/2$.
(*Hint:* $1 + 2 + 3 + \cdots + n = [n(n + 1)]/2$.)

VOCABULARY LIST

Be able to define each term. Pay special attention to the key terms, which are printed in red. In addition, describe in your own words, and give an example of, each term. Your examples should not be the ones given in class or in the textbook.

The bracketed numbers indicate the chapter in which the term first appeared, but you should define the terms again to show increased understanding of their meaning. Page numbers indicate the first appearance of the term in Chapter 5.

binomial coefficient (p. 270)
binomial experiment (p. 268)
binomial probability function (p. 269)
binomial random variable (p. 268)
constant function (p. 254)
continuous random variable (p. 251)
discrete random variable (p. 251)
experiment [1, 4] (p. 268)
failure (p. 268)
independent trials (p. 268)

mean of probability distribution (p. 260)
mutually exclusive events [4] (p. 253)
population parameter [1] (p. 260)
probability distribution (p. 253)
probability function (p. 254)
probability histogram (p. 256)
random variable (p. 250)
sample statistic [1] (p. 259)

(continued)

standard deviation of probability
distribution (p. 260)
success (p. 268)

trial (p. 268)
variance of probability distribution
(p. 260)

QUIZ A

Answer "True" if the statement is always true. If the statement is not always true, replace the words shown in bold with words that make the statement always true.

5.1   The number of hours you waited in line to register this semester is an example of a **discrete** random variable.

5.2   The number of automobile accidents you were involved in as a driver last year is an example of a **discrete** random variable.

5.3   The sum of all the probabilities in any probability distribution is always exactly **two**.

5.4   The various values of a random variable form a list of **mutually exclusive events**.

5.5   A binomial experiment always has **three or more** possible outcomes to each trial.

5.6   The formula $\mu = np$ may be used to compute the mean of a **discrete** population.

5.7   The binomial parameter p is the probability of **one success occurring in n trials** when a binomial experiment is performed.

5.8   A parameter is a statistical measure of some aspect of a **sample**.

5.9   **Sample statistics** are represented by letters from the Greek alphabet.

5.10   The probability of event A or B is equal to the sum of the probability of event A and the probability of event B when A and B are **mutually exclusive events**.

QUIZ B

5.1   a. Show that the following is a probability distribution:

x	$P(x)$
1	0.2
3	0.3
4	0.4
5	0.1

b. Find $P(x = 1)$.   c. Find $P(x = 2)$.
d. Find $P(x > 2)$.   e. Find the mean of x.
f. Find the standard deviation of x.

5.2 A T-shirt manufacturing company advertises that the probability of an individual T-shirt being irregular is 0.1. A box of 12 such T-shirts is randomly selected and inspected.

   a. What is the probability that exactly 2 of these 12 T-shirts are irregular?

   b. What is the probability that exactly 9 of these 12 T-shirts are *not* irregular?

   Let x be the number of T-shirts that are irregular in all such boxes of 12 T-shirts.

   c. Find the mean of x.   d. Find the standard deviation of x.

QUIZ C

5.1 What properties must an experiment possess in order for it to be a binomial probability experiment?

5.2 Student A uses a relative frequency distribution for a set of sample data and calculates the mean and standard deviation using formulas learned in Chapter 5. Student A justified her choice of formulas by saying that since relative frequencies are empirical probabilities, her sample is represented by a probability distribution and therefore her choice of formulas was correct. Student B argues that since the distribution represented a sample, then the mean and the standard deviation involved are known as $\bar{x}$ and s and must be calculated using the corresponding frequency distribution and formulas learned in Chapter 2. Who is correct, A or B? Justify your choice.

5.3 Student A and Student B were discussing one entry in a probability distribution chart.

x	$P(x)$
-2	0.1

Student B felt that this entry was okay since the $P(x)$ was a value between 0.0 and 1.0. Student A argued that this entry was impossible for a probability distribution since x was a -2, and negatives are not possible. Who is correct, A or B? Justify your choice.

NORMAL PROBABILITY DISTRIBUTIONS

Chapter Case Study

MEASURES OF INTELLIGENCE

Measuring intelligence is not a simple task; in fact, the I.Q. score discussed here is only one of many ways to measure a person's intelligence.

INTELLIGENCE MEASURES

Intelligence is a broad term referring to complex mental abilities of the individual. It is a term employed by lay persons to denote such qualities as quickness of mind, level of academic success, status on an occupational scale, or the attainment of eminence in a particular field of endeavor. Psychologists who measure intelligence have variously employed the term to indicate the amount of knowledge available and the rapidity with which new knowledge is acquired; the ability to adapt to new situations and to handle concepts, relationships, and abstract symbols; and even simply that phenomenon which intelligence tests measure. Validation studies of standardized tests of intelligence have demonstrated that they measure elements of mental abilities which appear to require intelligence, as the latter term is subjectively understood by scientists who study this human characteristic. I.Q. scores derived from clinically administered individual intelligence tests can predict academic achievement for the top 90% of the general population who proceed through school in regular classes, while identifying individuals in the bottom 10% with I.Q.'s below 80 who may require specialized educational, psychological, or medical assistance.

It would be erroneous to think that intelligence is completely measured by current standardized tests of I.Q. An I.Q. score is *not* the only measure of intelligence. It is merely the score earned by a person on a particular set of tasks or subtests on a test of *measured intelligence*, compared to the scores of those upon whom the test was normed (or standardized).

INTELLIGENCE TESTS

Arithmetic skills, information, reasoning, manipulation of objects, vocabulary, and memory functions constitute some of the tasks employed in tests of measured intelligence, whether they are called primary grade achievement tests, Scholastic Aptitude Tests (SAT), or intelligence tests. The objective of each of these types of tests is to appraise overall performance so as to obtain an estimate of general intellectual potential.

The Stanford-Binet and the Wechsler Scales are individually administered tests widely used today by educational psychologists, industrial psychologists, and clinical psychologists whose day-to-day work involves an intensive clinical, school, or job-related investigation of a person's intellectual functioning.

Prerequisite to a correct interpretation of an individual's I.Q. test score as one of the two components of intelligence is a knowledge of some basic characteristics of such I.Q. scores—specifically, knowledge of their norms, reliability, and validity. It is important in using an I.Q. test to be familiar with the complexities and subtleties of intelligence assessment through reading and clinical experience, so as to identify individual differences arising from any of a host of environmental circumstances, brain dysfunction, personality disturbance, and physical disability (Matarazzo, 1972, 1980; Matarazzo & Pankratz, 1980).

Norms are the scores obtained from a large number of subjects during the process of standardizing a test. These multiple person-derived scores serve as a standard for comparison against which a single individual may be evaluated. If such a client differs in sex, race, or socioeconomic status from the group upon whom a test's norms were derived, then the test is not valid for this individual. Test norms are generally reported as mental ages, standard scores, or percentiles. A child's mental age (MA) is the age of all other children whose test performance he or she equals. In one early form of its measurement, an I.Q. was obtained by dividing the mental age by the chronological age and multiplying by 100:

$$\text{I.Q.} = \frac{\text{MA}}{\text{CA}} \times 100$$

Thus an eight-year-old who scores as well as an average ten-year-old on the individually administered Stanford-Binet Test has an MA of 10 and a CA of 8. His I.Q. in this case is 125.

The measurement of intelligence as reflected by I.Q. test scores reveals that individuals vary by degrees along a continuous scale. The distribution of such scores is portrayed graphically in Figure 1. A percentile is a score that divides the sample population into 100 parts. It tells the percentage of individuals found below a given score. Standard scores show an individual's distance from the mean in standard deviation units. In Figure 1, the mean I.Q. is arbitrarily set at 100. That number is merely a convention. The standard deviation (SD) is a measure of the variability of the I.Q. scores of many individuals around that mean. One finds approximately 68% of the I.Q. scores of individual persons within one standard deviation above and below the mean for the age (sex, and so on) group of which they are a member. In Figure 1, 50% of all persons shown earned an I.Q. score between 90 and 110, with 25% obtaining I.Q.'s below 90 and the remaining 25% earning scores above 110.

Source: *Encyclopedia of Psychology*, 2nd Ed., Vol. 2, pp. 264–267 (New York: John Wiley & Sons, 1994).

Figure 1. The distribution of Wechsler Adult Intelligence Scale I.Q. categories. These categories of measured intelligence serve as only one of several indices of intelligence. (From J. D. Matarazzo, *Wechsler's measurement and appraisal of adult intelligence*, 5th ed. Baltimore: Williams and Wilkins, 1976), p. 124. © Oxford University Press, New York. Reprinted by permission.

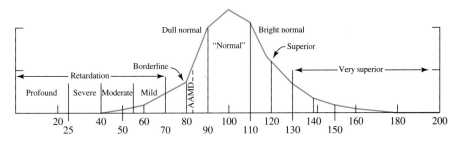

Is the I.Q. score a continuous variable? What type of distribution does the I.Q. score have? What is its mean and standard deviation? How are percentages obtained for intervals of scores? What percentage of the adult population has "superior" or higher intelligence? (See Exercise 6.95.)

● CHAPTER OBJECTIVES

Until now we have considered distributions of discrete variables only. In this chapter we examine one particular family of probability distributions of major importance whose domain is the set of all real numbers. These distributions are called the *normal*, the *bell-shaped*, or the *Gaussian distributions*. "Normal" is simply the traditional title of this particular type of distribution and is not a descriptive name meaning "typical." Although there are many other types of continuous distributions (rectangular, triangular, skewed, and so on), many variables have an approximately normal distribution. For example, several of the histograms drawn in Chapter 2 suggested a normal distribution. A mounded histogram that is approximately symmetric is an indication of such a distribution.

In addition to learning what a normal distribution is, we consider (1) how probabilities are found, (2) how they are represented, and (3) how normal distributions are used. Although there are other distributions of continuous variables, the normal distributions are the most important.

6.1 | NORMAL PROBABILITY DISTRIBUTIONS

Normal probability distribution

The **normal probability distribution** is considered to be the single most important probability distribution. There are an unlimited number of continuous random variables that have either a normal or an approximately normal distribution. There are also other probability distributions of both discrete and continuous random variables that are approximately normal under certain conditions.

Recall that in Chapter 5 we learned how to use a probability function to calculate probabilities associated with discrete random variables. The normal probability distribution has a continuous random variable and it uses two functions: one function to determine the ordinates (*y*-values) of the graph picturing the distribution, and a second to determine probabilities. Formula (6-1) expresses the ordinate (*y*-value) that corresponds to each abscissa (*x*-value).

Normal probability distribution function:

$$y = f(x) = \frac{e^{-1/2(x-\mu/\sigma)^2}}{\sigma\sqrt{2\pi}}, \text{ for all real } x \qquad \textbf{(6-1)}$$

Normal (bell-shaped) curve

When a graph of all such points is drawn, the **normal (bell-shaped) curve** will appear as shown in Figure 6-1.

FIGURE 6-1

The Normal
Probability
Distribution

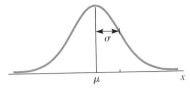

Formula (6-2) yields the probability associated with the interval from $x = a$ to $x = b$.

$$P(a \le x \le b) = \int_a^b f(x)dx \qquad (6\text{-}2)$$

The probability that x is within the interval from $x = a$ to $x = b$ is shown in Figure 6-2.

FIGURE 6-2

Shaded Area:
$P(a \le x \le b)$

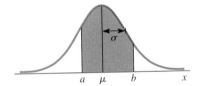

We will not be using the preceding formulas to calculate probabilities for normal distributions. The definite integral of formula (6-2) is a calculus topic and is mathematically beyond what is expected in elementary statistics. (These formulas often appear at the top of normal probability tables as identification.) *Instead of using formulas (6-1) and (6-2), we will use a table to find probabilities for normal distributions.* Before we learn to use the table, I must point out that the table is expressed in "standardized" form. It is standardized so that this one table can be used to find probabilities for all combinations of mean and standard deviation values. That is, the normal probability distribution with mean 38 and standard deviation 7 is very similar to the normal probability distribution with mean 123 and standard deviation 32. Recall the empirical rule and the percentage of the distribution that fall within certain intervals of the mean (see page 104). The three percentages held true for all normal distributions.

Percentage
Proportion
Probability
Area

NOTE   Percentage, proportion, and probability are basically the same concepts. **Percentage** (25%) is usually used when talking about a **proportion** (1/4) of a population. **Probability** is usually used when talking about the chance that the next individual item will possess a certain property. **Area** is the graphic representation of all three when we draw a picture to illustrate the situation.

The empirical rule is a fairly crude measuring device; with it we are able to find probabilities associated only with whole-number multiples of the standard deviation (within one, two, or three standard deviations of the mean). We will often be interested in the probabilities associated with fractional parts of the standard deviation. For example, we might want to know the probability that x is within 1.37 standard deviations of the mean. Therefore, we must refine the empirical rule so that we can deal with more precise measurements. This refinement is discussed in the next section.

6.2 | THE STANDARD NORMAL DISTRIBUTION

Standard normal
distribution

There are an unlimited number of normal probability distributions, but fortunately they are all related to one distribution, the **standard normal distribution**. The standard normal distribution is the normal distribution of the standard variable z [called "z-score"].

PROPERTIES OF THE STANDARD NORMAL DISTRIBUTION

1. The total area under the normal curve is equal to 1.
2. The distribution is mounded and symmetric; it extends indefinitely in both directions, approaching but never touching the horizontal axis.
3. The distribution has a mean of 0 and a standard deviation of 1.
4. The mean divides the area in half, 0.50 on each side.
5. Nearly all the area is between $z = -3.00$ and $z = 3.00$.

Table 3 in Appendix B lists the probabilities associated with the intervals from the mean (located at $z = 0.00$) to a specific value of z. Probabilities of other intervals will be found by using the table entries and the operations of addition and subtraction, in accordance with the preceding properties. Let's look at several illustrations demonstrating how to use Table 3 to find probabilities of the standard normal score, z.

 ILLUSTRATION 6-1

Find the area under the standard normal curve between $z = 0$ and $z = 1.52$.

FIGURE 6-3

Area from $z = 0$ to $z = 1.52$

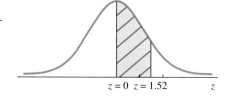

$z = 0$ $z = 1.52$ z

SOLUTION Table 3 is designed to read the area between $z = 0$ and $z = 1.52$ directly. The z-score is located on the margin, with the units and tenths digit located along the left side and the hundredths digit located along the top margin. For $z = 1.52$, locate the row labeled 1.5 and the column labeled 0.02; at their intersection you will find **0.4357**, the measure of the area or probability for the interval $z = 0.00$ to $z = 1.52$. Expressed as a probability: $P(0.00 < z < 1.52) = 0.4357$.

TABLE 6-1

A Portion of Table 3	z	0.00	0.01	0.02	...
	⋮				
	1.5			0.4357	
	⋮				

▲

EXERCISE 6.1

Find the area under the standard normal curve between $z = 0$ and $z = 1.37$.

Recall that one of the basic properties of probability is that the sum of all probabilities is exactly 1.0. Since the area under the normal curve represents the measure of probability, the total area under the bell-shaped curve is exactly 1 unit. This distribution is symmetric with respect to the vertical line drawn through $z = 0$, which cuts the area exactly in half at the mean. Can you verify this fact by inspecting formula (6-1)? That is, the area under the curve to the right of the mean is exactly one-half unit, 0.5, and the area to the left is also one-half unit, 0.5. Areas (probabilities) not given directly in the table can be found by relying on these facts.

Now let's look at some illustrations.

▼ **ILLUSTRATION 6-2**

Find the area under the normal curve to the right of $z = 1.52$; $P(z > 1.52)$.

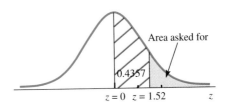

Area asked for

0.4357

$z = 0$ $z = 1.52$ z

SOLUTION The area to the right of the mean (all shading in the figure) is exactly 0.5000. The question asks for the shaded area that is not included in the 0.4357. Therefore, subtract 0.4357 from 0.5000.

$$P(z > 1.52) = 0.5000 - 0.4357 = \mathbf{0.0643}$$

▲

EXERCISE 6.2

Find the area to the right of $z = 2.03$, $P(z > 2.03)$.

SUGGESTION As we have done here, always draw and label a sketch. It is most helpful.

▼ | ILLUSTRATION   6 - 3

Find the area to the left of $z = 1.52$; $P(z < 1.52)$.

SOLUTION The total shaded area is made up of 0.4357 found in the table and the 0.5000 that is to the left of the mean. Therefore, add 0.4357 to 0.5000.

$$0.4357 + 0.5000 = \mathbf{0.9357}$$

▲ |

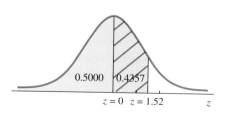

EXERCISE   6.3

Find the area to the left of $z = 1.73$, $P(z < 1.73)$.

NOTE   The addition and subtraction done in Illustrations 6-2 and 6-3 are correct because the "areas" represent mutually exclusive events (discussed in Section 4-5).

The symmetry of the normal distribution is a key factor in determining probabilities associated with values below (to the left of) the mean. The area between the mean and $z = -1.52$ is exactly the same as the area between the mean and $z = +1.52$. This fact allows us to find values related to the left side of the distribution.

▼ | ILLUSTRATION   6 - 4

EXERCISE   6.4

Find the area between -1.39 and the mean, $P(-1.39 < z < 0.00)$.

The area between the mean ($z = 0$) and $z = -2.1$ is the same as the area between $z = 0$ and $z = +2.1$; that is, $P(-2.1 < z < 0) = P(0 < z < 2.1)$.

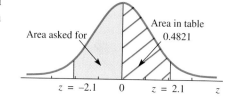

Thus, we have

$$P(-2.1 < z < 0) = \mathbf{0.4821}$$

▼ | ILLUSTRATION   6 - 5

EXERCISE   6.5

Find the area to the left of $z = -1.53$, $P(z < -1.53)$.

The area to the left of $z = -1.35$ is found by subtracting 0.4115 from 0.5000.

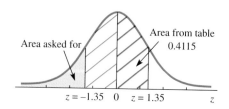

Therefore, we obtain

$$P(z < -1.35) = \mathbf{0.0885}$$

▲ |

▼| ILLUSTRATION 6-6

EXERCISE 6.6

Find the area between $z = -1.83$ and $z = 1.23$, $P(-1.83 < z < 1.23)$.

The area between $z = -1.5$ and $z = 2.1$, $P(-1.5 < z < 2.1)$, is found by adding the two areas together.

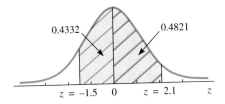

Therefore, we obtain

$$P(-1.5 < z < 0) + P(0 < z < 2.1) = 0.4332 + 0.4821 = \mathbf{0.9153}$$

▼| ILLUSTRATION 6-7

The area between $z = 0.7$ and $z = 2.1$, $P(0.7 < z < 2.1)$, is found by subtracting. The area between $z = 0$ and $z = 2.1$ contains all the area between $z = 0$ and $z = 0.7$. The area between $z = 0$ and $z = 0.7$ is therefore subtracted from the area between $z = 0$ and $z = 2.1$.

EXERCISE 6.7

Find the area between $z = 0.75$ and $z = 2.25$, $P(0.75 < z < 2.25)$.

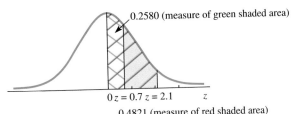

Thus, we have

$$P(0.7 < z < 2.1) = 0.4821 - 0.2580 = \mathbf{0.2241}$$

The normal distribution table can also be used to determine a z-score if we are given an area. The next illustration considers this idea.

▼| ILLUSTRATION 6-8

What is the z-score associated with the 75th percentile? (Assume the distribution is normal.) See Figure 6-4.

FIGURE 6-4

P_{75} and Its Associated z-Score

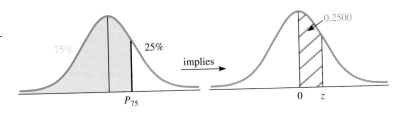

(continued)

EXERCISE   6.8

Find the *z*-score for the 80th percentile of a normal distribution.

SOLUTION To find this *z*-score, look in Table 3, Appendix B, and find the "area" entry that is closest to 0.2500; this area entry is 0.2486. Now read the *z*-score that corresponds to this area. From the table the *z*-score is found to be $z = \textbf{0.67}$. This says that the 75th percentile in a normal distribution is 0.67 $\left(\text{approximately } \frac{2}{3}\right)$ standard deviation above the mean.

z	...	0.07			0.08	...
:						
:						
0.6		0.2486	0.2500		0.2517	
:						

▼| ILLUSTRATION   6 - 9

What *z*-scores bound the middle 95% of a normal distribution? See Figure 6.5.

FIGURE 6-5

Middle 95% of Distribution and Its Associated *z*-score

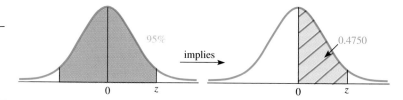

EXERCISE   6.9

Find the *z*-scores that bound the middle 75% of a normal distribution.

SOLUTION The 95% is split into two equal parts by the mean; 0.4750 is the area (percentage) between $z = 0$, the mean, and the *z*-score at the right boundary. Since we have the area, we look for the entry in Table 3 closest to 0.4750 (it happens to be exactly 0.4750) and read the *z*-score in the margin. We obtain $z = 1.96$. Therefore, $z = \textbf{-1.96 and } z = \textbf{1.96}$ bound the middle 95% of a normal distribution.

z	...	0.06	...
:			
:			
1.9		0.4750	
:			

▲|

EXERCISES

6.10   a. Describe the distribution of the standard normal score *z*.
            b. Why is this distribution called *standard normal*?

6.11   Find the area under the normal curve that lies between the following pairs of *z*-values.
            a. $z = 0$ to $z = 1.30$ b. $z = 0$ to $z = 1.28$
            c. $z = 0$ to $z = -3.20$ d. $z = 0$ to $z = -1.98$

6.12 Find the probability that a piece of data picked at random from a normal population will have a standard score (z) that lies between the following pairs of z-values.

a. $z = 0$ to $z = 2.10$ b. $z = 0$ to $z = 2.57$
c. $z = 0$ to $z = -1.20$ d. $z = 0$ to $z = -1.57$

6.13 Find the area under the standard normal curve that corresponds to the following z-values.

a. between 0 and 1.55   b. to the right of 1.55
c. to the left of 1.55   d. between -1.55 and 1.55

6.14 Find the probability that a piece of data picked at random from a normal population will have a standard score (z) that lies

a. between 0 and 0.84.   b. to the right of 0.84.
c. to the left of 0.84.   d. between -0.84 and 0.84.

6.15 Find the area under the normal curve that lies between the following pairs of z-values.

a. $z = -1.20$ to $z = 1.22$ b. $z = -1.75$ to $z = 1.54$
c. $z = -1.30$ to $z = 2.58$ d. $z = -3.5$ to $z = -0.35$

6.16 Find the probability that a piece of data picked at random from a normal population will have a standard score (z) that lies between the following pairs of z-values:

a. $z = -2.75$ to $z = 1.38$
b. $z = 0.67$ to $z = 2.95$
c. $z = -2.95$ to $z = -1.18$

6.17 Find the following areas under the normal curve.

a. to the right of $z = 0.00$ b. to the right of $z = 1.05$
c. to the right of $z = -2.30$ d. to the left of $z = 1.60$
e. to the left of $z = -1.60$

6.18 Find the probability that a piece of data picked at random from a normally distributed populaton will have a standard score that is

a. less than 3.00.   b. greater than -1.55.
c. less than -0.75.   d. less than 1.25.
e. greater than -1.25.

6.19 Find the following:

a. $P(0.00 < z < 2.35)$ b. $P(-2.10 < z < 2.34)$
c. $P(z > 0.13)$ d. $P(z < 1.48)$

6.20 Find the following:

a. $P(-2.05 < z < 0.00)$ b. $P(-1.83 < z < 2.07)$
c. $P(z < -1.52)$ d. $P(z < -0.43)$

6.21 Find the z-score for the standard normal distribution shown on each of the following diagrams.

a.

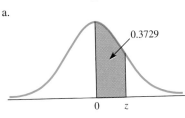

b.

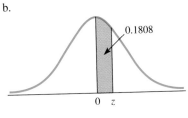

c.

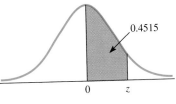

(continued)

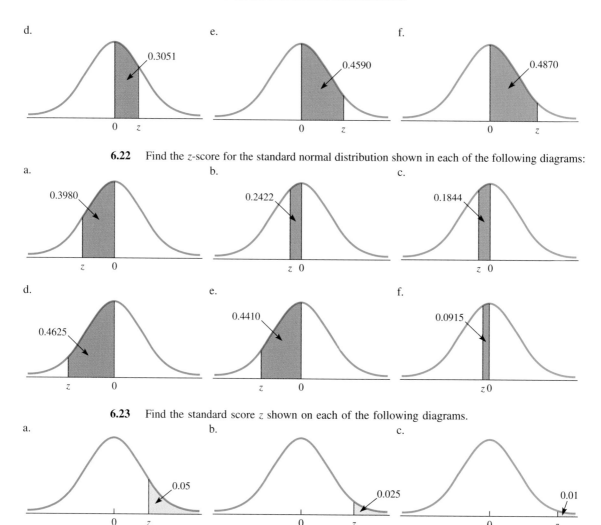

d.

e.

f.

0.3051

0.4590

0.4870

0   z   0   z   0   z

6.22   Find the z-score for the standard normal distribution shown in each of the following diagrams:

a.   b.   c.

0.3980   0.2422   0.1844

z 0 z 0 z 0

d.   e.   f.

0.4625   0.4410   0.0915

z 0 z 0 z0

6.23   Find the standard score z shown on each of the following diagrams.

a.   b.   c.

0.05   0.025   0.01

0   z 0 z 0 z

6.24   Find the standard score z shown on each of the following diagrams.

a.   b.   c.

0.7673   0.7190   0.1515

0   z z 0 z 0

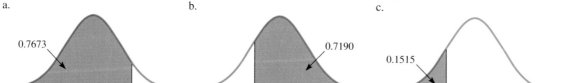

6.25   Find a value of z such that 40% of the distribution lies between it and the mean. (There are two possible answers.)

6.26   Find the standard z score such that

     a. 80% of the distribution is below (to the left of) this value.

     b. the area to the right of this value is 0.15.

6.27   Find the two z-scores that bound the middle 50% of a normal distribution.

6.28   Find the two standard scores z such that
   a. the middle 90% of a normal distribution is bounded by them.
   b. the middle 98% of a normal distribution is bounded by them.

6.29   Assuming a normal distribution, what is the z-score associated with the 90th percentile? the 95th percentile? the 99th percentile?

6.30   Assuming a normal distribution, what is the z-score associated with the 1 quartile? 2 quartile? 3 quartile?

6.3 | APPLICATIONS OF NORMAL DISTRIBUTIONS

In Section 6-2 we learned how to use Table 3 in Appendix B to convert information about the standard normal variable z into probability, or the opposite, to convert probability information about the standard normal distribution into z-scores. Now we are ready to apply this methodology to all normal distributions. The key is the standard score, z. The information associated with a normal distribution will be in terms of x-values or probabilities. We will use the z-score and Table 3 as the tools to "go between" the given information and the desired answer.

Recall that the standard score, z, was defined in Chapter 2.

$$\text{Standard score, } z: \quad z = \frac{x - (\text{mean of } x)}{(\text{standard deviation of } x)}$$

$$z = \frac{x - \mu}{\sigma} \tag{6-3}$$

(Note, when $x = \mu$, the standard score $z = 0$.)

▼ | ILLUSTRATION   6 - 10

Consider the intelligence quotient (I.Q.) scores for people. I.Q.s are normally distributed with a mean of 100 and a standard deviation of 16. If a person is picked at random, what is the probability that his or her I.Q. is between 100 and 115; that is, what is $P(100 < x < 115)$?

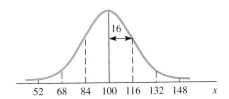

SOLUTION $P(100 < x < 115)$ is represented by the shaded area in the following figure.

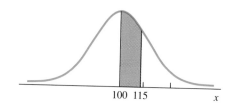

100 115 x

EXERCISE 6.31

Given $x = 58$, $\mu = 43$, and $\sigma = 5.2$, find z.

The variable x must be standardized by using formula (6-3). The z-values are shown on the next figure.

$$z = \frac{x - \mu}{\sigma}$$

EXERCISE 6.32

Find the probability that a randomly selected person will have an I.Q. score between 100 and 120.

When $x = 100$: $z = \dfrac{100 - 100}{16} = \mathbf{0.0}$

When $x = 115$: $z = \dfrac{115 - 100}{16} = 0.9375 = \mathbf{0.94}$

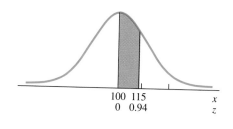

100 115
0 0.94
     x
     z

The value 0.3264 is found by using Table 3 in Appendix B.

Therefore,

$$P(100 < x < 115) = P(0.0 < z < 0.94) = \mathbf{0.3264}$$

Thus, the probability is 0.3264 that a person picked at random has an I.Q. between 100 and 115.

▼ ILLUSTRATION 6-11

Find the probability that a person selected at random will have an I.Q. greater than 90.

SOLUTION

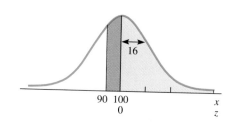

90 100
0
     x
     z

$$z = \frac{x - \mu}{\sigma} = \frac{90 - 100}{16} = \frac{-10}{16} = -0.625 = -0.63$$

$$P(x > 90) = P(z > -0.63)$$

$$= 0.2357 + 0.5000 = \mathbf{0.7357}$$

Thus, the probability is 0.7357 that a person selected at random will have an I.Q. greater than 90.

EXERCISE 6.33

Find the probability that a randomly selected person will have an I.Q. score above 80.

The normal table can be used to answer many kinds of questions that involve a normal distribution. Many times a problem will call for the location of a "cutoff point," that is, a particular value of x such that there is exactly a certain percentage in a specified area. The following illustrations concern some of these problems.

ILLUSTRATION 6-12

In a large class, suppose your instructor tells you that you need to obtain a grade in the top 10% of your class to get an A on a particular exam. From past experience she is able to estimate that the mean and standard deviation on this exam will be 72 and 13, respectively. What will be the minimum grade needed to obtain an A? (Assume that the grades will be approximately normally distributed.)

SOLUTION Start by converting the 10% to information that is compatible with Table 3 by subtracting:

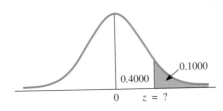

$$10\% = 0.1000 = 0.5000 - 0.4000$$

Look in Table 3 to find the value of z associated with the area entry closest to 0.4000; it is $z = 1.28$. Thus,

$$P(z > 1.28) = 0.10$$

EXERCISE 6.34

Suppose the instructor in Illustration 6-12 said the top 15% were to get A's. Find the minimum score to receive an A.

Now find the x-value that corresponds to $z = 1.28$ by using formula (6-3):

$$z = \frac{x - \mu}{\sigma}: 1.28 = \frac{x - 72}{13}$$

$$x - 72 = (13)(1.28)$$

$$x = 72 + (13)(1.28) = 72 + 16.64$$

$$= 88.64, \text{ or } \mathbf{89}$$

Thus, if you receive an 89 or higher, you can expect to be in the top 10% (which means an A).

Case Study

A Predictive Failure Cost Tool

6-1   *The standard normal probability distribution can be and is used in connection with many variables. This article shows its application in a predictive role.*

The normal curve is commonly used for analysis of process capability and percentage out of specification. Additionally, some engineers use it to calculate probability of occurrence of an event. But there is another use for the normal curve that is seldom taken advantage of—process cost analysis. Using a normal curve to estimate scrap and rework is an accurate method of estimating those costs, and the calculations are simple.

A factory, for example, makes a process change that is expected to reduce scrap and save the company $60,000 a year. Interestingly enough, neither the new or old process is capable of producing 100 percent of product to specifications, and the company has accepted sorting as a way of doing business.

The new process was started at the beginning of a week. Within three days, there was enough data for normal curve cost analysis to show that the process would not save $60,000. In fact, it would cost an additional $30,000 over the older process because, even though scrap had been reduced as predicted, the new process generated more rework at higher rework cost. But cost accounting didn't recognize the problem for a full month until it had enough data to identify the problem with the process' performance.

To use normal curve analysis for estimating cost, measurement error must be disregarded and the process must:

- Not be capable of meeting specifications.
- Be in statistical control.
- Produce a normal distribution.

A widget manufacturer, for example, shows how this statistical technique works. Table 1 shows the parameters for the widget process and cost information.

Table 1. Process parameters.	
Spec	0.5530 ± 0.0025
Process average	0.5535
Standard deviation	0.0015
Production rate	100,000 per year

Rework all undersize parts at a cost of $5.25 each.

Scrap all oversize parts at a cost of $15.34 each.

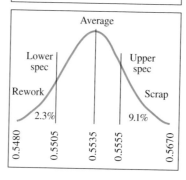

Figure 1. Distribution of widget process.

Figure 1 shows that the process is not capable of meeting specification. The process parameters indicate that all undersize parts are reworked and oversize parts are scrapped.

To determine rework costs, calculate the area under the normal curve (*z*-score) that represents the percent of undersized parts.

$$z\text{-score} = \frac{\text{lower spec} - \text{average}}{\text{std. dev.}}$$

where:

lower spec	= 0.5505
average	= 0.5535
std. dev.	= 0.0015
z-score	= −2.0.

A *z*-score of −2.0 represents 2.3 percent of parts undersized.

The total annual production of undersized parts is calculated by:

Parts reworked annually = (fraction defective)(annual production)

= (0.023)(100,000)

= 2,300 parts.

The annual cost of reworking parts is:

Annual rework cost = (parts reworked annually)(cost/part)

= (2,300)($5.25)

= $12,075.00.

The preceding calculations represent only the part production that can be reworked. Additionally, the cost of making scrap must also be considered. To determine the cost of making scrap, calculate the *z*-score for scrapped parts.

$$z\text{-score} = \frac{\text{upper spec} - \text{average}}{\text{std. dev.}}$$

where:

upper spec	= 0.5555
average	= 0.5535
std. dev.	= 0.0015
z-score	= 1.3.

A *z*-score of 1.3 represents 9.1 percent scrap.

The annual scrap production is calculated by:

Annual scrap production = (fraction defective)(annual production)

= (0.091)(100,000)

= 9,100.

The annual cost of scrap is:

Annual cost of scrap = (annual scrap production)(cost/scrapped part)

= (9,100)($15.34)

= $139,594.00.

The widget prediction shows that the process will generate $12,075.00 in rework and $139,594.00 in scrap or a total of $151,669.00 of failure cost. If the process is centered in the tolerance zone and the cost recalculated, the process will yield $25,090.00 in rework and $73,310.00 in scrap—a total failure cost of $98,400.00. This one step yields a $53,269.00 cost reduction over the previous process setting.

Obviously, some time needs to be spent to determine the process setting that will minimize scrap and rework cost. This requires a balance between scrap and rework to minimize cost impact. Shifting the process to eliminate scrap entirely will yield a substantial cost increase.

Source: Gregory Roth, "A Predictive Failure Cost Tool," *Quality*, October 1993.

▼ | I L L U S T R A T I O N   6 - 13

Find the 33rd percentile for I.Q. scores ($\mu = 100$ and $\sigma = 16$ from Illustration 6-10, p. 299).

EXERCISE 6.35

a. Verify the z-score for the lower spec and fraction of re-workable parts.

b. Verify the z-score for the upper spec and fraction of scrap parts.

S O L U T I O N

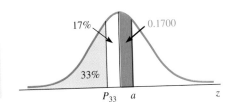

z	**0.04**
0.4	0.1700

$$P(0 < z < a) = 0.17$$
$$a = 0.44 \quad \text{(cutoff value of } z \text{ from Table 3)}$$
$$\text{33rd percentile of } z = -0.44 \quad \text{(below mean)}$$

Now we convert the 33rd percentile of the z-scores, -0.44, to an x-score:

$$z = \frac{x - \mu}{\sigma} \quad \text{[formula (6-3)]}$$

EXERCISE 6.36

Find the 25th percentile for I.Q. scores in Illustration 6-10.

$$-0.44 = \frac{x - 100}{16}$$
$$16(-0.44) = x - 100$$
$$100 - 7.04 = x$$
$$x = \mathbf{92.96}$$

▲ | Thus, 92.96 is the 33rd percentile for I.Q. scores.

Illustration 6-14 concerns a situation in which you are asked to find the mean μ when given the related information.

▼ | ILLUSTRATION   6 - 14

The incomes of junior executives in a large corporation are normally distributed with a standard deviation of $1,200. A cutback is pending, at which time those who earn less than $28,000 will be discharged. If such a cut represents 10% of the junior executives, what is the current mean salary of the group of junior executives?

SOLUTION If 10% of the salaries are below $28,000, then 40% (or 0.4000) are between $28,000 and the mean μ. Table 3 indicates that $z = -1.28$ is the standard score that occurs at $x = $28,000. Using formula (6-3) we can find the value of μ:

EXERCISE 6.37

If 20% of the salaries in Illustration 6-14 are below $28,000, find the current mean salary.

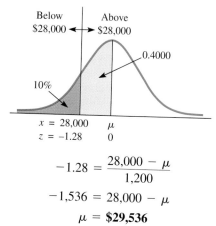

$$-1.28 = \frac{28,000 - \mu}{1,200}$$

$$-1,536 = 28,000 - \mu$$

$$\mu = \$29,536$$

▲ | That is, the current mean salary of junior executives is $29,536.

Referring again to the I.Q. scores, what is the probability that a person picked at random has an I.Q. of 125, $P(x = 125)$? (I.Q. scores are normally distributed with a mean of 100 and a standard deviation of 16.) This situation has two interpretations: (1) theoretical and (2) practical. Let's look at the theoretical interpretation first. Recall that the probability associated with a continuous random variable is represented by the area under the curve. That is, $P(a \le x \le b)$ is equal to the area between a and b under the curve. $P(x = 125)$ (that is, x is exactly 125) is then $P(125 \le x \le 125)$, or the area of the vertical line segment at $x = 125$. This area is zero. However, this is not the *practical* meaning of $x = 125$. It generally means 125 to the nearest integer value. Thus, $P(x = 125)$ would most likely be interpreted as

$$P(124.5 < x < 125.5)$$

The interval from 124.5 to 125.5 under the curve has a measurable area and is then nonzero. In situations of this nature, you must be sure of the meaning being used.

NOTE   A standard notation used to abbreviate "normal distribution with mean μ and standard deviation σ" is $N(\mu, \sigma)$. That is, $N(58, 7)$ represents "normal distribution, mean = 58 and standard deviation = 7."

EXERCISE   6.38

Generate a random
sample of 100 data
from a normal distri-
bution with mean 50
and standard devia-
tion 12.

MINITAB (Release 10) commands to generate n random data in C1
from a normal distribution with given mean A and standard
deviation B.

Session commands   *Menu commands*

Enter: RANDom n C1;   Choose: Calc > Random Data > Normal
        NORMal A B.   Enter: *Generate:*   n
                                     Store in: C1
                                     Mean: A
                                     Stand. dev.: B

If multiple samples (say, 12), all of same size, are wanted,
modify the above commands:

RANDom n C1-C12;   Store in: C1-C12

Note: To find descriptive statistics for each of these samples,
       use the session command DESCribe C1-C12, or any of the
       others listed in earlier chapters.

EXERCISE   6.39

Use the random sam-
ple of 100 data
found in Exercise
6.38 and find the
100 corresponding
y-values for the nor-
mal distribution
curve with mean 50
and standard devia-
tion 12.

MINITAB (Release 10) commands to find the ordinate values
(store them in C2) paired with the abscissas listed in C1 from
a normal distribution with a given mean A and standard
deviation B.

Session commands   *Menu commands*

Enter: PDF C1 C2;   Choose: Calc > Prob. Dist. > Normal
         NORMal A B.   Select: Probability Density
                                Enter: Mean: A
                                         Stand. dev.: B
                                         Input column: C1
                                         Optional Storage: C2

EXERCISE   6.40

Use the 100 ordered
pairs found in Exer-
cise 6.39 and draw
the curve for the nor-
mal distribution with
mean 50 and stan-
dard deviation 12.

To draw the graph of a normal probability curve with *x*-values
in C1 and *y*-values in C2:

Session commands   *Menu commands*

Enter: PLOT C2*C1;   Choose: Graph > Plot
        CONNect.   Enter: Graph variables: Y: C2 X: C1
                                   Data display: Display:
                                   Connect

MINITAB (Release 10) commands to find the cumulative
probabilities (store them in C3) paired with the abscissas
listed in C1 from a normal distribution with a given mean A and
standard deviation B.

(continued)

MINITAB (*continued*)

EXERCISE 6.41

Find the probability that a randomly selected value from a normal distribution with mean 50 and standard deviation 12 will be between 55 and 65. Verify your results by using Table 3.

Session commands

Enter: CDF C1 C3;
        NORMal A B.

Menu commands

Choose: Calc > Prob. Dist. > Normal
Select: Cumulative Probability
Enter: Mean: A
        Stand. dev.:   B
        Input column: C1
        Optional Storage: C3

Note 1. To find the probability between two *x*-values, enter the two values in C1, use the CDF, and subtract using numbers in C3.
     2. To draw a graph of the cumulative probability distribution (ogive), use PLOT C3*C1; and CONNect (similar to above with the distribution curve).

EXERCISES

6.42 Given that *x* is a normally distributed random variable with a mean of 60 and a standard deviation of 10, find the following probabilities.

   a. $P(x > 60)$ b. $P(60 < x < 72)$
   c. $P(57 < x < 83)$ d. $P(65 < x < 82)$
   e. $P(38 < x < 78)$ f. $P(x < 38)$

6.43 For a particular age group of adult males, the distribution of cholesterol readings, in mg/dl, is normally distributed with a mean of 210 and a standard deviation of 15.

   a. What percentage of this population would have readings exceeding 250?
   b. What percentage would have readings less than 150?

6.44 Computers are shut down for certain periods of time for routine maintenance, installation of new hardware, and so on. The down times for a particular computer are normally distributed with a mean of 1.5 hours and a standard deviation of 0.4 hour.

   a. What percentage of the down times exceed 3 hours?
   b. What percentage are between 1 and 2 hours?

6.45 According to the November 1993 issue of *Harper's* magazine, our kids spend from 1200 to 1800 hours a year in front of the television set. Suppose the time spent by kids in front of the television set per year is normally distributed with a mean equal to 1500 hours and a standard deviation equal to 100 hours.

   a. What percentage spend between 1400 and 1600 hours?
   b. What percentage spend between 1300 and 1700 hours?
   c. What percentage spend between 1200 and 1800 hours?
   d. Compare the results (a) through (c) with the empirical rule. Explain the relationship.

6.46 According to the August 15, 1994, issue of *National Review*, the average state and federal cigarette taxes per pack are 52 cents.

   a. Assuming that the state and federal cigarette taxes per pack are normally distributed with a standard deviation of 7 cents, what percentage of the states tax packages of cigarettes between 40 and 60 cents?
   b. The assumption of normality allows us to estimate a probability or percentage. How might this assumption affect the accuracy of the answer? Explain.

6.47   The length of useful life of a fluorescent tube used for indoor gardening is normally distributed. The useful life has a mean of 600 hr and a standard deviation of 40 hr. Determine the probability that

    a. a tube chosen at random will last between 620 and 680 hr.

    b. such a tube will last more than 740 hr.

6.48   At Pacific Freight Lines, bonuses are given to billing clerks when they complete 300 or more freight bills during an eight-hour day. The number of bills completed per clerk per eight-hour day is approximately normally distributed with a mean of 270 and a standard deviation of 16. What proportion of the time should a randomly selected billing clerk expect to receive a bonus?

6.49   The waiting time x at a certain bank is approximately normally distributed with a mean of 3.7 min and a standard deviation of 1.4 min.

    a. Find the probability that a randomly selected customer has to wait less than 2.0 min.

    b. Find the probability that a randomly selected customer has to wait more than 6 min.

    c. Find the value of the 75th percentile for x.

6.50   A brewery filling machine is adjusted to fill quart bottles with a mean of 32.0 oz of ale and a variance of 0.003. Periodically, a bottle is checked and the amount of ale is noted.

    a. Assuming the amount of fill is normally distributed, what is the probability that the next randomly checked bottle contains more than 32.02 oz?

    b. Let's say you buy 100 quart bottles of this ale for a party; how many bottles would you expect to find containing more than 32.02 oz of ale?

6.51   Final averages are typically approximately normally distributed with a mean of 72 and a standard deviation of 12.5. Your professor says that the top 8% of the class will receive an A; the next 20%, a B; the next 42%, a C; the next 18% a D; and the bottom 12% an F.

    a. What average must you exceed to obtain an A?

    b. What average must you exceed to receive a grade better than a C?

    c. What average must you obtain to pass the course? (You'll need a D or better.)

6.52   A radar unit is used to measure the speed of automobiles on an expressway during rush-hour traffic. The speeds of individual automobiles are normally distributed with a mean of 62 mph.

    a. Find the standard deviation of all speeds if 3% of the automobiles travel faster than 72 mph.

    b. Using the standard deviation found in (a), find the percentage of these cars that are traveling less than 55 mph.

    c. Using the standard deviation found in (a), find the 95th percentile for the variable "speed."

6.53   The weights of ripe watermelons grown at Mr. Smith's farm are normally distributed with a standard deviation of 2.8 lb. Find the mean weight of Mr. Smith's ripe watermelons if only 3% weigh less than 15 lb.

6.54   A machine fills containers with a mean weight per container of 16.0 oz. If no more than 5% of the containers are to weigh less than 15.8 oz, what must the standard deviation of the weights equal? (Assume normality.)

6.55   According to a USA Snapshot® (10-26-94), the average annual salary for a worker in the United States is $26,362. If we assume that the annual salaries for Americans are normally distributed with a standard deviation equal to $6,500, find the following:

    a. What percentage make below $15,000?

    b. What percentage make above $40,000?

6.56 According to the 1991 issue of *American Hospital Administration Hospital Statistics*, the average daily census total for 116 hospitals in Mississippi equals 10,872. Suppose the standard deviation of the daily census totals for these hospitals equals 1,505 patients. If the daily census totals are normally distributed:

    a. What percentage of the days does the daily census total less than 8500 patients in these hospitals? Approximately how often should we expect this to occur?

    b. What percentage of the days does the daily census total exceed 12,500 patients in these hospitals? Approximately how often should we expect this to occur?

6.57 Use a computer to find the probability that one randomly selected value of x from a normal distribution, mean 584.2 and standard deviation 37.3, will have a value

    a. less than 525.

    b. between 525 and 590.

    c. of at least 590.

    d. Verify the result using Table 3.

    e. Explain any differences you may find.

MINITAB commands

```
Enter 525, 590 into C1
CDF C1 C2;
  NORMal 584.2 37.3.
```

6.58

    a. Use a computer to generate your own abbreviated standard normal probability table (a short version of Table 3). Use z-values of 0.0 to 5.0 in intervals of 0.1.

    b. How are the values obtained related to Table 3 entries? Make the necessary adjustment and store the results in a column.

    c. Compare your results in (b) to the first column of Table 3. Comment on any differences you see.

MINITAB commands

```
a. SET C1
   0:5/.1.
   END
   CDF C1 C2;
     NORMal 0 1.
b. LET C3 = C2 − .5
   PRINt C1 C3
```

6.59 Use a computer to compare a random sample to the population from which the sample was drawn. Consider the normal population with mean 100 and standard deviation 16.

    a. List values of x from $\mu - 4\sigma$ to $\mu + 4\sigma$ in increments of half standard deviations and store them in a column.

    b. Find the ordinate (y-value) corresponding to each abscissa (x-value) for the normal distribution curve for $N(100, 16)$ and store them in a column.

    c. Graph the normal probability distribution curve for $N(100, 16)$.

    d. Generate 100 random data values from the $N(100, 16)$ distribution and store them in a column.

    e. Graph the histogram of the 100 data obtained in (d) using the numbers listed in (a) as class boundaries.

    f. Calculate other helpful descriptive statistics of the 100 data and compare the data to the expected distribution. Comment on the similarities and the differences you see.

MINITAB commands

```
a. Set C1
   36:164/8
   END
b. PDF C1 C2;
     NORMal 100 16.
c. PLOT C2*C1
d. RANDom 100 C3;
     NORM 100 16.
e. HIST C3;
     CUTP C1.
f. MEAN C3
   STDEv C3
```

6.60 Suppose you were to generate several random samples, all the same size, all from the same normal probability distribution. Will they all be the same? How will they differ? By how much will they differ?

a. Use a computer to generate 10 different samples, all of size 100, all from the normal probability distribution of mean 200 and standard deviation 25.
b. Draw histograms of all 10 samples using the same class boundaries.
c. Calculate several descriptive statistics for all 10 samples, separately.
d. Comment on the similarities and the differences you see.

See Helpful Hint on page 89.

MINITAB commands

```
a. RANDom 100 C1-C10;
   NORMal 200 25.
b. HISTogram C1-C10;
   CUTPoints 100:300/25.
c. DESCribe C1-C10
```

6.4 | NOTATION

The z-score is used throughout statistics in a variety of ways; however, the relationship between the numerical value of z and the area under the standard normal distribution curve does not change. Since z will be used with great frequency, we want a convenient notation to identify the necessary information. The convention that we will use as an "algebraic name" for a specific z-score is $z(\alpha)$, where α represents the "area to the right" of the z being named.

▼ | ILLUSTRATION 6 - 15

$z(0.05)$(read "z of 0.05") is the algebraic name for the z such that the area to the right and under the standard normal curve is exactly 0.05, as shown in Figure 6-6.

EXERCISE 6.61

Draw a figure showing $z(0.15)$ on the standard normal curve.

FIGURE 6-6

Area Associated with $z(0.05)$

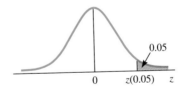

▼ | ILLUSTRATION 6 - 16

$z(0.60)$ (read "z of 0.60") is that value of z such that 0.60 of the area lies to its right, as shown in Figure 6-7.

EXERCISE 6.62

Draw a figure showing $z(0.82)$ on the standard normal curve.

FIGURE 6-7

Area Associated with $z(0.60)$

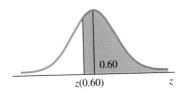

▲ |

Now let's find the numerical values of $z(0.05)$ and $z(0.60)$.

▼ | ILLUSTRATION   6 - 17

Find the numerical value of $z(0.05)$.

FIGURE 6-8

Find the Value of $z(0.05)$

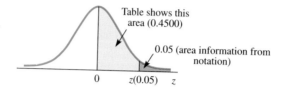

SOLUTION We must convert the area information in the notation into information that can be used with Table 3 in Appendix B; see the areas shown in Figure 6-8.

When we look in Table 3, we look for an area as close as possible to 0.4500.

EXERCISE 6.63

Find the value of $z(0.15)$.

z	...	0.04	0.05	...
⋮				
1.6	...	0.4495	0.4505	
⋮				

Therefore, $z(0.05) = $ **1.65**. (*Note:* We will use the z that corresponds to the area closest in value. If the value happens to be exactly halfway between the table entries, always use the larger value of z.)

▼ | ILLUSTRATION   6 - 18

Find the value of $z(0.60)$.

SOLUTION The value 0.60 is related to Table 3 by use of the area 0.1000, as shown in the following diagram. The closest values in Table 3 are 0.0987 and 0.1026.

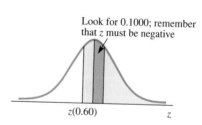

EXERCISE 6.64

Find the value of $z(0.82)$.

z	...	0.05	0.06	...
⋮				
0.2	...	0.0987	0.1026	
⋮				

▲ | Therefore, $z(0.60)$ is related to 0.25. Since $z(0.60)$ is below the mean, we conclude that $z(0.60) = $ **−0.25**.

In later chapters we will use this notation on a regular basis. The values of z that will be used regularly come from one of the following situations: (1) the z-score such that there is a specified area in one tail of the normal distribution, or (2) the z-scores that bound a specified middle proportion of the normal distribution.

Illustration 6-17 showed a commonly used one-tail situation; $z(0.05) = 1.65$ is located so that 0.05 of the area under the normal distribution curve is in the tail to the right.

▼ | ILLUSTRATION 6 - 19

Find $z(0.95)$.

SOLUTION $z(0.95)$ is located on the left-hand side of the normal distribution since the area to the right is 0.95. The area in the tail to the left then contains the other 0.05, as shown in Figure 6-9. Because of the symmetrical nature of the normal distribution, $z(0.95)$ is $-z(0.05)$, that is, $z(0.05)$ with its sign changed. Thus, $z(0.95) = -z(0.05) = -\mathbf{1.65}$.

FIGURE 6-9

Area Associated with $z(0.95)$

▲ |

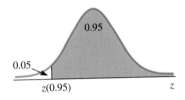

When the middle proportion of a normal distribution is specified, we can still use the "area to the right" notation to identify the specific z-score involved.

▼ | ILLUSTRATION 6 - 20

Find the z-scores that bound the middle 0.95 of the normal distribution.

SOLUTION Given 0.95 as the area in the middle (Figure 6-10), the two tails must contain a total of 0.05. Therefore, each tail contains 0.025, as shown in Figure 6-11.

FIGURE 6-10

Area Associated with Middle 0.95

FIGURE 6-11

Finding z-Scores for Middle 0.95

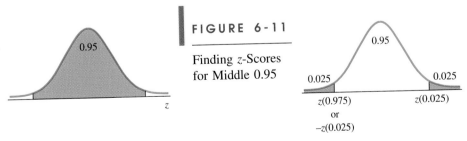

In order to find $z(0.025)$ in Table 3, we must determine the area between the mean and $z(0.025)$. It is 0.4750, as shown in Figure 6-12.

FIGURE 6-12

Finding the Value
of $z(0.025)$

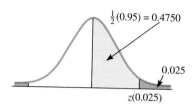

From Table 3,

	...	**0.06**	...
⋮			
1.9		0.4750	

Therefore, $z(0.025) = 1.96$ and $z(0.975) = -z(0.025) = -1.96$. The middle 0.95 of the normal distribution is bounded by **-1.96 and 1.96**.

EXERCISES

6.65 Using the $z(\alpha)$ notation (identify the value of α used within the parentheses), name each of the standard normal variable z's shown in the following diagrams.

a.

b.

c.

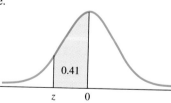

d.

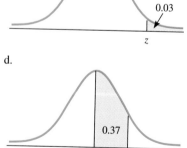

e.

f.

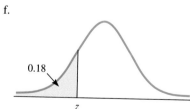

6.66 Using the $z(\sigma)$ notation (identify the value of α used within the parentheses), name each of the standard normal variable z's shown in the following diagrams.

a.

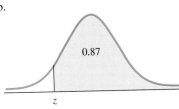

b.

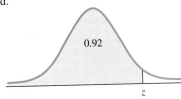

0.87

z

c.

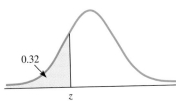

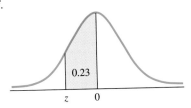

0.32

z

d.

0.92

z

e.

0.42

0   z

f.

0.23

z 0

6.67 We are often interested in finding the value of z that bounds a given area in the right-hand tail of the normal distribution, as shown in the accompanying figure. The notation $z(\alpha)$ represents the value of z such that $P(z > z(\alpha)) = \alpha$. Find the following:

a. $z(0.025)$

b. $z(0.05)$

c. $z(0.01)$

$z(\alpha)$

6.68 Use Table 3, Appendix B, to find the following values of z.
   a. $z(0.05)$ b. $z(0.01)$ c. $z(0.025)$ d. $z(0.975)$ e. $z(0.98)$

6.69 Complete the following charts of z-scores. The area A given in the tables is the area to the right under the normal distribution in the figures.
   a. z-scores associated with the right-hand tail: Given the area A, find $z(A)$.

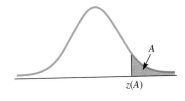

A

$z(A)$

A	0.10	0.05	0.025	0.02	0.01	0.005
$z(A)$						

(continued)

b. z-scores associated with the left-hand tail: Given the area A, find z(A).

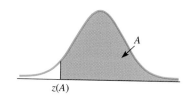

z(A)

A	0.995	0.99	0.98	0.975	0.95	0.90
z(A)						

6.70   a. Find the area under the normal curve for z between z(0.95) and z(0.025).
b. Find z(0.025) − z(0.95).

6.71   The z notation, z(α), combines two related concepts, the z-score and the *area to the right*, into a mathematical symbol. Identify the letter in each of the following as being a *z-score* or being an *area*, and then with the aid of a diagram explain what both the given number and the letter represent on the standard curve.
    a. z(A) = 0.10   b. z(0.10) = B   c. z(C) = −0.05   d. −z(0.05) = D

6.72   Understanding the z notation, z(α), requires us to know whether we have a z-score or an *area*. Each of the following expressions use the z notation in a variety of ways, some typical and some not so typical. Find the value asked for in each of the following, and then with the aid of a diagram explain what your answer represents.
    a. z(0.08)
    b. the area between z(0.98) and z(0.02)
    c. z(1.00 − 0.01)
    d. z(0.025) − z(0.975)

6.5 | NORMAL APPROXIMATION OF THE BINOMIAL

Binomial probability

In Chapter 5 we introduced the binomial distribution. Recall that the binomial distribution is a probability distribution of the discrete random variable x, the number of successes observed in n repeated independent trials. We will now see how **binomial probabilities**, that is, probabilities associated with a binomial distribution, can be reasonably estimated by using the normal probability distribution.

Let's look first at a few specific binomial distributions. Figures 6-13a, 6-13b, and 6-13c show the probabilities of x for 0 to n for three situations: n = 4, n = 8, and n = 24. For each of these distributions, the probability of success for one trial is 0.5. Notice that as n becomes larger, the distribution appears more and more like the normal distribution.

To make the desired approximation, we need to take into account one major difference between the binomial and the normal probability distributions. The binomial random variable is discrete, whereas the normal random variable is continuous. Recall that Chapter 5 demonstrated that the probability assigned to a particular value of x should be shown on a diagram by means of a straight-line segment whose length represents the probability (as in Figure 6-13). Chapter 5 suggested, however, that we can also use a histogram in which the area of each bar is equal to the probability of x.

FIGURE 6-13

Binomial
Distributions

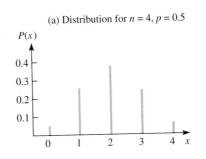

(a) Distribution for $n = 4$, $p = 0.5$

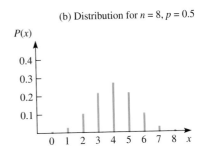

(b) Distribution for $n = 8$, $p = 0.5$

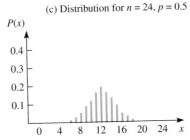

(c) Distribution for $n = 24$, $p = 0.5$

Let's look at the distribution of the binomial variable x, where $n = 14$ and $p = 0.5$. The probabilities for each x-value can be obtained from Table 2 in Appendix B. This distribution of x is shown in Figure 6-14. We see the very same distribution in Figure 6-15 in histogram form.

FIGURE 6-14

The Distribution of x
when $n = 14$, $p = 0.5$

FIGURE 6-15

Histogram for the Distribution
of x when $n = 14$, $p = 0.5$

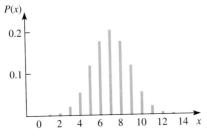

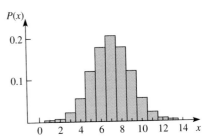

Let's examine $P(x = 4)$ for $n = 14$ and $p = 0.5$ to study the approximation technique. $P(x = 4)$ is equal to 0.061 (see Table 2 in Appendix B), the area of the bar above $x = 4$ in Figure 6-16. Area is the product of width and height. In this case

FIGURE 6-16

Area of Bar Above $x = 4$ Is 0.061. When $n = 14$, $p = 0.5$

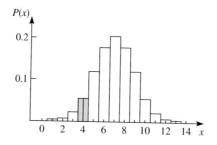

Continuity correction factor

the height is 0.061 and the width is 1.0; thus, the area is 0.061. Let's take a closer look at the width. For $x = 4$, the bar starts at 3.5 and ends at 4.5, so we are looking at an area bounded by $x = 3.5$ and $x = 4.5$. The addition and subtraction of 0.5 to the x-value is commonly called the **continuity correction factor**. It is our method of converting a discrete variable into a continuous variable.

Now let's look at the normal distribution related to this situation. We will first need a normal distribution with a mean and a standard deviation equal to those of the binomial distribution we are discussing. Formulas (5-7) and (5-8) give us these values.

$$\mu = np = (14)(0.5) = \mathbf{7.0}$$
$$\sigma = \sqrt{npq} = \sqrt{(14)(0.5)(0.5)}$$
$$= \sqrt{3.5} = \mathbf{1.87}$$

The probability that $x = 4$ is approximated by the area under the normal curve between $x = 3.5$ and $x = 4.5$, as shown in Figure 6-17. Figure 6-18 shows the entire distribution of the binomial variable x with a normal distribution of the same mean and standard deviation superimposed. Notice that the bars and the interval areas under the curve cover nearly the same area.

FIGURE 6-17

Probability That $x = 4$ Is Approximated by Shaded Area

FIGURE 6-18

Normal Distribution Superimposed over Distribution for Binomial Variable x

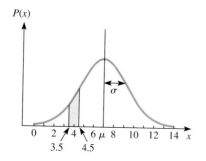

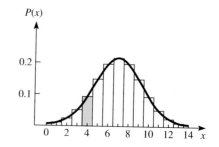

The probability that x is between 3.5 and 4.5 under this normal curve is found by using Table 3 and the methods outlined in Section 6.3.

$$P(3.5 < x < 4.5) = P\left(\frac{3.5 - 7.0}{1.87} < z < \frac{4.5 - 7.0}{1.87}\right)$$

$$= P(-1.87 < z < -1.34)$$

$$= 0.4693 - 0.4099 = \mathbf{0.0594}$$

Since the binomial probability of 0.061 and the normal probability of 0.0594 are reasonably close in value, the normal probability distribution seems to be a reasonable approximation of the binomial distribution.

The normal approximation of the binomial distribution is also useful for values of p that are not close to 0.5. The binomial probability distributions shown in Figures 6-19 and 6-20 suggest that binomial probabilities can be approximated by using the normal distribution. Notice that as n increases in size, the binomial distribution begins to look like the normal distribution. As the value of p moves away from 0.5, a larger n will be needed in order for the normal approximation to be reasonable. The "rule of thumb" on the next page is generally used as a guideline.

FIGURE 6-19   Binomial Distributions

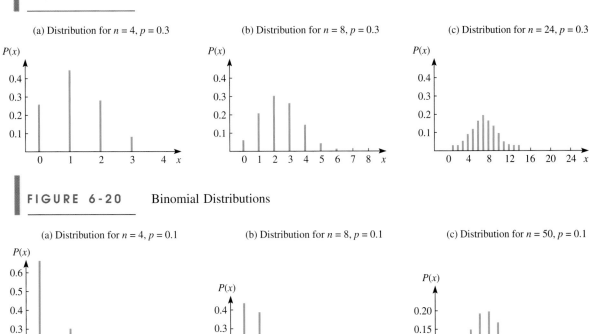

(a) Distribution for $n = 4, p = 0.3$ (b) Distribution for $n = 8, p = 0.3$ (c) Distribution for $n = 24, p = 0.3$

FIGURE 6-20   Binomial Distributions

(a) Distribution for $n = 4, p = 0.1$ (b) Distribution for $n = 8, p = 0.1$ (c) Distribution for $n = 50, p = 0.1$

> ### RULE
>
> The normal distribution provides a reasonable approximation to a binomial probability distribution whenever the values of np and $n(1 - p)$ both equal or exceed 5.

By now you may be thinking, "So what? I will just use the binomial table and find the probabilities directly and avoid all the extra work." But consider for a moment the situation presented in Illustration 6-21.

▼ **ILLUSTRATION   6 - 21**

An unnoticed mechanical failure has caused $\frac{1}{3}$ of a machine shop's production of 5000 rifle firing pins to be defective. What is the probability that an inspector will find no more than 3 defective firing pins in a random sample of 25?

EXERCISE 6.73

Find the values np and nq for a binomial experiment with $n = 100$ and $p = 0.02$. Does this binomial distribution satisfy the rule for normal approximation? Explain.

SOLUTION In this illustration of a binomial experiment, x is the number of defectives found in the sample, $n = 25$, and $p = P(\text{defective}) = \frac{1}{3}$. To answer the question by using the binomial distribution, we will need to use the binomial probability function [formula (5-5)]:

$$P(x) = \binom{25}{x}\left(\frac{1}{3}\right)^{x}\left(\frac{2}{3}\right)^{25-x} \quad \text{for} \quad x = 0, 1, 2, \ldots, \text{ or } 25$$

We must calculate the values for $P(0)$, $P(1)$, $P(2)$, and $P(3)$, since they do not appear in Table 2. This is a very tedious job because of the size of the exponent. In situations such as this, we can use the normal approximation method.

Now let's find $P(x \le 3)$ by using the normal approximation method. We first need to find the mean and standard deviation of x [formulas (5-7) and (5-8)]:

$$\mu = np = (25)\left(\tfrac{1}{3}\right) = \mathbf{8.333}$$

$$\sigma = \sqrt{npq} = \sqrt{(25)\left(\tfrac{1}{3}\right)\left(\tfrac{2}{3}\right)} = \mathbf{2.357}$$

EXERCISE 6.74

Calculate $P\left(x \le 3 \mid B\left(25, \tfrac{1}{3}\right)\right)$.

How good was the normal approximation? Explain. (*Hint:* If you use the computer, use the MINITAB commands on page 273.)

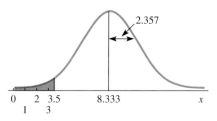

These values are shown in the figure above. The measure of the shaded area ($x < 3.5$) represents the probability of $x = 0, 1, 2,$ or 3. Remember that $x = 3$, the discrete binomial variable, covers the continuous interval from 2.5 to 3.5.

$$P(x \text{ is no more than 3}) = P(x \leq 3) \quad \text{(for discrete variable } x)$$
$$= P(x < 3.5) \quad \text{(using a continuous variable } x)$$

$$P(x < 3.5) = P\left(z < \frac{3.5 - 8.333}{2.357}\right) = P(z < -2.05)$$

$$= 0.5000 - 0.4798 = \textbf{0.0202}$$

▲ | Thus, P(no more than three defectives) is approximately 0.02.

EXERCISES

6.75 In which of the following binomial distributions does the normal distribution provide a reasonable approximation? Use Minitab commands to generate a graph of the distribution and compare the results to the "rule of thumb." State your conclusions.

    a. $n = 10$, $p = 0.3$
    b. $n = 100$, $p = 0.005$
    c. $n = 500$, $p = 0.1$
    d. $n = 50$, $p = 0.2$

```
MINITAB commands

Insert n and p as needed
SET C1
0:n
END
PDF C1 C2;
   BINOmial n p.
PLOT C2*C1;
   PROJect;
   GRID 2.
```

6.76 In order to see what happens when the normal approximation is improperly used, consider the binomial distribution with $n = 15$ and $p = 0.05$. Since $np = 0.75$, the rule of thumb ($np > 5$ and $nq > 5$) is not satisfied. Using the binomial tables, find the probability of one or fewer successes and compare this with the normal approximation.

6.77 Find the normal approximation for the binomial probability $P(x = 6)$, where $n = 12$ and $p = 0.6$. Compare this to the value of $P(x = 6)$ obtained from Table 2.

6.78 Find the normal approximation for the binomial probability $P(x = 4, 5)$, where $n = 14$ and $p = 0.5$. Compare this to the value of $P(x = 4, 5)$ obtained from Table 2.

6.79 Find the normal approximation for the binomial probability $P(x \leq 8)$, where $n = 14$ and $p = 0.4$. Compare this to the value of $P(x \leq 8)$ obtained from Table 2.

6.80 Find the normal approximation for the binomial probability $P(x \geq 9)$, where $n = 13$ and $p = 0.7$. Compare this to the value of $P(x \geq 9)$ obtained from Table 2.

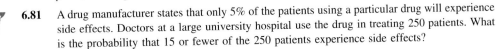 **6.81** A drug manufacturer states that only 5% of the patients using a particular drug will experience side effects. Doctors at a large university hospital use the drug in treating 250 patients. What is the probability that 15 or fewer of the 250 patients experience side effects?

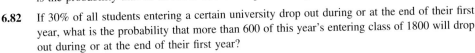 **6.82** If 30% of all students entering a certain university drop out during or at the end of their first year, what is the probability that more than 600 of this year's entering class of 1800 will drop out during or at the end of their first year?

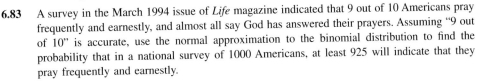 **6.83** A survey in the March 1994 issue of *Life* magazine indicated that 9 out of 10 Americans pray frequently and earnestly, and almost all say God has answered their prayers. Assuming "9 out of 10" is accurate, use the normal approximation to the binomial distribution to find the probability that in a national survey of 1000 Americans, at least 925 will indicate that they pray frequently and earnestly.

6.84 According to the August 1, 1994, *U.S. News & World Report*, 27% of America's kids are growing up with only one parent. If this report is accurate, what is the probability that in a national survey of 375 families at most 100 will have only one parent?

6.85 According to the *Bureau of Justice Statistics Sourcebook of Criminal Justice Statistics 1992*, 4.5% of young adults reported using alcohol daily for the past 30 days. Use the normal approximation to the binomial distribution to find the probability that, in a national poll of 1024 young adults, between 35 and 50 inclusive will indicate that they have used alcohol daily for the past 30 days.

CDF, NORM, and BINO will be helpful. See Exercises 5.74 and 6.57.

    a. Solve using normal approximation and Table 3.
    b. Solve using a computer and the normal approximation method.
    c. Solve using a computer and the binomial probability function.

6.86 According to the June 13, 1994, issue of *Time* magazine, the proportion of all workers who are union members equals 15.8%. Use the normal approximation to the binomial distribution to find the probability that, in a national survey of 2500 workers, at most 450 will be union members.

    a. Solve using normal approximation and Table 3.
    b. Solve using a computer and the normal approximation method.

IN RETROSPECT

We have learned about the standard normal probability distribution, the most important family of continuous random variables. We have learned to apply it to all other normal probability distributions, and how to use it to estimate probabilities of binomial distributions. We have seen a wide variety of problems (variables) that have this normal distribution or are reasonably well approximated by it.

In the next chapter we will examine sampling distributions and learn how to use the standard normal probability to solve additional applications.

CHAPTER EXERCISES

6.87 According to Chebyshev's theorem, at least how much area is there under the standard normal distribution between $z = -2$ and $z = +2$? What is the actual area under the standard normal distribution between $z = -2$ and $z = +2$?

6.88 The middle 60% of a normally distributed population lies between what two standard scores?

6.89 Find the standard score z such that the area above the mean and below z under the normal curve is
    a. 0.3962   b. 0.4846   c. 0.3712

6.90 Find the standard score z such that the area below the mean and above z under the normal curve is
    a. 0.3212   b. 0.4788   c. 0.2700

6.91 Given that z is the standard normal variable, find the value of k such that
    a. $P(|z| > 1.68) = k$
    b. $P(|z| < 2.15) = k$

6.92 Given that z is the standard normal variable, find the value of c such that
    a. $P(|z| > c) = 0.0384$
    b. $P(|z| < c) = 0.8740$

6.93 Find the following values of z:
    a. $z(0.12)$ b. $z(0.28)$ c. $z(0.85)$ d. $z(0.99)$

6.94 Find the area under the normal curve that lies between the following pairs of z-values:
    a. $z = -3.00$ and $z = 3.00$ b. $z(0.975)$ and $z(0.025)$
    c. $z(0.10)$ and $z(0.01)$

6.95 Reread the chapter case study on page 289 before continuing with this question.
    a. Is the I.Q. score a continuous variable? Explain.
    b. Find at least three bits of information about the I.Q. distribution that indicate the use of the normal distribution is appropriate.
    c. What interval of I.Q. score is classified as "superior"? What percentage of the adult population is classified as "superior"?
    d. What is the probability of randomly selecting one person from this population and their being classified as "above normal"?
    e. What proportion of this population is classified as "retarded"?

6.96 An article in *USA Today* (4-4-91) quotes a study involving 3365 people in Minneapolis–St. Paul between 1980 and 1982 and another 4545 between 1985 and 1987. It found that the average cholesterol level for males was 200. The authors of the study say the results of their study are probably similar nationwide. Assume that the cholesterol values for males in the United States are normally distributed with a mean equal to 200 and a standard deviation equal to 25.
    a. What percentage have readings between 150 and 225?
    b. What percentage have readings that exceed 250?

6.97 The length of life of a certain type of refrigerator is approximately normally distributed with a mean of 4.8 years and a standard deviation of 1.3 years.
    a. If this machine is guaranteed for 2 years, what is the probability that the machine you purchased will require replacement under the guarantee?
    b. What period of time should the manufacturer give as a guarantee if it is willing to replace only 0.5% of the machines?

6.98 The average length of time required for completing a certain academic achievement test is believed to be 150 min, and the standard deviation is 20 min. If we wish to allow sufficient time for only 80% to complete the test, when should the test be terminated? (Assume that the lengths of time required to complete this test are normally distributed.)

6.99 A machine is programmed to fill 10-oz containers with a cleanser. However, the variability inherent in any machine causes the actual amounts of fill to vary. The distribution is normal with a standard deviation of 0.02 oz. What must the mean amount μ be in order that only 5% of the containers receive less than 10 oz?

6.100 In a large industrial complex, the maintenance department has been instructed to replace light bulbs before they burn out. It is known that the life of light bulbs is normally distributed with a mean life of 900 hours of use and a standard deviation of 75 hours. When should the light bulbs be replaced so that no more than 10% of them will burn out while in use?

6.101 Suppose that x has a binomial distribution with $n = 25$ and $p = 0.3$.
   a. Explain why the normal approximation is reasonable.
   b. Find the mean and standard deviation of the normal distribution that is used in the approximation.

6.102 Let x be a binomial random variable for $n = 30$ and $p = 0.1$.
   a. Explain why the normal approximation is not reasonable.
   b. Find the function used to calculate the probability of any x from $x = 0$ to $x = 30$.
   c. Use a computer to list the probability distribution.

6.103
   a. Use a computer to list the binomial probabilities for the distribution where $n = 50$ and $p = 0.1$.
   b. Use the results from (a) and find $P(x \le 6)$.
   c. Find the normal approximation for $P(x \le 6)$, and compare the reuslts with those in (b).

6.104
   a. Use a computer to list both the probability distribution and the cumulative probability distribution for the binomial probability experiment with $n = 40$ and $p = 0.4$.
   b. Explain the relationship between the two distributions found in (a).
   c. If you could use only one of these lists when solving problems, which one would you prefer and why?

6.105 Consider the binomial experiment with $n = 300$ and $p = 0.2$.

Use MINITAB command CDF with a subcommand.

   a. Set up, but do not evaluate, the probability expression for 75 or fewer successes in the 300 trials.
   b. Use a computer and find $P(x \le 75)$ using the binomial probability function.
   c. Use a computer and find $P(x \le 75)$ using the normal approximation.
   d. Compare the answers in (a), (b), and (c).

6.106 The grades on an examination whose mean score is 525 and whose standard deviation is 80 are normally distributed.
   a. Anyone who scores below 350 will be retested. What percentage does this represent?
   b. The top 12% are to receive a special commendation. What score must be surpassed to receive this special commendation?
   c. The interquartile range of a distribution is the difference between Q_1 and Q_3, $Q_3 - Q_1$. Find the interquartile range for the grades on this examination.
   d. Find the grade such that only 1 out of 500 will score above it.

6.107 A soft-drink vending machine can be regulated so as to dispense an average of μ oz of soft drink per glass.
   a. If the ounces dispensed per glass are normally distributed with a standard deviation of 0.2 oz, find the setting for μ that will allow a 6-oz glass to hold (without overflowing) the amount dispensed 99% of the time.
   b. Use a computer and simulate drawing a sample of 40 glasses of soft drink from the machine [set using your answer to (a)].
   c. What percentage of your sample would have overflowed the cup?
   d. Does your sample seem to indicate the setting for μ is going to work? Explain.

Repeat (b) a few times. Try a different value for A and repeat (b). Observe how many would overflow in each set of 40.

```
MINITAB Commands

Insert A, the setting from (a)
RANDom 40 C1;
  NORMal A .2.
HISTogram C1;
  CUTPoint 5:6.2/.05.
```

6.108 A company asserts that 80% of the customers who purchase its special lawn mower will have no repairs during the first two years of ownership. Your personal study has shown that only 70 of the 100 in your sample lasted the two years without repair expenses. What is the probability of your sample outcome or less if the actual expenses-free percentage is 80%?

6.109 According to a report in *Time* (Oct. 8, 1990), 12% of the children below age 18 suffer from some form of psychological illness. If this claim is correct, use the normal approximation to the binomial distribution to compute the probability of finding 40 or more children in a sample of 250 under age 18 with some form of psychological illness.

6.110 A test-scoring machine is known to record an incorrect grade on 5% of the exams it grades. Find, by the appropriate method, the probability that the machine records
    a. 3 wrong grades in a set of 5 exams.
    b. no more than 3 wrong grades in a set of 5 exams.
    c. no more than 3 wrong grades in a set of 15 exams.
    d. no more than 3 wrong grades in a set of 150 exams.

6.111 It is believed that 58% of married couples with children agree on methods of disciplining their children. Assuming this to be the case, what is the probability that in a random survey of 200 married couples, we would find
    a. exactly 110 couples who agree?
    b. fewer than 110 couples who agree?
    c. more than 100 couples who agree?

6.112 A new drug is supposed to be 85% effective in treating a particular illness. (That is, 85% of the patients with this illness respond favorably to the drug.) Let x be the number of patients out of every group of 50 who respond favorably. Use the normal approximation method to find the following probabilities.
    a. $P(x > 45)$
    b. $P(40 < x < 50)$
    c. $P(x < 35)$

6.113 If 60% of the registered voters plan to vote for Ralph Brown for mayor of a large city, what is the probability that less than half of the voters, in a poll of 200 registered voters, plan to vote for Ralph Brown?

6.114 The following triangular distribution provides an approximation to the standard normal distribution. Line segment l_1 has the equation $y = x/9 + 1/3$, and segment l_2 has the equation $y = -x/9 + 1/3$.

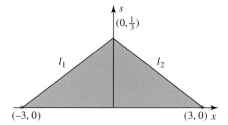

    a. Find the area under the entire triangular distribution.
    b. Find the area under the triangular distribution between 0 and 2.
    c. Find the area under the standard normal distribution between 0 and 2.

VOCABULARY LIST

Be able to define each term. Pay special attention to the key terms, which are printed in red. In addition, describe in your own words, and give an example of, each term. Your examples should not be the ones given in class or in the textbook.

The bracketed numbers indicate the chapters in which the term first appeared, but you should define the terms again to show increased understanding. Page numbers indicate the first appearance of the term in Chapter 6.

area representation for probability (p. 291)	normal distribution (p. 290)
bell-shaped curve (p. 290)	percentage (p. 291)
binomial distribution [5] (p. 314)	probability [4] (p. 291)
binomial probability (p. 314)	proportion (p. 291)
continuity correction factor (p. 316)	random variable [5] (p. 290)
continuous random variable (p. 290)	standard normal distribution
discrete random variable [1, 5] (p. 290)	(p. 292)
normal approximation of binomial (p. 314)	standard score [2] (p. 292)
normal curve (p. 290)	z-score [2] (p. 292)

QUIZ A

Answer "True" if the statement is always true. If the statement is not always true, replace the words shown in bold with words that make the statement always true.

6.1   The normal probability distribution is symmetric about **zero**.

6.2   The total area under the curve of any normal distribution is **1**.

6.3   The theoretical probability that a particular value of a **continuous** random variable will occur is exactly zero.

6.4   The unit of measure of the standard score is **the same as the unit of measure of the data**.

6.5   All **normal distributions** have the same general probability function and distribution.

6.6   When using the notation $z(0.05)$, the number in parentheses is the measure of the area to the **left** of the z-score.

6.7   Standard normal scores have a mean of **1** and a standard deviation of **zero**.

6.8   Probability distributions of **all** continuous random variables are normal.

6.9   We are able to add and subtract the areas under the curve because these areas represent the probabilities of **independent events**.

6.10   The most common distribution of a continuous random variable is the **binomial probability**.

QUIZ B

6.1   Find the following probabilities for z, the standard normal score:
   a. $P(0 < z < 2.42)$ b. $P(z < 1.38)$
   c. $P(z < -1.27)$ d. $P(-1.35 < z < 2.72)$

6.2 Find the value of the *z*-score indicated.
   a. $P(z > ?) = 0.2643$
   b. $P(z < ?) = 0.17$
   c. $z(0.04)$

6.3 Using the symbolic notation $z(\alpha)$, give the symbolic name for the *z*-score shown.
   a.   b.

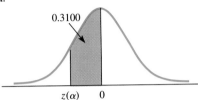

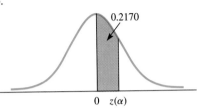

6.4 The lifetime of a flashlight battery is normally distributed about a mean of 35.6 hr with a standard deviation of 5.4 hr. Kevin selected one of these batteries at random and tested it. What is the probability that this one battery will last less than 40.0 hr?

6.5 The amount of time, *x*, spent commuting daily, one-way, to college by students is believed to have a mean of 22 min with a standard deviation of 9 min. If the length of time spent commuting is approximately normally distributed, find the time, *x*, that separates the 25% who spend the most time commuting from the rest of the commuters.

6.6 Thousands of high school students take the SAT each year. The scores attained by the students in a certain city are approximately normally distributed with a mean of 490 and a standard deviation of 70. Find
   a. the percentage of students who score between 600 and 700.
   b. the percentage of students who score less than 650.
   c. the 3rd quartile.
   d. the 15th percentile, P_{15}.
   e. the 95th percentile, P_{95}.

QUIZ C

6.1 In 50 words or less, describe the standard normal distribution.

6.2 Describe the meaning of the symbol $z(\alpha)$.

6.3 Explain why the standard normal distribution, as computed in Table 3, Appendix B, can be used to find probabilities for all normal distributions.

SAMPLE VARIABILITY

Chapter Case Study

GALLUPING ATTITUDES

*S*ince its founding in 1935, the American Institute of Public Opinion, better known as the Gallup Poll, has put approximately 20,000 questions to more than 2 million people.

One of the most interesting social changes to follow through the years in these pages is that of Americans' attitudes toward women at work and in politics. Herewith, an abbreviated reading of Gallup's progress report.

A WOMAN PRESIDENT

1937

Would you vote for a woman for President, if she qualified in every other respect?

		By Sex	Yes	No
Yes	34%	Men	27%	73%
No	66%	Women	41%	59%

1955

If the party whose candidate you most often support nominated a woman for President of the United States, would you vote for her if she seemed best qualified for the job?

		By Sex	Yes	No
Yes	52%	Men	47%	48%
No	44%	Women	57%	40%

1971

If your party nominated a woman for President, would you vote for her if she were qualified for the job?

		By Sex	Yes	No
Yes	66%	Men	65%	35%
No	29%	Women	67%	33%

1978

If your party nominated a woman for President, would you vote for her if she were qualified for the job?

		By Sex	Yes	No
Yes	76%	Men	76%	19%
No	19%	Women	77%	18%

1984

If your party nominated a woman for President, would you vote for her if she were qualified for the job?

		By Sex	Yes	No
Yes	78%	Men	77%	18%
No	17%	Women	78%	17%

1987

If your party nominated a woman for President, would you vote for her if she were qualified for the job?

		By Sex	Yes	No
Yes	82%	Men	81%	13%
No	12%	Women	83%	12%

Selected National Trend

Year	1937	1949	1955	1967	1969	1971	1975	1978	1983	1984	1987
Yes	34%	48%	52%	57%	54%	66%	73%	76%	80%	78%	82%
No	66%	48%	44%	39%	39%	29%	23%	19%	16%	17%	12%

Source: Copyright by the *Gallup Report*. Reprinted by permission.

There are many reasons for repeatedly sampling a population. "Galluping Attitudes" shows the Gallup Poll keeping track of America's attitude toward the possibility of having a woman President. In industry, products are constantly being resampled so that their quality can be monitored. In this chapter we talk about repeated samples drawn from a population in order to better understand the "behavior" of the sample mean. (See Exercise 7.54, p. 350.)

● CHAPTER OBJECTIVES

In Chapters 1 and 2 we discussed how to obtain and describe a sample. The description of the sample data is accomplished by using three basic concepts: (1) measures of central tendency (the mean is the most popularly used sample statistic), (2) measures of dispersion (the standard deviation is most commonly used), and (3) kind of distribution (normal, skewed, rectangular, and so on). The question that seems to follow is: What can be deduced about the statistical population from which a sample is taken?

To put this query at a more practical level, suppose that we have just taken a sample of 25 rivets made for the construction of airplanes. The rivets were tested for shearing strength, and the force required to break each rivet was the response variable. The various descriptive measures—mean, standard deviation, type of distribution—can be found for this sample. However, it is not the sample itself that we are interested in. The rivets that were tested were destroyed during the test, so they can no longer be used in the construction of airplanes. What we are trying to determine, from this sample, is information about the total population, and we certainly cannot test every rivet that is produced (there are too many, and there would be none left for building airplanes). Therefore, we must somehow deduce information, that is, make inferences about the population based on the results observed in the sample.

Suppose that we take another sample of 25 rivets from the same supply and test them by the same procedure. Do you think that we would obtain the same sample mean from the second sample that we obtained from the first? the same standard deviation?

After considering these questions we might suspect that we need to investigate the variability in the sample statistics from sample to sample. Thus, we need to find (1) measures of central tendency for the sample statistics of importance, (2) measures of dispersion for the sample statistics, and (3) the pattern of variability (distribution) of the sample statistics. Once we have this information, we will be better able to predict (make inferences about) the population parameters.

The objective of this chapter is to study the measures and the patterns of variability for the distribution formed by repeatedly observed values of a sample mean. (Look back at your results for Exercise 6.60. This exercise dealt with repeated samples taken from the same population. If you didn't complete that exercise before, it might be helpful to do so now.)

●

7.1 | SAMPLING DISTRIBUTIONS

To make inferences about a population, we need to discuss sample results a little more. A sample mean $\bar{x}$ is obtained from a sample. Do you expect that this value, $\bar{x}$, is exactly equal to the value of the population mean μ? Your answer should be "no." We do not expect that to happen, but we will be satisfied with our sample results if the sample mean is "close" to the value of the population mean. Let's consider a second question: If a second sample is taken, will the second sample have a mean equal to the population mean? Equal to the first sample mean? Again, no, we do not expect it to be equal to the population mean, nor do we expect the second sample mean to be a repeat of the first one. We do, however, again expect the values to be "close." (This argument should hold for any other sample statistic and its corresponding population value.)

The next questions should already have come to mind: What is "close"? How do we determine (and measure) this closeness? Just how would repeated sample statistics be distributed? To answer these questions we must look at a *sampling distribution*.

SAMPLING DISTRIBUTION OF A SAMPLE STATISTIC

The distribution of values for a sample statistic obtained from repeated samples, all of the same size and all drawn from the same population.

▼ | ILLUSTRATION 7 - 1

To illustrate the concept of a sampling distribution, let's consider a very small finite population, the set of even single-digit integers, {0, 2, 4, 6, 8}, and all possible samples of size 2; and look at two different sampling distributions that might be formed: (1) the sampling distribution of sample means and (2) the sampling distribution of sample ranges.

First we need to list all possible samples of size 2; there are 25 possible samples:	{0, 0}	{2, 0}	{4, 0}	{6, 0}	{8,0}
	{0, 2}	{2, 2}	{4, 2}	{6, 2}	{8,2}
	{0, 4}	{2, 4}	{4, 4}	{6, 4}	{8,4}
	{0, 6}	{2, 6}	{4, 6}	{6, 6}	{8,6}
	{0, 8}	{2, 8}	{4, 8}	{6, 8}	{8,8}

Each of these samples has a mean $\bar{x}$. These means are, respectively:	0	1	2	3	4
	1	2	3	4	5
	2	3	4	5	6
	3	4	5	6	7
	4	5	6	7	8

EXERCISE 7.1

Explain why the samples are equally likely; why $P(0) = 0.04$; and why $P(2) = 0.12$.

Each of these samples is equally likely, and thus each of the 25 sample means can be assigned a probability of $\frac{1}{25} = 0.04$ (Why? See Exercise 7.1.) The sampling distribution for sample means is shown in Table 7-1 as a probability distribution and shown in Figure 7-1 as a histogram.

TABLE 7-1

Sampling Distribution of
Sample Means

$\bar{x}$	$P(\bar{x})$
0	0.04
1	0.08
2	0.12
3	0.16
4	0.20
5	0.16
6	0.12
7	0.08
8	0.04

FIGURE 7-1

Histogram: Sampling Distribution
of Sample Means

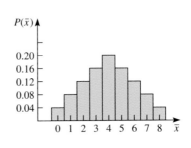

For the same set of all possible samples of size 2, let's find the sampling distribution for sample ranges. Each sample has a range R. These ranges are:

0	2	4	6	8
2	0	2	4	6
4	2	0	2	4
6	4	2	0	2
8	6	4	2	0

Again, each of these 25 sample ranges has a probability of 0.04. Table 7-2 shows the sampling distribution of sample ranges as a probability distribution, and Figure 7-2 shows the sampling distribution as a histogram.

TABLE 7-2

Sampling Distribution of
Sample Ranges

R	$P(R)$
0	0.20
2	0.32
4	0.24
6	0.16
8	0.08

FIGURE 7-2

Histogram: Sampling Distribution
of Sample Ranges

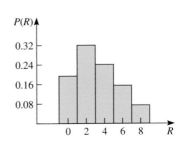

Illustration 7-1 is theoretical in nature and therefore expressed in probabilities. Since this population is quite small, it is easy to list all 25 possible samples of size 2 (a sample space) and assign probabilities. It is not always possible to do this.

Now, let's *empirically* (that is, by experimentation) investigate another sampling distribution.

▼ | ILLUSTRATION 7-2

Let's consider a population consisting of five equally likely integers: 1, 2, 3, 4, and 5. Let's observe a portion of the sampling distribution of sample means when 30 samples of size 5 are randomly selected. Figure 7-3 shows a histogram representation of the population.

FIGURE 7-3

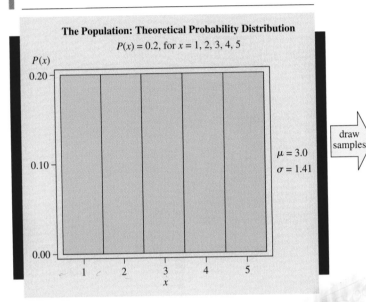

The Population: Theoretical Probability Distribution

$P(x) = 0.2$, for $x = 1, 2, 3, 4, 5$

$\mu = 3.0$
$\sigma = 1.41$

draw samples

TABLE 7-3

	30 Samples of Size 5				
No.	Sample	$\bar{x}$	No.	Sample	$\bar{x}$
1	4,5,1,4,5	3.8	16	4,5,5,3,5	4.4
2	1,1,3,5,1	2.2	17	3,3,1,2,1	2.0
3	2,5,1,5,1	2.8	18	2,1,3,2,2	2.0
4	4,3,3,1,1	2.4	19	4,3,4,2,1	2.8
5	1,2,5,2,4	2.8	20	5,3,1,4,2	3.0
6	4,2,2,5,4	3.4	21	4,4,2,2,5	3.4
7	1,4,5,5,2	3.4	22	3,3,5,3,5	3.8
8	4,5,3,1,2	3.0	23	3,4,4,2,2	3.0
9	5,3,3,3,5	3.8	24	3,3,4,5,3	3.6
10	5,2,1,1,2	2.2	25	5,1,5,2,3	3.2
11	2,1,4,1,3	2.2	26	3,3,3,5,2	3.2
12	5,4,3,1,1	2.8	27	3,4,4,4,4	3.8
13	1,3,1,5,5	3.0	28	2,3,2,4,1	2.4
14	3,4,5,1,1	2.8	29	2,1,1,2,4	2.0
15	3,1,5,3,1	2.6	30	5,3,3,2,5	3.6

using the 30 means

FIGURE 7-4

EXERCISE 7.2

Verify μ and σ for the population in Illustration 7-2.

EXERCISE 7.3

Table 7-3 lists 30 $\bar{x}$-values. Construct a group frequency distribution to verify the frequency distribution shown in Figure 7-4.

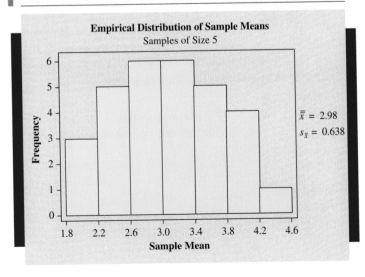

Empirical Distribution of Sample Means
Samples of Size 5

$\bar{\bar{x}} = 2.98$
$s_{\bar{x}} = 0.638$

Case Study   Average Aircraft Age

7-1

USA Today's *"Average aircraft age"* shows the average age of the aircraft that make up the fleets of the 12 biggest airline companies. Each company reported their own average fleet age. The information is shown in the form of a bar graph so that each company's average fleet age can be easily visualized and compared to the others. Using the variable "average fleet age," a frequency distribution of a histogram could have been used to present the data; however, that distribution would not be part of a sampling distribution.

EXERCISE   7.4

a. Construct a frequency histogram of average fleet age using class boundaries of 4.5, 6.5, ...

b. Explain why this distribution of "average fleet age" is not part of a sampling distribution.

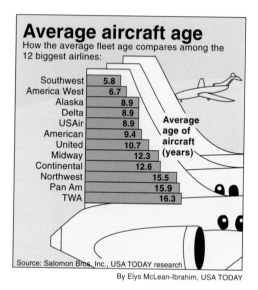

Average aircraft age
How the average fleet age compares among the 12 biggest airlines:

Airline	Average age of aircraft (years)
Southwest	5.8
America West	6.7
Alaska	8.9
Delta	8.9
USAir	8.9
American	9.4
United	10.7
Midway	12.3
Continental	12.6
Northwest	15.5
Pan Am	15.9
TWA	16.3

Source: Salomon Bros. Inc., USA TODAY research
By Elys McLean-Ibrahim, USA TODAY

Source: Copyright 1991, USA TODAY. Reprinted with permission.

Table 7-3 shows 30 samples and their means. The resulting sampling distribution, a frequency distribution, of sample means is shown in Figure 7-4. Notice that this distribution of sample means does not look like the population. Rather, it seems to display characteristics of a normal distribution; it is mounded and nearly symmetric about its mean (approximately 3.0).

▲

EXERCISE   7.5

Find the mean and standard deviation of the 30 $\bar{x}$-values in Table 7-3 to verify the values for $\bar{\bar{x}}$ and $s_{\bar{x}}$. Explain the meaning of the two symbols $\bar{\bar{x}}$ and $s_{\bar{x}}$.

The theory involved with sampling distributions that will be described in the remainder of this chapter requires *random sampling*.

RANDOM SAMPLE

A sample obtained in such a way that each possible sample of fixed size n has an equal probability of being selected. (See p. 25.)

EXERCISES

7.6   a. What is the sampling distribution of sample means?

b. A sample of size 3 is taken from a population and the sample mean found. Describe how this sample mean is related to the sampling distribution of sample means.

7.7   Consider the set of odd single-digit integers {1, 3, 5, 7, 9}.

a. Make a list of all samples of size 2 that can be drawn from this set of integers. (Sample with replacement; that is, the first number is drawn, observed, then replaced before the next drawing.)

b. Construct the sampling distribution of sample means for samples of size 2 selected from this set.

c. Construct the sampling distributions of sample ranges for samples of size 2.

7.8   Consider the set of even single-digit integers {0, 2, 4, 6, 8}.

a. Make a list of all the possible samples of size 3 that can be drawn from this set of integers. (Sample with replacement; that is, the first number is drawn, observed, then replaced before the next drawing.)

b. Construct the sampling distribution of the sample medians for samples of size 3.

c. Construct the sampling distribution of the sample means.

7.9   Using the telephone numbers listed in your local directory as your population, obtain randomly 20 samples of size 3. From each telephone number identified as a source, take the fourth, fifth, and sixth digits. (For example, for 245–8268, you would take the 8, the 2, and the 6 as your sample of size 3.)

a. Calculate the mean of the 20 samples.

b. Draw a histogram showing the 20 sample means. (Use classes −0.5 to 0.5, 0.5 to 1.5, 1.5 to 2.5, and so on.)

c. Describe the distribution of $\bar{x}$'s that you see in (b) (shape of distribution, center, amount of dispersion).

d. Draw 20 more samples and add the 20 new $\bar{x}$'s to the histogram in (b). Describe the distribution that seems to be developing.

7.10   Using a set of 5 dice, roll the dice and determine the mean number of dots showing on the five dice. Repeat the experiment until you have 25 sample means.

a. Draw a dotplot showing the distribution of the 25 sample means. (See Illustration 7-2, p. 331.)

b. Describe the distribution of $\bar{x}$'s in (a).

c. Repeat the experiment to obtain 25 more sample means and add these 25 $\bar{x}$'s to your dotplot. Describe the distribution of 50 means.

7.11   a. Simulate (using a computer or a random numbers table) the drawing of 100 samples, each size 5, from the uniform probability distribution of single-digit integers, 0 to 9.

b. Find the mean for each sample.

c. Construct a histogram of the sample means. (Use integer values as class marks.)

d. Describe the sampling distribution shown in the histogram.

```
MINITAB commands

a. RANDom 100 C1-C5;
     INTEger 0 9.
b. RMEAn C1-C5 C6.
c. HISTogram C6;
     MIDPoint 0:9.
```

7.12   a. Use a computer to draw 200 random samples, each of size 10, from the normal probability distribution with a mean 100 and standard deviation 20.
b. Find the mean for each sample.
c. Construct a frequency histogram of the 200 sample means.
d. Describe the sampling distribution shown in the histogram.

```
MINITAB commands

a. RANDom 200 C1-C10;
     NORMal 100 20.
b. RMEAn C1-C10 C11.
c. HISTogram C11;
     CUTPoint 74.8:125.2/6.3.
```

7.2 | THE CENTRAL LIMIT THEOREM

On the preceding pages we discussed the sampling distributions of two statistics, sample means and sample ranges. There are many others that could be discussed; however, the only sampling distribution of concern to us at this time is the *sampling distribution of sample means*.

Very useful information!

SAMPLING DISTRIBUTION OF SAMPLE MEANS

If all possible random samples, each of size n, are taken from any population with a mean μ and a standard deviation σ, the sampling distribution of sample means will

1. have a mean $\mu_{\bar{x}}$ equal to μ.
2. have a standard deviation $\sigma_{\bar{x}}$ equal to $\sigma/\sqrt{n}$.

Further, if the sampled population has a normal distribution, then the sampling distribution of $\bar{x}$ will also be normal for samples of all sizes.

This is a very interesting two-part statement. The first part tells us the relationship between the population mean and standard deviation, and the sampling distribution mean and standard deviation for all sampling distributions of sample means. The second part indicates that this information is not always useful. Stated differently it says that the mean value of only a few observations will be normally distributed when samples are drawn from a normally distributed population, but will not be normally distributed when the sampled population is uniform, skewed, or otherwise not normal. However, the *central limit theorem* gives us some additional and very important information about the sampling distribution of sample means.

CENTRAL LIMIT THEOREM

The sampling distribution of sample means will become normal as the sample size increases.

Truly amazing; $\bar{x}$ is normally distributed when n is large enough, no matter what shape the population is.

If the sampled distribution is normal, then the sampling distribution of sample means is normal, as stated above, and the central limit theorem (CLT) does not apply. Interestingly, if the sampled population is not normal, the sampling distribution will still be normally distributed under the right conditions. If the sampled distribution is nearly normal, the $\bar{x}$ distribution becomes normal for fairly small n (possibly ≤ 15). When the sampled distribution is far from normal in shape, n may have to be quite large (maybe 50 or more) before the normal distribution provides a satisfactory approximation.

By combining the preceding information, we can describe the sampling distribution of $\bar{x}$ completely: (1) the location of the center (mean), (2) a measure of spread indicating how widely it is dispersed (standard deviation), and (3) an indication of how it is distributed.

1. $\mu_{\bar{x}} = \mu$; the mean of the sampling distribution ($\mu_{\bar{x}}$) is equal to the mean of the population (μ).
2. $\sigma_{\bar{x}} = \sigma/\sqrt{n}$; the standard error of the mean ($\sigma_{\bar{x}}$; see the definition that follows) is equal to the standard deviation of the population (σ) divided by the square root of the sample size.

And the CLT tells us:

3. The distribution of sample means is normal when the parent population is normally distributed, or the distribution of sample means becomes approximately normal (regardless of the shape of the parent population), when the sample size is large enough.

NOTE   The n referred to is the size of each sample in the sampling distribution. (The number of repeated samples used in an empirical situation has no effect on the standard error.)

STANDARD ERROR OF THE MEAN ($\sigma_{\bar{x}}$)

The standard deviation of the sampling distribution of sample means.

We are unable to prove the above three facts at this time. However, their validity will be demonstrated by examining two illustrations. For the first illustration, let's consider a population for which the theoretical sampling distribution of all possible samples can be constructed.

▼ ILLUSTRATION   7 - 3

Let's consider all possible samples of size 2 that could be drawn from a population that contains the three numbers 2, 4, and 6. First let's look at the population itself: Construct a histogram to picture its distribution, Figure 7-5; calculate the mean μ and the standard deviation σ, Table 7-4. We must use the techniques from Chapter 5 for discrete probability distributions.

FIGURE 7-5

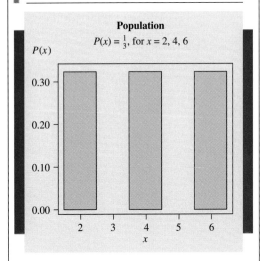

Population

$P(x) = \frac{1}{3}$, for $x = 2, 4, 6$

TABLE 7-4

Extensions Table for x

x	$P(x)$	$xP(x)$	$x^2P(x)$
2	$\frac{1}{3}$	$\frac{2}{3}$	$\frac{4}{3}$
4	$\frac{1}{3}$	$\frac{4}{3}$	$\frac{16}{3}$
6	$\frac{1}{3}$	$\frac{6}{3}$	$\frac{36}{3}$
Σ	$\frac{3}{3}$	$\frac{12}{3}$	$\frac{56}{3}$
	1.0	4.0	$18.6\overline{6}$

$\mu = \mathbf{4.0}$

$\sigma = \sqrt{18.6\overline{6} - (4.0)^2} = \sqrt{2.6\overline{6}} = \mathbf{1.63}$

draw samples

TABLE 7-5

All 9 possible samples of size 2

Sample	$\bar{x}$	Sample	$\bar{x}$	Sample	$\bar{x}$
2, 2	2	4, 2	3	6, 2	4
2, 4	3	4, 4	4	6, 4	5
2, 6	4	4, 6	5	6, 6	6

TABLE 7-6

Extensions Table for $\bar{x}$

$\bar{x}$	$P(\bar{x})$	$\bar{x}P(\bar{x})$	$\bar{x}^2P(\bar{x})$
2	$\frac{1}{9}$	$\frac{2}{9}$	$\frac{4}{9}$
3	$\frac{2}{9}$	$\frac{6}{9}$	$\frac{18}{9}$
4	$\frac{3}{9}$	$\frac{12}{9}$	$\frac{48}{9}$
5	$\frac{2}{9}$	$\frac{10}{9}$	$\frac{50}{9}$
6	$\frac{1}{9}$	$\frac{6}{9}$	$\frac{36}{9}$
Σ	$\frac{9}{9}$	$\frac{36}{9}$	$\frac{156}{9}$
	1.0	4.0	$17.3\overline{3}$

$\mu_{\bar{x}} = \mathbf{4.0}$

$\sigma_{\bar{x}} = \sqrt{17.3\overline{3} - (4.0)^2} = \sqrt{1.3\overline{3}} = \mathbf{1.15}$

FIGURE 7-6

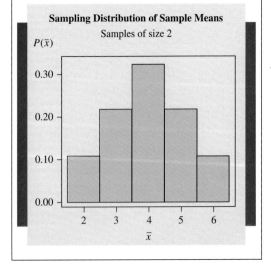

Sampling Distribution of Sample Means

Samples of size 2

Table 7-5 lists all the possible samples that could be drawn if samples of size 2 were to be drawn from this population. (One number is drawn, observed, and then returned to the population before the second number is drawn.) Table 7-5 also lists the means of these samples. The probability distribution for these means and the extensions are given in Table 7-6, along with the calculation of the mean and standard error of the mean for the sampling distribution. The histogram for the sampling distribution of sample means is shown in Figure 7-6.

Let's now check the truth of the three specific facts about the sampling distribution of sample means:

1. The mean $\mu_{\bar{x}}$ of the sampling distribution will equal the mean μ of the population: Both μ and $\mu_{\bar{x}}$ have the value **4.0**.
2. The standard error of the mean $\sigma_{\bar{x}}$ for the sampling distribution will equal the standard deviation σ of the population divided by the square root of the sample size n: $\sigma_{\bar{x}} = $ **1.15** and $\sigma = 1.63$, $n = 2$, $\sigma/\sqrt{n} = 1.63/\sqrt{2} = $ **1.15**; they are equal: $\sigma_{\bar{x}} = \sigma/\sqrt{n}$.
3. The distribution will become approximately normally distributed: The histogram (Figure 7-6) very strongly suggests normality.

Illustration 7-3, a theoretical situation, suggests that all three facts appear to hold true. Do these three facts occur when actual data are collected? Let's look back at Illustration 7-2 (p. 331) and see if all three facts are supported by the empirical sampling distribution that occurred there.

First, let's look at the population, the theoretical probability distribution from which the samples in Illustration 7-2 were taken. Figure 7-3 is the histogram showing the probability distribution for randomly selecting data from the population of equally likely integers 1, 2, 3, 4, 5. The population mean μ equals 3.0. The population standard deviation σ is $\sqrt{2}$, or 1.41. The population has a uniform distribution.

Now let's look at the empirical sampling distribution of the 30 sample means found in Illustration 7-2. Using the 30 values of $\bar{x}$ in Table 7-3, the observed mean of the $\bar{x}$'s, $\bar{\bar{x}}$, is 2.98 and the observed standard error of the mean, $s_{\bar{x}}$, is 0.638. The histogram of sampling distribution, Figure 7-4 (p. 331), appears to be mounded, approximately symmetrical, and centered near the value 3.0.

Now let's check the truth of the three specific properties.

1. $\mu_{\bar{x}}$ and μ will be equal. The mean of the population μ is **3.0**, and the observed sampling distribution mean $\bar{\bar{x}}$ is **2.98**; they are very close in value.
2. $\sigma_{\bar{x}}$ will equal $\sigma/\sqrt{n}$.
   $\sigma = 1.41$ and $n = 5$; therefore, $\sigma/\sqrt{n} = 1.41/\sqrt{5} = $ **0.632**; and $s_{\bar{x}} = $ **0.638**; they are very close in value. (Remember that we have taken only 30 samples, not all possible samples, of size 5.)
3. The population has a rectangular distribution; therefore, the $\bar{x}$ distribution should not be expected to be normal. However, the histogram in Figure 7-4 suggests that the $\bar{x}$ distribution has some of the properties of normality (mounded, symmetry).

Although Illustrations 7-2 and 7-3 do not constitute a proof, the evidence seems to strongly suggest that both statements, the *sampling distribution of sample means* and the *central limit theorem*, are true.

Having taken a look at these two specific illustrations, let's now look at four graphic illustrations that present the same information in slightly different form. In each of these graphic illustrations there are four distributions. The first graph shows the distribution of the parent population, the distribution of the individual x-values. Each of the other three graphs shows a sampling distribution of sample means, using three different sample sizes. In Figure 7-7 we have a uniform distribution, much like Figure 7-3 for the integer illustration, and the resulting distributions of sample means for samples of size 2, 5, and 30.

FIGURE 7-7

Uniform
Distribution

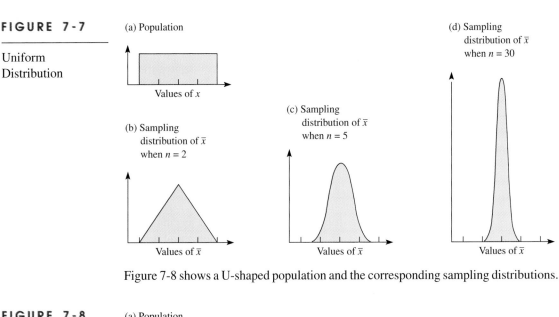

Figure 7-8 shows a U-shaped population and the corresponding sampling distributions.

FIGURE 7-8

U-Shaped
Distribution

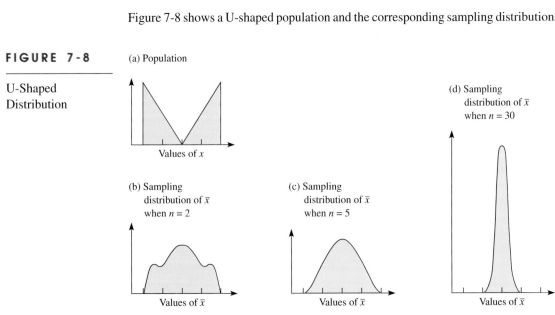

Figure 7-9 shows a J-shaped population and the three corresponding distributions.

FIGURE 7-9

J-Shaped
Distribution

EXERCISE   7.13

If a population has a
standard deviation σ
of 25 units, what is
the standard error of
the mean if samples
of size 16 are select-
ed? samples of size
36? size 100?

(a) Population

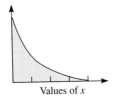

Values of x

(d) Sampling
distribution of $\bar{x}$
when $n = 30$

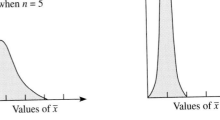

(b) Sampling
distribution of $\bar{x}$
when $n = 2$

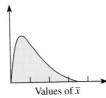

Values of $\bar{x}$

(c) Sampling
distribution of $\bar{x}$
when $n = 5$

Values of $\bar{x}$

Values of $\bar{x}$

Figure 7-10 shows a normally distributed population and the three sampling
distributions.

FIGURE 7-10

Normal Distribution

EXERCISE   7.14

a. What is the total
measure of the
area for any
probability
distribution?
b. Justify the state-
ment "$\bar{x}$ becomes
less variable as n
increases."

(a) Population

Values of x

(d) Sampling
distribution of $\bar{x}$
when $n = 30$

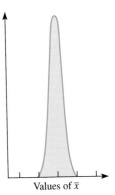

Values of $\bar{x}$

(b) Sampling
distribution of $\bar{x}$
when $n = 2$

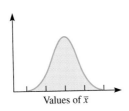

Values of $\bar{x}$

(c) Sampling
distribution of $\bar{x}$
when $n = 5$

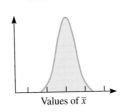

Values of $\bar{x}$

All three nonnormal distributions seem to verify the CLT; the sampling distributions of sample means appear to be approximately normal for all three when samples of size 30 were used. With the normal population (Figure 7-10), the sampling distributions for all sample sizes appear to be normal. Thus, you have seen an amazing phenomenon: No matter what the shape of a population, the sampling distribution of sample means either is normal or becomes approximately normal when n becomes sufficiently large.

You should notice one other point: The sample mean becomes less variable as the sample size increases. Notice that as n increases from 2 to 30, all the distributions become narrower and taller.

EXERCISES

7.15 A certain population has a mean of 500 and a standard deviation of 30. Many samples of size 36 are randomly selected and their mean calculated.
   a. What value would you expect to find for the mean of all these sample means?
   b. What value would you expect to find for the standard deviation of all these sample means?
   c. What shape would you expect the distribution of all these sample means to have?

7.16 Consider the experiment of taking a standardized mathematics test. The variable x is the raw score received. This exam has a mean of 720 and a standard deviation of 60. A group (sample) of 40 students takes the exam, and the sample mean $\bar{x}$ is 725.6. A sampling distribution of means is formed from the means of all such groups of 40 students.
   a. Determine the mean of this sampling distribution.
   b. Determine the standard deviation for this sampling distribution.

7.17 According to a USA Snapshot® (*USA Today*, October 21–23, 1994), the average amount spent per month for long-distance calls through the long-distance carrier is $31.65. If the standard deviation for long-distance calls through the long-distance carrier is $12.25 and a sample of 150 customers is selected, the mean of this sample belongs to a sampling distribution.
   a. What is the shape of this distribution?
   b. What is the mean of this sampling distribution?
   c. What is the standard deviation of this sampling distribution?

7.18 According to the 1993 *World Factbook*, the 1993 total fertility rate (mean number of children born per woman) for Madagascar is 6.75. Suppose the standard deviation of the total fertility rate is 2.5. The mean number of children for a sample of 200 randomly selected women is one value of many that form the sampling distribution of sample means.
   a. What is the mean value for this sampling distribution?
   b. What is the standard deviation of this sampling distribution?
   c. Describe the shape of this sampling distribution.

7.19   a. Use a computer to randomly select 100 samples of size 6 from a normal population with a mean $\mu = 20$ and standard deviation $\sigma = 4.5$.

b. Find mean $\bar{x}$ for each of the 100 samples.

c. Using the 100 sample means, construct a histogram, find mean $\bar{\bar{x}}$, and find the standard deviation $s_{\bar{x}}$.

d. Compare the results of part (c) with the three statements made in the CLT.

MINITAB commands

```
a. RANDom 100 C1-C6;
     NORMal 20 4.5.
b. RMEAn C1-C6 C7.
c. HISTogram C7;
     CUTPoint 12.8:27.2/1.8.
   MEAN C7
   STDEv C7
```

7.20   a. Use a computer to randomly select 200 samples of size 24 from a normal population with a mean $\mu = 20$ and standard deviation $\sigma = 4.5$.

b. Find mean $\bar{x}$ for each of the 200 samples.

c. Using the 200 sample means, construct a histogram, find mean $\bar{\bar{x}}$, and find the standard deviation $s_{\bar{x}}$.

d. Compare the results of part (c) with the three statements made in the CLT.

e. Compare these results to the results obtained in Exercise 7.19. Specifically: What effect did the increase in sample size from 6 to 24 have? What effect did the increase from 100 samples to 200 samples have?
(If you use MINITAB, see Exercise 7.19.)

7.3 | APPLICATION OF THE CENTRAL LIMIT THEOREM

When the sampling distribution of sample means is normally distributed, or approximately normally distributed, we will be able to answer probability questions with the aid of the standard normal distribution, Table 3 of Appendix B.

▼| ILLUSTRATION   7 - 4

Consider a normal population with $\mu = 100$ and $\sigma = 20$. If a sample of size 16 is selected at random, what is the probability that this sample will have a mean value between 90 and 110? That is, what is $P(90 < \bar{x} < 110)$?

SOLUTION Since the population is normally distributed, the sampling distribution of $\bar{x}$'s is normally distributed. To determine probabilities associated with a normal distribution, we will need to convert the statement $P(90 < \bar{x} < 110)$ to a probability statement concerning z in order to use Table 3, the standard normal distribution table. The sampling distribution is shown in the following figure, with $P(90 < \bar{x} < 110)$ represented by the shaded area:

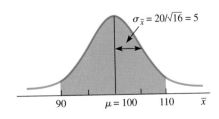

$\sigma_{\bar{x}} = 20/\sqrt{16} = 5$

90        $\mu = 100$ 110 $\bar{x}$

The formula for finding the z corresponding to a known value of $\bar{x}$ is

$$z = \frac{\bar{x} - \mu_{\bar{x}}}{\sigma_{\bar{x}}}$$ (7-1)

The mean and standard error of the mean are: $\mu_{\bar{x}} = \mu$ and $\sigma_{\bar{x}} = \sigma/\sqrt{n}$. Therefore, we will rewrite formula (7-1) in terms of μ, σ, and n:

$$z = \frac{\bar{x} - \mu}{\sigma/\sqrt{n}}$$ (7-2)

Returning to Illustration 7-4 and applying formula 7-2, we find:

the z-score for $\bar{x} = 90$: $z = \frac{\bar{x} - \mu}{\sigma/\sqrt{n}} = \frac{90 - 100}{20/\sqrt{16}} = \frac{-10}{5} = -2.00$

and

the z-score for $\bar{x} = 110$: $z = \frac{\bar{x} - \mu}{\sigma/\sqrt{n}} = \frac{110 - 100}{20/\sqrt{16}} = \frac{10}{5} = 2.00$

Therefore,

$$P(90 < \bar{x} < 110) = P(-2.00 < z < 2.00) = 2(0.4772) = 0.9544$$

EXERCISE 7.21

In Illustration 7-4, explain how 0.4772 was obtained and what it is.

Before we look at more illustrations, let's consider for a moment what is implied by saying that $\sigma_{\bar{x}} = \sigma/\sqrt{n}$. To demonstrate, let's suppose that $\sigma = 20$ and let's use a sampling distribution of samples of size 4. Now $\sigma_{\bar{x}}$ would be $20/\sqrt{4}$, or 10, and approximately 95% (0.9544) of all such sample means should be within the interval from 20 below to 20 above the population mean (within 2 standard deviations of the population mean). However, if the sample size were increased to 16, $\sigma_{\bar{x}}$ would become $20/\sqrt{16} = 5$ and approximately 95% of the sampling distribution would be within 10 units of the mean, and so on. As the sample size increases, the size of $\sigma_{\bar{x}}$ becomes smaller so that the distribution of sample means becomes much narrower. Figure 7-11 illustrates what happens to the distribution of $\bar{x}$'s as the size of the individual samples increases.

FIGURE 7-11

Distributions of Sample Means

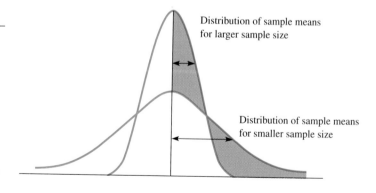

Distribution of sample means for larger sample size

Distribution of sample means for smaller sample size

Recall that the area (probability) under the normal curve is always exactly one. So as the width of the curve narrows, the height will have to increase in order to maintain this area.

▼ | ILLUSTRATION   7 - 5

Kindergarten children have heights that are approximately normally distributed about a mean of 39 in. and a standard deviation of 2 in. A random sample of size 25 is taken and the mean $\bar{x}$ is calculated. What is the probability that this mean value will be between 38.5 and 40.0 inches?

SOLUTION   We want to find: $P(38.5 < \bar{x} < 40.0)$. 38.5 and 40.0 are values of $\bar{x}$ and must be converted to z-scores (necessary for use of Table 3) as follows:

Using $z = \dfrac{\bar{x} - \mu}{\sigma/\sqrt{n}}$

EXERCISE   7.22

What is the probability that the sample of kindergarten children in Illustration 7-5 have a mean height less than 39.75 inches?

When $\bar{x} = 38.5$: $z = \dfrac{38.5 - 39.0}{2/\sqrt{25}}$

$= \dfrac{-0.5}{0.4} = \mathbf{-1.25}$

When $\bar{x} = 40.0$: $z = \dfrac{40.0 - 39.0}{2/\sqrt{25}}$

$= \dfrac{1.0}{0.4} = \mathbf{2.50}$

Therefore,

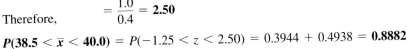

$$P(38.5 < \bar{x} < 40.0) = P(-1.25 < z < 2.50) = 0.3944 + 0.4938 = \mathbf{0.8882}$$

▼ | ILLUSTRATION   7 - 6

Referring to the heights of kindergarten children in Illustration 7-5, within what limits would the middle 90% of the sampling distribution of sample means for sample size 100 fall?

SOLUTION   The two tools we have to work with are formula (7-2) and Table 3. The formula relates the key values of the population to key values of the sampling distribution, and Table 3 relates area to z-scores. First, using Table 3, we find the middle 0.9000 is bounded by $z = \pm 1.65$.

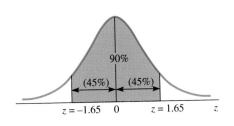

z	...	**0.04**		**0.05**
⋮		⋮		⋮
1.6	...	0.4495	0.4500	0.4505

Second, using formula (7-2), $z = \dfrac{\bar{x} - \mu}{\sigma/\sqrt{n}}$.

EXERCISE   7.23

Referring to Illustration 7-5, what height would bound the lower 25% of all samples of size 25?

If $z = -1.65$,

$$-1.65 = \dfrac{\bar{x} - 39}{2/\sqrt{100}}$$

$$-1.65 = \dfrac{\bar{x} - 39}{0.2}$$

$$\bar{x} - 39 = (-1.65)(0.2)$$

$$\bar{x} = 39 - 0.33$$

$$= 38.67$$

If $z = 1.65$,

$$1.65 = \dfrac{\bar{x} - 39}{2/\sqrt{100}}$$

$$1.65 = \dfrac{\bar{x} - 39}{0.2}$$

$$\bar{x} - 39 = (1.65)(0.2)$$

$$\bar{x} = 39 + 0.33$$

$$= 39.33$$

Thus,

$$P(38.67 < \bar{x} < 39.33) = 0.90$$

▲ Therefore, 38.67 in. and 39.33 in. are the limits that capture the middle 90% of the sample means.

▼ ## EXERCISES

7.24   A random sample of size 36 is to be selected from a population that has a mean μ of 50 and a standard deviation σ of 10.
   a. This sample of 36 has a mean value of $\bar{x}$ which belongs to a sampling distribution. Find the shape of this sampling distribution.
   b. Find the mean of this sampling distribution.
   c. Find the standard error of this sampling distribution.
   d. What is the probability that this sample mean will be between 45 and 55?
   e. What is the probability that the sample mean will have a value greater than 48?
   f. What is the probability that the sample mean will be within 3 units of the mean?

7.25   Consider the approximately normal population of heights of male college students with mean μ of 69 in. and standard deviation σ of 4 in. A random sample of 16 heights is obtained.
   a. Describe the distribution of x, height of male college student.
   b. Find the proportion of male college students whose height is greater than 70 in.
   c. Describe the distribution of $\bar{x}$, the mean of samples of size 16.
   d. Find the mean and standard error of the $\bar{x}$ distribution.
   e. Find $P(\bar{x} > 70)$.
   f. Find $P(\bar{x} < 67)$.

7.26   The amount of fill (weight of contents) put into a glass jar of spaghetti sauce is normally distributed with mean $\mu = 850$ g and standard deviation $\sigma = 8$ g.
   a. Describe the distribution of x, the amount of fill per jar.
   b. Find the probability that one jar selected at random contains between 848 and 855 g.
   c. Describe the distribution of $\bar{x}$, the mean weight for a sample of 24 such jars of sauce.
   d. Find the probability that a random sample of 24 jars has a mean weight between 848 and 855 g.

7.27   The heights of the kindergarten children mentioned in Illustration 7-5 (p. 343) are approximately normally distributed with $\mu = 39$ and $\sigma = 2$.

   a. If an individual kindergarten child is selected at random, what is the probability that he or she has a height between 38 and 40 inches?

   b. A classroom of 30 of these children is used as a sample. What is the probability that the class mean $\bar{x}$ is between 38 and 40 inches?

   c. If an individual kindergarten child is selected at random, what is the probability that he or she is taller than 40 inches?

   d. A classroom of 30 of these kindergarten children is used as a sample. What is the probability that the class mean $\bar{x}$ is greater than 40 inches?

7.28   Consider the experiment of taking a standardized mathematics test. The variable x is the raw score received. The exam has a mean of 720 and a standard deviation of 60. Assume that the variable x is normally distributed.

   a. Consider the experiment of an individual student taking this exam. Describe the distribution (shape of distribution, mean, and standard deviation) for the experiment.

   b. A group of 100 students takes the exam; the mean is reported as a result. Describe the sampling distribution (shape of distribution, mean, and standard deviation) for the experiment.

   c. What is the probability that the student in (a) scored less than 725.6?

   d. What is the probability that the mean of the group in (b) is less than 725.6?

7.29   According to the 1994 *World Almanac*, the average speed of winds in Honolulu, Hawaii, equals 11.4 miles per hour. Assume that wind speeds are approximately normally distributed with a standard deviation of 3.5 miles per hour.

   a. Find the probability that the wind speed on any one reading will exceed 13.5 miles per hour.

   b. Find the probability that the mean of a random sample of 9 readings exceeds 13.5 miles per hour.

   c. Do you think the assumption of normality is reasonable? Explain.

   d. What effect do you think the assumption of normality had on the answers to (a) and (b)? Explain.

7.30   According to an article in the *Omaha World Herald* (12-14-94), a questionnaire used by family doctors has been found to be very effective in diagnosing mental disorders. The questionnaire takes an average of 8.4 minutes for the doctor to administer. Suppose the administration times are normally distributed with $\sigma = 1.5$ minutes.

   a. If the questionnaire is administered to one randomly selected patient, what is the probability that the time required by that one patient exceeds 10 minutes?

   b. If the questionnaire is administered to 9 randomly selected patients, what is the probability that the average time required by the 9 patients exceeds 10 minutes?

   c. Do you think the assumption of normality is reasonable? Explain.

   d. What effect do you think the assumption of normality had on the answers to (a) and (b)? Explain.

7.31   According to the *World Almanac and Book of Facts–1994*, the median weekly earnings of full-time wage and salary women, age 16 years or older in 1992, equals $381. Assume that the wages and salaries are normally distributed with $\sigma = \$85$.

    a. Find the probability that the mean weekly earnings of a sample of 250 such women is between $375 and $385, if the mean equals $381.

    b. Do you think the assumption of normality is reasonable? Explain.

    c. What effect do you think the assumption of normality about the x distribution had on the answer to (a)? Explain.

    d. Do you think the assumption of mean equals $381 is reasonable? Explain.

    e. What effect do you think the assumption about the value of the mean had on the answer to (a)? Explain.

7.32   According to the August 1994 issue of *Employment and Earnings*, the June 1994 average weekly earnings for employees in general automotive repair shops was $406.15. Suppose the standard deviation for the weekly earnings for such employees is $55.50. Assuming that this mean and standard deviation are the current values, find the following probabilities for the mean of a sample of 100 such employees.

    a. The probability that the mean of the sample is less than $400.

    b. The probability that the mean of the sample is between $400 and $410.

    c. The probability that the mean of the sample is greater than $415.

    d. Explain why the assumption of normality about the x distribution was not involved in the solution to (a), (b), and (c).

7.33   a. Find $P(4 < \bar{x} < 6)$ for a random sample of size 4 drawn from a normal population with mean $\mu = 5$ and standard deviation $\sigma = 2$.

    b. Use a computer to randomly generate 100 samples, each of size 4, from a normal probability distribution with mean $\mu = 5$ and standard deviation $\sigma = 2$, and calculate the mean, $\bar{x}$, for each sample.

    c. How many of the sample means in (b) have values between 4 and 6? What percentage is that?

    d. Compare the answers to (a) and (c), and explain any differences that occurred.

```
MINITAB commands

a. SET C1
     DATA> 4 5 6
     DATA> END
     CDF C1 C2;
        NORMal 5 2.
     PRINt C1 C2
b. RANDom 100 C3-C6;
     NORMal 5 2.
     RMEAn C3-C6 C7
c. HISTogram C7;
     CUTPoint 1:9;
     BAR;
     SYMBol;
     TYPE 0;
     LABEl.
```

7.34   a. Find $P(46 < \bar{x} < 55)$ for a random sample size 16 drawn from a normal population with mean $\mu = 50$ and standard deviation $\sigma = 10$.

    b. Use a computer to randomly generate 200 samples, each of size 16, from a normal probability distribution with mean $\mu = 50$ and standard deviation $\sigma = 10$, and calculate the mean, $\bar{x}$, for each sample.

    c. How many of the sample means in (b) have values between 46 and 55? What percentage is that?

    d. Compare the answers to (a) and (c), and explain any differences that occurred. (If you use MINITAB, see Exercise 7.33.)

IN RETROSPECT

In Chapters 6 and 7 we have learned to use the standard normal probability distribution. We now have two formulas for calculating a z-score:

$$z = \frac{x - \mu}{\sigma} \quad \text{and} \quad z = \frac{\bar{x} - \mu}{\sigma/\sqrt{n}}$$

You must be careful to distinguish between these two formulas. The first gives the standard score when dealing with individual values from a normal distribution (x-values). The second formula deals with a sample mean ($\bar{x}$-value). The key to distinguishing between the formulas is to decide whether the problem deals with an individual x or a sample mean, $\bar{x}$. If it deals with the individual values of x, we use the first formula, as presented in Chapter 6. If, on the other hand, the problem deals with a sample mean, $\bar{x}$, we use the second formula and proceed as illustrated in this chapter.

The basic purpose for considering what happens under repeated sampling, as discussed in this chapter, is to form sampling distributions. The sampling distribution is then used to describe the variability that occurs from one sample to the next. Once this pattern of variability is known and understood for a specific sample statistic, we will be able to make predictions about the corresponding population parameter with a measure of how accurate the prediction is. The central limit theorem helps describe the distribution for sample means. We will begin to make inferences about population means in Chapter 8.

There are other reasons for repeated sampling. Repeated samples are commonly used in the field of production control, in which samples are taken to determine whether a product is of the proper size or quantity. When the sample statistic does not fit the standards, a mechanical adjustment of the machinery is necessary. The adjustment is then followed by another sampling to be sure the production process is in control.

The "standard error of the _____" is the name used for the standard deviation of the sampling distribution for whatever statistic is named in the blank. In this chapter we have been concerned with the standard error of the mean. However, we could also work with the standard error of proportion, median, or any other statistic.

You should now be familiar with the concept of a sampling distribution and, in particular, with the sampling distribution of sample means. In Chapter 8 we will begin to make predictions about the values of population parameters.

CHAPTER EXERCISES

7.35 The diameters of Red Delicious apples in a certain orchard are normally distributed with a mean of 2.63 in. and a standard deviation of 0.25 in.

  a. What percentage of the apples in this orchard have diameters less than 2.25 in.?

  b. What percentage of the apples in the orchard are larger than 2.56 in.?

A random sample of 100 apples is gathered and the mean diameter obtained is $\bar{x} = 2.56$.

  c. If another sample of size 100 is taken, what is the probability that its sample mean will be greater than 2.56 in.?

  d. Why is the z-score used in answering (a), (b), and (c)?

  e. Why is the formula for z used in (c) different from that used in (a) and (b)?

7.36   a. Find a value for e such that 95% of the apples in Exercise 7.35 are within e units of the mean 2.63. That is, find e such that $P(2.63 - e < x < 2.63 + e) = 0.95$.

b. Find a value for E such that 95% of the samples of 100 apples taken from the orchard in Exercise 7.35 will have mean values within E units of the mean 2.63. That is, find E such that $P(2.63 - E < \bar{x} < 2.63 + E) = 0.95$.

7.37   According to the 1992 issue of *Agricultural Statistics*, the 1991 per capita consumption of refined sugar is 64.9 pounds. If we assume the mean is still 64.9, find the probability that the mean per capita consumption for a sample of 100 individuals will be between 60 and 70 pounds. Assume that $\sigma = 17.5$ pounds.

7.38   According to an article in *Pharmaceutical News* (January 1991), a person age 65 or older will spend, on the average, $300 on personal-care products per year. If we assume that the amount spent on personal-care products by individuals 65 or older is normally distributed and has a standard deviation equal to $75, what is the probability that the mean amount spent by 25 randomly selected such individuals will fall between $250 and $350?

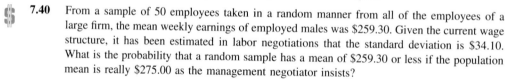

7.39   A shipment of steel bars will be accepted if the mean breaking strength of a random sample of 10 steel bars is greater than 250 pounds per square inch. In the past, the breaking strength of such bars has had a mean of 235 and a variance of 400.

a. What is the probability, assuming that the breaking strengths are normally distributed, that one randomly selected steel bar will have a breaking strength in the range from 245 to 255?

b. What is the probability that the shipment will be accepted?

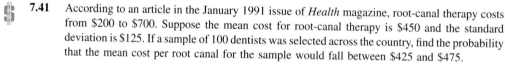

7.40   From a sample of 50 employees taken in a random manner from all of the employees of a large firm, the mean weekly earnings of employed males was $259.30. Given the current wage structure, it has been estimated in labor negotiations that the standard deviation is $34.10. What is the probability that a random sample has a mean of $259.30 or less if the population mean is really $275.00 as the management negotiator insists?

7.41   According to an article in the January 1991 issue of *Health* magazine, root-canal therapy costs from $200 to $700. Suppose the mean cost for root-canal therapy is $450 and the standard deviation is $125. If a sample of 100 dentists was selected across the country, find the probability that the mean cost per root canal for the sample would fall between $425 and $475.

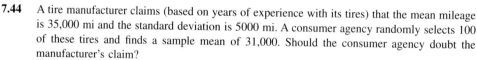

7.42   A report in *Newsweek* (November 12, 1990) stated that the day-care cost per week in Boston is $109. If this figure is taken as the mean cost per week and if the standard deviation were known to be $20, find the probability that a sample of 50 day-care centers would show a mean cost of $100 or less per week.

7.43   A manufacturer of light bulbs says that its light bulbs have a mean life of 700 hr and a standard deviation of 120 hr. You purchased 144 of these bulbs with the idea that you would purchase more if the mean life of your sample were more than 680 hr. What is the probability that you will not buy again from this manufacturer?

7.44   A tire manufacturer claims (based on years of experience with its tires) that the mean mileage is 35,000 mi and the standard deviation is 5000 mi. A consumer agency randomly selects 100 of these tires and finds a sample mean of 31,000. Should the consumer agency doubt the manufacturer's claim?

7.45 For large samples, the sample sum (Σx) has an approximately normal distribution. The mean of the sample sum is $n \cdot \mu$ and the standard deviation is $\sqrt{n} \cdot \sigma$. The distribution of savings per account for a savings and loan institution has a mean equal to $750 and a standard deviation equal to $25. For a sample of 50 such accounts, find the probability that the sum in the 50 accounts exceeds $38,000.

 7.46 The baggage weights for passengers using a particular airline are normally distributed with a mean of 20 lb and a standard deviation of 4 lb. If the limit on total luggage weight is 2125 lb, what is the probability that the limit will be exceeded for 100 passengers?

 7.47 A trucking firm delivers appliances for a large retail operation. The packages (or crates) have a mean weight of 300 lb and a variance of 2500 lb.
   a. If a truck can carry 4000 lb and 25 appliances need to be picked up, what is the probability that the 25 appliances will have an aggregate weight greater than the truck's capacity? (Assume that the 25 appliances represent a random sample.)
   b. If the truck has a capacity of 8000 lb, what is the probability that it will be able to carry the entire lot of 25 appliances?

 7.48 Every year, around Halloween, many street signs in a small city are defaced. The average repair cost per sign is $68.00 and the standard deviation is $12.40.
   a. If 300 signs are damaged this year, there is a 5% chance that the total repair costs for the 300 signs will exceed what value?
   b. You are about 68% certain that the total repair costs for 300 signs will fall within what interval centered around $20,400?

 7.49 A pop-music record firm wants the distribution of lengths of cuts on its records to have an average of 2 min 15 sec (135 sec) and a standard deviation of 10 sec so that disc jockeys will have plenty of time for commercials within each five-minute period. The population of times for cuts is approximately normally distributed with only a negligible skew to the right. You have just timed the cuts on a new release and have found that the 10 cuts average 140 sec.
   a. What percentage of the time will the average be 140 sec or longer, if the new release is randomly selected?
   b. If the music firm wants 10 cuts to average 140 sec less than 5% of the time, what must the population mean be given that the standard deviation remains at 10 sec?

7.50 Let's simulate the sampling distribution related to the disc jockey's concern for "length of cut" in Exercise 7.49.
   a. Use a computer to randomly generate 50 samples, each of size 10, from a normal distribution with mean 135 and standard deviation 10. Find the "sample total" and the sample mean for each sample.
   b. Using the 50 sample means, construct a histogram and find their mean and standard deviation.
   c. Using the 50 sample "totals," construct a histogram and find their mean and standard deviation.
   d. Compare the results obtained in (b) and (c). Explain any similarities and any differences observed.

```
MINITAB commands

a. RANDom 50 C1-C10;
     NORMal 135 10.
   RSUM C1-C10 C11
   RMEAn C1-C10 C12
b. HISTogram C12
   MEAN C12
   STDEv C12
c. HISTogram C11
   MEAN C11
   STDEv C11

extra: DESCribe C11 C12
```

7.51
a. Find the mean and standard deviation of x for a binomial probability distribution with $n = 16$ and $p = 0.5$.
b. Use a computer to construct the probability distribution and histogram for the binomial probability experiment with $n = 16$ and $p = 0.5$.
c. Use a computer to randomly generate 200 samples of size 25 from a binomial probability distribution with $n = 16$ and $p = 0.5$ and calculate the mean of each sample.
d. Construct a histogram and find the mean and standard deviation of the 200 sample means.
e. Compare the probability distribution of x found in (b) and the frequency distribution of $\bar{x}$ in (d). Does your information support the CLT? Explain.

```
MINITAB commands

b. SET C1
   0:16
   END
   PDF C1 C2;
   BINOmial 16 .5.
   PLOT C2*C1;
     AREA;
     STEP 0;
     COLOR 0.
c. RANDom 200 C3-C27;
     BINOmial 16 .5.
   RMEAn C3-C27 C28
d. HISTogram C28
   MEAN C28
   STDEv C28
```

7.52
a. Find the mean and standard deviation of x for a binomial probability distribution with $n = 200$ and $p = 0.3$.
b. Use a computer to construct the probability distribution and histogram for the random variable x of the binomial probability experiment with $n = 200$ and $p = 0.3$.
c. Use a computer to randomly generate 200 samples of size 25 from a binomial probability distribution with $n = 200$ and $p = 0.3$ and calculate the mean $\bar{x}$ of each sample.
d. Construct a histogram and find the mean and standard deviation of the 200 sample means.
e. Compare the probability distribution of x found in (b) and the frequency distribution of $\bar{x}$ in (d). Does your information support the CLT? Explain.

```
MINITAB commands

b. SET C1
   0:200
   END
   PDF C1 C2;
   BINOmial 200 .3.
   PLOT C2*C1;
   PROJECT.
c. RANDom 200 C3-C27;
     BINOmial 200 .3.
   RMEAn C3-C27 C28
d. HISTogram C28
   MEAN C28
   STDEv C28
```

7.53 A sample of 144 values is randomly selected from a population with a mean, μ, equal to 45 and a standard deviation, σ, equal to 18.

a. Determine the interval (smallest value to largest value) within which you would expect a sample mean to lie.
b. What is the amount of deviation from the mean for a sample mean of 46.3?
c. What is the maximum deviation you have allowed for in your answer to (a)?
d. How is this maximum deviation related to the standard error of the mean?

7.54 The Gallup Poll has been surveying the public for many years. The results summarized in the chapter case study (p. 327) list the findings for each of several years. When repeated sampling is used to track America's attitudes, the sample statistic reported, the percentage of yes responses, does not form a sampling distribution, but rather it forms a time series. Time series is a topic not covered in this text. To help recognize and understand the difference between repeated samples that belong to a sampling distribution and those that belong to a time series:

a. Plot a scatter diagram displaying the Selected National Trend information, using the year number as the input variable x and the percentage of yes responses as the output variable y.

b. On the scatter diagram drawn in (a), plot the percentages of no responses as a second output variable using the year number as the input variable x.

c. Do you see what could be called a trend? Explain.

d. Explain how the trends are shown and how they are different than a sampling distribution.

VOCABULARY LIST

Be able to define each term. Pay special attention to the key terms, which are printed in red. In addition, describe in your own words, and give an example of, each term. Your examples should not be ones given in class or in the textbook.

The bracketed numbers indicate the chapter in which the term first appeared, but you should define the terms again to show increased understanding of their meaning. Page numbers indicate the first appearance of the term in Chapter 7.

central limit theorem (p. 334)
frequency distribution [2] (p. 332)
probability distribution [5] (p. 329)
random sample [2] (p. 332)
repeated sampling (p. 329)

sampling distribution of sample means (p. 334)
standard error of the mean (p. 335)
z-score [2, 6] (p. 342)

QUIZ A

Answer "True" if the statement is always true. If the statement is not always true, replace the words shown in bold with words that make the statement always true.

7.1 A sampling distribution **is** a distribution listing all the sample statistics that describe a particular sample.

7.2 The histograms of **all** sampling distributions are symmetrically shaped.

7.3 The mean of the sampling distribution of $\bar{x}$'s is equal to the mean of the **sample**.

7.4 The standard error of the mean is the standard deviation of the population **from which the samples have been taken**.

7.5 The standard error of the mean **increases** as the sample size increases.

7.6 The shape of the distribution of sample means is always that of a **normal** distribution.

7.7 A **probability** distribution of a sample statistic is a distribution of all the values of that statistic that were obtained from all possible samples.

7.8 The central limit theorem provides us with a description of the three characteristics of a sampling distribution of sample **medians**.

7.9 A **frequency** sample is obtained in such a way that all possible samples of a given size have an equal chance of being selected.

7.10 We **do not need** to take repeated samples in order to use the concept of the sampling distribution.

QUIZ B

7.1 The lengths of the lake trout in Conesus Lake are believed to be normally distributed about a mean length of 15.6 in. with a standard deviation of 3.8 in.
    a. Kevin is going fishing at Conesus Lake tomorrow. If he catches one lake trout, what is the probability that it is less than 15.0 in.?
    b. If Captain Brian's fishing boat takes ten people fishing on Conesus Lake tomorrow and they catch a random sample of 16 fish, what is the probability that the mean length of their total catch is less than 15 in.?

7.2 Cigarette lighters manufactured by Easyvice Company are claimed to have a mean life of 20 months with a standard deviation of 6 months. The money-back guarantee allows you to return the lighter if it does not last at least 12 months from the date of purchase.
    a. If the length of life of these lighters is normally distributed, what percentage of the lighters will be returned to the company?
    b. If a random sample of 25 lighters is tested, what is the probability the sample mean will be more than 18 months?

7.3 Aluminum rivets produced by Rivets Forever, Inc., are believed to have shearing strengths that are distributed about a mean of 13.75 and have a standard deviation of 2.4. If this information is true and a sample of 64 such rivets is tested for shear strength, what is the probability that the mean strength will be between 13.6 and 14.2?

QUIZ C

7.1 "Two heads are better than one." If that's true, then "how good would several heads be?" To find out, a statistics instructor drew a line across the chalkboard and asked her class to estimate its length to the nearest inch. She collected their estimates, which ranged from 33 to 61 in., and calculated the mean value. She reported that the mean was 42.25 in. She then measured the line and found it to be 41.75 in. long. Does this show that "several heads are better than one"? What statistical theory supports this occurrence? Explain how.

7.2 The sampling distribution of sample means is more than just a distribution of the mean values that occur from many repeated samples taken from the same population. Describe what other specific condition must be met in order to have a sampling distribution of sample means.

7.3 Student A stated that "a sampling distribution of the standard deviation tells you how the standard deviation varies from sample to sample." Student B argues that "a population distribution tells you that." Who is right? Justify your answer.

7.4 Student A says that it is the "size of each sample used" and Student B says that it is the "number of samples used" that determines the spread of an empirical sampling distribution. Who is right? Justify your choice.

WORKING WITH YOUR OWN DATA

The central limit theorem is very important to the development of the rest of this course. Its proof, which requires the use of calculus, is beyond the intended level of this course. However, the truth of the CLT can be demonstrated both theoretically and by experimentation. The following series of questions will help to verify the central limit theorem both ways:

A | THE POPULATION

Consider the theoretical population that contains the three numbers 0, 3, and 6 in equal proportions.

1. a. Construct the theoretical probability distribution for the drawing of a single number, with replacement, from this population.
   b. Draw a histogram of this probability distribution.
   c. Calculate the mean μ and the standard deviation σ for this population.

B | THE SAMPLING DISTRIBUTION, THEORETICALLY

Let's study the theoretical sampling distribution formed by the means of all possible samples of size 3 that can be drawn from the given population.

2. Construct a list showing all the possible samples of size 3 that could be drawn from this population. (There are 27 possibilities.)
3. Find the mean for each of the 27 possible samples listed in answer to question 2.
4. Construct the probability distribution (the theoretical sampling distribution of sample means) for these 27 sample means.
5. Construct a histogram for this sampling distribution of sample means.
6. Calculate the mean $\mu_{\bar{x}}$ and the standard error of the mean $\sigma_{\bar{x}}$.
7. Show that the results found in answers 1(c), 5, and 6 support the three claims made by the central limit theorem. Cite specific values to support your conclusions.

Let's now see whether the central limit theorem can be verified empirically; that is, does it hold when the sampling distribution is formed by the sample means that result from several random samples?

8. Draw a random sample of size 3 from the given population. List your sample of three numbers and calculate the mean for this sample.

You may use a computer to generate your samples. You may take three identical "tags" numbered 0, 3, and 6, put them in a "hat," and draw your sample using replacement between each drawing. Or you may use dice; let 0 be represented by 1 and 2, let 3 be represented by 3 and 4, and 6 by 5 and 6. You may also use random numbers to simulate the drawing of your samples. Or you may draw your sample from the list of random samples at the bottom of the page. Describe the method you decide to use. (Ask your instructor for guidance.)

9. Repeat question 8 forty-nine (49) more times so that you have a total of fifty (50) sample means that have resulted from samples of size 3.
10. Construct a frequency distribution of the 50 sample means found in answering questions 8 and 9.
11. Construct a histogram of the frequency distribution of observed sample means.
12. Calculate the mean $\bar{x}$ and standard deviation $s_{\bar{x}}$ of the frequency distribution formed by the 50 sample means.
13. Compare the observed values of $\bar{x}$ and $s_{\bar{x}}$ with the values of $\mu_{\bar{x}}$ and $\sigma_{\bar{x}}$. Do they agree? Does the empirical distribution of $\bar{x}$ look like the theoretical one?

The following table contains 100 samples of size 3 that were generated randomly by computer:

6 3 0	0 3 0	6 6 0	3 3 6	6 6 3	6 3 3
0 0 3	3 0 6	3 3 0	3 6 6	0 3 0	6 6 3
6 6 6	0 3 0	6 3 6	0 6 3	6 0 3	6 3 3
6 0 0	3 0 6	6 3 3	3 3 0	3 3 0	3 3 3
3 3 3	3 0 0	6 6 6	3 3 6	0 0 6	0 6 3
6 6 6	0 0 6	3 3 0	0 6 6	0 0 3	6 6 3
0 0 6	0 0 6	6 6 6	6 3 6	6 6 0	3 0 0
3 6 6	6 3 0	3 6 3	3 0 0	3 3 6	0 6 0
3 0 0	0 3 6	6 3 3	6 0 6	3 3 6	6 0 3
0 3 6	3 6 3	6 6 3	6 6 0	3 3 3	3 0 0
6 3 0	6 6 0	0 3 0	6 6 0	3 6 6	0 3 6
6 3 3	0 3 0	6 6 0	6 6 3	6 6 0	3 0 3
3 6 3	3 6 0	0 0 6	0 3 3	3 6 6	0 3 6
0 6 0	6 0 0	0 6 0	0 6 6	0 3 3	0 3 6
3 3 6	3 3 3	3 3 6	6 3 6	3 3 3	3 6 6
6 3 3	3 0 0	3 0 6	6 0 3	3 6 6	6 0 3
0 3 3	6 3 0	0 3 6	0 3 6		

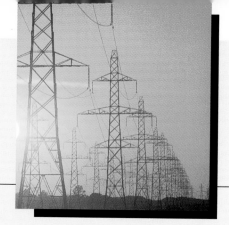

## PART THREE

INFERENTIAL STATISTICS

*T*he central limit theorem gave us some very important information about the sampling distribution of sample means. Specifically, it stated that in many realistic cases (when the random sample is large enough) a distribution of sample means is normally or approximately normally distributed about the mean of the population. With this information we were able to make probability statements about the likelihood of certain sample mean values occurring when samples are drawn from a population with a known mean and a known standard deviation. We are now ready to turn this situation around to the case in which the population mean is not known. We will draw one sample, calculate its mean value, and then make an inference about the value of the population mean based on the sample's mean value.

The objective of inferential statistics is to use the information contained in the sample data to increase our knowledge of the sampled population. In this part of the textbook we will learn about making two types of inferences: (1) the procedures for estimating the value of a population parameter and (2) the decision-making process known as a hypothesis test procedure. Specifically, we will learn about making these two types of inferences for the mean μ of a normal population, for the standard deviation σ of a normal population, and for the probability parameter p of a binomial population.

WILLIAM GOSSET

WILLIAM GOSSET ("Student"), a British industrial statistician, was born in Canterbury, England, on June 13, 1876, to Frederick and Agnes (Vidal) Gosset. William's educational background included studies at Winchester College, New College, and Oxford University. In 1899, upon leaving Oxford University, Gosset was employed as a brewer by the Arthur Guinness & Son Brewing Company. In 1906 Gosset married Marjory Surtees Philpotts, and they became the parents of two daughters and a son. William Gosset died in Beaconsfield, England, on October 16, 1937.

Guinness liked their employees to use pen names if publishing papers, so in 1908 Gosset adopted the pen name "Student" under which he published what was probably his most noted contribution to statistics, "The Probable Error of a Mean." Guinness sent Gosset to work under Karl Pearson, at the University of London, and eventually he took charge of the new Guinness Brewery in London.

In his paper "The Probable Error of a Mean," "Student" set out to find the distribution of the amount of error in the sample mean, $(\bar{x} - \mu)$, when divided by s, where s was the estimate of σ from a sample of any known size. The probable error of a mean, $\bar{x}$, could be calculated for any size sample by using this distribution of $(\bar{x} - \mu)/s$. Even though he was well aware of the insufficiency of a small sample to determine the form of the distribution of $\bar{x}$, he chose the normal distribution for simplicity, stating his opinion: "It appears probable that the deviation from normality must be very severe to lead to serious error."

Student's t-distribution did not immediately gain popularity. In September 1922, even 14 years after its publication, Student wrote to Fisher: "I am sending you a copy of Student's Tables as you are the only man that's ever likely to use them!" Today, Student's t-distribution is widely used and respected in statistical research.

INTRODUCTION TO STATISTICAL INFERENCES

CHATTANOOGA: CLEAN AND GREEN

We often read information describing the status of various aspects of life. In "'Tremendous' Strides Since 1st Earth Day" (USA Today, 4-21-95), Debbie Howlett reports: "And, as the 25th Earth Day approaches this weekend, Chattanooga, and much of the USA, shows remarkable environmental progress since 1970. In many places, the air is cleaner. So is the water. Factories have dramatically reduced toxic pollution and homeowners are recycling their trash at a record rate."

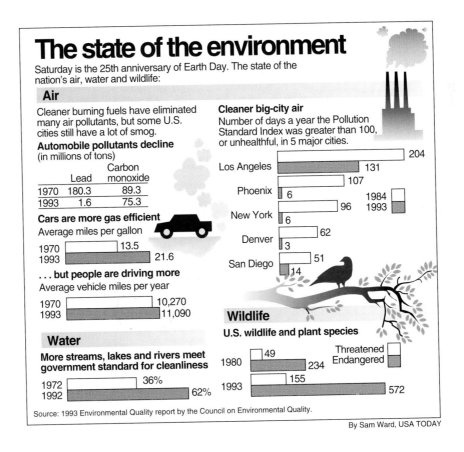

The state of the environment

Saturday is the 25th anniversary of Earth Day. The state of the nation's air, water and wildlife:

Air

Cleaner burning fuels have eliminated many air pollutants, but some U.S. cities still have a lot of smog.

Automobile pollutants decline
(in millions of tons)

	Lead	Carbon monoxide
1970	180.3	89.3
1993	1.6	75.3

Cars are more gas efficient
Average miles per gallon

| 1970 | 13.5 |
| 1993 | 21.6 |

...but people are driving more
Average vehicle miles per year

| 1970 | 10,270 |
| 1993 | 11,090 |

Cleaner big-city air
Number of days a year the Pollution Standard Index was greater than 100, or unhealthful, in 5 major cities.

1984 / 1993

Los Angeles	204 / 131
Phoenix	107 / 6
New York	96 / 6
Denver	62 / 3
San Diego	51 / 14

Water

More streams, lakes and rivers meet government standard for cleanliness

| 1972 | 36% |
| 1992 | 62% |

Wildlife

U.S. wildlife and plant species

Threatened / Endangered

| 1980 | 49 / 234 |
| 1993 | 155 / 572 |

Source: 1993 Environmental Quality report by the Council on Environmental Quality.

By Sam Ward, USA TODAY

The numbers reported in the "State of the environment" graphic are the results of an inference. The numerical values are estimates for many different population parameters: the average vehicle miles driven per year, the average miles per gallon our automobiles get, the millions of tons of air pollutants as a result of automobiles, and so on. These numerical values all appear to be results of statistical estimations, one of the topics of this chapter.

● CHAPTER OBJECTIVES

A random sample of 36 pieces of data yields a mean of 4.64. What can we deduce about the population from which the sample was taken? We will be asked to answer two types of questions:

1. What value or interval of values can we use to estimate the population mean?
2. Is the sample mean sufficiently different in value from the hypothesized mean value of 4.5 to justify arguing that 4.5 is not a correct value?

The first question requires us to *make an estimation*, whereas the second question requires us to *make a decision* about the population mean.

In this chapter we concentrate on learning about the basic concepts of estimation and hypothesis testing. We deal with questions about the population mean using two methods that assume the value of the population standard deviation is a known quantity. This assumption is seldom realized in real-life problems, but it will make our first look at the techniques of inference much simpler.

8.1 | THE NATURE OF ESTIMATION

A company manufactures rivets for use in building aircraft. One characteristic of extreme importance is the "shearing strength" of each rivet. The company's engineers must monitor their production so that they are certain the shearing strength of their rivets meets the required specs. To accomplish this, they take a sample and determine the mean shearing strength of the sample. Based on this sample information, the company can estimate the mean shearing strength for all the rivets it is manufacturing.

A sample of 36 rivets is randomly selected, and each rivet is tested for shearing strength. The resulting sample mean is $\bar{x} = 924.23$ lb. Based on this sample, we say, "We believe the mean shearing strength of all such rivets is 924.23 lb."

NOTE 1   Shearing strength is the force required to break a material in a "cutting" action. Obviously, the manufacturer is not going to test all rivets since the test destroys each rivet tested. Therefore, samples are tested and the information about the sample must be used to make inferences about the population of all such rivets.

NOTE 2   Throughout Chapter 8 we will treat the standard deviation σ as a known, or given, quantity, and concentrate on learning the procedures for making statistical inferences about the population mean μ. Therefore, to continue the explanation of statistical inferences, we will assume $\sigma = 18$ for the specific rivet described in our example.

POINT ESTIMATE FOR A PARAMETER

The value of the corresponding sample statistic.

That is, the sample mean, $\bar{x}$, is the point estimate (single number value) for the mean μ of the sampled population. For our rivet example, 924.23 is the point estimate for μ, the mean shearing strength of all rivets.

The quality of this point estimation should be questioned. Is it exact? Is it likely to be high? Or low? Would another sample yield the same result? Would another sample yield an estimate of nearly the same value? Or a value that is very different? How is "nearly the same" or "very different" measured? The quality of an estimation procedure (or method) is greatly enhanced if the sample statistic is both *less variable* and *unbiased*. The variability of a statistic is measured by the standard error of its sampling distribution. The sample mean can be made less variable by reducing its standard error, $\sigma/\sqrt{n}$. That requires using a larger sample, since as n increases, the standard error decreases.

EXERCISE 8.1

The use of a tremendously large sample does not solve the question of quality for an estimator. What problems do you anticipate with very large samples?

UNBIASED STATISTIC

A sample statistic whose sampling distribution has a mean value equal to the value of the population parameter being estimated. A statistic that is not unbiased is a *biased* statistic.

Figure 8-1 illustrates the concept of being unbiased and the effect of variability on the point estimate. The value A is the parameter being estimated, and the dots represent possible sample statistic values from the sampling distribution of the statistic. If A represents the true population mean μ, then the dots represent possible sample means from the $\bar{x}$ sampling distribution.

FIGURE 8-1   Effects of Variability and Bias

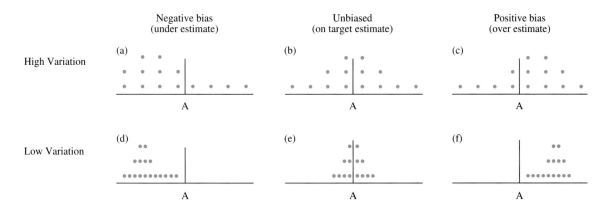

Figure 8-1(a), (c), (d), and (f) show biased statistics; (a) and (d) show sampling distributions whose mean values are less than the value of the parameter, while (c) and (f) show sampling distributions whose mean values are greater than the parameter. Figure 8-1(b) and (e) show sampling distributions that appear to have a mean value equal to the value of the parameter; therefore, they are unbiased. Figure 8-1 (a), (b), and (c) show more variability, while (d), (e), and (f) show less variability in the sampling distributions. Diagram (e) represents the best situation, an estimator that is unbiased (on target) and with low variability (all values close to the target).

The sample mean, $\bar{x}$, is an unbiased statistic because the mean value of the sampling distribution of sample means, $\mu_{\bar{x}}$, is equal to the population mean, μ. (Recall that the sampling distribution of sample means has a mean $\mu_{\bar{x}} = \mu$.) Therefore, the sample statistic $\bar{x} = 924.23$ is an unbiased point estimate for the mean strength of all rivets being manufactured in our example.

Sample means vary in value and form a sampling distribution in which not all samples result in $\bar{x}$-values equal to the population mean. If that is the case, we should not expect this sample of 36 rivets to produce a point estimate (sample mean) that is exactly equal to the mean μ of the sampled population. We should, however, expect the point estimate to be fairly close in value to the population mean. The sampling distribution and the central limit theorem provide the information needed to describe how close the point estimate, $\bar{x}$, is expected to be to the population mean, μ.

Recall that approximately 95% of a normal distribution is within two standard deviations of the mean and that the central limit theorem describes the sampling distribution of sample means as being nearly normal when samples are large enough. Samples of size 36 from populations of variables like rivet strength are generally considered large enough. Therefore, we should anticipate 95% of all samples randomly selected from a population with unknown mean μ and standard deviation $\sigma = 18$ will have means $\bar{x}$ between

$$\mu - 2\sigma_{\bar{x}} \quad \text{and} \quad \mu + 2\sigma_{\bar{x}}.$$

$$\mu - 2\left(\frac{\sigma}{\sqrt{n}}\right) \quad \text{and} \quad \mu + 2\left(\frac{\sigma}{\sqrt{n}}\right)$$

$$\mu - (2)\left(\frac{18}{\sqrt{36}}\right) \quad \text{and} \quad \mu + (2)\left(\frac{18}{\sqrt{36}}\right)$$

$$\mu - 6 \quad \text{and} \quad \mu + 6$$

EXERCISE 8.2

Explain why the standard error of sample means is 3 for the rivet example.

This suggests that 95% of all samples of size 36 randomly selected from the population of rivets would have a mean $\bar{x}$ between $\mu - 6$ and $\mu + 6$. Figure 8-2 shows the middle 95% of the distribution, the bounds of the interval covering the 95%, and the mean μ.

FIGURE 8-2

Sampling Distribution of $\bar{x}$'s, Unknown μ

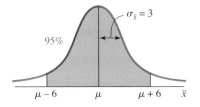

or expressed algebraically:
$$P(\mu - 6 < \bar{x} < \mu + 6) = 0.95$$

Now let's put all of this information together in the form of a *confidence interval.*

INTERVAL ESTIMATE

An interval bounded by two values and used to estimate the value of a population parameter. The values that bound this interval are statistics calculated from the sample that is being used as the basis for the estimation.

LEVEL OF CONFIDENCE $1 - \alpha$

The probability that the sample to be selected yields an interval that includes the parameter being estimated.

CONFIDENCE INTERVAL

An interval estimate with a specified level of confidence.

To construct the confidence interval, we will use the point estimate $\bar{x}$ as the central value of an interval in much the same way as we used the mean μ as the central value to find the interval that captures the middle 95% of the $\bar{x}$ distribution on page 362.

For our rivet example, we can find the bounds to an interval centered at $\bar{x}$:

$$\bar{x} - 2(\sigma_{\bar{x}}) \text{ to } \bar{x} + 2(\sigma_{\bar{x}})$$

$$924.23 - 6 \text{ to } 924.23 + 6$$

The resulting interval is: 918.23 to 930.23

EXERCISE 8.3

Verify that the level of confidence for a 2-standard-deviation interval is 95.44%.

The level of confidence assigned to this interval is approximately 95%, or 0.95. The bounds of the interval are 2 multiples ($z = 2.0$) of the standard error from the sample mean, and by looking at Table 3 in Appendix B, we can more accurately determine the level of confidence as **0.9544**. Putting all of this information together, we express the estimate as a confidence interval: **918.23 to 930.23** *is the 95.44% confidence interval for the mean shear strength of the rivets.* Or in an abbreviated form: **918.23 to 930.23**, *the 95.44% estimate for* μ.

Case Study

8-1

EXERCISE 8.4

a. What is the "average of the repeated samplings"?
b. What does "95 times out of 100" mean?
c. If a value of 52 has a ± 5 margin of error, how are these two numbers related to: point estimate, interval width, and interval bounds?

Gallup Report: Sampling Tolerances

We see survey results frequently in today's newspapers and magazines. A survey's results are often summarized using a statistic and a footnote that says, "Margin of error: ± 3." The statistic reported is the point estimate, and the margin of error is one-half of the interval width; therefore, 33 ± 3 represents the interval estimate 30 to 36 as Gallup describes in this article.

SAMPLING ERROR

In interpreting survey results, it should be borne in mind that all sample surveys are subject to sampling error, that is, the extent to which the results may differ from those that would be obtained if the whole population surveyed had been interviewed. The size of such sampling errors depends largely on the number of interviews.

A reported statistic is 33 and is subject to a sampling error of plus or minus 3. Another way of saying it is that very probably (95 times out of 100) the average of repeated samplings would be somewhere between 30 and 36, with the most likely figure the 33 obtained.

Source: Copyright 1986 by Gallup Report. Reprinted by permission.

EXERCISES

8.5   Identify each numerical value by "name" (mean, variance, etc.) and by symbol ($\bar{x}$, etc.):
   a. The mean height of 24 junior high school girls is 4'11".
   b. The standard deviation for I.Q. scores is 16.
   c. The variance among the test scores on last week's exam was 190.
   d. The mean height of all cadets who have ever entered West Point is 69 inches.

8.6   A random sample of the amount paid for taxi fare from downtown to the airport was obtained:

   15   19   17   23   21   17   16   18   12   18   20   22   15   18   20

   Use the data to find a point estimate for each of the following parameters.
      a. mean   b. variance   c. standard deviation

8.7   In each diagram below, I and II represent sampling distributions of two statistics that might be used to estimate a parameter. In each case, identify the statistic that you think would be the better estimator and describe why it is your choice.

   a.   b.   c.

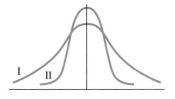
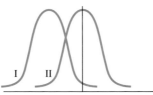
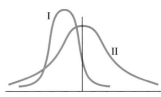

8.8 Suppose there are two statistics that will serve as an estimator for the same parameter. One of them is biased and the other unbiased.

  a. Everything else being equal, explain why you usually would prefer an unbiased estimator to a biased estimator.

  b. If a statistic is unbiased, does that ensure it is a good estimator? Why or why not? What other considerations must be taken into account?

  c. Describe a situation that might occur in which the biased statistic might be a better choice as an estimator than the unbiased statistic.

8.9 Being unbiased and having a small variability are two desirable characteristics of a statistic if it is going to be used as an estimator. Describe how the central limit theorem addresses both of these properties when estimating the mean of a population.

8.10 Find the level of confidence assigned to an interval estimate of the mean formed using the interval

  a. $\bar{x} - 1.3\sigma_{\bar{x}}$ to $\bar{x} + 1.3\sigma_{\bar{x}}$

  b. $\bar{x} - 1.65\sigma_{\bar{x}}$ to $\bar{x} + 1.65\sigma_{\bar{x}}$

  c. $\bar{x} - 1.96\sigma_{\bar{x}}$ to $\bar{x} + 1.96\sigma_{\bar{x}}$

  d. $\bar{x} - 2.3\sigma_{\bar{x}}$ to $\bar{x} + 2.3\sigma_{\bar{x}}$

  e. $\bar{x} - 2.6\sigma_{\bar{x}}$ to $\bar{x} + 2.6\sigma_{\bar{x}}$

8.11 Bill owns a collection of 25,000 baseball trading cards. A random sample of 100 of these cards reveals that 48 of them are worth $1, 27 are worth $5, 16—$10, 6—$25, and 3—$50. Estimate the total worth of Bill's collection.

8.12 Jackie has a stamp collection that she started when she was very young. She believes that she has approximately 7000 stamps. Devise a plan she might use to estimate the collection's worth.

8.2 | ESTIMATION OF MEAN μ (σ KNOWN)

In Section 8-1 we surveyed the basic ideas of estimation: point estimate, interval estimate, level of confidence, and confidence interval. These basic ideas are interrelated and used throughout statistics when an inference calls for an estimation. In this section we formalize the interval estimation process as it applies to estimating the population mean μ based on a random sample under the restriction that the population standard deviation σ is a known value.

The sampling distribution of sample means and the central limit theorem provide us with the needed information to ensure that the necessary **assumptions** are satisfied.

ASSUMPTIONS

The *conditions* that need to exist in order to correctly apply a statistical procedure.

NOTE The use of the word *assumptions* is somewhat a misnomer. It does not mean that we "assume" something to be the situation and continue, but that **we must be sure the conditions expressed by the assumptions do exist before applying a particular statistical method.**

> **The assumption for estimating mean μ using a known σ:** The sampling distribution of $\bar{x}$ has a normal distribution.

The information needed to ensure that this assumption (or condition) is satisfied is contained in the sampling distribution of sample means and in the central limit theorem (see Chapter 7, p. 334).

The sampling distribution for sample means, $\bar{x}$, is distributed about a mean equal to μ with a standard error equal to $\sigma/\sqrt{n}$; and (1) if the randomly sampled population is normally distributed, then $\bar{x}$ is normally distributed for all sample sizes, or (2) if the randomly sampled population is not normally distributed, then $\bar{x}$ is normally distributed for sufficiently large sample sizes.

Therefore, we can satisfy the required assumption by either (1) *knowing that the sampled population is normally distributed* or (2) *using a random sample containing a sufficiently large number of data.* The first possibility is obvious. We either know enough about the population to know it is normally distributed or we don't. The second way to satisfy the assumption is by applying the CLT. Inspection of various graphic displays of the sample data should yield an indication of the type of distribution the population possesses. The CLT can be applied to smaller samples (say, $n = 15$ or larger) when the data provides a strong indication of a *unimodal distribution* that is approximately *symmetric*. If there is evidence of some *skewness* present in the data, then the sample size will need to be much larger (perhaps $n \geq 50$). If the data contains evidence of an extremely skewed or *J*-shaped distribution, it is possible that the CLT does not apply.

The help of a professional statistician should be sought when treating extremely skewed data.

WARNING   There is no hard and fast rule defining "large enough"; the sample size that is "large enough" varies greatly according to the distribution of the population.

The $1 - \alpha$ confidence interval for the estimation of mean μ is found using the formula

$$\bar{x} - z\left(\frac{\alpha}{2}\right)\frac{\sigma}{\sqrt{n}} \quad \text{to} \quad \bar{x} + z\left(\frac{\alpha}{2}\right)\frac{\sigma}{\sqrt{n}} \tag{8-1}$$

The parts of the confidence interval formula are:

Point estimate
Confidence coefficient

1. $\bar{x}$ is the **point estimate** and the center point of the confidence interval.
2. $z(\alpha/2)$ is the **confidence coefficient**. It is the number of multiples of the standard error needed to formulate an interval estimate of the correct width to have a level of confidence of $1 - \alpha$. Figure 8-3 shows the relationship among (a) the level of confidence (the middle portion of the distribution), (b) the $\alpha/2$, the area used to identify the z-score ("area to the right" used with the critical-value notation), and (c) the confidence coefficient $z(\alpha/2)$, whose value is found using Table 4b of Appendix B.

FIGURE 8-3

Confidence
Coefficient, $z(\alpha/2)$

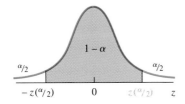

Standard error of the mean

Maximum error of estimate, E

Lower confidence limit

Upper confidence limit

3. $\sigma/\sqrt{n}$ is the standard error of the mean, the standard deviation of the sampling distribution of sample means.

4. $z(\alpha/2)(\sigma/\sqrt{n})$ is one-half the width of the confidence interval (the product of the confidence coefficient and the standard error) and is called the maximum error of estimate, E.

5. $\bar{x} - z(\alpha/2)(\sigma/\sqrt{n})$ is called the **lower confidence limit** (LCL), and $\bar{x} + z(\alpha/2)(\sigma/\sqrt{n})$ is called the **upper confidence limit** (UCL) for the confidence interval.

The estimation procedure is organized into a four-step process that will take into account all of the above and produce both the point estimate and the confidence interval.

Basically, the confidence interval is "point estimate ± maximum error."

THE CONFIDENCE INTERVAL: A FOUR-STEP MODEL

Step 1: Describe the population parameter of concern.
Step 2: Specify the confidence interval criteria.
    a. Check the assumptions.
    b. Determine the test statistic to be used.
    c. State the level of confidence, $1 - \alpha$.
Step 3: Collect and present sample evidence.
    a. Collect the sample information.
    b. Find the point estimate.
Step 4: Determine the confidence interval.
    a. Determine the confidence coefficients.
    b. Find the maximum error of estimate.
    c. Find the lower and upper confidence limits.
    d. Describe the results.

▼ ILLUSTRATION   8 - 1

The student body at many community colleges is considered a "commuter population." The student activities office wishes to obtain an answer to the question "How far (one way) does the average community-college student commute to college each day?" (Typically the "average student's commute distance" is meant to be the "mean distance" commuted by all students who commute.) A random sample of 100 commuting students was identified, and the one-way distance each commuted was obtained.

The resulting sample mean distance was 10.22 miles. Estimate the mean one-way distance commuted by all commuting students using: (a) a point estimate and (b) a 95% confidence interval. (Use $\sigma = 6$ miles.)

SOLUTION

STEP 1   Describe the population parameter of concern.
The mean μ of the one-way distances commuted by all commuting community-college students is the parameter of concern.

STEP 2   Specify the confidence interval criteria.
a. Check the assumptions.
The variable "distance commuted" most likely has a skewed distribution since the vast majority of the students will commute between 0 and 25 miles, with fewer commuting more than 25 miles. A sample size of 100 should be large enough for the CLT to satisfy the assumption, the $\bar{x}$ sampling distribution is normal.
b. Determine the test statistic to be used.
The test statistic used to calculate the confidence interval bounds will be the standard normal variable z using $\sigma = 6$.
c. State the level of confidence, $1 - \alpha$.
The question asks for 95% confidence, or $1 - \alpha = 0.95$.

STEP 3   Collect and present sample evidence.
a. Collect the sample information.
The sample information is given in the statement of the problem: $n = 100, \bar{x} = 10.22$.
b. Find the point estimate.
The point estimate for the mean one-way distance is **10.22** (the sample mean).

STEP 4   Determine the confidence interval.
a. Determine the confidence coefficient.
The confidence coefficient is found using Table 4b:

	A Portion of Table 4b			
Level of Confidence	α	. . .	0.05	**Confidence Coefficient**
$1 - \alpha = 0.95$	$z\left(\dfrac{\alpha}{2}\right)$	. . .	1.96 $\longrightarrow$	$z\left(\dfrac{\alpha}{2}\right) = 1.96$
	$1 - \alpha$	. . .	0.95	

b. Find the maximum error of estimate.
   Use the maximum error part of formula (8-1):

$$E = z\left(\frac{\alpha}{2}\right)\frac{\sigma}{\sqrt{n}} = 1.96\frac{6}{\sqrt{100}} = (1.96)(0.6) = \mathbf{1.176}$$

c. Find the lower and upper confidence limits.
   Using the point estimate ($\bar{x}$) from step 3b and the maximum error (E)
   from step 4b, the confidence interval limits are:

$$\bar{x} - z\left(\frac{\alpha}{2}\right)\frac{\sigma}{\sqrt{n}} \quad \text{to} \quad \bar{x} + z\left(\frac{\alpha}{2}\right)\frac{\sigma}{\sqrt{n}}$$

$$10.22 - 1.176 \quad \text{to} \quad 10.22 + 1.176$$

$$9.044 \quad \text{to} \quad 11.396$$

$$\mathbf{9.04} \quad \textbf{to} \quad \mathbf{11.40}$$

d. Describe the results.
   Therefore, with 95% confidence we can say, "The mean one-way distance is between 9.04 and 11.40 miles," and we abbreviate this as

9.04 to 11.40, the 95% confidence interval for μ

Let's look at another illustration of the estimation procedure.

ILLUSTRATION   8-2

"Particle size" is an important property of latex paint and is monitored during production as part of the quality-control process. Thirteen particle-size measurements were
taken using the Dwight P. Joyce Disc, and the sample mean was 3978.1 angstroms
[where 1 angstrom (1Å) $= 10^{-8}$ cm]. The particle size, x, is normally distributed
with a standard deviation $\sigma = 200$ angstroms. Find the 98% confidence interval for
the mean particle size for the sampled batch of paint.

SOLUTION

STEP 1   Parameter of concern: the mean particle size, μ, for the sampled batch of
paint.

STEP 2   Confidence interval criteria
   a. Assumptions: The variable "particle size" is normally distributed; there-
      fore, the sampling distribution of sample means is normal for all sample
      sizes.
   b. The test statistic: the standard normal variable z using $\sigma = 200$.
   c. The level of confidence: 98%, or $1 - \alpha = 0.98$.

Case Study

8-2

EXERCISE 8.15

"Rockies snow
brings little water"
lists "14.36 inches"
and "5.07 inches" as
statistics and uses
them as point esti-
mates. Describe why
these numbers are
statistics and why
they are also point
estimates.

Rockies Snow Brings Little Water

*When snow melts it becomes water, sometimes more water than others. This
newspaper article compares the water content of snow from two areas in the
United States that typically get about the same amount of snow annually. However,
the water content is very different. There are several point estimates for the average
included in the USA Today article.*

STEP 3   The sample
    a. Sample evidence: $n = 13$ and $\bar{x} = 3978.1$.
    b. The point estimate is $\bar{x} = \textbf{3978.1}$, the sample mean.

STEP 4   The confidence interval
    a. The confidence coefficient: $1 - \alpha = 0.98$, using Table 4b,
      $z(\alpha/2) = z(0.01) = \textbf{2.33}$.
    b. The maximum error: $E = z(\alpha/2)(\sigma/\sqrt{n}) = 2.33\,(200/\sqrt{13})$
                           $= (2.33)(55.47) = \textbf{129.2}$.

c. The lower and upper limits:

$$\bar{x} - z\!\left(\frac{\alpha}{2}\right)\frac{\sigma}{\sqrt{n}} \quad \text{to} \quad \bar{x} + z\!\left(\frac{\alpha}{2}\right)\frac{\sigma}{\sqrt{n}}$$

$$3978.1 - 129.2 = \mathbf{3848.9} \quad \text{to} \quad 3978.1 + 129.2 = \mathbf{4107.3}$$

d. The results: Therefore, with 98% confidence we can say, "The mean particle size is between 3848.9 and 4107.3 angstroms," and we can abbreviate this as

3848.9 to 4107.3, the 98% confidence interval for μ

Let's take another look at the concept "level of confidence." It was defined to be the probability that the selected sample will produce interval bounds that contain the parameter.

▼ | ILLUSTRATION   8 - 3

Single-digit random numbers, like the ones in Table 1 (Appendix B), have a mean value $\mu = 4.5$ and a standard deviation $\sigma = 2.87$ (see exercise 5.29, p. 263). Draw a sample of 40 single-digit numbers from Table 1 and construct the 90% confidence interval for the mean. Does the resulting interval contain the expected value of μ, 4.5? If we were to select another sample of 40 single-digit numbers from Table 1, would the same results occur? What might happen if we were to select a total of 15 different samples and construct the 90% confidence interval related to each? Would the expected value for μ, namely 4.5, be contained in all of them? Should we expect all 15 confidence intervals to contain 4.5? Think about the definition of "level of confidence"; it says that in the long run 90% of the samples will result in bounds that contain μ. In other words, 10% of the samples will *not* contain μ. Let's see what happens.

First we need to address the assumptions; if the assumptions are not satisfied, we cannot expect the 90% and the 10% to occur. We know that (1) the distribution of single-digit random numbers is rectangular (definitely not normal), (2) the distribution of single-digit random numbers is symmetric about their mean, (3) the $\bar{x}$ distribution for very small samples ($n = 5$) in Illustration 7-2, page 331, displayed a distribution that appeared to be approximately normal, and (4) there should be no skewness involved. Therefore, it seems reasonable to assume that $n = 40$ is large enough to allow the CLT to apply.

The first random sample was drawn from Table 1 in Appendix B.

The sample:	2	8	2	1	5	5	4	0	9	1
	0	4	6	1	5	1	1	3	8	0
	3	6	8	4	8	6	8	9	5	0
	1	4	1	2	1	7	1	7	9	3

Summary of sample: $n = 40$, $\Sigma x = 159$, $\bar{x} = 3.975$

Resulting 90% confidence interval:

$$\bar{x} \pm z\left(\frac{\alpha}{2}\right)\frac{\sigma}{\sqrt{n}}: \qquad 3.975 \pm 1.65\frac{2.87}{\sqrt{40}}$$

$$3.975 \pm (1.65)(0.454)$$

$$3.975 \pm 0.749$$

$$3.975 - 0.749 = \mathbf{3.23} \text{ to } 3.975 + 0.749 = \mathbf{4.72}$$

3.23 to 4.72, the 90% confidence interval for μ

Figure 8-4 shows this interval estimate, its bounds, and the expected mean μ.

FIGURE 8-4

The 90%
Confidence Interval

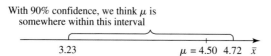

With 90% confidence, we think μ is
   somewhere within this interval

3.23 $\mu = 4.50$ 4.72 $\bar{x}$

The expected value for the mean, 4.5, does fall within the bounds of the confidence interval for this sample. Let's now select 14 more random samples from Table 1 in Appendix B each of size 40.

Table 8-1 lists the mean from the above sample and the means obtained from the 14 additional random samples of size 40. The 90% confidence intervals for the estimation of μ based on each of the 15 samples are listed in Table 8-1 and shown in Figure 8-5.

TABLE 8-1

Fifteen Samples of Size 40

Sample Number	Sample Mean $\bar{x}$	90% Confidence Interval Estimate for μ
1	3.975	3.23 to 4.72
2	4.64	3.89 to 5.39
3	4.56	3.81 to 5.31
4	3.96	3.21 to 4.71
5	5.12	4.37 to 5.87
6	4.24	3.49 to 4.99
7	3.44	2.69 to 4.19
8	4.60	3.85 to 5.35
9	4.08	3.33 to 4.83
10	5.20	4.45 to 5.95
11	4.88	4.13 to 5.63
12	5.36	4.61 to 6.11
13	4.18	3.43 to 4.93
14	4.90	4.15 to 5.65
15	4.48	3.73 to 5.23

FIGURE 8.5

Confidence Intervals from Table 8-1

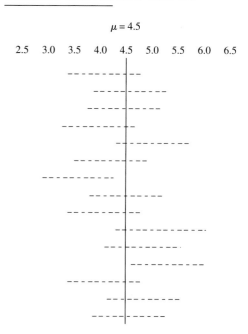

We see that 86.7% (13 of the 15) of the intervals contain μ and two of the 15 samples (sample 7 and sample 12) do not contain μ. The results here are "typical"; repeated experimentation might result in any number of intervals that contain 4.5. However, in the long run we should expect approximately $1 - \alpha = 0.90$ (or 90%) of the samples to result in bounds that contain 4.5 and approximately 10% that do not contain 4.5.

▲ |

Before using **MINITAB**, see Helpful Hint on page 89.

MINITAB (Release 10) commands to complete a $1 - \alpha = A$ confidence interval for mean μ with standard deviation $\sigma = B$ given and the sample data in C1.

Session commands

Enter:
  ZINTerval A $\sigma =$
  B in C1

Menu commands
Choose: Stat > Basic Stats > 1-Sample Z
  Enter: C1
  Select: Confidence interval
  Enter: $1 - \alpha$: A
    Sigma: B

EXERCISE 8.16

Using a computer, randomly select a sample of 40 single-digit numbers and find the 90% confidence interval for μ. Repeat several times, observing whether or not 4.5 is in the interval each time. Describe your results.

MINITAB commands

RANDom 40 C1;
  INTEger 0 9.
ZINTerval 90 2.87 C1

When the denominator increases, the value of the fraction decreases.

Sample Size

The confidence interval has two basic characteristics that determine its quality: its level of confidence and its width. It is preferred that the interval have a high level of confidence and be precise (narrow) at the same time. The higher the level of confidence, the more likely the interval is to contain the parameter, and the narrower the interval the more precise the estimation. However, these two properties seem to work against each other since it would seem that a narrower interval would tend to have a lower probability and a wider interval would be less precise. The maximum error part of the confidence interval formula specifies the relationship involved.

$$E = z\left(\frac{\alpha}{2}\right)\frac{\sigma}{\sqrt{n}} \qquad (8\text{-}2)$$

The components of this formula are: (a) maximum error E, one-half of the width of the confidence interval, (b) the confidence coefficient, $z(\alpha/2)$, which is determined by the level of confidence, (c) the sample size, n, and (d) the standard deviation, σ. The standard deviation σ is not a concern in this discussion because it is a constant (the standard deviation of a population does not change in value). That leaves three factors. Inspection of formula (8-2) indicates the following: Increasing the level of confidence will make the confidence coefficient larger and thereby force either the maximum error to increase or the sample size to increase; decreasing the maximum error will force the level of confidence to decrease or the sample size to increase; decreasing the sample size will force the maximum error to become larger or the level of confidence to decrease. We have a "three-way tug of war," as pictured in

Figure 8-6. An increase or decrease to any one of the three factors has an effect on one or both of the other two factors. The statistician's job is to "balance" the level of confidence, the sample size, and the maximum error so that an acceptable interval results.

FIGURE 8-6

The "Three-Way Tug of War" Between $1 - \alpha$, n, and E

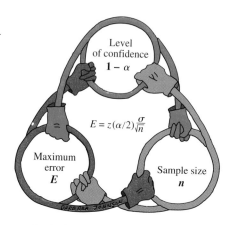

Let's look at an illustration of this relationship in action.

▼ ILLUSTRATION   8 - 4

Determine the size sample needed to estimate the mean weight of all second-grade boys if we want to be accurate within 1 lb with 95% confidence. Assume a normal distribution and that the standard deviation of such weights is 3 lb.

SOLUTION The desired level of confidence determines the confidence coefficient (using Table 4b):

Level of Confidence	A Portion of Table 4b			Confidence Coefficient
	α	. . .	0.05	
$1 - \alpha = 0.95$	$z\left(\dfrac{\alpha}{2}\right)$	. . .	1.96 ⟶	$z\left(\dfrac{\alpha}{2}\right) = 1.96$
	$1 - \alpha$	. . .	0.95	

The desired maximum error is $E = 1.0$. Now we are ready to use the maximum error formula:

$$E = z\left(\frac{\alpha}{2}\right)\frac{\sigma}{\sqrt{n}}: \qquad 1.0 = 1.96\frac{3}{\sqrt{n}}$$

Solve for n: $1.0 = \dfrac{5.88}{\sqrt{n}}$

$$\sqrt{n} = 5.88$$

$$n = (5.88)^2 = 34.57 = \mathbf{35}$$

Therefore, $n = 35$ is the sample size needed if you want a 95% confidence interval with a maximum error of no larger than 1.0 lb.

Using the maximum error formula (8-2) can be made a little easier by rewriting the formula into a form that expresses n in terms of the other values.

$$n = \left(\frac{z(\alpha/2) \cdot \sigma}{E}\right)^2 \tag{8-3}$$

If the maximum error is expressed as a multiple of the standard deviation σ, the actual value of σ is not needed in order to calculate the sample size.

ILLUSTRATION   8 - 5

EXERCISE 8.17

Find the sample size needed to estimate μ of a normal population with $\sigma = 9$ to within 2 units at the 90% level of confidence.

Find the sample size needed to estimate the population mean to within $\frac{1}{5}$ of a standard deviation with 99% confidence.

SOLUTION Determine the confidence coefficient (using Table 4b): $1 - \alpha = 0.99$, $z(\alpha/2) = \mathbf{2.58}$. The desired maximum error is $E = \sigma/5$. Now we are ready to use the sample size formula (8-3):

$$n = \left[\frac{z(\alpha/2) \cdot \sigma}{E}\right]^2 : \quad n = \left[\frac{(2.58) \cdot (\sigma)}{\sigma/5}\right]^2 = \left[\frac{(2.58\sigma)(5)}{\sigma}\right]^2 \; [(2.58)(5)]^2$$

$$n = 12.90^2 = 166.41 = \mathbf{167}$$

NOTE   When solving for the sample size n, it is customary to round up to the next larger integer, no matter what fraction (or decimal) results.

EXERCISES

8.18   Discuss the effect that each of the following have on the confidence interval.
   a. point estimate
   b. level of confidence
   c. sample size
   d. variability of the characteristic being measured

8.19   Discuss the conditions that we must be sure exist before we can estimate the population mean using the interval techniques of formula (8-1).

8.20   Determine the value of the confidence coefficient $z(\alpha/2)$ for each situation described:

    a. $1 - \alpha = 0.90$

    b. $1 - \alpha = 0.95$

    c. 98% confidence

    d. 99% confidence

8.21   Given the information: the sampled population is normally distributed, $n = 16$, $\bar{x} = 28.7$, and $\sigma = 6$.

    a. Find the 0.95 confidence interval estimate for μ.

    b. Are the assumptions satisfied? Explain why.

8.22   Given the information: the sampled population is normally distributed, $n = 55$, $\bar{x} = 78.2$, and $\sigma = 12$.

    a. Find the 0.98 confidence interval estimate for μ.

    b. Are the assumptions met? Explain.

8.23   Given the information: $n = 86$, $\bar{x} = 128.5$, and $\sigma = 16.4$.

    a. Find the 0.90 confidence interval estimate for μ.

    b. Are the assumptions satisfied? Explain how.

8.24   Given the information: $n = 45$, $\bar{x} = 72.3$, and $\sigma = 6.4$.

    a. Find the 0.99 confidence interval estimate for μ.

    b. Are the assumptions satisfied? Explain how.

8.25   In order to estimate the mean amount spent for textbooks per student during the fall semester at a large community college, a random sample of 75 students was surveyed. The sample mean was $158.30.

    a. What is the parameter of interest?

    b. Find the 90% confidence interval estimate for the mean cost per student for all students. (Use $\sigma = \$35$.)

8.26   A certain adjustment to a machine will change the length of the parts it is making but will not affect the standard deviation. The length of the parts is normally distributed, and the standard deviation is 0.5 mm. After an adjustment is made, a random sample is taken to determine the mean length of parts now being produced. The resulting lengths are:

75.3	76.0	75.0	77.0	75.4	76.3	77.0	74.9	76.5	75.8

    a. What is the parameter of interest?

    b. Find the point estimate for the mean length of all parts now being produced.

    c. Find the 0.99 confidence interval estimate for μ.

8.27   A sample of 60 night school students' ages is obtained in order to estimate the mean age of night school students. $\bar{x} = 25.3$ years. The population variance is 16.

    a. Give a point estimate for μ.

    b. Find the 95% confidence interval estimate for μ.

    c. Find the 99% confidence interval estimate for μ.

8.28   The lengths of 200 fish caught in Cayuga Lake had a mean of 14.3 in. The population standard deviation is 2.5 in.

    a. Find the 90% confidence interval for the population mean length.

    b. Find the 98% confidence interval for the population mean length.

8.29   An article titled "A Comparison of the Effects of Constant Co-operative Grouping versus Variable Co-operative Grouping on Mathematics Achievement Among Seventh Grade Students" (*International Journal of Mathematics Education in Science and Technology*, Vol. 24, No. 5, 1993) gives the mean percentile score on the California Achievement Test (CAT) for

20 students to be 55.20. Assume the population of CAT scores is normally distributed and that $\sigma = 19.5$.

    a. Make a point estimate for the mean of the population the sample represents.

    b. Find the maximum error of estimate for a level of confidence equal to 95%.

    c. Construct a 95% confidence interval estimate for the population mean.

    d. Explain the meaning of each of the above answers.

8.30   The 1994 *County and City Data Book* gives the number of workers 16 years and older for Nebraska as 775,085 with an average travel time to work as 15.8 minutes. A survey of 150 such workers in Lincoln, Nebraska, gave a sample mean equal to 27.5 minutes.

    a. Construct the 0.99 confidence interval estimate for the mean travel time to work of all such workers in Lincoln. Use $\sigma = 10.0$ minutes.

    b. What is the 15.8 given in the article?

    c. Based on the above results, what can you conclude about the mean travel time for workers in Lincoln compared to the workers for the whole state of Nebraska?

8.31   According to a USA Snapshot® (*USA Today*, 11-3-94), the annual teaching income for ski instructors in the Rocky Mountain and Sierra areas is $5600. (Assume $\sigma = \$1000$.)

    a. If this figure is based on a survey of 15 instructors and if the annual incomes are normally distributed, find a 90% confidence interval for μ, the mean annual teaching income for all ski instructors in the Rocky Mountain and Sierra areas.

    b. If the distribution of annual incomes is not normally distributed, what effect do you think that would have on the interval answer in part (a)? Explain.

8.32   How large a sample should be taken if the population mean is to be estimated with 99% confidence to within $75? The population has a standard deviation of $900.

8.33   A high-tech company wants to estimate the mean number of years of college education its employees have completed. A good estimate of the standard deviation for the number of years of college is 1.0. How large a sample needs to be taken to estimate μ to within 0.5 year with 99% confidence?

8.34   By measuring the amount of time it takes a component of a product to move from one work station to the next, an engineer has estimated that the standard deviation is 5 sec.

    a. How many measurements should be made in order to be 95% certain that the maximum error of estimation will not exceed 1 sec?

    b. What sample size is required for a maximum error of 2 sec?

8.35   An article titled "A Comparison of Computer-Assisted Instructional Methods" (*International Journal of Mathematics Education in Science and Technology*, Vol. 22, No. 6, 1991) gives the mean score on the Computer Anxiety Index (CAIN) to be 56.9 for 28 ninth-grade students in an algebra class that utilizes basic programming to solve problems. If the standard deviation for CAIN scores is known to equal 10.5, how large a sample would be needed to estimate μ, the mean for all such ninth-graders, with a maximum error of estimate equal to 2 with 95% confidence?

8.36   The weather page of the December 14, 1994, edition of the *Omaha World-Herald* in the listing of the Missouri River levels gave the following information: "The current rate of release from Gavins Point Dam is 17,500 cubic feet of water per second. The 1967–1993 average daily release for December is 19,500 CFS." If the mean CFS were not known, how many December days from 1967 to 1993 should be selected in order to estimate μ with a maximum error of estimate equal to 250 CFS with 95% confidence? Assume the CFS readings are normally distributed and that $\sigma = 500$ CFS.

8.3 | THE NATURE OF HYPOTHESIS TESTING

We all make decisions every day of our lives. Some of these decisions are of major importance, others are seemingly insignificant. All decisions follow the same basic pattern. We weigh the alternatives; then, based on our beliefs and preferences, and whatever evidence is available, we arrive at a decision and take the appropriate action. The statistical hypothesis test follows much the same process, except that it involves statistical information. In this section, we develop many of the concepts and attitudes of the hypothesis test while looking at several decision-making situations without any statistics.

A friend is having a party (Super Bowl party, home-from-college party, you know the situation, any excuse will do), and you have been invited. You must make a decision: attend or not attend. Simple decision; well maybe, except that you want to go only if you can be convinced the party is going to be more fun than your friend's typical party; further, you definitely do not want to go if it's going to be just another dud of a party. You have taken the position that "the party will be a dud" and you will not go unless you become convinced otherwise. Your friend assures you, "Guaranteed, the party will be a great time!" Do you go or not?

The decision-making process starts by identifying **something of concern** and then formulating two **hypotheses** about it.

HYPOTHESIS

A statement that something is true.

Your friend's statement, "The party will be a great time," is a hypothesis. Your position, "The party will be a dud," is also a hypothesis.

STATISTICAL HYPOTHESIS TEST

A process by which a decision is made between two opposing hypotheses. The two opposing hypotheses are formulated so that each hypothesis is the negation of the other. (That way one of them is always true, and the other one is always false.) Then one hypothesis is tested in hopes that it can be shown to be a very improbable occurrence, thereby implying the other hypothesis is the likely truth.

The two hypotheses involved in making a decision are known as the **null hypothesis** and the **alternative hypothesis**.

NULL HYPOTHESIS, H_0

The hypothesis we will test. Generally this is a statement that a population parameter has a specific value. The null hypothesis is so named because it is the "starting point" for the investigation. (The phrase "there is no difference" is often used in its interpretation.)

ALTERNATIVE HYPOTHESIS, H_a

A statement about the same population parameter that is used in the null hypothesis. Generally this is a statement that specifies the population parameter has a value different, in some way, from the value given in the null hypothesis. The rejection of the null hypothesis will imply the likely truth of this alternative hypothesis.

With regard to your friend's party, the two opposing viewpoints or hypotheses would be: "The party will be a great time" and "the party will be a dud." Which statement becomes the null hypothesis, and which becomes the alternative hypothesis?

Determining the statement of the null hypothesis and the statement of the alternative hypothesis is a very important step. The *basic idea* of the hypothesis test is for the evidence to have a chance to "disprove" the null hypothesis. The null hypothesis is the statement that the evidence might disprove. *Your concern* (belief or desired outcome), as the person doing the testing, is expressed in the alternative hypothesis. As the person making the decision, you believe that the evidence will demonstrate the feasibility of your "theory" by demonstrating the *unlikeliness* of the truth of the null hypothesis. The alternative hypothesis is sometimes referred to as the *research hypothesis* since it represents what the researcher hopes will be found to be "true." (If so, he/she will get a paper out of the research.)

Since the "evidence" (who's going to the party, what is going to be served, and so on) can only demonstrate the unlikeliness of the party being a dud, your initial position, "The party will be a dud," becomes the null hypothesis. Your friend's claim, "The party will be a great time," then becomes the alternative hypothesis.

H_0: "Party will be a dud"   vs. H_a: "Party will be a great time"

▼ | ILLUSTRATION   8 - 6

You are testing a new design for airbags used in automobiles, and you are concerned that they might not open properly. State the null and alternative hypotheses.

SOLUTION The two opposing possibilities are "bags open properly" or "bags do not open properly." Testing can only produce evidence that discredits the hypothesis "the bags open properly." Therefore, the null hypothesis is "they do open properly" and the alternative hypothesis is "they do not open properly."

▲ |

The alternative hypothesis can be the statement the experimenter wants to show to be true.

▼ ┃ I L L U S T R A T I O N   8 - 7

An engineer wishes to show that the new formula that was just developed results in a quicker drying paint. State the null and alternative hypotheses.

SOLUTION The two opposing possibilities are "does dry quicker" and "does not dry quicker." Since the engineer wishes to show "does dry quicker," the alternative hypothesis is "paint made with the new formula does dry quicker" and the null hypothesis is "paint made with the new formula does not dry quicker."

▲ ┃

Occasionally it might be reasonable to hope that the evidence does not lead to a rejection of the null hypothesis. Such is the case in Illustration 8-8.

▼ ┃ I L L U S T R A T I O N   8 - 8

EXERCISE   8.37

You are testing a new detonating system for explosives and you are concerned that the system is not reliable. State the null and alternative hypotheses.

You suspect that a brand-name detergent outperforms the store's brand of detergent and you wish to test the two detergents because you would prefer to buy the cheaper store brand. State the null and alternative hypotheses.

SOLUTION Your suspicion, "the brand-name detergent outperforms the store brand" is the reason for the test and therefore becomes the alternative hypothesis.

H_0: "There is no difference in detergent performance."
H_a: "The brand-name detergent performs better than the store brand."

▲ ┃ However, as a consumer, you are hoping not to reject the null hypothesis for budgetary reasons.

Before returning to our example about the party, we need to look at the four possible outcomes that could result from the null hypothesis being either true or false and the decision being either to "reject H_0" or to "fail to reject H_0." Table 8-2 shows these four possible outcomes.

TABLE 8-2

Four Possible Outcomes in a Hypothesis Test

Decision	Null Hypothesis	
	True	**False**
Fail to reject H_0	Type A correct decision	Type II error
Reject H_0	Type I error	Type B correct decision

Case Study

8-3

EXERCISE 8.38

State the instructor's hypothesis, the alternative hypothesis.

Evaluation of Teaching Techniques

Abstract: This study tests the effect of homework collection and quizzes on exam scores.

The hypothesis for this study is that an instructor can improve a student's performance (exam scores) through influencing the student's perceived effort-reward probability. An instructor accomplishes this by assigning tasks (teaching techniques) which are a part of a student's grade and are perceived by the student as a means of improving his or her grade in the class. The student is motivated to increase effort to complete those tasks which should also improve understanding of course material. The expected final result is improved exam scores.

The null hypothesis for this study is:

H_0: Teaching techniques have no significant effect on student's exam scores. . . .

Source: David R. Vruwink and Janon R. Otto, *The Accounting Review*, Vol. LXII, No. 2, April 1987. Reprinted by permission.

Type A correct decision
Type B correct decision
Type I error
Type II error

A **type A correct decision** occurs when the null hypothesis is true, and we decide in its favor. A **type B correct decision** occurs when the null hypothesis is false, and the decision is in opposition to the null hypothesis. A **type I error** will be committed when a true null hypothesis is rejected, that is, when the null hypothesis is true but we decided against it. A **type II error** is committed when we decide in favor of a null hypothesis that is actually false.

▼ ILLUSTRATION 8-9

Describe the four possible outcomes and the resulting actions that would occur for the hypothesis test in Illustration 8-8.

SOLUTION Recall H_0: "There is no difference in detergent performance"
H_a: "The brand-name detergent performs better than the store brand"

	Null Hypothesis Is True	Null Hypothesis Is False
Fail to Reject H_0	*Type A Correct Decision* Truth of situation: There is no difference between the detergents. Conclusion: It was determined that there was no difference. Action: The consumer bought the cheaper detergent, saving money and getting the same results.	*Type II Error* Truth of situation: The brand-name detergent is better. Conclusion: It was determined that there was no difference. Action: The consumer bought the cheaper detergent, saving money and getting inferior results.

(continued)

Table (*continued*)

	Null Hypothesis Is True	**Null Hypothesis Is False**
Reject H_0	*Type I Error* Truth of situation: There is no difference between the detergents. Conclusion: It was determined that the brand-name detergent was better. Action: The consumer bought the brand-name detergent, spending extra money to attain no better results.	*Type B Correct Decision* Truth of situation: The brand-name detergent is better. Conclusion: It was determined that the brand name detergent was better. Action: The consumer bought the brand-name detergent, spending more and getting better results.

EXERCISE 8.39

Describe the four possible decisions and the resulting actions with regard to your friend's party.

▲ ▎

EXERCISE 8.40

Describe how the type II error in the party example represents a "lost opportunity."

α (alpha)
β (beta)

NOTE   The type II error often results in what represents a "lost opportunity"; lost in this situation is the chance to use a product that yields better results.

When a decision is made, it would be nice to always make the correct decision. This, however, is not possible in statistics, since we make our decisions on the basis of sample information. The best we can hope for is to *control the probability* with which an error occurs. The probability assigned to the type I error is α (called "alpha"; α is the first letter of the Greek alphabet). The probability of the type II error is β (called "beta"; β is the second letter of the Greek alphabet). See Table 8-3.

TABLE 8-3

Probability with Which Decisions Occur

Error in Decision	Type	Probability	Correct Decision	Type	Probability
Rejection of a true H_0	I	α	Failure to reject a true H_0	A	$1 - \alpha$
Failure to reject a false H_0	II	β	Rejection of a false H_0	B	$1 - \beta$

EXERCISE 8.41

Explain why α is not always the probability of rejecting the null hypothesis.

EXERCISE 8.42

Explain how assigning a small probability to an error controls the likeliness of its occurrence.

To control these errors we will assign a small probability to each of them. The most frequently used probability values for α and β are 0.01 and 0.05. The probability assigned to each error depends on its seriousness. The more serious the error, the less willing we are to have it occur, and therefore a smaller probability will be assigned. α and β are each probabilities of errors, each under separate conditions, and they cannot be combined. Therefore, we cannot determine a single probability for making an incorrect decision. Likewise, the two correct decisions are distinctly separate and each has its own probability; $1 - \alpha$ is the probability of a correct decision when the null hypothesis is true, and $1 - \beta$ is the probability of a correct decision when the null hypothesis is false. $1 - \beta$ is called the **power** of the statistical test since it is the measure of the ability of a hypothesis test to reject a false null hypothesis, a very important characteristic.

Remember   Regardless of the outcome of a hypothesis test, you are never certain that a correct decision has been reached.

Let's look back at the two possible errors in decision that could occur in Illustration 8-9 while testing the detergents. Most people would become upset if they found out they were spending extra money for a detergent that performed no better than the cheaper brand. Likewise, most people would become upset if they found out they could have been buying a better detergent. Evaluating the relative seriousness of these errors requires knowing—whether this is your personal laundry or a professional laundry business—how much extra the brand-name detergent costs, and so on.

There is an interrelationship between the probability of the type I error (α), the probability of the type II error (β), and the sample size (n). This relationship is very much like the relationship between level of confidence, maximum error, and sample size discussed on pages 373 and 374. Figure 8-7 shows the "three-way tug of war" among α, β, and n. If any one of the three is increased or decreased, it has an effect on one or both of the others.

FIGURE 8-7

The "Three-Way Tug of War" Between α, β, and n

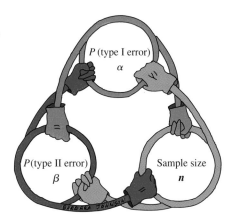

If α is reduced, then either β must increase or n must be increased; if β is decreased, then either α increases or n must be increased; if n is decreased, then either α increases or β increases. The choices for α, β, and n are definitely not arbitrary. At this time in our study of statistics, α will be given in the statement of the problem, as will the sample size n. Further discussion on the role of β, the P(type II error), is left for another time.

LEVEL OF SIGNIFICANCE α

The probability of committing the type I error.

Establishing the level of significance can be thought of as a "managerial decision." Typically, someone in charge determines the level of probability with which they are willing to risk a type I error.

At this point in the hypothesis test procedure, the evidence is collected, summarized, and the value of a **test statistic** is calculated.

TEST STATISTIC

A random variable whose value is calculated from the sample data and is used in making the decision "fail to reject H_0" or "reject H_0."

The value of the calculated test statistic is used in conjunction with a **decision rule** to determine either "reject H_0" or "fail to reject H_0." This decision rule must be established prior to collecting the data and specifies how you will reach the decision.

Back to your friend's party: You have to weigh the history of your friend giving parties, the time and place, others going, and so on, against your own criteria and then make your decision. As a result of the decision about the null hypothesis ("the party will be a dud"), you will take the appropriate action; you will either *go to* or *not go* to the party.

To complete a hypothesis test, you will need to write a **conclusion** that carefully describes the meaning of the decision relative to the intent of the hypothesis test.

THE CONCLUSION

a. If the decision is **"reject H_0,"** then the conclusion should be worded something like, **"There is sufficient evidence at the α level of significance to show that . . .** (*the meaning of the alternative hypothesis*)."

b. If the decision is **"fail to reject H_0,"** then the conclusion should be worded something like, **"There is not sufficient evidence at the α level of significance to show that . . .** (*the meaning of the alternative hypothesis*)."

When writing the decision and the conclusion, remember that (1) the decision is about H_0 and (2) the conclusion is a statement about whether or not the contention of H_a was upheld. This is consistent with the "attitude" of the whole hypothesis test procedure. The null hypothesis is the statement that is "on trial," and therefore the decision must be about it. The contention of the alternative hypothesis is the thought that brought about the need for a decision. Therefore, the question that led to the alternative hypothesis must be answered when the conclusion is written.

We must always remember that when the decision is made, nothing has been proved. Both decisions can lead to errors: "Fail to reject H_0" could be a type II error (the lack of sufficient evidence has led to great parties being missed more than once), and "reject H_0" could be a type I error (more than one person has decided to go to a party that was a dud).

EXERCISES

8.43   State the null and alternative hypotheses for each of the following:
  a. You are investigating a complaint that "special delivery mail takes too much time" to be delivered.
  b. You want to show that people find the new design for a recliner chair more comfortable than the old design.
  c. You are trying to show that cigarette smoke has an effect on the quality of a person's life.
  d. You are testing a new formula for hair conditioner hoping to show it is effective on "split ends."

8.44   When a parachute is inspected, the inspector is looking for anything that might indicate the parachute might not open.
  a. State the null and alternative hypotheses.
  b. Describe the four possible outcomes that can result depending on the truth of the null hypothesis and the decision reached.
  c. Describe the seriousness of the two possible errors.

8.45   When a medic at the scene of a serious accident inspects each victim, she administers the appropriate medical assistance to all victims, unless she is certain the victim is dead.
  a. State the null and alternative hypotheses.
  b. Describe the four possible outcomes that can result depending on the truth of the null hypothesis and the decision reached.
  c. Describe the seriousness of the two possible errors.

 8.46   A supplier of highway materials claims he can supply an asphalt mixture that will make roads paved with his materials less slippery when wet. A general contractor who builds roads wishes to test the supplier's claim. The null hypothesis is "roads paved with this asphalt mixture are no less slippery than roads paved with other asphalt." The alternative hypothesis is "roads paved with this asphalt mixture are less slippery than roads paved with other asphalt."
  a. Describe the meaning of the two possible types of errors that can occur in the decision when this hypothesis test is completed.
  b. Describe how the null hypothesis, as stated above, is a "starting point" for the decision to be made about the asphalt.

8.47   Describe the action that would result in a type I error and a type II error if each of the following null hypotheses were tested. (Remember, the alternative hypothesis is the negation of the null hypothesis.)
  a. H_0: The majority of Americans favor laws against assault weapons.
  b. H_0: This fast-food menu is not low salt.
  c. H_0: This building must not be demolished.
  d. H_0: There is no waste in government spending.

8.48   Describe the action that would result in a correct decision type A and a correct decision type B, if each of the null hypotheses in Exercise 8.47 were tested.

8.49   a. If the null hypothesis is true, what decision error could be made?
  b. If the null hypothesis is false, what decision error could be made?
  c. If the decision "reject H_0" is made, what decision error could have been made?
  d. If the decision "fail to reject H_0" is made, what decision error could have been made?

8.50 The director of an advertising agency is concerned with the effectiveness of a television commercial.
  a. What null hypothesis is she testing if she commits a type I error when she erroneously says that the commercial is effective?
  b. What null hypothesis is she testing if she commits a type II error when she erroneously says that the commercial is effective?

8.51 a. If α is assigned the value 0.001, what are we saying about the type I error?
  b. If α is assigned the value 0.05, what are we saying about the type I error?
  c. If α is assigned the value 0.10, what are we saying about the type I error?

8.52 a. If β is assigned the value 0.001, what are we saying about the type II error?
  b. If β is assigned the value 0.05, what are we saying about the type II error?
  c. If β is assigned the value 0.10, what are we saying about the type II error?

8.53 a. If the null hypothesis is true, the probability of a decision error is identified by what name?
  b. If the null hypothesis is false, the probability of a decision error is identified by what name?

8.54 Suppose that a hypothesis test is to be carried out by using $\alpha = 0.05$. What is the probability of committing a type I error?

8.55 The conclusion is part of the hypothesis test that communicates the findings of the test to the reader. As such, it needs special attention so that the reader receives an accurate picture of the findings.
  a. Carefully describe the "attitude" of the statistician and the statement of the conclusion when the decision is "reject H_0."
  b. Carefully describe the "attitude" and the statement of the conclusion when the decision is "fail to reject H_0."

8.56 Find the power of a test when the probability of the type II error is
  a. 0.01   b. 0.05   c. 0.10

8.57 A normally distributed population is known to have a standard deviation of 5, but its mean is in question. It has been argued to be either $\mu = 80$ or $\mu = 90$, and the following hypothesis test has been devised to settle the argument. The null hypothesis, $H_0: \mu = 80$, will be tested using one randomly selected data and comparing it to the critical value 86. If the data is greater than or equal to 86, the null hypothesis will be rejected.
  a. Find α, the probability of the type I error.
  b. Find β, the probability of the type II error.

8.58 Suppose the argument in Exercise 8.57 was to be settled using a sample of size 4; find α and β.

8.4 | HYPOTHESIS TEST OF MEAN μ (σ KNOWN): A PROBABILITY-VALUE APPROACH

In Section 8.3 we surveyed the concepts and much of the reasoning behind a hypothesis test while looking at nonstatistical illustrations. In this section we are going to formalize the hypothesis test procedure as it applies to statements concerning the mean μ of a population under the restriction that σ, the population standard deviation, is a known value.

> **The assumption for hypothesis tests about mean μ using a known σ:**
> The sampling distribution of $\bar{x}$ has a normal distribution.

The information needed to ensure that this assumption is satisfied is contained in the sampling distribution of sample means and in the central limit theorem (see Chapter 7, p. 334).

The sampling distribution for sample means, $\bar{x}$, is distributed about a mean equal to μ with a standard error equal to $\sigma/\sqrt{n}$, and is normally distributed when samples are randomly selected from a normal population or when the sample size is sufficiently large.

The hypothesis test is a well-organized, step-by-step procedure used to make a decision. Two different formats are commonly used for hypothesis testing. The **probability-value approach** is the hypothesis test process that has gained popularity in recent years, largely as a result of the convenience and the "number crunching" ability of the computer. This approach is organized as a five-step procedure.

Probability-value approach

THE PROBABILITY-VALUE HYPOTHESIS TEST: A FIVE-STEP MODEL

Step 1: 1Describe the population parameter of concern.
Step 2: State the null hypothesis (H_0) and the alternative hypothesis (H_a).
Step 3: Specify the test criteria.
    a. Check the assumptions.
    b. Identify the test statistic to be used.
    c. Determine the level of significance, α.
Step 4: Collect and present the sample evidence.
    a. Collect the sample information.
    b. Calculate the value of the test statistic.
    c. Calculate the p-value.
Step 5: Determine the results.
    a. Determine whether or not the p-value is smaller than α.
    b. Make a decision about H_0.
    c. Write a conclusion about H_a.

Think about the consequences of using weak rivets.

A commercial aircraft manufacturer buys rivets for use in assembling airliners. Each rivet supplier that wants to sell rivets to the aircraft manufacturer must demonstrate that its rivets meet the required specifications. One of the specs is: "The mean shearing strength of all such rivets, μ, is at least 925 lb." Each time the aircraft manufacturer buys rivets, it is concerned that the mean strength might be less than the 925-lb specification.

EXERCISE   8.59

In this example, the aircraft builder, the buyer of the rivets, is concerned that the rivets might not meet the mean-strength spec. State the aircraft manufacturer's null and alternative hypotheses.

NOTE 1   Each individual rivet has a shearing strength, which is determined by measuring the force required to shear ("break") the rivet. Clearly, not all the rivets can be tested. Therefore, a sample of rivets will be tested, and a decision about the mean strength of all the untested rivets will be based on the mean from those sampled and tested.

NOTE 2   We will use $\sigma = 18$ for our rivet example.

STEP 1   Describe the population parameter of concern.
The population parameter of interest is the mean μ, the mean shearing strength (or mean force required to shear) of the rivets being considered for purchase.

STEP 2   State the null hypothesis (H_0) and the alternative hypothesis (H_a).
The null hypothesis and the alternative hypothesis are formulated by inspecting the problem or statement to be investigated and first formulating two opposing statements concerning the mean μ. For our example, these two opposing statements are: (a) "The mean shearing strength is less than 925" ($\mu < 925$, the aircraft manufacturer's concern), and (b) "The mean shearing strength is at least 925" ($\mu \geq 925$, the rivet supplier's claim and the aircraft manufacturer's spec).

More specific instructions on p. 379.

NOTE   The *trichotomy law* from algebra states that two numerical values must be related in exactly one of three possible relationships: $<$, $=$, or $>$. All three of these possibilities must be accounted for between the two opposing hypotheses in order for the two hypotheses to be negations of each other. The three possible combinations of signs and hypotheses are shown in Table 8-4. Recall that the null hypothesis assigns a specific value to the parameter in question and therefore "equals" is always part of the null hypothesis.

TABLE 8-4

The Three Possible Statements of Null and Alternative Hypotheses

Null Hypothesis	Alternative Hypothesis
1. greater than or *equal* to ($\geq$)	less than ($<$)
2. less than or *equal* to ($\leq$)	greater than ($>$)
3. *equal* to ($=$)	not equal to ($\neq$)

The parameter of concern, the population mean μ, is related to the value 925.
Statement (a) becomes the alternative hypothesis:

$$H_a: \mu < 925 \text{ (the mean is less than 925)}$$

This statement represents the aircraft manufacturer's concern and says that "the rivets *do not* meet the required specs."

Statement (b) becomes the null hypothesis:

$$H_0: \mu = 925 \ (\geq) \ \text{(the mean is at least 925)}$$

This hypothesis represents the negation of the aircraft manufacturer's concern, and says that "the rivets *do* meet the required specs."

NOTE   The null hypothesis will be written with just the equal sign ("a value is assigned"). When "equal" is paired with "less than" or paired with "greater than," the combined symbol is written beside the null hypothesis as a reminder that all three signs have been accounted for in these two opposing statements.

Before continuing with our example, let's look at three illustrations that demonstrate formulating the statistical null and alternative hypotheses involving the population mean μ.

▼| I L L U S T R A T I O N   8 - 10

Suppose the EPA was suing the city of Rochester for noncompliance with carbon monoxide standards. Specifically, the EPA would want to show that the mean level of carbon monoxide in downtown Rochester's air is dangerously high, higher than 4.9 parts per million. State the null and alternative hypotheses.

S O L U T I O N   To state the two hypotheses, we first need to identify the population parameter in question: the "**mean** level of carbon monoxide pollution in Rochester." The parameter μ is being compared to the value **4.9** parts per million, the specific value of interest. The EPA is questioning the value of μ and wishes to show it to be **higher than** 4.9 (that is, $\mu > 4.9$). The three possible relationships (1) $\mu < 4.9$, (2) $\mu = 4.9$, or (3) $\mu > 4.9$ must be arranged to form two opposing statements: One states the EPA's position, "the mean level is higher than 4.9 ($\boldsymbol{\mu > 4.9}$)," and the other states the negation, "the mean level is **not higher than** 4.9 ($\boldsymbol{\mu \leq 4.9}$)." One of these two statements will become the null hypothesis H_0, and the other will become the alternative hypothesis H_a.

RECALL   There are two rules for forming the hypotheses: (1) The null hypothesis states that the parameter in question has a specified value ("H_0 must contain the equal sign"), and (2) the EPA's contention becomes the alternative hypothesis ("higher than"). Both rules indicate:

$$H_0: \mu = 4.9 \ (\leq) \quad \text{and} \quad H_a: \mu > 4.9$$

▼| I L L U S T R A T I O N   8 - 11

An engineer wants to show that applications of a paint made with the new formula dry ready for the next coat in a mean time of less than 30 minutes. State the null and alternative hypotheses for this test situation.

SOLUTION The parameter of interest is the **mean** drying time per application, and **30** minutes is the specified value. $\mu < 30$ corresponds to "the mean time is **less than** 30," whereas $\mu \geq 30$ corresponds to the negation, "the mean time is **not less than** 30." Therefore, the hypotheses are

▲

$$H_0: \mu = 30 \; (\geq) \quad \text{and} \quad H_a: \mu < 30$$

Illustration 8-12 demonstrates a "two-tailed" alternative hypothesis.

▼   ILLUSTRATION   8 - 12

Job satisfaction is very important when it comes to getting workers to produce. A standard job-satisfaction questionnaire was administered by union officers to a sample of assembly-line workers in a large plant in hopes of showing that the assembly workers' mean score on this questionnaire would be different than the established mean of 68. State the null and alternative hypotheses.

SOLUTION Either the **mean** job satisfaction score **is different than** 68 ($\boldsymbol{\mu \neq 68}$) or the mean **is equal to** 68 ($\boldsymbol{\mu = 68}$). Therefore,

▲

$$H_0: \mu = 68 \quad \text{and} \quad H_a: \mu \neq 68$$

EXERCISE   8.60

Professor Hart does not believe a statement he heard: "The mean weight of college women is 54.4 kg." State the null and alternative hypotheses he would use to challenge this statement.

NOTE 1   When "less than" appears together with "greater than," they become "not equal to."

NOTE 2   The alternative hypothesis is referred to as being "two-tailed" when H_a is "not equal."

The viewpoint of the experimenter greatly affects the way the hypotheses are formed. Generally, the experimenter is trying to show that the parameter value is different from the value specified. Thus, the experimenter is often hoping to be able to reject the null hypothesis so that the experimenter's theory has been substantiated. Illustrations 8-10, 8-11, and 8-12 also represent the three possible arrangements for the $<$, $=$, and $>$ relationships between the parameter μ and a specified value.

Table 8-5 lists some additional common phrases used in claims and indicates their negations and the hypothesis where each phrase will be used. Again, notice that "equals" is always in the null hypothesis. Also notice that the negation of "less than" is "greater than or equal to." Think of negation as "all the others" from the set of three signs.

TABLE 8-5

Common Phrases and Their Negations	$H_0: (\geq)$	$H_a: (<)$	$H_0: (\leq)$	$H_a: (>)$	$H_0: (=)$	$H_a: (\neq)$
	at least	less than	at most	more than	is	is not
	no less than	less than	no more than	more than	not different from	different from
	not less than	less than	not greater than	greater than	same as	not same as

Once the null and alternative hypotheses are established, we will work under the assumption that the null hypothesis is a true statement until there is sufficient evidence to reject it. This situation might be compared to a courtroom trial, where the accused is assumed to be innocent until sufficient evidence has been presented to show that innocence is totally unbelievable ("beyond a shadow of doubt"). At the conclusion of the hypothesis test, we will make one of two possible decisions. We will decide in opposition to the null hypothesis and say that we "reject H_0" (this corresponds to "conviction" of the accused in a trial), or we will decide in agreement with the null hypothesis and say that we "fail to reject H_0" (this corresponds to "fail to convict" or an "acquittal" of the accused in a trial).

Let's return to the rivet example. Recall:

$$H_0: \mu = 925 \; (\geq) \; \text{(at least 925)} \quad H_a: \mu < 925 \; \text{(less than 925)}$$

STEP 3   Specify the test criteria.

    a. Check the assumptions.

       Variables like shearing strength typically have a mounded distribution; therefore, a sample of size 50 should be large enough for the central limit theorem to apply. The sampling distribution of sample means can be expected to be normally distributed.

    b. Identify the test statistic to be used.

       For a hypothesis test of μ, we will want to compare the value of the sample mean to the value of the population mean as stated in the null hypothesis. We will use $\frac{\bar{x} - \mu}{\sigma/\sqrt{n}}$ as our calculated test statistic and call it $z^\star$ ("z star"). We will call it z since it is expected to have a standard normal distribution when the null hypothesis is true and the assumptions have been satisfied. The * ("star") is to remind us that this is the calculated value of the test statistic.

$$\text{test statistic:} \quad z^\star = \frac{\bar{x} - \mu}{\sigma/\sqrt{n}} \qquad \textbf{(8-4)}$$

    c. Determine the level of significance, α.

       Setting α was described as a managerial decision in Section 8.3. To see what is involved in determining α, the probability of the type I error for our rivet example, we start by identifying the four possible outcomes, their meaning, and the action related to each.

The type I error occurs when a true null hypothesis is rejected. This would occur when the manufacturer tested rivets that did meet the specs, and rejected them. Undoubtedly this would lead to the rivets not being purchased even though they did meet the specs. In order for the manager to set a level of significance, related information is needed—namely, how soon is the new supply of rivets needed? If they are needed tomorrow and this is the only vendor with an available supply, waiting a week to find acceptable rivets could be very expensive; therefore, rejecting good rivets could be considered a serious error. On the other hand, if the rivets

---

EXERCISE   8.61

State the null and alternative hypotheses used to test each of the following claims.

a. The mean reaction time is greater than 1.25 seconds.

b. The mean score on that qualifying exam is less than 335.

c. The mean selling price of homes in the area is not $230,000.

EXERCISE   8.62

Identify the four possible outcomes and describe the situation involved with each outcome with regard to the aircraft manufacturer's testing and buying rivets. Which is the more serious error, type I or type II? Explain.

There is more to this scenario, but hopefully you get the idea.

are not needed until next month, then this error may not be very serious. Only the manager will know all the ramifications, and therefore the manager's input is important here.

After much consideration, the manager assigns $\alpha = 0.05$ as the level of significance.

α will be assigned in the statement of exercises.

STEP 4   Collect and present the sample evidence.
　　　　a. Collect the sample information.

We are ready for the data. The sample must be a random sample drawn from the population whose mean μ is being questioned. A random sample of 50 rivets is selected, each rivet is tested, and the sample mean shearing strength is calculated.

$$\bar{x} = 921.18 \quad \text{and} \quad n = 50$$

　　　　b. Calculate the value of the test statistic.

The sample evidence ($\bar{x}$) is next converted into the calculated value of the test statistic, $z^{\star}$, using formula (8-4).

μ is 925 from H_0 and $\sigma = 18$ is the known quantity given in Note 2 on p. 388.

$$z^{\star} = \frac{\bar{x} - \mu}{\sigma/\sqrt{n}}: \; z^{\star} = \frac{921.18 - 925.0}{18/\sqrt{50}} = \frac{-3.82}{2.5456} = -1.5006 = \mathbf{-1.50}$$

　　　　c. Calculate the *p*-value.

EXERCISE   8.63

Calculate the test statistic $z^{\star}$, given H_0: $\mu = 56$, $\sigma = 7$, $\bar{x} = 54.3$, and $n = 36$.

PROBABILITY-VALUE, OR *p*-VALUE

The probability that the test statistic could be the value it is or a more extreme value (in the direction of the alternative hypothesis) when the null hypothesis is true. (*Note:* The symbol **P** will be used to represent the *p*-value, especially in algebraic situations.)

Draw a sketch of the standard normal distribution and locate $z^{\star}$ on it. To identify the area that represents the *p*-value, look at the sign in the alternative hypothesis. For this test, the alternative hypothesis indicates that we are interested in that part of the sampling distribution which is "less than" $z^{\star}$. Therefore, the *p*-value is the area that lies to the left of $z^{\star}$. Shade this area.

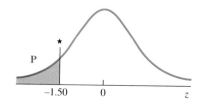

There is a 3rd option. Use a computer; see page 304.

To find its value, you have two options:

1. You may use Table 3 in Appendix B and calculate it:

$$\text{The } p\text{-value} = P(z < z^\star) = P(z < -1.50) = P(z > 1.50)$$
$$= 0.5000 - 0.4332 = \mathbf{0.0668}$$

or

EXERCISE 8.64

a. Calculate the p-value, given H_a: $\mu < 45$ and $z^\star = -2.3$.
b. Calculate the p-value, given H_a: $\mu > 58$ and $z^\star = 1.8$.

2. You may use Table 5 in Appendix B and the symmetry property. Table 5 is set up to allow you to read the p-value directly from the table, letting the table do the work the computer will typically do. Since $P(z < -1.50) = P(z > 1.50)$, simply look up $z^\star = 1.50$ on Table 5 and find the p-value is **0.0668**.

STEP 5   Determine the results.
   a. Determine whether or not the p-value is smaller than α.
      The p-value (0.0668) is **greater than** α (0.05).
   b. Make a decision about H_0.

Is the p-value small enough to indicate that the sample evidence is highly unlikely in the event that the null hypothesis is true? In order to make the decision, we need to know the **decision rule**.

EXERCISE 8.65

a. What decision is reached when the p-value is greater than α?
b. What decision is reached when α is greater than the p-value?

DECISION RULE

   a. If the p-value is **less than or equal to** the level of significance α, then the decision must be **reject H_0**.
   b. If the p-value is **greater than** the level of significance α, then the decision must be **fail to reject H_0**.

DECISION "Fail to reject H_0."
   c. Write a conclusion about H_a.

Specific information about writing the conclusion is on page 389.

CONCLUSION There is not sufficient evidence at the 0.05 level of significance to show that the mean shearing strength of the rivets is less than 925.
   "We failed to convict" the null hypothesis. In other words, a sample mean as small as 921.18 is likely to occur (as defined by α) when the true population mean value is 925.0 and $\bar{x}$ is normally distributed. The resulting action by the manager would be to buy the rivets.

NOTE   When the decision reached is "fail to reject H_0" (or "accept H_0," as many will say), it simply means "for the lack of better information, act as if the null hypothesis is true."
   Before looking at another example, let's look at the procedures for finding the p-value. The p-value is the area, under the curve of the probability distribution for the test statistic, that is more extreme than the calculated value of the test statistic.

There are three separate cases, and the direction (or sign) of the alternative hypothesis is the key. Table 8-6 outlines the procedure for all three cases.

TABLE 8-6   Finding p-Values

Case 1	p-value is the *area to the right of* $z^\star$	p-Value in Right Tail
H_a contains ">" "Right tail"	p-value $= P(z > z^\star)$	

Case 2	p-value is the *area to the left of* $z^\star$	p-Value in Left Tail
H_a contains "<" "Left tail"	The area of the left tail is the same as the area in the right tail bounded by the positive $z^\star$, therefore p-value $= P(z < z^\star) = P(z > \lvert z^\star \rvert)$	

Case 3	p-value is the *total area of both tails*	p-Value in Two Tails
H_a contains "≠" "Two-tailed"	p-value $= P(z < -\lvert z^\star \rvert) + P(z > \lvert z^\star \rvert)$ Since both areas are equal, find the probability of one tail and double it. Thus, p-value $= 2 \times P(z > \lvert z^\star \rvert)$	

Let's look at an illustration involving the two-tailed procedure.

ILLUSTRATION 8 - 13

Many of the large companies in a certain city have for years used the Kelley Employment Agency for testing prospective employees. The employment selection test used has historically resulted in scores normally distributed about a mean of 82 and a standard deviation of 8. The Brown Agency has developed a new test that is quicker

EXERCISE   8.66

Find the test statistic $z^\star$ and the p-value for each of the following situations.

a. H_0: $\mu = 22.5$,
H_a: $\mu > 22.5$,
$\bar{x} = 24.5$, $\sigma = 6$,
$n = 36$

b. H_0: $\mu = 200$,
H_a: $\mu < 200$,
$\bar{x} = 192.5$,
$\sigma = 40$, $n = 50$

c. H_0: $\mu = 12.4$,
H_a: $\mu \neq 12.4$,
$\bar{x} = 11.52$,
$\sigma = 2.2$, $n = 16$

and easier to administer and therefore less expensive. Brown claims that its test results are the same as those obtained on the Kelley test. Many of the companies are considering a change from the Kelley Agency to the Brown Agency in order to cut costs. However, they are unwilling to make the change if the Brown test results have a different mean value. An independent testing firm tested 36 prospective employees. A sample mean of 79 resulted. Determine the p-value associated with this hypothesis test. (Assume $\sigma = 8$.)

SOLUTION

STEP 1   Parameter of concern: the mean of all test scores using the Brown Agency test.

STEP 2   The hypotheses: The Brown Agency's test results will be **different** (the concern) if the mean test score **is not equal to** 82. They will be the **same** if the mean **is equal to** 82. Therefore,

$$H_0: \mu = 82 \text{ (test results have the same mean)}$$

$$H_a: \mu \neq 82 \text{ (test results have a different mean)}$$

STEP 3   The test criteria
   a. The assumptions: If the Brown test scores are distributed the same as the Kelley scores, they will be normally distributed and the sampling distribution will be normal for all sample sizes.
   b. The test statistic: The test statistic will be $z^\star$, formula (8-4).
   c. The level of significance is omitted when a question asks for the p-value and not a decision.

STEP 4   The sample evidence
   a. Sample information: $n = 36$, $\bar{x} = 79$.
   b. Calculated test statistic:

$$\text{formula (8-4): } z^\star = \frac{\bar{x} - \mu}{\sigma/\sqrt{n}}: \quad z^\star = \frac{79 - 82}{8/\sqrt{36}} = \frac{-3}{1.3333} = \mathbf{-2.25}$$

   c. The p-value: Since the alternative hypothesis indicates a two-tailed test, we must find the probability associated with both tails. The p-value is found by doubling the area of one tail.
   Since $z^\star = -2.25$, the value of $\left|z^\star\right| = 2.25$.

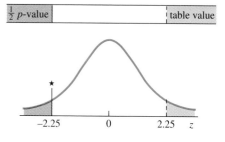

Thus, the p-value $= 2 \times P(z > |z^\star|) = 2 \times P(z > 2.25)$
Using Table 3: p-value $= 2 \times P(z > 2.25) = 2 \times (0.5000 - 0.4878)$
$$= 2 \times (0.0122) = \mathbf{0.0244}$$
Using Table 5: p-value $= 2 \times P(z > 2.25) \approx 2 \times (0.012)$ or **approx.** 0.024

STEP 5   The results. When the question asks for the p-value and not a decision, the result is the p-value.

The p-value for this hypothesis test is **0.0244**. Each individual company now will make a decision whether to (a) continue to use Kelley's services or (b) change to the Brown Agency. Each will need to establish the level of significance that best fits its own situation and then make a decision using the decision rule described previously.

▲

The idea of a p-value is to express a degree of belief in the null hypothesis: (a) When the p-value is minuscule (something like 0.0003), the null hypothesis would be rejected by everybody because the sample results are very unlikely for a true H_0; (b) when the p-value is fairly small (like 0.012), the evidence against H_0 is quite strong and H_0 will be rejected by many; (c) when the p-value begins to get larger (say, 0.02 to 0.08), there is too much probability that data like the sample involved could have occurred even if H_0 were true, and the rejection of H_0 is not an easy decision; and (d) when the p-value gets large (like 0.15 or more), the data are not at all unlikely if the H_0 is true, and no one will reject H_0.

The advantages of the p-value approach are: (1) The results of the test procedure are expressed in terms of a continuous probability scale from 0.0 to 1.0, rather than simply on a "reject" or "fail to reject" basis. (2) A p-value can be reported and the user of the information can decide on the strength of the evidence as it applies to his/her own situation. (3) Computers can do all the calculations and report the p-value, thus eliminating the need for tables.

Do your opponents show you their poker hands before you bet?

The disadvantage of the p-value approach is the tendency for people to put off determining the level of significance. This should not be allowed to happen, as it is then possible for someone to set the level of significance after the fact, leaving open the possibility that their "preferred" decision will result. This is probably only important when the reported p-value falls in the "hard choice" range (say, 0.02 to 0.08), case (c) as described above.

▼

ILLUSTRATION   8 - 14

According to the results of Exercise 5.29, page 263, the mean of single-digit random numbers is 4.5 and the standard deviation is $\sigma = 2.87$. Draw a random sample of 40 single-digit numbers from Table 1 in Appendix B and test the hypothesis, "the mean of the single-digit numbers in Table 1 is 4.5." Use $\alpha = 0.10$.

SOLUTION

STEP 1   The parameter of interest is the mean μ of the population of single-digit numbers in Table 1 of Appendix B.

STEP 2   The null and alternative hypotheses:

$$H_0: \mu = 4.5 \text{ (mean is 4.5)}$$
$$H_a: \mu \neq 4.5 \text{ (mean is not 4.5)}$$

STEP 3   a. Assumptions: Samples of size 40 should be large enough to satisfy the CLT; see discussion of this on page 371.
   b. The test statistic: $z^{\star}$, calculated using formula (8-4).
   c. The level of significance: $\alpha = 0.10$ (given in the statement of the problem).

STEP 4   The sample evidence
   a. The sample: The following random sample was drawn from Table 1.

2	8	2	1	5	5	4	0	9	1
0	4	6	1	5	1	1	3	8	0
3	6	8	4	8	6	8	9	5	0
1	4	1	2	1	7	1	7	9	3

From the sample: $\bar{x} = 3.975$, $n = 40$.
   b. The calculated test statistic:

formula (8-4): $z^{\star} = \dfrac{\bar{x} - \mu}{\sigma/\sqrt{n}}$: $z^{\star} = \dfrac{3.975 - 4.50}{2.87/\sqrt{40}} = \dfrac{-0.525}{0.454} = -1.156 = \mathbf{-1.16}$

   c. The p-value: Since the alternative hypothesis indicates a two-tailed test, we must find the probability associated with both tails. The p-value is found by doubling the area of one tail. Since $z^{\star} = -1.16$, the value of $|z^{\star}| = 1.16$.

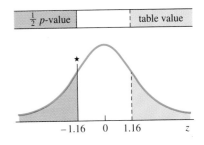

EXERCISE 8.67

Calculate the p-value, given H_a: $\mu \neq 245$ and $z^{\star} = 1.1$.

p-value $= 2 \times P(z > |z^{\star}|) = 2 \times P(z > 1.16)$
$= 2 \times (0.5000 - 0.3770) = 2 \times (0.1230) = \mathbf{0.2460}$

STEP 5   The results: The p-value is **larger than** the level of significance.

DECISION   "Fail to reject H_0."

CONCLUSION   The observed sample mean is not significantly different from 4.5 at the 0.10 level of significance.

Suppose that we were to take another sample of size 40 from Table 1. Would we obtain the same results? Suppose that we took a third sample and a fourth. What results might we expect? What does the p-value in Illustration 8-14 measure? Table 8-7 lists (a) the means obtained from 50 different random samples of size 40 that were taken from Table 1 in Appendix B, (b) the 50 values of $z^\star$ corresponding to the 50 $\bar{x}$'s, and (c) their 50 corresponding p-values. Figure 8-8 shows a histogram of the 50 $z^\star$-values.

TABLE 8-7   a. The Means of 50 Random Samples Taken from Table 1 in Appendix B

3.850	5.075	4.375	4.675	5.200	4.250	3.775	4.075	5.800	4.975
4.225	4.125	4.350	4.925	5.100	4.175	4.300	4.400	4.775	4.525
4.225	5.075	4.325	5.025	4.725	4.600	4.525	4.800	4.550	3.875
4.750	4.675	4.700	4.400	5.150	4.725	4.350	3.950	4.300	4.725
4.975	4.325	4.700	4.325	4.175	3.800	3.775	4.525	5.375	4.225

b. The $z^\star$-Values Corresponding to the 50 Means

−1.432	1.267	−0.275	0.386	1.543	−0.551	−1.598	−0.937	2.865	1.047
−0.606	−0.826	−0.331	0.937	1.322	−0.716	−0.441	−0.220	0.606	0.055
−0.606	1.267	−0.386	1.157	0.496	0.220	0.055	0.661	0.110	−1.377
0.551	0.386	0.441	−0.220	1.432	0.496	−0.331	−1.212	−0.441	0.496
1.047	−0.386	0.441	−0.386	−0.716	−1.543	−1.598	0.055	1.928	−0.606

c. The p-Values Corresponding to the 50 Means

0.152	0.205	0.783	0.700	0.123	0.582	0.110	0.349	0.004	0.295
0.545	0.409	0.741	0.349	0.186	0.474	0.659	0.826	0.545	0.956
0.545	0.205	0.700	0.247	0.620	0.826	0.956	0.509	0.912	0.168
0.582	0.700	0.659	0.826	0.152	0.620	0.741	0.226	0.659	0.620
0.295	0.700	0.659	0.700	0.474	0.123	0.110	0.956	0.054	0.545

FIGURE 8-8

The 50 Values of $z^\star$ from Table 8-7

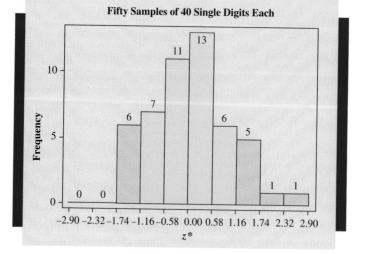

EXERCISE 8.68

Describe in your own words what the p-value measures.

The histogram shows that 6 values of $z^{\star}$ were less than -1.16 and 7 values were greater than 1.16. That means 13 of the 50 samples, or 26%, have mean values more extreme than the mean ($\bar{x} = 3.975$) in Illustration 8-14. This observed relative frequency of 0.26 represents an empirical look at the p-value. Notice that the empirical value for the p-value (0.26) is very similar to the calculated p-value of 0.2460. Check the list of p-values; do you find 13 of the 50 p-values are less than 0.2460? Which samples resulted in a $\left|z^{\star}\right| \geq 1.16$? Which samples resulted in a p-value greater than 0.2460? How do they compare?

The p-value approach was "made" for the computer!

▲ |

EXERCISE 8.69

Use a computer to select 40 random single-digit numbers. Find the sample mean, $z^{\star}$, and p-value. Repeat several times as in Table 8-7. Describe your findings.

MINITAB commands
RANDom 40 C1;
  INTEger 0 9.
ZTESt 4.5 2.87 C1;
  ALTErnative 0.

MINITAB (Release 10) commands to complete a hypothesis test for mean μ = A with standard deviation σ = B given and the sample data in C1 against the alternative hypothesis coded as K: -1 = left-tailed, 0 = two-tailed, and +1 = right-tailed.

Session commands

Enter:
  ZTESt μ = A σ = B C1;
  ALTErnative K.

Menu commands

Choose:
  Stat > Basic Stat > 1-Sample Z
  Enter: C1
  Select: Test mean Enter: A
  Enter: Alternative: -1 or 0 or 1
  Enter: Sigma: B

The solution to the rivet example, used in this section, looks like the following when solved on MINITAB. With the 50 data values in C1 on MINITAB's worksheet, the following two commands would result in the output listed below.

To use a set of four MINITAB commands repeatedly, see the Helpful Hint on page 89.

ZTESt 925 18 C1;   ⟵ You tell the computer the key information:
ALTErnative -1. ⟵ Z test of mu = 925, sigma = 18, data in C1
                    ⟵ Alternative hypothesis is "less than"

Z-Test

Test of mu = 925.00 vs mu < 925.00
The assumed sigma = 18.0

Variable	N	Mean	StDev	SE Mean	Z	P-Value
C1	50	921.18	17.58	2.546	-1.50	0.0668

EXERCISE 8.70

Describe how MINI-TAB found each of the six numerical values it reported as results.

All that is left is for you to make the decision and write the conclusion.

EXERCISES

8.71   State the null hypothesis H_0 and the alternative hypothesis H_a that would be used for a hypothesis test related to each of the following statements.

  a. The mean age of the students enrolled in evening classes at a certain college is greater than 26 years.

  b. The mean weight of packages shipped on Air Express during the past month was less than 36.7 lb.

  c. The mean life of fluorescent light bulbs is at least 1600 hr.

  d. The mean weight of college football players is no more than 210 lb.

  e. The mean strength of welds by a new process is different from 570 lb per unit area, the mean strength of welds by the old process.

8.72   A paint manufacturer wishes to test the hypothesis that "the addition of an additive will increase the average coverage of the company's paint." The average coverage has been 450 sq ft/gal. Let μ be the average coverage with the additive included. The null hypothesis is "the average coverage will not increase with the addition of the additive," ($\mu = 450$). The alternative hypothesis is "the average coverage will increase with the addition of the additive," ($\mu > 450$). Describe the meaning of the two possible types of errors that can occur in the decision when this test of the hypothesis is conducted.

8.73   Suppose we want to test the hypothesis that the mean hourly charge for automobile repairs is at least $50 per hour at the repair shops in a nearby city. Explain the conditions that would exist if we make an error in decision by committing a

  a. type I error   b. type II error

8.74   Describe how the null hypothesis, as stated in Illustration 8-11, is a "starting point" for the decision to be made about the drying time for paint made with the new formula.

8.75   Assume that z is the test statistic and calculate the value of $z^{\star}$ for each of the following:

  a. $H_0: \mu = 10$, $\sigma = 3$, $n = 40$, $\bar{x} = 10.6$

  b. $H_0: \mu = 120$, $\sigma = 23$, $n = 25$, $\bar{x} = 126.2$

  c. $H_0: \mu = 18.2$, $\sigma = 3.7$, $n = 140$, $\bar{x} = 18.93$

  d. $H_0: \mu = 81$, $\sigma = 13.3$, $n = 50$, $\bar{x} = 79.6$

8.76   Assume that z is the test statistic and calculate the value of $z^{\star}$ for each of the following:

  a. $H_0: \mu = 51$, $\sigma = 4.5$, $n = 40$, $\bar{x} = 49.6$

  b. $H_0: \mu = 20$, $\sigma = 4.3$, $n = 75$, $\bar{x} = 21.2$

  c. $H_0: \mu = 138.5$, $\sigma = 3.7$, $n = 14$, $\bar{x} = 142.93$

  d. $H_0: \mu = 815$, $\sigma = 43.3$, $n = 60$, $\bar{x} = 799.6$

8.77   There are only two possible decisions as a result of a hypothesis test.

  a. State the two possible decisions.

  b. Describe the conditions that will lead to each of the two decisions identified in (a).

8.78   For each of the following pairs of values, state the decision that will occur and state why.

  a. p-value $= 0.014$, $\alpha = 0.02$ b. p-value $= 0.118$, $\alpha = 0.05$

  c. p-value $= 0.048$, $\alpha = 0.05$ d. p-value $= 0.064$, $\alpha = 0.10$

8.79   The calculated p-value for a hypothesis test is p-value $= 0.084$. What decision about the null hypothesis would occur if

  a. the hypothesis test is completed at the 0.05 level of significance?

  b. the hypothesis test is completed at the 0.10 level of significance?

8.80 a. A one-tailed hypothesis test is to be completed at the 0.05 level of significance. What calculated values of p will cause a rejection of H_0?

b. A two-tailed hypothesis test is to be completed at the 0.02 level of significance. What calculated values of p will cause a "fail to reject H_0" decision?

8.81 Calculate the p-value for each of the following:

a. H_0: $\mu = 10$ $z^\star = 1.48$
   H_a: $\mu > 10$
b. H_0: $\mu = 105$ $z^\star = -0.85$
   H_a: $\mu < 105$
c. H_0: $\mu = 13.4$ $z^\star = 1.17$
   H_a: $\mu \neq 13.4$
d. H_0: $\mu = 8.56$ $z^\star = -2.11$
   H_a: $\mu < 8.56$
e. H_0: $\mu = 110$ $z^\star = -0.93$
   H_a: $\mu \neq 110$

8.82 Calculate the p-value for each of the following:

a. H_0: $\mu = 20$ ($\bar{x} = 17.8$, $\sigma = 9$, $n = 36$)
   H_a: $\mu < 20$
b. H_0: $\mu = 78.5$ ($\bar{x} = 79.8$, $\sigma = 15$, $n = 100$)
   H_a: $\mu > 78.5$
c. H_0: $\mu = 1.587$ ($\bar{x} = 1.602$, $\sigma = 0.15$, $n = 50$)
   H_a: $\mu \neq 1.587$

8.83 Find the value of $z^\star$ for each of the following:

a. H_0: $\mu = 35$ versus H_a: $\mu > 35$ when p-value $= 0.0582$
b. H_0: $\mu = 35$ versus H_a: $\mu < 35$ when p-value $= 0.0166$
c. H_0: $\mu = 35$ versus H_a: $\mu \neq 35$ when p-value $= 0.0042$

8.84 The null hypothesis, H_0: $\mu = 48$, was tested against the alternative hypothesis, H_a: $\mu > 48$. A sample of 75 resulted in a calculated p-value of 0.102. If $\sigma = 3.5$, find the value of the sample mean, $\bar{x}$.

8.85 The following MINITAB output was used to complete a hypothesis test.

```
MTB > ZTEST OF MU = 525 SIGMA = 60 DATA IN C1;
SUBC> ALTERNATIVE = -1.
TEST OF MU = 525.00 VS MU L.T. 525.00
THE ASSUMED SIGMA = 60.0
```

N	MEAN	STDEV	SE MEAN	Z	P VALUE
38	512.14	64.78	9.733	-1.32	0.093

a. State the null and alternative hypotheses.
b. If the test is completed using $\alpha = 0.05$, what decision and conclusion are reached?
c. Verify the value of the standard error of the mean.

8.86   The following MINITAB output was used to complete a hypothesis test.

```
MTB > ZTEST OF MU = 6.25 SIGMA = 1.4 DATA IN C1;
SUBC> ALTERNATIVE = 0.
TEST OF MU = 6.250 VS MU N.E. 6.250
THE ASSUMED SIGMA = 1.40

           N     MEAN    STDEV    SE MEAN     Z     P VALUE
          78    6.596    1.273    0.1585    2.18    0.029
```

    a. State the null and alternative hypotheses.
    b. If the test is completed using $\alpha = 0.05$, what decision and conclusion are reached?
    c. Verify the value of the standard error of the mean.
    d. Find the values for $\Sigma\, x$ and $\Sigma\, x^2$.

8.87   An article titled "Comparisons of Mathematical Competencies and Attitudes of Elementary Education Majors with Established Norms of a General College Population" (*School Science and Mathematics*, Vol. 93, No. 3, March 1993) reported the mean score on a test of mathematical competency for 165 elementary education majors to be 32.63. Test the null hypothesis that μ, the mean for the population of elementary education majors, is 35.70 (the established norm of the general college population) versus the alternative that $\mu < 35.70$. Assume that $\sigma = 6.73$.
    a. Describe the parameter of interest.
    b. State the null and alternative hypotheses.
    c. Calculate the value for $z^{\star}$ and find the p-value.
    d. State your decision and conclusion using $\alpha = 0.001$.

8.88   The marketing director of A & B Cola is worried that the product is not attracting enough young consumers. To test this hypothesis, she randomly surveys 100 A & B Cola consumers. The mean age of an individual in the community is 32 years, and the standard deviation is 10 years. The surveyed consumers of A & B Cola have a mean age of 35. At the 0.01 level of significance, is this sufficient evidence to conclude that A & B Cola consumers are, on the average, older than the average person living in the community? Complete this hypothesis test using the p-value approach.

8.89   An article titled "Too Many Cesareans" appeared in *Consumer Reports* (February 1991). Figures from 1988 indicated only 12.6 vaginal births per 100 women with a previous cesarean. Many experts believe that at least half of the women with prior cesareans could have a successful vaginal birth. Consider a study involving 125 hospitals. The average number per 100 women of vaginal births, following a previous cesarean, was 17.3. From prior experience, the standard deviation, σ, is believed to be 5. Complete the hypothesis test of H_0: $\mu = 12.6$ vs. H_a: $\mu > 12.6$ at the 0.02 level of significance using the p-value approach.

8.90   According to an article in *Good Housekeeping* (February 1991) a 128-lb woman who walks for 30 minutes four times a week at a steady, 4 mi/hr pace can lose up to 10 pounds over a span of a year. Suppose 50 women with weights between 125 and 130 lb performed the four walks per week for a year and at the end of the year the average weight loss for the 50 was 9.1 lb. Assuming that the standard deviation, σ, is 5, complete the hypothesis test of H_0: $\mu = 10.0$ vs. H_a: $\mu \neq 10.0$ at the 0.05 level of significance using the p-value approach.

8.5 | HYPOTHESIS TEST OF MEAN μ (σ KNOWN): A CLASSICAL APPROACH

In Section 8.3 we surveyed the concepts and much of the reasoning behind a hypothesis test while looking at nonstatistical illustrations. In this section we are going to formalize the hypothesis test procedure as it applies to statements concerning the mean μ of a population under the restriction that σ, the population standard deviation, is a known value.

> **The assumption for hypothesis tests about mean μ using a known σ:**
> The sampling distribution of $\bar{x}$ has a normal distribution.

The information needed to ensure that this assumption is satisfied is contained in the sampling distribution of sample means and in the central limit theorem (see Chapter 7, p. 334).

The sampling distribution for sample means, $\bar{x}$, is distributed about a mean equal to μ with a standard error equal to $\sigma/\sqrt{n}$, and is normally distributed when samples are randomly selected from a normal population or when the sample size is sufficiently large.

Classical approach

The hypothesis test is a well-organized, step-by-step procedure used to make a decision. Two different formats are commonly used for hypothesis testing. The **classical approach** is the hypothesis test process that has enjoyed popularity for many years. This approach is organized as a five-step procedure.

THE CLASSICAL HYPOTHESIS TEST: A FIVE-STEP MODEL

Step 1: Describe the population parameter of concern.
Step 2: State the null hypothesis (H_0) and the alternative hypothesis (H_a).
Step 3: Specify the test criteria.
    a. Check the assumptions.
    b. Identify the test statistic to be used.
    c. Determine the level of significance, α.
    d. Determine the critical region(s) and critical value(s).
Step 4: Collect and present the sample evidence.
    a. Collect the sample information.
    b. Calculate the value of the test statistic.

(Model continued)

Step 5: Determine the results.
   a. Determine whether or not the calculated test statistic is in the critical region.
   b. Make a decision about H_0.
   c. Write a conclusion about H_a.

Using weak rivets could have terrible consequences.

A commercial aircraft manufacturer buys rivets for use in assembling airliners. Each rivet supplier who wants to sell rivets to the aircraft manufacturer must demonstrate that its rivets meet the required specifications. One of the specs is: "The mean shearing strength of all such rivets, μ, is at least 925 lb." Each time the aircraft manufacturer buys rivets, it is concerned that the mean strength might be less than the 925-lb specification.

EXERCISE   8.91

In this example, the aircraft builder, the buyer of the rivets, is concerned that the rivets might not meet the mean-strength spec. State the aircraft manufacturer's null and alternative hypotheses.

NOTE 1   Each individual rivet has a shearing strength, which is determined by measuring the force required to shear ("break") the rivet. Clearly, not all the rivets can be tested. Therefore, a sample of rivets will be tested, and a decision about the mean strength of all the untested rivets will be based on the mean from those sampled and tested.

NOTE 2   We will use $\sigma = 18$ for our rivet example.

STEP 1   Describe the population parameter of concern.
The population parameter of interest is the mean μ, the mean shearing strength (or mean force required to shear) of the rivets being considered for purchase.

More specific instructions on p. 379.

STEP 2   State the null hypothesis (H_0) and the alternative hypothesis (H_a).
The null hypothesis and the alternative hypothesis are formulated by inspecting the problem or statement to be investigated and first formulating two opposing statements concerning the mean μ. For our example, these two opposing statements are: (a) "The mean shearing strength is less than 925" ($\mu < 925$, the aircraft manufacturer's concern), and (b) "The mean shearing strength is at least 925" ($\mu \geq 925$, the rivet supplier's claim and the aircraft manufacturer's spec).

NOTE   The *trichotomy law* from algebra states that two numerical values must be related in exactly one of three possible relationships: $<$, $=$, or $>$. All three of these possibilities must be accounted for between the two opposing hypotheses in order for the two hypotheses to be negations of each other. The three possible combinations of signs and hypotheses are shown in Table 8-8. Recall that the null hypothesis assigns a specific value to the parameter in question and therefore "equals" is always part of the null hypothesis.

TABLE 8-8

The Three Possible Statements of Null and Alternative Hypotheses

Null Hypothesis	Alternative Hypothesis
1. greater than or *equal* to ($\geq$)	less than ($<$)
2. less than or *equal* to ($\leq$)	greater than ($>$)
3. *equal* to ($=$)	not equal to ($\neq$)

The parameter of concern, the population mean, μ, is related to the value 925.

Statement (a) becomes the alternative hypothesis:

$$H_a: \mu < 925 \text{ (the mean is less than 925)}$$

This statement represents the aircraft manufacturer's concern and says that "the rivets *do not* meet the required specs."

Statement (b) becomes the null hypothesis:

$$H_0: \mu = 925 \ (\geq) \text{ (the mean is at least 925)}$$

This hypothesis represents the negation of the aircraft manufacturer's concern, and says that "the rivets *do* meet the required specs."

NOTE   The null hypothesis will be written with just the equal sign ("a value is assigned"). When "equal" is paired with "less than" or paired with "greater than," the combined symbol is written beside the null hypothesis as a reminder that all three signs have been accounted for in these two opposing statements.

Before continuing with our illustration, let's look at three illustrations that demonstrate formulating the statistical null and alternative hypotheses involving population mean μ.

▼ ┃ I L L U S T R A T I O N   8 - 15

A consumer advocate group would like to disprove a car manufacturer's claim that a specific model will average 24 miles per gallon of gasoline. Specifically, the group would like to show that the mean miles per gallon is considerably lower than 24 miles per gallon. State the null and alternative hypotheses.

S O L U T I O N   To state the two hypotheses, we first need to identify the population parameter in question: the "**mean** mileage attained by this car model." The parameter μ is being compared to the value **24** miles per gallon, the specific value of interest. The advocates are questioning the value of μ and wish to show it to be **lower than 24** (that is, $\mu < 24$). There are three possible relationships: (1) $\mu < 24$, (2) $\mu = 24$, or (3) $\mu > 24$. These three statements must be arranged to form two opposing statements: one that states what the advocates are trying to show, "the mean level is lower than 24 ($\mu < 24$)," whereas the "**negation**" is "the mean level is **not lower** than 24 ($\mu \geq 24$)." One of these two statements will become the null hypothesis, and the other will become the alternative hypothesis.

RECALL   There are two rules for forming the hypotheses: (1) The null hypothesis states that the parameter in question has a specified value ("H_0 must contain the equal sign"), and (2) the consumer advocate group's contention becomes the alternative hypothesis ("lower than"). Both rules indicate:

$$H_0: \mu = 24 \ (\geq) \quad \text{and} \quad H_a: \mu < 24$$

▼ | ILLUSTRATION   8 - 16

Suppose the EPA was suing a large manufacturing company for not meeting federal emissions guidelines. Specifically, the EPA is claiming that the mean amount of sulfur dioxide in the air is dangerously high, higher than 0.09 part per million. State the null and alternative hypotheses for this test situation.

SOLUTION The parameter of interest is the **mean** amount of sulfur dioxide in the air, and **0.09** part per million is the specified value. $\mu > 0.09$ corresponds to "the mean amount is **greater than** 0.09," whereas $\mu \leq 0.09$ corresponds to the negation, "the mean time is **not greater than** 0.09." Therefore, the hypotheses are

▲ |
$$H_0: \mu = 0.09 \ (\leq) \quad \text{and} \quad H_a: \mu > 0.09$$

Illustration 8-17 demonstrates a "two-tailed" alternative hypothesis.

▼ | ILLUSTRATION   8 - 17

Job satisfaction is very important when it comes to getting workers to produce. A standard job satisfaction questionnaire was administered by union officers to a sample of assembly line workers in a large plant in hopes of showing that the assembly workers' mean score on this questionnaire would be different than the established mean of 68. State the null and alternative hypotheses.

SOLUTION Either the **mean** job satisfaction score **is different than** 68 ($\mu \neq 68$) or the mean score **is equal to** 68 ($\mu = 68$). Therefore,

▲ |
$$H_0: \mu = 68 \quad \text{and} \quad H_a: \mu \neq 68$$

NOTE 1   When "less than" appears together with "greater than," they become "not equal to."

EXERCISE 8.92

Professor Hart does not believe the statement "the mean distance commuted daily by the nonresident students at our college is no more than 9 miles." State the null and alternative hypotheses he would use to challenge this statement.

NOTE 2   The alternative hypothesis is referred to as being "two-tailed" when H_a is "not equal."

The viewpoint of the experimenter greatly affects the way the hypotheses are formed. Generally, the experimenter is trying to show that the parameter value is different from the value specified. Thus, the experimenter is often hoping to be able to reject the null hypothesis so that the experimenter's theory has been substantiated. Illustrations 8-15, 8-16, and 8-17 also represent the three possible arrangements for the $<$, $=$, and $>$ relationships between the parameter μ and a specified value.

Table 8-9 lists some additional common phrases used in claims and indicates the phrase of its negation and the hypothesis where each phrase will be used. Again, notice that "equals" is always in the null hypothesis. Also notice that the negation of "less than" is "not less than," which is equivalent to "greater than or equal to." Think of negation of "one sign" as the "other two signs combined."

TABLE 8-9 Common Phrases and Their Negations

H_0: ($\geq$)	H_a: ($<$)	H_0: ($\leq$)	H_a: ($>$)	H_0: ($=$)	H_a: ($\neq$)
at least	less than	at most	more than	is	is not
no less than	less than	no more than	more than	not different from	different from
not less than	less than	not greater than	greater than	same as	not same as

---

EXERCISE 8.93

State the null and alternative hypotheses used to test each of the following claims.

a. The mean reaction time is less than 1.25 seconds.

b. The mean score on that qualifying exam is different than 335.

c. The mean selling price of homes in the area is no more than $230,000.

Once the null and alternative hypotheses are established, we will work under the assumption that the null hypothesis is a true statement until there is sufficient evidence to reject it. This situation might be compared to a courtroom trial, where the accused is assumed to be innocent until sufficient evidence has been presented to show that innocence is totally unbelievable ("beyond a shadow of doubt"). At the conclusion of the hypothesis test, we will make one of two possible decisions. We will decide in opposition to the null hypothesis and say that we "reject H_0" (this corresponds to "conviction" of the accused in a trial), or we will decide in agreement with the null hypothesis and say that we "fail to reject H_0" (this corresponds to "fail to convict" or an "acquittal" of the accused in a trial).

Let's return to the rivet example. Recall:

$$H_0: \mu = 925 \; (\geq) \; \text{(at least 925)} \qquad H_a: \mu < 925 \; \text{(less than 925)}$$

STEP 3 Specify the test criteria.

a. Check the assumptions.

Variables like shearing strength typically have a mounded distribution; therefore, a sample of size 50 should be large enough for the central limit theorem to apply. The sampling distribution of sample means can be expected to be normally distributed.

b. Identify the test statistic to be used.

For a hypothesis test of μ, we will want to compare the value of the sample mean to the value of the population mean as stated in the null hypothesis. We will use $\frac{\bar{x} - \mu}{\sigma/\sqrt{n}}$ as our calculated test statistic and call it $z^\star$ ("z star"). We will call it z since it is expected to have a standard normal distribution when the null hypothesis is true and the assumptions have been satisfied. The $\star$ ("star") is to remind us that this is the calculated value of the test statistic.

$$\text{test statistic:} \quad z^\star = \frac{\bar{x} - \mu}{\sigma/\sqrt{n}} \qquad \text{(8-4)}$$

c. Determine the level of significance, α.

Setting α was described as a managerial decision in Section 8.3. To see what is involved in determining α, the probability of the type I error for our rivet example, we start by identifying the four possible outcomes, their meaning, and the action related to each.

EXERCISE   8.94

Identify the four possible outcomes and describe the situation involved with each outcome with regard to the aircraft manufacturer's testing and buying rivets. Which is the more serious error, type I or type II? Explain.

There is more to this scenario, but hopefully you get the idea.

α will be assigned in the statement of exercises.

The type I error occurs when a true null hypothesis is rejected. This would occur when the manufacturer tested rivets that in truth did meet the specs, and rejected them. Undoubtedly this would lead to the rivets not being purchased even though they did meet the specs. In order for the manager to set a level of significance, related information is needed—namely, how soon is the new supply of rivets needed? If they are needed tomorrow and this is the only vendor with an available supply, waiting a week to find acceptable rivets could be very expensive; therefore, rejecting good rivets could be considered a serious error. On the other hand, if the rivets are not needed until next month, then this error may not be very serious. Only the manager will know all the ramifications, and therefore the manager's input is important here.

After much consideration, the manager assigns $\alpha = 0.05$ as the level of significance.

  d. Determine the critical region(s) and critical value(s).

The standard normal variable z is our test statistic for this hypothesis test; draw a sketch of the standard normal distribution, label the scale as z, and locate its mean value, 0.

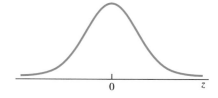

CRITICAL REGION

The set of values for the test statistic that will cause us to reject the null hypothesis. The set of values that are not in the critical region is called the *noncritical region* (sometimes called the *acceptance region*).

Recall that we are working under the assumption that the null hypothesis is true. Thus, we are assuming that the mean shearing strength of all rivets in the sampled population is 925. If this is the case, then when we select a random sample of 50 rivets, we can expect this sample mean, $\bar{x}$, to be part of a normal distribution that is centered at 925 and

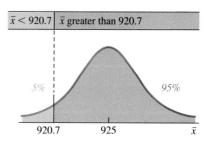

to have a standard error of $18/\sqrt{50}$, or approximately 2.55. Approximately 95% of the sample mean values will be greater than 920.7 (a value 1.65 standard errors below the mean, $925 - (1.65)(2.55) = 920.7$). Thus, if H_0 is true and $\mu = 925$, then we expect $\bar{x}$ to be greater than 920.7 approximately 95% of the time and less than 920.7 only 5% of the time.

If, however, the value of $\bar{x}$ that we obtain from our sample is less than 920.7, say 919.5, we will have to make a choice. It could be that either such an $\bar{x}$-value (919.5) is a member of the sampling distribution with mean 925 although it has a very low probability of occurrence (less than 0.05), or $\bar{x} = 919.5$ is a member of a sampling distribution whose mean is less than 925, which would make it a value that is more likely to occur.

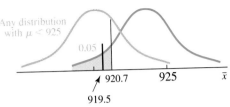

In statistics, we "bet" on the "more probable to occur" and consider the second choice to be the right one. Thus, the left-hand tail of the z distribution becomes the critical region. And the level of significance α becomes the measure of its area.

CRITICAL VALUE(S)

The "first" or "boundary" value(s) of the critical region(s).

Information about critical value notation is on pages 309–312.

The critical value for our illustration is $-z(0.05)$ and has the value of -1.65, as found in Table 4a, Appendix B.

Having completed the ground rules for the test (steps 1, 2, and 3), we are now ready for data.

critical region	noncritical region

0.05

-1.65 0 z

STEP 4   Collect and present the sample evidence.
a. Collect the sample information.
   The sample must be a random sample drawn from the population whose mean μ is being questioned. A random sample of 50 rivets is selected, each rivet is tested, and the sample mean shearing strength is calculated.

$$\bar{x} = 921.18 \text{ and } n = 50$$

b. Calculate the value of the test statistic.
   The sample evidence ($\bar{x}$) is next converted into the calculated value of the test statistic, $z^{\star}$, using formula (8-4).
      μ is 925 from H_0, and $\sigma = 18$ is the known quantity given in Note 2 on p. 404.

$$z^{\star} = \frac{\bar{x} - \mu}{\sigma/\sqrt{n}}: \quad z^{\star} = \frac{921.18 - 925.0}{18/\sqrt{50}} = \frac{-3.82}{2.5456} = -1.5006 = \mathbf{-1.50}$$

STEP 5   Determine the results.

a. Determine whether or not the calculated test statistic is in the critical region. Graphically this determination is shown by locating the value for $z^\star$ on the sketch in Step 3(d). The calculated value of z, $z^\star = -1.50$, is in the noncritical region.

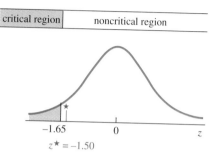

b. Make a decision about H_0. In order to make the decision, we need to know the *decision rule*.

DECISION RULE

a. If the test statistic **falls within the critical region**, we will **reject H_0**. (The critical value is part of the critical region.)
b. If the test statistic is **in the noncritical region**, we will **fail to reject H_0**.

Specific information about writing the conclusion is on page 384.

DECISION   "Fail to reject H_0."

c. Write a conclusion about H_a.

CONCLUSION There is not sufficient evidence at the 0.05 level of significance to show that the rivets have a mean shearing strength less than 925. "We failed to convict" the null hypothesis. In other words, a sample mean as small as 921.18 is likely to occur (as defined by α) when the true population mean value is 925.0. Therefore, the resulting action would be to buy the rivets.

▲ |

Before we look at another illustration, let's summarize briefly some of the details we have seen thus far:

1. The null hypothesis specifies a particular value of a population parameter.
2. The alternative hypothesis can take three forms. Each form dictates a specific location of the critical region(s), as shown in the following table.
3. For many hypothesis tests, the sign in the alternative hypothesis "points" in the direction in which the critical region is located. [Think of the not equal to sign ($\neq$) as being both less than ($<$) and greater than ($>$), thus pointing in both directions.]

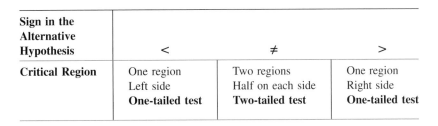

Sign in the Alternative Hypothesis	<	≠	>
Critical Region	One region Left side **One-tailed test**	Two regions Half on each side **Two-tailed test**	One region Right side **One-tailed test**

EXERCISE   8.97

Find the critical value(s) and region(s) when H_a: $\mu < 19$ and $\alpha = 0.01$.

Significance level

EXERCISE   8.98

Find the critical value(s) and region(s) when H_a: $\mu > 34$ and $\alpha = 0.02$.

The value assigned to α is called the **significance level** of the hypothesis test. Alpha cannot be interpreted to be anything other than the risk (or probability) of rejecting the null hypothesis when it is actually true. We will seldom be able to determine whether the null hypothesis is true or false; we will decide only to "reject H_0" or to "fail to reject H_0." The relative frequency with which we reject a true hypothesis is α, but we will never know the relative frequency with which we make an error in decision. The two ideas are quite different; that is, a type I error and an error in decision are two different things altogether. Remember, there are two types of errors, type I and type II.

Let's look at another hypothesis test.

▼| ILLUSTRATION   8 - 18

It has been claimed that the mean weight of women students at a college is 54.4 kg. Professor Hart does not believe the statement and sets out to show the mean weight is not 54.4 kg. To test the claim he collects a random sample of 100 weights from among the women students. A sample mean of 53.75 kg results. Is this sufficient evidence for Professor Hart to reject the statement? Use $\alpha = 0.05$ and $\sigma = 5.4$ kg.

EXERCISE   8.99

How many pounds is 54.4 kilograms?

SOLUTION

STEP 1   Describe the parameter of concern: the mean weight of all women students at the college.

STEP 2   State the null and alternative hypotheses.
The mean weight is **equal** to 54.4 kg, or the mean weight is **not equal** to 54.4 kg.

H_0: $\mu = 54.4$ (mean weight is 54.4)
H_a: $\mu \neq 54.4$ (mean weight is not 54.4) (Remember: $\neq$ is $<$ and $>$ together.)

STEP 3   Specify the test criteria.
   a. Check the assumptions: Weights of an adult group of women are generally approximately normally distributed; therefore, a sample of $n = 100$ is large enough to allow the CLT to apply.
   b. Identify the test statistic. The test statistic will be $z^\star$ and formula (8-3).
   c. Determine the level of significance: $\alpha = 0.05$ (given in the statement of problem).

d. Determine the critical regions and critical values: The critical region is both the left tail and the right tail because both smaller and larger values of the sample mean suggest the null hypothesis is wrong. The level of significance will be split in half, with

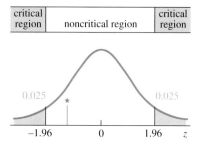

0.025 being the measure of each tail. The critical values are found in Table 4b, Appendix B.

$$\pm z(0.025) = \pm 1.96$$

STEP 4   Collect and present the sample evidence.
   a. The sample information: $\bar{x} = 53.75$, $n = 100$.
   b. Calculate the value of the test statistic. Using formula (8-4) and information from H_0: $\mu = 54.4$ and $\sigma = 5.4$ (given).

$$z^\star = \frac{\bar{x} - \mu}{\sigma/\sqrt{n}}; \quad z^\star = \frac{53.75 - 54.4}{5.4/\sqrt{100}} = \frac{-0.65}{0.54} = -1.204 = \mathbf{-1.20}$$

STEP 5   Determine the results.
   a. Determine whether or not the calculated test statistic is in the critical region. The calculated value of z, $z^\star = -1.20$, is in the noncritical region (shown in red on figure above).
   b. Make a decision.

   DECISION "Fail to reject H_0."
   c. Write a conclusion.

   CONCLUSION There is not sufficient evidence at the 0.05 level of significance to show that the women students have a mean weight different than the 54.4 kg claimed. In other words, there is no statistical evidence to support Professor Hart's contentions.

▼| ILLUSTRATION   8 - 19

The student body at many community colleges is considered a "commuter population." The following question was asked of the Student Affairs Office at one such college: "How far (one way) does the average community-college student commute to college daily?" The office answered: "No more than 9.0 mi." The inquirer was not convinced of the truth of this and decided to test the statement. She took a random sample of 50 students and found a mean commuting distance of 10.22 mi. Test the hypothesis stated above at a significance level of $\alpha = 0.05$, using $\sigma = 5$ mi.

SOLUTION

STEP 1   The parameter of concern: μ, the mean one-way distance traveled by all commuting students.

STEP 2   The hypotheses: The claim "no more than 9.0 mi" implies that the three possible relationships should be grouped "no more than 9.0" ($\leq$) versus "more than 9.0" ($>$).

$$H_0: \mu = 9.0 \ (\leq) \ (\text{no more than 9.0 mi})$$
$$H_a: \mu > 9.0 \ (\text{more than 9.0 mi})$$

STEP 3   The test criteria:

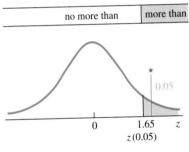

a. Commuting distances is a mounded and skewed distribution, but with a sample size $n = 50$, the CLT will hold; therefore, the assumptions are satisfied.
b. The test statistic is $z^\star$ and is calculated using formula (8-3).
c. The level of significance is given: $\alpha = 0.05$
d. The critical region is the right tail because the alternative hypothesis is "greater than." The critical value is $z(0.05) = 1.65$.

STEP 4   The sample evidence: From the sample: $\bar{x} = 10.22$, $n = 50$; from H_0: $\mu = 9.0$; and $\sigma = 5.0$ was given. Using formula (8-4):

$$z^\star = \frac{\bar{x} - \mu}{\sigma/\sqrt{n}}: \quad z^\star = \frac{10.22 - 9.0}{5.0/\sqrt{50}}$$
$$= \frac{1.22}{0.707} = \mathbf{1.73}$$

STEP 5   The results: The calculated value of z, $z^\star = 1.73$, is in the critical region (shown in red on figure above).

DECISION "Reject H_0."

CONCLUSION There is sufficient evidence at the 0.05 level of significance to conclude that the average commuting community-college student probably travels more than 9.0 miles one-way to college.

▼| ILLUSTRATION   8 - 20

According to the results of Exercise 5.29, page 263, the mean of single-digit random numbers is 4.5 and the standard deviation is $\sigma = 2.87$. Draw a random sample of 40 single-digit numbers from Table 1 in Appendix B and test the hypothesis, "the mean of the single-digit numbers in Table 1 is 4.5." Use $\alpha = 0.10$.

SOLUTION

STEP 1   The parameter of interest is the mean μ of the population of single-digit numbers in Table 1 of Appendix B.

STEP 2   The null and alternative hypotheses:

$$H_0: \mu = 4.5 \text{ (mean is 4.5)}$$
$$H_a: \mu \neq 4.5 \text{ (mean is not 4.5)}$$

STEP 3   a. Assumptions: Samples of size 40 should be large enough to satisfy the CLT; see discussion of this on page 371.
b. The test statistic: $z^\star$, calculated using formula (8-4).
c. The level of significance: $\alpha = 0.10$ (given in the statement of the problem).
d. A two-tailed critical region will be used, and 0.05 will be the area of each tail. The critical values are $\pm z(0.05) = \pm 1.65$.

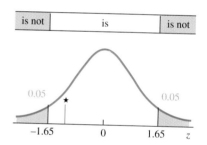

STEP 4   The sample evidence
a. The sample: The following random sample was drawn from Table 1.

2	8	2	1	5	5	4	0	9	1
0	4	6	1	5	1	1	3	8	0
3	6	8	4	8	6	8	9	5	0
1	4	1	2	1	7	1	7	9	3

From the sample: $\bar{x} = 3.975$, $n = 40$.
b. The calculated test statistic:

formula (8-4): $z^\star = \dfrac{\bar{x} - \mu}{\sigma/\sqrt{n}}$: $z^\star = \dfrac{3.975 - 4.50}{2.87/\sqrt{40}} = \dfrac{-0.525}{0.454} = -1.156 = \mathbf{-1.16}$

STEP 5   The results: $z^\star = -1.16$ is in the noncritical region (shown in red of figure above).

DECISION "Fail to reject H_0."

CONCLUSION The observed sample mean is not significantly different from 4.5 at the 0.10 level of significance.

▲ |

Suppose that we were to take another sample of size 40 from the table of random digits. Would we obtain the same results? Suppose that we took a third sample or a fourth? What results might we expect? What is the level of significance? Yes, its value is 0.10, but what does it measure? Table 8-10 lists the means obtained from

TABLE 8-10

Twenty Random Sample of Size 40 Taken from Table 1 in Appendix B

Sample Number	Sample Mean (x)	Calculated z ($z^\star$)	Decision Reached
1	4.62	+0.26	Fail to reject H_0
2	4.55	+0.11	Fail to reject H_0
3	4.08	−0.93	Fail to reject H_0
4	5.00	+1.10	Fail to reject H_0
5	4.30	−0.44	Fail to reject H_0
6	3.65	−1.87	Reject H_0
7	4.60	+0.22	Fail to reject H_0
8	4.15	−0.77	Fail to reject H_0
9	5.05	+1.21	Fail to reject H_0
10	4.80	+0.66	Fail to reject H_0
11	4.70	+0.44	Fail to reject H_0
12	4.88	+0.83	Fail to reject H_0
13	4.45	−0.11	Fail to reject H_0
14	3.93	−1.27	Fail to reject H_0
15	5.28	+1.71	Reject H_0
16	4.20	−0.66	Fail to reject H_0
17	3.48	−2.26	Reject H_0
18	4.78	+0.61	Fail to reject H_0
19	4.28	−0.50	Fail to reject H_0
20	4.23	−0.61	Fail to reject H_0

EXERCISE 8.100

Use a computer to select 40 random single-digit numbers. Repeat it several times as in Table 8-10. Find the sample mean and $z^\star$. Describe your findings after several tries.

MINITAB commands

```
RANDom 40 C1;
  INTEger 0 9.
ZTESt 4.5 2.87 C1;
  ALTErnative 0.
```

20 different random samples of size 40 that were taken from Table 1 in Appendix B. The calculated value of $z^\star$ that corresponds to each $\bar{x}$ and the decision each would dictate are also listed. The 20 calculated z-scores are shown in Figure 8-9. Note that 3 of the 20 samples (or 15%) caused us to reject the null hypothesis, even though we know the null hypothesis is true for this situation. Can you explain this?

FIGURE 8-9

z-Scores from
Table 8-10;
$\alpha = 0.10$

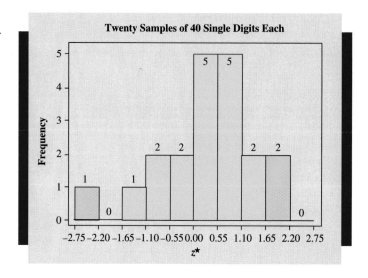

 *MINITAB commands
for hypothesis
testing are on
page 399.*

REMEMBER α is the probability that we "reject H_0" when it is actually a true statement. Therefore, we can anticipate that a type I error will occur α of the time when testing a true null hypothesis. In the above empirical situation, we observed a 15% rejection rate. If we were to repeat this experiment many times, the proportion of samples that would lead to a rejection would vary, but the observed relative frequency of rejection should be approximately α or 10%.

EXERCISES

8.101   State the null hypothesis, H_0, and the alternative hypothesis, H_a, that would be used for a hypothesis test related to each of the following statements.
   a. The mean age of the youths who hang out at the mall is less than 16 years.
   b. The mean height of professional basketball players is more than 6 ft 6 in.
   c. The mean elevation drop for ski trails at eastern ski centers is at least 285 ft.
   d. The mean diameter of the rivets is no more than 0.375 in.
   e. The mean cholesterol level of male college students is different from 200 units.

8.102   Suppose you want to test the hypothesis that "the mean salt content of frozen 'lite' dinners is more than 350 mg per serving." An average of 350 mg is an acceptable amount of salt per serving; therefore, you use it as standard. The null hypothesis is "the average content is not more than 350 mg" ($\mu = 350$). The alternative hypothesis is "the average content is more than 350 mg" ($\mu > 350$).
   a. .Describe the conditions that would exist if your decision results in a type I error.
   b. Describe the conditions that would exist if your decision results in a type II error.

8.103   Suppose you wanted to test the hypothesis that the mean minimum home service call charge for plumbers is at most $85 in your area. Explain the conditions that would exist if you make an error in decision by committing a
    a. type I error
    b. type II error

8.104   Describe how the null hypothesis in Exercise 8.102 is a "starting point" for the decision to be made about the salt content.

8.105   a. What is the critical region?
    b. What is the critical value?

8.106   Since the size of the type I error can always be made smaller by reducing the size of the critical region, why don't we always choose critical regions that make α extremely small?

8.107   Determine the test criteria (critical values and critical region for z) that would be used to test the null hypothesis at the given level of significance, as described in each of the following:
    a. H_0: $\mu = 20$ ($\alpha = 0.10$)
      H_a: $\mu \neq 20$
    b. H_0: $\mu = 24$ ($\leq$)   ($\alpha = 0.01$)
      H_a: $\mu > 24$
    c. H_0: $\mu = 10.5$ ($\geq$)   ($\alpha = 0.05$)
      H_a: $\mu < 10.5$
    d. H_0: $\mu = 35$ ($\alpha = 0.01$)
      H_a: $\mu \neq 35$

8.108   Determine the test criteria used to test the following null hypotheses.
    a. H_0: $\mu = 55$ ($\geq$)   ($\alpha = 0.02$)
      H_a: $\mu < 55$
    b. H_0: $\mu = -86$ ($\geq$)   ($\alpha = 0.01$)
      H_a: $\mu < -86$
    c. H_0: $\mu = 107$ ($\alpha = 0.05$)
      H_a: $\mu \neq 107$
    d. H_0: $\mu = 17.4$ ($\leq$)   ($\alpha = 0.10$)
      H_a: $\mu > 17.4$

8.109   The null hypothesis, H_0: $\mu = 250$, was tested against the alternative hypothesis, H_a: $\mu < 250$. A sample of $n = 85$ resulted in a calculated test statistic of $z^\star = -1.18$. If $\sigma = 22.6$, find the value of the sample mean, $\bar{x}$. Find the sum of the sample data, Σx.

8.110   Find the value of $\bar{x}$ for each of the following:
    a. H_0: $\mu = 580$, $z^\star = 2.10$, $\sigma = 26$, $n = 55$
    b. H_0: $\mu = 75$, $z^\star = -0.87$, $\sigma = 9.2$, $n = 35$

8.111   The calculated value of the test statistic is actually the number of standard errors that the sample mean differs from the hypothesized value of μ in the null hypothesis. Suppose that the null hypothesis is H_0: $\mu = 4.5$, σ is known to be 1.0, and a sample of size 100 results in $\bar{x} = 4.8$.
    a. How many standard errors is $\bar{x}$ above 4.5?
    b. If the alternative hypothesis is H_a: $\mu > 4.5$ and $\alpha = 0.01$, would you reject H_0?

8.112   Consider the hypothesis test where the hypotheses are

$$H_0: \mu = 26.4$$
$$H_a: \mu < 26.4$$

A sample of size 64 is randomly selected and yields a sample mean of 23.6.
   a. If it is known that $\sigma = 12$, how many standard errors below $\mu = 26.4$ is the sample mean, $\bar{x} = 23.6$?
   b. If $\alpha = 0.05$, would you reject H_0? Explain.

8.113   There are only two possible decisions as a result of a hypothesis test.
   a. State the two possible decisions.
   b. Describe the conditions that will lead to each of the two decisions identified in (a).

8.114   a. What proportion of the probability distribution is in the critical region, provided the null hypothesis is correct?
   b. What error could be made if the test statistic falls in the critical region?
   c. What proportion of the probability distribution is in the noncritical region, provided the null hypothesis is not correct?
   d. What error could be made if the test statistic falls in the noncritical region?

8.115   The following MINITAB output was used to complete a hypothesis test.

```
MTB > ZTEST OF MU = 15.0, SIGMA = .5, DATA IN C1
TEST OF MU = 15.0000 VS MU N.E. 15.0000
THE ASSUMED SIGMA = 0.50

                N       MEAN      STDEV     SE MEAN      Z
C1              30     15.6333    0.4270    0.0913     6.94
MTB>STOP
```

   a. State the null and alternative hypotheses.
   b. If the test is completed using $\alpha = 0.01$, what decision and conclusion are reached?
   c. Verify the value of the standard error of the mean.

8.116   The following MINITAB output was used to complete a hypothesis test.

```
MTB > ZTEST OF MU = 72, SIGMA = 12, DATA IN C1;
SUBC> ALTERNATIVE = 1.

TEST OF MU = 72.00 VS MU G.T. 72.00
THE ASSUMED SIGMA = 12.0

           N     MEAN     STDEV    SE MEAN      Z
C1         36     75.2    11.87     2.00      1.60
```

   a. State the null and alternative hypotheses.
   b. If the test is completed using $\alpha = 0.05$, what decision and conclusion are reached?
   c. Verify the value of the standard error of the mean.

8.117 The manager at Air Express feels that the weights of packages shipped recently are less than in the past. Records show that in the past packages have had a mean weight of 36.7 lb and a standard deviation of 14.2 lb. A random sample of last month's shipping records yielded a mean weight of 32.1 lb for 64 packages. Is this sufficient evidence to reject the null hypothesis in favor of the manager's claim? Use $\alpha = 0.01$.

8.118 A fire insurance company felt that the mean distance from a home to the nearest fire department in a suburb of Chicago was at least 4.7 mi. It set its fire insurance rates accordingly. Members of the community set out to show that the mean distance was less than 4.7 mi. This, they felt, would convince the insurance company to lower its rates. They randomly identified 64 homes and measured the distance to the nearest fire department for each. The resulting sample mean was 4.4. If $\sigma = 2.4$ mi, does the sample show sufficient evidence to support the community's claim at the $\alpha = 0.05$ level of significance?

8.119 The 1993–1994 edition of the *American Hospital Association Hospital Statistics* gives data compiled from the American Hospital Association 1992 annual survey of hospitals. The average stay in days for 3057 nongovernment not-for-profit hospitals is given to be 7.0 days. A sample of 40 such hospitals was selected in 3 midwestern states to test the hypothesis that the average stay is less than the national average. The sample mean equals 6.1 days. Is this sufficient evidence to reject the null hypothesis? Use $\alpha = 0.05$ and $\sigma = 1.5$ days.

8.120 An article titled "The Trouble with Margarine" (*Consumer Reports*, March 1991) discusses a Dutch research study which shows that trans-fatty acids tend to cause a rise in LDL cholesterol (which increases the risk of coronary heart disease) and tend to cause a lowering of HDL cholesterol (which lowers the risk of coronary heart disease). Margarine is much higher in trans-fatty acids than butter. The estimated U.S. intake of trans-fatty acids is 8 g per day. Consider a research project involving 150 individuals in which their daily intake of trans-fatty acids was measured. Suppose the sample mean intake was 12.5 g. Assuming that $\sigma = 8.0$, test the research hypothesis that $\mu > 8$ at $\alpha = 0.05$.

IN RETROSPECT

Two forms of inference were studied in this chapter: estimation and hypothesis testing. They may be, and often are, used separately. It would, however, seem natural for the rejection of a null hypothesis to be followed by a confidence interval estimate. (If the value claimed is wrong, we will often want to estimate the true value.)

These two forms of inference are quite different but they are related. There is a certain amount of crossover between the use of the two inferences. For example, suppose that you had sampled and calculated a 90% confidence interval for the mean of a population. The interval was 10.5 to 15.6. Following this someone claims that the true mean is 15.2. Your confidence interval estimate can be compared to this claim. If the claimed value falls within your interval estimate, you would fail to reject the null hypothesis that $\mu = 15.2$ at a 10% level of significance in a two-tailed test. If the claimed value (say, 16.0) falls outside the interval, you would then reject the null hypothesis that $\mu = 16.0$ at $\alpha = 0.10$ in a two-tailed test. If a one-tailed test is required, or if you prefer a different value of α, a separate hypothesis test must be used.

Many users of statistics (especially those marketing a product) will claim that their statistical results prove that their product is superior. But remember, the hypothesis test does not prove or disprove anything. The decision reached in a hypothesis test has probabilities associated with the four various situations. If "fail to reject H_0" is the decision, it is possible that an error has occurred. Further, if "reject H_0" is the decision reached, it is possible for this to be an error. Both errors have probabilities greater than zero.

In this chapter we have restricted our discussion of inferences to the mean of a population for which the standard deviation is known. In Chapters 9 and 10, we will discuss inferences about the population mean and remove the restriction about the known value for standard deviation. We will also look at inferences about the parameters proportion, variance, and standard deviation.

CHAPTER EXERCISES

8.121 A sample of 64 measurements is taken from a continuous population, and the sample mean is found to be 32.0. The standard deviation of the population is known to be 2.4. An interval estimation is to be made of the mean with a level of confidence of 90%. State or calculate the following items.

a. $\bar{x}$

b. σ

c. n

d. $1 - \alpha$

e. $z(\alpha/2)$

f. $\sigma_{\bar{x}}$

g. E (maximum error of estimate)

h. upper confidence limit

i. lower confidence limit

8.122 Suppose that a confidence interval is assigned a level of confidence of $1 - \alpha = 95\%$. How is the 95% used in constructing the confidence interval? If $1 - \alpha$ was changed to 90%, what effect would this have on the confidence interval?

8.123 The expected mean of a continuous population is 100, and its standard deviation is 12. A sample of 50 measurements gives a sample mean of 96. Using a 0.01 level of significance, a test is to be made to decide between "the population mean is 100" or "the population mean is different than 100." State or find each of the following:

a. H_0 b. H_a c. α d. μ (based on H_0)   e. $\bar{x}$

f. σ g. $\sigma_{\bar{x}}$ h. $z^{\star}$, z-score for $\bar{x}$ i. p-value   j. decision

k. Sketch the standard normal curve and locate $z^{\star}$, p-value, and α.

8.124 The expected mean of a continuous population is 200, and its standard deviation is 15. A sample of 80 measurements gives a sample mean of 205. Using a 0.01 level of significance, a test is to be made to decide between "the population mean is 200" or "the population mean is different than 200." State or find each of the following:

a. H_0 b. H_a c. α d. $z(\alpha/2)$ e. μ (based on H_0)

f. $\bar{x}$ g. σ h. $\sigma_{\bar{x}}$ i. $z^{\star}$, z-score for $\bar{x}$ j. decision

k. Sketch the standard normal curve and locate $\alpha/2$, $z(\alpha/2)$, the critical region, and $z^{\star}$.

8.125   Suppose that a hypothesis test is assigned a level of significance of $\alpha = 0.01$. How is the 0.01 used in completing the hypothesis test? If α were changed to 0.05, what effect would this have on the test procedure?

8.126   From a population of unknown mean μ and a standard deviation $\sigma = 5.0$, a sample of $n = 100$ is selected and the sample mean 40.6 is found. Compare the concepts of confidence interval estimation and hypothesis testing by answering the following:
   a. Determine the 95% confidence interval estimate for μ.
   b. Complete the hypothesis test involving H_a: $\mu \neq 40$ using the *p*-value approach and $\alpha = 0.05$.
   c. Complete the hypothesis test involving H_a: $\mu \neq 40$ using the classical approach and $\alpha = 0.05$.
   d. On one sketch of the standard normal curve, locate: the interval representing the confidence interval estimate from (a); the $z^\star$, *p*-value, and α from (b); and the $z^\star$ and critical regions from (c). Describe the relationship between these three separate procedures.

8.127   From a population of unknown mean μ and a standard deviation $\sigma = 5.0$, a sample of $n = 100$ is selected and the sample mean 41.5 is found. Compare the concepts of confidence interval estimation and hypothesis testing by answering the following:
   a. Determine the 95% confidence interval estimation for μ.
   b. Complete the hypothesis test involving H_a: $\mu \neq 40$ using the *p*-value approach and $\alpha = 0.05$.
   c. Complete the hypothesis test involving H_a: $\mu \neq 40$ using the classical approach and $\alpha = 0.05$.
   d. On one sketch of the standard normal curve, locate: the interval representing the confidence interval estimate from (a); the $z^\star$, *p*-value, and α from (b); and the $z^\star$ and critical regions from (c). Describe the relationship between these three separate procedures.

8.128   From a population of unknown mean μ and a standard deviation $\sigma = 5.0$, a sample of $n = 100$ is selected and the sample mean 40.9 is found. Compare the concepts of confidence interval estimation and hypothesis testing by answering the following:
   a. Determine the 95% confidence interval estimation for μ.
   b. Complete the hypothesis test involving H_a: $\mu > 40$ using the *p*-value approach and $\alpha = 0.05$.
   c. Complete the hypothesis test involving H_a: $\mu > 40$ using the classical approach and $\alpha = 0.05$.
   d. On one sketch of the standard normal curve, locate: the interval representing the confidence interval estimate from (a); the $z^\star$, *p*-value, and α from (b); and the $z^\star$ and critical regions from (c). Describe the relationship between these three separate procedures.

8.129   The weights of full boxes of a certain kind of cereal are normally distributed with a standard deviation of 0.27 oz. A sample of 18 randomly selected boxes produced a mean weight of 9.87 oz.
   a. Find the 95% confidence interval for the true mean weight of a box of this cereal.
   b. Find the 99% confidence interval for the true mean weight of a box of this cereal.
   c. What effect did the increase in the level of confidence have on the width of the confidence interval?

8.130 The standard deviation of a normally distributed population is equal to 10. A sample size of 25 is selected, and its mean is found to be 95.
    a. Find an 80% confidence interval estimate for μ.
    b. If the sample size were 100, what would be the 80% confidence interval?
    c. If the sample size were 25 but the standard deviation were 5 (instead of 10), what would be the 80% confidence interval?

8.131 A random sample of the scores of 100 applicants for clerk-typist positions at a large insurance company showed a mean score of 72.6. The preparer of the test maintained that qualified applicants should average 75.0.
    a. Determine the 99% confidence interval estimate for the mean score of all applicants at the insurance company. Assume that the standard deviation of test scores is 10.5.
    b. Can the insurance company conclude that it is getting qualified applicants (as measured by this test)?

8.132 Waiting times (in hours) at a popular restaurant are believed to be approximately normally distributed with a variance of 2.25 hr during busy periods.
    a. A sample of 20 customers revealed a mean waiting time of 1.52 hr. Construct the 95% confidence interval for the estimate of the population mean.
    b. Suppose that the mean of 1.52 hr had resulted from a sample of 32 customers. Find the 95% confidence interval.
    c. What effect does a larger sample size have on the confidence interval?

8.133 An article titled "Evaluation of a Self-Efficacy Scale for Preoperative Patients (*AORN Journal*, Vol. 60, No. 1, July 1994) describes a 32-item rating scale used to determine efficacy (measure of effectiveness) expectations as well as outcome expectations. The 32-item scale was administered to 200 preoperative patients. The mean efficacy expectation score for the ambulating item equaled 4.00. Construct the 0.95 confidence interval estimate for the mean of all such preoperative patients. Use $\sigma = 0.94$.

8.134 The June 18, 1990, issue of *Insight* describes the beneficial results of a wellness program begun in 1988 in Wellsburg, W. Va. One thousand of the town's 4000 residents participated in the program. Five hundred of the 1000 participants noted a reduction in their average cholesterol by 15 points. Consider another wellness program involving 500 participants. Suppose their cholesterol was lowered by 20 points on the average. If we regard these results as representative of what such a program can accomplish, estimate the mean reduction in cholesterol for all participants in such programs with a 95% confidence interval. (Assume the standard deviation , σ, is 7.5 units.)

8.135 A new paint has recently been developed by a research laboratory. Twenty-five gallons are tested, and the mean coverage is found to be 515 sq ft/gal. Assuming the standard deviation to be 50 sq ft/gal, find a 99% confidence interval estimate for μ, the mean coverage per gallon of all such paint.

8.136 An automobile manufacturer wants to estimate the mean gasoline mileage that its customers will obtain with its new compact model. How many sample runs must be performed in order that the estimate be accurate to within 0.3 mpg at 95% confidence? (Assume that $\sigma = 1.5$.)

8.137 A fish hatchery manager wants to estimate the mean length of her three-year-old hatchery-raised trout. She wants to make a 99% confidence interval estimate accurate to within $\frac{1}{3}$ of a standard deviation. How large a sample does she need to take?

8.138 We are interested in estimating the mean life of a new product. How large a sample do we need to take in order to estimate the mean to within $\frac{1}{10}$ of a standard deviation with 90% confidence?

8.139 In "Much Ado About Probability" (*Physical Therapy*, September 1990), the editor of the journal discusses the difference between statistical significance and practical significance. Suppose the mean value for a home in a residential area of a large city is listed as $150,000. A current sample of homes from the area finds a mean of $153,500 with a *p*-value equal to 0.005. Does this statistical significance necessarily have practical significance?

 8.140 According to an article in *Health* magazine (March 1991) supplementation with potassium reduced the blood pressure readings in a group of mild hypertensive patients from an average of 158/100 to 143/84.5. Consider a study involving the use of potassium supplementation to reduce the systolic blood pressure for mild hypertensive patients. Suppose 75 patients with mild hypertension were placed on potassium for six weeks. The response measured was the systolic reading at the beginning of the study minus the systolic reading at the end of the study. The mean drop in the systolic reading was 12.5 units. Assume the population standard deviation, σ, to be 7.5 units. Calculate the value of the test statistic $z^\star$ and the *p*-value for testing H_0: $\mu = 10.0$ vs. H_a: $\mu > 10.0$.

 8.141 The college bookstore tells prospective students that the average cost of its textbooks is $32 per book with a standard deviation of $4.50. The engineering science students think that the average cost of their books is higher than the average for all students. To test the bookstore's claim against their alternative, the engineering students collect a random sample of size 45.
   a. If they use $\alpha = 0.05$, what is the critical value of the test statistic?
   b. The engineering students' sample data are summarized by $n = 45$ and $\Sigma x = 1470.25$. Is this sufficient evidence to support their contention?

 8.142 A rope manufacturer, after conducting a large number of tests over a long period of time, has found that the rope has a mean breaking strength of 300 lb and a standard deviation of 24 lb. Assume that these values are μ and σ. It is believed that by using a recently developed high-speed process, the mean breaking strength has been decreased.
   a. Design a null and alternative hypothesis such that rejection of the null hypothesis will imply that the mean breaking strength has decreased.
   b. Using the decision rule established in (a), what is the *p*-value associated with rejecting the null hypothesis when 45 tests result in a sample mean of 295?
   c. If the decision rule in (a) is used with $\alpha = 0.01$, what is the critical value for the test statistic and what value of $\bar{x}$ corresponds to it if a sample of size 45 is used?

 8.143 A manufacturing process produces ball bearings with diameters having a normal distribution and a standard deviation of $\sigma = 0.04$ cm. Ball bearings that have diameters that are too small or too large are undesirable. To test the null hypothesis that $\mu = 0.50$ cm, a sample of 25 is randomly selected and the sample mean is found to be 0.51.
   a. Design a null and alternative hypothesis such that rejection of the null hypothesis will imply that the ball bearings are undesirable.
   b. Using the decision rule established in (a), what is the *p*-value for the sample results?
   c. If the decision rule in (a) is used with $\alpha = 0.02$, what is the critical value for the test statistic?

 8.144 The admissions office at Memorial Hospital recently stated that the mean age of its patients was 42 years. A random sample of 120 ages was obtained from the admissions office records in an attempt to disprove the claim. Is a sample mean of 44.2 years significantly larger than the claimed 42 years at the $\alpha = 0.05$ level? Use $\sigma = 20$ years.
   a. Solve using the *p*-value approach.
   b. Solve using the classical approach.

8.145   In a large supermarket the customer's waiting time to check out is approximately normally distributed with a standard deviation of 2.5 min. A sample of 24 customer waiting times produced a mean of 10.6 min. Is this evidence sufficient to reject the supermarket's claim that its customer checkout time averages no more than 9 min? Complete this hypothesis test using the 0.02 level of significance.

  a. Solve using the p-value approach.

  b. Solve using the classical approach.

8.146   At a very large firm, the clerk-typists were sampled to see whether the salaries differed among departments for workers in similar categories. In a sample of 50 of the firm's accounting clerks, the average annual salary was $16,010. The firm's personnel office insists that the average salary paid to all clerk-typists in the firm is $15,650 and that the standard deviation is $1,800. At the 0.05 level of significance, can we conclude that the accounting clerks receive, on the average, a different salary from that of the clerk-typists?

  a. Solve using the p-value approach.

  b. Solve using the classical approach.

8.147   A random sample of 280 rivets is selected from a very large shipment. The rivets are to have a mean diameter of no more than $\frac{5}{16}$ in. Does a sample mean of 0.3126 show sufficient reason to reject the null hypothesis that the mean diameter is no more than $\frac{5}{16}$ in. at the $\alpha = 0.01$ level of significance? Use $\sigma = 0.0006$.

  a. Solve using the p-value approach.

  b. Solve using the classical approach.

8.148   A manufacturer of automobile tires believes it has developed a new rubber compound that has superior wearing qualities. It produced a test run of tires made with this new compound and had them road tested. The data recorded was the amount of tread wear per 10,000 miles. In the past, the mean amount of tread wear per 10,000 miles, for tires of this quality, has been 0.0625 inches.

The null hypothesis to be tested here is "the mean amount of wear on the tires made with the new compound is the same mean amount of wear with the old compound, 0.0625 inches per 10,000 miles," H_0: $\mu = 0.0625$. Three possible alternative hypotheses could be used: (1) H_a: $\mu < 0.0625$, (2) H_a: $\mu \neq 0.0625$, (3) H_a: $\mu > 0.0625$.

  a. Explain the meaning of each of these three alternatives.

  b. Which one of the possible alternative hypotheses should the manufacturer use if it hopes to conclude that "use of the new compound does yield superior wear"?

8.149   All drugs must be approved by the Food and Drug Administration (FDA) before they can be marketed by a drug manufacturer. The FDA must weigh the error of marketing an ineffective drug, with the usual risks of side effects, against the consequences of not allowing an effective drug to be sold. Suppose, using standard medical treatment, that the mortality rate (r) of a certain disease is known to be A. A manufacturer submits for approval a drug that is supposed to treat this disease. The FDA sets up the hypothesis to test the mortality rate for the drug as (1) H_0: $r = A$, H_a: $r < A$, $\alpha = 0.005$; or (2) H_0: $r = A$, H_a: $r > A$, $\alpha = 0.005$.

  a. If $A = 0.95$, which test do you think the FDA would use?

  b. If $A = 0.05$, which test do you think the FDA would use? Explain.

8.150   The drug manufacturer in Exercise 8.149 has a different viewpoint of the matter. It wants to market the new drug starting as soon as possible so that it can beat its competitors to the marketplace and make money. Its position is "market the drug unless the drug is totally ineffective."

  a. How do you think the drug company will set up the alternative hypothesis if it is doing the testing? H_a: $r < A$, H_a: $r \neq A$, or H_a: $r > A$.

b. Does the mortality rate of the existing treatment affect the alternative hypothesis? $A = 0.95$ or $A = 0.05$.

8.151 The following MINITAB output shows a simulated sample of size 25 randomly generated from a normal population with $\mu = 130$ and $\sigma = 10$. ZINTerval was then used to set a 95% confidence interval for μ.

```
MTB > RANDOM 25 OBSERVATIONS INTO C1;
SUBC> NORMAL MU = 130 SIGMA = 10.
MTB > PRINT C1
C1
116.187   119.832   121.782   122.320   141.436   129.197   119.172
120.713   135.765   131.153   122.307   126.155   137.545   141.154
123.405   143.331   121.767   109.742   140.524   150.600   121.655
127.992   136.434   139.768   125.594

MTB > ZINTERVAL 95 PERCENT CONFIDENCE, SIGMA = 10, DATA IN C1

THE ASSUMED SIGMA = 10.0

           N      MEAN     STDEV    SE MEAN    95.0 PERCENT C.I.
C1        25     129.02    10.18     2.00      (125.10, 132.95)
```

a. State the confidence interval estimate that resulted.

b. Verify the values reported for the standard error of mean and the interval bounds.

8.152 Use a computer and generate 50 random samples, each of size $n = 25$, from a normal probability distribution with $\mu = 130$ and standard deviation $\sigma = 10$.

a. Calculate the 95% confidence interval based on each sample mean.

b. What proportion of these confidence intervals contain $\mu = 130$?

c. Explain what the proportion found in (b) represents.

8.153 The following MINITAB output shows a simulated sample of size 28 randomly generated from a normal population with $\mu = 18$ and $\sigma = 4$. ZTESt was then used to complete a hypothesis test for $\mu = 18$ against a two-tailed alternative.

a. State the alternative hypothesis, the decision, and the conclusion that resulted.

b. Verify the values reported for the standard error of mean, $z^\star$, and the p-value.

```
MTB > RANDOM 28 OBSERVATIONS INTO C1;
SUBC> NORMAL MU = 18 SIGMA = 4.
MTB > PRINT C1

C1

18.7734   21.4352   15.5438   20.2764   23.2434   15.7222   13.9368
14.4112   15.7403   19.0970   19.0032   20.0688   12.2466   10.4158
 8.9755   18.0094   20.0112   23.2721   16.6458   24.6146   17.8078
16.5922   16.1385   12.3115   12.5674   18.9141   22.9315   13.3658
```

(continued)

MINITAB (*continued*)
```
MTB > ZTEST MU = 18 SIGMA = 4 C1

TEST OF MU = 18.000 VS MU N.E. 18.000
THE ASSUMED SIGMA = 4.00

            N     MEAN    STDEV   SE MEAN      Z    P VALUE
C1         28   17.217    4.053    0.756    -1.04     0.30
```

8.154   Use a computer and generate 50 random samples, each of size $n = 28$, from a normal probability distribution with $\mu = 18$ and standard deviation $\sigma = 4$.
   a. Calculate the $z^{\star}$ corresponding to each sample mean.
   b. In regard to the *p*-value approach, find the proportion of 50 $z^{\star}$-values that are "more extreme" than the $z = -1.04$ that occurred in Exercise 8.153 (H_a: $\mu \neq 18$). Explain what this proportion represents.
   c. In regard to the classical approach, find the critical values for a two-tailed test using $\alpha = 0.01$; find the proportion of 50 $z^{\star}$-values that fall in the critical region. Explain what this proportion represents.

8.155   Use a computer and generate 50 random samples, each of size $n = 28$, from a normal probability distribution with $\mu = 19$ and standard deviation $\sigma = 4$.
   a. Calculate the $z^{\star}$ corresponding to each sample mean that would result when testing the null hypothesis $\mu = 18$.
   b. In regard to the *p*-value approach, find the proportion of 50 $z^{\star}$-values that are "more extreme" than the $z = -1.04$ that occurred in Exercise 8.153 (H_a: $\mu \neq 18$). Explain what this proportion represents.
   c. In regard to the classical approach, find the critical values for a two-tailed test using $\alpha = 0.01$; find the proportion of 50 $z^{\star}$-values that fall in the noncritical region. Explain what this proportion represents.

VOCABULARY LIST

Be able to define each term. Pay special attention to the key terms, which are printed in red. In addition, describe in your own words, and give an example of, each term. Your examples should not be ones given in class or in the textbook.

The bracketed numbers indicate the chapters in which the terms first appeared, but you should define the terms again to show increased understanding of their meaning. Page numbers indicate the first appearance of the term in Chapter 8.

acceptance region (p. 408)
alpha (α) (p. 382)
alternative hypothesis (p. 379)
assumptions (p. 365)
beta (β) (p. 382)
biased statistics (p. 361)
calculated value (*) (p. 409)
conclusion (p. 384)
confidence coefficient (p. 366)
confidence interval (p. 363)

confidence interval model (p. 367)
critical region (p. 408)
critical value (p. 409)
decision (p. 378)
decision rule (pp. 393, 410)
estimation (p. 360)
hypothesis (p. 378)
hypothesis test (p. 378)
hypothesis test, classic model (p. 403)
hypothesis test, *p*-value model (p. 387)

interval estimate (p. 363)
level of confidence (p. 363)
level of significance (p. 383)
lower confidence limit (p. 367)
maximum error of estimate (p. 367)
noncritical region (p. 408)
null hypothesis (p. 379)
parameter [1] (pp. 361, 379)
point estimate (p. 361)
p-value (p. 392)
sample size (p. 373)

sample statistic [1, 2] (p. 361)
standard error of mean [7] (p. 367)
test criteria (pp. 391, 407)
test statistic (pp. 384, 407, 391)
type A correct decision (p. 381)
type B correct decision (p. 381)
type I error (p. 381)
type II error (p. 381)
unbiased statistic (p. 361)
upper confidence limit (p. 367)
$z(\alpha)$ [6] (p. 366)

QUIZ A

Answer "True" if the statement is always true. If the statement is not always true, replace the words shown in bold with words that make the statement always true.

8.1 **Beta** is the probability of a type I error.

8.2 $1 - \alpha$ is known as the level of significance of a hypothesis test.

8.3 The standard error of the mean is the standard deviation of the **sample selected**.

8.4 The maximum error of estimate is controlled by three factors: **level of confidence, sample size, and standard deviation**.

8.5 Alpha is the measure of the area under the curve of the standard score that lies in the **rejection region** for H_0.

8.6 The risk of making a **type I error** is directly controlled in a hypothesis test by establishing a level for α.

8.7 Failing to reject the null hypothesis when it is false is a **correct decision**.

8.8 If the noncritical region in a hypothesis test is made wider (assuming σ and n remain fixed), α becomes larger.

8.9 Rejection of a null hypothesis that is false is a **type II error**.

8.10 To conclude that the mean is higher (or lower) than a claimed value, the value of the test statistic must fall in the **acceptance region**.

QUIZ B

Answer all questions, showing all formulas, substitutions, and work.

8.1 An unhappy post office customer is disturbed with the waiting time to buy stamps. Upon registering his complaint he was told, "The average waiting time in the past has been about 4 min with a standard deviation of 2 min." The customer collected a sample of $n = 45$ customers and found the mean wait was 5.3 min. Find the 95% confidence interval estimate for the mean waiting time.

8.2   State the null (H_0) and the alternative (H_a) hypotheses that would be used to test each of the following claims.
   a. The mean weight of professional football players is more than 245 lb.
   b. The mean monthly amount of rainfall in Monroe County is less than 4.5 in.
   c. The mean weight of the baseball bats used by major league players is not equal to 35 oz.

8.3   Determine the test criteria [level of significance, test statistic, critical region(s), and critical value(s)] that would be used in completing each hypothesis test using $\alpha = 0.05$.
   a. H_0: $\mu = 43$ b. H_0: $\mu = 0.80$ c. H_0: $\mu = 95$
      H_a: $\mu < 43$ H_a: $\mu > 0.80$ H_a: $\mu \neq 95$
      (given $\sigma = 6$)   (given $\sigma = 0.13$)   (given $\sigma = 12$)

8.4   Find each of the following:
   a. $z(0.05)$ b. $z(0.01)$ c. $z(0.12)$
   d. $z(0.95)$ e. $z(0.98)$ f. $z(0.75)$

8.5   In the past, the grapefruits grown in a particular orchard have had a mean diameter of 5.50 in. and a standard deviation of 0.6 in. The owner believes this year's crop is larger than in the past. He collected a random sample of 100 and found a sample mean diameter of 5.65.
   a. Find the value of the test statistic, $z^\star$, that corresponds to $\bar{x} = 5.65$.
   b. Calculate the p-value for the owner's hypothesis.

8.6   A manufacturer of light bulbs claims that its light bulbs have a mean life of 1520 hr with a standard deviation of 85 hr. A random sample of 40 such bulbs is selected for testing. If the sample produces a mean value of 1498.3 hr, is there sufficient evidence to claim that the mean life is significantly less than the manufacturer claimed? Use $\alpha = 0.01$.

QUIZ C

8.1   The noise level in a hospital may be a critical factor influencing a patient's rate of recovery. Suppose for the sake of discussion that a research commission has recommended a maximum mean noise level of 30 db with a standard deviation of 10 db. The staff of a hospital intends to sample one of its wards to determine if the noise level is significantly higher than the recommended level. The following hypothesis test will be completed.

$$H_0: \mu = 30 \quad (\leq)$$
$$H_a: \mu > 30$$
$$\alpha = 0.05$$

   a. Identify the correct interpretation for the meaning of each hypothesis with regard to the recommendation and justify your choice.

   H_0: (a) noise level is not significantly higher than the recommended level
       or
       (b) noise level is significantly higher than the recommended level
   H_a: (a) noise level is not significantly higher than the recommended level
       or
       (b) noise level is significantly higher than the recommended level

b. Which statement below best describes the type I error?
   (1) Decision reached was, noise level is within level recommended, when in fact it actually was within in.
   (2) Decision reached was, noise level is within level recommended, when in fact it actually exceeded it.
   (3) Decision reached was, noise level exceeds the level recommended, when it fact it actually exceeded it.
   (4) Decision reached was, noise level exceeds the level recommended, when in fact it actually was within it.

c. Which statement above best describes the type II error?

d. If alpha were changed from 0.05 to 0.01, identify and justify the effect (increases, decreases, or remains the same) on each of the following: P(type I error), P(type II error).

INFERENCES INVOLVING ONE POPULATION

THE AMERICAN GENDER EVOLUTION:
GETTING WHAT WE WANT

THE NEW WOMAN SURVEY

The results reported here are based on a telephone survey of 1,201 adults supervised by the polling firm of Yankelovich Clancy Shulman and conducted during March 1990. It has a sampling error of ± 3 percent and all statistics exclude "don't know" answers.

THE CHANGING FACE OF MARRIAGE

Marriage is as important as ever, but in the past 20 years, the roles of husband and wife—and their expectations of marriage—have changed dramatically. Today, couples often fall into one of two categories—Traditional and Egalitarian—which reflect opposing ideals of marriage. Other couples, struggling to integrate old and new ideals, fall somewhere in between, in marriages that could be called Transitional.

In the Traditional marriage, the wife generally derives her identity through her family and her role as wife and mother. The husband defines himself through his work and his role as breadwinner.

Many older Americans—those over age 55—have Traditional marriages. They are most likely to say, for instance, that the ideal woman is a good homemaker (see Chart 1).

In an Egalitarian marriage, the couple views the union as an equal partnership. Both husband and wife gain a sense of identity through work and family; they share the burden of breadwinning as well as the pleasure of nurturing children.

Egalitarian marriages are most likely to be found among younger Americans—those under age 45. . . .

. . . Most American women and men under age 45 believe that both partners should be responsible for earning a living (see Chart 2). . . .

It's inevitable that over the next decade more couples will be hoping to find an Egalitarian marriage. This year, 65 percent of American wives under 65 are working outside the home, according to the U.S. Bureau of Labor Statistics. As more wives remain in the work force, fewer Americans will expect men to be the sole breadwinner. . . .

Perhaps that is why some men go even further and insist that wives are obligated to work. Almost half of American men today actually disapprove of a wife with no children who doesn't work; so do about one third of women (see Chart 3). . . .

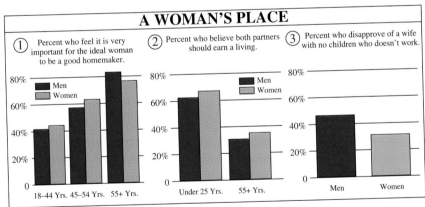

A WOMAN'S PLACE

(1) Percent who feel it is very important for the ideal woman to be a good homemaker.

(2) Percent who believe both partners should earn a living.

(3) Percent who disapprove of a wife with no children who doesn't work.

Source: "The American Gender Evolution: Getting What We Want," *New Woman* (November 1990). Reprinted by permission.

Today's newspapers and magazines often report the findings of survey polls about various aspects of life. When an article reports 62% or a graph shows 62%, did you ever wonder what is this 62%: Is it a parameter or a statistic? Somewhere in the article you might find bits of information like: "telephone survey of 1201 adults" and "has a sampling error of ± 3 percent." What does this information tell us? How does this information relate to the statistical inferences that we have been learning about? How do terms like point estimate, level of confidence, maximum error of estimate, *and* confidence interval estimation *relate to the values reported in the article as it applies to this situation involving percentage of people? Is percentage of people a population parameter, and if so, is it related to any of the parameters that we have studied? We will return to these questions at the end of the chapter.*

● CHAPTER OBJECTIVES

In Chapter 8 we learned about two forms of statistical inference: confidence intervals and hypothesis testing. The study of these two types of inference was restricted to the unlikely circumstance that the population parameter σ was known so that we could focus our attention on learning the basic procedures. The standard deviation or variance of a population is seldom known in a real-world application, so now we will remove that restriction and find out how both types of inference about the population mean μ are treated in a more realistic manner. We will also learn how to perform both types of inference when our concern is about the population parameters p, the binomial probability of success, and, σ, standard deviation.

9.1 | INFERENCES ABOUT MEAN μ (σ UNKNOWN)

Inferences about the population mean μ are based on the sample mean $\bar{x}$ and information obtained from the sampling distribution of sample means. Recall the sampling distribution of sample means has a mean μ and a standard error of $\sigma/\sqrt{n}$ for all samples of size n, and it is normally distributed when the sampled population has a normal distribution or when the sample size is sufficiently large. This means the test statistic $z^\star = (\bar{x} - \mu)/(\sigma/\sqrt{n})$ has a standard normal distribution. However, when σ is unknown, the standard error $\sigma/\sqrt{n}$ is also unknown. Therefore, the sample standard deviation s will be used as the point estimate for σ. As a result, $s/\sqrt{n}$ will be used as an estimate for the standard error of the mean and our test statistic will become $(\bar{x} - \mu)/(s/\sqrt{n})$.

When a known σ is being used to make an inference about mean μ, a sample provides one value for use in the formulas; that one value is $\bar{x}$. When s is also used, the sample provides two values: the sample mean $\bar{x}$ and the estimated standard error $s/\sqrt{n}$. As a result, the z-statistic will be replaced with a statistic that accounts for the use of an estimated standard error. This new statistic is known as the **Student's t-statistic**.

Student's *t*-statistic

In 1908 W. S. Gosset, an Irish brewery employee, published a paper about this t-distribution under the pseudonym "Student." In deriving the t-distribution, Gosset assumed that the samples were taken from normal populations. Although this might seem to be quite restrictive, satisfactory results are obtained when selecting large samples from many nonnormal populations.

Figure 9-1 presents a diagrammatic organization for the inferences about the population mean as discussed in Chapter 8 and in this first section of Chapter 9. Two situations exist: σ is known, or σ is unknown. As stated before, σ is almost never a known quantity in real-world problems; therefore, the standard error will almost always be estimated using $s/\sqrt{n}$. The use of an estimated standard error of the mean requires the use of the t-distribution. Almost all real-world inferences about the population mean will be completed using the Student's t-statistic.

FIGURE 9-1   Do I Use the z-Statistic or the t-Statistic?

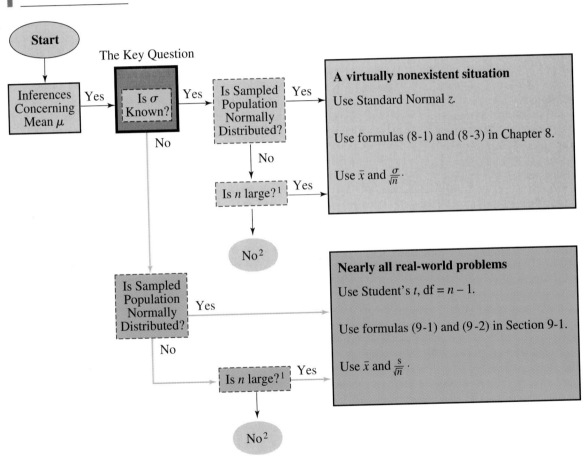

1. Is n large? — Samples as small as $n \geq 20$ may be considered large enough for the central limit theorem to hold if the sample data are unimodal, nearly symmetrical, short-tailed, and without outliers. Samples that are not symmetrical require larger sample sizes, with 50 sufficing, except for extremely skewed samples. See discussion on page 366.
2. Requires the use of a nonparametric technique; see Chapter 14.

The *t*-distribution has the following properties (see also Figure 9-2):

PROPERTIES OF THE *t*-DISTRIBUTION (df > 2)[1]

1. *t* is distributed with a mean of 0.
2. *t* is distributed symmetrically about its mean.
3. *t* is distributed so as to form a family of distributions, a separate distribution for each different number of *degrees of freedom* (df ≥ 1).
4. The *t*-distribution approaches the normal distribution as the number of degrees of freedom increases.
5. *t* is distributed with a variance greater than 1, but as the degrees of freedom increases, the variance approaches 1.
6. *t* is distributed so as to be less peaked at the mean and thicker at the tails than the normal distribution.

_____

1. Not all of the properties hold for df = 1 and df = 2. Since we will not encounter situations where df = 1, 2, these special cases are not discussed further.

FIGURE 9-2

Student's
t-Distributions

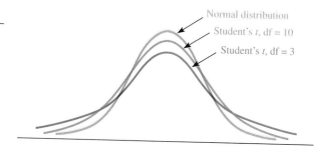

DEGREES OF FREEDOM, df

A parameter that identifies each different distribution of Student's *t*-distribution. For the methods presented in this chapter, the value of df will be the sample size minus 1, df = *n* − 1.

The *number of degrees of freedom associated with* s^2 is the divisor ($n − 1$) used to define sample variance s^2 [formula (2-6), p. 79]. That is, **df = *n* − 1**. Sample variance is the mean of the squared deviations. The number of degrees of freedom is the "number of unrelated deviations" available for use in estimating σ^2. Recall that the sum of the deviations, $\Sigma (x − \bar{x})$, must be zero. From a sample of size *n*, only the first *n* − 1 of these deviations has freedom of value. That is, the last, or

nth, value of $(x - \bar{x})$ must make the sum of the n deviations total exactly zero. As a result, variance is said to average $n - 1$ unrelated squared deviation values, and this number, $n - 1$, was named "degrees of freedom."

EXERCISE 9.1

Make a list of four numbers that total "zero." How many numbers were you able to pick without restriction? Explain how this demonstrates "degrees of freedom."

Although there is a separate t-distribution for each degrees of freedom, df = 1, df = 2, ..., df = 20, ..., df = 40, and so on, only certain key critical values of t will be necessary for our work. Consequently, the table for Student's t-distribution (Table 6 in Appendix B) is a table of critical values rather than a complete table, such as Table 3 for the standard normal distribution for z. As you look at Table 6, you will note that the left side of the table is identified by "df," degrees of freedom. This left-hand column starts at 1 at the top and lists consecutive df values to 30, then jumps to 35, ..., to "df > 100" at the bottom. As previously stated, as the degrees of freedom increases, the t-distribution approaches the characteristics of the standard normal z-distribution. Once df is "greater than 100," the critical values of the t-distribution are the same as the corresponding critical values of the standard normal distribution.

EXERCISE 9.2

Explain the relationship between the critical values found in the bottom row of Table 6 and the critical values of z given in Table 4.

The critical values of the Student's t-distribution to be used for both constructing a confidence interval and for hypothesis testing will be obtained from Table 6 in Appendix B. To obtain the value of t, you will need to know two identifying values: (1) df, the number of degrees of freedom (identifying the distribution of interest), and (2) α, the area under the curve to the right of the right-hand critical value. A notation much like that used with z will be used to identify a critical value. $t(\text{df}, \alpha)$, read as "t of df, α," is the symbol for the value of t with df degrees of freedom and an area of α in the right-hand tail, as shown in Figure 9-3.

FIGURE 9-3

t-Distribution
Showing $t(\text{df}, \alpha)$

I L L U S T R A T I O N   9 - 1

Find the value of $t(10, 0.05)$
(see the diagram).

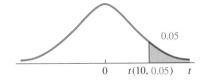

EXERCISE 9.3

Find the value of:

a. $t(12, 0.01)$,
b. $t(22, 0.025)$.

S O L U T I O N   There are 10 degrees of freedom, and 0.05 is the area to the right. In Table 6, Appendix B, we look for the row df = 10 and column marked $\alpha = 0.05$. From the table we see that $t(10, 0.05) = $ **1.81**.

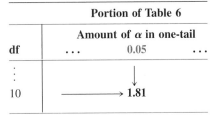

	Portion of Table 6		
	Amount of α in one-tail		
df	...	0.05	...
$\vdots$			
10		$\longrightarrow$ 1.81	

For the values of t on the left side of the mean, we can use one of two notations. The t-value shown in Figure 9-4 could be named $t(df, 0.95)$, since the area to the right of it is 0.95, or it could be identified by $-t(df, 0.05)$, since the t-distribution is symmetric about its mean, zero.

FIGURE 9-4

t-Value of Left Side

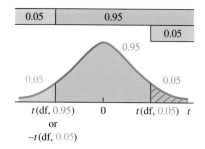

 ILLUSTRATION 9-2

Find the value of $t(15, 0.95)$.

EXERCISE 9.4

Find the value of:

a. $t(18, 0.90)$,
b. $t(9, 0.99)$.

SOLUTION There are 15 degrees of freedom. In Table 6 we look for the column marked $\alpha = 0.05$ and come down to row $df = 15$. The table gives us $t(15, 0.05) = 1.75$; therefore, $t(15, 0.95) = -t(15, 0.05) = \mathbf{-1.75}$ (the value is negative because it is to the left of the mean; see the following figure).

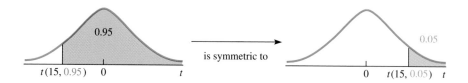

 ILLUSTRATION 9-3

Find the values of the t-distribution that bound the middle 0.90 of the area under the curve for the distribution with $df = 17$.

SOLUTION The middle 0.90 leaves 0.05 for the area of each tail. The value of t that bounds the right-hand tail is $t(17, 0.05) = 1.74$, as found in Table 6. The value bounding the left-hand tail is -1.74, since the t-distribution is symmetric about its mean zero.

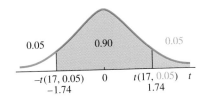

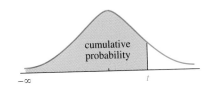

EXERCISE 9.5

Find the values of t that bound the middle 0.95 of the distribution for df = 12.

If the df needed is not listed in the left-hand column of Table 6, then use the next smaller value of df that is listed. For example, $t(72, 0.05)$ will be estimated using $t(70, 0.05) = 1.67$.

Most computer software packages will calculate either the area related to a specified t-value or the t-value that bounds a specified area. The accompanying figure shows the relationship between the cumulative distribution probability and a specific t-value for a t-distribution with df degrees of freedom. If the t-value is known and the cumulative area from $-\infty$ to t is wanted, then the **cumulative distribution function (CDF)** is used. If the area from $-\infty$ to t is known and the value of t is wanted, then the **inverse cumulative distribution function (INVCDF)** is used.

MINITAB (Release 10) commands used to determine the cumulative probability from $-\infty$ to a specified value of t = A for the t-distribution with **df** degrees of freedom:

Session commands

Enter: CDF A;
       T df.

Menu commands

Choose: Calc > Prob. Dist. > t
Select: Cumulative Probability
Enter: Degrees of freedom: df
        Input constant: A

MINITAB (Release 10) commands used to determine the value of t bounding a specified cumulative probability B from $-\infty$ to t for the t-distribution with **df** degrees of freedom:

Session commands

Enter: INVCdf B;
       T df.

Menu commands

Choose: Calc > Prob. Dist. > t
Select: Inverse Cumulative Probability
Enter: Degrees of freedom: df
        Input constant: B

In Illustration 9-3 the critical value of t with df = 17 and 0.05 as the area in the left-hand tail was found to be $-t(17, 0.05) = -1.74$.

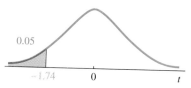

If we have the value of $t = -1.74$ and we want to find the area to the left, then we use MINITAB's CDF command.

```
CDF  -1.74;
T 17.
```

```
Cumulative Distribution Function
Student's t distribution with 17 d.f.

      X       P(X <= x)
  -1.7400      0.0500
```

Area to the left of $t = -1.74$, with df = 17.

If we know the area to the left of a particular t is 0.05, then we could use MINITAB's INVCdf command to find the value of t.

```
INVCdf .05;
T 17.
```

Use MINITAB to find the answers to Exercises 9.3, 9.4, and 9.5.

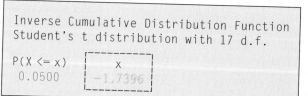

```
Inverse Cumulative Distribution Function
Student's t distribution with 17 d.f.

P(X <= x)          X
 0.0500        -1.7396
```

The t bounding the given area is -1.74.

We are now ready to make inferences about the population mean μ using the sample standard deviation. As mentioned earlier, use of the t-distribution has a condition.

> **The assumption for inferences about mean μ when σ is unknown:** The sampled population is normally distributed.

Confidence Interval Procedure

The procedure used to make confidence interval estimations using the sample standard deviation is very similar to that used when σ is known (see pp. 365–370). The difference is the use of Student's t in place of the standard normal z and the use of s, the sample standard deviation, as an estimate of σ. The central limit theorem implies that this technique can also be applied to nonnormal populations when the sample size is sufficiently large.

The formula for the $1 - \alpha$ confidence interval of estimation for μ is

$$\bar{x} - t\left(\text{df}, \frac{\alpha}{2}\right)\frac{s}{\sqrt{n}} \quad \text{to } \bar{x} + t\left(\text{df}, \frac{\alpha}{2}\right)\frac{s}{\sqrt{n}}, \quad \text{where df} = n - 1 \qquad \textbf{(9-1)}$$

▼ | ILLUSTRATION   9 - 4

A random sample of size 20 is taken from the weights of babies born at Northside Hospital during the year 1994. A mean of 6.87 lb and a standard deviation of 1.76 lb were found for the sample. Estimate, with 95% confidence, the mean weight of all babies born in this hospital in 1994. Based on past information, it is assumed that weights of newborns are normally distributed.

The four-step confidence interval model is on page 367.

SOLUTION

STEP 1   Describe the population parameter of interest: the mean weight of newborns at Northside Hospital.

STEP 2   Specify the confidence interval criteria.
   a. Check assumptions: Past information indicates that sampled population is normal and σ is unknown.
   b. The test statistic: t will be the test statistic.
   c. Confidence level: $1 - \alpha = 0.95$.

STEP 3   Collect and present sample evidence.
   a. Sample evidence: $n = 20$, $\bar{x} = 6.87$, and $s = 1.76$.
   b. Point estimate: $\bar{x} = 6.87$.

df is used to find the confidence coefficient in Table 6, while n is used in the formula.

STEP 4   Determine the confidence interval limits.

a. Confidence coefficients: two-tailed situation with $\alpha/2 = 0.025$, $n = 20$; therefore, df = 19 and from Table 6, we find $t(\text{df}, \alpha/2) = t(19, 0.025) = 2.09$. See the figure.

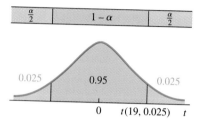

b. Maximum error: Using part of formula (9-1),

$$E = t\left(\text{df}, \frac{\alpha}{2}\right)\frac{s}{\sqrt{n}} = t(19, 0.025)\frac{s}{\sqrt{n}} = 2.09 \cdot \frac{1.76}{\sqrt{20}} = (2.09)(0.394) = \mathbf{0.82}$$

c. Confidence limits: Using formula (9-1),

$$\bar{x} - E \text{ to } \bar{x} + E$$
$$6.87 - 0.82 \text{ to } 6.87 + 0.82$$
$$\mathbf{6.05} \text{ to } \mathbf{7.69}$$

6.05, 7.69, the 95% confidence interval for μ

That is, with 95% confidence we estimate the mean weight to be between 6.05 and 7.69 lb.

▲|

EXERCISE 9.6

Construct a 95% confidence interval estimate for the mean μ using the sample information $n = 24, \bar{x} = 16.7$, and $s = 2.6$.

MINITAB (Release 10) commands to complete a $1 - \alpha = K$ confidence interval estimate for mean μ with unknown standard deviation and the sample data in C1.

Session commands   *Menu commands*

Enter: Choose:
  TINTerval K   Stat > Basic Stats > 1-Sample t
  data in C1   Enter: C1
                           Select: Confidence interval
                           Enter: $1 - \alpha$: K

The solution to Illustration 9-4 looks like this when solved on MINITAB: With the 20 weights in C1 on MINITAB's worksheet, the following command would result in the output listed below.

```
TINTerval 95 C1
```

Compare the MINITAB output to the solution of Illustration 9-4.

```
Confidence Interval

Variable    N    Mean    StDev    SE Mean    95.0% C.I.
   C1      20   6.870   1.760    0.394     (6.047, 7.693)
```

Hypothesis-Testing Procedure

The t-statistic is used to complete a hypothesis test about the population mean μ in much the same manner as z was used in Chapter 8. In hypothesis-testing situations, we will use formula (9-2) to calculate the value of the test statistic $t^{\star}$.

$$t^{\star} = \frac{\bar{x} - \mu}{s/\sqrt{n}}, \text{ with df} = n - 1 \qquad \text{(9-2)}$$

The calculated t, like z, is the number of estimated standard errors $\bar{x}$ is from the hypothesized mean μ. As with confidence intervals, the central limit theorem indicates that the t-distribution can also be applied to nonnormal populations when the sample size is sufficiently large.

Probability-Value Approach

▼ | ILLUSTRATION 9 - 5

Let's return to the hypothesis of Illustration 8-10 (p. 389) where the EPA wanted to show that "the mean carbon monoxide level of air pollution is higher than 4.9." Does a random sample of 22 readings (sample results: $\bar{x} = 5.1$ and $s = 1.17$) present sufficient evidence to support the EPA's claim? Use $\alpha = 0.05$. Previous studies have indicated that such readings have an approximately normal distribution.

The five-step p-value hypothesis test model is on page 387.

SOLUTION

STEP 1   Describe the population parameter of concern: The parameter of concern is the mean pollution level of air in downtown Rochester.

Procedures for writing H_0 and H_a are on pages 388–390.

STEP 2   State the null and alternative hypotheses:

H_0: $\mu = 4.9$ ($\leq$) (no higher than)

H_a: $\mu > 4.9$ (higher than)

STEP 3   Specify the test criteria.
   a. Check the assumptions: The assumptions are satisfied since the population is approximately normal and the sample size is large enough for the CLT to apply.
   b. Identify the test statistic: The test statistic is $t^{\star}$, formula (9-2), with df $= n - 1 = 21$.
   c. The level of significance: $\alpha = 0.05$.

STEP 4 Collect and present the sample evidence.
    a. Collect the sample information: $n = 22$, $\bar{x} = 5.1$, and $s = 1.17$.
    b. Calculate the value of the test statistic: Using formula (9-2),

$$t^\star = \frac{\bar{x} - \mu}{s/\sqrt{n}}; \quad t^\star = \frac{5.1 - 4.9}{1.17/\sqrt{22}} = \frac{0.20}{0.2494} = 0.8018 = \mathbf{0.80}$$

Instructions for finding p-value are on page 394.

    c. Calculate the p-value:

$$\mathbf{P} = P(t^\star > 0.80, \text{ with df} = 21) \text{ as shown on the figure.}$$

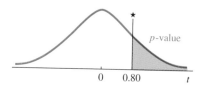

If you are doing this hypothesis test with the aid of a computer, most likely the computer will calculate the p-value for you. If you are not using a computer, we will "simulate" the computer's calculation of the p-value by using either Table 6 or Table 7. If you use Table 6 in Appendix B, you will "place bounds" on the p-value: By inspecting the df = 21 row of Table 6, you can place bounds on the p-value. Locate the $t^\star$ along the row opposite df = 21. If $t^\star$ is not listed, locate the two values $t^\star$ falls between, and then read the bounds for the p-value from the top of the table. In this case, $t^\star = 0.80$ is between 0.69 and 1.32; therefore, **P** is between 0.10 and 0.25. (Use the one-tailed heading when H_a is one-tailed.)

Portion of Table 6

df	Amount of α in one-tail		
	0.25	P	0.10
⋮	⋮	↑	⋮
21	0.69	0.80	1.32

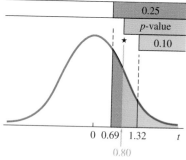

Separately these three areas and t-values look like this:

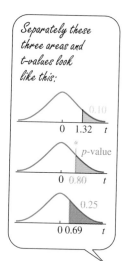

The 0.69 entry in the table tells us that $P(t > 0.69) = 0.25$, as shown on the above figure in purple. The 1.32 entry in the table tells us that $P(t > 1.32) = 0.10$, as shown in orange. You can see that the p-value **P** (shown in red) is between 0.10 and 0.25. Therefore, $\mathbf{0.10 < P < 0.25}$, and we say that **0.10 and 0.25 are the "bounds" for the p-value**.

    Table 7 in Appendix B is designed so that you will be able to read the p-value directly from the table for many situations: Simply locate the p-value at the intersection of the $t^\star$ row and the df column. The p-value for $t^\star = 0.80$ with df = 21 is **0.216**.

EXERCISE 9.7

Calculate the value of $t^\star$ and **P**, and state the decision that would occur for the following hypothesis test:
H_0: $\mu = 32$,
H_a: $\mu > 32$,
$\alpha = 0.05$,
$n = 16$, $\bar{x} = 32.93$,
$s = 3.1$.

Portion of Table 7		
$t^\star$	df	21
$\vdots$		
0.80	$\longrightarrow$	0.216

$\mathbf{P} = P(t^\star > 0.80,\text{ with df} = 21) = \mathbf{0.216}$

STEP 5   Determine the results: The p-value is greater than the level of significance, $\alpha = 0.05$.

DECISION Fail to reject H_0.

CONCLUSION At the 0.05 level of significance, we do not have sufficient evidence to show that "the mean carbon monoxide level is higher than 4.9."

▼ | **ILLUSTRATION   9 - 6**

Determine the p-value for the following hypothesis test.

$$H_0: \mu = 55 \quad \text{vs.} \quad H_a: \mu \neq 55, \text{ with df} = 15 \text{ and } t^\star = -1.80$$

EXERCISE 9.8

Calculate the value of $t^\star$ and **P**, and state the decision that would occur for the following hypothesis test:
H_0: $\mu = 73$,
H_a: $\mu \neq 73$,
$\alpha = 0.05$,
$n = 12$, $\bar{x} = 71.46$,
$s = 4.1$.

SOLUTION $\mathbf{P} = P(t < -1.80) + P(t > 1.80) = 2P(t > 1.80)$

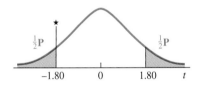

Using Table 6, find 1.80 in row df $= 15$, read the value or bounds for **P** from the two-tailed heading at the top of the table: $\mathbf{0.05 < P < 0.10}$.
   Using Table 7, $P(t > 1.80) = 0.046$; $\mathbf{P} = 2P(t > 1.80) = 2(0.046) = \mathbf{0.092}$.

▼ | **ILLUSTRATION   9 - 7**

Instructions for finding a two-tailed p-value are on page 394.

Determine the p-value for the following hypothesis test.

$$H_0: \mu = 525$$
$$H_a: \mu < 525$$
$$\text{df} = 23 \text{ and } t^\star = -0.84$$

SOLUTION $\mathbf{P} = P(t < -0.84,\text{ with df} = 23)$ (See figure on next page.)
   Due to symmetry, $P(t < -0.84) = P(t > 0.84)$. Using Table 6, $\mathbf{0.10 < P < 0.25}$. Using Table 7: Because Table 7 is not a complete table, occasionally bounds will result. Generally, these bounds will be narrower than the bounds from using Table 6. (Results are shown in the table on the next page.)

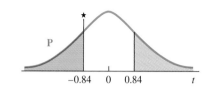

EXERCISE   9.9

Place bounds on the p-value for the following hypothesis test: H_0: $\mu = 125$, H_a: $\mu > 125$, df = 16, and $t^\star = 1.08$.

▲ |

Portion of Table 7

Degrees of Freedom

$t^\star$	21	**23**	25
⋮			⋮
0.80	0.216		0.216
0.84		P	
0.90	**0.189**		0.188

$0.188 < P < 0.216$

There is a third choice: Use MINI-TAB to calculate the p-value; see page 437.

MINITAB (Release 10) commands to complete a hypothesis test for mean $\mu = A$ with unknown standard deviation and the sample data in C1 against the alternative hypothesis coded as K: −1 = left-tailed, 0 = two-tailed, and +1 = right-tailed.

Session commands   *Menu commands*

Enter: Choose:
   TTESt μ = A   Stat > Basic Stats > 1-Sample t
   data in C1;   Enter: C1
   ALTErnative K.   Select: Test mean
                                  Enter: A
                                  Enter: Alternative: −1 or 0 or 1

The solution to Illustration 9-7 looks like this when solved on MINITAB: With the 24 data values in C1 on MINITAB's worksheet, the following commands result in the output listed below.

```
TTESt 525 C1;
ALTErnative −1.
```

Compare MINITAB results to the solution found in Illustration 9-7.

```
T-Test of the Mean
Test of mu = 525 vs mu < 525

Variable      N      Mean     StDev    SE Mean       T    P-Value
     C1      24    521.84     18.45      3.766    -0.84     0.2051
```

Classical Approach

Let's look at another hypothesis-testing situation and complete it using the classical approach.

▼ | ILLUSTRATION 9 - 8

On a popular self-image test, which results in normally distributed scores, the mean score for public-assistance recipients is expected to be 65. A random sample of 28 public-assistance recipients in Emerson County are given the test. They achieve a mean score of 62.1, and their scores have a standard deviation of 5.83. Do the Emerson County assistance recipients test lower, on the average, than what is expected, at the 0.01 level of significance?

The five-step classical hypothesis test model is on page 403.

SOLUTION

STEP 1 Describe the population parameter of concern: The parameter of interest is the mean self-image test score, μ, for all Emerson County public-assistance recipients.

STEP 2 State the null and alternative hypotheses:

$$H_0: \mu = 65 \ (\geq) \text{ (mean is not less than 65)}$$
$$H_a: \mu < 65 \text{ (mean is less than 65)}$$

STEP 3 Specify the test criteria.
   a. Check the assumptions: The test is expected to produce normally distributed scores; therefore, the assumption has been satisfied.
   b. Identify the test statistic: The test statistic will be t, since the null hypothesis is about mean μ and the standard deviation σ is not given. (The standard deviation mentioned in the problem is from the sample.)
   c. Determine the level of significance: $\alpha = 0.01$ (given in statement of problem).
   d. Determine the critical region and critical value: df $= n - 1 = 27$. The critical region is the left-hand tail since the H_a expresses concern for values related to "less than." The critical value is obtained from Table 6: $-t(27, 0.01) = -2.47$.

Portion of Table 6	
	α **in one-tail**
df	0.01
⋮	↓
27	⟶ 2.47

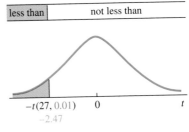

STEP 4   Collect and present the sample evidence.
    a. Sample information: $n = 28$, $\bar{x} = 62.1$, and $s = 5.83$.
    b. Calculate the value of the test statistic: Using formula (9-2),

$$t^\star = \frac{\bar{x} - \mu}{s/\sqrt{n}}; \quad t^\star = \frac{62.1 - 65.0}{5.83/\sqrt{28}} = \frac{-2.9}{1.1018} = -2.632 = \mathbf{-2.63}$$

STEP 5   Determine the results: Compare the calculated $t^\star$ to the critical region
(see the accompanying figure); $t^\star = -2.63$ falls in the critical region.

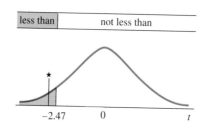

DECISION Reject H_0.

CONCLUSION At the 0.01 level of significance, we do have sufficient evidence to
conclude that the Emerson County assistance recipients test significantly lower, on
the average, than the expected 65.

▼ | ILLUSTRATION   9 - 9

Determine the critical values and region for the following hypothesis test.

$$H_0: \mu = 35$$
$$H_a: \mu \neq 35$$
$$df = 45 \text{ and } \alpha = 0.02$$

SOLUTION The critical region is split between both tails since the H_a expresses
concern for values "not equal to." The critical value for the right-hand tail is obtained
from Table 6:

$$t(44, 0.01) \approx t(40, 0.01) = 2.42$$

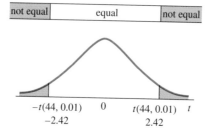

EXERCISE 9.10

Find the critical region, $t^\star$, and state the decision for the hypothesis test:
$H_0: \mu = 52$,
$H_a: \mu < 52$,
$\alpha = 0.01$, $n = 10$,
$\bar{x} = 50.47$, $s = 3.9$.

When df is not in the table, use the next smaller df listed.

EXERCISE 9.11

Determine the critical region for a two-tailed test at 0.02 level of significance when the sample size is 55.

▲ |

Case Study

9.1

Mother's Use of Personal Pronouns When Talking with Toddlers

The calculated t-value and the probability value for five different hypothesis tests are given in the following article. The expression t(44) = 1.92 means t* = 1.92 with df = 44 and is significant with p-value < 0.05. Can you verify the p-values? Explain.

EXERCISE   9.12

a. Verify that
$t(44) = 1.92$ is
significant at the
0.05 level.
b. Verify that
$t(44) = 3.41$ is
significant at the
0.01 level.
c. Explain why
$t(44) = 1.81$,
$p < .10$, makes
sense only if the
hypothesis test is
two-tailed. If the
test is one-tailed,
what level would
be reported?

ABSTRACT. The verbal interaction of 2-year-old children (N = 46; 16 girls, 30 boys) and their mothers was audiotaped, transcribed, and analyzed for the use of personal pronouns, the total number of utterances, the child's mean length of utterance, and the mother's responsiveness to her child's utterances. Mothers' use of the personal pronoun we was significantly related to their children's performance on the Stanford-Binet at age 5 and the Wechsler Intelligence Scale for Children at age 8. Mothers' use of we in social–vocal interchange, indicating a system for establishing a shared relationship with the child, was closely connected with their verbal responsiveness to their children. The total amount of maternal talking, the number of personal pronouns used by mothers, and their verbal responsiveness to their children were not related to mothers' social class or years of education.

Mothers tended to use more first person singular pronouns (*I* and *me*), $t(44) = 1.81$, $p < .10$, and used significantly more first person plural pronouns (*we*), $t(44) = 1.92$, $p < .05$, with female children than with male children. The mothers also were more verbally responsive to their female children, $t(44) = 2.0$, $p < .06$.

In general, mothers talked more to their first born children, $t(44) = 3.41$, $p < .001$, and were more responsive to their first born children, $t(44) = 3.71$, $p < .001$. Yet, the proportion of personal pronouns used when speaking to first born children was not different from that used when speaking to later born children.

Source: Dan R. Laks, Leila Beckwith, Sarale E. Cohen, THE JOURNAL OF GENETIC PSYCHOLOGY, 151(1), 25–32, 1990. Reprinted with permission of the Helen Dwight Reid Educational Foundation. Published by Heldref Publications, 4000 Albemarle St., N.W., Washington, D.C., 20016. Copyright © 1990.

EXERCISES

9.13   Find these critical values using Table 6 in Appendix B:

a. $t(25, 0.05)$ b. $t(10, 0.10)$

c. $t(15, 0.01)$ d. $t(21, 0.025)$

e. $t(21, 0.95)$ f. $t(26, 0.975)$

g. $t(27, 0.99)$ h. $t(60, 0.025)$

9.14   Using the notation of Exercise 9.13, name and find the following critical values of t:

a.   b.   c.

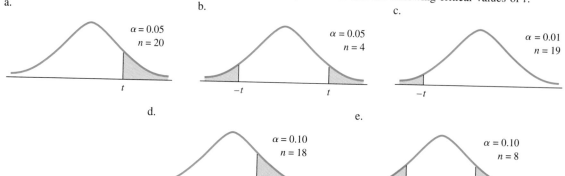

d.   e.

9.15   Ninety percent of Student's t-distribution lies between $t = -1.89$ and $t = 1.89$ for how many degrees of freedom?

9.16   Ninety percent of Student's t-distribution lies to the right of $t = -1.37$ for how many degrees of freedom?

9.17   a. Find the first percentile of Student's t-distribution with 24 degrees of freedom.
b. Find the 95th percentile of Student's t-distribution with 24 degrees of freedom.
c. Find the first quartile of Student's t-distribution with 24 degrees of freedom.

9.18   Find the percent of the Student's t-distribution that lies between the following values.
a. df $= 12$ and t ranges from -1.36 to 2.68
b. df $= 15$ and t ranges from -1.75 to 2.95

9.19   a. State two ways in which the standard normal distribution and the Student's t-distribution are alike.
b. State two ways in which they are different.

9.20   The variance for each of the Student's t-distributions is equal to df/(df $- 2$). Find the standard deviation for a Student's t-distribution with each of the following degrees of freedom.
a. 10   b. 20   c. 30

9.21   The data from a study reported in "White-Collar Crime and Criminal Careers: Some Preliminary Findings" (*Crime and Delinquency*, July 1990) indicate that white-collar criminals are likely to be older and to show a lower frequency of offending than street criminals. For example, the mean age of onset of offending for those convicted of antitrust offenses was 54 years with $n = 35$. If the standard deviation is estimated to be 7.5 years, set a 90% confidence interval on the true mean age.

9.22   Taking a random sample of 25 individuals registering for classes at a particular college, we find that the mean waiting time in the registration line was 22.6 min and the standard deviation was 8.0 min. Waiting time has a mounded distribution. Using a 90% confidence interval, estimate the mean waiting time for all individuals registering.

9.23   A study in the *Journal of Obstetric, Gynecologic, and Neonatal Nursing* (October 1994) reported on 80 new mothers who met 7 eligibility requirements. The gestation period for these 80 participants ranged from 37.0 to 42.5 weeks. Suppose that in a similar study 20 new mothers, who met the 7 eligibility requirements, had a mean gestation period equal to 40 weeks and a standard deviation of 1.5 weeks. Estimate, with 95% confidence, the mean gestation period of all new mothers who meet the 7 eligibility requirements based on the study of 20. Assume that the gestation periods are normally distributed.

9.24   Ten randomly selected shut-ins were each asked to list how many hours of television they watched per week. The results are

82	66	90	84	75	88	80	94	110	91

Determine the 90% confidence interval estimate for the mean number of hours of television watched per week by shut-ins. Assume the number of hours is normally distributed.

9.25   While doing an article on the high cost of college education, a reporter took a random sample of the cost of textbooks for a semester. The random variable x is the cost of one book. Her sample data can be summarized by $n = 41$, $\Sigma x = 550.22$, and $\Sigma(x - \bar{x})^2 = 1617.984$.
   a. Find the sample mean, $\bar{x}$.
   b. Find the sample standard deviation, s.
   c. Find the 90% confidence interval to estimate the true mean textbook cost for the semester based on this sample.

9.26   The addition of a new accelerator is claimed to decrease the drying time of latex paint by more than 4%. Several test samples were conducted with the following percentage decrease in drying time.

5.2	6.4	3.8	6.3	4.1	2.8	3.2	4.7

If we assume that the percentage decrease in drying time is normally distributed:
   a. Find the 95% confidence interval for the true mean decrease in the drying time based on this sample. (The sample mean and standard deviation were found in answering Exercise 2.147).
   b. Did the interval estimate reached in (a) result in the same conclusion as you expressed in answering part (c) of Exercise 2.147 for this same data?

9.27   The pulse rates for 13 adult women were

83	58	70	56	76	64	80	76	70	97	68	78	108

Verify the results shown on the last line of the MINITAB output below.

```
MTB > T INTERVAL 90 PERCENT CONFIDENCE INTERVAL FOR DATA IN C1

              N     MEAN     STDEV    SE MEAN    90.0 PERCENT C.I.
    C1        13    75.69    14.54     4.03       (68.50, 82.88)
```

9.28 The weights of the drained fruit found in 21 randomly selected cans of peaches packed by Sunny Fruit Cannery were (in ounces)

11.0	11.6	10.9	12.0	11.5	12.0	11.2
10.5	12.2	11.8	12.1	11.6	11.7	11.6
11.2	12.0	11.4	10.8	11.8	10.9	11.4

Using a computer or a calculator,
  a. Calculate the sample mean and standard deviation.
  b. Assume normality and construct the 98% confidence interval for the estimate of the mean weight of drained peaches per can.
(If you use MINITAB, use TINTerval.)

9.29 State the null hypothesis, H_0, and the alternative hypothesis, H_a, that would be used to test each of the following claims.
  a. The mean weight of honey bees is at least 11 g.
  b. The mean age of patients at Memorial Hospital is no more than 54 years.
  c. The mean amount of salt in granola snack bars is different from 75 mg.

9.30 Determine the p-value for the following hypothesis tests involving the Student's t-distribution with 10 degrees of freedom.
  a. $H_0: \mu = 15.5$ $H_a: \mu < 15.5$ $t^\star = -2.01$
  b. $H_0: \mu = 15.5$ $H_a: \mu > 15.5$ $t^\star = 2.01$
  c. $H_0: \mu = 15.5$ $H_a: \mu \neq 15.5$ $t^\star = 2.01$
  d. $H_0: \mu = 15.5$ $H_a: \mu \neq 15.5$ $t^\star = -2.01$

9.31 Determine the test criteria that would be used to test the null hypotheses below.
  a. $H_0: \mu = 10$ ($\alpha = 0.05, n = 15$)
     $H_a: \mu \neq 10$
  b. $H_0: \mu = 37.2$ ($\alpha = 0.01, n = 25$)
     $H_a: \mu > 37.2$
  c. $H_0: \mu = -20.5$ ($\alpha = 0.05, n = 18$)
     $H_a: \mu < -20.5$
  d. $H_0: \mu = 32.0$ ($\alpha = 0.01, n = 42$)
     $H_a: \mu > 32.0$

9.32 Compare the p-value and classical approaches to hypothesis testing by comparing the p-value and decision of the p-value approach to the critical values and decision of the classical approach for each of the following situations. Use $\alpha = 0.05$.
  a. $H_0: \mu = 128$, $H_a: \mu \neq 128$, $n = 15$, $t^\star = 1.60$
  b. $H_0: \mu = 18$, $H_a: \mu > 18$, $n = 25$, $t^\star = 2.16$
  c. $H_0: \mu = 38$, $H_a: \mu < 38$, $n = 45$, $t^\star = -1.73$
  d. Compare the results of the two techniques for each case.

9.33 A student group maintains that the average student must travel for at least 25 minutes in order to reach college each day. The college admissions office obtained a random sample of 22 one-way travel times from students. The sample had a mean of 19.4 min and a standard deviation of 9.6 min. Does the admissions office have sufficient evidence to reject the students' claim? Use $\alpha = 0.01$.
  a. Solve using the p-value approach.
  b. Solve using the classical approach.

9.34 Homes in a nearby college town have a mean value of $88,950. It is assumed that homes in the vicinity of the college have a higher value. To test this theory, a random sample of 12 homes is chosen from the college area. Their mean valuation is $92,460 and the standard deviation is $5,200. Complete a hypothesis test using $\alpha = 0.05$. Assume prices are normally distributed.

 a. Solve using the *p*-value approach.
 b. Solve using the classical approach.

9.35 An article in the *American Journal of Public Health* (March 1994) describes a large study involving 20,143 individuals. The article states that the mean percentage intake of kilocalories from fat was 38.4% with a range from 6% to 71.6%. A small sample study was conducted at a university hospital to determine if the mean intake of patients at that hospital was different from 38.4%. A sample of 15 patients had a mean intake of 40.5% with a standard deviation equal to 7.5%. Assuming that the sample is from a normally distributed population, test the hypothesis of "different from" using a level of significance equal to 0.05.

 a. What evidence do you have that the assumption of normality is reasonable? Explain.
 b. Complete the test using the *p*-value approach. Include $t^\star$, *p*-value, and your conclusion.
 c. Complete the test using the classical approach. Include the critical values, $t^\star$, and your conclusion.

9.36 A study in the journal *PAIN*, October 1994, reported on six patients with chronic myofascial pain syndrome. The mean duration of pain had been 3.0 years for the 6 patients and the standard deviation had been 0.5 year. Test the hypothesis that the mean pain duration of all patients who might have been selected for this study was greater than 2.5 years. Use $\alpha = 0.05$ and assume that the durations are normally distributed.

 a. What role does the assumption of normality play in this solution? Explain.
 b. Complete the test using the *p*-value approach. Include $t^\star$, *p*-value, and your conclusion.
 c. Complete the test using the classical approach. Include the critical values, $t^\star$, and your conclusion.

9.37 It is claimed that the students at a certain university will score an average of 35 on a given test. Is the claim reasonable if a random sample of test scores from this university yields 33, 42, 38, 37, 30, 42? Complete a hypothesis test using $\alpha = 0.05$. Assume test results are normally distributed.

 a. Solve using the *p*-value approach.
 b. Solve using the classical approach.

9.38 Gasoline pumped from a supplier's pipeline is supposed to have an octane rating of 87.5. On 13 consecutive days a sample was taken and analyzed with the following results.

| 88.6 | 86.4 | 87.2 | 88.4 | 87.2 | 87.6 | 86.8 |
| 86.1 | 87.4 | 87.3 | 86.4 | 86.6 | 87.1 | |

Many of these data sets are on your data disk.

 a. If the octane ratings have a normal distribution, is there sufficient evidence to show that these octane readings were taken from gasoline with a mean octane significantly less than 87.5 at the 0.05 level? (The sample mean and standard deviation were found in answering Exercise 2.148.)
 b. Did the statistical decision reached in (a) result in the same conclusion as you expressed in answering part (c) of Exercise 2.148 for this same data?

9.39   In order to test the null hypothesis "the mean weight for adult males equals 160 lb" against the alternative "the mean weight for adult males exceeds 160 lb," the weights of 16 males were determined with the following results.

| 173 | 178 | 145 | 146 | 157 | 175 | 173 | 137 |
| 152 | 171 | 163 | 170 | 135 | 159 | 199 | 131 |

Assume normality and verify the results shown in the following MINITAB analysis (*Note:* ALT = −1, 0, +1 represents lower-tail, two-tail, and upper-tail tests, respectively.)

```
MTB > TTEST OF MU = 160 DATA IN C1;
SUBC> ALTERNATIVE = 1.
TEST OF MU = 160.00 VS MU G.T. 160.00

        N      MEAN     STDEV    SE MEAN     T     P VALUE
C1      16    160.25    18.49     4.62      0.05     0.48
```

9.40   "Obesity raises heart-attack risk" according to a study published in the March 1990 issue of the *New England Journal of Medicine.* "Those about 15 to 25 percent above desirable weight had twice the heart disease rate." Suppose the data listed below are the percentages above desired weight for a sample of patients involved in a similar study.

18.3	19.7	22.1	19.2	17.5	12.7	22.0	17.2
21.1	16.2	15.4	19.9	21.5	19.8	22.5	16.5
13.0	22.1	27.7	17.9	22.2	19.7	18.1	22.4
17.3	13.3	22.1	16.3	21.9	16.9	15.4	19.3

Use a computer to test the null hypothesis, $\mu = 18\%$, versus the alternative hypothesis, $\mu \neq 18\%$. Use $\alpha = 0.05$.

9.41   Use a computer to complete the calculations and the hypothesis test for this exercise. (If you use MINITAB, see Exercise 9.39 for program commands.) Delco Products, a division of General Motors, produces commutators designed to be 18.810 mm in overall length. (A commutator is a device used in the electrical system of an automobile.) The following sample of 35 commutators was taken while monitoring the manufacturing process.

This data is saved on disk as EX2-020.

18.802	18.810	18.780	18.757	18.824
18.827	18.825	18.809	18.794	18.787
18.844	18.824	18.829	18.817	18.785
18.747	18.802	18.826	18.810	18.802
18.780	18.830	18.874	18.836	18.758
18.813	18.844	18.861	18.824	18.835
18.794	18.853	18.823	18.863	18.808

Source: With permission of Delco Products Division, GMC.

Is there sufficient evidence to reject the claim that these parts meet the design requirements "mean length is 18.810" at the $\alpha = 0.01$ level of significance?

9.42 How important is the assumption "the sampled population is normally distributed" to the use of the Student's t-distribution? Use a computer and the three sets of MINITAB commands to simulate drawing 100 samples of size 10 from each of three different types of population distributions. The first four commands will generate 1000 data and construct a histogram so that you can see what the population looks like. The next two commands will generate 100 samples of size 10 from the same population; each row represents a sample. (The 10 columns of 100 data each represent 100 samples of 10 data each.) The next two commands calculate the mean and standard deviation for each of the 100 row samples. The LET command calculates the $t^\star$ for each of the 100 samples. The last four commands construct histograms of the 100 sample means and the 100 $t^\star$-values, using the subcommand CUTPOINT and class boundaries starting at the mean and going in both directions by approximately "half" standard deviations. (See Helpful Hint on page 89.)

```
NORMAL POPULATION

RANDom 1000 C1;
  NORMal 100 50.
HISTogram C1;
  CUTPoint -100:300/25.
RANDom 100 C2-C11;
  NORMal 100 50.
RMEAn C2-C11 C12
RSTDev C2-C11 C13
# calculate t* for each x-bar
LET C14 = (C12-100)/(C13/SQRT(10))
HISTogram C12;
  CUTPoints 52:148/8.
HISTogram C14;
  CUTPoints -5:5/0.5.
```

```
RECTANGULAR POPULATION

RANDom 1000 C1;
  UNIForm 0 200.
HISTogram C1;
  CUTPoint 0:200/25.
RANDom 100 C2-C11;
  UNIForm 0 200.
RMEAn C2-C11 C12
RSTDev C2-C11 C13
# calculate t* for each x-bar
LET C14 = (C12-100)/(C13/SQRT(10))
HISTogram C12;
  CUTPoints 44:156/8.
HISTogram C14;
  CUTPoints -5:5/0.5.
```

```
SKEWED POPULATION

RANDom 1000 C1;
  EXPO 100.
HISTogram C1;
  CUTPoint 0:700/25.
RANDom 100 C2-C11;
  EXPO 100.
RMEAn C2-C11 C12
RSTDev C2-C11 C13
# calculate t* for each x-bar
LET C14 = (C12-100)/(C13/SQRT(10))
HISTogram C12;
  CUTPoints 20:180/8.
HISTogram C14;
  CUTPoints -5:5/0.5.
```

For the samples from the normal population:
   a. Does the x-bar distribution appear to be normal? Find percentages for intervals and compare to the normal distribution.
   b. Does the $t^\star$-distribution appear to have a t-distribution with df $= 9$? Find percentages for intervals and compare them to the t-distribution.

For the samples from the rectangular population:
   c. Does the x-bar distribution appear to be normal? Find percentages for intervals and compare them to the normal distribution.
   d. Does the $t^\star$- distribution appear to have a t-distribution with df $= 9$? Find percentages for intervals and compare them to the t-distribution.

For the samples from the skewed population:
   e. Does the x-bar distribution appear to be normal? Find percentages for intervals and compare them to the normal distribution.
   f. Does the $t^\star$-distribution appear to have a t-distribution with df $= 9$? Find percentages for intervals and compare them to the t-distribution.

In summary:
   g. In each of the preceding three situations, the sampling distribution for x-bar appears to be slightly different than the $t^\star$-distribution. Explain why.
   h. Does the normality condition appear to be necessary in order for the calculated test statistic $t^\star$ to have a Student's t-distribution? Explain.

9.2 | INFERENCES ABOUT THE BINOMIAL PROBABILITY OF SUCCESS

Perhaps the most common inference of all is an inference involving the binomial parameter p, the "probability of success." Yes, every one of us uses this inference, even if only very casually. There are thousands of examples of situations in which we are concerned about something either "happening" or "not happening." There are only two possible outcomes of concern, and that is the fundamental property of a binomial experiment. The other needed ingredient is for multiple independent trials to exist. Asking 5 people whether they are "for" or "against" some issue can create 5 independent trials; if 200 people are asked the same question, 200 independent trials may be involved; if 30 items are inspected to see if each "exhibits a particular property" or "not," there will be 30 repeated trials—these are the makings of a binomial inference.

Binomial parameter p
Observed or sample binomial probability (p')

The **binomial parameter p** is defined to be the probability of success on a single trial in a binomial experiment. We define p', **the observed or sample binomial probability**, to be

$$p' = \frac{x}{n} \tag{9-3}$$

where the random variable x represents the number of successes that occur in a sample consisting of n trials. Also recall that the mean and standard deviation of the binomial random variable x are found by using formula (5-7), $\mu = np$, and formula (5-8), $\sigma = \sqrt{npq}$, where $q = 1 - p$. The distribution of x is considered to be approximately normal if n is larger than 20 and if np and nq are both larger than 5. This commonly accepted *rule of thumb* allows us to use the normal distribution to estimate probabilities for the binomial random variable x, the number of successes in n trials, and to make inferences concerning the binomial parameter p, the probability of success while performing an individual trial.

Generally, it is easier and more meaningful to work with the distribution of p' (the observed probability of occurrence) than with x (the number of occurrences). Consequently, we will convert formulas (5-7) and (5-8) from the units of x (integers) to units of proportions (percentages expressed as decimals) by dividing each formula by n, as shown in Table 9-1.

TABLE 9-1

Formulas (9-4) and (9-5)

Variable	Mean		Standard Deviation	
x	$\mu_x = np$	(5-7)	$\sigma_x = \sqrt{npq}$	(5-8)
$\dfrac{x}{n}$	$\dfrac{np}{n}$		$\dfrac{\sqrt{npq}}{n}$	
p'	$\mu_{p'} = p$	(9-4)	$\sigma_{p'} = \sqrt{\dfrac{pq}{n}}$	(9-5)

EXERCISE   9.43

a. Does it seem reasonable that the mean of the sampling distribution of observed values of p' should be p, the true proportion? Explain.

b. Explain why p' is an unbiased estimator for the population p.

EXERCISE   9.44

Show that $(\sqrt{npq})/n$ simplifies to $\sqrt{(pq)/n}$.

The standard deviation of a sampling distribution is called "standard error."

Since $\mu_{p'} = p$, the sample statistic p' is an unbiased estimator for p.

The information about the sampling distribution of p' is summarized as follows:

> If a sample of size n is randomly selected from a large population with $p = P(\text{success})$, then the sampling distribution of p' has
>
> 1. a mean $\mu_{p'}$ equal to p,
> 2. a standard error $\sigma_{p'}$ equal to $\sqrt{(pq)/n}$, and
> 3. an approximately normal distribution if n is sufficiently large.

In practice, use of the following guidelines will ensure normality:

1. The sample size is greater than 20.
2. The sample consists of less than 10% of the population.
3. The products np and nq are both larger than 5.

> **The assumptions for inferences about the binomial parameter p:** The n random observations forming the sample are selected independently from a population that is not changing during the sampling.

Confidence Interval Procedure

Inferences concerning the population binomial parameter p, $P(\text{success})$, are made using procedures that closely parallel the inference procedures for the population mean μ. When we estimate the population proportion p, we will base our estimations on the unbiased sample statistic p'. The point estimate, p', becomes the center of the confidence interval, and the maximum error of estimate is a multiple of the standard error. The level of confidence determines the confidence coefficient, the number of multiples of the standard error.

$$p' - z\left(\frac{\alpha}{2}\right) \cdot \sqrt{\frac{p'q'}{n}} \quad \text{to} \quad p' + z\left(\frac{\alpha}{2}\right) \cdot \sqrt{\frac{p'q'}{n}} \tag{9-6}$$

where $p' = x/n$ and $q' = 1 - p'$. Notice that the standard error, $\sqrt{(pq)/n}$, has been replaced by $\sqrt{(p'q')/n}$. Since we are estimating p, we do not know its value and therefore must use the best replacement available. That replacement is p', the observed value or the point estimate for p. This replacement will cause little change in the standard error or the width of our confidence interval provided n is sufficiently large.

▼ | ILLUSTRATION   9 - 10

While talking about the cars that fellow students drive, several statements were made about types, ages, makes, colors, and so on. Dana decided he wanted to estimate the proportion of convertibles students drove, so he randomly identified 200 cars in the student parking lot, of which he found 17 to be convertibles. Find the 90% confidence interval for the proportion of convertibles driven by students.

The four-step confidence interval model is on page 367.

SOLUTION

STEP 1   Describe the population parameter of concern: The parameter of interest is the proportion (percentage) of convertibles driven by students.

STEP 2   Specify the confidence interval criteria.
     a. Check the assumptions: The sample was randomly selected, and each subject's response was independent of those of the others surveyed.
     b. Determine the test statistic to be used: z will be used as the test statistic. p' is expected to be approximately normal since $n = 200$ is greater than 20 and both np [approximated by $np' = 200(17/200) = 17$] and nq [approximated by $nq' = 200(183/200) = 183$] are larger than 5.
     c. State the level of confidence, $1 - \alpha$: $1 - \alpha = \textbf{0.90}$.

STEP 3   Collect and present sample evidence.
     a. Collect the sample information: $n = 200$ cars were identified, and $x = 17$ were convertibles.
     b. Find the point estimate: $p' = x/n = 17/200 = \textbf{0.085}$.

STEP 4   Determine the confidence interval limits.
     a. Determine the confidence coefficients: This is the z-score identifying the number of standard errors needed to attain the level of confidence and is found using Table 4 in Appendix B, $z(\alpha/2) = z(0.05) = \textbf{1.65}$ (see diagram.)

EXERCISE 9.45

Find the 95% confidence interval for proportion of convertibles in Illustration 9-10 based on a sample of $n = 400$ with $x = 92$.

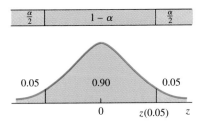

     b. The maximum error of estimate: Using the maximum error part of formula (9-6),

$$E = z\left(\frac{\alpha}{2}\right) \cdot \sqrt{\frac{p'q'}{n}} = 1.65\sqrt{\frac{(0.085)(0.915)}{200}} = (1.65)\sqrt{0.000389} = (1.65)(0.020) = \textbf{0.033}$$

     c. Find the lower and upper confidence limits:

$$p' - E \quad \text{to} \quad p' + E$$
$$0.085 - 0.033 \quad \text{to} \quad 0.085 + 0.033$$
$$\textbf{0.052} \quad \textbf{to} \quad \textbf{0.118}$$

d. Results: **0.052 to 0.118, 90% confidence interval for p = P(drives convertible)**.

That is, the true proportion of students who drive convertibles is between 0.052 and 0.118, with 90% confidence.

By using the maximum-error part of the confidence interval formula, it is possible to determine the size of the sample that must be taken in order to estimate p with a desired accuracy. The **maximum error of estimate** for a proportion is

Maximum error of estimate

$$E = z\left(\frac{\alpha}{2}\right) \cdot \sqrt{\frac{pq}{n}} \qquad\qquad \textbf{(9-7)}$$

In order to determine the sample size from this formula, we must decide on the quality we want for our final confidence interval. This quality is measured in two ways: the level of confidence and the preciseness (narrowness) of the interval. The level of confidence we establish will in turn determine the confidence coefficient, $z(\alpha/2)$. The desired preciseness will determine the maximum error of estimate, E. (Remember that we are estimating p, the binomial probability; therefore, E will typically be expressed in hundredths.)

Remember:
$q = 1 - p$.

For ease of use, formula (9-7) can be solved for n as follows:

$$n = \frac{[z(\alpha/2)]^2 \cdot p^* \cdot q^*}{E^2} \qquad\qquad \textbf{(9-8)}$$

where p^* and q^* are provisional values of p and q used for planning.

By inspecting formula (9-8), we can observe that there are three components determining the sample size: (1) level of confidence ($1 - \alpha$, which in turn determines the confidence coefficient), (2) the provisional value of p (p^* determines the value of q^*), and (3) the maximum error, E. An increase or decrease in one of these three components affects the sample size. If the level of confidence is increased or decreased (while the other components are held constant), then the sample size will increase or decrease, respectively. If the product of p^* and q^* is increased or decreased (other components held constant), then the sample size will increase or decrease, respectively. (The product p^*q^* is largest when $p^* = 0.5$ and decreases as the value of p^* becomes further from 0.5.) An increase or decrease in the desired maximum error will have the opposite effect on the sample size since E appears in the denominator of the formula. If no provisional values for p and q are available, then use $p^* = 0.5$ and $q^* = 0.5$. Using $p^* = 0.5$ is safe because it gives the largest sample size of any possible value of p. Using $p^* = 0.5$ works reasonably well when the true value is "near 0.5" (say, between 0.3 and 0.7); however, as p gets nearer to either 0 or 1, a sizable overestimate in sample size will occur.

▼ | ILLUSTRATION   9 - 11

Determine the sample size that is required to estimate the true proportion of blue-eyed community college students if you want your estimate to be within 0.02 with 90% confidence.

SOLUTION

STEP 1   The level of confidence is $1 - \alpha = 0.90$; therefore, the confidence coefficient is $z(\alpha/2) = z(0.05) = \mathbf{1.65}$ from Table 4 in Appendix B; see diagram.

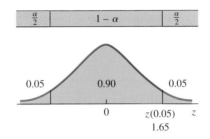

STEP 2   The desired maximum error is $E = 0.02$.

STEP 3   Since no estimate was given for p, use $p^* = 0.5$, and $q^* = 1 - p^* = 0.5$.

STEP 4   Use formula (9-8) to find n:

$$n = \frac{[z(\alpha/2)]^2 \cdot p^* \cdot q^*}{E^2} = \frac{(1.65)^2 \cdot (0.5) \cdot (0.5)}{(0.02)^2} = \frac{0.680625}{0.0004} = 1701.56 = \mathbf{1702}$$

▼ | ILLUSTRATION   9 - 12

EXERCISE   9.46

Find the sample size n needed for a 95% interval estimate in Illustration 9-11.

An automobile manufacturer purchases bolts from a supplier who claims the bolts to be approximately 5% defective. Determine the sample size that will be required to estimate the true proportion of defective bolts if we want our estimate to be within ± 0.02 with 90% confidence.

SOLUTION

STEP 1   The level of confidence is $1 - \alpha = 0.90$; the confidence coefficient is

$$z\left(\frac{\alpha}{2}\right) = z(0.05) = \mathbf{1.65}$$

Always round up to the next larger integer no matter how small the decimal.

STEP 2   The desired maximum error is $E = 0.02$.

STEP 3   Since there is an estimate for p (supplier's claim is "5% defective"), use $p^* = 0.05$ and $q^* = 1 - p^* = 0.95$.

STEP 4   Use formula (9-8) to find n:

$$n = \frac{[z(\alpha/2)]^2}{E^2} = \frac{(1.65)^2 \cdot (0.05) \cdot (0.95)}{(0.02)^2} = \frac{0.12931875}{0.0004} = 323.3 = \mathbf{324}$$

EXERCISE   9.47

Find *n* for a 90% interval estimate of
p with $E = 0.02$
using an estimate of
$p = 0.25$.

Notice the difference in the sample size required in Illustrations 9-11 and 9-12. The only mathematical difference between the problems is the value that was used for $p*$. In Illustration 9-11 we used $p* = 0.5$, and in Illustration 9-12 we used $p* = 0.05$. Recall that the use of the provisional value $p* = 0.5$ gives the maximum sample size. As you can see, it will be an advantage to have an indication of the value expected for p, especially as p becomes increasingly further from 0.5.

Hypothesis-Testing Procedure

When the binomial parameter p is to be tested using a hypothesis-testing procedure, we will use a test statistic that represents the difference between the observed proportion and the hypothesized proportion divided by the standard error. This test statistic is assumed to be normally distributed when the null hypothesis is true, when the assumptions for the test have been satisfied, and when n is sufficiently large ($n > 20$, $np > 5$, and $nq > 5$).

The value of the test statistic $z^\star$ is calculated using formula (9-9):

$$z^\star = \frac{p' - p}{\sqrt{\dfrac{pq}{n}}} \quad \text{where } p' = \frac{x}{n} \qquad \textbf{(9-9)}$$

Probability-Value Approach

▼ ILLUSTRATION   9 - 13

Many people sleep-in on the weekends to make up for "short nights" during the work week. The Better Sleep Council reports that 61% of us get more than seven hours of sleep per night on the weekend. A random sample of 350 adults found that 235 had more than seven hours each night last weekend. At the 0.05 level of significance, does this evidence show that more than 61% get seven or more hours per night on the weekend?

SOLUTION

STEP 1   Describe the parameter of concern: The proportion of adults who get more than seven hours of sleep per night on weekends.

STEP 2   State the null and alternative hypotheses:

H_0: $p = P(7 +$ hours of sleep$) = 0.61$ (no more than 61%)

H_a: $p > 0.61$ (more than 61%)

STEP 3   Specify the test criteria.
a. Assumptions: The random sample of 350 adults were independently surveyed.

b. The test statistic: The standard normal z will be used. Since $n = 350$ is larger than 20 and both $np = (350)(0.61) = 213.5$ and $nq = (350)(0.39) = 136.5$ are larger than 5, p' is expected to be normally distributed.
c. Level of significance: $\alpha = 0.05$.

STEP 4   Collect and present the sample evidence.
a. Sample information: $n = 350$ and $x = 235$; $p' = x/n = 235/350 = $ **0.671**.
b. Calculate the test statistic: Using formula (9-9),

$$z^\star = \frac{p' - p}{\sqrt{\dfrac{pq}{n}}}; \quad z^\star = \frac{0.671 - 0.61}{\sqrt{\dfrac{(0.61)(0.39)}{350}}} = \frac{0.061}{\sqrt{0.0006797}} = \frac{0.061}{0.0261} = \mathbf{2.34}$$

c. Calculate the p-value: p-value $= P(z > 2.34)$.

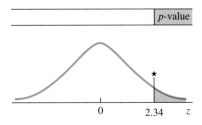

There are three ways to find the p-value: (1) Use Table 3 in Appendix B and calculate it; (2) use Table 5 in Appendix B and look it up; or (3) use a computer and have it calculated.

Method 1: Use Table 3 in Appendix B:

$$\text{p-value} = P(z > 2.34) = 0.5000 - 0.4904 = \mathbf{0.0096}$$

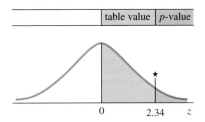

Method 2: Use Table 5 in Appendix B: Simply look up $z^\star$ and read the p-value from Table 5. If the $z^\star$ is between table values, place bounds on the p-value.

$$\mathbf{0.0094} < \text{p-value} < \mathbf{0.0107}$$

Portion of Table 5	
$z^\star$	p-value
2.30	0.0107
2.34	p-value
2.35	0.0094

Method 3: Use a computer and the cumulative distribution function (CDF): MINITAB session commands **CDF 2.34**; and **NORMal 0.0 1.0**. will produce the following results.

```
Cumulative Distribution Function
Normal with mean = 0 and stand. dev. = 1.00

   x         P(X <= x)
2.3400        0.9904
```

cumulative probability	p-value

p-value = $1 - 0.9904 = 0.0096$

EXERCISE 9.48

Test H_0: $p = 0.50$ against H_a: p < 0.50 using the p-value approach, $\alpha = 0.05$, and the sample $n = 250$ and $x = 113$.

STEP 5 Determine the results: Compare the p-value to α. The p-value is smaller than α.

DECISION Reject H_0.

CONCLUSION There is sufficient reason to conclude that the proportion of adults in the sampled population who are getting more than seven hours of sleep nightly on weekends is significantly higher than 61%.

▲

Classical Procedure

▼ ILLUSTRATION 9 - 14

While talking about the cars that fellow students drive (Illustration 9-10), Tom made the claim that 15% of the students drive convertibles. Jody finds this hard to believe and she wants to check the validity of Tom's claim, using Dana's random sample. At a level of significance of 0.10, is there sufficient evidence to reject Tom's claim if there were 17 convertibles in his sample of 200 cars?

SOLUTION

STEP 1 Describe the parameter of concern: $p = P$(student drives convertible).

EXERCISE 9.49

Test H_0: $p = 0.70$
against H_a: $p > 0.70$
using the classical
approach, $\alpha = 0.05$,
and the sample
$n = 300$ and
$x = 224$.

EXERCISE 9.50

Show that the
hypothesis test
completed as Illustra-
tion 9-14 was un-
necessary since the
confidence interval
had already been
completed in Illustra-
tion 9-10.

STEP 2   State the null and alternative hypotheses:

H_0: $p = 0.15$ (15% do drive convertibles)

H_a: $p \neq 0.15$ (the percent is different than 15)

STEP 3   Specify the test criteria.
   a. Check the assumptions: The sample was randomly selected, and each subject's response was independent of other responses.
   b. Test statistic to be used: The standard normal z will be used. Since $n = 200$ is larger than 20 and both np and nq are larger than 5, p' is expected to be normally distributed.
   c. The level of significance: $\alpha = 0.10$.
   d. Determine the critical region(s) and critical value(s): This test is two-tailed; therefore, $\alpha = 0.10$ will be split equally, 0.05 in each tail. $z(0.05) = 1.65$. See the figure and Table 4 in Appendix B.

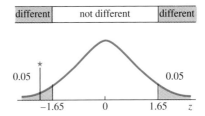

STEP 4   Collect and present the sample evidence.
   a. Collect the sample information: $n = 200$, $x = 17$, $p' = x/n = 17/200 = $ **0.085**
   b. Calculate the value of the test statistic:

$$z^\star = \frac{p' - p}{\sqrt{\dfrac{pq}{n}}}: \quad z^\star = \frac{0.085 - 0.150}{\sqrt{[(0.15)(0.85)]/200}} = \frac{-0.065}{\sqrt{0.00064}} = \frac{-0.065}{0.02525} = \mathbf{-2.57}$$

STEP 5   Determine the results: Compare the calculated test statistic to the critical region. $z^\star$ is in the critical region (shown in red on figure above).

DECISION Reject H_0.

CONCLUSION There is sufficient evidence to reject Tom's claim and conclude that the percentage of students who drive convertibles is significantly different than 15%.

There is a relationship between confidence intervals and two-tailed hypothesis tests when the level of confidence and the level of significance add up to 1. The confidence coefficients and the critical values are the same, meaning the width of

Case Study   The Methods Behind the Polling Madness

9.2

EXERCISE 9.51

a. Find the standard error of proportion when $n = 751$ using $p^* = 0.5$.

b. What level of confidence is related to the sampling error of 0.036 reported in Case Study 9-2?

The "sampling error" graph shows the margin of error that results from using samples of various sizes. It specifically points out that a sample of size $n = 751$ yields a 3.6% (or 0.036) sampling error. This sampling error is also called the "maximum error" of estimate and is a multiple of the standard error of proportion.

The "applying sampling error" chart shows how the estimates we read as headlines can be interpreted and also helps explain why the various polls seem to report what appear at first glance to be inconsistent findings. Notice that when the sampling error of 3.6% is applied to the 44%, the proportion of voters favoring Clinton on that day could have been any percentage from 40.4% to 47.6% and for Bush the percentage could have been any value from 29.4% to 36.6%; thus, the margin Clinton had that day could have been anywhere from 3.8% to 18.2%. If you had taken your own random sample of 751 voters on that day, do you think your sample results would have been the same 33% and 44%?

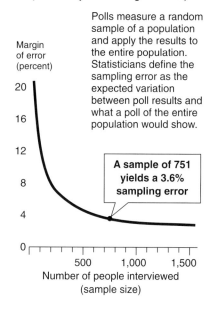

Sampling error
(Commonly called margin of error)

Polls measure a random sample of a population and apply the results to the entire population. Statisticians define the sampling error as the expected variation between poll results and what a poll of the entire population would show.

A sample of 751 yields a 3.6% sampling error

Margin of error (percent)

Number of people interviewed (sample size)

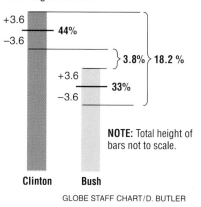

Applying sampling error

In a *Wall Street Journal*/ NBC poll of 751 registered voters released Thursday, 44% preferred Clinton, 33% preferred Bush, and 17% were for Perot. Assuming there were no other sources of error, Clinton's lead over Bush could be as small as 3.8% or as big as 18.2%.

+3.6
−3.6
44%

3.8% } 18.2 %

+3.6
−3.6
33%

NOTE: Total height of bars not to scale.

Clinton   Bush

GLOBE STAFF CHART/D. BUTLER

Source: The Boston Globe, November 2, 1992.

the confidence interval and the width of the noncritical region are the same. The point estimate is the center of the confidence interval, and the hypothesized mean is the center of the noncritical region. Therefore, if the hypothesized value of p is contained in the confidence interval, then the test statistic will be in the noncritical

Case Study

9.3

EXERCISE 9.52

a. Find the standard error of proportion that Mr. Brown used in Case Study 9-3.
b. Explain why the null hypothesis would be $p = 0.00$.
c. Is the alternative hypothesis $p > 0$ or $p < 0$? Explain.
d. Using the point estimate and the standard error found in (a), calculate the test statistic $z^\star$.
e. What do you find for a p-value?
f. Calculate the 95% confidence interval estimate for p; the 98% interval estimate.
g. Explain how it is possible for these intervals to show a reduction in p due to passive smoke exposure.

Statisticians Occupy Front Lines in Battles Over Passive Smoking

In the controversy over passive smoking, the difference between 90% and 95% has become a matter of life and death.

The U.S. Environmental Protection Agency says there is a 90% probability that the risk of lung cancer for passive smokers is somewhere between 4% and 35% higher than for those who aren't exposed to environmental smoke. To statisticians, this calculation is called the "90% confidence interval."

And that, say tobacco-company statisticians, is the rub. "Ninety-nine percent of all epidemiological studies use a 95% confidence interval," says Gio B. Gori.

These five percentage points will haunt the coming battle in a North Carolina courtroom where tobacco interests have sued the EPA.

When statisticians on both sides go at it, calculator-to-calculator, in the coming trial, they will present a series of arcane arguments about how these unknowns affect the study's reliability.

When the 19%-higher-risk figure was first calculated, a statistical test determined its "statistical significance," that is, the odds that the answer was the result of chance instead of reality, explains Kenneth G. Brown, an independent statistician who did the risk calculations.

This latter calculation showed that there were only two chances out of 100—a probability of 0.02—that the 19% figure was a matter of happenstance. This more than meets the standard of 0.05, at which most scientific studies are considered statistically significant.

Mr. Brown says that it was during the reviews of the final drafts that this controversial 90% confidence interval was added. The reason the EPA didn't use the standard 95% confidence interval, Dr. Gori says, is that it would be so wide it might even hint that passive smoking actually reduced the risk of cancer.

Source: The Wall Street Journal, July 23, 1993, by Jerry E. Bishop.

Relating the numbers in this article to the terms studied in this section, the 19% is the point estimate, the 4% is the lower limit, and the 35% is the upper limit to the 90% confidence interval estimate for a parameter p. The probability of 0.02 that Mr. Brown mentioned appears to be a p-value for a hypothesis test. Most likely, the null hypothesis was p = 0.0 and the alternative was one-tailed. Dr. Gori's comment about the confidence interval being so wide that it might hint at a reduction seems rather flip. What was meant by it?

region (see Figure 9-5). Further, if the hypothesized probability p does not fall within the confidence interval, then the test statistic will be in the critical region (see Figure 9-6). This comparison should be used only when the hypothesis test is two-tailed and when the same value of α is involved for both procedures.

FIGURE 9-5

Confidence Interval Contains p

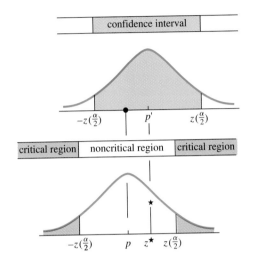

FIGURE 9.6

Confidence Interval Does Not Contain p

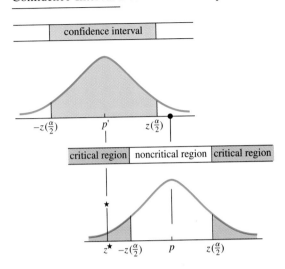

E XERCISES

9.53 Forty-five (45) of the 150 elements in a random sample are classified as "success."
   a. Explain why x and n are assigned the values 45 and 150, respectively.
   b. Determine the value of p'. Explain how p' is found and the meaning of p'.

9.54 For each of the following situations, find p'.
   a. $x = 24$ and $n = 250$
   b. $x = 640$ and $n = 2050$
   c. 892 of 1280 responded "yes"

9.55 a. What is the relationship between $p = P(\text{success})$ and $q = P(\text{failure})$? Explain.
   b. Explain why the relationship between p and q can be expressed by the formula $q = 1 - p$.
   c. If $p = 0.6$, what is the value of q?
   d. If the value of $q' = 0.273$, what is the value of p'?

9.56 Find (1) α, (2) area of one-tail, and (3) the confidence coefficients of z that are used with each of the following levels of confidence.
   a. $1 - \alpha = 0.90$ b. $1 - \alpha = 0.95$ c. $1 - \alpha = 0.98$ d. $1 - \alpha = 0.99$

9.57 A telephone survey was conducted to estimate the proportion of households with a personal computer. Of the 350 households surveyed, 75 had a personal computer.
   a. Give a point estimate for the proportion in the population who have a personal computer.
   b. Give the maximum error of estimate with 95% confidence.

9.58   A bank randomly selected 250 checking-account customers and found that 110 of them also had savings accounts at this same bank. Construct a 95% confidence interval for the true proportion of checking-account customers who also have savings accounts.

9.59   In a sample of 60 randomly selected students, only 22 favored the amount being budgeted for next year's intramural and interscholastic sports. Construct the 99% confidence interval for the proportion of all students who support the proposed budget amount.

9.60   *USA Today* (2-27-95) reported on a poll of 750 children and teenagers, aged 10 to 16. Two of the findings reported were: "4 out of 5 kids say Entertainment television should teach youngsters right from wrong," and "2 out of 3 say shows like *The Simpsons* and *Married . . . With Children* encourage kids to disrespect their parents." The poll (margin of error: plus/minus 3%) was sponsored by Children Now, an advocacy group.
  a. Find the point estimate, the maximum error of estimate, and the 95% confidence interval that results from the "4 out of 5" sample summary.
  b. Find the point estimate, the maximum error of estimate, and the 95% confidence interval that results from the "2 out of 3" sample summary.
  c. Compare the maximum error calculated in parts (a) and (b) to the margin of error mentioned in the article. What level of confidence do you believe the margin of error "plus/minus 3%" represents? Explain.
  d. In Case Study 9-2, the sampling error for a sample of 751 is reported as being 3.6%. The article mentioned above reports that a sample of size 750 has a margin of error of 3%. Explain why these two polls have different sampling errors. Assume both are using 95% confidence.

9.61   An article titled "Why Don't Women Buy CDs?" appeared in the September 1994 issue of *Music* magazine. Yehuda Shapiro, marketing director of Virgin Retail Europe, found that across Europe 40% of his customers who buy classical records are women. Determine a 90% confidence interval for the true value of p if the 40% estimate is based on 1000 randomly selected buyers.

9.62   An article in *World Health Statistics Quarterly* (Vol. 45, 1992) discusses Chagas disease. According to the article, "There are two stages of the human disease: the acute stage, which appears shortly after the infection, and the chronic stage, which may last several years; the latter irreversibly affects internal organs namely the heart, esophagus and colon, and the peripheral nervous system." In Brasilia, Brazil, 14.6% of 2413 samples tested from blood banks were positive for the parasite *T. cruzi*, which causes Chagas disease. Find a 99% confidence interval for p, the proportion who would test positive for *T. cruzi* in all blood banks in Brasilia.

9.63   "Parents should spank children when they think it is necessary, said 51% of adult respondents to a survey—though most child-development experts say spanking is not appropriate. The survey of 7225 adults . . . was co-sponsored by *Working Mother* magazine and Epcot Center at Walt Disney World." This statement appeared in the Rochester *Democrat & Chronicle* (12-20-90). Find the 99% confidence maximum error of estimate for the parameter p, P(should spank when necessary), for the adult population.

9.64   In a survey of 12,000 adults aged 19 to 74, National Cancer Institute researchers found that 9% in the survey ate at least the recommended two servings of fruit or juice and three servings of vegetables per day (*Ladies Home Journal*, April 1991). Use this information to determine a 95% confidence interval for the true proportion in the population who follow the recommendation.

9.65   Construct 90% confidence intervals for the binomial parameter p for each of the following pairs of values. Write your answers on the chart.

Observed Proportion $p' = x/n$	Sample Size	Lower Limit	Upper Limit
a. $p' = 0.3$	$n = 30$		
b. $p' = 0.7$	$n = 30$		
c. $p' = 0.5$	$n = 10$		
d. $p' = 0.5$	$n = 100$		
e. $p' = 0.5$	$n = 1000$		

    f. Explain the relationship between answers (a) and (b).
    g. Explain the relationship between answers (c), (d), and (e).

9.66   a. Calculate the maximum error of estimate for p for the 95% confidence interval for each of the situations listed in the table.

	Approximate Value of p				
Sample Size n	**0.1**	**0.3**	**0.5**	**0.7**	**0.9**
100					
500					
1000					
1500					

    b. Explain the relationship between answers in column 0.1 and 0.9; 0.3 and 0.7.
    c. Explain the relationship between answers in column 0.5 and the values that can be read from the sampling-error graph shown in Case Study 9-2 on page 463.

9.67   Karl Pearson once tossed a coin 24,000 times and recorded 12,012 heads.
    a. Calculate the point estimate for $p = P(\text{head})$ based on Pearson's results.
    b. Determine the standard error of proportion.
    c. Determine the 95% confidence interval estimate for $p = P(\text{head})$.
    d. It must have taken Mr. Pearson many hours to toss a coin 24,000 times. You can simulate 24,000 coin tosses using the MINITAB commands listed below.

(*Note:* A Bernoulli experiment is like a "single" trial binomial experiment. That is, one toss of a coin is one Bernoulli experiment with $p = 0.5$; and 24,000 tosses of a coin either is a binomial experiment with $n = 24,000$ or is 24,000 Bernoulli experiments. Code: $0 = $ tail, $1 = $ head. The sum of the 1's will be the number of heads in the 24,000 tosses.)

```
RANDOM 24000 C1;
  BERNoulli 0.5.
SUM C1 K1
LET K2 = K1/24000
PRINt K2
```

    e. How do your simulated results compare to Pearson's?
    f. Use these six commands and generate another set of 24,000 coin tosses. Compare these results to those above. Explain what you can conclude from these results.

9.68 The "rule of thumb" stated on page 454 indicated that we could expect the sampling distribution of p' to be approximately normal when "$n > 20$ and both np and nq were greater than 5." What happens when these guidelines are not followed?

    a. Use the following set of MINITAB commands and your computer will show you what happens. Try $n = 15$ and $p = 0.1$ ($K1 = n$ and $K2 = p$). Do the distributions look normal? Explain what causes the "gaps." Why do the histograms look alike? Try some different combinations of n ($K1$) and p ($K2$):

```
LET K1 = 15
LET K2 = 0.1
RANDom 1000 C1;
  BINOmial K1 K2.
LET C2 = C1/K1
LET C3 = (C2 - K2)/SQRT(K2*(1-K2)/K1)
HISTogram C2
HISTogram C3
```

    b. Try $n = 15$ and $p = 0.01$.
    c. Try $n = 50$ and $p = 0.03$.
    d. Try $n = 20$ and $p = 0.2$.
    e. Try $n = 20$ and $p = 0.8$.
    f. What happens when the rule of thumb is not followed? (See Helpful Hint on page 89.)

9.69 According to the June 1994 issue of *Bicycling*, only 16% of all bicyclists own helmets. You wish to conduct a survey in your city to determine what percent of the bicyclists own helmets. Use the national figure of 16% for your initial estimate of p.

    a. Find the sample size if you want your estimate to be within 0.02 with 90% confidence.
    b. Find the sample size if you want your estimate to be within 0.04 with 90% confidence.
    c. Find the sample size if you want your estimate to be within 0.02 with 98% confidence.
    d. What effect does changing the level of confidence have on the sample size? Explain.
    e. What effect does changing the maximum error have on the sample size? Explain.

9.70 A bank believes that approximately $\frac{2}{5}$ of its checking-account customers have used at least one other service provided by the bank within the last six months. How large a sample will be needed to estimate the true proportion to within 5% at the 98% level of confidence?

9.71 A hospital administrator wants to conduct a telephone survey to determine the proportion of people in a city who have been hospitalized for at least three days in the past five years. How large a sample must she take to be 95% confident that the sample proportion will be within 0.03 of the true proportion of the city?

9.72 According to the May 1990 issue of *Good Housekeeping*, only about 14% of lung cancer patients survive for five years after diagnosis. Suppose you wanted to see if this survival rate were still true. How large a sample would you need to take to estimate the true proportion surviving for five years after diagnosis to within 1% with 95% confidence? (Use the 14% as the value of p.)

9.73 State the null hypothesis, H_0, and the alternative hypothesis, H_a, that would be used to test these claims:

    a. More than 60% of all students at our college work part-time jobs during the academic year.
    b. The probability of our team winning tonight is less than 0.50.
    c. No more than one-third of cigarette smokers are interested in quitting.
    d. At least 50% of all parents believe in spanking their children when appropriate.
    e. A majority of the voters will vote for the school budget this year.
    f. At least three-quarters of the trees in our county were seriously damaged by the storm.
    g. The results show the coin was not tossed fairly.
    h. The single-digit numbers generated by the computer do not seem to be equally likely with regard to being odd or even.

9.74   Determine the test criteria that would be used to test the following hypotheses when z is used as the test statistic and the classical approach is used.
   a. $H_0: p = 0.5$ and $H_a: p > 0.5$, with $\alpha = 0.05$
   b. $H_0: p = 0.5$ and $H_a: p \neq 0.5$, with $\alpha = 0.05$
   c. $H_0: p = 0.4$ and $H_a: p < 0.4$, with $\alpha = 0.10$
   d. $H_0: p = 0.7$ and $H_a: p > 0.7$, with $\alpha = 0.01$

9.75   Determine the p-value for each of the following hypothesis-testing situations.
   a. $H_0: p = 0.5, H_a: p \neq 0.5, z^\star = 1.48$
   b. $H_0: p = 0.7, H_a: p \neq 0.7, z^\star = -2.26$
   c. $H_0: p = 0.4, H_a: p > 0.4, z^\star = 0.98$
   d. $H_0: p = 0.2, H_a: p < 0.2, z^\star = -1.59$

9.76   The binomial random variable, x, may be used as the test statistic when testing hypotheses about the binomial parameter, p, when n is small (say, 15 or less). (Use Table 2 in Appendix B and determine the p-value for each of the following situations.
   a. $H_0: p = 0.5, H_a: p \neq 0.5$, where $n = 15$ and $x = 12$
   b. $H_0: p = 0.8, H_a: p \neq 0.8$, where $n = 12$ and $x = 4$
   c. $H_0: p = 0.3, H_a: p > 0.3$, where $n = 14$ and $x = 7$
   d. $H_0: p = 0.9, H_a: p < 0.9$, where $n = 13$ and $x = 9$

9.77   The binomial random variable, x, may be used as the test statistic when testing hypotheses about the binomial parameter, p. When n is small (say, 15 or less), Table 2 in Appendix B provides the probabilities for each value of x separately, thereby making it unnecessary to estimate probabilities of the discrete binomial random variable with the continuous standard normal variable z. Use Table 2 and determine the value of α for each of the following:
   a. $H_0: p = 0.5$ and $H_a: p > 0.5$, where $n = 15$ and the critical region is $x = 12, 13, 14, 15$
   b. $H_0: p = 0.3$ and $H_a: p < 0.3$, where $n = 12$ and the critical region is $x = 0, 1$
   c. $H_0: p = 0.6$ and $H_a: p \neq 0.6$, where $n = 10$ and the critical region is $x = 0, 1, 2, 3, 9, 10$
   d. $H_0: p = 0.05$ and $H_a: p > 0.05$, where $n = 14$ and the critical region is $x = 4, 5, 6, 7, \ldots, 14$

9.78   Use Table 2 in Appendix B and determine the critical region used in testing each of the following hypotheses. (*Note:* Since x is discrete, choose critical regions that do not exceed the value of α given.)
   a. $H_0: p = 0.5$ and $H_a: p > 0.5$, where $n = 15$ and $\alpha = 0.05$
   b. $H_0: p = 0.5$ and $H_a: p \neq 0.5$, where $n = 14$ and $\alpha = 0.05$
   c. $H_0: p = 0.4$ and $H_a: p < 0.4$, where $n = 10$ and $\alpha = 0.10$
   d. $H_0: p = 0.7$ and $H_a: p > 0.7$, where $n = 13$ and $\alpha = 0.01$

9.79   You are testing the hypothesis $p = 0.7$ and have decided to reject this hypothesis if after 15 trials you observe 14 or more successes.
   a. If the null hypothesis is true and you observe 13 successes, then which of the following will you do? (1) Correctly fail to reject H_0? (2) Correctly reject H_0? (3) Commit a type I error? (4) Commit a type II error?
   b. Find the significance level of your test.
   c. If the true probability of success is $\frac{1}{2}$ and you observe 13 successes, then which of the following will you do: (1) correctly fail to reject H_0? (2) correctly reject H_0? (3) commit a type I error? (4) commit a type II error?
   d. Calculate the p-value for your hypothesis test after 13 successes are observed.

9.80 You are testing the null hypothesis $p = 0.4$ and will reject this hypothesis if $z^\star$ is less than -2.05.

    a. If the null hypothesis is true and you observe $z^\star$ equal to -2.12, then which of the following results: (1) correctly fail to reject H_0, (2) correctly reject H_0, (3) commit a type I error, (4) commit a type II error?

    b. What is the significance level for this test?

    c. What is the p-value for $z^\star = -2.12$?

9.81 An insurance company states that 90% of its claims are settled within 30 days. A consumer group selected a random sample of 75 of the company's claims to test this statement. If the consumer group found that 55 of the claims were settled within 30 days, do they have sufficient reason to support their contention that fewer than 90% of the claims are settled within 30 days? Use $\alpha = 0.05$.

    a. Solve using the p-value approach.

    b. Solve using the classical approach.

9.82 A county judge has agreed that he will give up his county judgeship and run for a state judgeship unless there is evidence that more than 25% of his party is in opposition. A random sample of 800 party members included 217 who opposed him. Does this sample suggest that he should run for the state judgeship in accordance with his agreement? Carry out this hypothesis test by using $\alpha = 0.10$.

    a. Solve using the p-value approach.

    b. Solve using the classical approach.

9.83 A politician claims that she will receive 60% of the vote in an upcoming election. The results of a properly designed random sample of 100 voters showed that 50 of those sampled will vote for her. Is it likely that her assertion is correct at the 0.05 level of significance?

    a. Solve using the p-value approach.

    b. Solve using the classical approach.

9.84 The full-time student body of a college is composed of 50% males and 50% females. Does a random sample of students (30 male, 20 female) from an introductory chemistry course show sufficient evidence to reject the hypothesis that the proportion of male and of female students who take this course is the same as that of the whole student body? Use $\alpha = 0.05$.

    a. Solve using the p-value approach.

    b. Solve using the classical approach.

9.85 The January 14, 1991, issue of *Newsweek* reported that in a telephone survey of 759 adults, 27% were worried about maintaining their mortgage payments. Suppose you conducted a similar survey and found that 150 out of 759 were worried about maintaining their mortgage payments. Does your evidence (150 of 759) show that the percentage of worriers is significantly less than the 27% reported in *Newsweek*? Test using $\alpha = 0.05$.

9.86 The article "Making Up for Lost Time" (*U.S. News & World Report*, July 30, 1990) reported that more than half of the country's workers aged 45 to 64 want to quit work before they reach age 65. Suppose you conduct a survey of 1000 randomly chosen workers in order to test H_0: $p = 0.5$ versus H_a: $p < 0.5$, where p represents the proportion who want to quit before they reach age 65. If 460 of the 1000 respond that they want to quit work before age 65,

    a. calculate the value of the test statistic and the p-value.

    b. complete the hypothesis test using $\alpha = 0.01$.

9.3 | INFERENCES ABOUT VARIANCE AND STANDARD DEVIATION

Problems often arise that require us to make inferences about variability. For example, a soft-drink bottling company has a machine that fills 16-oz bottles. It needs to control the standard deviation σ (or variance σ^2) in the amount of soft drink, x, put into each bottle. The mean amount placed in each bottle is important, but a correct mean amount does not ensure that the filling machine is working correctly. If the variance is too large, there will be many bottles that are overfilled and many that are underfilled. Thus, the bottling company will want to maintain as small a standard deviation (or variance) as possible.

When discussing inferences about the spread of data, it is customary to talk about variance instead of the standard deviation, because the techniques (the formulas used) employ the sample variance rather than the standard deviation. However, remember that the standard deviation is the positive square root of the variance; thus, to talk about the variance of a population is comparable to talking about the standard deviation.

Inferences about the variance of a normally distributed population use the **chi-square**, χ^2, distributions. ("ki-square"; that's "ki" as in kite; χ is the Greek lowercase letter chi.) The chi-square distributions, like the Student t-distributions, are a family of probability distributions, each one being identified by the parameter, number of degrees of freedom. In order to use the chi-square distribution, we must be aware of its properties (see Figure 9-7 also).

Chi-square

PROPERTIES OF THE CHI-SQUARE DISTRIBUTION

1. χ^2 is nonnegative in value; it is zero or positively valued.
2. χ^2 is not symmetrical; it is skewed to the right.
3. χ^2 is distributed so as to form a family of distributions, a separate distribution for each different number of degrees of freedom.

FIGURE 9-7

Various Chi-Square Distributions

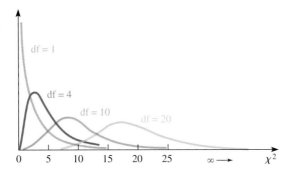

The critical values for chi-square are obtained from Table 8 in Appendix B. Each critical value is identified by two pieces of information: degrees of freedom (df) and the area under the curve to the right of the critical value being sought. Thus, $\chi^2(\text{df}, \alpha)$ is the symbol used to identify the critical value of chi-square with df degrees of freedom and with α area to the right, as shown in Figure 9-8. Since the chi-square distribution is not symmetrical, the critical values associated with right and left tails are given separately in Table 8.

FIGURE 9-8

Chi-Square
Distribution
Showing $\chi^2(\text{df}, \alpha)$

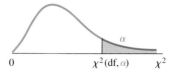

▼ | ILLUSTRATION   9 - 15

Find $\chi^2(20, 0.05)$.

SOLUTION   See the figure. Use Table 8 in Appendix B to find the value, as shown in the table: $\chi^2(20, 0.05) = \mathbf{31.4}$.

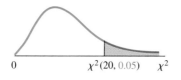

	Portion of Table 8	
	Area to the right	
df	...	0.05	...
⋮		
20	⟶	**31.4**

▲ |

EXERCISE   9.87

Find:

a. $\chi^2(10, 0.01)$,
b. $\chi^2(12, 0.025)$.

NOTE   When df > 2, the mean value of the chi-square distribution is df. The mean is located to the right of the mode (the value where the curve reaches its high point) and just to the right of the median (the value that splits the distribution, 50% on either side). By locating the zero at the left extreme and the value of df on your sketch of the χ^2 distribution, you will establish an approximate scale so that other values can be located in their respective positions. See Figure 9-9.

FIGURE 9-9

Location of Mean,
Median, and Mode
for χ^2 Distribution

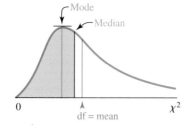

▼ ┃ ILLUSTRATION   9 - 16

Find $\chi^2(14, 0.90)$.

SOLUTION See the figure. Use Table 8 in Appendix B to find the value, as shown in the table: $\chi^2(14, 0.90) = $ **7.79**.

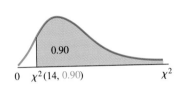

Portion of Table 8		
	Area to the right	
df	... 0.90 ...	
⋮	↓	
14	⟶ **7.79**	

▲ ┃

EXERCISE   9.88

Find:

a. $\chi^2(10, 0.95)$,
b. $\chi^2(22, 0.995)$.

We are ready to use chi-square to make inferences about the population variance or standard deviation.

> **The assumptions for inferences about the variance σ^2 or standard deviation σ:** The sampled population is normally distributed.

The t procedures for inferences about the mean (Section 9.1) were based on the assumption of normality, but they are generally very useful even when the sampled population is nonnormal, especially for larger samples. However, the same is not true about the inference procedures for standard deviation. The statistical procedures for standard deviation are very sensitive to nonnormal distributions (skewness, in particular), and this makes it very difficult to determine whether an apparent significant result is the result of the sample evidence or a violation of the assumptions. Therefore, the only inference procedure to be presented here will be the hypothesis test for the standard deviation of a normal population.

Hypothesis-Testing Procedure

Let's return to our illustration about the bottling company that wishes to detect when the variability in the amount of soft drink placed in each bottle gets out of control. A variance of 0.0004 is considered acceptable, and the company will want to adjust the bottle-filling machine when the variance, σ^2, becomes larger than this value. The decision will be made by using the hypothesis-testing procedure. The null hypothesis is "the variance is no larger than the specified value 0.0004"; the alternative hypothesis is "the variance is larger than 0.0004."

$$H_0: \sigma^2 = 0.0004 \; (\leq) \; \text{(variance is not larger than 0.0004)}$$
$$H_a: \sigma^2 > 0.0004 \; \text{(variance is larger than 0.0004)}$$

The test statistic that will be used in testing hypotheses about the population variance or standard deviation is obtained by using the formula

$$\chi^{2\star} = \frac{(n-1)s^2}{\sigma^2}, \quad \text{with df} = n - 1 \qquad \text{(9-10)}$$

When random samples are drawn from a normal population of a known variance σ^2, the quantity $\frac{(n-1)s^2}{\sigma^2}$ possesses a probability distribution that is known as the chi-square distribution with $n - 1$ degrees of freedom.

Probability-Value Approach

▼ | ILLUSTRATION   9 - 17

The soft-drink bottling company wants to control the variability in the amount of fill by not allowing the variance to exceed 0.0004. Does a sample of size 28 with a variance of 0.0007 indicate that the bottling process is out of control (with regard to variance) at the 0.05 level of significance?

SOLUTION

STEP 1   Parameter of concern: The variance σ^2 for the amount of fill is of concern.

STEP 2   Null and alternative hypotheses:

H_0: $\sigma^2 = 0.0004$ ($\leq$) (is not larger than 0.0004)
H_a: $\sigma^2 > 0.0004$ (is larger than 0.0004)

STEP 3   Test criteria:
   a. Assumptions: The amount of fill put into a bottle is generally normally distributed. By checking the distribution of the sample, we could verify this.
   b. Test statistic: chi-square, $\chi^{2\star}$, formula (9-10), with df $= n - 1 = 27$.
   c. Level of significance: $\alpha = 0.05$.

STEP 4   Sample evidence:
   a. Sample information: $n = 28$ and $s^2 = 0.0007$.
   b. Calculate the test statistic:

$$\chi^{2\star} = \frac{(n-1)s^2}{\sigma^2}: \quad \chi^{2\star} = \frac{(28-1)(0.0007)}{0.0004} = \frac{(27)(0.0007)}{0.0004} = \mathbf{47.25}$$

Specific instructions are on page 394.

   c. Calculate p-value: Since the concern is for larger values, the p-value is the area to the right of $\chi^{2\star} = 47.25$ as shown in the figure.

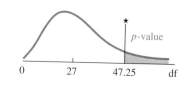

$$\mathbf{P} = P(\chi^2 > 47.25, \text{df} = 27)$$

There are two ways to find the p-value. (1) Use Table 8 in Appendix B and estimate it or (2) use a computer and have it calculated.

Method 1:   Use Table 8 in Appendix B to place "bounds" on the p-value. By inspecting the df $= 27$ row and locating $\chi^{2\star} = 47.25$, we can determine an interval within which the p-value must lie. See the portion of Table 8.

$$0.005 < P(\chi^2 > 47.25, \text{df} = 27) < 0.01; \; \mathbf{0.005 < P < 0.01}$$

	Portion of Table 8		
	Area to the right		
df	... 0.01	P	0.005
⋮		↑	
27	... 47.0	47.25	49.6

Method 2:   Use the computer and the cumulative distribution function (CDF) to find the cumulative probability up to $\chi^{2\star} = 47.25$ and then subtract from 1.0 to find the p-value.

MINITAB session commands **CDF 47.25;** and **CHISquare 27.** will produce the following results.

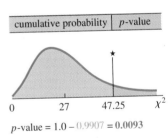

```
Cumulative Distribution Function
Chisquare with 27 d.f.

    X         P(X <= x)
47.2500       0.9907
```

cumulative probability | p-value

p-value $= 1.0 - 0.9907 = 0.0093$

STEP 5   The results: The p-value is smaller than the specified level of significance, $\alpha = 0.05$.

DECISION Reject H_0.

CONCLUSION At the 0.05 level of significance, we conclude that the bottling process is out of control with regard to the variance.

▼ | ILLUSTRATION   9 - 18

EXERCISE   9.89

Complete the following hypothesis test using the *p*-value approach:
H_0: $\sigma^2 = 532$ vs.
H_a: $\sigma^2 > 532$ using
$\alpha = 0.05$ and
sample information
$n = 18$ and
$s^2 = 785$. Assume
that the sample was
taken from a normal
population.

Find the *p*-value for the following hypothesis test.

$$H_0: \sigma^2 = 12 \ (\geq)$$
$$H_a: \sigma^2 < 12$$
$$df = 15 \quad \text{and} \quad \chi^{2\star} = 7.88$$

SOLUTION Since the concern is for smaller values, the *p*-value is the area to the left of $\chi^{2\star} = 7.88$ as shown in the figure.

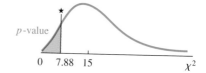

$$\mathbf{P} = P(\chi^2 < 7.88, df = 15)$$

There are two ways to find the *p*-value. Use Table 8 in Appendix B, or use a computer.

Method 1: We will place "bounds" on the *p*-value, **P**, by using Table 8. By inspecting the df = 15 row and locating $\chi^{2\star} = 7.88$, we can determine an interval within which **P** must lie. See the portion of Table 8.

$$0.05 < P(\chi^2 < 7.88, df = 15) < 0.10; \ \mathbf{0.05 < P < 0.10}$$

Portion of Table 8

		Area in left-hand tail		
df	...	**0.05**	**P**	**0.10**
15	...	7.26	7.88	8.55

Method 2: Use the computer and the cumulative distribution function (CDF) to find the cumulative probability up to $\chi^{2\star} = 7.88$, which will be the *p*-value.

MINITAB session commands **CDF 7.88**; and **CHISquare 15**. will produce the following results.

```
Cumulative Distribution Function
Chisquare with 15 d.f.

   X        P(X <= x)
7.8800      0.0715
```

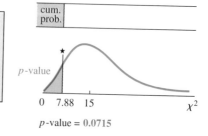

p-value = 0.0715

▲ |

▼ | ILLUSTRATION   9 - 19

Find the *p*-value for the following hypothesis test.

$$H_0: \sigma^2 = 32 \quad \text{vs.} \quad H_a: \sigma^2 \neq 32 \text{ with df} = 18 \text{ and when}$$
$$\text{a. } \chi^{2\star} = 7.38 \qquad \text{b. } \chi^{2\star} = 29.3$$

SOLUTION a. Since the concern is for values different than 32, the *p*-value is the
area in both tails. The chi-square distribution is not symmetri-
cal; therefore, we find the area of one tail and double it. Since
$\chi^{2\star} = 7.38$ is in the left tail, follow the procedure used in Illustration 9-18. See the figure.

$$\tfrac{1}{2}P = P(\chi^2 < 7.38, \text{ df} = 18)$$

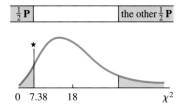

Method 1: Use Table 8 in Appendix B to place "bounds" on the *p*-value. $\tfrac{1}{2}P$ is
found using Table 8. By inspecting the df = 18 row and locating
$\chi^{2\star} = 7.38$, we can determine an interval within which $\tfrac{1}{2}P$ must lie.
See the portion of Table 8.

$$0.01 < P(\chi^2 < 7.38, \text{ df} = 18) < 0.025;$$

by doubling, **0.02 < P < 0.05**

	Portion of Table 8		
	Area in left-hand tail		
df	...	0.01	$\tfrac{1}{2}$P	0.025
⋮				
18	...	7.01	7.38	8.23

Method 2: Use the computer and the cumulative distribution function (CDF) to
find the cumulative probability up to $\chi^{2\star} = 7.38$, which will be one-
half of the *p*-value.

MINITAB session commands **CDF 7.38**; and **CHISquare 18.** will produce the following results.

```
Cumulative Distribution Function
Chisquare with 18 d.f.

    x         P(X <= x)
7.3800        0.0135
```

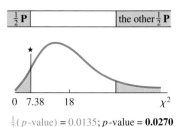

$\frac{1}{2}(p\text{-value}) = 0.0135; p\text{-value} = \textbf{0.0270}$

b. The concern (H_a) is for values different than 32; the p-value is the area in both tails. The chi-square distribution is not symmetrical; therefore, we find the area of one tail and double it. Since $\chi^{2\star} = 29.3$ is in the right tail, follow the procedure used in Illustration 9-17 to find $\frac{1}{2}\mathbf{P}$. See the figure.

$$\tfrac{1}{2}\mathbf{P} = P(\chi^2 > 29.3,\ df = 18)$$

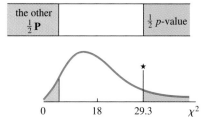

Method 1: Use Table 8 of Appendix B to place "bounds" on one-half of the p-value, $\frac{1}{2}\mathbf{P}$. By inspecting the df = 18 row and locating $\chi^{2\star} = 29.3$, we can determine an interval within which $\frac{1}{2}\mathbf{P}$ must lie. See the portion of Table 8.

$$0.025 < P(\chi^2 > 29.3,\ df = 18) < 0.05;$$

by doubling, $\mathbf{0.05 < P < 0.10}$

Portion of Table 8			
Area in right-hand tail			
df	0.05	$\frac{1}{2}$P	0.025
⋮			
18	28.9	29.3	31.5

EXERCISE   9.90

Find the p-value for the following hypothesis test:
H_0: $\sigma^2 = 52$ vs.
H_a: $\sigma^2 \neq 52$ using sample information $n = 41$ and $s^2 = 78.2$.

Method 2: Use the computer and the cumulative distribution function (CDF) to find the cumulative probability up to $\chi^{2\star} = 29.3$; then one-half of the p-value will be found by subtracting the cumulative probability from 1.0.

MINITAB session commands **CDF 29.3**; and **CHISquare 18**. will produce the following results.

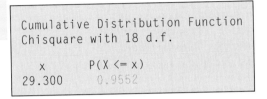

```
Cumulative Distribution Function
Chisquare with 18 d.f.

      x        P(X <= x)
  29.300        0.9552
```

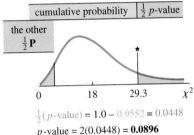

$\frac{1}{2}(p\text{-value}) = 1.0 - 0.9552 = 0.0448$

$p\text{-value} = 2(0.0448) = \textbf{0.0896}$

▲ |

Classical Approach

▼ | ILLUSTRATION   9 - 20

A photographic chemical is claimed by its manufacturer to have a shelf life that is normally distributed about a mean of 180 days with a standard deviation of no more than 10 days. As a user of this chemical, Fast Photo is concerned that the standard deviation might be greater than 10 days; otherwise, they will buy a larger quantity while the chemical is part of a special promotion. Twelve samples were randomly selected and tested, with a standard deviation of 14 days resulting. At the 0.05 level of significance, does this sample present sufficient evidence to show the standard deviation is greater than 10 days?

SOLUTION

STEP 1   Parameter of concern: The standard deviation σ for the shelf life of the chemical.

STEP 2   Null and alternative hypotheses:

H_0: $\sigma = 10$ ($\leq$) (st. dev. is not greater than 10)

H_a: $\sigma > 10$ (st. dev. is greater than 10)

The null and alternative hypotheses could be expressed equivalently in terms of variance:

H_0: $\sigma^2 = 100$ ($\leq$) (variance is not greater than 100)

H_a: $\sigma^2 > 100$ (variance is greater than 100)

Choosing to express the hypotheses in terms of standard deviation or variance has no effect on the rest of the hypothesis-testing procedure.

STEP 3   Test criteria:

a. Check assumptions: The manufacturer claims "shelf life" is normally distributed; this could be verified by checking the distribution of the sample.

b. Test statistic: chi-square, $\chi^{2\star}$, formula (9-10), with df $= n - 1 = 11$.

c. Level of significance: $\alpha = 0.05$.

d. Critical region: Critical region is the right-hand tail since concern expressed in alternative hypothesis is "greater than." See the figure. From Table 8: $\chi^2 (11, 0.05) = 19.7$.

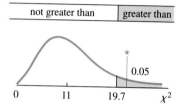

STEP 4   Sample evidence:

a. Sample information: $n = 12$ and $s = 14$.

b. Test statistic:

$$\chi^{2\star} = \frac{(n-1)s^2}{\sigma^2}: \quad \chi^{2\star} = \frac{(12-1)(14)^2}{(10)^2} = \frac{2156}{100} = \mathbf{21.56}$$

STEP 5   Results: The calculated $\chi^{2\star}$ is in the critical region (shown in red on the figure above).

DECISION   Reject H_0.

CONCLUSION   There is sufficient evidence at the 0.05 significance level to conclude that the shelf life of this chemical does have a standard deviation greater than 10 days. Therefore, Fast Photo should purchase the chemical accordingly.

NOTE   When sample data are skewed, just one outlier can greatly affect the standard deviation. It is very important, especially when using small samples, that the sampled population be normal; otherwise, the procedures are not reliable.

MINITAB (Release 10) does not have commands that perform the inferences for population variance and standard deviation. This list of session commands will help you complete a hypothesis test about variance and standard deviation based on a sample of data in C1. Enter your choice for the hypothesized variance as K1. These commands will find the area under the chi-square curve both to the left of $\chi^{\star}$ and to the right of $\chi^{\star}$. If test

---

EXERCISE 9.91

Complete the following hypothesis test using the classical approach:
$H_0: \sigma^2 = 0.25$ vs.
$H_a: \sigma^2 < 0.25$ using
$\alpha = 0.05$ and
sample information
$n = 10$ and
$s = 0.31$.

▲

EXERCISE 9.92

Calculate the standard deviation for these data:

a. 5, 6, 7, 7, 8, 10,
b. 5, 6, 7, 7, 8, 15.
c. What effect did the largest value changing from 10 to 15 have on the standard deviation?
d. Why do you think 15 might be called an outlier?

Use the Helpful Hint on page 89 when using a series of session commands.

is one-tailed to the left side, C3 is the p-value; if test is one-tailed to the right side, C4 is the p-value; if test is two-tailed, the p-value is equal to 2 times the smaller value, C3 or C4.

Session commands

```
LET K1 = variance from null hypothesis
LET K2 = COUNt(C1)-1
LET K3 = STDEv(C1)
LET C2 = (K2*(K3**2))/K1
PRINt K2 K3 C2
CDF C2 C3;
  CHISquare K2.
LET C4 = 1 - C3
PRINt C3 C4
```

EXERCISE   9.93

Describe what each of the MINITAB session commands does as it finds the two areas under the chi-square curve. Why do the two areas sum to 1.0?

EXERCISES

9.94 Find these critical values by using Table 8 of Appendix B.

  a. $\chi^2(18, 0.01)$ b. $\chi^2(16, 0.025)$
  c. $\chi^2(8, 0.10)$ d. $\chi^2(28, 0.01)$
  e. $\chi^2(22, 0.95)$ f. $\chi^2(10, 0.975)$
  g. $\chi^2(50, 0.90)$ h. $\chi^2(24, 0.99)$

9.95 Using the notation of Exercise 9.94, name and find the critical values of χ^2.

  a.   b.

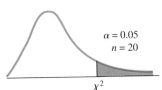

$\alpha = 0.05$
$n = 20$
χ^2

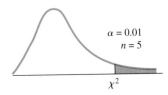

$\alpha = 0.01$
$n = 5$
χ^2

  c.   d.

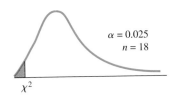

$\alpha = 0.025$
$n = 18$
χ^2

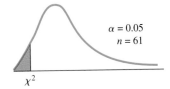

$\alpha = 0.05$
$n = 61$
χ^2

9.96   Using the notation of Exercise 9.94, name and find the critical values of χ^2.

a.

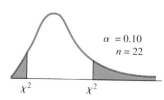

$\alpha = 0.10$
$n = 22$

$\chi^2 \qquad \chi^2$

b.

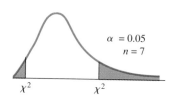

$\alpha = 0.05$
$n = 7$

$\chi^2 \qquad \chi^2$

c.

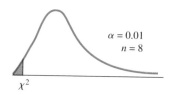

$\alpha = 0.01$
$n = 8$

χ^2

d.

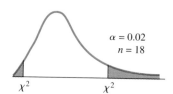

$\alpha = 0.02$
$n = 18$

$\chi^2 \qquad \chi^2$

9.97   a. What value of chi-square for 5 degrees of freedom subdivides the area under the distribution curve such that 5% is to the right and 95% is to the left?

b. What is the value of the 95th percentile for the chi-square distribution with 5 degrees of freedom?

c. What is the value of the 90th percentile for the chi-square distribution with 5 degrees of freedom?

9.98   a. The central 90% of the chi-square distribution with 11 degrees of freedom lies between what values?

b. The central 95% of the chi-square distribution with 11 degrees of freedom lies between what values?

c. The central 99% of the chi-square distribution with 11 degrees of freedom lies between what values?

9.99   For a chi-square distribution having 12 degrees of freedom, find the area under the curve for chi-square values ranging from 3.57 to 21.0.

9.100   For a chi-square distribution having 35 degrees of freedom, find the area under the curve between $\chi^2(35, 0.96)$ and $\chi^2(35, 0.15)$.

9.101   State the null hypothesis, H_0, and the alternative hypothesis, H_a, that would be used to test these claims:

a. The standard deviation has increased from its previous value of 24.

b. The standard deviation is no larger than 0.5 oz.

c. The standard deviation is not equal to 10.

d. The variance is no less than 18.

e. The variance is different from the value of 0.025, the value called for in the specs.

f. The variance has increased from 34.5.

9.102   Calculate the value for the test statistic, $\chi^{2\star}$, for each of these situations:

a. $H_0: \sigma^2 = 20$, $n = 15$, $s^2 = 17.8$

b. $H_0: \sigma^2 = 30$, $n = 18$, $s = 5.7$

c. $H_0: \sigma = 42$, $n = 25$, $s = 37.8$

d. $H_0: \sigma = 12$, $n = 37$, $s^2 = 163$

9.103   Calculate the p-value for each of the following hypothesis tests.
  a. H_a: $\sigma^2 \neq 20$, $n = 15$, $\chi^{2\star} = 27.8$
  b. H_a: $\sigma^2 > 30$, $n = 18$, $\chi^{2\star} = 33.4$
  c. H_a: $\sigma \neq 42$, df $= 25$, $\chi^{2\star} = 37.9$
  d. H_a: $\sigma < 12$, df $= 40$, $\chi^{2\star} = 26.3$

9.104   Determine the test criteria that would be used to test the following using the classical approach:
  a. H_0: $\sigma = 0.5$ and H_a: $\sigma > 0.5$, with $n = 18$ and $\alpha = 0.05$
  b. H_0: $\sigma^2 = 8.5$ and H_a: $\sigma^2 < 8.5$, with $n = 15$ and $\alpha = 0.01$
  c. H_0: $\sigma = 20.3$ and H_a: $\sigma \neq 20.3$, with $n = 10$ and $\alpha = 0.10$
  d. H_0: $\sigma^2 = 0.05$ and H_a: $\sigma^2 \neq 0.05$, with $n = 8$ and $\alpha = 0.02$
  e. H_0: $\sigma = 0.5$ and H_a: $\sigma < 0.5$, with $n = 12$ and $\alpha = 0.10$

9.105   A random sample of 51 observations was selected from a normally distributed population. The sample mean was $\bar{x} = 98.2$, and the sample variance was $s^2 = 37.5$. Does this sample show sufficient reason to conclude that the population standard deviation is not equal to 8 at the 0.05 level of significance?

9.106   In the past the standard deviation of weights of certain 32.0-oz packages filled by a machine was 0.25 oz. A random sample of 20 packages showed a standard deviation of 0.35 oz. Is the apparent increase in variability significant at the 0.10 level of significance? Assume package weight is normally distributed.
  a. Solve using the p-value approach.
  b. Solve using the classical approach.

9.107   According to the September 1994 issue of *Money* magazine, the average loan rate for a 30-year fixed mortgage equaled 8.71%, the average rate for a 15-year fixed mortgage equaled 8.22%, and the average for a 1-year adjustable was 5.63%. The average for a home-equity line was 8.83%. Many times the variability of loan rates is also of interest. Test the hypothesis that the standard deviation for home-equity lines differs from 0.5% using a sample of 15 loan institutions that gave a standard deviation of 1.1%. Give the critical values, $\chi^{2\star}$, and your conclusion. Assume that the mortgage rates are normally distributed, and use a level of significance equal to 0.05.
  a. What role does the assumption of normality play in this solution? Explain.
  b. Describe how you might attempt to determine whether or not it is realistic to assume that mortgage rates are normally distributed.
  c. Complete the test using the p-value approach. Include $\chi^{2\star}$, p-value, and your conclusion.
  d. Complete the test using the classical approach. Include the critical values, $\chi^{2\star}$, and your conclusion.

9.108   A commercial farmer harvests his entire field of a vegetable crop at one time. Therefore, he would like to plant a variety of green beans that mature all at one time (small standard deviation between maturity times of individual plants). A seed company has developed a new hybrid strain of green beans that it believes to be better for the commercial farmer. The maturity time of the standard variety has an average of 50 days and a standard deviation of 2.1 days. A random sample of 30 plants of the new hybrid showed a standard deviation of 1.65 days. Does this sample show a significant lowering of the standard deviation at the 0.05 level of significance? Assume that maturity time is normally distributed.
  a. Solve using the p-value approach.
  b. Solve using the classical approach.

9.109 A car manufacturer claims that the miles per gallon for a certain model has a mean equal to 40.5 mi with a standard deviation equal to 3.5 mi. Use the following data, obtained from a random sample of 15 such cars, to test the hypothesis that the standard deviation differs from 3.5. Use $\alpha = 0.05$. Assume normality.

| 37.0 | 38.0 | 42.5 | 45.0 | 34.0 | 32.0 | 36.0 | 35.5 |
| 38.0 | 42.5 | 40.0 | 42.5 | 36.0 | 30.0 | 37.5 | |

    a. Solve using the *p*-value approach.
    b. Solve using the classical approach.

9.110 The manager at Quik Delivery is concerned that the fees currently being charged for delivery of items classified as small packages are too varied. Company policy does not allow the standard deviation of the fees for items of this class to exceed $1. Using the set of data given in Exercise 2.45 (p. 67), calculate the *p*-value needed to test the manager's concern.

9.111 The chi-square distribution was described on page 471 as a family of distributions. Let's investigate these distributions and observe some of their properties.

    a. Use the list of MINITAB session commands and generate several large random samples of data from various chi-square distributions. Enter each different df value at **K** of the second command. Use df values of 1, 2, 3, 5, 10, 20, and 80 (others if you wish).

    b. What appears to be the relationship between the mean of the sample and the number of degrees of freedom?

See Helpful Hint on page 89.

    c. How do the values of the mean, median, and mode appear to be related? Do your results agree with the information on page 472?

    d. Have the computer generate samples for two additional degrees of freedom, df = 120 and 150. Describe how these distributions seem to be changing as df increases.

```
RANDom 1000 C1;
  CHISquare K.
MEAN C1
MEDIan C1
HISTogram C1
```

9.112 How important is the assumption "the sampled population is normally distributed" for the use of the chi-square distributions? Use a computer and the two sets of MINITAB commands that follow to simulate drawing 200 samples of size 10 from each of two different types of population distributions. The first four commands will generate 2000 data and construct a histogram so that you can see what the population looks like. The next two commands will generate 200 samples of size 10 from the same population; each row represents a sample. The next three commands calculate the standard deviation and $\chi^{2\star}$ for each of the 200 samples. The last three commands construct histograms of the 200 sample standard deviations and the 200 $\chi^{2\star}$ values.

```
RANDom 2000 C1;
  NORMal 100 50.
HISTogram C1;
  CUTPoint -100:300/25.
RANDom 200 C2-C11;
  NORMal 100 50.
RSTDev C2-C11 C12
# calculating chi-sq* for each sample stdev
LET C13 = (9*C12**2)/(50**2)
HISTogram C12
HISTogram C13;
  CUTPoint -1:41/2.
```

```
RANDom 2000 C1;
  EXPO 100.
HISTogram C1;
  CUTPoint 0:700/25.
RANDom 200 C2-C11;
  EXPO 100.
RSTDev C2-C11 C12
# calculating chi-sq* for each sample stdev
LET C13 = (9*C12**2)/(100**2)
HISTogram C12
HISTogram C13;
  CUTPoint 0:40/1.
```

For the samples from the normal population:
    a. Does the sampling distribution of sample standard deviations appear to be normal? Describe the distribution.
    b. Does the $\chi^{2\star}$-distribution appear to have a chi-square distribution with df = 9? Find percentages for intervals (less than 2, less than 4, . . . , more than 15, more than 20, etc.), and compare them to the percentages expected as estimated using Table 8, "Critical Values of the Chi-Square Distribution," in Appendix B.

For the samples from the skewed population:
    c. Does the sampling distribution of sample standard deviations appear to be normal? Describe the distribution.
    d. Does the $\chi^{2\star}$-distribution appear to have a chi-square distribution with df = 9? Find percentages for intervals (less than 2, less than 4, . . . , more than 15, more than 20, etc.), and compare them to the percentages expected as estimated using Table 8.

In summary:
    e. Does the normality condition appear to be necessary in order for the calculated test statistic $\chi^{2\star}$ to have a χ^{2} distribution? Explain.

IN RETROSPECT

We have been studying inferences, both confidence intervals and hypothesis tests, for the three basic population parameters (mean μ, proportion p, and standard deviation σ) of a single population. Most inferences about a single population are concerned with one of these three parameters. Figure 9-10 presents a visual organization of the techniques studied throughout Chapters 8 and 9 along with the key questions that you must ask yourself as you are deciding which test statistic and which formula to use.

In this chapter we also used the maximum error of estimate, formula (9-7), to determine the size of sample required to make estimates about the population proportion with the desired accuracy. Case Study 9-2 presents a graph showing the "margin of error" (maximum error of estimate) for sample sizes up to 1500. These are for 95% confidence interval estimates as you found out in Exercise 9.51. By combining a point estimate with its corresponding maximum error of estimate (sampling error), we can construct an interval estimate based on the information reported. In Exercise 9.52 you did some calculating to verify the 95% level of confidence. Most polls and surveys use the 95% confidence level, even though they do not report the 95%.

In the next chapter we will discuss inferences about two populations whose respective means, proportions, and standard deviations are to be compared.

FIGURE 9-10   Choosing the Right Inference Technique

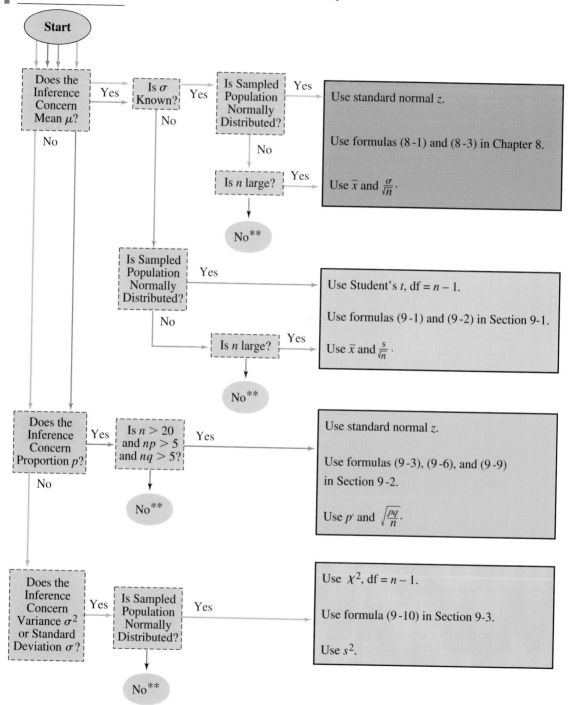

No** means that a nonparametric technique (normally not required) is used; see Chapter 14.

CHAPTER EXERCISES

 9.113 One of the objectives of a large medical study was to estimate the mean physician fee for cataract removal. For 25 randomly selected cases the mean fee was found to be $1550 with a standard deviation of $125. Set a 99% confidence interval on μ, the mean fee for all physicians. Assume fees are normally distributed.

 9.114 A natural-gas utility is considering a contract for purchasing tires for its fleet of service trucks. The decision will be based on expected mileage. For a sample of 100 tires tested, the mean mileage was 36,000 and the standard deviation was 2000 miles. Estimate the mean mileage that the utility should expect from these tires using a 96% confidence interval.

 9.115 According to an article in *Changing Times* (March 1991), financial assets in the United States fell by $6700 per household between the end of 1989 and the end of 1990. Suppose that in order to check this statement, a survey of 500 households is conducted and that the decrease in financial assets was $6300 with a standard deviation equal to $3250. Calculate the value of the test statistic and the p-value for the research hypothesis $\mu < \$6700$.

 9.116 It has been suggested that abnormal male children tend to occur more in children born to older-than-average parents. Case histories of 20 abnormal males were obtained, and the ages of the 20 mothers were

31	21	29	28	34	45	21	41	27	31
43	21	39	38	32	28	37	28	16	39

The mean age at which mothers in the general population give birth is 28.0 years.
   a. Calculate the sample mean and standard deviation.
   b. Does the sample give sufficient evidence to support the claim that abnormal male children have older-than-average mothers? Use $\alpha = 0.05$. Assume ages have a normal distribution.
      (1) Solve using the p-value approach.
      (2) Solve using the classical approach.

 9.117 The water pollution readings at State Park Beach seem to be lower than last year. A sample of 12 readings was randomly selected from the records of this year's daily readings:

3.5	3.9	2.8	3.1	3.1	3.4
4.8	3.2	2.5	3.5	4.4	3.1

Does this sample provide sufficient evidence to conclude that the mean of this year's pollution readings is significantly lower than last year's mean of 3.8 at the 0.05 level? Assume that all such readings have a normal distribution.

 9.118 In a large cherry orchard the average yield has been 4.35 tons per acre for the last several years. A new fertilizer was tested on 15 randomly selected one-acre plots. The yields from these plots follow:

3.56	5.00	4.88	4.93	3.92
4.25	5.12	5.13	4.79	4.45
5.35	4.81	3.48	4.45	4.72

At the 0.05 level of significance, do we have sufficient evidence to claim that there was a significant increase in production? Assume yield per acre is normally distributed.
a. Solve using the *p*-value approach.
b. Solve using the classical approach.

9.119 The scores for a standardized reading test used in the state of Nebraska are normally distributed and have a mean value of 80. A school district wishes to test the hypothesis that its mean is different from that of the state. Twenty randomly selected students were tested and the results summarized by $\bar{x} = 77.5$ and $s = 2.5$. Complete the hypothesis test using $\alpha = 0.05$.
a. Solve using the *p*-value approach.
b. Solve using the classical approach.

9.120 A manufacturer of television sets claims that the maintenance expenditures for its product will average no more than $50 during the first year following the expiration of the warranty. A consumer group has asked you to substantiate or discredit the claim. The results of a random sample of 50 owners of such television sets showed that the mean expenditure was $61.60 and the standard deviation was $32.46. At the 0.01 level of significance, should you conclude that the producer's claim is true or not likely to be true?
a. Solve using the *p*-value approach.
b. Solve using the classical approach.

9.121 *The LEXIS*, a national law journal, found from a survey conducted on April 6–7, 1991, that nearly two-thirds of the 800 people surveyed said doctors should not be prosecuted for helping people with terminal illnesses commit suicide. The poll carries a margin of error of plus or minus 3.5%.
a. Describe how this survey of 800 people fits the properties of a binomial experiment. Specifically identify: *n*, a trial, success, *p*, and *x*.
b. Exactly what is the "two-thirds" reported? How was it obtained? Is it a parameter or a statistic?
c. Calculate the 95% confidence maximum error of estimate for the population proportion of all people who believe doctors should not be prosecuted.
d. How is the maximum error, found in (c), related to the 3.5% mentioned in the survey report?

9.122 The marketing research department of an instant-coffee company conducted a survey of married men to determine the proportion of married men who preferred their brand. Twenty of the 100 in the random sample preferred the company's brand. Use a 95% confidence interval to estimate the proportion of all married men who prefer this company's brand of instant coffee. Interpret your answer.

9.123 A company is drafting an advertising campaign that will involve endorsements by noted athletes. In order for the campaign to succeed, the endorser must be both highly respected and easily recognized. A random sample of 100 prospective customers are shown photos of various athletes. If the customer recognizes an athlete, then the customer is asked whether he or she respects the athlete. In the case of a top woman golfer, 16 of the 100 respondents recognized her picture and indicated that they also respected her. At the 95% level of confidence, what is the true proportion with which this woman golfer is both recognized and respected?

9.124 A local auto dealership advertises that 90% of customers whose autos were serviced by their service department are pleased with the results. As a researcher, you take exception to this statement because you are aware that many people are reluctant to express dissatisfaction even if they are not pleased. A research experiment was set up in which those in the sample had received service by this dealer within the past two weeks. During the interview, the individuals were led to believe that the interviewer was new in town and was considering taking his car to this dealer's service department. Of the 60 sampled, 14 said that they were dissatisfied and would not recommend the department.

    a. Estimate the proportion of dissatisfied customers using a 95% confidence interval.

    b. Given your answer to (a), what can be concluded about the dealer's claim?

9.125 In obtaining the sample size to estimate a proportion, the formula $n = [z(\alpha/2)]^2 pq/E^2$ is used. If a reasonable estimate of p is not available, then it is suggested that $p = 0.5$ be used because this will give the maximum value for n. Calculate the value of $pq = p(1 - p)$ for $p = 0.1, 0.2, 0.3, \ldots, 0.8, 0.9$ in order to obtain some idea about the behavior of the quantity pq.

9.126 A consumer group was interested in determining the proportion of dentists who would accept the patient's insurance payment as the full payment for a routine exam. Determine the sample size needed to estimate the true proportion to within 0.01 with 95% confidence.

9.127 *Prevention* magazine reported in its latest survey that 64% of adult Americans, or 98 million people, were overweight. The telephone survey of 1254 randomly selected adults was conducted November 8–29, 1990, and had a margin of error of three percentage points.

    a. Calculate the maximum error of estimate for 0.95 confidence with $p' = 0.64$.

    b. How is the margin of error of three percentage points related to answer (a)?

    c. How large a sample would be needed to reduce the maximum error to 0.02 with 95% confidence?

9.128 "Two of five Americans believe the country should rely on nuclear power more than other energy sources for energy in the 1990s, according to a poll released yesterday. . . . The telephone poll, taken April 10–11, has a margin of error of plus or minus 3 points." This statement appeared in the Rochester *Democrat & Chronicle* on April 21, 1991. Forty percent plus or minus three points sounds like a confidence interval.

    a. What is another name for the "margin of error of plus or minus 3 points"?

    b. If we assume a 95% level of confidence, how large a sample is needed for a maximum error of 0.03?

9.129 A machine is considered to be operating in an acceptable manner if it produces 0.5% or fewer defective parts. It is not performing in an acceptable manner if more than 0.5% of its production is defective. The hypothesis $H_0: p = 0.005$ is tested against the hypothesis $H_a: p > 0.005$ by taking a random sample of 50 parts produced by the machine. The null hypothesis is rejected if two or more defective parts are found in the sample. Find the probability of the type I error.

9.130 You are interested in comparing the hypothesis $p = 0.8$ against the alternative $p < 0.8$. In 100 trials you observe 73 successes. Calculate the p-value associated with this result.

9.131 An instructor asks each of the 54 members of his class to write down "at random" one of the numbers 1, 2, 3, . . . , 13, 14, 15. Since the instructor believes that students like gambling, he considers that 7 and 11 are lucky numbers. He counts the number of students, x, who selected 7 or 11. How large must x be before the hypothesis of randomness can be rejected at the 0.05 level?

9.132   Today's newspapers and magazines often report the findings of survey polls about various aspects of life. *The American Gender Evolution* (chapter case study, p. 431) reports "62% of the men believe both partners should earn a living." Other bits of information given in the article are: "telephone survey of 1201 adults" and "has a sampling error of ± 3 percent." Relate this information to the statistical inferences you have been studying in this chapter.

     a. Is a percentage of people a population parameter, and if so, how is it related to any of the parameters that we have studied?

     b. Based on the information given, find the 95% confidence interval for the true proportion of men who believe both partners should earn a living.

     c. Explain how the terms "point estimate," "level of confidence," "maximum error of estimate," and "confidence interval" relate to the values reported in the article and to your answers in (b).

9.133   In order to test the hypothesis that the standard deviation on a standard test is 12, a sample of 40 randomly selected students was tested. The sample variance was found to be 155. Does this sample provide sufficient evidence to show that the standard deviation differs from 12 at the 0.05 level of significance?

9.134   Bright-lite claims that its 60-watt light bulb burns with a length of life that is approximately normally distributed with a standard deviation of 81 hours. A sample of 101 bulbs had a variance of 8075. Is this sufficient evidence to reject Bright-lite's claim in favor of the alternative, "the standard deviation is larger than 81 hours," at the 0.05 level of significance?

     a. Solve using the *p*-value approach.

     b. Solve using the classical approach.

9.135   A production process is considered to be out of control if the produced parts have a mean length different from 27.5 mm or a standard deviation that is greater than 0.5 mm. A sample of 30 parts yields a sample mean of 27.63 mm and a sample standard deviation of 0.87 mm. If we assume part length is a normally distributed variable:

     a. At the 0.05 level of significance, does this sample indicate that the process should be adjusted in order to correct the standard deviation of the product?

     b. At the 0.05 level of significance, does this sample indicate that the process should be adjusted in order to correct the mean value of the product?

9.136   Oranges are selected at random from a large shipment that just arrived. A sample is taken to estimate the size (circumference, in inches) of the oranges. The sample data are summarized as follows: $n = 100$, $\Sigma x = 878.2$, and $\Sigma(x - \bar{x})^2 = 49.91$.

     a. Determine the mean and standard deviation for this sample.

     b. What is the point estimate for μ, the mean circumference of oranges in this shipment?

     c. What is the 95% confidence interval estimate for μ?

9.137   According to the January 28, 1991, issue of the *Christian Science Monitor*, the cost of replacing light bulbs shot out by drug dealers was $7200 per public housing complex last year. Suppose a study was conducted to determine the average maintenance costs for public-housing complexes. For a sample of 20 such complexes in the Northeast, the mean maintenance cost was $75,000 with a standard deviation of $15,500. Determine a 95% confidence interval estimate for μ, the mean, cost per complex in the Northeast. Assume normality.

9.138   Suppose that a sample of size 30 was used to test H_0: $\sigma^2 = 17$ versus H_a: $\sigma^2 > 17$ at $\alpha = 0.05$. How large would s^2 need to be before the null hypothesis would be rejected?

VOCABULARY LIST

Be able to define each term. Pay special attention to the key terms, which are printed in red. In addition, describe in your own words, and give an example of, each term. Your examples should not be ones given in class or in the textbook.

The bracketed numbers indicate the chapter(s) in which the term appeared previously, but you should define the terms again to show increased understanding of their meaning. Page numbers indicate the first appearance of the term in Chapter 9.

calculated value [8] (p. 441)
chi-square (p. 474)
chi-square distribution (p. 471)
conclusion [8] (p. 443)
confidence interval [8] (p. 439)
critical region [8] (p. 445)
critical value [8] (p. 443)
decision [8] (p. 443)
degrees of freedom (p. 434)
hypothesis test [8] (p. 441)
inference [8] (p. 432)
level of confidence [8] (p. 439)
level of significance [8] (p. 441)
maximum error of estimate [8] (p. 440)
observed binomial probability (p') (p. 454)
one-tailed test [8] (p. 442)

parameter [1, 8] (p. 437)
p-value [8] (p. 442)
proportion [6] (p. 454)
random variable [5, 6] (p. 454)
response variable [1] (p. 454)
sample size [8] (p. 434)
sample statistic [1, 2] (p. 455)
σ known [8] (p. 432)
σ unknown (p. 432)
standard error [7, 8] (p. 432)
standard normal, z [2, 6, 8] (p. 459)
Student's t (p. 432)
test statistic [8] (p. 441)
two-tailed test [8] (p. 442)

QUIZ A

Answer "True" if the statement is always true. If the statement is not always true, replace the words shown in bold with words that make the statement always true.

9.1 The Student's t-distribution is an approximately normal distribution but is more **dispersed** than the normal distribution.

9.2 The **chi-square** distribution is used for inferences about the mean when σ is unknown.

9.3 The **Student's t**-distribution is used for all inferences about a population's variance.

9.4 If the test statistic falls in the critical region, the null hypothesis has **been proved true**.

9.5 When the test statistic is t and the number of degrees of freedom gets very large, the critical value of t is very close to that of z.

9.6 When making inferences about one mean when the value of σ is not known, the **z-score** is the test statistic.

9.7 The chi-square distribution is a skewed distribution whose mean is **2** for df > 2.

9.8 Often the concern with testing the variance (or standard deviation) is to keep its size under control or relatively small. Therefore, many of the hypothesis tests with chi-square will be **one-tailed**.

9.9 $\sqrt{npq}$ is the standard error of proportion.

9.10 The sampling distribution of p' is approximately distributed as **chi-square**.

QUIZ B

Answer all questions, showing all formulas, substitutions, and work.

9.1   Find each of the following:
   a. $z(0.02)$ b. $t(18, 0.95)$ c. $\chi^2(25, 0.95)$

9.2   A random sample of 25 was selected from a normally distributed population for the purpose of estimating the population mean, μ. The sample statistics are $n = 25$, $\bar{x} = 28.6$, $s = 3.50$.
   a. Find the point estimate for μ.
   b. Find the maximum error of estimate for the 0.95 confidence interval estimate.
   c. Find the lower confidence limit (LCL) and the upper confidence limit (UCL) for the 0.95 confidence interval estimate for μ.

9.3   Thousands of area elementary school students were recently given a nationwide standardized exam testing their composition skills. If 64 of a random sample of 100 students passed this exam, construct the 0.98 confidence interval estimate for the true proportion of all area students who passed the exam.

9.4   State the null (H_0) and the alternative (H_a) hypotheses that would be used to test each of the following claims.
   a. The mean weight of professional basketball players is no more than 225 lb.
   b. The standard deviation for the monthly amounts of rainfall in Monroe County is less than 3.7 in.
   c. Approximately 40% of MCC's daytime students own their own car.

9.5   Determine the test criteria [level of significance, test statistic, critical region(s), and critical value(s)] that would be used in completing each hypothesis test using the classic approach and $\alpha = 0.05$.
   a. H_0: $\mu = 43$ b. H_0: $p = 0.80$ c. H_0: $\mu = 95$
      H_a: $\mu < 43$ H_a: $p > 0.80$ H_a: $\neq 95$
      (given $\sigma = 6$)   (σ unknown, $n = 22$)
   d. H_0: $\sigma = 12$
      H_a $\sigma \neq 12$
      ($n = 28$)

9.6   The manufacturer of a new model car, called Orion, claims the typical Orion will average 26 mpg of gasoline. An independent consumer group is somewhat skeptical of this claim and thinks the mean gas mileage is less than the 26 claimed. A sample of 24 randomly selected Orions produced the following results:

Sample statistics: mean $= 24.15$,   st. dev. $= 4.87$

At the 0.05 level of significance, does the consumer group have sufficient evidence to refute the manufacturer's claim?

9.7   A coffee machine is supposed to dispense 6 fluid ounces of coffee into a paper cup. In reality, the amount dispensed varies from cup to cup. However, if the machine is operating properly, the standard deviation of the amounts dispensed should be 0.1 or less of an ounce. A random sample of 15 cups produced a standard deviation of 0.13 oz. Does this represent sufficient evidence, at the 0.10 level of significance, to conclude that "the machine is not operating properly"?

9.8 An unhappy customer is disturbed with the waiting time at the post office when buying stamps. Upon registering his complaint, he was told, "You wait more than one minute for service no more than half of the time when only buying stamps." Not believing this to be the case, the customer collected some data from people who had just purchased stamps only.

$$\text{Sample statistics: } n = 60, \quad x = n(\text{more than one minute}) = 35$$

At the 0.02 level of significance, does our unhappy customer have sufficient evidence to refute the post office's claim?

QUIZ C

9.1 Student B says the range of a set of data may be used to obtain a crude estimate for the standard deviation of the population. Student A is not sure. How will Student B correctly explain how and under what circumstances his statement is true?

9.2 Is it (a) the null hypothesis or (b) the alternative hypothesis that the researcher usually believes to be true? Explain.

9.3 When you reject a null hypothesis, student A says that you are expressing disbelief in the value of the parameter as claimed in the null hypothesis, whereas student B says that you are expressing the belief that the sample statistic came from a population other than one related to the parameter claimed in the null hypothesis. Who is correct? Explain.

9.4 "The Student t-distribution must be used when making inferences about the population mean, μ, when the population standard deviation, σ, is not known" is a true statement. Student A states that the z-score plays a role sometimes when using the t-distribution. Explain the conditions that exist and the role played by z that make student A's statement correct.

9.5 Student A says that the percentage of the sample means that fall outside the critical values of the sampling distribution determined by a true null hypothesis is the p-value for the test. Student B says that the percentage student A is describing is the level of significance. Who is correct? Explain.

9.6 Student A carries out a study in which he is willing to run a 1% risk of making a type I error. He rejects the null hypothesis and claims that his statistic is significant at the 0.99 level of confidence. Student B argues that student A's claim is not properly worded. Who is right? Explain.

9.7 Student A claims that when you employ a 95% confidence interval to determine an estimation, you do not know for sure whether your inference is correct (the parameter is contained within the interval) or not. Student B claims that you do; you have shown that the parameter cannot be less than the lower limit or greater than the upper limit of the interval. Who is right? Explain.

9.8 Student A says that the best way to improve a confidence interval estimate is to increase the level of confidence. Student B argues that using a high confidence level does not really improve the desirability of the resulting interval estimate. Who is right? Explain.

10

INFERENCES INVOLVING
TWO POPULATIONS

Chapter Case Study

YOUTHFUL IDEAS ABOUT OLD AGE: AN ANALYSIS OF CHILDREN'S DRAWINGS

An interesting application of two sample statistics is discussed in the following article. The "heights" and the "quality" of the children's artwork were measured and analyzed, and hypothesis tests were completed using both independent and dependent samples. All of these are included in our study of Chapter 10.

ABSTRACT

Youthful ideas about old age are investigated by analyzing children's drawings of young and old people. The age of the children ranged between ten and a half years and eleven and a half years of age, and each child drew four pictures (a young woman, an old woman, a young man, and an old man). Analyses consisted of a content analysis of drawings; the calculation of standard scores by means of the Harris-Goodenough Draw-A-Person (DAP) test scoring procedure (in order to effect comparisons of scores within each child's set of drawings); and measurement of the height of each drawing.

HEIGHT COMPARISONS

Inspection of drawings suggested that there were regular differences in sizes of the different classes of people depicted. Measurement supported this view. Mean heights of figures in the four groups are shown in Table 3. The old people were invariably smaller than the young ones, women were larger than men, and drawings by girls were bigger than those by boys.

The means of the DAP standard scores were calculated for all categories of drawing. These values are shown in Table 4. Given that the higher the standard score, the more "sophisticated," detailed and complex the drawing, it appears that DAP scores echo some of the findings of the mean height calculations: namely, that drawings of young people were "better" than those of old people, and that drawings of women, on the whole, were "better" than those of men.

TESTS FOR THE SIGNIFICANCE OF DIFFERENCES

t-tests for the significance of the difference between two means for correlated samples were carried out, in order to compare scores for drawings of young and old people and for scores of drawings of female and male people. Scores of girl and boy artists were also treated separately. Similarly, *t*-tests for the significance of the difference between two means for independent samples enabled comparisons of drawings of girls and boys.

In the correlated samples, all differences between drawings of old and young people were found to be significant. Differences between boys' drawings of old and young women were found to be highly significant at the .001 level ($t = 7.42$) as were those of girls ($t = 6.55$). Similarly, boys' drawings of old and young men were also significantly different at the .001 level ($t = 5.02$). Girls' drawings of young and old men differed at the .05 level ($t = 2.3225$). Tests for independent samples carried out on drawings of girls and boys in all four categories (young woman, old woman, young man, old man) produced no significant differences.

Source: Nancy Falchikov, *International Journal of Aging and Human Development*, 3 (no. 2), 1990, 79–99. Reprinted by permission.

What are the correlated samples? What are independent samples? How do you compare the means for independent samples? (See Exercise 10.121 at the end of the chapter.)

FIGURE 1

A Young Woman (boy artist)

FIGURE 2

An Old Woman (girl artist)

TABLE 3 Mean Height of Figures Measured in Centimeters (sample standard deviations in parentheses)

	Young Woman	Old Woman	Young Man	Old Man
Drawings by girls	14.01 (2.29)	12.46 (2.67)	12.24 (3.44)	11.69 (3.26)
Drawings by boys	11.88 (2.90)	10.36 (2.21)	9.13 (2.32)	8.29 (2.61)

TABLE 4 Mean Standard Scores (standard deviations in parentheses)

	Young Woman	Old Woman	Young Man	Old Man
Drawings by girls	110.6 (8.25)	98.9 (7.84)	104.9 (9.00)	100.2 (8.42)
Drawings by boys	113.6 (11.66)	102.4 (9.26)	108.3 (9.44)	100.1 (9.47)

● CHAPTER OBJECTIVES

In Chapters 8 and 9 you were introduced to the basic concepts of estimation and hypothesis testing in connection with inferences about one population and the parameters mean, proportion, and variance (or standard deviation). In this chapter we continue to investigate the inferences about these same three parameters, but we will now use these parameters as a basis for comparing two populations.

When sampling two different populations, two types of samples are used: independent samples and dependent samples. After learning about the two types of samples in Section 10.1, we will learn the techniques for using the mean of the paired differences, the difference between two means, the difference between two proportions, and the ratio of two variances for comparing two populations. Since we will be studying several different situations, Figure 10-1 is offered as a "road map" to help you organize these various inference techniques.

FIGURE 10-1   "Road Map" to Two Population Inferences

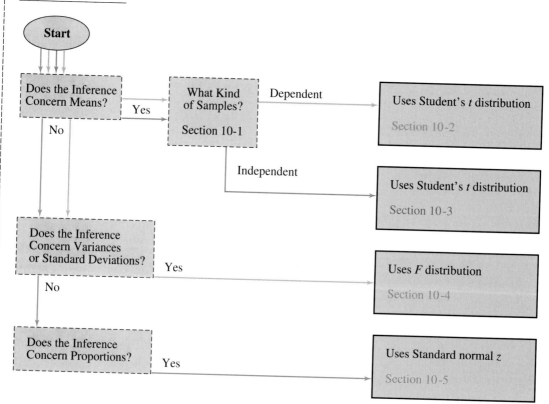

10.1 | INDEPENDENT AND DEPENDENT SAMPLES

Source

Dependent sampling
Independent sampling

In this chapter we are going to study the procedures for making inferences about two populations. When comparing two populations, we need two samples, one from each population. Two basic kinds of samples can be used: independent and dependent. The dependence or independence of two samples is determined by the sources used for the data. A **source** can be a person, an object, or anything that yields a piece of data. If the same set of sources or related sets are used to obtain the data representing both populations, we have **dependent sampling**. If two unrelated sets of sources are used, one set from each population, we have **independent sampling**. The following illustrations should amplify these ideas.

▼ | ILLUSTRATION   10 - 1

A test will be conducted to see whether the participants in a physical fitness class actually improve in their level of fitness. It is anticipated that approximately 500 people will sign up for this course, and the instructor decides that she will give 50 of the participants a set of tests before the course begins (a pretest) and then will give another set of tests to 50 participants at the end of the course (a posttest). The following two sampling procedures are proposed:

> ▼ *Plan A:* Randomly select 50 participants from the list of those enrolled and give them the pretest. At the end of the course, make another or second random selection of size 50 and give them the posttest.
> ▼ *Plan B:* Randomly select 50 participants and give them the pretest; give the same set of 50 the posttest upon completion of the course.

Plan A illustrates independent sampling—the sources (the class participants) used for each sample (pretest and posttest) were selected separately. Plan B illustrates dependent sampling—the sources used for both samples (pretest and posttest) are the same.

▲ |

Typically, when both a pretest and a posttest are used, the same subjects are used in the study. Thus, pretest versus posttest (before versus after) studies are usually dependent samples.

EXERCISE   10.1

Describe how you could select two independent samples from among your classmates to compare the heights of female and male students.

EXERCISE   10.2

Describe how you could select two dependent samples from among your classmates to compare their heights as they entered high school to when they entered college.

ILLUSTRATION   10 - 2

A test is being designed to compare the wearing quality of two brands of automobile tires. The automobiles will be selected and equipped with the new tires and then driven under "normal" conditions for one month. Then a measurement will be taken to determine how much wear took place. Two plans are proposed:

> ▼ *Plan C:* A sample of cars will be selected randomly and equipped with brand A and driven for the month, and another sample of cars will be selected and equipped with brand B and driven for the month.

Case Study   ## Exploring the Traits of Twins

10.1

Studies that involve identical twins are a natural for the dependent sampling technique discussed in this section.

A NEW STUDY SHOWS THAT KEY CHARACTERISTICS MAY BE INHERITED

EXERCISE   10.3

Explain why studies involving identical twins result in dependent samples of data.

Like many identical twins reared apart, Jim Lewis and Jim Springer found they had been leading eerily similar lives. Separated four weeks after birth in 1940, the Jim twins grew up 45 miles apart in Ohio and were reunited in 1979. Eventually they discovered that both drove the same model blue Chevrolet, chain-smoked Salems, chewed their fingernails and owned dogs named Toy. Each had spent a good deal of time vacationing at the same three-block strip of beach in Florida. More important, when tested for such personality traits as flexibility, self-control and sociability, the twins responded almost exactly alike.

The project is considered the most comprehensive of its kind. The Minnesota researchers report the results of six-day tests of their subjects, including 44 pairs of identical twins who were brought up apart. Well-being, alienation, aggression and the shunning of risk or danger were found to owe as much or more to nature as to nurture. Of eleven key traits or clusters of traits analyzed in the study, researchers estimated that a high of 61 percent of what they call "society potency" (a tendency toward leadership or dominance) is inherited, while "social closeness" (the need for intimacy, comfort and help) was lowest, at 33 percent.

Source: Copyright 1987 by Time Inc. All rights reserved. Reprinted by permission of TIME.

▼ *Plan D:* A sample of cars will be selected randomly, equipped with one tire of brand A and one tire of brand B (the other two tires are not part of the test), and driven for the month.

In this illustration we might suspect that many other factors must be taken into account when testing automobile tires—such as age, weight, and mechanical condition of the car; driving habits of drivers; location of the tire on the car; and where and how much the car is driven. However, at this time we are trying only to illustrate dependent and independent samples. Plan C is independent (unrelated sources), and plan D is dependent (common sources).

Independent and dependent samples each have their advantages; these will be emphasized later. Both methods of sampling are often used.

EXERCISES

10.4 In trying to estimate the amount of growth that took place in the trees planted by the County Parks Commission recently, 36 trees were randomly selected from the 4000 planted. The heights of these trees were measured and recorded. One year later another set of 42 trees was randomly selected and measured. Do the two sets of data (36 heights, 42 heights) represent dependent or independent samples? Explain.

10.5 Twenty people were selected to participate in a psychology experiment. They answered a short multiple-choice quiz about their attitudes on a particular subject and then viewed a 45-minute film. The following day the same 20 people were asked to answer a follow-up questionnaire about their attitudes. At the completion of the experiment, the experimenter will have two sets of scores. Do these two samples represent dependent or independent samples? Explain.

10.6 An experiment is designed to study the effect diet has on the uric acid level. Twenty white rats are used for the study. Ten rats are randomly selected and given a junk-food diet. The other ten receive a high-fiber, low-fat diet. Uric acid levels of the two groups are determined. Do the resulting sets of data represent dependent or independent samples? Explain.

10.7 Two different types of disc centrifuges are used to measure the particle size in latex paint. A gallon of paint is randomly selected, and 10 specimens are taken from it for testing on each of the centrifuges. There will be two sets of data, 10 data each, as a result of the testing. Do the two sets of data represent dependent or independent samples? Explain.

10.8 An insurance company is concerned that garage A charges more for repair work than garage B charges. It plans to send 25 cars to each garage and obtain separate estimates for the repairs needed for each car.
    a. How can the company do this and obtain independent samples? Explain in detail.
    b. How can the company do this and obtain dependent samples? Explain in detail.

10.9 A study is being designed to determine the reasons why adults choose to follow a healthy diet plan. One thousand men and 1000 women will be surveyed. Upon completion, the reasons that men choose a healthy diet will be compared to the reasons that women choose a healthy diet.
    a. How can the data be collected if independent samples are to be obtained? Explain in detail.
    b. How can the data be collected if dependent samples are to be obtained? Explain in detail.

10.10 Suppose that 400 students in a certain college are taking elementary statistics this semester. Two samples of size 25 are needed in order to test some precourse skill against the same skill after the students complete the course.
    a. Describe how you would obtain your samples if you were to use dependent samples.
    b. Describe how you would obtain your samples if you were to use independent samples.

10.2 | INFERENCES CONCERNING THE MEAN DIFFERENCE USING TWO DEPENDENT SAMPLES

The procedures for comparing two population means are based on the relationship between two sets of sample data, one sample from each population. When dependent samples are involved, the data are thought of as "**paired data**." The data may be

Paired data

paired as a result of the data being obtained from "before" and "after" studies; from pairs of identical twins as in Case Study 10-1; from a "common" source, as with the amounts of tire wear for each brand from a car in plan D of Illustration 10-2; or from matching two subjects of similar traits to form "matched pairs." The pairs of data values are compared directly to each other by using the difference in their numerical values. The resulting difference is called a **paired difference**.

Paired difference

$$\text{Paired difference:} \quad d = x_1 - x_2 \tag{10-1}$$

The concept of using paired data this way has a built-in ability to remove the effect of otherwise uncontrolled factors. The tire-wear problem (Illustration 10-2) is an excellent example of such additional factors. The wearing ability of a tire is greatly affected by a multitude of factors: the size and weight of the car, the age and condition of the car, the driving habits of the driver, the number of miles driven, the condition and types of roads driven on, the quality of the material used to make the tire, and so on. We have created paired data by placing one tire from each brand on a car. Since one tire of each brand will be tested under the same conditions, same car, same driver, and so on, the extraneous causes of wear are neutralized.

A test was conducted to compare the wearing quality of the tires produced by two tire companies using plan D, as described in Illustration 10-2. All the afore-mentioned factors will have an equal effect on both brands of tires, car by car. Table 10-1 gives the amount of wear (in thousandths of an inch) that resulted from the test. One tire of each brand was placed on each of six test cars. The position (left or right side, front or back) was determined with the aid of a random numbers table.

TABLE 10-1

Amount of Tire Wear

Car	1	2	3	4	5	6
Brand A	125	64	94	38	90	106
Brand B	133	65	103	37	102	115

Since the various cars, drivers, and conditions are the same for each tire of a paired set of data, it would make sense to use a third variable, the paired difference d. Our two dependent samples of data will be combined into one set of d values, where $d = B - A$.

EXERCISE 10.11

Find the paired differences, $d = A - B$, for this set of data:

Pairs	1	2	3	4	5
Sample A	3	6	1	4	7
Sample B	2	5	1	2	8

Car	1	2	3	4	5	6
$d = B - A$	8	1	9	−1	12	9

The sample statistics needed will be the mean of the sample differences, $\bar{d}$

$$\bar{d} = \frac{\sum d}{n} \tag{10-2}$$

Formulas (10-2) and (10-3) are adaptations of formulas (2-1) and (2-10)

and the standard deviation of the sample differences, s_d

$$s_d = \sqrt{\frac{\Sigma d^2 - \left[\frac{(\Sigma d)^2}{n}\right]}{n-1}} \tag{10-3}$$

▼ | ILLUSTRATION 10 - 3

EXERCISE 10.12

Find the mean $\bar{d}$ and the standard deviation, s_d of the paired differences in Exercise 10.11.

Find the mean and standard deviation of the paired differences in Table 10-1.

SOLUTION The summary of data: $n = 6$, $\Sigma d = 38$, and $\Sigma d^2 = 372$.

$$\text{The mean:} \quad \frac{\Sigma d}{n} = \frac{38}{6} = 6.333 = \mathbf{6.3}$$

The standard deviation:

$$s_d = \sqrt{\frac{\Sigma d^2 - \left[\frac{(\Sigma d)^2}{n}\right]}{n-1}} = \sqrt{\frac{372 - \left[\frac{(38)^2}{6}\right]}{6-1}} = \sqrt{26.27} = 5.13 = \mathbf{5.1}$$

▲ |

The difference between the two population means, when dependent samples are used, is equivalent to the mean of the paired differences. Therefore, when an inference is to be made about the difference of two means and paired differences are being used, the inference will in fact be about the mean of the paired differences. The mean of the sample paired differences will be used as the point estimate for these inferences.

In order to make inferences about the mean of all possible paired differences μ_d, we need to know about the sampling distribution of $\bar{d}$.

When paired observations are randomly selected from normal populations, the paired difference, $d = x_1 - x_2$, will be approximately normally distributed about a mean μ_d with a standard deviation of σ_d.

This is the situation in which the t-test for one mean is applied; namely, we wish to make inferences about an unknown mean (μ_d) where the random variable (d) involved has an approximately normal distribution with an unknown standard deviation (σ_d).

Inferences about the mean of all possible paired differences μ_d are based on samples of n dependent pairs of data and the t distribution with $n - 1$ degrees of freedom, under the following assumption.

The assumption for inferences about the mean of paired differences μ_d:
The paired data are randomly selected from normally distributed populations.

Confidence Interval

Formula (10-4) is an adaptation of formula (9-1).

The $1 - \alpha$ confidence interval for estimating the mean difference μ_d is found using this formula:

$$\bar{d} - t\left(df, \frac{\alpha}{2}\right)\frac{s_d}{\sqrt{n}} \quad \text{to} \quad \bar{d} + t\left(df, \frac{\alpha}{2}\right)\frac{s_d}{\sqrt{n}}, \quad \text{where df} = n - 1 \quad \textbf{(10-4)}$$

▼ | ILLUSTRATION   10 - 4

Construct the 95% confidence interval for the mean difference in the paired data on tire wear, as reported in Table 10-1. The sample information is $n = 6$ pieces of paired data, $\bar{d} = 6.3$, and $s_d = 5.1$ (calculated in Illustration 10-3). Assume the amount of wear is approximately normally distributed for both brands of tires.

SOLUTION

STEP 1   Population parameter of interest: The mean difference in the amount of wear that occurred between the two brands of tires.

STEP 2   Confidence interval criteria:
   a. Check the assumptions: Both sampled populations are approximately normal.
   b. Test statistic: t with df $= 6 - 1 = 5$.
   c. Confidence level: $1 - \alpha = 0.95$.

STEP 3   Sample evidence:
   a. Sample information: $n = 6$, $\bar{d} = 6.3$, and $s_d = 5.1$.
   b. Point estimate: $\bar{d} = \textbf{6.3}$.

STEP 4   Confidence interval limits:
   a. Confidence coefficients: Two-tailed situation with $\alpha/2 = 0.025$ in one tail. From Table 6 in Appendix B, $t(df, \alpha/2) = t(5, 0.025) = 2.57$.

EXERCISE   10.13

Find $t(15, 0.025)$. Describe the role this number plays in the confidence interval.

   b. Maximum error: Using the maximum error part of formula (10-4),

$$E = t\left(df, \frac{\alpha}{2}\right)\frac{s_d}{\sqrt{n}} = 2.57\left(\frac{5.1}{\sqrt{6}}\right) = (2.57)(2.082) = 5.351 = \textbf{5.4}$$

   c. Confidence limits:

$$\bar{d} \pm E$$
$$6.3 \pm 5.4$$
$$6.3 - 5.4 = 0.9 \quad \text{to} \quad 6.3 + 5.4 = 11.7$$

   d. The results: **0.9 to 11.7, the 95% confidence interval for μ_d**

▲ | That is, with 95% confidence we can say that the mean difference in the amount of wear is between 0.9 and 11.7.

NOTE   This confidence interval is quite wide, due, in part, to the small sample size. Recall from the central limit theorem that as the sample size increases, the standard error (estimated by $s_d/\sqrt{n}$) decreases.

EXERCISE   10.14

a. Find the 95% confidence interval for μ_d given $n = 26$, $\overline{d} = 6.3$, and $s_d = 5.1$.

b. Compare your interval to the interval found in Illustration 10-4.

MINITAB (Release 10) commands to complete a $1 - \alpha = K\%$ confidence interval for mean μ_d with unknown standard deviation and with two dependent sets of sample data in C1 and C2. You must first use the session command LET to find the paired differences and store the differences in C3.

Enter: LET C3 = C1 - C2   *Menu commands*

Session command

Enter:
   TINTerval K data in C3

Choose:
   Stat > Basic Stats > 1-Sample t
Enter: C3
Select: Confidence interval
Enter: $1 - \alpha$: K

With Brand A data in C1 and Brand B data in C2 on MINITAB's worksheet, the following commands were used to solve Illustration 10-4; the resulting output is shown below.

EXERCISE   10.15

Use a computer to find the 95% confidence interval for estimating μ_d based on these paired data:

Before	75	68	40	30	43	65
After	70	69	32	30	39	63

```
LET C3 = C2 - C1
TINTerval 95 C3
```

```
Confidence Intervals

Variable    N    Mean    StDev    SE Mean      95.0% C.I.
      C3    6    6.33     5.13       2.09    (0.95, 11.71)
```

Hypothesis Testing

When testing a null hypothesis about the mean difference, the test statistic used will be the difference between the sample mean $\overline{d}$ and the hypothesized value of μ_d divided by the estimated standard error. This statistic is assumed to have a t distribution when the null hypothesis is true and the assumptions for the test are satisfied. The value of the test statistic $t^\star$ is calculated using formula (10-5):

Formula (10-5) is an adaptation of formula (9-2).

$$t^\star = \frac{\overline{d} - \mu_d}{s_d/\sqrt{n}}, \quad \text{where df} = n - 1 \qquad \textbf{(10-5)}$$

p-Value Approach

▼ | ILLUSTRATION   10 - 5

In a study dealing with high blood pressure and the drugs used to aid in controlling it, the effect of calcium channel blockers on pulse rate was one of many specific concerns. Twenty-six patients were randomly selected from a large pool of potential subjects, and their pulse rates were established. A calcium channel blocker was administered to each patient for a fixed period of time, and then each patient's pulse rate was again determined. The two resulting sets of data appeared to have an approximately normal distribution, and resulting statistics were $\bar{d} = 1.07$ and $s_d = 1.74$ ($d = $ before $-$ after). Does the resulting sample information provide sufficient evidence to show that this calcium channel blocker did lower the pulse rate? Use $\alpha = 0.05$.

SOLUTION

STEP 1   Parameter of concern: The mean difference (reduction) in pulse rate from before to after having used the calcium channel blocker for the time period of the test.

STEP 2   Null and alternative hypothesis:

> H_0: $\mu_d = 0$ ($\leq$) (did not lower rate)   Remember: $d = $ before $-$ after.
> H_a: $\mu_d > 0$ (did lower rate)

"Lower rate" means "after" is less than "before" and "before − after" is positive.

STEP 3   Test criteria:
  a. Assumptions: Since the data in both sets were approximately normal, it seems reasonable to assume that the two populations are approximately normally distributed.
  b. Test statistic:

$$t^\star = \frac{\bar{d} - \mu_d}{s_d/\sqrt{n}}, \quad \text{with df} = n - 1 = 25$$

  c. Level of significance: $\alpha = 0.05$.

STEP 4   Sample evidence:
  a. Sample information: $n = 26$, $\bar{d} = 1.07$, and $s_d = 1.74$.
  b. Calculated test statistic:

$$t^\star = \frac{\bar{d} - \mu_d}{s_d/\sqrt{n}}: \quad t^\star = \frac{1.07 - 0.0}{1.74/\sqrt{26}} = \frac{1.07}{0.34} = \textbf{3.14}$$

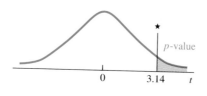

EXERCISE   10.16

Use the *p*-value approach to complete the hypothesis test with alternative hypothesis $\mu_d > 0$ based on the paired data listed below and $d = B - A$. Use $\alpha = 0.05$. Assume normality.

A 700 830 860 1080 930
B 720 820 890 1100 960

---

c. The p-value: $\mathbf{P} = P(t^{\star} > 3.14$, with df $= 25) = \mathbf{0.002}$ (from Table 7 in Appendix B).

STEP 5   The results: The p-value is smaller than α.

DECISION   Reject H_0.

CONCLUSION   At the 0.05 level of significance, we can conclude that the calcium channel blocker does lower the pulse rate.

▲

"Statistical significance" does not always have the same meaning when the "practical" application of the results is considered. In the preceding detailed hypothesis test, the results showed a statistical significance with a p-value of 0.002, that is, 2 chances in 1000. However, a more practical question might be, Is lowering the pulse rate by this small average amount worth the risks involving possible side effects of this medication? Actually the whole issue is far more reaching than just this one issue of pulse rate.

MINITAB (Release 10) commands to complete a hypothesis test for mean difference μ_d = A, with unknown standard deviation and two dependent sets of the sample data in C1 and C2, against the alternative hypothesis coded as K: −1 = left-tailed, 0 = two-tailed, and +1 = right-tailed. You must first use the session command LET to find the paired differences and store them in C3.

First enter:

    LET C3 = C2 − C1

Session commands

Enter:

    TTESt μ = A data in C3;
    ALTErnative K.

Menu commands

Choose:
    Stat > Basic Stats > 1-Sample t
Enter: C3
Select: Test mean
Enter: A
Select Alternative:
    not equal, less than or greater than

EXERCISE   10.17

Use a computer to complete the hypothesis test with alternative hypothesis $\mu_d < 0$ based on the paired data listed below and $d = M - N$. Use $\alpha = 0.02$. Assume normality.

M	58	78	45	38	49	62
N	62	86	42	39	47	68

With the before-data in C1 and the after-data in C2 on MINITAB's worksheet, the following commands are used to solve Illustration 10-5; the output is shown below.

```
LET C3 = C1 − C2
TTESt 0 C3;
ALTErnative 1.
```

```
T-Test of the Mean
Test of mu = 0.00 vs mu > 0.00

Variable    N    Mean    StDev    SE Mean     T     P-Value
   C3      26    1.07    1.74      0.34      3.14    0.002
```

Classical Approach

▼ ILLUSTRATION   10 - 6

Suppose the sample data in Table 10-1 had been collected with the hope of showing that "the two brands do not wear equally." Do the data provide sufficient evidence for us to conclude that the two brands show unequal wear, at the 0.05 level of significance? Assume the amount of wear is approximately normally distributed for both brands of tires.

SOLUTION

STEP 1   Parameter of concern: The mean difference in the amount of wear between the two brands.

STEP 2   Null and alternative hypotheses:

H_0: $\mu_d = 0$ (no difference)   Remember: $d = B - A$.
H_a: $\mu_d \neq 0$ (difference)

STEP 3   Test criteria:
    a. Assumptions: The assumption of normality is included in the statement of the problem.
    b. Test statistic:

$$t^\star = \frac{\bar{d} - \mu_d}{s_d/\sqrt{n}}, \text{ with df} = n - 1 = 6 - 1 = 5$$

    c. Level of significance: $\alpha = 0.05$.
    d. Critical values: Critical region is split between two tails since the alternative hypothesis is "not equal." From Table 6 in Appendix B, $t(5, 0.025) = 2.57$. (See the figure.)

Difference	No difference	Difference

−2.57   0   2.57 t

EXERCISE 10.18

Use the classical approach to complete the hypothesis test with alternative hypothesis $\mu_d > 0$ based on the paired data listed below and $d = 0 - Y$. Use $\alpha = 0.01$. Assume normality.

| **Oldest** | 199 | 162 | 174 | 159 | 173 |
| **Youngest** | 194 | 162 | 167 | 156 | 176 |

STEP 4   Sample evidence:
Sample information: $n = 6$, $\bar{d} = 6.3$, and $s_d = 5.1$.
Calculate the test statistic:

$$t^\star = \frac{\bar{d} - \mu_d}{s_d/\sqrt{n}}: \quad t^\star = \frac{6.3 - 0.0}{5.1/\sqrt{6}} = \frac{6.3}{2.08} = \mathbf{3.03}$$

STEP 5   Results: $t^\star$ is in the critical region as shown in red on the figure above.

DECISION Reject H_0.

CONCLUSION There is a significant difference in the mean amount of wear at the 0.05 level of significance.

▲

EXERCISES

10.19   Salt-free diets are often prescribed to people with high blood pressure. The following data were obtained from an experiment designed to estimate the reduction in diastolic blood pressure as a result of following a salt-free diet for two weeks. Assume diastolic readings to be normally distributed.

Before	93	106	87	92	102	95	88	110
After	92	102	89	92	101	96	88	105

a. What is the point estimate for the mean reduction in the diastolic reading after two weeks on this diet?
b. Find the 98% confidence interval for the mean reduction.

10.20   All students who enroll in a certain memory course are given a pretest before the course begins. A random sample of ten students who finished the course were given a posttest, and their scores are listed below.

Student	1	2	3	4	5	6	7	8	9	10
Before	93	86	72	54	92	65	80	81	62	73
After	98	92	80	62	91	78	89	78	71	80

MINITAB was used to find the 95% confidence interval for the mean improvement in memory resulting from taking the memory course, as measured by the difference in test scores (d = after − before). Verify the results shown on the output by calculating them.

```
LET C3 = C2 - C1
TINTerval 95 C3

Confidence Intervals

Variable      N    Mean    StDev    SE Mean      95% C.I.
      C3     10    6.10     4.79       1.52    (2.67, 9.53)
```

10.21   An experiment was designed to estimate the mean difference in weight gain for pigs fed ration A as compared to those fed ration B. Eight pairs of pigs were used. The pigs within each pair were littermates. The rations were assigned at random to the two animals within each pair. The gains (in pounds) after 45 days are shown in the following table.

Litter	1	2	3	4	5	6	7	8
Ration A	65	37	40	47	49	65	53	59
Ration B	58	39	31	45	47	55	59	51

Assuming weight gain is normal, find the 95% confidence interval estimate for the mean of the differences μ_d, where d = ration A − ration B.

10.22   A sociologist is studying the effects of a certain motion picture film on the attitudes of black men toward white men. Twelve black men were randomly selected and asked to fill out a questionnaire before and after viewing the film. The scores received by the 12 men are shown in the following table. Assuming the questionnaire scores are normal, construct a 95% confidence interval for the mean shift in score that takes place when this film is viewed.

Before	10	13	18	12	9	8	14	12	17	20	7	11
After	5	9	13	17	4	5	11	14	13	18	7	12

10.23   Two men, A and B, who usually commute to work together, decide to conduct an experiment to see whether one route is faster than the other. The men feel that their driving habits are approximately the same, and therefore they decide on the following procedure. Each morning for two weeks A will drive to work on one route and B will use the other route. On the first morning, A will toss a coin. If heads appear, he will use route I; if tails appear, he will use route II. On the second morning, B will toss the coin: heads, route I; tails, route II. The times, recorded to the nearest minute, are shown in the following table. Assume commute times are normal and estimate the mean of the differences with a 95% confidence interval.

					Day					
Route	**M**	**Tu**	**W**	**Th**	**F**	**M**	**Tu**	**W**	**Th**	**F**
I	29	26	25	25	25	24	26	26	30	31
II	25	26	25	25	24	23	27	25	29	30

10.24 We want to know which of two types of filters should be used over an oscilloscope to help the operator pick out the image on the screen of the cathode-ray tube. A test was designed in which the strength of a signal could be varied from zero to the point where the operator first detects the image. At this point, the intensity setting is read. The lower the setting, the better the filter. Twenty operators were asked to make one reading for each filter.

	Filter			Filter	
Operator	1	2	Operator	1	2
1	96	92	11	88	88
2	83	84	12	89	89
3	97	92	13	85	86
4	93	90	14	94	91
5	99	93	15	90	89
6	95	91	16	92	90
7	97	92	17	91	90
8	91	90	18	78	80
9	100	93	19	77	80
10	92	90	20	93	90

Assuming the intensity readings are normally distributed, estimate the mean difference between the two readings using a 90% confidence interval.

10.25 State the null hypothesis, H_0, and the alternative hypothesis, H_a, that would be used to test these claims:

a. The mean of the differences between the posttest and the pretest scores is greater than 10.

b. The mean weight gain, due to the change in diet for the laboratory animals, is at least 10 oz.

c. The mean weight loss experienced by people on a new diet plan was no less than 12 lb.

d. As a result of a special training session, it is believed that the mean of the difference in performance scores will not be zero.

10.26 Determine the test criteria that would be used with the classical approach to test the following hypotheses when t is used as the test statistic.

a. H_0: $\mu_d = 0.5$ and H_a: $\mu_d > 0.5$, with $n = 15$ and $\alpha = 0.05$

b. H_0: $\mu_d = 4.5$ and H_a: $\mu_d \neq 4.5$, with $n = 25$ and $\alpha = 0.05$

c. H_0: $\mu_d = 1.45$ and H_a: $\mu_d < 1.45$, with $n = 12$ and $\alpha = 0.10$

d. H_0: $\mu_d = 0.75$ and H_a: $\mu_d > 0.75$, with $n = 18$ and $\alpha = 0.01$

10.27 The corrosive effects of various soils on coated and uncoated steel pipe were tested by using a dependent sampling plan. The data collected are summarized by

$$n = 40 \qquad \Sigma d = 220 \qquad \Sigma d^2 = 6222$$

where d is the amount of corrosion on the coated portion subtracted from the amount of corrosion on the uncoated portion. Does this sample provide sufficient reason to conclude that the coating is beneficial? Use $\alpha = 0.01$.

a. Solve using the p-value approach.

b. Solve using the classical approach.

10.28   To test the effect of a physical fitness course on one's physical ability, the number of sit-ups that a person could do in one minute, both before and after the course, was recorded. Ten randomly selected participants scored as shown in the following table. Can you conclude that a significant amount of improvement took place? Use $\alpha = 0.01$.

Before	29	22	25	29	26	24	31	46	34	28
After	30	26	25	35	33	36	32	54	50	43

    a. Solve using the *p*-value approach.
    b. Solve using the classical approach.

10.29   A group of ten recently diagnosed diabetics were tested to determine whether an educational program was effective in increasing their knowledge of diabetes. They were given a test, before and after the educational program, concerning self-care aspects of diabetes. The scores on the test were as follows:

Patient	1	2	3	4	5	6	7	8	9	10
Before	75	62	67	70	55	59	60	64	72	59
After	77	65	68	72	62	61	60	67	75	68

The following MINITAB output may be used to determine whether the scores improved as a result of the program. Verify the following as shown on the output: mean difference (MEAN), standard deviation (STDEV), standard error of the difference (SE MEAN), $t^\star$ (T), and *p*-value.

```
MTB > LET C3 = C2 - C1
MTB > TTEST OF MU = 0 DIFFERENCES IN C3;
SUBC> ALTERNATIVE = 1.

TEST OF MU = 0.000 VS MU G.T. 0.000
        N     MEAN    STDEV    SE MEAN     T      P VALUE
C3      10    3.200    2.741    0.867     3.69     0.0025
```

10.30   As metal parts experience wear the metal is displaced. The table lists displacement measurements (mm) on metal parts that have undergone durability cycling for the equivalent of 100,000-plus miles. The first column is the serial number for the part, the second column lists the before test (BT) displacement measurement of the new part, the third column lists the end of test (EOT) measurements, and the fourth column lists the change (i.e., wear) in the parts.

Serial	Displacement (mm)		
Number	BT	EOT	Difference
1	4.609	4.604	−0.005
2	5.227	5.208	−0.019
3	5.255	5.193	−0.062
4	4.622	4.601	−0.021
5	4.630	4.589	−0.041
6	5.207	5.188	−0.019
7	5.239	5.198	−0.041
8	4.605	4.596	−0.009
9	4.622	4.576	−0.046
10	4.753	4.736	−0.017
11	5.226	5.218	−0.008
12	5.094	5.057	−0.037
13	4.702	4.683	−0.019
14	5.152	5.111	−0.041

Source: Problem data provided by AC Rochester Division, General Motors, Rochester, N.Y.

a. Does it seem right that all the difference values are negative? Explain.

Use a computer to complete the following:

b. Find the sample mean, variance, and standard deviation for the before test (BT) data.
c. Find the sample mean, variance, and standard deviation for the end of test (EOT) data.
d. Find the sample mean, variance, and standard deviation for the difference data.
e. How is the sample mean difference related to the means of BT and EOT? Are the variances and standard deviations related in the same way?
f. At the 0.01 level, do the data show that a significant amount of wear took place?

 10.31 An article titled "Fuel and Energy Metabolism in Fasting Humans" (*American Journal of Clinical Nutrition*, July 1994) compared the fasting glucose levels (mmol/L) of 6 volunteers after 12 hours of fasting and after 60 hours of fasting. The mean difference was found significant with a *p*-value less than 0.001. Suppose a similar experiment with 6 volunteers gave the following results.

Subject	After 12 Hours of Fasting	After 60 Hours of Fasting
1	6.5	5.9
2	7.1	6.9
3	5.5	5.2
4	8.0	6.9
5	6.3	5.9
6	7.1	6.5

Let $d = 12$ hours fasting $- 60$ hours fasting.

Test the null hypothesis that the mean difference equals zero versus the alternative that the mean difference is positive at $\alpha = 0.05$.

a. Complete the hypothesis test using the p-value approach.

b. Complete the hypothesis test using the classical approach.

 10.32   An article titled "Influencing Diet and Health Through Project LEAN" (*Journal of Nutrition Education*, July/August 1994) compared 28 individuals with borderline-high or high cholesterol levels before and after a nutrition education session. The participants' cholesterol levels were significantly lowered, and the p-value was reported to be less than 0.001. A similar study involving 10 subjects was performed with the following results.

Subject	Presession Cholesterol	Postsession Cholesterol
1	295	265
2	279	266
3	250	245
4	235	240
5	255	230
6	290	230
7	310	235
8	260	250
9	275	250
10	240	215

Let $d = $ presession cholesterol $-$ postsession cholesterol.

Test the null hypothesis that the mean difference equals zero versus the alternative that the mean difference is positive at $\alpha = 0.05$. Assume normality.

a. Complete the hypothesis test using the p-value approach.

b. Complete the hypothesis test using the classical approach.

10.3 | INFERENCES CONCERNING THE DIFFERENCE BETWEEN MEANS USING TWO INDEPENDENT SAMPLES

When comparing the means of two populations, we typically consider the difference between their means, $\mu_1 - \mu_2$. The inferences about $\mu_1 - \mu_2$ will be based on the difference between the observed sample means, $\bar{x}_1 - \bar{x}_2$. This observed difference, $\bar{x}_1 - \bar{x}_2$, belongs to a sampling distribution, the characteristics of which are described in the following statement.

Why is $\bar{x}_1 - \bar{x}_2$ an unbiased estimator of $\mu_1 - \mu_2$?

If independent samples of sizes n_1 and n_2 are drawn randomly from large populations with means μ_1 and μ_2 and variances σ_1^2 and σ_2^2, respectively, the sampling distribution of $\bar{x}_1 - \bar{x}_2$, the difference between the sample means, has

1. a mean, $\mu_{\bar{x}_1 - \bar{x}_2} = \mu_1 - \mu_2$

$$2. \text{ a standard error, } \sigma_{\bar{x}_1 - \bar{x}_2} = \sqrt{\left(\frac{\sigma_1^2}{n_1}\right) + \left(\frac{\sigma_2^2}{n_2}\right)} \qquad \textbf{(10-6)}$$

If both populations have normal distributions, then the sampling distribution of $\bar{x}_1 - \bar{x}_2$ will also be normally distributed.

The preceding statement is true for all sample sizes given that the populations involved are normal and the population variances σ_1^2 and σ_2^2 are known quantities. However, as with inferences about one mean, the variance of a population is generally an unknown quantity. Therefore, it will be necessary to estimate the standard error by replacing the variances, σ_1^2 and σ_2^2, in formula (10-6) with the best estimates available, namely, the sample variances, s_1^2 and s_2^2. The estimated standard error will be found using formula (10-7).

$$\text{Estimated standard error} = \sqrt{\left(\frac{s_1^2}{n_1}\right) + \left(\frac{s_2^2}{n_2}\right)} \qquad \textbf{(10-7)}$$

Inferences about the difference between two population means $\mu_1 - \mu_2$ will be based on the following assumptions.

> ### THE ASSUMPTIONS FOR INFERENCES ABOUT THE DIFFERENCE BETWEEN TWO MEANS $\mu_1 - \mu_2$:
>
> The samples are randomly selected from normally distributed populations, and the samples are selected in an independent manner.

No assumptions are made about the population variances.

Since the samples provide the information for determining the standard error, the t distribution will be used as the test statistic. The inferences are divided into two cases.

▼ *Case 1:* The t distribution will be used, and the number of degrees of freedom will be calculated.

▼ *Case 2:* The t distribution will be used, and the number of degrees of freedom will be approximated.

Case 1 will occur when you are completing the inferences *using a computer and the statistical software calculates the number of degrees of freedom* for you. The calculated value for df is a function of both sample sizes and their relative sizes, and both sample variances and their relative sizes. The value of df will be a number between the smaller of $\text{df}_1 = n_1 - 1$ or $\text{df}_2 = n_2 - 1$, and the sum of the degrees of freedom, $\text{df}_1 + \text{df}_2 = [(n_1 - 1) + (n_2 - 1)] = n_1 + n_2 - 2$.

Case 2 will occur when you are completing the inference *without the aid of a computer and its statistical software package.* Use of the t distribution with the smaller of $\text{df}_1 = n_1 - 1$ or $\text{df}_2 = n_2 - 1$ degrees of freedom will give conservative

---

EXERCISE   10.33

Two independent random samples resulted in the following:

Sample 1:
$n_1 = 12$ and
$s_1^2 = 190$

Sample 2:
$n_2 = 18$ and
$s_2^2 = 150$

Find the estimate for the standard error for the difference of two means.

EXERCISE 10.34

Two independent random samples of sizes 18 and 24 were obtained to make inferences about the difference between two means. What is the number of degrees of freedom? Discuss both cases.

Would you say the difference between 5 and 8 is −3? How would you express the difference? Explain.

results. Because of this approximation, the true level of confidence for an interval estimate will be slightly higher than the reported level of confidence; or the true *p*-value and the true level of significance for a hypothesis test will be slightly less than reported. The gap between these reported values and the true values will be quite small, unless the sample sizes are quite small and unequal or the sample variances are very different in value. The gap will decrease as the samples increase in size or as the sample variances are more alike in value.

Since the only difference between the two cases is the number of degrees of freedom used to identify the *t* distribution involved, we will study case 2 first.

NOTE $A > B$ ("A is greater than B") is equivalent to $B < A$ ("B is less than A"). When the difference between A and B is being discussed, it is customary to express the difference as "larger subtract smaller" so that the resulting difference is positive, $A - B > 0$. To express the difference as "smaller subtract larger" results in $B - A < 0$ ("the difference is negative") and is at best "clever," and is usually unnecessarily confusing. Therefore, it is recommended that the difference be expressed "larger subtract smaller."

Confidence Intervals

We will use formula (10-8) for calculating the endpoints of the $1 - \alpha$ confidence interval.

$$(\bar{x}_1 - \bar{x}_2) - t\left(df, \frac{\alpha}{2}\right)\sqrt{\left(\frac{s_1^2}{n_1}\right) + \left(\frac{s_2^2}{n_2}\right)} \quad \text{to} \quad (\bar{x}_1 - \bar{x}_2) + t\left(df, \frac{\alpha}{2}\right)\sqrt{\left(\frac{s_1^2}{n_1}\right) + \left(\frac{s_2^2}{n_2}\right)}, \qquad \textbf{(10-8)}$$

where df equals the smaller of df_1 or df_2 when calculating the confidence interval without the aid of a computer and its statistical software

▼| ILLUSTRATION   10 - 7

The heights (measured in inches) of 20 randomly selected women and 30 randomly selected men were independently obtained from the student body of a certain college in order to estimate the difference in their mean heights. The sample information is given in Table 10-2. Assume that heights are approximately normally distributed for both populations.

TABLE 10-2

Sample Information for Illustration 10-7

Sample	Number	Mean	Standard Deviation
Female (*f*)	20	63.8	2.18
Male (*m*)	30	69.8	1.92

Find the 95% confidence interval for the difference between the mean heights, $\mu_m - \mu_f$.

SOLUTION

STEP 1   Population parameter of interest: The difference between the mean height of male students and the mean height of female students, $\mu_m - \mu_f$.

STEP 2   Confidence interval criteria:
     a. Check the assumptions: Both populations are approximately normal, and the samples were random and independently selected.
     b. Test statistic: t with df $= 19$; the smaller of $n_m - 1 = 30 - 1 = 29$ or $n_f - 1 = 20 - 1 = 19$.
     c. Confidence level: $1 - \alpha = 0.95$.

STEP 3   Sample evidence:
     a. Sample information: See Table 10-2.
     b. Point estimate: For $\mu_m - \mu_f$ it is $\bar{x}_m - \bar{x}_f = 69.8 - 63.8 =$ **6.0** inches.

STEP 4   Confidence interval limits:
     a. Confidence coefficients: We have a two-tailed situation with $\alpha/2 = 0.025$ in one tail and df $= 19$. From Table 6 in Appendix B, $t(\text{df}, \alpha/2) = t(19, 0.025) = 2.09$.

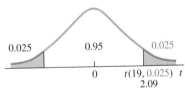

     b. Maximum error: Using the maximum error part of formula (10-8),

$$E = t\left(\text{df}, \frac{\alpha}{2}\right)\sqrt{\left(\frac{s_m^2}{n_m}\right) + \left(\frac{s_f^2}{n_f}\right)} = 2.09\sqrt{\left(\frac{1.92^2}{30}\right) + \left(\frac{2.18^2}{20}\right)} = (2.09)(0.60) = \mathbf{1.25}$$

     c. Confidence limits:

$$(\bar{x}_m - \bar{x}_f) \pm E$$
$$6.00 \pm 1.25$$
$$6.00 - 1.25 = 4.75 \text{ to } 6.00 + 1.25 = 7.25$$

4.75 to 7.25, the 95% confidence interval for $\mu_m - \mu_f$

That is, with 95% confidence, we can say that the difference between the mean heights of male and female students is between 4.75 and 7.25 inches.

▲

EXERCISE 10.35

Find the 90% confidence interval for the difference between two means based on this information about two samples. Assume independent samples from normal populations.

Sample Number	Mean	St. Dev.
1	20	35	22
2	15	30	16

Hypothesis Tests

When testing a null hypothesis about the difference between two population means, the test statistic used will be the difference between the observed difference of sample means and the hypothesized difference of the population means, divided by the estimated standard error. The test statistic is assumed to have a t distribution when the null hypothesis is true and the normality assumption has been satisfied. The calculated value of the test statistic is found using formula (10-9).

$$t^\star = \frac{(\bar{x}_1 - \bar{x}_2) - (\mu_1 - \mu_2)}{\sqrt{\left(\dfrac{s_1^2}{n_1}\right) + \left(\dfrac{s_2^2}{n_2}\right)}}, \qquad (10\text{-}9)$$

where df is the smaller of df_1 or df_2 when calculating $t^\star$ without the aid of a computer and its statistical software

NOTE   The hypothesized difference between the two population means $\mu_1 - \mu_2$ can be any specified value. The most common value specified is zero; however, the difference can be nonzero.

p-Value Approach

▼ I L L U S T R A T I O N   10 - 8

Suppose that we are interested in comparing the academic success of college students who belong to fraternal organizations with the academic success of those who do not belong to fraternal organizations. The reason for the comparison centers on the recent concern that the fraternity members, on the average, are achieving academically at a lower level than the nonfraternal students. (Cumulative grade-point average is used to measure academic success.) Random samples of size 40 are taken from each population. The sample results are listed below.

EXERCISE   10.36

Find the value of $t^\star$ for the difference between two means based on this information about two samples:

Sample	Number	Mean	St. Dev.
1	18	38.2	14.2
2	25	43.1	10.6

TABLE 10-3

Sample Information for Illustration 10-8

Sample	Number	Mean	Standard Deviation
Fraternity members (f)	40	2.03	0.68
Nonmembers (n)	40	2.21	0.59

Complete a hypothesis test using $\alpha = 0.05$. Assume that grade-point averages for both groups are approximately normally distributed.

SOLUTION

STEP 1   Parameter of concern: The difference between the mean grade-point average for the fraternity members and that for the nonfraternity members.

STEP 2 Null and alternative hypotheses:

$$H_0: \mu_n - \mu_f = 0 \text{ (fraternity averages are no lower)}$$
$$H_a: \mu_n - \mu_f > 0 \text{ (fraternity averages are lower)}$$

Remember: "Larger subtract smaller" results in a positive difference.

STEP 3 Test criteria:
   a. Assumptions: Both populations are approximately normal, and random samples were selected. Since the two populations are separate, the samples are also independent.
   b. Test statistic: $t^{\star}$ calculated using formula (10-9) with df = the smaller of df_n or df_f; since both n's are 40, df = 40 − 1 = 39.
   c. Level of significance: $\alpha = 0.05$.

STEP 4 Sample evidence:
   a. Sample information: See Table 10-3.
   b. Test statistic:

$$t^{\star} = \frac{(\bar{x}_n - \bar{x}_f) - (\mu_n - \mu_f)}{\sqrt{\left(\dfrac{s_n^2}{n_n}\right) + \left(\dfrac{s_f^2}{n_f}\right)}} = \frac{(2.21 - 2.03) - (0.00)}{\sqrt{\left(\dfrac{0.59^2}{40}\right) + \left(\dfrac{0.68^2}{40}\right)}}$$

$$t^{\star} = \frac{0.18}{\sqrt{0.00870 + 0.01156}} = \frac{0.18}{0.1423} = 1.26$$

   c. The p-value: Using Table 7 in Appendix B, **P** $= P(t^{\star} > 1.26$, with df $= 39)$; $0.100 < \mathbf{P} < 0.119$.
      Using MINITAB (CDF 1.26; T 39.), the cumulative probability is 0.8924; therefore, p-value $= 1 - 0.8924 = \mathbf{0.1076}$.

See page 437 for info on MTB commands.

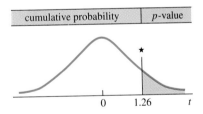

cumulative probability	p-value

STEP 5 Results: **P** is greater than α.

DECISION Fail to reject H_0.

CONCLUSION At the 0.05 level of significance, the claim that the fraternity members achieve at a level lower than nonmembers is not supported by the sample data.

▼| ILLUSTRATION   10 - 9

Two independent random samples were selected from two normally distributed populations for the purpose of comparing their means. The sample results are given in Table 10-4.

TABLE 10-4

Sample Information for Illustration 10-9

Sample	Number	Mean	Standard Deviation
A	15	53.8	5.9
B	12	56.9	10.3

The hypothesis "the mean of population A is different than the mean of B" is to be tested. Find the p-value.

SOLUTION

STEP 1   Parameter of concern: The difference between the mean of population A and the mean of population B, $\mu_B - \mu_A$.

STEP 2   Null and alternative hypotheses:

$$H_0: \mu_B - \mu_A = 0 \text{ (same)}$$
$$H_a: \mu_B - \mu_A \neq 0 \text{ (different)}$$

STEP 3   Test criteria:
  a. Assumptions: Both populations are given to be approximately normal, and the samples were random and independently selected.
  b. Test statistic: $t^\star$ calculated using formula (10-9) with df = the smaller of $n_A - 1 = 15 - 1 = 14$ or $n_B - 1 = 12 - 1 = 11$; df = 11.
  c. Level of significance: Not needed since the question asks for the p-value.

STEP 4   Sample evidence:
  a. Sample information: See Table 10-4.
  b. Test statistic:

$$t^\star = \frac{(\bar{x}_B - \bar{x}_A) - (\mu_B - \mu_A)}{\sqrt{\left(\frac{s_B^2}{n_B}\right) + \left(\frac{s_A^2}{n_A}\right)}} = \frac{(56.9 - 53.8) - (0.00)}{\sqrt{\left(\frac{10.3^2}{12}\right) + \left(\frac{5.9^2}{15}\right)}}$$

$$t^\star = \frac{3.1}{\sqrt{8.8408 + 2.3207}} = \frac{3.1}{3.3409} = 0.93$$

EXERCISE   10.37

Find the p-value
for the hypothesis
test involving
H_a: $\mu_B - \mu_A \neq 0$
with df = 18 and
$t^{\star} = 1.3$.

Calculate **P** *using
a computer.*

▲

c. The p-value: $\frac{1}{2}\mathbf{P} = P(t^{\star} > 0.93$, with df = 11), from Table 7 in Appendix B. (See the figure.)

$$0.169 < \tfrac{1}{2}\mathbf{P} < 0.195$$

$$\mathbf{0.34 < P < 0.39}$$

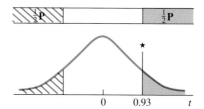

The p-value, the probability that these sample results or more extreme results will occur provided the null hypothesis is true and the assumptions are correctly satisfied, is approximately one chance in three.

Classical Approach

▼

ILLUSTRATION   10 - 10

Many students have complained that the soft-drink vending machine A (in the student recreation room) dispenses less drink than machine B (in the faculty lounge). To test this belief, a student randomly sampled several servings from each machine and carefully measured them, with the results as shown in Table 10-5.

TABLE 10-5

Machine	Number	Mean	Standard Deviation
A	10	5.38	1.59
B	12	5.92	0.83

Sample Information
for Illustration 10-10

Does this evidence support the hypothesis that the mean amount dispensed by A is less than the amount dispensed by B? Assume the amounts dispensed by both machines are normally distributed and complete the test using $\alpha = 0.05$.

SOLUTION

*Remember: "Larger
subtract smaller"
results in a positive
difference.*

STEP 1   Parameter of concern: The difference between the mean amount dispensed by machine A and the mean amount dispensed by machine B.

STEP 2   Null and alternative hypotheses:

$$H_0: \mu_B - \mu_A = 0 \text{ (no less)}$$
$$H_a: \mu_B - \mu_A > 0 \text{ (A dispenses less than B)}$$

STEP 3   Test criteria:

a. **Assumptions:** Both populations are assumed to be approximately normal, and the samples were random and independently selected.

b. **Test statistic:** $t^\star$ calculated using formula (10-9) with df = the smaller of $n_a - 1 = 10 - 1 = 9$ or $n_B - 1 = 12 - 1 = 11$; df = 9.

c. **Level of significance:** $\alpha = 0.05$.

d. **Critical value:** Critical region is right-hand tail, since alternative hypothesis is "greater than." From Table 6, $t(9, 0.05) = 1.83$. (See the figure.)

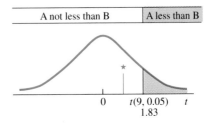

STEP 4   Sample evidence:

a. **Sample information:** See Table 10-5.

b. **Test statistic:**

$$t^\star = \frac{(\bar{x}_B - \bar{x}_A) - (\mu_B - \mu_A)}{\sqrt{\left(\dfrac{s_B^2}{n_B}\right) + \left(\dfrac{s_A^2}{n_A}\right)}} = \frac{(5.92 - 5.38) - (0.00)}{\sqrt{\left(\dfrac{0.83^2}{12}\right) + \left(\dfrac{1.59^2}{10}\right)}}$$

$$t^\star = \frac{0.54}{\sqrt{0.0574 + 0.2528}} = \frac{0.54}{0.557} = 0.97$$

STEP 5   **Results:** $t^\star$ is in the noncritical region as shown in red on the figure above.

DECISION Fail to reject H_0.

CONCLUSION The evidence is not sufficient to show that machine A dispenses less soft drink than machine B, at the 0.05 level of significance. Thus, for lack of evidence we will proceed as though the two machines dispense, on average, the same amount.

▲

EXERCISE   10.38

Suppose the calculated $t^\star$ had been 1.80 in Illustration 10-10. Using df = 9 or using df = 20 results in different answers. Explain how the word *conservative* (p. 513) applies here.

Most computer statistical packages will complete the inferences for the difference between two means by calculating the number of degrees of freedom. MINITAB's TWOSAMPLE command performs both the confidence interval and the hypothesis test at the same time.

Case Study

10.2

EXERCISE 10.39

a. Explain why the samples discussed in this article are independent samples.

b. Many of the reported *p*-values are 0.000, meaning that the *p*-value is less than 0.0005. Explain what is implied by these *p*-values.

c. The *p*-value for "activity in professional societies" is 0.394. Estimate the calculated t^*-value. Explain what is implied by this *p*-value.

d. Explain why there is reason to question the use of the two-sample *t*-test as studied in this chapter to analyze the data Tong and Bures collected. (*Hint:* It has to do with assumptions.)

An Empirical Study of Faculty Evaluation Systems: Business Faculty Perceptions

Hsin-Min Tong and Allen L. Bures report a whole series of t-test results from comparing various factors of perceptions between AACSB accredited schools and non-AACSB accredited schools. For each t-test they report a two-tailed probability (p-value).

Table 1 shows how respondents from American Assembly of Collegiate Schools of Business (AACSB)-accredited and non-AACSB-accredited schools rated the importance of the ten faculty evaluation factors. Clearly AACSB-accredited schools and non-AACSB-accredited schools are different, with the AACSB-accredited institutions placing greater weight on research and publication, and with the non-accredited institutions putting more emphasis on classroom teaching, campus committee work, student advising, public service, advisor to student organizations, and consultation (business, government). There is no significant difference between AACSB-accredited and non-AACSB-accredited schools in the importance given to a faculty member's activity in professional societies.

TABLE 1   *t*-Test Results of Importance of Faculty Evaluation Factors, AACSB-Accredited vs. Non-AACSB-Accredited Schools

Factor	AACSB-Accredited Schools ($n = 176$)	Non-AACSB-Accredited Schools ($n = 74$)	Two-Tailed *t*-Test Probability
Articles in professional journals	4.49	2.87	0.000
Classroom teaching	3.34	4.31	0.000
Books as author or editor	3.23	2.45	0.000
Papers at professional meetings	3.09	2.65	0.001
Activity in professional societies	2.55	2.65	0.394
Campus committee work	2.25	3.16	0.000
Student advising	1.80	3.36	0.000
Public service	1.98	2.47	0.000
Advisor to student organizations	1.71	2.30	0.000
Consultation (business, government)	1.73	2.30	0.001

Note: A five-point scale with 1 = "not at all important" to 5 = "extremely important" was used in this analysis.

Source: Hsin-Min Tong and Allen L. Bures in *Journal of Education for Business*, April 1987. Reprinted with permission of the Helen Dwight Reid Educational Foundation. Published by Heldref Publications, 4000 Albemarle St., N.W., Washington, D.C. 20016. Copyright © 1987.

MINITAB (Release 10) commands to complete a $1 - \alpha = K\%$ confidence interval or a hypothesis test for the difference between two population means with unknown standard deviation and with two independent sets of sample data in C1 and C2.

Session commands   *Menu commands*

Enter: Choose:
  TWOSample K data C1 C2;   Stat > Basic Stats > 2-Sample t
  ALTErnative −1 or 0 or +1.   Select: Different columns
                            Enter: C1
                            Enter: C2
                            Select Alternative:
                              less than or not equal to or
                                    greater than
                        Enter: $1 - \alpha$: K

Illustration 10-8 was solved using MINITAB. With 40 cumulative grade-point averages for nonmembers in C1 and 40 averages for fraternity members in C2, the following commands resulted in the output shown below. Compare these results to the solution of Illustration 10-8. Notice the difference in **P** and df values. Explain.

```
TWOSample 95 with data in C1 C2;
ALTErnative +1.
```

```
Two Sample T-Test and Confidence Interval

Twosample T for C1 vs C2

         N     Mean    StDev    SE Mean
C1      40     2.21    0.59     0.09
C2      40     2.03    0.68     0.11

95% C.I. for mu C1 − mu C2: (−0.10, 0.46)
T-Test mu C1 = mu C2 (vs >): T= 1.26 P= 0.106 DF= 75
```

Sample Summaries
Nonmembers
Fraternity members

95% confidence interval
Hypothesis test summary

EXERCISES

10.40   Find the confidence coefficient, $t(df, \alpha/2)$, that would be used to find the maximum error for each of the following situations when estimating the difference between two means, $\mu_1 - \mu_2$.

    a. $1 - \alpha = 0.95$, $n_1 = 25$, $n_2 = 15$ b. $1 - \alpha = 0.98$, $n_1 = 43$, $n_2 = 32$
    c. $1 - \alpha = 0.99$, $n_1 = 19$, $n_2 = 45$

10.41 An experiment was conducted to compare the mean absorptions of two drugs in specimens of muscle tissue. Seventy-two tissue specimens were randomly divided into two equal groups. Each group was tested with one of the two drugs. The sample results were $\bar{x}_A = 7.9$, $\bar{x}_B = 8.5$, $s_A = 0.11$, and $s_B = 0.10$. Assume both populations are normal. Construct the 98% confidence interval for the difference in the mean absorption rates.

10.42 Experimentation with a new rocket nozzle has led to two slightly different designs. The following data summaries resulted from testing these two designs.

	n	Σx	Σx^2
Design 1	36	278.4	2163.76
Design 2	42	310.8	2332.26

Determine the 99% confidence interval for the difference in the means for these two rocket nozzles.

10.43 A study was designed to estimate the difference in diastolic blood pressures between men and women. MINITAB was used to construct a 99% confidence interval for the difference between the means based on the following sample data.

Males	76	76	74	70	80	68	90	70	90	72	76	80	68	72	96	80
Females	76	70	82	90	68	60	62	68	80	74	60	62	72			

```
MTB > TWOSample 99.0 C1 C2
Twosample T for males vs females

              N    Mean    StDev   SE Mean
males        16    77.37    8.35     2.1
females      13    71.08    9.22     2.6

99% C.I. for mu males − mu females: ( −2.9, 15.5)
```

Verify the results by calculating the two sample means and standard deviations, and the confidence interval bounds.

10.44 A study comparing attitudes toward death was conducted in which organ donors (individuals who had signed organ donor cards) were compared with nondonors. The study is reported in the journal *Death Studies* (Vol. 14, No. 3, 1990). *Templer's Death Anxiety Scale* (DAS) was administered to both groups. On this scale, high scores indicate high anxiety concerning death. The results were reported as follows:

	n	Mean	St. Dev.
Organ Donors	25	5.36	2.91
Nonorgan Donors	69	7.62	3.45

Construct the 95% confidence interval for difference between the means, $\mu_{non} - \mu_{donor}$.

10.45 The two independent samples shown in the following table were obtained in order to estimate the difference between the two population means. Construct the 98% confidence interval.

Sample A	6	7	7	6	6	5	6	8	5	4
Sample B	7	2	4	3	3	5	4	6	4	2

 10.46 "Is the length of a steel bar affected by the heat treatment technique used?" This was the question being tested when the following data were collected.

Heat Treatment	Lengths (to the nearest inch)							
1	156	159	151	153	157	159	155	155
	151	152	158	154	156	156	157	155
	156	159	153	157	157	159	158	155
	159	152	150	154	156	156	157	160
2	154	156	150	151	156	155	153	154
	149	150	150	151	154	155	155	154
	154	156	150	151	156	154	153	154
	149	150	150	151	154	148	155	158

a. Find the means and standard deviations for the two sets of data.
b. Find evidence about the sample data (both graphic and numeric) that supports the assumption of normality for the two sampled populations.
c. Find the 95% confidence interval for $\mu_1 - \mu_2$.

 10.47 A USA Snapshot® (*USA Today*, 10-12-94) reported the longest average workweeks for non-supervisory employees in private industry to be mining (45.4 hours) and manufacturing (42.3 hours). The same article reported the shortest average workweeks to be retail trade (29 hours) and services (32.4 hours). A study conducted in Missouri found the following results for a similar study.

Industry	n	Average Hours/Week	Standard Deviation
Mining	15	47.5	5.5
Manufacturing	10	43.5	4.9

Set a 95% confidence interval on the difference in average length of workweek between mining and manufacturing. Assume normality for the sampled populations and that the samples were selected randomly.

10.48 An article titled "Stages of Change for Reducing Dietary Fat to 30% of Energy or Less" (*Journal of the American Dietetic Association*, Vol. 94, No. 10, October 1994) measured the energy from fat (expressed as a percent) for two different groups. Sample 1 was a random sample of 614 adults who responded to mailed questionnaires, and sample 2 was a convenience sample of 130 faculty, staff, and graduate students. The following table gives the percent of energy from fat for the two groups.

Group	n	Mean	Standard Deviation
1	614	35.0	6.3
2	130	32.0	9.1

a. Construct the 95% confidence interval on $\mu_1 - \mu_2$.
b. Do these samples satisfy the assumptions for this confidence interval? Explain.

10.49 State the null and alternative hypotheses that would be used to test the following claims.
a. There is a difference between the mean age of employees at the two different large companies.
b. The mean of population 1 is greater than the mean of population 2.
c. The difference between the means of the two populations is more than 20 pounds.
d. The mean of population A is less than 50 more than the mean of population B.

10.50 Calculate the estimate for the standard error of difference between two independent means for each of the following cases.
a. $s_1^2 = 12,$ $s_2^2 = 15,$ $n_1 = 16,$ and $n_2 = 21$
b. $s_1^2 = 0.054,$ $s_2^2 = 0.087,$ $n_1 = 8,$ and $n_2 = 10$
c. $s_1 = 2.8,$ $s_2 = 6.4,$ $n_1 = 16,$ and $n_2 = 21$

10.51 Determine the p-value for the following hypothesis tests for the difference between two means with population variances unknown.
a. $H_a: \mu_1 - \mu_2 > 0,$ $n_1 = 6,$ $n_2 = 10,$ $t^\star = 1.3$
b. $H_a: \mu_1 - \mu_2 < 0,$ $n_1 = 16,$ $n_2 = 9,$ $t^\star = -2.8$
c. $H_a: \mu_1 - \mu_2 \neq 0,$ $n_1 = 26,$ $n_2 = 16,$ $t^\star = 1.8$
d. $H_a: \mu_1 - \mu_2 \neq 5,$ $n_1 = 26,$ $n_2 = 35,$ $t^\star = 1.8$

10.52 Determine the test criteria that would be used for the following hypothesis tests (using the classical approach) about the difference between two means with population variance unknown.
a. $H_a: \mu_1 - \mu_2 \neq 0,$ $n_1 = 26,$ $n_2 = 16,$ $\alpha = 0.05$
b. $H_a: \mu_1 - \mu_2 < 0,$ $n_1 = 36,$ $n_2 = 27,$ $\alpha = 0.01$
c. $H_a: \mu_1 - \mu_2 > 0,$ $n_1 = 8,$ $n_2 = 11,$ $\alpha = 0.10$
d. $H_a: \mu_1 - \mu_2 \neq 10,$ $n_1 = 14,$ $n_2 = 15,$ $\alpha = 0.05$

10.53 A study was designed to compare the attitudes of two groups of nursing students toward computers. Group 1 had previously taken a statistical methods course that involved significant computer interaction through the use of statistical packages. Group 2 had taken a statistical methods course that did not use computers. The students' attitudes were measured by administering the Computer Anxiety Index (CAIN). The results were as follows:

$$\text{Group 1 (with computers):} \quad n = 10 \quad \bar{x} = 60.3 \quad s = 7.5$$
$$\text{Group 2 (without computers):} \quad n = 15 \quad \bar{x} = 67.2 \quad s = 2.1$$

Do the data show that the mean score for those with computer experience was significantly less than the mean score for those without computer experience? Use $\alpha = 0.05$.

10.54 If a random sample of 18 homes south of Center Street in Provo has a mean selling price of $125,000 and a standard deviation of $2,400, and a random sample of 18 homes north of Center Street has a mean selling price of $126,000 and a standard deviation of $4,800, can you conclude that there is a significant difference between the selling price of homes in these two areas of Provo at the 0.05 level?
  a. Solve using the *p*-value approach.
  b. Solve using the classical approach.

10.55 A study was designed to investigate the effect of a calcium-deficient diet on lead consumption in rats. One hundred rats were randomly divided into 2 groups of 50 each. One group served as a control group, and the other was the experimental, or calcium-deficient group. The response recorded was the amount of lead consumed per rat. The results were summarized by:

$$\text{Control group:} \quad n = 50 \quad \bar{x} = 5.2 \quad s = 1.1$$
$$\text{Experimental:} \quad n = 50 \quad \bar{x} = 7.6 \quad s = 1.3$$

Test H_0: $\mu_E - \mu_C = 0$ versus H_a: $\mu_E - \mu_C > 0$ at $\alpha = 0.05$.
  a. Solve using the *p*-value approach.
  b. Solve using the classical approach.

10.56 The purchasing department for a regional supermarket chain is considering two sources from which to purchase 10-lb bags of potatoes. A random sample taken from each source shows the following results.

	Idaho Supers	Idaho Best
Number of Bags Weighed	100	100
Mean Weight	10.2 lb	10.4 lb
Sample Variance	0.36	0.25

At the 0.05 level of significance, is there a difference between the mean weights of the 10-lb bags of potatoes?
  a. Solve using the *p*-value approach.
  b. Solve using the classical approach.

10.57   MINITAB was used to complete a *t*-test of the difference between the two means using the following two independent samples.

Sample 1	33.7	21.6	32.1	38.2	33.2	35.9	34.1	39.8	23.5	21.2	23.3	18.9	30.3		
Sample 2	28.0	59.9	22.3	43.3	43.6	24.1	6.9	14.1	30.2	3.1	13.9	19.7	16.6	13.8	62.1	28.1

```
MTB > TWOSample C1 C2
Twosample T for sample 1 vs sample 2

              N     Mean    StDev    SE Mean
sample1      13    29.68     7.07      2.0
sample2      16    26.9     17.4       4.4

T-Test mu sample1 = mu sample2 (vs not =) : T=0.59 P=0.56 DF=20
```

a. Verify the results by calculating the two sample means and standard deviations, and the calculated $t^\star$.
b. Use Table 7 in Appendix B to verify the *p*-value based on the calculated df.
c. Find the *p*-value using the smaller number of degrees of freedom. Compare the two *p*-values.

 10.58   A study was conducted to assess the safety and efficiency of receiving nitroglycerin from a transdermal system (i.e., a patch worn on the skin), which intermittently delivers the medication, versus oral medication (pills). Twenty patients who suffer from angina (chest pain) due to physical effort were enrolled in trials. All received patches, some ($n = 8$) contained nitroglycerin, the others ($n = 12$) contained a placebo. Suppose the resulting "time to angina" data were summarized:

	Mean Time to Angina (sec)				
	Active	**Placebo**	**Difference**	**SE**	**p-Value**[a]
Day 1 AM	320.00	287.00	33.00	9.68	0.0029
Day 7 PM	314.00	285.25	28.75	13.74	0.0500

[a]For treatment difference.

a. Determine the value of *t* for the difference between two independent means given the difference and the standard error (SE) for the day 1 AM data.
b. Verify the *p*-value.
c. Determine the value of *t* for the difference between two independent means given the difference and the standard error (SE) for the day 7 PM data.
d. Verify the *p*-value.

10.59   Twenty laboratory mice were randomly divided into two groups of 10. Each group was fed according to a prescribed diet. At the end of three weeks, the weight gained by each animal was recorded. Do the data in the following table justify the conclusion that the mean weight gained on diet B was greater than the mean weight gained on diet A, at the $\alpha = 0.05$ level of significance?

Diet A	5	14	7	9	11	7	13	14	12	8
Diet B	5	21	16	23	4	16	13	19	9	21

10.60   The quality of latex paint is monitored by measuring different characteristics of the paint. One characteristic of interest is the particle size. Two different types of disc centrifuges (JLDC, Joyce Loebl Disc Centrifuge, and the DPJ, Dwight P. Joyce disc) are used to measure the particle size. It is thought that these two methods yield different measurements. Thirteen readings were taken from the same batch of latex paint using both the JLDC and the DPJ discs.

JLDC	DPJ
4714	4295
4601	4271
4696	4326
4896	4530
4905	4618
4870	4779
4987	4752
5144	4744
3962	3764
4006	3797
4561	4401
4626	4339
4924	4700

Source: With permission of SCM Corporation.

Assuming particle size is normally distributed:
   a. Determine whether there is a significant difference between the readings at the 0.10 level of significance.
   b. What is your estimate for the difference between the two readings?

10.61   The material used in making parts affects not only how long the part lasts but also how difficult it is to take apart to repair. The following measurements are for screw torque removal for a specific screw after several operations of use. The first column lists the part number, the second column lists the screw torque removal measurements for assemblies made with material A, and the third column lists the screw torque removal measurements for assemblies made with material B. Assume torque measurements are normally distributed.

Part Number	Removal Torque (NM, Newton-meters)	
	Material A	Material B
1	16	11
2	14	14
3	13	13
4	17	13
5	18	10
6	15	15
7	17	14
8	16	12
9	14	11
10	16	14
11	15	13
12	17	12
13	14	11
14	16	13
15	15	12

Source: Problem data provided by AC Rochester Division, General Motors, Rochester, NY.

a. Find the sample mean, variance, and standard deviation for the material A data.
b. Find the sample mean, variance, and standard deviation for the material B data.
c. At the 0.01 level, do these data show a significant difference in the mean torque required to remove the screws from the two different materials?

 10.62 A study in *Holistic Nursing Practice* (July 1992) compares the sleep–wake frequency for two groups of neonates (newborns). The mean sleep–wake frequency for neonates less than 12 hours old was not significantly different from the mean sleep–wake frequency for neonates greater than 12 hours old. Suppose the following results were obtained for the sleep–wake frequencies for two such groups.

Sample	n	Mean	s
Neonates < 12 hours old	40	7.38	6.30
Neonates ≥ 12 hours old	40	6.86	5.27

a. Find the value for $t^{\star}$ and the corresponding p-value for testing $H_0: \mu_1 = \mu_2$ versus $H_a: \mu_1 \neq \mu_2$.
b. What evidence is given that would indicate that the sampled populations are not normally distributed? What effect might this have on the hypothesis test in (a)?

10.63 Use a computer to demonstrate the truth of the statement describing the sampling distribution of $\bar{x}_1 - \bar{x}_2$. Use two theoretical normal populations: $N_1(100, 20)$ and $N_2(120, 20)$.

a. To get acquainted with the two theoretical populations, randomly select a very large sample from each. Generate 2000 data values; calculate mean and standard deviation; and construct a histogram using class boundaries that are multiples of one-half of a standard deviation (10) starting at the mean for each population.

b. If samples of size 8 are randomly selected from each population, what do you expect the distribution of $\bar{x}_1 - \bar{x}_2$ to be like (shape of distribution, mean, standard error)?

c. Randomly draw a sample of size 8 from each population, and find the mean of each sample. Find the difference between the sample means. Repeat 99 more times.

d. The set of 100 $(\bar{x}_1 - \bar{x}_2)$ values form an empirical sampling distribution of $\bar{x}_1 - \bar{x}_2$. Describe the empirical distribution: shape (histogram), mean, and standard error. (Use class boundaries that are multiples of standard error from mean for easy comparison to the expected.)

e. Using the information found above, verify the statement about the $\bar{x}_1 - \bar{x}_2$ sampling distribution made on page 512.

f. Repeat the experiment a few times and compare the results.

```
MINITAB commands

a. Populations
RANDom 2000  C1;
  NORMal 100 20.
RANDom 2000 C2;
  NORMal 120 20.
DESCribe C1 C2
HISTogram C1;
  CUTPoint 20:180/10.
HISTogram C2;
  CUTPoint 40:200/10.
c. Samples
RANDom 100 C3-C10;
  NORMal 100 20.
RMEAn C3-C10 C11
RANDom 100 C12-C19;
  NORMal 120 20.
RMEAn C12-C19 C20
LET C21 = C20 - C11
d. Sampling Distribution
DESCribe C21
HISTogram C21;
  GRID 2;
  CUTPoint -20:60/10.
```

The MINITAB commands needed for these exercises are on page 531.

10.64 One of the reasons for being conservative when determining the number of degrees of freedom to use with the t distribution was the possibility that the population variances might be unequal. Extremely different values cause a lowering in the number of df used. Repeat Exercise 10.63 using theoretical normal distributions of $N(100, 9)$ and $N(120, 27)$ and both sample sizes of 8. Check all three properties of the sampling distribution: normality, its mean value, and its standard error. Describe in detail what you discover. Do you think we should be concerned about the choice of df? Explain.

10.65 Unbalanced sample sizes is a factor in determining the number of degrees of freedom for inferences about the difference between two means. Repeat Exercise 10.63 using theoretical normal distributions of $N(100, 20)$ and $N(120, 20)$ and sample sizes of 5 and 20. Check all three properties of the sampling distribution: normality, its mean value, and its standard error. Describe in detail what you discover. Do you think we should be concerned when using unbalanced sample sizes? Explain.

10.66 One part of the assumptions for the two-sample t-test is the "sampled populations are to be normally distributed." What happens when they are not normally distributed? Repeat Exercise 10.63 using two theoretical populations that are not normal and using samples of size 10. The exponential distribution has a continuous random variable, it has a J-shaped distribution, and its mean and standard deviation are the same value. Use the two exponential distributions with means of 50 and 80: Exp(50) and Exp(80). Check all three properties of the sampling distribution: normality, its mean value, and its standard error. Describe in detail what you discover. Do you think we should be concerned when sampling nonnormal populations? Explain.

10.64 MINITAB commands	10.65 MINITAB commands	10.66 MINITAB commands

```
# Populations
RANDom 2000 C1;
  NORMal 100 9.
RANDom 2000 C2;
  NORMal 120 27.
DESCribe C1 C2
HISTogram C1;
  CUTPoint 64:136/4.5.
HISTogram C2;
  CUTPoint 12:228/13.5.

# Samples
RANDom 100 C3-C10;
  NORMal 100 9.
RMEAn C3-C10 C11
RANDom 100 C12-C19;
  NORMal 120 27.
RMEAn C12-C19 C20

# Sampling
Distribution
LET C21 = C20 - C11
DESCribe C21
HISTogram C21;
  GRID 2;
  CUTPoint -20:60/10.
```

```
# Populations
RANDom 2000 C1;
  NORMal 100 20.
RANDom 2000 C2;
  NORMal 120 20.
DESCribe C1 C2
HISTogram C1;
  CUTPoint 20:180/10.
HISTogram C2;
  CUTPoint 40:200/10.

# Samples
RANDom 100 C3-C7;
  NORMal 100 20.
RMEAn C3-C7 C8
RANDom 100 C9-C28;
  NORMal 120 20.
RMEAn C9-C28 C29

# Sampling
Distribution
LET C30 = C29 - C8
DESCribe C30
HISTogram C30;
  GRID 2;
  CUTPoint -20:60/10.
```

```
# Populations
RANDom 2000 C1;
  EXPOnential 50.
HISTogram C1;
  CUTPoint 0:350/25.
RANDom 2000 C2;
  EXPOnential 80.
HISTogram C2;
  CUTPoint 0:560/40.
DESCribe C1 C2

# Samples
RANDom 100 C3-C12;
  EXPOnential 50.
RMEAn C3-C12 C13
RANDom 100 C14-C23;
  EXPOnential 80.
RMEAn C14-C23 C24

# Sampling
Distribution
LET C25 = C24 - C13
DESCribe C25
HISTogram C25;
  GRID 2;
  CUTPoint -90:180/30.
```

10.4 | INFERENCES CONCERNING THE RATIO OF VARIANCES USING TWO INDEPENDENT SAMPLES

When comparing two populations, it is natural that we compare their two most fundamental distribution characteristics, their "center" and their "spread," by comparing their means and standard deviations. We have learned, in the two previous sections, how to use the t distribution to make inferences comparing two population means using either dependent or independent samples. These procedures were intended for uses with normal populations but work quite well even when the populations are not exactly normally distributed.

The next logical step in comparing two populations is to compare their standard deviations, the most often used measure of spread. However, sampling distributions dealing with sample standard deviations (or variances) are very sensitive to slight departures from the assumptions. Therefore, the only inference procedure to be presented here will be the hypothesis test for the equality of standard deviations (or variances) for two normal populations.

The soft-drink bottling company discussed in Section 9-3 (page 471) is trying to decide whether to install a modern, high-speed bottling machine. There are, of course, many concerns in making this decision, and one of them is the concern that the increased speed may result in an increased variability in the amount of fill placed in each bottle; and such an increase would not be acceptable. To this, the manufacturer of the new system responded that the variance in fills will be no larger with the new machine than with the old. (The new system will fill several bottles in the same amount of time as the old system fills one bottle—the reason why the change is being considered.) A test is set up to statistically test the bottling company's concern "standard deviation of new machine is greater than standard deviation of old" against the manufacturer's claim "standard deviation of new is no greater than standard deviation of old."

▼ | ILLUSTRATION   10 - 11

State the null and alternative hypotheses to be used for comparing the variances of the two soft-drink bottling machines.

SOLUTION There are several equivalent ways to express the null and alternative hypotheses, but since the test procedure uses the ratio of variances, the recommended convention is to express the null and alternative hypotheses as ratios of the population variances. Further, it is recommended that the "larger" or "expected to be larger" variance be used as the numerator. The concern of the soft-drink company is that the new modern machine (m) will result in a larger standard deviation in the amount of fill than its present machine (p); $\sigma_m > \sigma_p$, or equivalently, $\sigma_m^2 > \sigma_p^2$, which becomes $\sigma_m^2/\sigma_p^2 > 1$ when the inequality is expressed with a ratio. The manufacturer of the new machine claims the variance of the new machine is no larger; that becomes $\sigma_m^2 \le \sigma_p^2$, or equivalently, $\sigma_m^2/\sigma_p^2 \le 1$. We want to test the manufacturer's claim (the null hypothesis) against the company's concern (the alternative hypothesis).

$$H_0: \frac{\sigma_m^2}{\sigma_p^2} = 1 \ (\le) \ \text{(no more variable)}$$

▲ |

$$H_a: \frac{\sigma_m^2}{\sigma_p^2} > 1 \quad \text{(new is more variable)}$$

F distribution

Inferences about the ratio of variances for two normally distributed populations use the **F distribution**. The F distribution, similar to the Student's t distribution and the χ^2 distribution, is a family of probability distributions. Each F distribution is identified by two numbers of degrees of freedom, one for each of the two samples involved.

Before continuing with the details of the hypothesis test procedure, let's learn about the F distribution.

PROPERTIES OF THE *F* DISTRIBUTION

1. *F* is nonnegative in value; it is zero or positively valued.
2. *F* is nonsymmetrical; it is skewed to the right.
3. *F* is distributed so as to form a family of distributions; there is a separate distribution for each pair of numbers of degrees of freedom.

For inferences discussed in this section, the number of degrees of freedom for each sample is $df_1 = n_1 - 1$ and $df_2 = n_2 - 1$. Each different combination of degrees of freedom results in a different *F* distribution; and each *F* distribution looks approximately like the distribution shown in Figure 10-2.

FIGURE 10-2

F Distribution

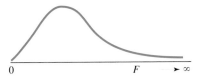

The critical values for the *F* distribution are identified using three values: $\mathbf{df_n}$, the degrees of freedom associated with the sample whose variance is in the numerator of the calculated $F^\star$; $\mathbf{df_d}$, the degrees of freedom associated with the sample whose variance is in the denominator; and $\boldsymbol{\alpha}$, the area under the distribution curve to the right of the critical value being sought. Therefore, the symbolic name for a critical value of *F* will be $F(df_n, df_d, \alpha)$, as shown in Figure 10-3.

FIGURE 10-3

A Critical Value of *F*

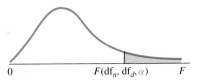

Since it takes three values to identify a single critical value of *F*, tabling *F* is not as simple as with previously studied distributions. The tables presented in this textbook are organized so as to have a different table for each different value of α, the "area to the right." Table 9a in Appendix B shows the critical values for $F(df_n, df_d, \alpha)$, when $\alpha = 0.05$; Table 9b gives the critical values when $\alpha = 0.025$; Table 9c gives the values when $\alpha = 0.01$.

▼ | ILLUSTRATION   10 - 12

Find $F(5, 8, 0.05)$, the critical *F*-value for samples of size 6 and size 9 with 5% of the area in the right-hand tail.

SOLUTION From Table 9a ($\alpha = 0.05$), we obtain the value shown in the accompanying partial table. Therefore, $F(5, 8, 0.05) = 3.69$.

EXERCISE 10.68

Find the values of $F(12, 24, 0.01)$ and $F(24, 12, 0.01)$.

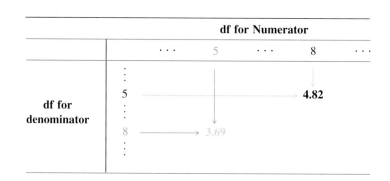

Notice that $F(8, 5, 0.05)$ is 4.82. The degrees of freedom associated with the numerator and with the denominator must be kept in the correct order; 3.69 is quite different from 4.82. Check some other pairs to verify that interchanging the degrees of freedom numbers will result in different F-values.

We are ready to use F to complete a hypothesis test about the ratio of two population variances.

> **The assumptions for inferences about the ratio of two variances,**
> σ_1^2/σ_2^2: The samples are randomly selected from normally distributed populations, and the two samples are selected in an independent manner.

▼ | ILLUSTRATION   10 - 13

Recall that our soft-drink bottling company was to make a decision about the equality of the variances of amounts of fill between its present machine and a modern high-speed outfit. Does the sample information in Table 10-6 present sufficient evidence to reject the null hypothesis (the manufacturer's claim) that the modern high-speed bottle-filling machine fills bottles with no more variance than the company's present machine? Assume the amount of fill is normally distributed for both machines, and complete the test using $\alpha = 0.01$.

TABLE 10-6

Sample Information for Illustration 10-13

Sample	n	s^2
Present machine (p)	22	0.0008
Modern high-speed machine (m)	25	0.0018

SOLUTION

STEP 1 Parameter of concern: The ratio of the variances in the amount of fill placed in bottles by the new machine and the company's present machine.

STEP 2 Null and alternative hypotheses: (established in Illustration 10-11).

$$H_0: \frac{\sigma_m^2}{\sigma_p^2} = 1 \ (\leq) \text{ (no more variable)}$$

$$H_a: \frac{\sigma_m^2}{\sigma_p^2} > 1 \quad \text{(more variable)}$$

EXERCISE 10.69

Express the H_0 and H_a of Illustration 10-13 equivalently in terms of standard deviations.

NOTE When the "expected to be largest" variance is in the numerator for a one-tail test, the alternative hypothesis will state, "the ratio of the variances is greater than 1."

STEP 3 Test criteria:
    a. Check the assumptions: The assumptions are two parts: the sampled populations are normally distributed (given in statement of problem) and the samples are independently selected (drawn from two separate populations).
    b. Test statistic: $F^\star$, the ratio of the sample variances.

The test statistic to be used is the ratio of the sample variances, formula (10-10).

$$F^\star = \frac{s_1^2}{s_2^2}, \quad \text{with} \quad df_1 = n_1 - 1 \quad \text{and} \quad df_2 = n_2 - 1 \qquad \textbf{(10-10)}$$

The sample variances are assigned to the numerator and denominator in the order established by the null and alternative hypotheses for one-tail tests. The calculated ratio, $F^\star$, will have an F distribution with $df_n = n_n - 1$ (numerator) and $df_d = n_d - 1$ (denominator) when the assumptions and the null hypothesis are true.

    c. Level of significance: $\alpha = 0.01$.

Both hypothesis test models (p-value and classical) share the same first three steps, namely, steps 1–3c. Illustration 10-13 is completed below, using the p-value approach (steps 4 and 5), and then on page 537, using the classical approach (steps 3d, 4, and 5).

Probability-Value Approach The solution of Illustration 10-13 is completed here using the p-value approach.

STEP 4 Collect and present the sample evidence:
    a. Sample information: See Table 10-6 on page 534.
    b. Calculate the test statistic:

$$F^\star = \frac{s_m^2}{s_p^2} = \frac{0.0018}{0.0008} = \textbf{2.25}$$

c. Calculate the *p*-value: *p*-value $= P(F^\star > 2.25$, with $\mathrm{df}_n = 24$ and $\mathrm{df}_d = 21$).

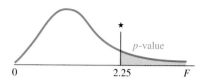

Using Tables 9a, b, and c in Appendix B to estimate the *p*-value is very limited. However, for Illustration 10-13, the *p*-value can be estimated. By inspecting Tables 9a and 9b, you will find that $F(24, 21, 0.025) = 2.37$ and $F(24, 21, 0.05) = 2.05$. $F^\star = 2.25$ is between the values 2.37 and 2.05; therefore, the *p*-value is between 0.025 and 0.05; **$0.025 < P < 0.05$**.

If you are working with a computer, the computer may be able to calculate the *p*-value for you by using the cumulative probability distribution for *F*. The MINITAB instructions for finding the cumulative probability (area under the curve from zero to $F^\star$) are given below. The *p*-value is the area to the right of $F^\star$ and is found by subtraction.

MINITAB (Release 10) commands to determine cumulative
probability from 0 up to the value of $F^\star$ = K for the
probability distribution $F(\mathrm{df}_n, \mathrm{df}_d)$:

Session commands   Menu commands

Enter: CDF K;   Choose: Calc > Prob. Dist. > F
       F df_n df_d.   Select: Cumulative Probability
                            Enter: df_n
                            Enter: df_d
                            Select: Input Constant
                            Enter: K

For Illustration 10-13, the cumulative probability is found using the CDF command.

```
CDF 2.25;
F 24 21.
```

```
Cumulative Distribution Function
F distribution with 24 d.f. in numerator and 21 d.f. in denominator

     X        P(X <= x)
  2.2500       0.9677
```

The *p*-value is then found by subtracting the cumulative probability from 1. See the figure.

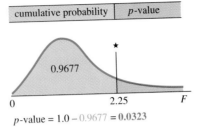

p-value = 1.0 − 0.9677 = **0.0323**

STEP 5   Results: *p*-value is larger than α.

DECISION Fail to reject H_0.

CONCLUSION At the 0.01 level of significance, the samples do not present suffi-cient evidence to indicate an increase in variance.

Classical Approach   The solution of Illustration 10-13 is completed here using the classical approach (steps 1–3c begin on page 535).

STEP 3   Test criteria:
d. Determine the critical region and critical value: The test is one-tailed and the critical region is the right-hand tail, since the alternative hypoth-esis says "greater than." The number of degrees of freedom for the numerator, df_n, = 24 (25 − 1) since the sample from the modern high-speed machine is associated with the numerator, as specified by the null hypothesis. df_d = 21 since the sample associated with the denomi-nator has size 22. The critical value is $F(24, 21, 0.01)$ = 2.80. (See the figure and Table 9c, Appendix B.)

EXERCISE 10.71

Find the critical value for the hypothesis test with H_a: $\sigma_1 > \sigma_2$, with n_1 = 7 and n_2 = 10, α = 0.05.

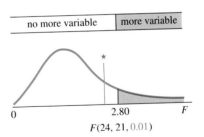

$F(24, 21, 0.01)$

STEP 4   Sample evidence:
a. Sample information: See Table 10-6 on page 534.
b. Calculate the test statistic:

$$F^{\star} = \frac{s_m^2}{s_p^2} = \frac{0.0018}{0.0008} = \mathbf{2.25}$$

STEP 5   Result: $F^{\star}$ is in the noncritical region (as shown in red on the figure above).

Decision Fail to reject H_0.

▲   **Conclusion** At the 0.01 level of significance, the samples do not present sufficient evidence to indicate an increase in variance.

The tables of critical values for the F distribution give only the right-hand critical values. This will not be a problem since the right-hand critical value is the only critical value that will be needed. You can adjust the numerator–denominator order so that all the "activity" is in the right-hand tail. There are two cases: one-tailed tests and two-tailed tests.

▼ *One-tailed tests:* Arrange the null and alternative hypotheses so that the alternative is always "greater than." The $F^\star$-value is calculated using the same order as specified in the null hypothesis (as in Illustration 10-13; also, see Illustration 10-14).

▼ *Two-tailed tests:* When the value of $F^\star$ is calculated, always use the sample with the largest variance for the "numerator"; this will make $F^\star$ larger than 1 and place it in the right-hand tail of the distribution. Thus, you will need only the critical value for the right-hand tail (see Illustration 10-15).

α must still be split between the two tails.

▼   **Illustration   10 - 14**

Reorganize the following alternative hypothesis so that the critical region will be the right-hand tail.

$$H_a: \sigma_1^2 < \sigma_2^2 \quad \text{or} \quad \frac{\sigma_1^2}{\sigma_2^2} < 1 \qquad \text{(Population 1 is less variable)}$$

Solution Reverse the direction of the inequality, and reverse the roles of the numerator and denominator.

$$H_a: \sigma_2^2 > \sigma_1^2 \quad \text{or} \quad \frac{\sigma_2^2}{\sigma_1^2} > 1 \qquad \text{(Population 2 is more variable)}$$

The calculated test statistic $F^\star$ will be s_2^2 / s_1^2.

▼   **Illustration   10 - 15**

Find $F^\star$ and the critical values for the following hypothesis test so that only the right-hand critical value is needed. Use $\alpha = 0.05$ and the sample information: $n_1 = 10$, $n_2 = 8$, $s_1 = 5.4$, $s_2 = 3.8$.

$$H_0: \sigma_2^2 = \sigma_1^2 \quad \text{or} \quad \frac{\sigma_2^2}{\sigma_1^2} = 1 \qquad \text{vs.} \qquad H_a: \sigma_2^2 \neq \sigma_1^2 \quad \text{or} \quad \frac{\sigma_2^2}{\sigma_1^2} \neq 1$$

Case Study

10.3

Personality Characteristics of Police Applicants: Comparisons Across Subgroups and with Other Populations

Bruce N. Carpenter and Susan M. Raza concluded that "police applicants are somewhat more like each other than are those in the normative population" when the F test of homogeneity of variance resulted in a p-value of less than 0.005. Homogeneity means that the group's scores are less variable than the scores for the normative population.

EXERCISE 10.72

a. What null and alternative hypotheses did Carpenter and Raza test?
b. What does "$p < 0.005$" mean?
c. Use MINITAB to calculate the p-value for $F(237, 305) = 1.36$.

To determine whether police applicants are a more homogeneous group than the normative population, the *F* test of homogeneity of variance was used. With the exception of scales *F*, *K*, and 6, where the differences are nonsignificant, the results indicate that the police applicants form a somewhat more homogeneous group than the normative population ($F(237, 305) = 1.36$, $p < .005$). Thus, police applicants are somewhat more like each other than are individuals in the normative population.

Source: Reproduced from the Journal of Police Science and Administration, Vol. 15, no. 1, pp. 10–17, with permission of the International Association of Chiefs of Police, P.O. Box 6010, 13 Firstfield Road, Gaithersburg, Maryland 20878.

EXERCISE 10.73

What would be the value of $F^\star$ in Illustration 10-15 if $F^\star = s_2^2/s_1^2$ were used? Why is it less than 1?

SOLUTION When the alternative hypothesis is two-tailed ($\neq$), the calculated $F^\star$ can be either $F^\star = s_1^2/s_2^2$ or $F^\star = s_2^2/s_1^2$. The choice is ours; we only need to make sure that we keep the df_n and df_d in the correct order. We make the choice by looking at the sample information and using the sample with the larger standard deviation or variance as the sample for the numerator. Therefore, in this illustration, $F^\star = s_1^2/s_2^2 = (5.4)^2/(3.8)^2 = 29.16/14.44 = \mathbf{2.02}$.

The critical values for this test are: left tail, $F(9, 7, 0.975)$, and right tail, $F(9, 7, 0.025)$, as shown in the figure.

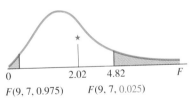

$F(9, 7, 0.975)$ $F(9, 7, 0.025)$

EXERCISE 10.74

When the hypothesis test is two-tailed and MINITAB is used to calculate the p-value, as shown above, what additional step must be taken?

Since we chose the sample with the larger standard deviation (or variance) for the numerator, the value of $F^\star$ will be greater than 1 and will be in the right-hand tail; therefore, only the right-hand critical value is needed. (All critical values for left-hand tails will be values between 0 and 1.)

▲

EXERCISES

10.75 State the null hypothesis, H_0, and the alternative hypothesis, H_a, that would be used to test the following claims.
   a. The variances of populations A and B are not equal.
   b. The standard deviation of population I is larger than the standard deviation of population II.
   c. The ratio of the variances for populations A and B is different from 1.
   d. The variability within population C is less than the variability within population D.

10.76 Using the $F(df_1, df_2, \alpha)$ notation, name each of the critical values shown on the following figures.

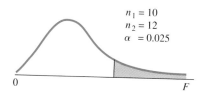

$n_1 = 10$
$n_2 = 12$
$\alpha = 0.025$

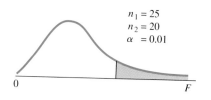

$n_1 = 25$
$n_2 = 20$
$\alpha = 0.01$

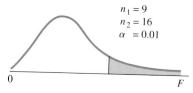

$n_1 = 9$
$n_2 = 16$
$\alpha = 0.01$

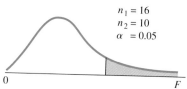

$n_1 = 16$
$n_2 = 10$
$\alpha = 0.05$

10.77 Find the following critical values for F from Tables 9a, 9b, and 9c in Appendix B.
   a. $F(24, 12, 0.05)$ b. $F(30, 40, 0.01)$
   c. $F(12, 10, 0.05)$ d. $F(5, 20, 0.01)$
   e. $F(15, 18, 0.025)$ f. $F(15, 9, 0.025)$
   g. $F(40, 30, 0.05)$ h. $F(8, 40, 0.01)$

10.78 Determine the test criteria that would be used to test the following hypotheses (classical model) when F is used as the test statistic.
   a. $H_0: \sigma_1^2 = \sigma_2^2$ and $H_a: \sigma_1^2 > \sigma_2^2$, with $n_1 = 10$, $n_2 = 16$, and $\alpha = 0.05$
   b. $H_0: \sigma_1^2/\sigma_2^2 = 1$ and $H_a: \sigma_1^2/\sigma_2^2 \neq 1$, with $n_1 = 25$, $n_2 = 31$, and $\alpha = 0.05$
   c. $H_0: \sigma_1^2/\sigma_2^2 = 1$ and $H_a: \sigma_1^2/\sigma_2^2 > 1$, with $n_1 = 10$, $n_2 = 10$, and $\alpha = 0.01$
   d. $H_0: \sigma_1 = \sigma_2$ and $H_a: \sigma_1 < \sigma_2$, with $n_1 = 16$, $n_2 = 16$, and $\alpha = 0.01$

10.79 A bakery is considering buying one of two gas ovens. The bakery requires that the temperature remain constant during a baking operation. A study was conducted to measure the variance in temperature of the ovens during the baking process. The variance in temperature before the thermostat restarted the flame for the Monarch oven was 2.4 for 16 measurements. The variance for the Kraft oven was 3.2 for 12 measurements. Does this information provide sufficient reason to conclude that there is a difference in the variances for the two ovens? Assume measurements are normally distributed and use a 0.01 level of significance.

10.80 "Body Composition in Paraplegic Male Athletes" (*Medicine and Science in Sports and Exercise*, 1987, Vol. 19) compares the body composition and anthropometric characteristics of male paraplegic athletes with able-bodied athletes. One comparison was the circumference of the neck. The data may be summarized as follows:

	n	Mean	St. Dev.
Paraplegic	22	40.3 cm	2.1 cm
Able-bodied	30	41.5 cm	3.1 cm

Test the null hypothesis of equal variances against the alternative of unequal variances. Use $\alpha = 0.05$. (Assume normality.)

10.81 A study in *Pediatric Emergency Care* (June 1994) compared the injury severity between younger and older children. One measure reported was the Injury Severity Score (ISS). The standard deviation of ISS scores for 37 children eight years or younger was 23.9, and the standard deviation for 36 children older than eight years was 6.8. Assume that ISS scores are normally distributed for both age groups.

    a. At the 0.01 level of significance, is there sufficient reason to conclude that the standard deviation of ISS scores for younger children is larger than the standard deviation of ISS scores for older children?

    b. Use a computer and calculate the *p*-value for this hypothesis test.

10.82   a. Two independent samples, each of size 3, are drawn from a normally distributed population. Find the probability that one of the sample variances is at least 19 times larger than the other one.

    b. Two independent samples, each of size 6, are drawn from a normally distributed population. Find the probability that one of the sample variances is no more than 11 times larger than the other one.

10.83 A study was conducted to determine whether or not there was equal variability in male and female systolic blood pressures. Random samples of 16 men and 13 women were used to test the experimenter's claim that the variances were unequal. MINITAB was used to calculate the observed value of F, $F^\star$ and the *p*-value.

Men	120	120	118	112	120	114	130	114
	124	125	130	100	120	108	112	122
Women	122	102	118	126	108	130	104	
	116	102	122	120	118	130		

```
LET K1 = STDEv(C1)
LET K2 = STDEv(C2)
LET C3 = (STDEv(C2)**2)/(STDEv(C1)**2)
CDF C3 C4;
  F 12 15.
LET C5 = 1 - C4
NAME K1 'STDEV1' K2 'STDEV2' C3 'F*' C4 'CUMPROB' C5 'P-VALUE'
PRINT K1 K2 C3-C5
```

Data Display

```
STDEV1    7.88643
STDEV2    9.91761

Row       F*         CUMPROB      P-VALUE
 1      1.58144     0.801026     0.198974
```

Verify the $F^\star$ and p-value given by MINITAB by calculating the values yourself.

10.84 The quality of the end product is somewhat determined by the quality of the materials used. Textile mills monitor the tensile strength of the fibers used in weaving their yard goods. The following data are tensile strengths of cotton fibers from two suppliers.

Supplier A	78	82	85	83	77	84	90	82	93	82
	80	82	77	80	80					
Supplier B	76	79	83	78	72	73	69	80	74	77
	78	78	73	76	78	79				

Calculate the observed value of F, $F^\star$ for comparing the variances of these two sets of data.

10.85 Use a computer to demonstrate the truth of the theory presented in this section.

a. The underlying assumptions are "the populations are normally distributed" and while conducting a hypothesis test for the equality of two standard deviations, it is assumed that the standard deviations are equal. Generate very large samples of two theoretical populations: $N(100, 20)$ and $N(120, 20)$. Find graphic and numerical evidence that the populations satisfy the assumptions.

b. Randomly select 100 samples, each of size 8, from both populations and find the standard deviation of each sample.

MINITAB commands

```
a. Populations
RANDom 2000 C1;
  NORMal 100 20.
RANDom 2000 C2;
  NORMal 120 20.
DESCribe C1 C2
HISTogram C1;
  CUTPoint 20:180/10.
HISTogram C2;
  CUTPoint 40:200/10.

b. 100 samples
RANDom 100 C3-C10;
  NORMal 100 20.
RSTDev C3-C10 C11
RANDom 100 C12-C19;
  NORMal 120 20.
RSTDev C12-C19 C20
```

c. Using the first sample drawn from each population as a pair, calculate the $F^\star$-statistic. Repeat for all samples. Describe the sampling distribution of the 100 $F^\star$-values using both graphic and numerical statistics.

d. Generate the probability distribution for $F(7, 7)$, and compare it to the observed distribution of $F^\star$. Do the two graphs agree? Explain.

```
c. Sampling distribution
LET C21 = (C20/C11)**2
DESCribe C21
HISTogram C21;
  GRID 2;
  CUTPoint 0:5/.2.

d. Cum. probabilities
SET C22
DATA> 0:5/.2
DATA> END
PDF C22 C23;
  F 7 7.
PLOT C23*C22;
CONNect.
```

10.86 It was stated in this section that the F test was very sensitive to minor digressions from the assumptions. Repeat Exercise 10.85 using $N(100, 20)$ and $N(120, 30)$. Notice that the only change from Exercise 10.85 is the seemingly slight increase in the standard deviation of the second population. Answer the same questions using the same kind of information and you will see very different results.

MINITAB commands

```
a. Populations
RANDom 2000 C1;
  NORMal 100 20.
RANDom 2000 C2;
  NORMal 120 25.
DESCribe C1 C2
HISTogram C1;
  CUTPoint 20:180/10.
HISTogram C2;
  CUTPoint 20:220/12.5.

c. Sampling distribution
LET C21 = (C20/C11)**2
DESCribe C21
HISTogram C21;
  GRID 2;
  CUTPoint 0:5/.2.
```

```
b. Samples
RANDom 100 C3-C10;
  NORMal 100 20.
RSTDev C3-C10 C11
RANDom 100 C12-C19;
  NORMal 120 25.
RSTDev C12-C19 C20

d. Cum. probabilities
SET C22
DATA> 0:5/.2
DATA> END
PDF C22 C23;
  F 7 7.
PLOT C23*C22;
CONNect.
```

10.5 | INFERENCES CONCERNING THE DIFFERENCE BETWEEN PROPORTIONS USING TWO INDEPENDENT SAMPLES

We are often interested in making statistical comparisons between the proportions, percentages, or probabilities associated with two populations. The following questions

The 3 "p" words (proportion, percentage, probability) are all the binomial parameter p, "P(success)."

ask for such comparisons: Is the proportion of homeowners who favor a certain tax proposal different from the proportion of renters who favor it? Did a larger percentage of this semester's class than of last semester's class pass statistics? Is the probability of a Democratic candidate winning in New York greater than the probability of a Republican candidate winning in Texas? Do students' opinions about the new code of conduct differ from those of the faculty? You have probably asked similar questions.

RECALL   The properties of a binomial experiment are as follows:

Binomial experiments are completely defined on page 268.

1. The observed probability is $p' = x/n$, where x is the number of observed successes in n trials.
2. $q' = 1 - p'$.
3. p is the probability of success on an individual trial in a binomial probability experiment of n repeated independent trials.

In this section, we will compare two population proportions by using the difference between the observed proportions, $p_1' - p_2'$, of two independent samples. The observed difference, $p_1' - p_2'$, belongs to a sampling distribution, the characteristics of which are described in the following statement.

EXERCISE 10.87

Only 75 of the 250 people interviewed were able to name the vice president of the United States. Find the values for x, n, p', and q'.

If independent samples of sizes n_1 and n_2 are drawn randomly from large populations with $p_1 = P_1$(success) and $p_2 = P_2$(success), respectively, the sampling distribution of $p_1' - p_2'$ has these properties:

1. a mean $\mu_{p_1' - p_2'} = p_1 - p_2$

2. a standard error $\sigma_{p_1' - p_2'} = \sqrt{\dfrac{p_1 q_1}{n_1} + \dfrac{p_2 q_2}{n_2}}$ **(10-11)**

3. an approximately normal distribution if n_1 and n_2 are sufficiently large

In practice, use the following guidelines to ensure normality.

1. The sample sizes are both larger than 20.
2. The products $n_1 p_1$, $n_1 q_1$, $n_2 p_2$, and $n_2 q_2$ are all larger than 5.
3. The samples consist of less than 10% of respective populations.

EXERCISE 10.88

If $n_1 = 40$, $p_1' = 0.9$, $n_2 = 50$, and $p_2' = 0.9$:

a. Find the estimated values for both np's and both nq's.

b. Would this situation satisfy the guidelines for approximately normal? Explain.

NOTE p_1 and p_2 are unknown; therefore, these products will be estimated by $n_1 p_1'$, $n_1 q_1'$, $n_2 p_2'$, and $n_2 q_2'$.

Inferences about the difference between two population proportions, $p_1 - p_2$, will be based on the following assumptions.

The assumptions for inferences about the difference between two proportions $p_1 - p_2$: The n_1 random observations and the n_2 random observations forming the two samples are selected independently from two populations that are not changing during the sampling.

Confidence Intervals

When we estimate the difference between two proportions $p_1 - p_2$, we will base our estimates on the unbiased sample statistic $p'_1 - p'_2$. The point estimate, $p'_1 - p'_2$, becomes the center of the confidence interval and the confidence limits are found using formula (10-12).

$$(p'_1 - p'_2) - z(\alpha/2) \cdot \sqrt{\frac{p'_1 q'_1}{n_1} + \frac{p'_2 q'_2}{n_2}} \quad \text{to} \quad (p'_1 - p'_2) + z(\alpha/2) \cdot \sqrt{\frac{p'_1 q'_1}{n_1} + \frac{p'_2 q'_2}{n_2}} \quad \textbf{(10-12)}$$

▼| ILLUSTRATION   10 - 16

In studying his campaign plans, Mr. Morris wishes to estimate the difference between men's and women's views regarding his appeal as a candidate. He asks his campaign manager to take two samples and find the 99% confidence interval for the difference. A sample of 1000 voters was taken from each population, with 388 men and 459 women favoring Mr. Morris.

SOLUTION The campaign manager determined the confidence interval as follows:

STEP 1   Population parameter of interest: The difference between the proportion of men voters and the proportion of women voters who plan to vote for Mr. Morris.

STEP 2   Confidence interval criteria:
   a. Check the assumptions: The sample sizes are larger than 20, and the estimated values for $n_m p_m$, $n_m q_m$, $n_f p_f$, and $n_f q_f$ are all larger than 5; therefore, the sampling distribution of $p'_f - p'_m$ should have an approximately normal distribution.
   b. Test statistic: $z^\star$ calculated using formula (10-12).
   c. Confidence level: $1 - \alpha = 0.99$.

STEP 3   Sample evidence:
   a. Sample information: $n_m = 1000$, $x_m = 388$, $n_f = 1000$, $x_f = 459$.

$$p'_m = \frac{x_m}{n_m} = \frac{388}{1000} = 0.388; \quad \text{and} \quad q'_m = 1 - 0.388 = 0.612$$

$$p'_f = \frac{x_f}{n_f} = \frac{459}{1000} = 0.459; \quad \text{and} \quad q'_f = 1 - 0.459 = 0.541$$

   b. Point estimate: $p'_f - p'_m = 0.459 - 0.388 = \textbf{0.071}$. (It is customary to place the larger value first; that way, the point estimate for the difference is a positive value.)

STEP 4   Confidence interval limits:

a. Confidence coefficients: Two-tailed situation, with $\alpha/2$ in each tail: $z(\alpha/2) = z(0.005) = \mathbf{2.58}$.

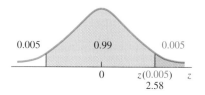

EXERCISE 10.89

Find the 95% confidence interval for $p_A - p_B$.

Sample	n	x
A	125	45
B	150	48

b. Maximum error: Using part of formula (10-12),

$$E = z(\alpha/2) \cdot \sqrt{\frac{p'_f q'_f}{n_f} + \frac{p'_m q'_m}{n_m}} = 2.58 \sqrt{\frac{(0.459)(0.541)}{1000} + \frac{(0.388)(0.612)}{1000}}$$

$$E = 2.58\sqrt{0.000248 + 0.000237} = (2.58)(0.022) = \mathbf{0.057}$$

c. Confidence limits:

$$(p'_f - p'_m) \pm E$$

$$0.071 \pm 0.057$$

$$0.071 - 0.057 = 0.014 \quad \text{to} \quad 0.071 + 0.057 = 0.128$$

0.014 to 0.128, the 99% confidence interval for $p_f - p_m$

d. Results: With 99% confidence, we can say that there is a difference of from 1.4% to 12.8% in Mr. Morris's voter appeal. That is, a larger proportion of women than men favor Mr. Morris, and the difference in proportion is between 1.4% and 12.8%.

Hypothesis Tests

When the null hypothesis, "there is no difference between two proportions," is being tested, the test statistic will be the difference between the observed proportions divided by the standard error, and it is found using formula (10-13).

$$z^\star = \frac{p'_1 - p'_2}{\sqrt{pq\left[\left(\dfrac{1}{n_1}\right) + \left(\dfrac{1}{n_2}\right)\right]}} \qquad \text{(10-13)}$$

NOTES

1. The null hypothesis is $p_1 = p_2$, or $p_1 - p_2 = 0$ ("difference is zero").
2. Nonzero differences between proportions are not discussed in this section.
3. The numerator of formula (10-13) could be written as $(p'_1 - p'_2) - (p_1 - p_2)$, but since the null hypothesis is assumed to be true during the test, $p_1 - p_2 = 0$. By substitution the numerator becomes simply $p'_1 - p'_2$.

4. Since the null hypothesis is $p_1 = p_2$, the standard error of $p'_1 - p'_2$,

$$\sqrt{\left(\frac{p_1 q_1}{n_1}\right) + \left(\frac{p_2 q_2}{n_2}\right)}, \quad \text{can be written as} \quad \sqrt{pq\left[\left(\frac{1}{n_1}\right) + \left(\frac{1}{n_2}\right)\right]}, \quad \text{where}$$

$p = p_1 = p_2$ and $q = 1 - p$.

5. When the null hypothesis states $p_1 = p_2$ and does not specify the values of either p_1 or p_2, the two sets of sample data will be pooled to obtain the estimate for p. This pooled probability, known as p'_p, is the total number of successes divided by the total number of observations for the two samples combined, and it is found using formula (10-14).

EXERCISE   10.90

Find the values of p'_p and q'_p for these samples:

Sample	x	n
E	15	250
R	25	275

$$p'_p = \frac{x_1 + x_2}{n_1 + n_2} \qquad \text{(10-14)}$$

and q'_p is its complement,

$$q'_p = 1 - p'_p \qquad \text{(10-15)}$$

When the pooled estimate, p'_p, is being used, formula (10-13) becomes formula (10-16).

$$z^\star = \frac{p'_1 - p'_2}{\sqrt{(p'_p)(q'_p)\left[\left(\frac{1}{n_1}\right) + \left(\frac{1}{n_2}\right)\right]}} \qquad \text{(10-16)}$$

p-Value Approach

 ILLUSTRATION   10 - 17

A salesman for a new manufacturer of walkie-talkies claims not only that they cost the retailer less but also that the percentage of defective walkie-talkies found among his products will be no higher than the percentage of defectives found in a competitor's line. To test this statement, the retailer took random samples of each manufacturer's product. The sample summaries are given in Table 10-7. Can we reject the salesman's claim at the 0.05 level of significance?

TABLE 10-7

Walkie-Talkie
Sample Information

Product	Number Defective	Number Checked
Salesman's	15	150
Competitor's	6	150

SOLUTION

STEP 1   Population parameter of interest: The difference between the proportion of defectives of the salesman's product and the proportion of defectives of the competitor's product.

STEP 2   State the hypotheses: The concern of the retailer is that the salesman's less expensive product may be of a poorer quality, meaning a greater proportion of defectives. By using the difference, the "suspected larger proportion minus the smaller proportion," the alternative hypothesis is "the difference is positive (greater than zero)."

H_0: $p_s - p_c = 0$ ($\leq$) (salesman's product defective rate is no higher than competitor's)
H_a: $p_s - p_c > 0$ (salesman's product defective rate is higher than competitor's)

STEP 3   Test criteria:
   a. Assumptions: Populations are very large (all walkie-talkies produced); the samples are larger than 20; the estimated products $n_s p'_s$, $n_s q'_s$, $n_c p'_c$, and $n_c q'_c$ are all larger than 5. Therefore, the sampling distribution should have an approximately normal distribution.
   b. Test statistic: $z^\star$ calculated using formula (10-16).
   c. Level of significance: $\alpha = 0.05$.

STEP 4   Sample evidence:
   a. Sample information:

$$p'_s = \frac{x_s}{n_s} = \frac{15}{150} = \textbf{0.10}; \quad p'_c = \frac{x_c}{n_c} = \frac{6}{150} = \textbf{0.04}$$

$$p'_p = \frac{x_s + x_c}{n_s + n_c} = \frac{15 + 6}{150 + 150} = \frac{21}{300} = \textbf{0.07}; \quad q'_p = 1 - p'_p = 1 - 0.07 = \textbf{0.93}$$

   b. Test statistic:

EXERCISE 10.91

Find the p-value for the test with alternative hypothesis $p_E < p_R$ using data in Exercise 10.90.

$$z^\star = \frac{p'_s - p'_c}{\sqrt{(p'_p)(q'_p)\left[\left(\frac{1}{n_s}\right) + \left(\frac{1}{n_c}\right)\right]}} = \frac{0.10 - 0.04}{\sqrt{(0.07)(0.93)\left[\left(\frac{1}{150}\right) + \left(\frac{1}{150}\right)\right]}}$$

$$z^\star = \frac{0.06}{\sqrt{0.000868}} = \frac{0.06}{0.02946} = \textbf{2.04}$$

   c. Find the p-value: $\mathbf{P} = P(z^\star > 2.04)$. Using Table 5 in Appendix B, **0.0202 < P < 0.0228**. Using Table 3 in Appendix B, we can calculate $\mathbf{P}$; $\mathbf{P} = P(z^\star > 2.04) = 0.5000 - 0.4793 = \textbf{0.0207}$.

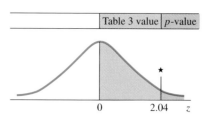

STEP 5   Results: **P** is less than $\alpha = 0.05$.

DECISION Reject H_0.

CONCLUSION At the 0.05 level of significance, there is sufficient evidence to reject the salesman's claim; the proportion of his company's walkie-talkies that are defective is higher than the proportion of his competitor's walkie-talkies that are defective.

▲

Classical Approach

▼ ILLUSTRATION 10 - 18

Joe claimed the probability that a commuting college student has car trouble of some type (car wouldn't start, flat tire, accident) on the way to college in the morning is greater than the probability that the student will have car trouble on the way to work or home after class. He has several theories to explain his claim. The Sports Car Club thinks that the ideas of "before class" and "after class" have nothing to do with whether or not a student has car trouble. The club decides to challenge Joe's claim, and two large samples are gathered in order to test the theory. The resulting sample statistics are given in Table 10-8. Complete the hypothesis test with $\alpha = 0.02$.

TABLE 10-8

Car Troubles
Sample Information

Sample	Number with Car Trouble	Number Sampled
Before (b)	30	500
After (a)	28	600

SOLUTION

STEP 1 **Population parameter of interest:** The difference in the probability of car trouble before and after classes, on the day of the sampling, for commuting college students.

STEP 2 **Hypotheses:**

The positive difference, "larger minus smaller," is more understandable.

H_0: $p_b = p_a$ or $p_b - p_a = 0$ (not greater than)
H_a: $p_b > p_a$ or $p_b - p_a > 0$ (Joe's claim, p_b greater than p_a)

STEP 3 **Test criteria:**
  a. **Assumptions:** Populations are large and samples are larger than 20; the products $n_b p_b'$, $n_b q_b'$, $n_a p_a'$, and $n_a q_a'$ are all larger than 5; therefore, the sampling distribution is expected to be approximately normally distributed.
  b. **Test statistic:** $z^\star$, found using formula (10-16).

Case Study   Smokers Need More Time in Recovery Room

10.4

The following report says that a study involving 327 patients in Long Beach, New Jersey, showed that 38% of the nonsmokers spent less than one hour in recovery after surgery, compared to 23% of smokers. Further, 19% of the smokers spent more than two hours, compared to 7% of the nonsmokers.

EXERCISE 10.92

"38% of the non-smokers spent less than one hour in re-covery vs. 23% of smokers" was report-ed in "Smokers Need More Time in Recovery Room." As-suming the patients were equally divided between the two samples, find the *p*-value associated with the hypothesis test using this infor-mation. Do you be-lieve this information is significant? Explain.

Smokers take longer to recover from anesthesia after surgery than nonsmokers do, new research shows.

A study reported over the weekend at the American Society of Anesthesiolo-gists meeting in Las Vegas found that smokers were nearly three times more likely than nonsmokers to spend two or more hours in the recovery room. "It's a very strong difference," says researcher Dr. David Handlin, of Monmouth Medical Cen-ter, Long Branch, N.J.

Of the 327 patients studied:

▼ 38% of nonsmokers spent less than one hour in the recovery room, vs. 23% of smokers.

▼ 19% of smokers spent more than two hours in recovery, vs. 7% of nonsmokers.

Longer recovery room stays demand more nursing care and drive up health care costs, says Handlin. Other studies have shown that smokers tend to have a higher risk of complications from surgery. Handlin says this is the first to show that smokers take longer to recover from anesthesia.

Longer recovery stays may be due to mild lung disease or a lowered ability of the blood to carry oxygen, says Handlin. Some experts say quitting smoking weeks before surgery may cut recovery time.

Source: Copyright 1990, USA TODAY. Reprinted with permission.

EXERCISE 10.93

Find the value of $z\star$ that would be used to test the difference between the propor-tions, given the following:

Sample	*n*	*x*
G	380	323
H	420	332

c. Level of significance: $\alpha = 0.02$.
d. Critical values: One-tail test with critical region in right-hand tail since the alternative hypothesis states that the difference is "greater than." $z(0.02) = 2.05$. (See the figure.)

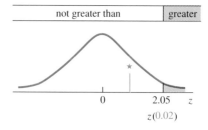

Case Study   **One Too Many**

10.5

"One too many" reports that 36% (overall) admit to sometimes drinking too much. The graphic indicates that a sample of women and a sample of men were polled separately, yielding results of 46% and 26%. The "overall" statistic appears to be a pooled statistic.

EXERCISE 10.94

"One too many" shows the percentage of people who admit to sometimes drinking too much.

a. What is the point estimate for the proportion of all men who sometimes drink too much? Women?

b. In this section, a pooled observed probability, p'_p, was introduced. Explain its relationship to the overall percentage reported.

c. Assuming the 340 men and 340 women were polled, construct the 95% confidence interval for the difference between the proportion of men and women who drink too much.

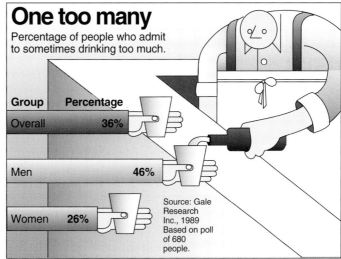

USA SNAPSHOTS®

A look at statistics that shape the nation

One too many

Percentage of people who admit to sometimes drinking too much.

Group	Percentage
Overall	36%
Men	46%
Women	26%

Source: Gale Research Inc., 1989 Based on poll of 680 people.

By Marcia Staimer, USA TODAY

Source: Copyright 1990, USA TODAY. Reprinted with permission.

STEP 4   Sample evidence:

     a. Sample information: $n_b = 500$, $x_b = 30$, $n_a = 600$, $x_a = 28$.
       Sample statistics:

$$p'_b = \frac{x_b}{n_b} = \frac{30}{500} = \mathbf{0.060}; \quad p'_a = \frac{x_a}{n_a} = \frac{28}{600} = 0.04666\overline{6} = \mathbf{0.047}$$

$$p'_p = \frac{x_b + x_a}{n_b + n_a} = \frac{30 + 28}{500 + 600} = \frac{58}{1100} = 0.0527 = \mathbf{0.053}; \quad q'_p = 1 - p'_p = 1 - 0.053 = \mathbf{0.947}$$

b. Test statistic:

$$z^\star = \frac{p_b' - p_a'}{\sqrt{(p_p')(q_p')\left[\left(\dfrac{1}{n_b}\right) + \left(\dfrac{1}{n_a}\right)\right]}} = \frac{0.060 - 0.047}{\sqrt{(0.053)(0.947)\left[\left(\dfrac{1}{500}\right) + \left(\dfrac{1}{600}\right)\right]}}$$

$$z^\star = \frac{0.013}{\sqrt{(0.05019)(0.002 + 0.0017)}} = \frac{0.013}{\sqrt{0.000186}} = \frac{0.013}{0.0136} = \mathbf{0.96}$$

STEP 5   Results: $z^\star$ is in the noncritical region (shown in red on the figure on page 550).

DECISION Fail to reject H_0.

CONCLUSION At the 0.02 level of significance, there is no evidence to indicate that the probability of car trouble is greater before class than after class.

▲

EXERCISES

▼

10.95   According to the publication *Housing Economics* (March 1991), 10.6% of all renters have an IRA or Keogh retirement plan compared to 24.2% of all households and 32.0% of homeowners. Suppose you conducted a survey of 200 renters and 300 homeowners. If 15% of the renters and 25% of the homeowners had an IRA or Keogh plan, find a 99% confidence interval estimating the difference in proportions for the two groups.

10.96   In a random sample of 40 brown-haired individuals, 22 indicated that they used hair coloring. In another random sample of 40 blonde individuals, 26 indicated that they used hair coloring. Use a 92% confidence interval to estimate the difference in the proportion of these groups that use hair coloring.

10.97   An article titled "Nurse Executive Turnover" (*Nursing Economics*, January/February 1993) compared two groups of nurse executives. One group had participated in a unique program for nurse executives called the Wharton Fellows Program, and the other group had not participated in the program. Eighty-seven of 341 Wharton Fellows had experienced one change in position, and 9 of 40 non-Wharton Fellows had experienced one change in position. Find a 99% confidence interval for the difference in population proportions.

10.98   In a survey of 300 people from city A, 128 preferred New Spring soap to all other brands of deodorant soap. In city B, 149 of 400 people preferred New Spring. Find the 98% confidence interval for the difference in the two proportions.

10.99   The proportions of defective parts produced by two machines were compared, and the following data were collected.

Machine 1: $n = 150$; number of defective parts $= 12$

Machine 2: $n = 150$; number of defective parts $= 6$

Determine a 90% confidence interval for $p_1 - p_2$.

10.100   Show that the standard error of $p_1' - p_2'$, which is $\sqrt{\left(\frac{p_1 q_1}{n_1}\right) + \left(\frac{p_2 q_2}{n_2}\right)}$, reduces to $\sqrt{pq\left[\left(\frac{1}{n_1}\right) + \left(\frac{1}{n_2}\right)\right]}$ when $p_1 = p_2 = p$.

10.101   State the null hypothesis, H_0, and the alternative hypothesis, H_a, that would be used to test these claims:

    a. There is no difference between the proportions of men and women who will vote for the incumbent in next month's election.

    b. The percentage of boys who cut classes is greater than the percentage of girls who cut classes.

    c. The percentage of college students who drive old cars is higher than the percentage of noncollege people of the same age who drive old cars.

10.102   Determine the test criteria that would be used to test (classical model) the following hypotheses when z is used as the test statistic.

    a. H_0: $p_1 = p_2$ and H_a: $p_1 > p_2$, with $\alpha = 0.05$

    b. H_0: $p_A = p_B$ and H_a: $p_A \neq p_B$, with $\alpha = 0.05$

    c. H_0: $p_1 - p_2 = 0$ and H_a: $p_1 - p_2 < 0$, with $\alpha = 0.04$

    d. H_0: $p_m - p_f = 0$ and H_a: $p_m - p_f > 0$, with $\alpha = 0.01$

10.103   The December 2, 1994, edition of the *Omaha World Herald* reported on a study involving 4462 Vietnam War Army veterans who were between the ages of 31 and 49 in 1985. Of that number, 1162 said they had never smoked, 1292 said they were former smokers, and 2008 said they smoked. Among nonsmokers, 2.2% said they suffered persistent impotence, compared with 2.0% of former smokers and 3.7% of current smokers. Researchers said the difference in the rate of impotence reported by nonsmokers and former smokers was statistically insignificant. Find the value of $z^\star$ and the p-value for testing the null hypothesis of no difference in the rate of impotence in nonsmokers and former smokers (that is $p_1 = p_2$). The alternative hypothesis is that a difference exists.

10.104   In a survey of working parents (both parents working), one of the questions asked was "Have you refused a job, promotion, or transfer because it would mean less time with your family?" Two hundred men and 200 women were asked this question. Twenty-nine percent of the men and 24% of the women responded "yes." Based on this survey, can we conclude that there is a difference in the proportion of men and women responding "yes" at the 0.05 level of significance?

10.105   Two randomly selected groups of citizens were exposed to different media campaigns that dealt with the image of a political candidate. One week later the citizen groups were surveyed to see whether they would vote for the candidate. The results were as follows:

	Exposed to Conservative Image	Exposed to Moderate Image
Number in Sample	100	100
Proportion for the Candidate	0.40	0.50

Is there sufficient evidence to show a difference in the effectiveness of the two image campaigns at the 0.05 level of significance?

    a. Solve using the p-value approach.

    b. Solve using the classical approach.

10.106 U.S. military active duty personnel and their dependents are provided with free medical care. A survey was conducted to compare obstetrical care between military and civilian (pay for medical care) families ("Use of Obstetrical Care Compared Among Military and Civilian Families," *Public Health Reports*, May/June 1989). The number of women who began prenatal care by the second trimester were reported as follows:

	Military	Civilian
Prenatal Care Began by 2nd Trimester	358	6786
Total Sample	407	7363

Is there a significant difference between the proportion of military and civilian women who begin prenatal care by the second trimester? Use $\alpha = 0.02$.

10.107 The July 28, 1990, issue of *Science News* reported that smoking boosts death risk for diabetics. The death risk is increased more for women than men. Suppose as a follow-up study we investigated the smoking rates for male and female diabetics and obtained the following data.

Gender	*n*	Number Who Smoke
Male	500	215
Female	500	170

  a. Test the research hypothesis that the smoking rate (proportion of smokers) is higher for males than for females. Calculate the *p*-value.
  b. What decision and conclusion would be reached at the 0.05 level of significance?

10.108 The guidelines to ensure the sampling distribution of $p'_1 - p'_2$ is normal include several conditions about the size of several values. The two binomial distributions $B(100, 0.3)$ and $B(100, 0.4)$ satisfy all of those guidelines.
  a. Verify that $B(100, 0.3)$ and $B(100, 0.4)$ satisfy all of the guidelines.
  b. Use a computer to randomly generate 200 random samples from each of the binomial populations. Find the observed proportion for each sample and the value of the 200 differences between two proportions.
  c. Describe the observed sampling distribution using both graphic and numerical statistics.
  d. Does the empirical sampling distribution appear to have an approximately normal distribution? Explain.

```
MINITAB commands

b. RANDom 200 C1;
     BINOmial 100 .3.
   RANDom 200 C2;
     BINOmial 100 .2.
   LET C3 = C1/100
   LET C4 = C2/100
   LET C5 = C3 - C4

c. DESCribe C5
   HIST C5;
     GRID 2;
     PERCent;
     CUTP -.14:.34/.06.
```

IN RETROSPECT

In this chapter we began the comparisons of two populations by first distinguishing between independent and dependent samples, which are statistically very important and useful sampling procedures. We then proceeded to examine the inferences concerning the comparison of means, variances, and proportions for two populations.

The use of confidence intervals and hypothesis tests can sometimes be interchanged; that is, a confidence interval can often be used in place of a hypothesis test. For example, Illustration 10-16 called for a confidence interval. Now suppose that Mr. Morris asked: "Is there a difference in my voter appeal to men voters as opposed to women voters?" To answer his question, you would not need to complete a hypothesis test if you chose to test at $\alpha = 0.01$ using a two-tailed test. "No difference" would mean a difference of zero, which is not included in the interval from 0.014 to 0.128 (the interval determined in Illustration 10-16). Therefore, a null hypothesis of "no difference" would be rejected, thereby substantiating the conclusion that a significant difference exists in voter appeal between the two groups.

We are always making comparisons between two groups: We compare means and we compare proportions. In this chapter we have learned how to statistically compare two populations by making inferences about their means, proportions, or variances. For convenience, Table 10-9 identifies the formulas to use when making inferences about comparisons between two populations. The chapter case study at the beginning of this chapter compares children's drawings of people and illustrates the use of topics studied in three sections of this chapter: both dependent and independent samples and the difference between two means using both types of samples.

In Chapters 8 through 10 we have learned how to use confidence intervals and hypothesis tests to answer questions about means, proportions, and standard deviations for one or two populations. From here we can expand our techniques to include inferences about more than two populations as well as inferences of different types.

TABLE 10-9   Formulas to Use for Inferences Involving Two Populations

		Formula to Be Used	
Situations	**Test Statistic**	**Confidence Interval**	**Hypothesis Test**
Difference between two means			
Using dependent samples	t	Formula (10-4) (p. 502)	Formula (10-5) (p. 503)
Using independent samples	t	Formula (10-8) (p. 514)	Formula (10-9) (p. 516)
Ratio of two variances or standard			
deviations	F		Formula (10-10) (p. 535)
Difference between two proportions	z	Formula (10-12) (p. 545)	Formula (10-16) (p. 547)

CHAPTER EXERCISES

10.109   Using a 95% confidence interval, estimate the difference in I.Q. between the oldest and the youngest members (brothers and sisters) of a family based on the following random sample of I.Q.s. Assume normality.

Oldest	145	133	116	128	85	100	105	150	97	110	120	130
Youngest	131	119	103	93	108	100	111	130	135	113	108	125

10.110   The diastolic blood pressures for 15 patients were determined using two techniques: the standard method used by medical personnel and a method using an electronic device with a digital readout. The results were as follows:

Patient	1	2	3	4	5	6	7	8	9	10	11	12	13	14	15
Standard method	72	80	88	80	80	75	92	77	80	65	69	96	77	75	60
Digital method	70	76	87	77	81	75	90	75	82	64	72	95	80	70	61

Assuming blood pressure is normally distributed, determine the 90% confidence interval for the mean difference in the two readings, where d = standard method − digital readout.

10.111   Ten new recruits participated in a rifle-shooting competition at the end of their first day at training camp. The same ten competed again at the end of a full week of training and practice. Their resulting scores are shown in the following table.

Time of Competition	Recruit									
	1	2	3	4	5	6	7	8	9	10
First day	72	29	62	60	68	59	61	73	38	48
One week later	75	43	63	63	61	72	73	82	47	43

Does this set of ten pairs of data show that there was a significant amount of improvement in the recruits' shooting abilities during the week? Use $\alpha = 0.05$.

10.112   Twelve automobiles were selected at random to test two new mixtures of unleaded gasolines. Each car was given a measured allotment of the first mixture, x, and driven; then the distance traveled was recorded. The second mixture, y, was immediately tested in the same manner. The order in which the x and y mixtures were tested was also randomly assigned. The results are given in the following table.

Mixture						Car						
	1	2	3	4	5	6	7	8	9	10	11	12
x	7.9	5.6	9.2	6.7	8.1	7.3	8.1	5.4	6.9	6.1	7.1	8.1
y	7.7	6.1	8.9	7.1	7.9	6.7	8.2	5.0	6.2	5.7	6.2	7.5

Can you conclude that there is no real difference in mileage obtained by these two gasoline mixtures at the 0.10 level of significance? Assume mileage is normal.

    a. Solve using the p-value approach.

    b. Solve using the classical approach.

 10.113 A test that measures math anxiety was given to 50 male and 50 female students. The results were as follows:

$$\text{Males:} \quad \bar{x} = 70.5 \quad s = 13.2$$
$$\text{Females:} \quad \bar{x} = 75.7 \quad s = 13.6$$

Construct a 95% confidence interval for the difference between the mean anxiety scores.

 10.114 The same achievement test is given to soldiers selected at random from two units. The scores they attained are summarized as follows:

$$\text{Unit 1:} \quad n_1 = 70 \quad \bar{x}_1 = 73.2 \quad s_1 = 6.1$$
$$\text{Unit 2:} \quad n_2 = 60 \quad \bar{x}_2 = 70.5 \quad s_2 = 5.5$$

Construct a 90% confidence interval for the difference in the mean level of the two units.

 10.115 The score on a certain psychological test is used as an index of status frustration. The scale ranges from 0 (low frustration) to 10 (high frustration). The test was administered to independent random samples of seven radical rightists and eight Peace Corps volunteers.

Radical Rightists	6	10	3	8	8	7	9	
Peace Corps Volunteers	3	5	2	0	3	1	0	4

Construct a 95% confidence interval for the difference between the mean scores of the two groups.

 10.116 The performance on an achievement test in a beginning computer science course was administered to two groups. One group had a previous computer science course in high school; the other group did not. The test results are below. Assuming test scores are normal, construct a 98% confidence interval for the difference between the two means.

Group 1 (had high school course)	17	18	27	19	24	36	27	26
	35	22	18	29	29	26	33	
Group 2 (no high school course)	19	25	28	27	21	24	18	14	28
	21	22	20	21	14	29	28	25	17
	20	28	31	27					

10.117   Two methods were used to study the latent heat of ice fusion. Both method A (an electrical method) and method B (a method of mixtures) were conducted with the specimens cooled to $-0.72°C$. The data in the following table represent the change in total heat from $-0.72°C$ to water at $0°C$ in calories per gram of mass.

Method A	Method B
79.98	80.02
80.04	79.94
80.02	79.98
80.04	79.97
80.03	79.97
80.03	80.03
80.04	79.95
79.97	79.97
80.05	
80.03	
80.02	
80.00	
80.02	

Construct a 95% confidence interval for the difference between the means.

10.118   Ten soldiers were selected at random from each of two companies to participate in a rifle-shooting competition. Their scores are shown in the following table.

Company A	72	29	62	60	68	59	61	73	38	48
Company B	75	43	63	63	61	72	73	82	47	43

Construct a 95% confidence interval for the difference between the mean scores for the two companies.

10.119   In the article "Management of Family and Employment Responsibilities by Mexican-American and Anglo-American Women" (*Social Work*, May 1990), Marlow reports the following results.

Age	Mexican-American Women	Anglo-American Women
18–24	15	11
25–34	31	29
35–44	15	15
45–54	6	16
55–64	3	1

Source: Copyright 1990, National Association of Social Workers, Inc. *Social Work.*

For the purposes of this exercise, use 15–25, 25–35, . . . , as the class boundaries for the classification in place of 18–24, 25–34, ...

    a. Draw histograms of these two distributions.

    b. Calculate the mean and standard deviation for the Mexican-American distribution.

    c. Calculate the mean and standard deviation for the Anglo-American distribution.

    d. Is the difference between these two sample means significant at the 0.05 level?

    e. Does it appear that these two samples have the same distributions? Justify your conclusion.

10.120 A test concerning some of the fundamental facts about AIDS was administered to two groups, one consisting of college graduates and the other consisting of high school graduates. A summary of the test results follows:

College graduates:	$n = 75$	$\bar{x} = 77.5$	$s = 6.2$
High school graduates:	$n = 75$	$\bar{x} = 50.4$	$s = 9.4$

Do these data show that the college graduates, on the average, score significantly higher on the test? Use $\alpha = 0.05$.

10.121 The article that forms the chapter case study describes using statistics to analyze children's artwork and demonstrates several concepts studied in this chapter.

    a. Explain why dependent (correlated) samples were used when pictures of young subjects were compared to pictures of old subjects.

    b. Explain why independent samples were involved when drawings by girls were compared to drawings by boys.

    c. Were independent or dependent techniques used to obtain the four t-values that are reported?

    d. Explain what it means to say, "$t = 7.42$ is highly significant."

    e. Why were both independent and dependent samples involved? Why do the t-values reported show significance?

10.122 To compare the merits of two short-range rockets, 8 of the first kind and 10 of the second kind are fired at a target. If the first kind has a mean target error of 36 ft and a standard deviation of 15 ft, while the second kind has a mean target error of 52 ft and a standard deviation of 18 ft, does this indicate that the second kind of rocket is less accurate than the first? Use $\alpha = 0.01$ and assume normal distribution for target error.

10.123 The following data were collected concerning waist sizes of men and women. Do these data present sufficient evidence to conclude that men have larger mean waist sizes than women at the 0.05 level of significance? Assume waist sizes are normally distributed.

Men	33	33	30	34	34	40	35	35	32
	34	32	35	32	32	34	36	30	38
Women	22	29	27	24	28	28			
	27	26	27	26	25				

10.124   The women's health section of *Good Housekeeping* (Feb. 1991) mentions a study connecting a woman's athletic ability and her menstrual cycle. Researchers selected 16 runners who logged at least 35 miles a week. Half had ceased having menstrual periods, and half had normal cycles. The researchers found no significant difference in the exercise capacities between the two groups. Suppose the following data were obtained:

Group	x	Mean Exercise Capacity	St. Dev.
No period	8	30.5 min	9.5
Normal cycle	8	31.3	8.0

a. Test the research hypothesis that the athletic capacities differ, and give the *p*-value.
b. If the test is being completed at the 0.05 level of significance, what decision and conclusion are reached?

10.125   A group of 17 students participated in an evaluation of a special training session that claimed to improve memory. The students were randomly assigned to two groups: group A, the test group, and group B, the control group. All 17 students were tested for the ability to remember certain material. Group A was given the special training; group B was not. After one month both groups were tested again, with the results as shown in the following table. Do these data support the alternative hypothesis that the special training is effective at the $\alpha = 0.01$ level of significance?

Group A

Time of Test	\multicolumn{9}{c}{Student}								
	1	**2**	**3**	**4**	**5**	**6**	**7**	**8**	**9**
Before	23	22	20	21	23	18	17	20	23
After	28	29	26	23	31	25	22	26	26

Group B

Time of Test	\multicolumn{8}{c}{Student}							
	10	**11**	**12**	**13**	**14**	**15**	**16**	**17**
Before	22	20	23	17	21	19	20	20
After	23	25	26	18	21	17	18	20

10.126   A study was conducted to investigate the effectiveness of two different teaching methods used in physical education: the task method and the command method. Both were used to teach tennis to college students. A group of size 30 was used for each method. The students in both groups were instructed for the same length of time and then tested on the forehand and backhand tennis strokes.

Results	Command Group	Task Group	Calculated t
Forehand	$\bar{x} = 55.6$	$\bar{x} = 59.8$	$t^{\star} = 2.39$
Backhand	$\bar{x} = 35.87$	$\bar{x} = 41.83$	$t^{\star} = 3.59$

    a. Is one method more effective than the other in teaching tennis to college students?

    b. If so, for which stroke and with what level of significance?

 10.127 A manufacturer designed an experiment to compare the difference between men and women with respect to the times they required to assemble a product. Fifteen men and 15 women were tested to determine the time they required, on the average, to assemble the product. The times required by the men had a standard deviation of 4.5 min, and the times required by the women had a standard deviation of 2.8 min. Do these data show that the amount of time needed by men is more variable than the time needed by women? Use $\alpha = 0.05$ and assume the times are approximately normally distributed.

 10.128 A soft-drink distributor is considering two new models of dispensing machines. Both the Harvard Company machine and the Fizzit machine can be adjusted to fill the cups to a certain mean amount. However, the variation in the amount dispensed from cup to cup is a primary concern. Ten cups dispensed from the Harvard machine showed a variance of 0.065, whereas 15 cups dispensed from the Fizzit machine showed a variance of 0.033. The factory representative from the Harvard Company maintains that his machine had no more variability than the Fizzit machine. Assume amount dispensed is normally distributed.

    At the 0.05 level of significance, does the sample refute the representative's assertion?

 10.129 A survey was conducted to determine the proportion of Democrats as well as Republicans who support a "get tough" policy in South America. The results of the survey were as follows:

        Democrats: $n = 250$ number in support $= 120$

        Republicans: $n = 200$ number in support $= 105$

Construct the 98% confidence interval for the difference between the proportions of support.

 10.130 According to a report in *Science News* (Vol. 137, No. 8), the percentage of seniors who used an illicit drug during the previous month was 19.7% in 1989. The figure was 21.3% in 1988. The annual survey of 17,000 seniors is conducted by researchers at the University of Michigan in Ann Arbor.

    a. Set a 95% confidence interval on the true decrease in usage.

    b. Does the interval found in part (a) suggest that there has been a significant decrease in the usage of illicit drugs by seniors? Explain.

 10.131 Of a random sample of 100 stocks on the New York Stock Exchange, 32 made a gain today. A random sample of 100 stocks on the American Stock Exchange showed 27 stocks making a gain.

    a. Construct a 99% confidence interval, estimating the difference in the proportion of stocks making a gain.

    b. Does the answer to (a) suggest that there is a significant difference between the proportions of stocks making gains on the two stock exchanges?

10.132 A consumer group compared the reliability of two comparable microcomputers from two different manufacturers. The proportion requiring service within the first year after purchase was determined for samples from each of two manufacturers.

Manufacturer	Sample Size	Proportion Needing Service
1	75	0.15
2	75	0.09

Find a 0.95 confidence interval for $p_1 - p_2$.

10.133 In determining the "goodness" of a test question, a teacher will often compare the percentage of better students who answer it correctly to the percentage of the poorer students who answer it correctly. One expects that the better students will answer the question correctly more frequently than the poorer students. On the last test, 35 of the students with the top 60 grades and 27 with the bottom 60 answered a certain question correctly. Did the students with the top grades do significantly better on this question? Use $\alpha = 0.05$.
   a. Solve using the p-value approach.
   b. Solve using the classical approach.

10.134 The April 4, 1991, issue of *USA Today* reported results from the *New England Journal of Medicine*. In a study of 987 deaths in southern California, the average right-hander died at age 75 and the average left-hander died at age 66. In addition, it was found that 7.9% of the lefties died from accident-related injuries, excluding vehicles, versus 1.5% for the right-handers; and 5.3% of the left-handers died while driving vehicles versus 1.4% of the right-handers.

Suppose you examine 1000 randomly selected death certificates of which 100 were left-handers and 900 were right-handers. If you found that 5 of the left-handers and 18 of the right-handers died while driving a vehicle, would you have evidence to show that the proportion of left-handers who die at the wheel is significantly higher than the proportion of right-handers? Calculate the p-value and interpret its meaning.

10.135 The December 31, 1991, issue of *Newsweek* summarized results obtained at the University of Utah Medical Center concerning breast cancer. One hundred and three healthy women who each had two close relatives with breast cancer were compared with 31 women with no such family history. Both sets were examined for proliferative breast disease (PBD), benign but multiplying breast lesions. The results can be summarized as follows:

Group	Number	Percent with PBD
1. Family history of breast cancer	103	35
2. No family history	31	13

   a. Calculate the test statistic for the research hypothesis that $p_1 > p_2$ and the p-value.
   b. State the decision and the conclusion if this hypothesis test is completed at the 0.02 level of significance.

10.136 "It's a draw, according to two Australian researchers. By age 25, up to 29% of all men and up to 34% of all women have some gray hair, but this difference is so small that it's considered insignificant" ("Silver Threads Among the Gold: Who'll Find Them First, a Man or a Woman?" *Family Circle*, June 26, 1990). If 1000 men and 1000 women were involved in this research, would the 5% difference mentioned be significant at the 0.01 level?

VOCABULARY LIST

Be able to define each term. Pay special attention to the key terms, which are printed in red. In addition, describe in your own words, and give an example of, each term. Your examples should not be ones given in class or in the textbook.

The bracketed numbers indicate the chapters in which the term previously appeared, but you should define the terms again to show increased understanding of their meaning. Page numbers indicate the first appearance of the term in Chapter 10.

binomial *p* [5, 9] (p. 544)	percentage (p. 543)
binomial experiment [5] (p. 544)	probability (p. 543)
confidence interval [8, 9] (p. 502)	proportion (p. 543)
dependent means (p. 499)	pooled observed probability (p. 547)
dependent samples (p. 497)	*p*-value [8, 9] (p. 504)
F **distribution (p. 532)**	ratio of variances (p. 532)
F-statistic (p. 535)	source (of data) (p. 497)
hypothesis test [8, 9] (p. 503)	standard error [8, 9] (pp. 513, 544)
independent means (p. 512)	*t* **distribution [9] (p. 501)**
independent samples (p. 497)	test statistic (p. 503)
mean difference (p. 501)	*t*-**statistic [9] (p. 503)**
paired difference (p. 500)	*z*-statistic [8, 9] (p. 546)

QUIZ A

Answer "True" if the statement is always true. If the statement is not always true, replace the words shown in bold with words that make the statement always true.

10.1 When the means of two unrelated samples are used to compare two populations, we are dealing with **two dependent means**.

10.2 The use of **paired data (dependent means)** often allows for the control of unmeasurable or confounding variables because each pair is subjected to these confounding effects equally.

10.3 The **chi-square distribution** is used for making inferences about the ratio of the variances of two populations.

10.4 The *F* **distribution** is used when two dependent means are to be compared.

10.5 In comparing two independent means when the σ's are unknown, we need to use the **standard normal** distribution.

10.6 The **standard normal score** is used for all inferences concerning population proportions.

10.7 The *F* distribution is a **symmetric** distribution.

10.8 The number of degrees of freedom for the critical value of t is equal to **the smaller of $n_1 - 1$ or $n_2 - 1$** when making inferences about the difference between two independent means for the case when the degrees of freedom is estimated.

10.9 In constructing a confidence interval for the mean difference in paired data, the interval **increases** in width when the sample size is increased.

10.10 A **pooled estimate** for any statistic in a problem dealing with two populations is a value arrived at by combining the two separate sample statistics so as to achieve the best possible point estimate.

QUIZ B

Answer all questions, showing all formulas, substitutions, and work.

10.1 State the null (H_0) and the alternative (H_a) hypotheses that would be used to test each of these claims:

     a. There is no significant difference in the mean batting averages for the baseball players of the two major leagues.

     b. The standard deviation for the monthly amounts of rainfall in Monroe County is less than the standard deviation for the monthly amounts of rainfall in Orange County.

     c. There is a significant difference between the percentage of male and female college students who own their own car.

10.2 Determine the test criteria [test statistic, critical region(s), and critical value(s)] that would be used in completing each hypothesis test (classical model) using $\alpha = 0.05$.

     a. $H_0: \mu_1 - \mu_2 = 43$ b. $H_0: \sigma_1/\sigma_2 = 1.0$ c. $H_0: p_1 - p_2 = 0.2$
        $H_a: \mu_1 - \mu_2 > 43$ $H_a: \sigma_1/\sigma_2 > 1.0$ $H_a: p_1 - p_2 \neq 0.2$
        ($n_1 = 50, n_2 = 40$)   ($n_1 = 16, n_2 = 25$)
     d. $H_0: \mu_d = 12$ e. $H_0: \mu_1 - \mu_2 = 17$
        $H_a: \mu_d \neq 12$ $H_a: \mu_1 - \mu_2 > 17$
        ($n = 28$)   ($n_1 = 8, n_2 = 10$)

10.3 Find each of the following:

     a. $z(0.04)$ b. $t(15, 0.025)$
     c. $t(45, 0.01)$ d. $F(8, 12, 0.01)$

10.4 Twenty college freshmen were randomly divided into two groups. The members of one group were assigned to a statistics section using programmed materials only. Members of the other group were assigned to a section in which the professor lectured. At the end of the semester, all were given the same final exam. The results were as follows:

Programmed	76	60	85	58	91	44	82	64	79	88
Lecture	81	62	87	70	86	77	90	63	85	83

At the 5% level of significance, do these data show sufficient evidence to conclude that on the average the students in the lecture sections performed significantly better on the final exam?

10.5 The weights of eight people before they stopped smoking and five weeks after they stopped smoking are as follows:

Person	1	2	3	4	5	6	7	8
Before	148	176	153	116	129	128	120	132
After	154	179	151	121	130	136	125	128

At the 0.05 level of significance, does this sample present enough evidence to justify the conclusion that weight increases if one quits smoking?

10.6 In a nationwide sample of 600 school-age boys and 500 school-age girls, 288 boys and 175 girls admitted to having committed a destruction-of-property offense. Use this sample data and construct a 95% confidence interval for the difference between the proportion of boys and girls who have committed this offense.

QUIZ C

10.1 To compare the accuracy of two short-range missiles, 8 of the first kind and 10 of the second kind are fired at a target. Let x be the distance by which the missile missed the target. Do these two sets of data (8 distances and 10 distances) represent dependent or independent samples? Explain.

10.2 Let's assume that there are 400 students in our college taking elementary statistics this semester. Describe how you could obtain two dependent samples of size 20 from these students in order to test some precourse skill against the same skill after completing the course. Be very specific.

10.3 Student A says he "doesn't see what all the fuss about the difference between independent and dependent means is all about; the results are almost the same regardless of the method used." Professor C suggests, "Student A should compare the procedures a bit more carefully." Help student A discover that there is a substantial difference between the procedures.

10.4 Suppose you are testing H_0: $\mu_d = 0$ versus H_a: $\mu_d < 0$ and the sample paired differences are all negative. Does this mean there is sufficient evidence to reject the null hypothesis? How can it not be significant? Explain.

10.5 Truancy is very disruptive to the educational system. A group of high school teachers and counselors have developed a group counseling program that they hope will help improve the truancy situation in their school. They have selected the 80 students in their school with the worst truancy records and have randomly assigned half of them to the group counseling program. At the end of the school year, the 80 students will be rated with regard to their truancy. When the scores have been collected, they will be turned over to you for evaluation. Explain what you will do to complete the study.

10.6 You wish to estimate and compare the proportion of Catholic families whose children attend a private school to the proportion of non-Catholic families whose children attend private schools. How would you go about estimating the two proportions and the difference between them?

WORKING WITH YOUR OWN DATA

As consumers, we all purchase many bottled, boxed, canned, or packaged products. Seldom, if ever, do we question whether or not the content amount stated on the container is accurate.

Here are a few typical content listings found on various containers:

28 FL OZ (1 PT 12 OZ)
750 ml
5 FL OZ (148 ml)
32 FL OZ (1 QT) 0.951
NET WT 10 OZ 283 GRAMS
NET WT $3\frac{3}{4}$ OZ 106 g—48 tea bags
140 1-PLY NAPKINS
77 SQ FT—92 TWO-PLY SHEETS—11 × 11 IN.

Have you ever thought, "I wonder if I am getting the amount that I am paying for?" And if this thought did cross your mind, did you attempt to check the validity of the content claim? The following article appeared in the *Times Union* of Rochester, New York, on February 16, 1972. (This kind of situation occurs frequently.)

Milk Firm Accused of Short Measure*
The processing manager of Dairylea Cooperative, Inc., has been named in a warrant charging that the cooperative is distributing cartons of milk in the Rochester area containing less than the quantity represented.

. . . an investigator found shortages in four quarts of Dairylea milk purchased Friday.
Asst. Dist. Atty. Howard R. Relin, who issued the warrant, said the shortages ranged from $1\frac{1}{8}$ to $1\frac{1}{4}$ ounces per quart. A quart of milk contains 32 fluid ounces.
. . . the state Agriculture and Markets Law . . . provides that a seller of a commodity shall not sell or deliver less of the commodity than the quantity represented to be sold.
. . . the purpose of the law under which . . . the dairy is charged is to ensure honest, accurate, and fair dealing with the public. There is no requirement that intent to violate the law be proved, he said.

This situation poses a very interesting legal problem: There is no need to show intent to "short the customer." If caught, the violators are fined automatically and the fines are often quite severe.

A | A HIGH-SPEED FILLING OPERATION

A high-speed piston-type machine used to fill cans with hot tomato juice was sold to a canning company. The guarantee stated that the machine would fill 48-oz cans

---

*From *The Times-Union*, Rochester, N.Y., Feb. 16, 1972.

with a mean amount of 49.5 oz, a standard deviation of 0.072 oz, and a maximum spread of 0.282 oz while operating at a rate of filling 150 to 170 cans per minute. On August 12, 1994, a sample of 42 cans was gathered and the following weights were recorded. The weights, measured to the nearest $\frac{1}{8}$ oz, are recorded as variations from 49.5 oz.

$-\frac{1}{8}$	0	$-\frac{1}{8}$	0	0	0	$-\frac{1}{8}$	0	0	0
0	0	$-\frac{1}{8}$	0	$\frac{1}{8}$	0	$\frac{1}{8}$	0	$\frac{1}{8}$	0
0	0	$-\frac{1}{8}$	0	0	0	0	$-\frac{1}{8}$	0	0
0	0	0	0	0	0	0	0	0	0
0	0								

1. Calculate the mean $\bar{x}$, the standard deviation s, and the range of the sample data.
2. Construct a histogram picturing the sample data.
3. Does the amount of fill differ from the prescribed 49.5 oz at the $\alpha = 0.05$ level? Test the hypothesis that $\mu = 49.5$ against an appropriate alternative.
4. Does the amount of variation, as measured by the standard deviation, satisfy the guarantee at the $\alpha = 0.05$ level?
5. Assuming that the filling machine continues to fill cans with an amount of tomato juice that is distributed normally and the mean and standard deviation are equal to the values found in question 1, what is the probability that a randomly selected can will contain less than the 48 oz claimed on the label?
6. If the amount of fill per can is normally distributed and the standard deviation can be maintained, find the setting for the mean value that would allow only 1 can in every 10,000 to contain less than 48 oz.

B | YOUR OWN INVESTIGATION

Select a packaged product that has a quantity of fill per package that you can and would like to investigate.

1. Describe your selected product, including the quantity per package, and describe how you plan to obtain your data.
2. Collect your sample of data. (Consult your instructor for advice on size of sample.)
3. Calculate the mean $\bar{x}$ and standard deviation s for your sample data.
4. Construct a histogram or stem-and-leaf diagram picturing the sample data.
5. Does the mean amount of fill meet the amount prescribed on the label? Test using $\alpha = 0.05$.
6. Assume that the item you selected is filled continually. The amount of fill is normally distributed, and the mean and standard deviation are equal to the values found in question 3. What is the probability that one randomly selected package contains less than the prescribed amount?

MORE INFERENTIAL STATISTICS

*I*n Part Three we studied the two inferential statistical techniques: confidence intervals and hypothesis tests, for one and two populations, and the three population parameters: mean, μ; standard deviation, σ; and proportion, p. In Part Four we will learn how to use the estimation and the hypothesis test techniques in other statistical situations. Some of these situations will be extensions of previous methods. (For example, analysis of variance will be used to deal with more than two populations when the question involves the mean, and the binomial experiment will be expanded to a multinomial experiment.) Other situations will be alternatives to methods previously studied (such as the nonparametric techniques) or will deal with new inferences (such as the nonparametric methods and the regression and correlation inferential techniques).

Sir Ronald A. Fisher

Sir Ronald Alymer Fisher, British statistician, was born in London on February 17, 1890. In 1912, after earning a B.A. from Gonville and Caius College in Cambridge, Fisher worked as a statistician for an investment company until 1915, at which time he became a public school teacher. In 1917 he married Ruth Eillean Gralton Guiness with whom he had eight children. Fisher received his M.A. in 1920 and his Sc.D. in 1926. Many years later, he retired to Australia, where he died at the age of 72 on July 29, 1962.

In 1919 Fisher was hired by Rothamstead Experimental Station to do statistical work with its plant breeding experiments. It was there that he pioneered the application of statistical procedures to the design of scientific experiments. During his employment at Rothamstead, Fisher introduced the principle of randomization and originated the concept of the analysis of variance (an even more important achievement). In 1925 Fisher wrote "Statistical Methods for Research," a work that remained in print for more than 50 years. Fisher has played a major role in the development and application of statistics.

ADDITIONAL APPLICATIONS OF CHI-SQUARE

Chapter Case Study

TRIAL BY NUMBER

Karl Pearson's chi-square test measured the fit between theory and reality, ushering in a new sort of decision making.

Our world is inundated with statistics. Every medical fear or triumph is charted by a complex analysis of chances. Think of cancer, heart disease, AIDS: The less we know, the more we hear of probabilities. This daily barrage is not a matter of mere counting but of inference and decision in the face of uncertainty. No committee changes our schools or our prisons without studies on the effects of busing or early parole. Money markets, drunken driving, family life, high energy physics, and deviant human cells are all subject to tests of significance and data analysis.

This all began in 1900 when Karl Pearson published his chi-square test of goodness of fit, a formula for measuring how well a theoretical hypothesis fits some observations. The basic idea is simple enough. Suppose that you think a die will fall equally often on each of its six faces. You roll it 600 times. It seems to come up six all too frequently. Could this simply be chance? How well does the hypothesis—that the die is fair—fit the data? The result that would best fit your theory would be that each face came up just 100 times in 600. In practice the ratios are almost always different, even with a fair die, because even for many throws there will always be the factor of chance. How different should they be to make you suspect a poor fit between your theory and your 600 observations? Pearson's chi-square test gives one measure of how well theory and data correspond.

The chi-square test can be used for hypotheses and data where observations naturally fall into discrete categories that statisticians call cells. If, for example, you are testing to find whether a certain treatment for cholera is worthless, then the patients divide among four cells: treated and recovered, treated and died, untreated and recovered, and untreated and died. If the treatment is worthless, you expect no difference in recovery rate between treated and untreated patients. But chance and uncontrollable variables dictate that there will almost always be some difference. Pearson's test takes this into account, telling you how well your hypothesis—that the treatment is worthless—fits your observations. . . .

The chi-square test was a tiny event in itself, but it was the signal for a sweeping transformation in the ways we interpret our numerical world. Now there could be a standard way in which ideas were presented to policy makers and to the general public. . . .

Statistical talk has now invaded many aspects of daily life, creating sometimes spurious impressions of objectivity. Policy makers demand numbers to measure every risk and hope. Thus the 1975 *Reactor Safety Study* of the Nuclear Regulatory Commission attached probabilities to various kinds of danger. Some critics suggest that this is just a way of escaping the responsibility of admitting ignorance. Other critics point to the increasing use of complex models generated to "fit" reams of data in some statistical sense—models that can become a fantasyland without any connection to the real world.

For better or worse, statistical inference has provided an entirely new style of reasoning. The quiet statisticians have changed our world—not by discovering new facts or technical developments but by changing the ways we reason, experiment, and form our opinions about it.

Source: Ian Hacking. Reprinted by permission from the November issue of SCIENCE '84. Copyright © 1984 by the American Association for the Advancement of Science.

If you were to roll a die 600 times, how different from 100 could the observed frequencies for each face be before the results would become significantly different from equally likely at the 0.05 level? (See Exercise 11.49 on p. 607.)

● CHAPTER OBJECTIVES

Previously we discussed and used the chi-square distribution to test the value of the variance (or standard deviation) of a single population. The chi-square distribution may also be used for tests in other types of situations. In this chapter we investigate some of these uses. Specifically, we will look at two tests: a multinomial experiment and the contingency table. These two types of tests are to compare experimental results with expected results in order to determine (1) preferences, (2) independence, and (3) homogeneity.

Enumerative data

    The information that we will use in these techniques will be **enumerative**; that is, the data will be placed in categories and counted. These counts become the enumerative information used.

●

11.1 │ CHI-SQUARE STATISTIC

There are many problems for which the data are categorized and the results shown by way of counts. For example, a set of final exam scores can be displayed as a frequency distribution. These frequency numbers are counts, the number of data that fall in each cell. A survey asks voters whether they are registered as Republican, Democrat, or other, and whether or not they support a particular candidate. The results are usually displayed on a chart that shows the number of voters for each possible category. There were numerous illustrations of this way of presenting data throughout the previous ten chapters.

Cell

    Suppose that we have a number of **cells** into which n observations have been sorted. (The term *cell* is synonymous with the term *class*; the terms *class* and *frequency* were defined and first used in earlier chapters. Before continuing, a brief review of Sections 2.1, 2.2, and 3.1 might be beneficial.) The **observed frequencies** in each

Observed frequency

cell are denoted by $O_1, O_2, O_3, \ldots, O_k$ (see Table 11-1). Note that the sum of all observed frequencies is equal to

$$O_1 + O_2 + \ldots + O_k = n$$

where n is the sample size. What we would like to do is to compare the observed

Expected frequency

frequencies with some **expected**, or **theoretical, frequencies**, denoted by $E_1, E_2, E_3, \ldots, E_k$ (see Table 11-1), for each of these cells. Again, the sum of these expected frequencies must be exactly

$$E_1 + E_2 + \ldots + E_k = n$$

TABLE 11-1

Observed Frequencies

| | *k* Categories | | | | | |
	1st	2nd	3rd	· · ·	*k*th	Total
Observed Frequency	O_1	O_2	O_3	· · ·	O_k	n
Expected Frequency	E_1	E_2	E_3	· · ·	E_k	n

We will then decide whether the observed frequencies seem to agree or seem to disagree with the expected frequencies. This will be accomplished by a hypothesis test using **chi-square**, χ^2.

Chi-square

The calculated value of the test statistic will be

$$\chi^{2\star} = \sum_{\text{all cells}} \frac{(O - E)^2}{E} \tag{11-1}$$

This calculated value for chi-square will be the sum of several nonnegative numbers, one from each cell (or category). The numerator of each term in the formula for χ^2 is the square of the difference between the values of the observed and the expected frequencies. The closer together these values are, the smaller the value of $(O - E)^2$; the farther apart, the larger the value of $(O - E)^2$. The denominator for each cell puts the size of the numerator into perspective. That is, a difference $(O - E)$ of 10 resulting from frequencies of 110 (O) and 100 (E) seems quite different from a difference of 10 resulting from 15 (O) and 5 (E).

These ideas suggest that small values of chi-square indicate agreement between the two sets of frequencies, whereas larger values indicate disagreement. Therefore, it is customary for these tests to be one-tailed, with the critical region on the right.

In repeated sampling, the calculated value of $\chi^{2\star}$ [formula (11-1)] will have a sampling distribution that can be approximated by the chi-square probability distribution when n is large. This approximation is generally considered adequate when all the expected frequencies are equal to or greater than 5. Recall that there is a separate chi-square distribution for each degree of freedom, df. The appropriate value of df will be described with each specific test. The critical value of chi-square, $\chi^2(\text{df}, \alpha)$, can be found in Table 8 of Appendix B.

The underlying assumption for this chi-square test is that the sample information is obtained from a random sample drawn from a population in which each individual is classified according to the categorical variable(s) involved in the test. A **categorical**

Categorical variable

variable is a variable that classifies or categorizes each individual into exactly one of several cells or classes; these cells or classes are all inclusive and mutually exclusive. The resulting face from a rolled die is a categorical variable: The list of outcomes $\{1, 2, 3, 4, 5, 6\}$ form a set of all-inclusive and mutually exclusive categories.

In this chapter we permit a certain amount of "liberalization" with respect to the null hypothesis and its testing. In previous chapters the null hypothesis has always been a statement about a population parameter (μ, σ, or p). However, there are other types of hypotheses that can be tested, such as "this die is fair" or "height and weight of individuals are independent." Notice that these hypotheses are not claims about a parameter, although sometimes they could be stated with parameter values specified.

Suppose that I claim that "this die is fair," $p = P(\text{any one number}) = \frac{1}{6}$, and you want to test it. What would you do? Was your answer something like "Roll this die many times, recording the results"? Suppose that you decide to roll the die 60 times. If the die is fair, what do you expect will happen? Each number $(1, 2, \ldots, 6)$ should appear approximately $\frac{1}{6}$ of the time (that is, 10 times). If it happens that approximately 10 of each number occur, you will certainly accept the claim of fairness $\left(p = \frac{1}{6}\right)$. If it happens that the die seems to favor some numbers, you will reject the claim. (The test statistic χ^2 will have a large value in this case, as we will soon see.)

11.2 | Inferences Concerning Multinomial Experiments

Multinomial
experiment

The preceding die problem is a good illustration of a **multinomial experiment.** Let's consider this problem again. Suppose that we want to test this die (at $\alpha = 0.05$) and decide whether to fail to reject or reject the claim "this die is fair." (The probability of each number is $\frac{1}{6}$.) The die is rolled from a cup onto a smooth flat surface 60 times, with the observed frequencies shown in the following table.

Number	1	2	3	4	5	6
Observed freq.	7	12	10	12	8	11

The null hypothesis that the die is fair is assumed to be true. This allows us to calculate the expected frequencies. If the die was fair, we certainly would expect 10 occurrences for each number.

Now let's calculate an observed value of χ^2. These calculations are shown in Table 11-2. The calculated value is $\chi^{2\star} = 2.2$.

TABLE 11-2

Computations for
Calculating χ^2

Number	Observed (O)	Expected (E)	$O - E$	$(O - E)^2$	$(O - E)^2/E$
1	7	10	-3	9	0.9
2	12	10	2	4	0.4
3	10	10	0	0	0.0
4	12	10	2	4	0.4
5	8	10	-2	4	0.4
6	11	10	1	1	0.1
Total	60	60	0 (ck)		2.2

Note $\Sigma(O - E)$ must equal zero since $\Sigma O = \Sigma E = n$. You can use this fact as a check, as shown in Table 11-2.

Before continuing, let's set up the hypothesis-testing format.

STEP 1   Parameters of concern: The probability with which each face lands on top; $P(1)$, $P(2)$, $P(3)$, $P(4)$, $P(5)$, $P(6)$.

STEP 2   Null and alternative hypothesis:

H_0: The die is fair (each $p = \frac{1}{6}$).

H_a: The die is not fair (at least one p is different).

STEP 3   Test criteria:
a. Assumptions: The data were collected in a random manner.
b. Test statistic to be used: $\chi^{2\star} = \Sigma(O - E)^2/E$, with df degrees of freedom

In a multinomial experiment, df $= k - 1$, where k is the number of cells. For our illustration, df $= k - 1 = 6 - 1 = 5$.

c. Level of significance: $\alpha = 0.05$.

STEP 4   Sample evidence:

a. Calculated value of test statistic: $\chi^{2\star} = 2.2$ (calculations shown in Table 11-2)

b. Calculate the p-value: p-value $= P(\chi^{2\star} > 2.2 \mid$ df $= 5)$.

MINITAB commands to find p-value:
CDF 2.2;
CHISquare 5.
See page 475.

Using computer: p-value $= 0.821$

Using Table 8 in Appendix B: $0.75 < p$-value < 0.90

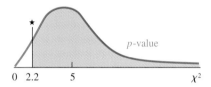

STEP 5   Results: p-value is greater than α.

DECISION Fail to reject H_0.

CONCLUSION At the 0.05 level of significance, the observed frequencies are not significantly different from those expected of a fair die.

If you use the classical approach, the preceding hypothesis test is completed using steps 1 and 2 from page 574 and steps 3–5 as follows:

STEP 3   Test criteria:

Assumptions: The data were collected randomly.

Test statistic: $\chi^2 = \Sigma(O - E)^2/E$, with df $= k - 1 = 6 - 1 = 5$.

Level of significance: $\alpha = 0.05$.

Critical value and region:

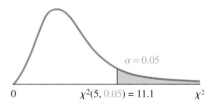

STEP 4   Sample evidence:

$\chi^{2\star} = 2.2$ (calculations shown on Table 11-2)

STEP 5   Results: $\chi^{2\star}$ is in the noncritical region.

DECISION Fail to reject H_0.

(*continued*)

CONCLUSION At the 0.05 level of significance, the observed frequencies are not significantly different from those expected of a fair die.

Before we look at other illustrations, we must define the term *multinomial experiment* and we must state the guidelines for completing the chi-square test for it.

MULTINOMIAL EXPERIMENT

An experiment with the following characteristics:

1. It consists of n identical independent trials.
2. The outcome of each trial fits into exactly one of k possible cells.
3. There is a probability associated with each particular cell, and these individual probabilities remain constant during the experiment. (It must be true that $p_1 + p_2 + \cdots + p_k = 1$.)
4. The experiment will result in a set of observed frequencies, O_1, $O_2, \ldots, O_k$, where each O_i is the number of times a trial outcome falls into that particular cell. (It must be the case that $O_1 + O_2 + \cdots + O_k = n$.)

The die example meets the definition of a multinomial experiment because it has all four of the characteristics described in the definition.

1. The die was rolled n (60) times in an identical fashion, and these trials were independent of each other. (The result of each trial was unaffected by the results of other trials.)
2. Each time the die was rolled, one of six numbers resulted, and each number was associated with a cell.
3. The probability associated with each cell was $\frac{1}{6}$, and this was constant from trial to trial. (Six values of $\frac{1}{6}$ sum to 1.0.)
4. When the experiment was complete, we had a list of frequencies (7, 12, 10, 12, 8, and 11) that summed to 60, indicating that each of the outcomes was taken into account.

The testing procedure for multinomial experiments is very much as it was in previous chapters. The biggest change comes with the statement of the null hypothesis. It may be a verbal statement, such as in the die example: "This die is fair." Often the alternative to the null hypothesis is not stated. However, in this book the alternative hypothesis will be shown, since it aids in organizing and understanding the problem. However, it will not be used to determine the location of the critical region, as was the case in previous chapters. *For multinomial experiments we will always use a one-tailed critical region, and it will be the right-hand tail of the χ^2 distribution, because larger deviations from the expected values (positive or negative) lead to an increase in the calculated χ^2 value.*

The critical value will be determined by the level of significance assigned (α) and the number of degrees of freedom. The number of degrees of freedom (df) will be 1 less than the number of cells (k) into which the data are divided:

$$\text{df} = k - 1 \qquad\qquad \textbf{(11-2)}$$

Each expected frequency, E_i, will be determined by multiplying the total number of trials n by the corresponding probability (p_i) for that cell. That is,

$$E_i = n \cdot p_i \qquad\qquad \textbf{(11-3)}$$

One guideline should be met to ensure a good approximation to the chi-square distribution: Each expected frequency should be at least 5 (that is, each $E_i \geq 5$). Sometimes it is possible to combine "smaller" cells to meet this guideline. If this guideline cannot be met, corrective measures to ensure a good approximation should be used. These corrective measures are not covered in this book but are discussed in many other sources.

▼| ILLUSTRATION   11 - 1

College students have regularly insisted on freedom of choice when registering for courses. This semester there were seven sections of a particular mathematics course. They were scheduled to meet at various times with a variety of instructors. Table 11-3 shows the number of students who selected each of the seven sections. Do the data indicate that the students had a preference for certain sections, or do they indicate that each section was equally likely to be chosen?

TABLE 11-3

Data for
Illustration 11-1

	Section							
	1	**2**	**3**	**4**	**5**	**6**	**7**	**Total**
Number of Students	18	12	25	23	8	19	14	119

SOLUTION If no preference was shown in the selection of sections, we would expect the 119 students to be equally distributed among the seven classes. Thus, if no preference was the case, then we would expect 17 students to register for each section. The test is completed as shown in steps 1–5, at the 5% level of significance.

STEP 1 Population parameter of concern: Preference of each section, the probability that a particular section is selected at registration.

STEP 2 The hypothesis:

H_0: There was no preference shown (equally distributed).

H_a: There was a preference shown (not equally distributed).

STEP 3   Test criteria:

a. The 119 students represent a random sample of the population of all students who register for this particular course. Since there were no new regulations related to the selection of courses, this or others, and registration seemed to proceed in its usual patterns, there is no reason to believe this is other than a random sample.

b. Test statistic to be used: chi-square, using formula (11-1).

c. Level of significance: $\alpha = 0.05$.

STEP 4   Sample evidence:

a. The calculated value:

We use formula (11-1) to calculate the test statistic, $\chi^{2\star}$:

$$\chi^2 = \frac{(18 - 17)^2}{17} + \frac{(12 - 17)^2}{17} + \frac{(25 - 17)^2}{17}$$

$$+ \frac{(23 - 17)^2}{17} + \frac{(8 - 17)^2}{17} + \frac{(19 - 17)^2}{17} + \frac{(14 - 17)^2}{17}$$

$$= \frac{(1)^2 + (-5)^2 + (8)^2 + (6)^2 + (-9)^2 + (2)^2 + (-3)^2}{17}$$

$$= \frac{1 + 25 + 64 + 36 + 81 + 4 + 9}{17} = \frac{220}{17}$$

$$= 12.9411$$

$$\chi^{2\star} = \mathbf{12.94}$$

b. Find the p-value: p-value $= P(\chi^{2\star} > 12.94 | df = 6)$.

Using computer: p-value $= 0.044$

Using Table 8: $0.025 < p$-value < 0.05

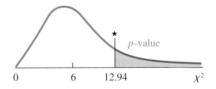

STEP 5   Results: p-value is less than α.

DECISION Reject H_0.

CONCLUSION At the 0.05 level of significance, there does seem to be a preference shown. We cannot determine, from the given information, what the preference is. It could be teacher preference, time preference, or a case of schedule conflict. Conclusions must be worded carefully to avoid suggesting conclusions that we cannot support.

Not all multinomial experiments result in equal expected frequencies, as we will see in Illustration 11-2.

▼ | ILLUSTRATION   11 - 2

The Mendelian theory of inheritance claims that the frequencies of round and yellow, wrinkled and yellow, round and green, and wrinkled and green peas will occur in the ratio $9 : 3 : 3 : 1$ when two specific varieties of peas are crossed. In testing this theory, Mendel obtained frequencies of 315, 101, 108, and 32, respectively. Do these sample data provide sufficient evidence to reject this theory at the 0.05 level of significance?

SOLUTION

STEP 1   Parameters of concern: The proportions: P(round and yellow), P(wrinkled and yellow), P(round and green), P(wrinkled and green).

STEP 2   Hypotheses:

H_0: $9 : 3 : 3 : 1$ is the ratio of inheritance.

H_a: $9 : 3 : 3 : 1$ is not the ratio of inheritance.

STEP 3   Test criteria:
   a. Assumptions: We will assume that Mendel's results form a random sample.
   b. Test statistic: χ^2, using formula (11-1) with 3 degrees of freedom.
   c. Significance level: $\alpha = 0.05$.

EXERCISE 11.1

Explain how $9 : 3 : 3 : 1$ became $\frac{9}{16}, \frac{3}{16}, \frac{3}{16}$, and $\frac{1}{16}$.

STEP 4   Sample evidence:
   a. The calculated value: The ratio $9 : 3 : 3 : 1$ indicates probabilities of $\frac{9}{16}, \frac{3}{16}, \frac{3}{16}$, and $\frac{1}{16}$. Therefore, the expected frequencies are $9n/16, 3n/16, 3n/16$, and $n/16$, where

$$n = \Sigma O_i = 315 + 101 + 108 + 32 = 556$$

The computations necessary for calculating χ^2 are given in Table 11-4. Thus, $\chi^{2\star} = \mathbf{0.47}$.

TABLE 11-4   Computations Needed to Calculate χ^2

O	E	$O - E$	$(O - E)^2/E$
315	312.75	2.25	0.0162
101	104.25	−3.25	0.1013
108	104.25	3.75	0.1349
32	34.75	−2.75	0.2176
556	556.00	0	0.4700

EXERCISE 11.2

Explain how 312.75, 2.25, and 0.0162 were obtained in Table 11-4.

Case Study

11.1

Why We Rearrange Furniture

The data collected from a survey of 300 adults and used to create "Why We Rearrange Furniture" is actually that of a multinomial experiment.

EXERCISE   11.3

Verify that "Why we rearrange furniture" is a multinomial experiment.

a. What is one trial?
b. What is the variable?
c. What are the many levels of results from each trial?

USA SNAPSHOTS®

A look at statistics that shape our lives

Why we rearrange furniture

Other
Purchasing new furniture 15%
Bored with arrangement 36%
14%
16%
19%
Redecorating
Moving to new residence

Source: Southwestern Bell Freedom Phone survey of 300 adults

By Marcy E. Mullins, USA TODAY

Source: Copyright 1990, USA TODAY. Reprinted with permission.

Case Study

11.2

Methods Used to Fall Asleep

"Methods Used to Fall Asleep" displays the results of surveying 1000 adults about what method they use to fall asleep.

EXERCISE   11.4

Why is the information shown in "Methods used to fall asleep" not that of a multinomial experiment? Be specific.

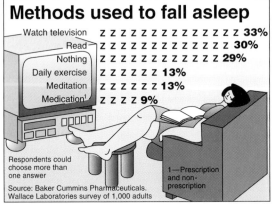

USA SNAPSHOTS®

A look at statistics that shape our lives

Methods used to fall asleep

Watch television   Z Z Z Z Z Z Z Z Z Z Z Z Z **33%**
Read   Z Z Z Z Z Z Z Z Z Z Z Z **30%**
Nothing   Z Z Z Z Z Z Z Z Z Z Z **29%**
Daily exercise   Z Z Z Z Z Z **13%**
Meditation   Z Z Z Z Z Z **13%**
Medication[1]   Z Z Z Z **9%**

Respondents could choose more than one answer

Source: Baker Cummins Pharmaceuticals. Wallace Laboratories survey of 1,000 adults

1—Prescription and non-prescription

By Julie Stacey, USA TODAY

Source: Copyright 1990, USA TODAY. Reprinted with permission.

b. The p-value: p-value $= P(\chi^{2\star} > 0.47 | \mathrm{df} = 3)$.

Using computer: p-value $= 0.925$

Using Table 8: $0.90 < p$-value < 0.95

STEP 5   Results: p-value $> \alpha$.

DECISION Fail to reject H_0.

CONCLUSION At the 0.05 level of significance, there is not sufficient evidence to reject Mendel's theory.

▲ |

MINITAB (Release 10) commands to find the cumulative probability, for the chi-square distribution with A degrees of freedom, from zero up to the $\chi^{2\star}$ listed in C1 and store the cumulative probability in C2. To find the p-value, subtract the cumulative probability from 1.0 or use the session command LET and the p-value will be in C3.

Session commands

```
Enter: CDF C1 C2;
          CHISquare A.

LET C3 = C1 - C2
PRINt C3
```

Menu commands

```
Choose: Calc > Prob. Dist. > ChiSquare
Select: Cumulative Probability
Enter:  Degrees of freedom: A
        Input column:  C1
        Output column: C2
```

Also see illustration on page 475.

EXERCISES

▼

11.5 State the null hypothesis H_0 and the alternative hypothesis H_a that would be used to test the following statements.
   a. The five numbers, 1, 2, 3, 4, and 5, are equally likely to be drawn.
   b. That multiple-choice question has a history of students selecting answers in the ratio of $2 : 3 : 2 : 1$.
   c. The poll will show a distribution of 16%, 38%, 41%, and 5% for the possible ratings of excellent, good, fair, and poor on that issue.

11.6   Determine the test criteria that would be used to test the null hypothesis for the following multinomial experiments.
   a. H_0: $P(1) = P(2) = P(3) = P(4) = 0.25$, with $\alpha = 0.05$.
   b. H_0: $P(I) = 0.25$, $P(II) = 0.40$, $P(III) = 0.35$, with $\alpha = 0.01$.

11.7   A manufacturer of floor polish conducted a consumer-preference experiment to determine which of five different floor polishes was the most appealing in appearance. A sample of 100 consumers viewed five patches of flooring that had each received one of the five polishes. Each consumer indicated the patch he or she preferred. The lighting and background were approximately the same for all patches. The results were as follow:

Polish	A	B	C	D	E	Total
Frequency	27	17	15	22	19	100

   a. State the hypothesis for "no preference" in statistical terminology.
   b. What test statistic will be used in testing this null hypothesis?
   c. Complete the hypothesis test using $\alpha = 0.10$.
      (1) Solve using the p-value approach.
      (2) Solve using the classical approach.

11.8   A certain type of flower seed will produce magenta, chartreuse, and ochre flowers in the ratio $6 : 3 : 1$ (one flower per seed). A total of 100 seeds are planted and all germinate, yielding the following results.

Magenta	Chartreuse	Ochre
52	36	12

   a. If the null hypothesis $(6 : 3 : 1)$ is true, what is the expected number of magenta flowers?
   b. How many degrees of freedom are associated with chi-square?
   c. Complete the hypothesis test using $\alpha = 0.10$.
      (1) Solve using the p-value approach.
      (2) Solve using the classical approach.

11.9   "Looking Up to Athletes" (*USA Today*, 5-7-91) reported: "Here's how sports team members say athletes do as role models for children: Excellent—16%, Good—38%, Fair—41%, Poor—5%." Suppose you took a poll of 350 members within your community and obtained the following results: Excellent—44, Good—145, Fair—133, Poor—28. Do your results show that your community has a significantly different idea about athletes as role models than the sports team members?

   MINITAB was used to analyze this problem. With the percentages entered into C1 as probabilities (i.e., in decimal form), and the observed frequencies entered into C2, the following commands were used and the following results occurred.

```
NAME C1 'P' C2 'OBS' C3 'EXP' C4 'CHI-SQ'
NAME C5 'SUM(P)' C6 'SUM(OBS)' C7 'SUM(EXP)' C8 'CHI-SQ★'
NAME K1 'DF' K2 'P-VALUE'
LET K1 = COUNT(C2)
LET C3 = C1*SUM(C2)
LET C4 = ((C2-C3)**2)/C3
SUM C1 C5
SUM C2 C6
SUM C3 C7
SUM C4 C8
CDF C8 C9;
  CHISquare 3.
LET K2 = 1 - C9
PRINt C1-C4
PRINt C5-C8
PRINt K1 K2

Row      P      OBS      EXP      CHI-SQ
 1     0.16      44     56.0     2.57143
 2     0.38     145    133.0     1.08271
 3     0.41     133    143.5     0.76829
 4     0.05      28     17.5     6.30000

Row    SUM(P)     SUM(OBS)     SUM(EXP)     CHI-SQ★
 1        1          350          350       10.7224

DF          3
P-VALUE     0.0133
```

Verify the results by:
    a. Calculating the expected values, cell chi-square values, and chi-square★.
    b. Finding df.
    c. Checking the p-value using Table 8.

11.10 A large supermarket carries four qualities of ground beef. Customers are believed to purchase these four varieties with probabilities of 0.10, 0.30, 0.35, and 0.25, respectively, from the least to most expensive variety. A sample of 500 purchases resulted in sales of 46, 162, 191, and 101 of the respective qualities. Does this sample contradict the expected proportions? Use $\alpha = 0.05$. Use a computer to complete this exercise. (If you use MINITAB, see Exercise 11.9.)

11.11 An article titled "Aging Baby Boomers Give Boost to Vacation Home Market" (*Sunday World-Herald*, Omaha, Nebraska, 9-4-94) reported the following preferred unit sizes from a nationwide telephone survey.

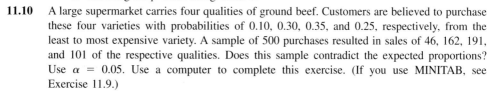

Unit Size	Percent Preferring
Studio/efficiency	18.2%
1 bedroom	18.2
2 bedrooms	40.4
3 bedrooms	18.2
Over 3 bedrooms	5.0

(*continued*)

The following results were obtained from a survey of 300 individuals in Nebraska who do not own vacation property. Does the distribution of preferences in Nebraska differ from the distribution obtained nationally? Test at the 0.05 level of significance.

Unit Size	Number Preferring
Studio/efficiency	75
1 bedroom	60
2 bedrooms	105
3 bedrooms	45
Over 3 bedrooms	15

11.12 The 1993 edition of the *Digest of Educational Statistics* gives the following distribution of the ages of persons 18 and over who hold a bachelor's or higher degree.

Age	18–24	25–34	35–44	45–54	55–64	65 or over
Percent	5	29	30	16	10	10

A survey of 500 randomly chosen persons age 18 and over who hold a bachelor's or higher degree in Alaska gave the following distribution.

Age	18–24	25–34	35–44	45–54	55–64	65 or over
Number	30	150	155	75	35	55

Test the null hypothesis that the age distribution is the same in Alaska as it is nationally at a level of significance equal to 0.05. Include in your answer the critical value, $\chi^{2\star}$, decision, and conclusion.

11.13 A program for generating random numbers on a computer is to be tested. The program is instructed to generate 100 single-digit integers between 0 and 9. The frequencies of the observed integers were as follows:

Integer	0	1	2	3	4	5	6	7	8	9
Frequency	11	8	7	7	10	10	8	11	14	14

At the 0.05 level of significance, is there sufficient reason to believe that the integers are not being generated uniformly?

11.14 *Nursing Magazine* (March 1991) reports results of a survey of over 1800 nurses across the country concerning job satisfaction and retention. Nurses from magnet hospitals (hospitals that successfully attract and retain nurses) describe the staffing situation in their units as follows:

Staffing Situation	Percent
1. Desperately short of help—patient care has suffered	12%
2. Short, but patient care hasn't suffered	32
3. Adequate	38
4. More than adequate	12
5. Excellent	6

A survey of 500 nurses from nonmagnet hospitals gave the following responses to the staffing situation.

Staffing Situation	1	2	3	4	5
Number	165	140	125	50	20

Do the data indicate that the nurses from the nonmagnet hospitals have a different distribution of opinions? Use $\alpha = 0.05$.

11.15 Why is this chi-square test typically a one-tail test with the critical region in the right tail?

   a. What kind of value would result if the observed frequencies and the expected frequencies were very close in value? Explain how you would interpret this situation.

   b. Suppose you had to roll a die 60 times as an experiment to test the fairness of the die as discussed in the example on page 574, but instead of rolling the die yourself, you paid your little brother $1 to roll it 60 times and keep a tally of the numbers. He agreed to perform this deed for you and ran off to his room with the die, returning in a few minutes with his resulting frequencies. He demanded his $1. You, of course, pay him before he hands over his results, which were: 10, 10, 10, 10, 10, and 10. The observed results are exactly what you had "expected." Right? Explain your reactions. What value of $\chi^{2\star}$ will result? What do you think happened? What do you demand of your little brother and why? What possible role might the left tail have in the hypothesis test?

   c. Why is the left tail not typically of concern?

11.16 According to the September 20, 1994, issue of *Family Circle*, 39% of Americans own guns for hunting, 30% for protection, 19% for both hunting and protection, and 12% for other reasons. A survey in Memphis, Tennessee, of 2000 individuals gave the following results.

Why Do You Own a Gun?	Number Responding
Hunting	740
Protection	600
Both	360
Other	300

(*continued*)

a. Test the null hypothesis that the distribution of reasons for owning a gun is the same in Memphis as it is nationally as reported by *Family Circle*. Use a level of significance equal to 0.05.

b. What caused the calculated value of $\chi^{2\star}$ to be so large? Does it seem right that one cell should have this much effect on the results? How could this test be completed differently (hopefully, more meaningfully) so that the results might not be affected as they were in (a)? Be specific.

11.3 | INFERENCES CONCERNING CONTINGENCY TABLES

Contingency table

A **contingency table** is an arrangement of data into a two-way classification. The data are sorted into cells, and the number of data in each cell is reported. The contingency table involves two factors (or variables), and the usual question concerning such tables is whether the data indicate that the two variables are independent or dependent.

Test of Independence

To illustrate the use and analysis of a contingency table, let's consider the gender of liberal-arts college students and their favorite academic area.

▼| ILLUSTRATION   11 - 3

Each person in a group of 300 students was identified as male or female and then asked whether he or she preferred taking liberal arts courses in the area of math–science, social science, or humanities. Table 11-5 is a contingency table that shows the frequencies found for these categories. Does this sample present sufficient evidence to reject the null hypothesis "preference for math–science, social science, or humanities is independent of the gender of a college student" at the 0.05 level of significance?

TABLE 11-5

Contingency Table
Showing Sample
Results for
Illustration 11-3

	Favorite Subject Area			
Gender	**Math–Science (M–S)**	**Social Science (SS)**	**Humanities (H)**	**Total**
Male (*m*)	37	41	44	122
Female (*f*)	35	72	71	178
Total	72	113	115	300

SOLUTION

STEP 1   Population parameters of concern: The independence of variables "gender" and "favorite course" require us to discuss the probability of the various answers and the effect that answers of one variable have on the probability of answers related to the other variable. Independence, as defined in Chapter 4, requires $P(M\text{–}S|M) = P(M\text{–}S|F) = P(M\text{–}S)$; that is, gender has no effect on the probability of a person's choice being math–science; and so on.

STEP 2   Hypotheses:

H_0: Preference for math–science, social science, or humanities is independent of the gender of a college student.

H_a: Subject preference is not independent of the gender of the student.

STEP 3   Level of significance: $\alpha = 0.05$.

STEP 4a   The sample evidence: In the case of contingency tables, the number of degrees of freedom is exactly the same number as the number of cells in the table that may be filled in freely when you are given the marginal totals. The totals in this illustration are shown in the following table.

				122
				178
72	113	115		300

Given these totals, you can fill in only two cells before the others are all determined. (The totals must, of course, be the same.) For example, once we pick two arbitrary values (say, 50 and 60) for the first two cells of the first row, the other four cell values are fixed (see the following table).

50	60	C		122
D	E	F		178
72	113	115		300

They have to be $C = 12$, $D = 22$, $E = 53$, and $F = 103$. Otherwise, the totals will not be correct. Therefore, for this problem there are two free choices. Each free choice corresponds to 1 degree of freedom. Hence the number of degrees of freedom for our example is 2 (df = 2).

Before the calculated value of chi-square can be found, we need to determine the expected values E for each cell. To do this we must recall the null hypothesis, which asserts that these factors are independent. Therefore, we would expect the values to be distributed in proportion to the **marginal totals**. There are 122 males; we would expect them to be distributed among M–S, SS, and H proportionally to the 72, 113, and 115 totals. Thus, the expected cell counts for males are

$$\frac{72}{300} \cdot 122 \qquad \frac{113}{300} \cdot 122 \qquad \frac{115}{300} \cdot 122$$

Marginal totals

Similarly, we would expect for the females

$$\frac{72}{300} \cdot 178 \qquad \frac{113}{300} \cdot 178 \qquad \frac{115}{300} \cdot 178$$

Thus, the expected values are as shown in Table 11-6. (Always check the new totals against the old totals.)

TABLE 11-6

Expected Values

	M–S	SS	H	Total
	29.28	45.95	46.77	122.00
	42.72	67.05	68.23	178.00
Total	72.00	113.00	115.00	300.00

EXERCISE   11.17

Find the expected value for the cell shown.

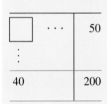

NOTE   We can think of the computation of the expected values in a second way. Recall that we assume the null hypothesis to be true until there is evidence to reject it. Having made this assumption in our example, in effect we are saying that the event that a student picked at random is male and the event that a student picked at random prefers math–science courses are independent. Our point estimate for the probability that a student is male is 122/300, and the point estimate for the probability that the student prefers math–science courses is 72/300. Therefore, the probability that both events occur is the product of the probabilities. [Refer to formula (4-8a), p. 221.] Thus, (122/300) · (72/300) is the probability of a selected student being male and preferring math–science. Therefore, the number of students out of 300 that are expected to be male and prefer math–science is found by multiplying the probability (or proportion) by the total number of students (300). Thus, the expected number of males who prefer math–science is (122/300)(72/300)(300) = (122/300)(72) = 29.28. The other expected values can be determined in the same manner.

Typically the contingency table is written so that it contains all this information (see Table 11-7).

TABLE 11-7

Contingency Table Showing Sample Results and Expected Values

Gender of Student	Favorite Subject Area			Total
	Math–Science (M–S)	Social Science (SS)	Humanities (H)	
Male	37 (29.28)	41 (45.95)	44 (46.77)	122
Female	35 (42.72)	72 (67.05)	71 (68.23)	178
Total	72	113	115	300

The calculated chi-square is

$$\chi^2 = \sum \frac{(O - E)^2}{E}$$

$$= \frac{(37 - 29.28)^2}{29.28} + \frac{(41 - 45.95)^2}{45.95} + \frac{(44 - 46.77)^2}{46.77}$$

$$+ \frac{(35 - 42.72)^2}{42.72} + \frac{(72 - 67.05)^2}{67.05} + \frac{(71 - 68.23)^2}{68.23}$$

$$= 2.035 + 0.533 + 0.164 + 1.395 + 0.365 + 0.112$$

$$= 4.604$$

$$\chi^{2\star} = \mathbf{4.604}$$

STEP 4b   Calculate the p-value: p-value = $P(\chi^{2\star} > 4.604 \,|\, df = 2)$.

Using computer: p-value = 0.1001

Using Table 8: $0.10 < p\text{-value} < 0.25$

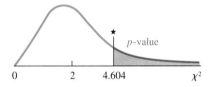

STEP 5   Results: p-value is greater than $\alpha = 0.05$.

DECISION Fail to reject H_0.

CONCLUSION At the 0.05 level of significance, the evidence does not allow us to reject the idea of independence between the gender of a student and the student's preferred academic subject area.

If you use the classical approach, the preceding hypothesis test is completed using steps 1 and 2 from pages 586 and 587 and steps 3–5 as follows:

STEP 3   Test criteria:

Test statistic: $\chi^{2\star} = \sum \frac{(O - E)^2}{E}$ with 2 degrees of freedom.

Critical region: The critical value is found in Table 8.

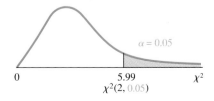

STEP 4   Sample evidence:

$$\chi^{2\star} = 4.604 \text{ (calculations shown on p. 589)}$$

STEP 5   Results: $\chi^{2\star}$ is in a noncritical region.

DECISION Fail to reject H_0.

CONCLUSION At the 0.05 level of significance, the evidence does not allow us to reject the idea of independence between the gender of a student and the student's preferred academic subject area.

▲|

r × c contingency table

In general, the **r × c contingency table** (*r* is the number of rows; *c* is the number of columns) will be used to test the independence of the row factor and the column factor. The number of degrees of freedom will be determined by

$$df = (r - 1) \cdot (c - 1) \tag{11-4}$$

where *r* and *c* are both greater than 1. (This value for df should agree with the number of cells counted according to the general description on page 587.)

The expected frequencies for an *r × c* contingency table will be found by means of the formulas in each cell in Table 11-8, where *n* = grand total. In general, the **expected frequency** at the intersection of the ***i*th row and the *j*th column** is given by

Expected frequency

$$E_{i,j} = \frac{Row\ total \times Column\ total}{Grand\ total} = \frac{R_i \times C_j}{n} \tag{11-5}$$

TABLE 11-8

Expected Frequencies for an *r × c* Contingency Table

Rows	1	2	· · ·	*j*th Column	· · ·	*c*	Total
			Columns				
1	$\dfrac{R_1 \times C_1}{n}$	$\dfrac{R_1 \times C_2}{n}$		$\dfrac{R_1 \times C_j}{n}$			R_1
2	$\dfrac{R_2 \times C_1}{n}$						R_2
⋮	⋮	⋮	⋮	⋮			⋮
*i*th Row	$\dfrac{R_i \times C_1}{n}$			$\dfrac{R_i \times C_j}{n}$			R_i
⋮	⋮			⋮			⋮
r							
Total	C_1	C_2	· · ·	C_j	· · ·		*n*

We should again observe the previously mentioned guideline: Each $E_{i,j}$ should be at least 5.

EXERCISE 11.18

Identify these values
from Table 11-7.

a. C_2
b. R_1
c. n
d. $E_{2,3}$

NOTE   The notation used in Table 11-8 and formula (11-5) may be unfamiliar to you. For convenience in referring to cells or entries in a table, $E_{i,j}$ (or E_{ij}) can be used to denote the entry in the ith row and the jth column. That is, the first letter in the subscript corresponds to the row number and the second letter corresponds to the column number. Thus, $E_{1,2}$ is the entry in the first row, second column, and $E_{2,1}$ is the entry in the second row, first column. Referring to Table 11-6, $E_{1,2}$ for that table is 45.95 and $E_{2,1}$ is 42.72. The notation used in Table 11-8 is interpreted in a similar manner; that is, R_1 corresponds to the total from row 1, and C_1 corresponds to the total from column 1.

Test of Homogeneity

Homogeneity

There is another type of contingency table problem. It is called a **test of homogeneity**. This test is used when one of the two variables is controlled by the experimenter so that the row (or column) totals are predetermined.

For example, suppose that we were to poll registered voters about a piece of legislation proposed by the governor. In the poll, 200 urban, 200 suburban, and 100 rural residents will be randomly selected and asked whether they favor or oppose the governor's proposal. That is, a simple random sample is taken for each of these three groups. A total of 500 voters are to be polled. But notice that it has been predetermined (before the sample is taken) just how many are to fall within each row category, as shown in Table 11-9, and each category is sampled separately.

TABLE 11-9

Registered Voter
Poll with
Predetermined
Row Totals

Type of Residence	Governor's Proposal		
	Favor	Oppose	Total
Urban			200
Suburban			200
Rural			100
Total			500

In a test of this nature, we are actually testing the hypothesis "the distribution of proportions within the rows is the same for all rows." That is, the distribution of proportions in row 1 is the same as in row 2, is the same as in row 3, and so on. The alternative to this is "the distribution of proportions within the rows is not the same for all rows." This type of example may be thought of as a comparison of several multinomial experiments.

Beyond this conceptual difference, the actual testing for independence and homogeneity with contingency tables is the same. Let's demonstrate this by completing the polling illustration.

▼ ┃   ILLUSTRATION   11 - 4

Each person in a random sample of 500 registered voters (200 urban, 200 suburban, and 100 rural residents) was asked his or her opinion about the governor's proposed legislation. Does the sample evidence shown in Table 11-10 support the hypothesis that "voters within the different residence groups have different opinions about the governor's proposal"? Use $\alpha = 0.05$.

TABLE 11-10

Sample Results for Illustration 11-4

Type of Residence	Governor's Proposal		
	Favor	Oppose	Total
Urban	143	57	200
Suburban	98	102	200
Rural	13	87	100
Total	254	246	500

SOLUTION

STEP 1   Population parameters of concern: The proportion of voters who favor or oppose—that is, the proportion of urban voters who favor, the proportion of suburban voters who favor, the proportion of rural voters who favor, and the proportion of all three groups, separately, who oppose.

STEP 2   Hypotheses:

H_0: The proportion of voters favoring the proposed legislation is the same in all three groups.

H_a: The proportion of voters favoring the proposed legislation is not the same in all three groups. (That is, in at least one group the proportions are different from the others.)

STEP 3   Level of significance: $\alpha = 0.05$.

STEP 4a   The evidence: The expected values are found by using formula (11-5) and are given in Table 11-11.

TABLE 11-11

Sample Results and
Expected Values

Type of Residence	Governor's Proposal		Total
	Favor	**Oppose**	**Total**
Urban	143	57	200
	(101.6)	(98.4)	
Suburban	98	102	200
	(101.6)	(98.4)	
Rural	13	87	100
	(50.8)	(49.2)	
Total	254	246	500

NOTE   Each expected value is used twice in the calculation of $\chi^{2\star}$; therefore, it is a good idea to keep extra decimal places while doing the calculations.

$$\chi^2 = \frac{(143 - 101.6)^2}{101.6} + \frac{(57 - 98.4)^2}{98.4} + \frac{(98 - 101.6)^2}{101.6}$$

$$+ \frac{(102 - 98.4)^2}{98.4} + \frac{(13 - 50.8)^2}{50.8} + \frac{(87 - 49.2)^2}{49.2}$$

$$= 16.87 + 17.42 + 0.13 + 0.13 + 28.13 + 29.04$$

$$= 91.72$$

$$\chi^{2\star} = \mathbf{91.72}$$

$$df = (r - 1)(c - 1) = (3 - 1)(2 - 1) = \mathbf{2}$$

STEP 4b   The *p*-value:

$$p\text{-value} = P(\chi^{2\star} > 91.72 | df = 2) < 0.005.$$

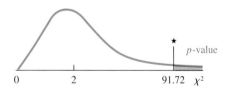

STEP 5   Results: *p*-value is less than $\alpha = 0.05$.

DECISION Reject H_0.

CONCLUSION The three groups of voters do not all have the same proportions favoring the proposed legislation.

Case Study   50% Opposed to Building Nuclear Plants

11.3

The results of a poll by the National Center for Telephone Research of New York for Gannett News Service asked registered voters, "Do you favor or oppose building more nuclear power plants in New York State?" Percentages of voters who answered "yes," "no," or "not sure" for each of four geographic regions of the state as well as for the whole state are shown in the table. If the actual number of voters in each category were given, we would have a contingency table and we would be able to complete a hypothesis test about the homogeneity of the four regions.

NEW YORK STATE POLL

The New York Poll was conducted Dec. 15–18 by the National Center for Telephone Research of New York for Gannett News Service. Trained interviewers surveyed 1,000 registered voters. The calls were placed at random, but in proportion to past voter turnouts around the state.

EXERCISE   11.19

Refer to Case Study 11-3.

a. Explain why the following question could be tested using the chi-square statistic: "Is the support for building more nuclear power plants the same in all four regions of New York State?"

b. Explain why this is a test of homogeneity.

Half of all New Yorkers oppose construction of nuclear power plants in the state, the New York State Poll shows.

And women, far more than men, account for the opposition.

Exactly 50 percent of New Yorkers surveyed said they opposed more nuclear plants in the state. That matches the percentage of voters who said they were against more plants two years ago in a New York State poll.

The latest poll, however, shows a drop since 1977 in the percentage that favors nuclear plants—down to 32 percent from 38 percent. There is a corresponding increase of 6 percent among those who say they're not sure.

Resistance to nuclear plants, regionally, ran highest in western New York, and lowest in the New York City north suburbs.

There is a great deal of uncertainty about the safety of nuclear plants, the poll found. But slightly more people seem to think they're safe than think they are unsafe.

Thus, it seems that most New Yorkers agree with Gov. Carey's decision last year during the campaign, to not allow more nuclear plants to be built until the radioactive waste problem is solved.

Here are the figures on the main question. "Do you favor or oppose building more nuclear power plants in New York State?":

	Yes	No	Not Sure
Total	32	50	18
Upstate	35	48	17
NY City	30	54	16
No. suburbs	32	45	23
Western NY	28	56	16

Source: John Omicinski, Chief, GNS Albany Bureau. Copyright 1979 by the Gannett News Service, Albany Bureau. Reprinted by permission.

Case Study

11.4

Why We Don't Exercise

The USA Snapshots below show circle graphs that represent relative frequency distributions for six categories of responses by men and women separately. Does the distribution of responses given by men appear to be significantly different from the distribution of responses given by women? This question calls for a test of homogeneity. Can the sample information given here be expressed on a contingency table?

EXERCISE 11.20

The USA Snapshots® "Why we don't exercise" and "Why women don't exercise" in Case Study 11-4 show the responses given by 1000 men and 1000 women who do not exercise.

a. Express the information on these circle graphs as a contingency table showing relative frequencies. Use two columns (M, W) and six rows (the six different responses).

b. Change the relative frequencies to frequencies and construct another 6 × 2 contingency table.

c. Complete a hypothesis test to determine if there is a significant difference between the distribution of responses given by men and women. Use $\alpha = 0.05$.

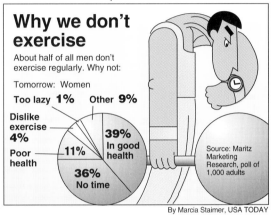

USA SNAPSHOTS®

A look at statistics that shape our lives

Why we don't exercise

About half of all men don't exercise regularly. Why not:

Tomorrow: Women

Too lazy **1%**   Other **9%**

Dislike exercise **4%**

Poor health   **11%**   **39%** In good health

36% No time

Source: Maritz Marketing Research, poll of 1,000 adults

By Marcia Staimer, USA TODAY

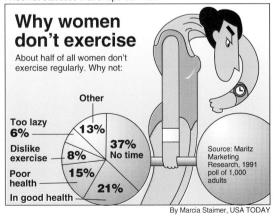

USA SNAPSHOTS®

A look at statistics that shape our lives

Why women don't exercise

About half of all women don't exercise regularly. Why not:

Other

Too lazy **6%**   **13%**

Dislike exercise **8%**   **37%** No time

Poor health **15%**

In good health   **21%**

Source: Maritz Marketing Research, 1991 poll of 1,000 adults

By Marcia Staimer, USA TODAY

MINITAB (Release 10) commands to calculate the chi-square value for a contingency table listed in **C1** and **C2**. Use a column for each column of observed frequencies in the table.

Session commands

Menu commands

Enter: CHISquare C1 C2

Choose: **Stat > Tables > ChiSq. Test**
Enter: Columns containing table: **C1 C2**

C O M P U T E R S O L U T I O N MINITAB Printout for Illustration 11-4:

Information given to computer

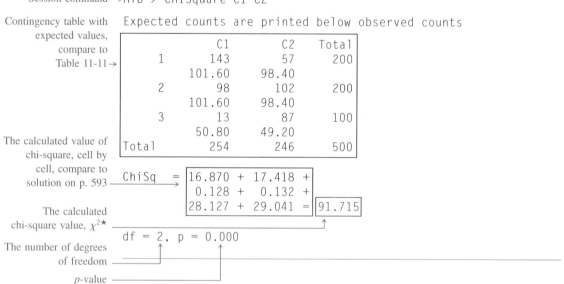

```
                          C1        C2
                        143.       57.
                         98.      102.
                         13.       87.
```

Session command → MTB > CHISquare C1 C2

Contingency table with expected values, compare to Table 11-11 →

Expected counts are printed below observed counts

```
              C1       C2    Total
    1        143       57      200
          101.60    98.40
    2         98      102      200
          101.60    98.40
    3         13       87      100
           50.80    49.20
Total       254      246      500
```

The calculated value of chi-square, cell by cell, compare to solution on p. 593

```
ChiSq  =  16.870 + 17.418 +
           0.128 +  0.132 +
          28.127 + 29.041 = 91.715
```

The calculated chi-square value, $\chi^{2\star}$

df = 2, p = 0.000

The number of degrees of freedom

p-value

EXERCISES

11.21 State the null hypothesis H_0 and the alternative hypothesis H_a that would be used to test the following statements.

a. The voters expressed preferences that were not independent of their party affiliations.

b. The distribution of opinions is the same for all three communities.

c. The proportion of "yes" responses was the same for all categories surveyed.

11.22 The "test of independence" and the "test of homogeneity" are completed in identical fashion, using the contingency table to display and organize the calculations. Explain how these two hypothesis tests differ.

11.23 The manager of an assembly process wants to determine whether the number of defective articles manufactured depends on the day of the week the articles are produced. She collected the following information.

Day of Week	Mon.	Tues.	Wed.	Thurs.	Fri.
Nondefective	85	90	95	95	90
Defective	15	10	5	5	10

Is there sufficient evidence to reject the hypothesis that the number of defective articles is independent of the day of the week on which they are produced? Use $\alpha = 0.05$.

11.24 The following table is from the July 1993 publication of *Vital and Health Statistics* from the Centers for Disease Control and Prevention/National Center for Health Statistics. The individuals in the following table have an eye irritation, or a nose irritation, or a throat irritation. They have only one of the three.

| Type of Irritation | Age (years) | | | |
	18–29	30–44	45–64	65 and over
Eye	440	567	349	59
Nose	924	1311	794	102
Throat	253	311	157	19

Test the null hypothesis that the type of ENT irritation is independent of the age group at a level of significance equal to 0.05.

11.25 A paper titled "Research Skills and the Research Environment: A Needs Assessment of Allied Health Faculty" (*Journal of Allied Health*, May 1988) reported a survey of allied health faculty. One question asked was "Do you need help with statistical analysis?" Suppose the data, with the respondents broken down by degree, were as follows:

| | Degree | | |
	BS	Master's	Doctoral
n	513	953	525
Percentage Needing Help with Statistics	76.4	77.2	52.8

The paper gives the calculated χ^2 value as 108.94. Verify this value.

11.26   Students use many kinds of criteria when selecting courses. "Teacher who is a very easy grader" is often one criterion. Three teachers are scheduled to teach statistics next semester. A sample of the grade distributions for these three teachers is shown below.

	Professor		
Grades	#1	#2	#3
A	12	11	27
B	16	29	25
C	35	30	15
Other	27	40	23

Using the 0.01 level of significance, test the hypothesis "Student's grades are independent of the student's professor."
  a. State the null and alternative hypotheses.
  b. Determine the test criteria: level of significance, test statistic, critical region, critical value(s).
  c. Use a computer to complete the calculation of $\chi^{2\star}$.
  d. State the decision and conclusion.
  e. Which professor is the easiest grader? Explain, citing specific supporting evidence.

11.27   A psychologist is investigating how a person reacts to a certain situation. He feels the reaction may be influenced by how ethnically pure the person's neighborhood is. He collects data on 500 people and finds the results shown in the following table.

Ethnically Pure Neighborhood	Reaction			
	Mild	Medium	Strong	Total
Yes	170	100	30	300
No	70	100	30	200
Total	240	200	60	500

Does there appear to be a relationship between neighborhood and reaction at the 0.10 level of significance?
  a. Solve using the *p*-value approach.
  b. Solve using the classical approach.

11.28   Fear of darkness is a common emotion. The following data were obtained by asking 200 individuals in each age group whether they had serious fears of darkness. At $\alpha = 0.01$, do we have sufficient evidence to reject the hypothesis that "the same proportion of each age group has serious fears of darkness"? (*Hint*: The contingency table must account for all 1000 people.)

Age Group	Elementary	Jr. High	Sr. High	College	Adult
No. Who Fear Darkness	83	72	49	36	114

a. Solve using the *p*-value approach.
b. Solve using the classical approach.

11.29 The following table is taken from the publication *1992 Uniform Crime Reports for the United States*. It gives the percent distribution by region for 1992 motor vehicle thefts.

Note: Two rows do not add up to 100.0 due to round-off error.

		Type of Vehicle		
Region	**Total**	**Autos**	**Trucks/Buses**	**Other Vehicles**
Total	100.0	79.6	15.1	5.4
Northeastern	100.0	92.4	4.8	2.9
Midwestern	100.0	83.7	11.1	5.2
Southern	100.0	74.2	18.7	7.1
Western	100.0	72.8	21.6	5.6

Suppose a study involving vehicle thefts and region of the country gave the following results. One hundred records were randomly selected from each region.

	Type of Vehicle		
Region	**Autos**	**Trucks/Buses**	**Other Vehicles**
Northeastern	85	10	5
Midwestern	80	15	5
Southern	65	25	10
Western	60	30	10

Test the null hypothesis that the distribution of types of vehicle thefts is homogeneous across regions at a level of significance equal to 0.05.

11.30 The following table is found in *Vital and Health Statistics* (Series 23, Number 16). The table gives the percent distribution by months pregnant when prenatal care began for mothers who had never married as well as mothers who had ever married.

Mother's Marital Status at Time of Birth	Months Pregnant When Prenatal Care Began		
	Less Than 3 Months	**3 or 4 Months**	**5 Months or More or No Care**
Never married	43.8	39.1	17.2
Ever married	70.6	22.7	6.7

(continued)

A similar study gave the following observed numbers.

Mother's Marital Status at Time of Birth	Months Pregnant When Prenatal Care Began		
	Less Than 3 Months	3 or 4 Months	5 Months or More or No Care
Never married	430	380	190
Ever married	700	200	100

Test the null hypothesis that months pregnant when prenatal care began is independent of the mother's marital status at time of birth at a level of significance equal to 0.01.

11.31   "Doctor's Age Key in Decision on Cesareans" (*USA Today*, 8-29-90) reports on a survey of 2213 obstetricians, "98% of doctors under 40 encouraged vaginal delivery in patients with a previous C-section vs. 84% over 55." How can a doctor's age be a key to the decision as to whether a woman has a cesarean childbirth or not? We know that people's ages affect their opinions, so why not expect a doctor's age to affect his or her opinion on method of childbirth? If you were commissioned to carry out a study regarding the effect that a doctor's age had on preference for or against C-sections, what information would you want to collect and how would you chart and analyze it?

IN RETROSPECT

In this chapter we have been concerned with tests of hypotheses using chi-square, with the cell probabilities associated with the multinominal experiment, and with the simple contingency table. In each case the basic assumptions are that a large number of observations have been made and that the resulting test statistic, $\Sigma[(O - E)^2/E]$, is approximately distributed as chi-square. In general, if n is large and the minimum allowable expected cell size is 5, this assumption is satisfied.

The article in Case Study 11-3 shows a contingency table with the results listed as relative frequencies. This is a common practice when reporting the results of public opinion polls. However, to statistically interpret the results, the actual frequencies must be used.

The contingency table can also be used to test homogeneity. The test for homogeneity and the test for independence look very similar and, in fact, are carried out in exactly the same way. The concepts being tested, however, equal distributions and independence, are quite different. The two tests are easily distinguished from one another, for the test of homogeneity has predetermined marginal totals in one direction in the table. That is, before the data are collected, the experimenter determines how many subjects will be observed in each category. The only predetermined number in the test of independence is the grand total.

A few words of caution: The correct number of degrees of freedom is critical if the test results are to be meaningful. The degrees of freedom determine, in part, the critical region, and its size is important. Like other tests of hypothesis, failure to reject H_0 does not mean outright acceptance of the null hypothesis.

CHAPTER EXERCISES

11.32 The psychology department at a certain college claims that the grades in its introductory course are distributed as follows: 10% A's, 20% B's, 40% C's, 20% D's, and 10% F's. In a poll of 200 randomly selected students who had completed this course, it was found that 16 had received A's, 43 B's, 65 C's, 48 D's, and 28 F's. Does this sample contradict the department's claim at the 0.05 level?

  a. Solve using the *p*-value approach.
  b. Solve using the classical approach.

11.33 When interbreeding two strains of roses, we expect the hybrid to appear in three genetic classes in the ratio $1 : 3 : 4$. If the results of an experiment yield 80 hybrids of the first type, 340 of the second type, and 380 of the third type, do we have sufficient evidence to reject the hypothesized genetic ratio at the 0.05 level of significance?

  a. Solve using the *p*-value approach.
  b. Solve using the classical approach.

11.34 A sample of 200 individuals are tested for their blood type, and the results are used to test the hypothesized distribution of blood types:

Blood Type	A	B	O	AB
Percent	0.41	0.09	0.46	0.04

The observed results were as follows:

Blood Type	A	B	O	AB
Number	75	20	95	10

At the 0.05 level of significance, is there sufficient evidence to show that the stated distribution is incorrect?

11.35 According to the *1993 Statistical Abstract of the United States*, the distribution of AIDS deaths by age (1982–1992) is as follows:

Age (years)	Percent of Deaths
Under 12	1
13–29	18
30–39	45
40–49	24
50–59	8
60 and over	4

(continued)

Suppose a study of 1000 AIDS deaths over the past year gave the following observed numbers in the six age groups.

Age (years)	Number observed
Under 12	8
13–29	205
30–39	390
40–49	270
50–59	120
60 and over	7

Test the null hypothesis that the distribution of AIDS deaths over the past year is the same as over the time period 1982–1992 at the level of significance 0.05.

11.36 The weights (x) of 300 adult males were determined and used to test the hypothesis that the weights were normally distributed with a mean of 160 lb and a standard deviation of 15 lb. The data were grouped into the following classes:

Weight (x)	Observed Frequency
$x < 130$	7
$130 \leq x < 145$	38
$145 \leq x < 160$	100
$160 \leq x < 175$	102
$175 \leq x < 190$	40
190 and over	13

Using the normal tables, the percentages for the classes are 2.28%, 13.59%, 34.13%, 34.13%, 13.59%, and 2.28%, respectively. Do the observed data show significant reason to discredit the hypothesis that the weights are normally distributed with a mean of 160 lb and a standard deviation of 15 lb? Use the *p*-value approach. Use $\alpha = 0.05$.

11.37 An article titled "Human Papillomavirus Infection and Its Relationship to Recent and Distant Sexual Partners" (*Obstetrics & Gynecology*, November 1994) gave the following results concerning age and the percent who were HPV-positive among the 290 participants in the study.

Age	N	HPV-Positive (%)
≤20	27	40.7
21–25	81	37.0
26–30	108	31.5
31–35	74	24.3

Complete the test of the hypothesis that the same proportion of each age group is HPV-positive for the population this sample represents. Use $\alpha = 0.05$.

11.38 A survey report in the March 1991 issue of *McCall's* magazine presented findings concerning job stress affecting women. One question concerned women getting verbal abuse in offices. Suppose the results were reported as follows:

Country	Percent Who Often Get Verbal Abuse in Office
USA	11
Australia	5
Brazil	17
Germany	4
Japan	10

If there were 4400 respondents in each country, would the data above indicate a true difference in the rate of verbal office abuse in the five countries?
   a. Calculate the value of chi-square for these data.
   b. Complete the hypothesis test at the 0.01 level of significance.
   c. Find the *p*-value.

11.39 Four brands of popcorn were tested for popping. One hundred kernels of each brand were popped, and the number of kernels not popped was recorded in each test (see the following table). Can we reject the null hypothesis that all four brands pop equally? Test at $\alpha = 0.05$.

Brand	A	B	C	D
No. Not Popped	14	8	11	15

   a Solve using the *p*-value approach.
   b. Solve using the classical approach.

11.40 In an article in the *Journal of Geography* (January/February 1991), Brothers reports a relationship between the presence or absence of beech trees and the classification of uplands (above 550 ft elevation) and bottomlands (below 550 ft elevation).

	Uplands	Bottomlands
Beech Present	18	6
Beech Absent	36	40

Use the data to test the independence of these two factors. Use $\alpha = 0.05$.

11.41   In the May 1990 issue of *Social Work*, Marlow reported the following results.

Number of Children Living at Home	Mexican-American Women	Anglo-American Women
0	23	38
1	22	9
2	17	15
3	7	9
4	1	1

Source: Copyright 1990, National Association of Social Workers, Inc. *Social Work.*

a. Calculate the expected values for each of the cells on the contingency table.
b. Calculate the chi-square for the data above.
c. State the null and alternative hypotheses used to test the claim "The distribution for number of children living at home is different for Mexican-American and Anglo-American women."
d. State the test criteria using $\alpha = 0.05$.
e. State the decision and conclusion.
f. Did the statistical decision above result in the same conclusion as you expressed in answering part (d) of Exercise 2.162, p. 120, for the same data?

11.42   In the May 1990 issue of *Social Work*, Marlow reported the following results:

Choice of Working	Mexican-American Women	Anglo-American Women
Yes	23	13
No, woman would not if she did not have to	21	12
No, but woman still would work	26	47

Source: Copyright 1990, National Association of Social Workers, Inc. *Social Work.*

a. Calculate the expected values for each of the cells on the contingency table.
b. Calculate the chi-square for the data above.

11.43   a. Case Study 11-3 (p. 594) reports the findings of a sample of 1000 people. Suppose that the percentages reported in this news article had come from the following contingency table. Can the null hypothesis "the distribution of yes, no, and not sure opinions is the same for all four geographic areas of New York State" be true? Use $\alpha = 0.05$.
   b. Suppose that there had been only 400 people in the sample, 100 from each geographic location. Calculate the observed value of chi-square, $\chi^{2\star}$. Compare this value to that found in (a). Explain the difference.

	Yes	No	Not Sure
Upstate	96	150	54
NY City	105	144	51
No. suburbs	64	90	46
Western NY	56	112	32

11.44 "Attitudes in Conflict" (*USA Today*, 3-11-91) reported, "We love our doctors . . . and hospitals . . . but dislike the system." Below are the ratings for each from a Gallup Poll of 1000 adults.

Rating	Physician	Hospital	Health-Care System
Excellent	53%	47%	6%
Good	39	35	32
Fair	6	12	32
Poor	1	5	24
Don't know	1	1	4
Total	100%	100%	98%[1]

[1]Doesn't total 100% due to rounding

a. Change the two 32s to 33 and convert the table of relative frequencies above to frequencies, assuming that each column represents 1000 responses.

b. Is there sufficient reason to reject the claim that "we love our doctors and hospitals" with the same distribution of attitudes? Use $\alpha = 0.05$.

c. Is there sufficient reason to claim that "our love for the doctors, the hospitals, and the system" is not all distributed the same? Use $\alpha = 0.05$.

d. Does it seem fair to conclude, as the article did, that "we love doctors and hospitals but dislike the system"? How do you justify your answer?

11.45 Does test failure reduce academic aspirations and thereby contribute to the decision to drop out of school? These were the concerns of a study titled "Standards and School Dropouts: A National Study of Tests Required for High School Graduation." The table reports the responses of 283 students selected from schools with low graduation rates to the question "Do tests required for graduation discourage some students from staying in school?"

	Urban	Suburban	Rural	Total
Yes	57	27	47	131
No	23	16	12	51
Unsure	45	25	31	101
Total	125	68	90	283

Source: American Journal of Education (November 1989), University of Chicago Press. © 1989 by The University of Chicago. All Rights Reserved.

Does there appear to be a relationship at the 0.05 level of significance between a student's response and the school's location?

11.46 The following table shows the number of reported crimes committed last year in the inner part of a large city. The crimes were classified according to type of crime and district of the inner city where it occurred. Do these data show sufficient evidence to reject the hypothesis that the type of crime and the district in which it occurred are independent? Use $\alpha = 0.01$.

			Crime		
District	Robbery	Assault	Burglary	Larceny	Stolen Vehicle
1	54	331	227	1090	41
2	42	274	220	488	71
3	50	306	206	422	83
4	48	184	148	480	42
5	31	102	94	596	56
6	10	53	92	236	45

11.47 Last year's work record for absenteeism in each of four categories for 100 randomly selected employees is compiled in the following table. Do these data provide sufficient evidence to reject the hypothesis that the rate of absenteeism is the same for all categories of employees? Use $\alpha = 0.01$ and 240 work days for the year.

	Married Male	Single Male	Married Female	Single Female
Number of Employees	40	14	16	30
Days Absent	180	110	75	135

    a. Solve using the *p*-value approach.
    b. Solve using the classical approach.

11.48 A survey of employees at an insurance firm was concerned with worker–supervisor relationships. One statement for evaluation was "I am not sure what my supervisor expects." The results of the survey are found in the following contingency table.

Years of Employment	I Am Not Sure What My Supervisor Expects		
	True	Not True	Totals
Less than 1 year	18	13	31
1–3 years	20	8	28
3–10 years	28	9	37
More than 10 years	26	8	34
Total	92	38	130

Can we reject the null hypothesis that "the responses to the statement and years of employment are independent" at the 0.10 level of significance?

    a. Solve using the p-value approach.

    b. Solve using the classical approach.

11.49 If you were to roll a die 600 times, how different from 100 could the observed frequencies for each face be before the results would become significantly different from equally likely at the 0.05 level?

11.50 Consider the following set of data.

	Response		
	Yes	No	Total
Group 1	75	25	100
Group 2	70	30	100
Total	145	55	200

    a. Compute the value of the test statistic $z^\star$ that would be used to test the null hypothesis that $p_1 = p_2$, where p_1 and p_2 are the proportions of "yes" responses in the respective groups.

    b. Compute the value of the test statistic $\chi^{2\star}$ that would be used to test the hypothesis that "response is independent of group."

    c. Show that $\chi^{2\star} = (z^\star)^2$.

VOCABULARY LIST

Be able to define each term. Pay special attention to the key terms, which are printed in red. In addition, describe in your own words, and give an example of, each term. Your examples should not be ones given in class or in the textbook.

The bracketed numbers indicate the chapters in which the term previously appeared, but you should define the terms again to show increased understanding of their meaning. Page numbers indicate the first appearance of the term in Chapter 11.

cell (p. 572)
chi-square [9] (p. 573)
column [3] (p. 590)
contingency table (p. 586)
degrees of freedom [9, 10] (p. 577)
enumerative data (p. 572)
expected frequency (p. 572)
homogeneity (p. 591)

hypothesis test [8, 9, 10] (p. 574)
independence [4] (p. 586)
marginal totals [3] (p. 587)
multinomial experiment (p. 574)
observed frequency [2, 4] (p. 572)
$r \times c$ contingency table (p. 590)
rows [3] (p. 590)
statistic [1, 2, 8] (p. 573)

QUIZ A

Answer "True" if the statement is always true. If the statement is not always true, replace the words shown in bold with words that make the statement always true.

11.1   The number of degrees of freedom for a test of a multinomial experiment is **equal to** the number of cells in the experimental data.

11.2   The **expected frequency** in a chi-square test is found by multiplying the hypothesized probability of a cell by the number of pieces of data in the sample.

11.3   The **observed** frequency of a cell should not be allowed to be smaller than 5 when a chi-square test is being conducted.

11.4   In the **multinomial experiment** we have $(r - 1)$ times $(c - 1)$ degrees of freedom (r is the number of rows, and c is the number of columns).

11.5   A multinomial experiment consists of n **identical, independent trials**.

11.6   A **multinomial experiment** arranges the data into a two-way classification such that the totals in one direction are predetermined.

11.7   The charts for both the multinomial experiment and the contingency table **must** be set up in such a way that each piece of data will fall into exactly one of the categories.

11.8   The test statistic $\Sigma[(O - E)^2/E]$ has a distribution that is **approximately normal**.

11.9   The data used in a chi-square multinomial test are always **enumerative** in nature.

11.10 The null hypothesis being tested by a test of **homogeneity** is that the distribution of proportions is the same for each of the subpopulations.

QUIZ B

Answer all questions. Show formulas, substitutions, and work.

11.1   State the null and alternative hypotheses that would be used to test each of the following claims:
   a. The single-digit numerals generated by a certain random number generator were not equally likely.
   b. The results of the last election in our city suggest that the votes cast were not independent of the voter's registered party.
   c. The distributions of types of crimes committed against society are the same in the four largest U.S. cities.

11.2   Find each of the following:
   a. $\chi^2(12, 0.975)$ b. $\chi^2(17, 0.005)$

11.3   Three hundred consumers were each asked to identify which one of three different items they found to be the most appealing. The table shows the number that preferred each item.

Item	1	2	3
Number	85	103	112

Do these data present sufficient evidence at the 0.05 level of significance to indicate that the three items are not equally preferred?

11.4 To study the effect of the type of soil on the amount of growth attained by a new hybrid plant, saplings were planted in three different types of soil and their subsequent amounts of growth classified into three categories:

Growth	Clay	Sand	Loam
		Soil Type	
Poor	16	8	14
Average	31	16	21
Good	18	36	25
Total	65	60	60

Does the quality of growth appear to be distributed differently for the tested soil types at the 0.05 level?
  a. State the null and alternative hypotheses.
  b. Find the expected value for the cell containing 36.
  c. Calculate the value of chi-square for these data.
  d. Find the p-value.
  e. Find the test criteria [level of significance, test statistic, its distribution, critical region, and critical value(s)].
  f. State the decision and the conclusion for this hypothesis test.

QUIZ C

11.1 Explain how a multinomial experiment and a binomial experiment are similar and also how they are different.

11.2 Explain the distinction between a test for independence and a test for homogeneity.

11.3 Student A says that the tests for independence and homogeneity are the same, and student B says that they are not at all alike because they are tests of different concepts. Both students are partially right and partially wrong. Explain.

11.4 You are interpreting the results of an opinion poll on the role of recycling in your town. A random sample of 400 people were asked to respond strongly in favor, slightly in favor, neutral, slightly against, or strongly against on each of several questions. There are four key questions that concern you and you plan to analyze their results.
  a. How do you calculate the expected probabilities for each answer?
  b. How would you decide if the four questions were answered the same?

ANALYSIS OF VARIANCE

Chapter Case Study

TILLAGE TEST PLOTS REVISITED

The 18 plot yields can be analyzed using the one-way analysis of variance technique with 3 replicates at each of 6 different treatment levels (tillage method).

All tillage tools are not created equal although the weather can certainly make them look that way. That's our analysis after putting six of your favorite primary tillage tools through the paces last season.

We staked out 18 plots on a gently sloping field with a Chalmers silty clay loam soil type.

How good is a tie? The yields from individual tillage treatments showed less variability than we had hoped, probably due to the late planting. That's why university researchers conduct these studies over many years. The only yield differences worth noting are the 7.4 bu/A spreads between the standard chisel and light disk at 115.2 bu. and the moldboard plow at 122.6 bu. The coulter-chisel and heavy disk, and surprisingly enough the V-chisel, yielded virtually the same as the moldboard plow. However, we think there are other factors that add even more merit to heavy-duty conservation tillage.

Tillage Plot Yields Bu A @ 15.5% Moisture

| | Rep | | | Tillage |
Tillage	I	II	III	Average
Plow	118.3	125.6	123.8	122.6
V-chisel	115.8	122.5	118.9	119.1
Coulter chisel	124.1	118.5	113.3	118.6
Std. chisel	109.2	114.0	122.5	115.2
Hvy. disk	118.1	117.5	121.4	119.0
Lt. disk	118.3	113.7	113.7	115.2

Source: Larry Reichenberger, Associate Machine Editor. Reprinted with permission from *Successful Farming*, February 1979. Copyright 1979 by Meredith Corporation. All rights reserved.

Draw a dotplot showing these 6 sets of data side by side on the scale. Is there much variation in the data? Is more of the variation between or within the levels? Does the method of tillage seem to have an effect on the mean yield? (See Exercise 12.29, p. 637.)

● CHAPTER OBJECTIVES

Analysis of variance

Previously, we have tested hypotheses about two means. In this chapter we are concerned with testing a hypothesis about several means. The **analysis of variance technique (ANOVA)**, which we are about to explore, will be used to test a hypothesis about several means, for example,

$$H_0: \mu_1 = \mu_2 = \mu_3 = \mu_4 = \mu_5$$

By using our former technique for hypotheses about two means, we could test several hypotheses if each stated a comparison of two means. For example, we could test

$$
\begin{array}{ll}
H_1: \mu_1 = \mu_2 & H_2: \mu_1 = \mu_3 \\
H_3: \mu_1 = \mu_4 & H_4: \mu_1 = \mu_5 \\
H_5: \mu_2 = \mu_3 & H_6: \mu_2 = \mu_4 \\
H_7: \mu_2 = \mu_5 & H_8: \mu_3 = \mu_4 \\
H_9: \mu_3 = \mu_5 & H_{10}: \mu_4 = \mu_5
\end{array}
$$

In order to test the null hypothesis, H_0, that all five means are equal, we would have to test each of these ten hypotheses using our former technique for two means. Rejection of any one of the ten hypotheses about two means would cause us to reject the null hypotheses that all five means are equal. If we failed to reject all ten hypotheses about the means, we would fail to reject the main null hypothesis. Suppose we tested a null hypothesis that dealt with several means by testing all the possible pairs of two means; the overall type-I error rate would become much larger than the value of α associated with a single test. The ANOVA techniques allow us to test the null hypothesis (all means are equal) against the alternative hypothesis (at least one mean value is different) with a specified value of α.

In this chapter we introduce ANOVA. ANOVA experiments can be very complex, depending on the situation. We will restrict our discussion to the most basic experimental design, the single-factor ANOVA.

12.1 | INTRODUCTION TO THE ANALYSIS OF VARIANCE TECHNIQUE

We will begin our discussion of the analysis of variance technique by looking at an illustration.

▼ ILLUSTRATION   12 - 1

The temperature at which a plant is maintained is believed to affect the rate of production in the plant. The data in Table 12-1 are the number, x, of units produced in one hour for randomly selected one-hour periods when the production process in

Level

Replicate

the plant was operating at each of three temperature **levels**. The data values from repeated samplings are called **replicates**. Four replicates, or data values, were obtained for two of the temperatures and five were obtained for the third temperature. Do these data suggest that temperature has a significant effect on the production level at the 0.05 level?

TABLE 12-1

Sample Results for
Illustration 12-1

	Temperature Levels		
	Sample from 68°F ($i = 1$)	Sample from 72°F ($i = 2$)	Sample from 76°F ($i = 3$)
	10	7	3
	12	6	3
	10	7	5
	9	8	4
		7	
Column Totals	$C_1 = 41$ $\bar{x}_1 = 10.25$	$C_2 = 35$ $\bar{x}_2 = 7.0$	$C_3 = 15$ $\bar{x}_3 = 3.75$

EXERCISE 12.1

Draw a dotplot of data in Table 12-1. Represent the data using integers 1, 2, and 3 indicating the level of test factor the data are from. Do you see a "difference" between the levels?

The level of production is measured by the mean value; $\bar{x}_i$ indicates the observed production mean at level i, where $i = 1$, 2, and 3 corresponds to temperatures of 68°F, 72°F, and 76°F, respectively. There is a certain amount of variation among these means. Since sample means do not necessarily repeat when repeated samples are taken from a population, some variation can be expected, even if all three population means are equal. We will next pursue the question "Is this variation among the $\bar{x}$'s due to chance, or is it due to the effect that temperature has on the production rate?"

SOLUTION The null hypothesis that we will test is

$$H_0: \mu_{68} = \mu_{72} = \mu_{76}$$

That is, the true production mean is the same at each temperature level tested. In other words, the temperature does not have a significant effect on the production rate. The alternative to the null hypothesis is

$$H_a: \text{not all temperature level means are equal}$$

EXERCISE 12.2

State the hypotheses used to test "the mean rating is the same in all three departments."

Thus, we will want to reject the null hypothesis if the data show that one or more of the means are significantly different from the others.

We will make the decision to reject H_0 or fail to reject H_0 by using the F distribution and the F test statistic. Recall from Chapter 10 that the calculated value of F is the ratio of two variances. The analysis of variance procedure will separate the variation among the entire set of data into two categories. To accomplish this separation, we first work with the numerator of the fraction used to define sample variance, formula (2-6):

$$s^2 = \frac{\sum(x - \bar{x})^2}{n - 1}$$

Sum of squares

The numerator of this fraction is called the **sum of squares**:

$$\text{sum of squares} = \sum(x - \bar{x})^2 \qquad \textbf{(12-1)}$$

Total sum of squares, SS(total)

We calculate the **total sum of squares, SS(total)**, for the total set of data by using a formula that is equivalent to formula (12-1) but does not require the use of $\bar{x}$. This equivalent formula is

$$\text{SS(total)} = \sum(x^2) - \frac{\left(\sum x\right)^2}{n} \qquad \textbf{(12-2)}$$

Now we can find the SS(total) for our illustration by using formula (12-2):

$$\sum(x^2) = 10^2 + 12^2 + 10^2 + 9^2 + 7^2 + 6^2 + 7^2$$
$$+ 8^2 + 7^2 + 3^2 + 3^2 + 5^2 + 4^2$$
$$= 731$$
$$\sum x = 10 + 12 + 10 + 9 + 7 + 6 + 7$$
$$+ 8 + 7 + 3 + 3 + 5 + 4$$
$$= 91$$
$$\text{SS(total)} = 731 - \frac{(91)^2}{13}$$
$$= 731 - 637 = \textbf{94}$$

Next, 94, the SS(total), must be separated into two parts: the sum of squares, SS(temperature), due to temperature levels; and the sum of squares, SS(error), due to experimental error of replication. This splitting is often called **partitioning**, since SS(temperature) + SS(error) = SS(total); that is, in our illustration SS(temperature) + SS(error) = 94.

Partitioning

Variation between the factor levels

The sum of squares, SS(factor) [SS(temperature) for our illustration] that measures the **variation between the factor levels** (temperatures) is found by using formula (12-3):

$$\text{SS(factor)} = \left(\frac{C_1^2}{k_1} + \frac{C_2^2}{k_2} + \frac{C_3^2}{k_3} + \cdots\right) - \frac{\left(\sum x\right)^2}{n} \qquad \textbf{(12-3)}$$

where C_i represents the column total, k_i represents the number of replicates at each level of the factor, and n represents the total sample size ($n = \sum k_i$).

NOTE   The data have been arranged so that each column represents a different level of the factor being tested.

Now we can find the SS(temperature) for our illustration by using formula (12-3):

$$\text{SS(temperature)} = \left(\frac{41^2}{4} + \frac{35^2}{5} + \frac{15^2}{4}\right) - \frac{(91)^2}{13}$$
$$= (420.25 + 245.00 + 56.25) - 637.0$$
$$= 721.5 - 637.0 = \textbf{84.5}$$

Variation within rows

The sum of squares SS(error) that measures the **variation within the rows** is found by using formula (12-4):

$$SS(\text{error}) = \sum(x^2) - \left(\frac{C_1^2}{k_1} + \frac{C_2^2}{k_2} + \frac{C_3^2}{k_3} + \cdots\right) \tag{12-4}$$

The SS(error) for our illustration can now be found by using formula (12-4):

$$\sum(x^2) = 731 \text{ (found previously)}$$

$$\left(\frac{C_1^2}{k_1} + \frac{C_2^2}{k_2} + \frac{C_3^2}{k_3}\right) = 721.5 \text{ (found previously)}$$

$$SS(\text{error}) = 731.0 - 721.5 = \mathbf{9.5}$$

NOTE SS(total) = SS(factor) + SS(error). Inspection of formulas (12-2), (12-3), and (12-4) will verify this.

For convenience we will use an ANOVA table to record the sums of squares and to organize the rest of the calculations. The format of an ANOVA table is shown in Table 12-2.

TABLE 12-2

Format for
ANOVA Table

Source	df	SS	MS
Factor			
Error			
Total			

We have calculated the three sums of squares for our illustration. The degrees of freedom, df, associated with each of the three sources are determined as follows:

1. df(factor) is one less than the number of levels (columns) for which the factor is tested:

$$df(\text{factor}) = c - 1 \tag{12-5}$$

where c represents the number of levels for which the factor is being tested (number of columns on the data table).

2. df(total) is one less than the total number of data:

$$df(\text{total}) = n - 1 \tag{12-6}$$

where n represents the number of data in the total sample (that is, $n = k_1 + k_2 + k_3 + \cdots$, where k_i is the number of replicates at each level tested).

3. df(error) is the sum of the degrees of freedom for all the levels tested (columns in the data table). Each column has $k_i - 1$ degrees of freedom; therefore,

$$df(\text{error}) = (k_1 - 1) + (k_2 - 1) + (k_3 - 1) + \cdots$$

or

$$df(\text{error}) = n - c \tag{12-7}$$

The degrees of freedom for our illustration are

$$df(temperature) = c - 1 = 3 - 1 = 2$$
$$df(total) = n - 1 = 13 - 1 = 12$$
$$df(error) = n - c = 13 - 3 = 10$$

The sums of squares and the degrees of freedom must check. That is,

$$SS(factor) + SS(error) = SS(total) \qquad \text{(12-8)}$$
$$df(factor) + df(error) = df(total) \qquad \text{(12-9)}$$

Mean square, MS(factor), MS(error)

The **mean square** for the factor being tested, the **MS(factor)**, and for error, the **MS(error)**, will be obtained by dividing the sum-of-square value by the corresponding number of degrees of freedom. That is,

$$MS(factor) = \frac{SS(factor)}{df(factor)} \qquad \text{(12-10)}$$

$$MS(error) = \frac{SS(error)}{df(error)} \qquad \text{(12-11)}$$

The mean squares for our illustration are

$$MS(temperature) = \frac{SS(temperature)}{df(temperature)} = \frac{84.5}{2}$$
$$= \textbf{42.25}$$

$$MS(error) = \frac{SS(error)}{df(error)} = \frac{9.5}{10}$$
$$= \textbf{0.95}$$

The completed ANOVA table appears as shown in Table 12-3. The hypothesis test is now completed using the two mean squares as measures of variance. The calculated value of the **test statistic, $F^\star$**, is found by dividing the MS(factor) by the MS(error):

Test statistic, $F^\star$

$$F^\star = \frac{MS(factor)}{MS(error)} \qquad \text{(12-12)}$$

TABLE 12-3

ANOVA Table for Illustration 12-1

Source	df	SS	MS
Temperature	2	84.5	42.25
Error	10	9.5	0.95
Total	12	94.0	

The calculated value of F for our illustration is found by using formula (12-12):

$$F^\star = \frac{MS(temperature)}{MS(error)} = \frac{42.25}{0.95}$$
$$F^\star = \textbf{44.47}$$

The hypothesis test may be completed by using either (a) the p-value approach or (b) the classical approach.

a. The p-value approach requires the calculation of the p-value. The p-value is represented by the area under the F distribution to the right of $F^\star$. For our illustration, p-value = $P(F > 44.47 \mid df_n = 2, df_d = 10)$.

MINITAB will calculate p-values. Use CDF C1; F df$_n$ df$_d$.

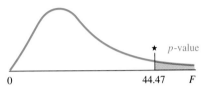

Using the computer: p-value = 0.00001

Using Table 9c in Appendix B: p-value < 0.01

EXERCISE 12.3

Use a computer to find the p-value when $F^\star$ = 4.572, df_{Factor} = 5, and df_{Error} = 22.

b. The classical approach requires a critical region. The critical value, F(2, 10, 0.05), may be obtained from Table 9a in Appendix B. The critical value, $F(2, 10, 0.05) = 4.10$, and the critical region are shown on the accompanying figure.

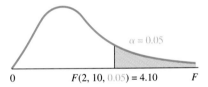

NOTE   Since the calculated value of F, $F^\star$, is found by dividing MS(temperature) by MS(error), the number of degrees of freedom for the numerator is df(temperature) = 2 and the number of degrees of freedom for the denominator is df(error) = 10.

We rejected H_0 because the p-value is smaller than α, or equivalently, because the value $F^\star$ fell in the critical region. Therefore, we conclude that at least one of the room temperatures does have a significant effect on the production rate. The differences in the mean production rates at the tested temperature levels were found to be significant. The mean at 68°F is certainly different from the mean at 76°F, since the sample means for these levels are the largest and smallest, respectively. Whether any other pairs of means are significantly different cannot be determined from the ANOVA procedure alone. ▲

In this section we saw how the ANOVA technique separated the variance among the sample data into two measures of variance: (1) MS(factor), the measure of variance between the levels tested, and (2) MS(error), the measure of variance within the levels being tested. Then these measures of variance can be compared. For our illustration, the between-level variance was found to be significantly larger than the within-level variance (experimental error). This led us to the conclusion that temperature did have a significant effect on the variable x, the number of units of production completed per hour.

In the next section we will use several illustrations to demonstrate the logic of the analysis of variance technique.

12.2 | THE LOGIC BEHIND ANOVA

Many experiments are conducted to determine the effect that different levels of some test factor have on a response variable. The test factor may be temperature (as in Illustration 12-1), the manufacturer of a product, the day of the week, or any number of other things. In this chapter we are investigating the *single-factor* analysis of variance. Basically, the design for the single-factor ANOVA is to obtain independent random samples at each of the several levels of the factor being tested. We will then make a statistical decision concerning the effect that the levels of the test factors have on the response (observed) variable.

Illustrations 12-2 and 12-3 demonstrate the logic of the analysis of variance technique. Briefly, the reasoning behind the technique proceeds like this: In order to compare the means of the levels of the test factor, a measure of the **variation between the levels** (between columns on the data table), the **MS(factor)**, will be compared to a measure of the **variation within the levels** (within the columns on the data table), the **MS(error)**. If the MS(factor) is significantly larger than the MS(error), then we will conclude that the means for each of the factor levels being tested are not all the same. This implies that the factor being tested does have a significant effect on the response variable. If, however, the MS(factor) is not significantly larger than the MS(error), we will not be able to reject the null hypothesis that all means are equal.

Variation between levels
Variation within levels

▼| ILLUSTRATION   12 - 2

Do the data in Table 12-4 show sufficient evidence to conclude that there is a difference in the three population means μ_F, μ_G, and μ_H?

TABLE 12-4

Sample Results for Illustration 12-2

	Factor Levels		
	Sample from Level F	**Sample from Level G**	**Sample from Level H**
	3	5	8
	2	6	7
	3	5	7
	4	5	8
Column Totals	$C_F = 12$ $\bar{x} = 3.00$	$C_G = 21$ $\bar{x} = 5.25$	$C_H = 30$ $\bar{x} = 7.50$

FIGURE 12-1

Data from
Table 12-4

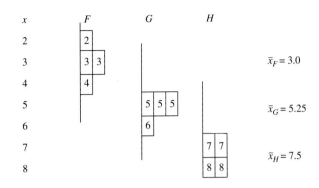

Figure 12-1 shows the relative relationship among the three samples. A quick look at the figure suggests that the three sample means are different from each other, implying that the sampled populations have different mean values.

▲

Within-sample and
between-sample
variation

These three samples demonstrate relatively little **within-sample variation**, although there is a relatively large amount of **between-sample variation**. Let's look at another illustration.

▼ ILLUSTRATION 12 - 3

Do the data in Table 12-5 show sufficient evidence to conclude that there is a difference in the three population means μ_J, μ_K, and μ_L?

TABLE 12-5

Sample Results for
Illustration 12-3

	Factor Levels		
	Sample from Level J	Sample from Level K	Sample from Level L
	3	5	6
	8	4	2
	6	3	7
	4	7	5
Column Totals	$C_J = 21$	$C_K = 19$	$C_L = 20$
	$\bar{x}_J = 5.25$	$\bar{x}_K = 4.75$	$\bar{x}_L = 5.00$

Figure 12-2 (p. 620) shows the relative relationship among the three samples. A quick look at the figure *does not suggest* that the three sample means are different from each other.

Case Study

Waiting for a Verdict

12.1

This USA Snapshot® *reports the average amount of time it takes for each of several types of felony charges to move through trial and to a verdict. Does the type of felony appear to have an effect on the average amount of time required?*

EXERCISE 12.4

Referring to "waiting for a verdict,"

a. Does the type of felony charge appear to have an effect on the average amount of time it takes to move through the trial to a verdict?

b. What additional information would be needed in order to determine whether the type of felony has a significant effect on the waiting time?

USA SNAPSHOTS®
A look at statistics that shape the nation

Waiting for a verdict
Average time it takes for a felony charge to move through trial court to a verdict, according to new survey:

6.2 months — Homicide
4.2 months — Sexual assault
3.5 months — Robbery
3.2 months — Larcenies/burglaries

Source: Justice Department

By Dale Glasgow, USA TODAY

Source: Copyright 1986, USA TODAY. Reprinted with permission.

FIGURE 12-2

Data from
Table 12-5

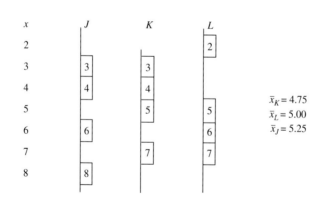

x	J	K	L
2			2
3	3	3	
4	4	4	
5		5	5
6	6		6
7		7	7
8	8		

$\bar{x}_K = 4.75$
$\bar{x}_L = 5.00$
$\bar{x}_J = 5.25$

There is little between-sample variation for these three samples (that is, the sample means are relatively close in value), whereas the within-sample variation is relatively large (that is, the data values within each sample cover a relatively wide range of values).

Case Study

12.2

Teens Are Big Spenders

The average spent per week by teenagers in each age group seems to increase substantially as the teenager gets older. Data on the amounts spent by each of 2110 teenagers were collected and divided by age group. This is similar to how one-way analysis of variance treats data of this type.

EXERCISE 12.5

The amount of money spent varies from teenager to teenager, just as it varies from person to person at all ages. "Teens are big spenders" suggests that the average amount spent is affected by the age category the teenager belongs to. Describe how the data presented are organized in the same way as data are organized for one-way analysis of variance.

USA SNAPSHOTS®

A look at statistics that shape your finances

Teens are big spenders

Teen-agers spent $71 billion last year. Females spent an average $55.50 a week, while males spent an average $48.40 a week.

Age	Avg. spent per week
12¹–15	$35.20
16–17	$53.90
18–19	$78.89

1—12-year-olds are traditionally included in teen studies
Source: Teenage Research Unlimited's survey of 2,110 people aged 12–19

By Sam Ward, USA TODAY

Source: Copyright 1990, USA TODAY. Reprinted with permission.

Basic assumptions

To complete a hypothesis test for analysis of variance, we must agree on some ground rules, or assumptions. In this chapter we will use the following three **basic assumptions**:

1. Our goal is to investigate the effect that various levels of the factor under test have on the response variable. Typically, we want to find the level that yields the most advantageous values of the response variable. This, of course, means that we probably will want to reject the null hypothesis in favor of the alternative. Then a follow-up study could determine the "best" level of the factor.
2. We must assume that the effects due to chance and due to untested factors are normally distributed and that the variance caused by these effects is constant throughout the experiment.
3. We must assume independence among all observations of the experiment. (Recall that independence means the results of one observation of the experiment do not affect the results of any other observation.) We will usually conduct the tests in a randomly assigned order to ensure independence. This technique also helps to avoid data contamination.

12.3 | APPLICATIONS OF SINGLE-FACTOR ANOVA

Before continuing our ANOVA discussion, let's identify the notation, particularly the subscripts that are used (see Table 12-6). Notice that each piece of data has two subscripts; the first subscript indicates the column number (test factor level), and the second subscript identifies the replicate (row) number. The column totals, C_i, are listed across the bottom of the table. The grand total, T, is equal to the sum of all x's and is found by adding the column totals. Row totals can be used as a cross-check but serve no other purpose.

TABLE 12-6

Notation Used
in ANOVA

	Factor Levels				
Replication	**Sample from Level 1**	**Sample from Level 2**	**Sample from Level 3**	$\cdots$	**Sample from Level C**
$k = 1$	$x_{1,1}$	$x_{2,1}$	$x_{3,1}$		$x_{c,1}$
$k = 2$	$x_{1,2}$	$x_{2,2}$	$x_{3,2}$		$x_{c,2}$
$k = 3$	$x_{1,3}$	$x_{2,3}$	$x_{3,3}$		$x_{c,3}$
$\vdots$					
Column	C_1	C_2	C_3	$\cdots$	C_c	T
Totals	$T = \text{grand total} = \text{sum of all } x\text{'s} = \Sigma x = \Sigma C_i$				

A mathematical model (equation) is often used to express a particular situation. In Chapter 3 we used a mathematical model to help explain the relationship between the values of bivariate data. The equation $\hat{y} = b_0 + b_1 x$ served as the model when we believed that a straight-line relationship existed. The probability functions studied in Chapter 5 are also examples of mathematical models. For the single-factor ANOVA,

Mathematical model   the **mathematical model**, formula (12-13), is an expression of the composition of each piece of data entered in our data table.

$$x_{c,k} = \mu + F_c + \epsilon_{k(c)} \tag{12-13}$$

We interpret each term of this model as follows:

1. μ is the mean value for all the data without respect to the test factor.
2. F_c is the effect that the factor being tested has on the response variable at each different level c.

Experimental error   3. $\epsilon_{k(c)}$ (ϵ is the lowercase Greek letter epsilon) is the **experimental error** that occurs among the k replicates in each of the c columns.

Let's look at another hypothesis test concerning an analysis of variance.

▼| ILLUSTRATION   12 - 4

A rifle club performed an experiment on a randomly selected group of beginning shooters. The purpose of the experiment was to determine whether shooting accuracy is affected by the method of sighting used: only the right eye open, only the left eye open, or both eyes open. Fifteen beginning shooters were selected and split into three groups. Each group experienced the same training and practicing procedures with one exception: the method of sighting used. After completing training, each student was given the same number of rounds and asked to shoot at a target. Their scores appear in Table 12-7.

TABLE 12-7

Sample Results for Illustration 12-4

	Method of Sighting		
	Right Eye	**Left Eye**	**Both Eyes**
	12	10	16
	10	17	14
	18	16	16
	12	13	11
	14		20
			21

EXERCISE 12.6

Consider the following table for a single-factor ANOVA. Find the following:

a. $x_{1,2}$
b. $x_{2,1}$
c. C_1
d. Σx
e. $\Sigma (C_i)^2$

	Level of Factor		
Replicates	1	2	3
1	3	2	7
2	0	5	4
3	1	4	5

At the 0.05 level of significance, is there sufficient evidence to reject the claim that these methods of sighting are equally effective?

SOLUTION In this experiment the factor is "method sighting" and the levels are the three different methods of sighting (right eye, left eye, and both eyes open). The replicates are the scores received by the students in each group. The null hypothesis to be tested is "the three methods of sighting are equally effective," or "the mean scores attained using each of the three methods are the same."

STEP 1 Parameters of concern: the "mean" at each level of the test factor is of concern: the mean score using right eye μ_R, the mean score using left eye μ_L, and the mean score using both eyes μ_B. The factor being tested, "method of sighting," has three levels—right, left, or both.

STEP 2 Hypotheses:

H_0: $\mu_R = \mu_L = \mu_B$

H_a: The means are not all equal (that is, at least one mean is different).

STEP 3 Level of significance: $\alpha = 0.05$.

STEP 4   The sample evidence:
a. Calculate test statistic:

$$df(numerator) = df(method) = 3 - 1 = 2 \text{ [using formula (12-5)]}$$
$$df(denominator) = df(error) = 15 - 3 = 12 \text{ [using formula (12-7)]}$$

Calculate the test statistic $F^\star$: Table 12-8 is used to find column totals.

TABLE 12-8

Sample Results for Illustration 12-4

| Replicates | Factor Levels: Method of Sighting | | |
	Right Eye	Left Eye	Both Eyes
$k = 1$	12	10	16
$k = 2$	10	17	14
$k = 3$	18	16	16
$k = 4$	12	13	11
$k = 5$	14		20
$k = 6$			21
Totals	$C_R = 66$	$C_L = 56$	$C_B = 98$

First, the summations Σx and Σx^2 need to be calculated:

$$\Sigma x = 12 + 10 + 18 + 12 + 14 + 10 + 17 + \cdots + 21 = \mathbf{220}$$
$$\Sigma x^2 = 12^2 + 10^2 + 18^2 + 12^2 + 14^2 + 10^2 + \cdots + 21^2 = \mathbf{3392}$$

Using formula (12-2), we find

$$SS(total) = 3392 - \frac{(220)^2}{15} = 3392 - 3226.67 = \mathbf{165.33}$$

Using formula (12-3), we find

$$SS(method) = \left(\frac{66^2}{5} + \frac{56^2}{4} + \frac{98^2}{6}\right) - 3226.67$$
$$= 3255.87 - 3226.67 = \mathbf{29.20}$$

Using formula (12-4), we find

$$SS(error) = 3392 - 3255.87 = \mathbf{136.13}$$

Use formula (12-8) to check the sum of squares.

$$SS(method) + SS(error) = SS(total)$$
$$29.20 \quad + \quad 136.13 \quad = \quad 165.33$$

The number of degrees of freedom is found using formulas (12-5), (12-6), and (12-7):

$$df(\text{method}) = 3 - 1 = \mathbf{2}$$
$$df(\text{total}) = 15 - 1 = \mathbf{14}$$
$$df(\text{error}) = 15 - 3 = \mathbf{12}$$

Using formulas (12-10) and (12-11), we find

$$MS(\text{method}) = \frac{29.20}{2} = \mathbf{14.60}$$

$$MS(\text{error}) = \frac{136.13}{12} = \mathbf{11.34}$$

The results of these computations are combined in the ANOVA table shown in Table 12-9.

TABLE 12-9

ANOVA Table for Illustration 12-4

Source	df	SS	MS
Method	2	29.20	14.60
Error	12	136.13	11.34
Total	14	165.33	

The calculated value of the test statistic is then found using formula (12-12).

$$F^\star = \frac{14.60}{11.34} = \mathbf{1.287}$$

Use MINITAB commands
CDF 1.287;
F 2 12.

b. Find the *p*-value: *p*-value = $P(F > 1.287 | df_n = 2, df_d = 12)$.

Using the computer: *p*-value = 0.312
Using Table 9a: *p*-value > 0.05

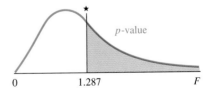

STEP 5 Results: *p*-value is greater than α.

DECISION Fail to reject H_0.

CONCLUSION The data show no evidence that would give reason to reject the null hypothesis that the three methods are equally effective.

If you use the classical approach, the preceding hypothesis test is completed using steps 3–5 as follows:

STEP 3   Test criteria:
Test statistic: $F^\star$ with $df_n = 2$ and $df_d = 12$.
Critical region: The critical value is found in Table 9a.

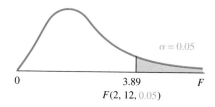

$$\alpha = 0.05$$

$$0 \qquad 3.89 \qquad F$$

$$F(2, 12, 0.05)$$

STEP 4   Sample evidence:
$F^\star = 1.287$ (calculation shown on pages 624 and 625)

STEP 5   Results: $F^\star$ is in noncritical region.

DECISION Fail to reject H_0.

CONCLUSION The data show no evidence that would give reason to reject the null hypothesis that the three methods are equally effective.

▲

MINITAB (Release 10) commands to calculate the one-way analysis of variance table when data are listed in one column, C1, and the level of the test factor is listed in a second column, C2.

Session commands   *Menu commands*

Enter: ONEWay C1 C2   Choose: Stat > ANOVA > Oneway
                                 Enter: Columns containing data: C1
                                          Column containing levels: C2

MINITAB (Release 10) commands to calculate the one-way analysis of variance table when the data for each level of the test factor are listed in separate columns, C1, C2, C3.

Session commands   *Menu commands*

Enter: AOVOneway C1 C2 C3   Choose:
                            Stat > ANOVA > Oneway(unstacked)
                            Enter:
                              Columns containing data: C1 C2 C3

NOTE MINITAB's DOTPLOT command produces side-by-side dotplots that are very useful in visualizing the within-sample variation, the between-sample variation, and the relationship between them.

COMPUTER SOLUTION MINITAB Printout for Illustration 12-4:

Data is entered on the MINITAB worksheet with all data values in C1 and test factor level identifiers in C2 ("stacked").

Information given to
computer→

ROW	C1	C2
1	12.	1.
2	10.	1.
3	18.	1.
4	12.	1.
5	14.	1.
6	10.	2.
7	17.	2.
8	16.	2.
9	13.	2.
10	16.	3.
11	14.	3.
12	16.	3.
13	11.	3.
14	20.	3.
15	21.	3.

MTB > ONEWAY C1 C2

ANALYSIS OF VARIANCE

The ANOVA table ⟶
compare to Table 12-9

SOURCE	DF	SS	MS	F	P
FACTOR	2	29.2	14.6	1.29	0.312
ERROR	12	136.1	11.3		
TOTAL	14	165.3			

The calculated value of
F, $F^\star$

Sample statistics for
each factor level→

LEVEL	N	MEAN	ST.DEV.
1	5	13.200	3.033
2	4	14.000	3.162
3	6	16.333	3.724

The calculated p-value

MTB > DOTPLOT C1;
SUBC > BY C2.

(continued)

MINITAB (*continued*)

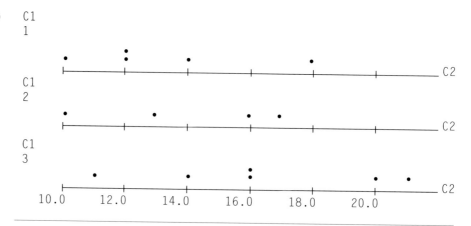

Recall the null hypothesis, that there is no difference between the levels of the factor being tested. A "fail to reject H_0" decision must be interpreted as the conclusion that there is no evidence of a difference due to the levels of the tested factor, whereas the rejection of H_0 implies that there is a difference between the levels. That is, at least one level is different from the others. If there is a difference, the next problem is to locate the level or levels that are different. Locating this difference may be the main object of the analysis. In order to find the difference, the only method that is appropriate at this stage is to inspect the data. It may be obvious which level(s) caused the rejection of H_0. In Illustration 12-1 it seems quite obvious that at least one of the levels [level 1 (68°F) or level 3 (76°F), because they have the largest and smallest sample means] is different from the other two. If the higher values are more desirable for finding the "best" level to use, we would choose that corresponding level of the factor.

Thus far we have discussed analysis of variance for data dealing with one factor. It is not unusual for problems to have several factors of concern. The ANOVA techniques presented in this chapter can be developed further and applied to more complex cases.

EXERCISES

12.7   State the null hypothesis H_0 and the alternative hypothesis H_a that would be used to test the following statements.
   a. The mean value of x is the same at all five levels of the experiment.
   b. The scores are the same at all four locations.
   c. The four levels of the test factor do not significantly affect the data.
   d. The three different methods of treatment do affect the variable.

12.8   Find the *p*-value for each of the following situations.
   a. $F^\star = 3.852$, $df_{Factor} = 3$, $df_{Error} = 12$
   b. $F^\star = 4.152$, $df_{Factor} = 5$, $df_{Error} = 18$

12.9   Determine the test criteria that would be used to test the null hypothesis for the following multinomial experiments.

    a. H_0: $\mu_1 = \mu_2 = \mu_3 = \mu_4$ with $n = 18$, $\alpha = 0.05$
    b. H_0: $\mu_1 = \mu_2 = \mu_3 = \mu_4 = \mu_5$ with $n = 15$, $\alpha = 0.01$
    c. H_0: $\mu_1 = \mu_2 = \mu_3$ with $n = 25$, $\alpha = 0.05$

12.10   Why does df(factor), the number of degrees of freedom associated with the factor, always appear first in the critical value notation F [df(factor), df(error), α]?

12.11   Suppose that an F test (as described in this chapter) has a p-value of 0.04.

    a. What is the interpretation of p-value $= 0.04$?
    b. What is the interpretation of the situation if you had previously decided on a 0.05 level of significance?
    c. What is the interpretation of the situation if you had previously decided on a 0.02 level of significance?

12.12   Suppose that an F test (as described in this chapter) has a critical value of 2.2, as shown in this figure:

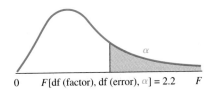

$$0 \qquad F[\text{df (factor), df (error)}, \alpha] = 2.2 \qquad F$$

    a. What is the interpretation of a calculated value of F larger than 2.2?
    b. What is the interpretation of a calculated value of F smaller than 2.2?
    c. What is the interpretation if the calculated F were 0.1? 0.01?

12.13   a. State the null hypothesis, in a general form, for the one-way ANOVA.
    b. State the alternative hypothesis, in a general form, for the one-way ANOVA.
    c. What must happen in order to "reject H_0"?
    d. How would a decision of "reject H_0" be interpreted?
    e. What must happen in order to "fail to reject H_0"?
    f. How would a decision of "fail to reject H_0" be interpreted?

12.14   The following table of data is to be used for single-factor ANOVA. Find each of the following:

    a. $x_{3,2}$ b. $x_{4,3}$ c. C_3 d. Σx e. $\Sigma(C_i)^2$

	Level of Factor			
Replicates	**1**	**2**	**3**	**4**
1	13	12	16	14
2	17	8	18	11
3	9	15	10	19

12.15 A new operator was recently assigned to a crew of workers who perform a certain job. From the records of the number of units of work completed by each worker each day last month, a sample of size five was randomly selected for each of the two experienced workers and the new worker. At the 0.05 level of significance, does the evidence provide sufficient reason to reject the claim that there is no difference in the amount of work done by the three workers?

	Workers		
	New	**A**	**B**
Units of Work (replicates)	8	11	10
	10	12	13
	9	10	9
	11	12	12
	8	13	13

12.16 An employment agency wants to see which of three types of ads in the help-wanted section of local newspapers is the most effective. Three types of ads (big headline, straightforward, and bold print) were randomly alternated over a period of weeks, and the number of people responding to the ads was noted each week. Do these data support the null hypothesis that there is no difference in the effectiveness of the ads, as measured by the mean number responding, at the 0.01 level of significance?

	Type of Advertisement		
	Big Headline	**Straightforward**	**Bold Print**
Number of Responses (replicates)	23	19	28
	42	31	33
	36	18	46
	48	24	29
	33	26	34
	26		34

12.17 An article titled "Can Aspirin Prevent and Treat Pre-Eclampsia?" (*American Journal of Maternal/Child Nursing*, September/October 1994) compares the plasma volume (mL/cm) of nonpregnant women, of normal pregnant women, and of women with pregnancy-induced hypertension (PIH). Consider a small-scale study with five women in each group. The plasma volume for each subject is given.

Nonpregnant Women	**Normal Pregnant Women**	**Women with PIH**
20	25	20
18	22	19
25	25	22
17	29	19
26	30	23

Complete an ANOVA table for the above data, and test the hypothesis "the mean plasma volume is the same for all groups." Use $\alpha = 0.05$.

12.18 An article titled "A Contrast in Images: Nursing and Nonnursing College Students" (*Nursing Education*, March 1994) described a seven-point Likert scale questionnaire to compare nursing students, business majors, engineering majors, human service majors, and social science majors with respect to several attitudes towards nursing. The nonnursing students found nursing a significantly less dangerous profession than did the nursing students. A sample of five students from each of the five majors were asked to respond to the danger associated with the nursing profession. The scores for each of the 25 students are given below.

Nursing	Business	Engineering	Human Service	Social Science
6	4	5	4	3
5	4	4	4	3
5	3	4	5	2
7	3	6	4	4
4	2	2	6	2

Complete an ANOVA table for the above data. Test the null hypothesis that the mean score is the same for each of the five groups. Use the 0.01 level of significance.

12.19 Students with three different high school academic backgrounds were compared with respect to their aptitude in computer science. The students were classified as having excellent, above-average, or average or below-average high school academic backgrounds. Each student was given the KSW computer-science aptitude test, and the score was recorded. The results were as follows:

High School Academic Performance		
Excellent	**Above Average**	**Average or Below**
16	21	4
22	19	20
15	16	13
20	17	18
23	5	8
16	20	6
21	18	11
	19	
	14	
	22	
	13	

(*continued*)

Refer to the following MINITAB analysis of the experiment and verify the analysis of variance table and test for the null hypothesis $\mu_E = \mu_{AA} = \mu_{AB}$ at $\alpha = 0.05$.

```
MTB > ONEWAY ANOVA, DATA IN C1, LEVELS IN C2
ANALYSIS OF VARIANCE ON C1
SOURCE     DF      SS       MS       F        P
  CS        2    214.7    107.4     4.65    0.021
ERROR      22    507.9     23.1
TOTAL      24    722.6
```

12.20 A study was conducted to assess the effectiveness of treating vertigo (motion sickness) with the Transdermal Therapeutic System (TTS—patch worn on skin). Two other treatments, both oral (one pill containing a drug and one a placebo) were used. The age and the gender of the patients for each treatment are listed below.

TTS		Antivert		Placebo	
47-f	53-m	51-f	43-f	67-f	38-m
41-f	58-f	53-f	56-f	52-m	59-m
63-m	62-f	27-m	48-m	47-m	33-f
59-f	34-f	29-f	52-f	35-f	32-f
62-f	47-f	31-f	19-f	37-f	26-f
24-m	35-f	25-f	31-f	40-f	37-m
43-m	34-f	52-f	48-f	31-f	49-f
20-m	63-m	55-f	53-m	45-f	49-m
55-f	46-f	32-f	63-m	41-f	38-f
		51-f	54-m	49-m	
		21-f			

Is there a significant difference between the mean age of the three test groups? Use $\alpha = 0.05$. Use a computer to complete this exercise.

12.21 The article "An Investigation of High School Preparation As Predictors of the Cultural Literacy of Developmental, Nondevelopmental and ESL College Students" (*RTDE*, Fall 1990) reported on a study that examined the cultural literacy of developmental, nondevelopmental, and ESL (English as a Second Language) college freshmen.
  a. How many student scores were in the samples?
  b. The students were divided into how many groups?

Analysis of Variance by Group for Total Score	Source	df	SS	MS	F	P
	Group	2	4062.06	2031.03	14.49	0.0001
	Error	117	16394.53	140.12		
	Total	119	20456.59			

Source	df	SS	MS	F	P
Group	2	0.95	0.475	1.93	0.1493
Error	117	28.75	0.246		
Total	119	29.70			

Analysis of Variance by Group for Foreign Language Preparation

c. Given the sum of square (SS) and degree of freedom (df) values, verify the mean square (MS), the calculated F-value, and the p-value for each figure.
d. Do the statistics in the first table show that the total scores were different for the groups involved? Explain.
e. Do the statistics in the second table show that the foreign-language preparation scores were different for the groups involved? Explain.

12.22 An article titled "The Effectiveness of Biofeedback and Home Relaxation Training on Reduction of Borderline Hypertension" (*Health Education*, October/November 1988) compared different methods of reducing blood pressure. Biofeedback ($n = 13$ subjects), Biofeedback/Relaxation ($n = 15$), and Relaxation ($n = 14$) were three methods compared. There were no differences among the three groups on pretest diastolic or systolic blood pressures. There was a significant posttest difference between groups on the systolic measure, $F(2, 39) = 4.14$, $p < 0.025$, and diastolic measure, $F(2, 39) = 5.56$, $p < 0.008$.
   a. Verify that df(method) = 2 and df(error) = 39.
   b. Use Tables 9a, 9b, and 9c in Appendix B to verify that for systolic, $p < 0.025$ and for diastolic, $p < 0.01$.

IN RETROSPECT

In this chapter we have presented an introduction to the statistical techniques known as analysis of variance. The techniques studied here were restricted to the test of a hypothesis that dealt with questions about means from several populations. We were restricted to normal populations and populations with homogeneous (equal) variances. The test of multiple means is accomplished by partitioning the sum of squares into two segments: (1) the sum of squares due to variation between the levels of the factor being tested and (2) the sum of squares due to the variation between the replicates within each level. The null hypothesis about means is then tested by using the appropriate variance measurements.

Note that we restricted our development to one-factor experiments. This one-factor technique represents only a beginning to the study of analysis of variance techniques.

Refer back to the Chapter Case Study, on page 611, and you will see that the data given use the method of tillage (plow, and so on) as the factor and the yield from each field plot as the replicate of a one-factor analysis of variance experiment. (See Exercise 12.29.)

CHAPTER EXERCISES

12.23   Samples of peanut butter produced by three different manufacturers were tested for salt content with the following results.

Brand 1:	2.5	8.3	3.1	4.7	7.5	6.3
Brand 2:	4.5	3.8	5.6	7.2	3.2	2.7
Brand 3:	5.3	3.5	2.4	6.8	4.2	3.0

Is there a significant difference in the mean amount of salt in these samples? Use $\alpha = 0.05$.
   a. State the null and alternative hypotheses.
   b. Determine the test criteria: level of significance, test statistic, critical region, critical value(s).
   c. Using the information on the computer printout below, state the decision and conclusion to the hypothesis test.
   d. What does the p-value tell you? Explain.
Each level of data is entered into a separate column.

```
MTB > AOVONEWAY C1 C2 C3

ANALYSIS OF VARIANCE
SOURCE     DF      SS       MS       F       P
FACTOR      2     4.68     2.34    0.64    0.541
ERROR      15    54.88     3.66
TOTAL      17    59.56

                              INDIVIDUAL 95 PCT CI'S FOR MEAN
                              BASED ON POOLED STDEV
LEVEL    N     MEAN     STDEV   ---+--------|--------+--------|--
  C1     6    5.400     2.359         (--------.*--------)
  C2     6    4.500     1.669      (--------*--------)
  C3     6    4.200     1.621      (--------*--------)
                                ---+--------|--------+--------|--
POOLED STDEV = 1.913             3.0      4.5      6.0      7.5
```

12.24   A new all-purpose cleaner is being test-marketed by placing sales displays in three different locations within various supermarkets. The number of bottles sold from each location within each of the supermarkets tested is reported below.

	I	40	35	44	38
Locations	**II**	32	38	30	35
	III	45	48	50	52

a. State the null and alternative hypotheses for testing "the location of the sales display had no effect on the number of bottles sold."
b. Using $\alpha = 0.01$, determine the test criteria: level of significance, test statistic, critical region, critical value(s).
c. Using the information on the computer printout below, state the decision and conclusion to the hypothesis test.
d. What does the p-value tell you? Explain.

Data are entered into C1, location identifier is entered into C2.

```
MTB > ONEWAY C1 C2

ANALYSIS OF VARIANCE ON C1
SOURCE     DF     SS       MS       F        P
C2          2    460.7    230.3    19.51    0.001
ERROR       9    106.2     11.8
TOTAL      11    566.9

                                    INDIVIDUAL 95 PCT CI'S FOR MEAN
                                    BASED ON POOLED STDEV
                                    - - - - - +- - - - - - - +- - - - - - - +- - - - - .
LEVEL    N     MEAN     STDEV
1        4    39.250    3.775              (- - - ★ - - -)
2        4    33.750    3.500       (- - - ★ - - -)
3        4    48.750    2.986                       (- - - ★ - - - )
                                    - - - - - +- - - - - - - +- - - - - - - +- - - - -
POOLED STDEV = 3.436                 35.0        42.0        49.0
```

 12.25 An experiment was designed to compare the lengths of time that four different drugs provided pain relief following surgery. The results (in hours) are shown in the following table.

Drug			
A	B	C	D
8	6	8	4
6	6	10	4
4	4	10	2
2	4	10	
		12	

Is there enough evidence to reject the null hypothesis that there is no significant difference in the length of pain relief for the four drugs at $\alpha = 0.05$?

12.26   The distance required to stop a vehicle on wet pavement was measured to compare the stopping power of four major brands of tires A tire of each brand was tested on the same vehicle on a controlled wet pavement. The resulting distances are shown in the following table. At $\alpha = 0.05$, is there sufficient evidence to conclude that there is a difference in the mean stopping distance?

	Brand of Tire			
	A	B	C	D
Distance (replicate)	37	37	33	41
	34	40	34	41
	38	37	38	40
	36	42	35	39
	40	38	42	41
	32		34	43

12.27   A certain vending company's soft-drink dispensing machines are supposed to serve six ounces of beverage. Various machines were sampled and the resulting amounts of dispensed drink were recorded, as shown in the following table. Does this sample evidence provide sufficient reason to reject the null hypothesis that all five machines dispense the same average amount of soft drink? Use $\alpha = 0.01$.

	Machines				
	A	B	C	D	E
Amounts of Soft	3.8	6.8	4.4	6.5	6.2
Drink Dispensed	4.2	7.1	4.1	6.4	4.5
	4.1	6.7	3.9	6.2	5.3
	4.4		4.5		5.8

12.28   Suburbs, each with its own attributes, are located around every metropolitan area. There is always the "rich" one (the most expensive one), the least expensive one, and so on. Does the suburb affect the transfer value of its homes? x = transfer value, the amount on which county transfer taxes are paid.

Suburb A	Suburb B	Suburb C	Suburb D	Suburb E
105	101	95	74	79
114	88	107	135	89
85	105	101	165	140
177	100	92	114	114
104	161	91	80	80
135	113	89	115	86
	94			94
				102

a. Do the sample data show sufficient evidence to conclude that the suburbs represented do have a significant effect on the transfer value of their homes? Use $\alpha = 0.01$.

b. Construct a graph that demonstrates the conclusion reached in (a).

12.29 "Tillage Test Plots Revisited" (Chapter Case Study, p. 611) presents the yield per plot obtained in an experiment designed to compare six different methods of tilling the ground.

a. Construct a dotplot showing the six samples separately and side by side. Using one-way ANOVA, test the claim that "all tillage tools are not created equal."

b. State the null and alternative hypotheses and describe the meaning of each.

c. Complete the hypothesis test using $\alpha = 0.05$.

12.30 The August 1994 issue of *Employment and Earnings* gives the mean duration of unemployment in July 1994 for three groups of women, age 16 years and over. The mean for "married with spouse present" equals 15.4 weeks, the mean for "widowed, divorced, or separated" equals 17.1 weeks, and the mean for "single (never married)" equals 14.1 weeks.

A similar small survey is conducted in a midwestern city. The durations are given below.

Married, Spouse Present	Widowed/Div./Separated	Single (never married)
14	18	13
16	17	12
15	20	14
14	21	15
17	20	15
19	22	16
18	18	
19		

Complete an ANOVA for the above data, and test the null hypothesis that the mean duration of unemployment is the same for all three groups at a 0.05 level of significance.

12.31 An experiment compared typing speeds for clerk-typists on a standard electric typewriter and on a Teletype Model 43 computer terminal. Twelve typists were randomly assigned to type on the two machines. The scores are shown in the following table.

Standard Electric	Terminal
62	52
78	60
48	47
63	48
55	52
	40
	51

Is there sufficient evidence to conclude that there is a difference in the population means for the two types of machines? Use $\alpha = 0.05$.

12.32 To compare the effectiveness of three different methods of teaching reading, 26 children of equal reading aptitude were divided into three groups. Each group was instructed for a given period of time using one of the three methods. After completing the instruction period, all students were tested. The test results are shown in the following table. Is the evidence sufficient to reject the hypothesis that all three instruction methods are equally effective? Use $\alpha = 0.05$.

	Method I	Method II	Method III
	45	45	44
	51	44	50
	48	46	45
	50	44	55
Test Scores (replicates)	46	41	51
	48	43	51
	45	46	45
	48	49	47
	47	44	

12.33 The *1994 County and City Data Book* gives the median family income for the following three counties in Nebraska: Lancaster $36,467, Hall $30,822, and Sarpy $38,315. The following data represent the family incomes (in thousands) for nine randomly selected individuals from each of the three counties.

Lancaster	Hall	Sarpy
45	32	40
39.5	30	42
42	37	45
35	35	39.5
40	28.5	40
37	37.5	38
44	31	51
48.5	37.6	47.5
50	25	41

Complete an ANOVA for the above data, and test the null hypothesis that the mean family income is the same for each of the three counties. Use a 0.05 level of significance.

12.34 An article titled "Few Feel Economic Rebound" (*USA Today*, 10-7-94) gives the median household income for Asians as $38,347, whites as $32,960, Hispanics as $22,886, and blacks as $19,532. Suppose the following sample data were collected for the incomes (in thousands of dollars) for these four groups in Seattle, Washington.

Asians	Whites	Hispanics	Blacks
40.0	45.0	40.0	35.0
35.5	29.0	24.5	29.5
37.5	33.5	29.5	25.0
42.0	35.0	30.0	31.0
29.5	40.5	35.0	37.5
50.0	27.5	42.5	22.5
44.5	28.0	27.5	37.5

Complete an ANOVA for the above data. Test the null hypothesis that the mean household income is the same for the four groups. Use a 0.025 level of significance.

 12.35 The May 16, 1994, issue of *Fortune* magazine gave the percent change in home prices during 1993 for the top five markets as follows: Denver 12.7%, Salt Lake City–Ogden 9.7%, Miami–Hialeah 8.3%, Nashville 7.0%, and Portland 7.0%. The following data represent the percent change in home prices for randomly selected homes in St. Louis, Kansas City, and Oklahoma City.

St. Louis	Kansas City	Oklahoma City
3.0	4.5	1.0
2.5	2.5	-2.5
-1.5	7.0	-3.5
4.0	9.0	2.0
-1.0	1.5	4.6
5.5	2.0	0.5

Complete an ANOVA table for the above data. Test the null hypothesis that the mean percent change is equal for the three cities. Use a 0.01 level of significance.

 12.36 The following table shows the number of arrests made last year for violations of the narcotic drug laws in 24 communities. The data given are rates of arrest per 10,000 inhabitants. At $\alpha = 0.05$, is there sufficient evidence to reject the hypothesis that the mean rates of arrests are the same in all four sizes of communities?

Cities (over 250,000)	Cities (under 250,000)	Suburban Communities	Rural Communities
45	23	25	8
34	18	17	16
41	27	19	14
42	21	28	17
37	26	31	10
28	34	37	23

12.37   For the following data, find SS(error) and show that

$$SS(error) = [(k_1 - 1)s_1^2 + (k_2 - 1)s_2^2 + (k_3 - 1)s_3^2]$$

where s_i is the variance for the ith factor level.

Factor Level		
1	**2**	**3**
8	6	10
4	6	12
2	4	14

12.38   For the following data, show that

$$SS(factor) = k_1 (\bar{x}_1 - \bar{x})^2 + k_2(\bar{x}_2 - \bar{x})^2 + k_3(\bar{x}_3 - \bar{x})^2$$

where $\bar{x}_1, \bar{x}_2, \bar{x}_3$ are the means for the three factor levels and $\bar{x}$ is the overall mean.

Factor Level		
1	**2**	**3**
6	13	9
8	12	11
10	14	7

 12.39   An article in the *Journal of Pharmaceutical Sciences* (Dec. 1987) discusses the change of plasma protein binding of Diazepam at various concentrations of Imipramine. Suppose the results were reported as follows:

Diazepam Alone (1.25 mg/mL)	Diazepam with Imipramine		
	1.25	2.50	5.00
97.99	97.68	96.29	93.92

The values given represent mean plasma protein binding and $n = 8$ for each of the four groups. Find the sum of squares among the four groups.

 12.40   A study reported in the *Journal of Research and Development in Education* (Summer 1989) evaluates the effectiveness of social skills training and cross-age tutoring for improving academic skills and social communication behaviors among boys with learning disabilities. Twenty boys were divided into three groups, and their scores on the Test of Written Spelling (TWS) may be summarized as follows:

Use the information in Exercises 12.37 and 12.38.

Group	n	TWS Mean	St. Dev.
Social skills training and tutoring components	7	21.43	9.48
Social skills training only	7	20.00	8.91
Neither component	6	20.83	9.06

Calculate the entries of the ANOVA table using these results.

12.41 Seven golf balls from each of six manufacturers were randomly selected and tested for durability. Each ball was hit 300 times or until failure occurred, whichever came first. Do these sample data show sufficient reason to reject the null hypothesis that the six different brands tested withstood the durability test equally well? Use $\alpha = 0.05$.

	Manufacturer Brand					
	A	B	C	D	E	F
	300	190	228	276	162	264
	300	164	300	296	175	168
	300	238	268	62	157	254
Number of Hits (replicate)	260	200	280	300	262	216
	300	221	300	230	200	257
	261	132	300	175	256	183
	300	156	300	211	92	93

VOCABULARY LIST

Be able to define each term. Pay special attention to the key terms, which are printed in red. In addition, describe in your own words, and give an example of, each term. Your examples should not be ones given in class or in the textbook.

The bracketed numbers indicate the chapters in which the term previously appeared, but you should define the terms again to show increased understanding of their meaning. Page numbers indicate the first appearance of the term in Chapter 12.

analysis of variance (ANOVA) (p. 612)
between-sample variation (p. 619)
degrees of freedom [9, 10, 11] (p. 616)
experimental error (p. 622)
levels of the tested factor (p. 613)
mathematical model (p. 622)
mean square, MS(factor), MS(error) (p. 616)
partitioning (p. 614)
randomize [2] (p. 618)
replicate (p. 613)

response variable [1] (p. 618)
sum of squares (p. 614)
test statistic, $F^{\star}$ (p. 616)
total sum of squares, SS(total) (p. 614)
variance [2, 9, 10] (p. 613)
variation between levels, MS(factor) (p. 614)
variation within a level, MS(error) (p. 615)
within-sample variation (p. 619)

QUIZ A

Answer "True" if the statement is always true. If the statement is not always true, replace the words shown in bold with words that make the statement always true.

12.1   To partition the sum of squares for the total is to separate the numerical value of SS(total) into two values such that the **sum** of these two values is equal to SS(total).

12.2   A **sum of squares** is actually a measure of variance.

12.3   **Experimental error** is the name given to the variability that takes place between the levels of the test factor.

12.4   **Experimental error** is the name given to the variability that takes place among the replicates of an experiment as it is repeated under constant conditions.

12.5   **Fail to reject H_0** is the desired decision when the means for the levels of the factor being tested are all different.

12.6   The **mathematical model** for a particular problem is an equational statement showing the anticipated makeup of an individual piece of data.

12.7   The degrees of freedom for the factor are equal to **the number of factors tested.**

12.8   The measure of a specific level of a factor being tested in an ANOVA is the **variance** of that factor level.

12.9   We **need not** assume that the observations are independent to do analysis of variance.

12.10 The rejection of H_0 **indicates** that you have identified the level(s) of the factor that is (are) different from the others.

QUIZ B

12.1   Determine the truth (T/F) for each statement with regard to the one-factor analysis of variance technique.

_____ a. The mean squares are measures of variance.

_____ b. "There is no difference between the mean values of the random variable at the various levels of the test factor" is a possible interpretation of the null hypothesis.

_____ c. "The factor being tested has no effect on the random variable x" is a possible interpretation of the alternative hypothesis.

_____ d. "There is no variance among the mean values of x for each of the different factor levels" is a possible interpretation of the null hypothesis.

_____ e. The "partitioning" of the variance occurs when the SS(total) is separated into SS(factor) and SS(error).

_____ f. We will want to reject the null hypothesis and conclude that the factor has an effect on the variable when the amount of variance assigned to the factor is significantly larger than the variance assigned to error.

_____ g. In order to apply the F-test, the sample size from each factor level must be the same.

_____ h. In order to apply the *F*-test, the sample standard deviation from each factor level must be the same.

_____ i. If 20 is subtracted from every data value, then the calculated value of the *F*-statistic is also reduced by 20.

When the calculated value of *F*, $F^\star$, is greater than the table value for *F*,

_____ j. the decision will be: "Fail to reject H_0."

_____ k. the conclusion will be: "The factor being tested does have an effect on the variable."

Independent samples were collected in order to test the effect a factor had on a variable. The data are summarized in the following ANOVA table:

	SS	df
Factor	810	2
Error	720	8
Total	1530	10

Is there sufficient evidence to reject the null hypothesis that all levels of the test factor have the same effect on the variable?

_____ l. The null hypothesis could be: $\mu_A = \mu_B = \mu_C = \mu_D$

_____ m. The calculated value of *F* is 1.125.

_____ n. The critical value of *F* for $\alpha = 0.05$ is 6.06.

_____ o. The null hypothesis can be rejected at $\alpha = 0.05$.

12.2 Determine the values *A*, *B*, *C*, *D*, and *E* missing in the table.

	SS	df	MS	$F^\star$
Factor	A	4	18	E
Error	B	18	D	
Total	144	C		

Find the values of

   a. *A*   b. *B*   c. *C*   d. *D*   e. *E*

QUIZ C

12.1 In 50 words or less, explain what a single-factor ANOVA experiment is.

12.2 A state environmental agency tested three different scrubbers used to reduce the resulting air pollution in the generation of electricity. The primary concern was the emission of particulate matter. Several trials were run with each scrubber. The amount of particulate emission was recorded for each trial.

			Replicates Amounts of Emission				
	I	11	10	12	9	13	12
Scrubbers Tested	II	12	10	12	8	9	
	III	9	11	10	7	8	

a. State the mathematical model for this experiment.
b. State the null and alternative hypotheses.
c. Calculate and form the ANOVA table.
d. Complete the testing of H_0 using a 0.05 level of significance. State the decision and conclusion clearly.
e. Construct a graph representing the data that is helpful in picturing the results of the hypothesis test.

13

LINEAR CORRELATION AND REGRESSION ANALYSIS

Chapter Case Study

CHARTING THE POUNDS

*H*eight and weight charts, similar to the one below, have been around for many years. They are used in research, by insurance companies, by doctors, and by each of us to determine the "ideal" weight for our height.

This is the standard chart used in a major new study of women's weight and heart attacks. The study found that women weighing at least 6 percent below these "ideals" had 23 percent less of a risk of heart attacks.

Height	Women's Weight (lb.)		
	Small Frame	**Medium Frame**	**Large Frame**
4′10″	102–111	109–121	118–131
4′11″	103–113	111–123	120–134
5′0″	104–115	113–126	122–137
5′1″	106–118	115–129	125–140
5′2″	108–121	118–132	128–143
5′3″	111–124	121–135	131–147
5′4″	114–127	124–138	134–151
5′5″	117–130	127–141	137–155
5′6″	120–133	130–144	140–159
5′7″	123–136	133–147	143–163
5′8″	126–139	136–150	146–167
5′9″	129–142	139–153	149–170
5′10″	132–145	142–156	152–173
5′11″	135–148	145–159	155–176
6′0″	138–151	148–162	158–179

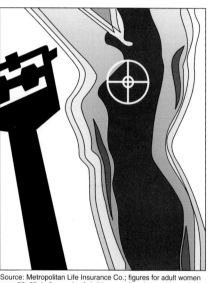

Source: Metropolitan Life Insurance Co.; figures for adult women ages 25–59, in 3 pounds of clothing, wearing shoes with one-inch heels

By Suzy Parker, USA TODAY

Source: Copyright 1990, USA TODAY. Reprinted with permission.

Draw a scatter diagram showing the information for each frame size, using height as the input variable. Each inch of height seems to add how many pounds to the ideal weight? (See Exercise 13.68, p. 692.)

● CHAPTER OBJECTIVES

In Chapter 3 the basic ideas of regression and linear correlation analysis were introduced. (If these concepts are not fresh in your mind, review Chapter 3 before beginning this chapter.) Chapter 3 was only a first look—a presentation of the basic graphic (the scatter diagram) and descriptive statistical aspects of linear correlation and regression analysis. In this chapter we take a second, more detailed look at linear correlation and regression analysis.

Previously we used the linear correlation coefficient to measure the strength of the linear relationship between two variables. Now we will determine whether there is a linear relationship by using a hypothesis test in which the probability of a type I error is fixed by the value assigned to α. In Chapter 3 we introduced a set of formulas for finding the equation of the straight line of best fit. Now we wish to ask, "Is the linear equation of any real use?" Previously we used the equation of the line of best fit to make point predictions. Now we will make confidence interval estimations. In short, this second look at linear correlation and regression analysis will be a much more complete presentation than that in Chapter 3.

Bivariate data   Recall that **bivariate data** are ordered pairs of numerical values. The two values are each response variables. They are paired with each other as a result of a common bond (see p. 135).

13.1 | LINEAR CORRELATION ANALYSIS

In Chapter 3 the linear correlation coefficient was presented as a quantity that measures the strength of a linear relationship (dependency). Now let's take a second look at this concept and see how r, the coefficient of linear correlation, works. Intuitively, we want to think about how to measure the mathematical linear dependency of one variable on another. As x increases, does y tend to increase (decrease)? How strong (consistent) is this tendency? We are going to use two measures of dependence—covariance and the coefficient of linear correlation—to measure the relationship between two variables. We'll begin our discussion by examining a set of bivariate data and identifying some related facts as we prepare to define covariance.

FIGURE 13-1

Graph of Data for
Illustration 13-1

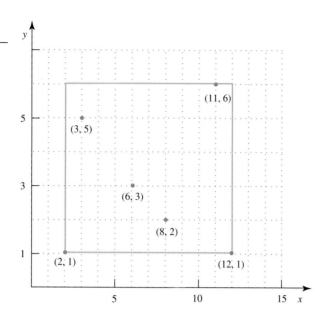

▼ ■ I L L U S T R A T I O N   13 - 1

Let's consider the following sample of six pieces of bivariate data: (2, 1), (3, 5), (6, 3), (8, 2), (11, 6), (12, 1). (See Figure 13-1.) The mean of the six *x*-values (2, 3, 6, 8, 11, 12) is $\bar{x} = 7$. The mean of the six *y*-values (1, 5, 3, 2, 6, 1) is $\bar{y} = 3$.

Centroid

The point $(\bar{x}, \bar{y})$, which is (7, 3), is located as shown on the graph of the sample points in Figure 13-2. The point $(\bar{x}, \bar{y})$ is called the **centroid** of the data. If a vertical and a horizontal line are drawn through the centroid, the graph is divided into four sections, as shown in Figure 13-2. Each point (x, y) lies a certain distance from each of these two lines. $(x - \bar{x})$ is the horizontal distance from (x, y) to the vertical line passing through the centroid. $(y - \bar{y})$ is the vertical distance from (x, y) to the horizontal line passing through the centroid. Both the horizontal and vertical distances of each data point from the centroid can be measured, as shown in Figure 13-3. The distances may be positive, negative, or zero, depending on the position of the point (x, y) in reference to $(\bar{x}, \bar{y})$. $[(x - \bar{x})$ and $(y - \bar{y})$ are represented by means of braces, with positive or negative signs, as shown in Figure 13-3.]

▲ ■

Covariance

One measure of linear dependency is the covariance. The **covariance of *x* and *y*** is defined as the sum of the products of the distances of all values of *x* and *y* from the centroid, $\Sigma[(x - \bar{x})(y - \bar{y})]$, divided by $n - 1$:

FIGURE 13.2

The Point (7, 3) Is
the Centroid

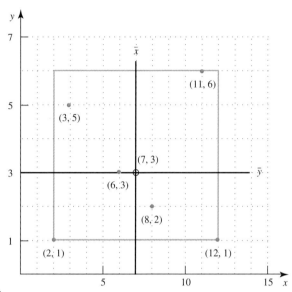

FIGURE 13.3

Measuring the Distance of Each
Data Point from the Centroid

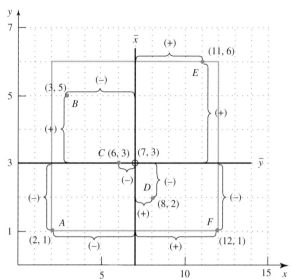

TABLE 13-1

Calculations for
Finding covar (x, y)
for the Data of
Illustration 13-1

Points	$x - \bar{x}$	$y - \bar{y}$	$(x - \bar{x})(y - \bar{y})$
A(2, 1)	−5	−2	10
B(3, 5)	−4	2	−8
C(6, 3)	−1	0	0
D(8, 2)	1	−1	−1
E(11, 6)	4	3	12
F(12, 1)	5	−2	−10
Total	0	0	3

$$\text{covar}(x, y) = \frac{\sum_{i=1}^{n}(x_i - \bar{x})(y_i - \bar{y})}{n - 1} \qquad \textbf{(13-1)}$$

The covariance for the data given in Illustration 13-1 is calculated in Table 13-1.
The covariance, written as covar(x, y), of the data is $+\frac{3}{5} = 0.6$.

NOTE $\Sigma(x - \bar{x}) = 0$ and $\Sigma(y - \bar{y}) = 0$. This will always happen. Why? (See
p. 78.)

The covariance is positive if the graph is dominated by points to the upper right and to the lower left of the centroid. The products of $(x - \bar{x})$ and $(y - \bar{y})$ are positive in these two sections. If the majority of the points are in the upper-left and lower-right sections relative to the centroid, the sum of the products is negative. Figure 13-4 shows data that represent a positive dependency (a), a negative dependency (b), and little or no dependency (c). The covariances for these three situations would definitely be positive in part (a), negative in (b), and near zero in (c). (The sign of the covariance is always the same as the sign of the slope of the regression line.)

FIGURE 13-4   Data and Covariance

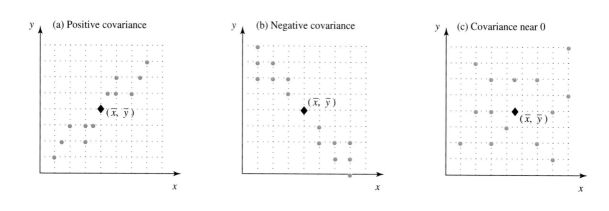

The biggest disadvantage of covariance as a measure of linear dependency is that it does not have a standardized unit of measure. One reason for this is that the spread of the data is a strong factor in the size of the covariance. For example, if we were to multiply each data point in Illustration 13-1 by 10, we would have (20, 10), (30, 50), (60, 30), (80, 20), (110, 60), and (120, 10). The relationship of the points to each other would be changed only in that they would be much more spread out. However, the covariance for this new set of data is 60. Does this mean that the amount of dependency between the x and y variables is stronger than in the original case? No, it does not; the relationship is the same even though each data point was multiplied by 10. This is the trouble with covariance as a measure. We must find some way to eliminate the effect of the spread of the data when we measure dependency.

If we standardize x and y by dividing the distance of each from the respective mean by the respective standard deviation,

$$x' = \frac{x - \bar{x}}{s_x} \quad \text{and} \quad y' = \frac{y - \bar{y}}{s_y}$$

This calculation is assigned in Exercise 13.9, page 655.

and then compute the covariance of x' and y', we will have a covariance that is *not* affected by the spread of the data. This is exactly what is accomplished by the linear correlation coefficient. It divides the covariance of x and y by a measure of the spread of x and by a measure of the spread of y (the standard deviations of x and of y are used as measures of spread). Therefore, by definition, the **coefficient of linear correlation** is

Coefficient of linear correlation

$$r = \text{covar}(x', y') = \frac{\text{covar}(x, y)}{s_x \cdot s_y} \tag{13-2}$$

The coefficient of linear correlation standardizes the measure of dependency and allows us to compare the relative strengths of dependency of different sets of data. [Formula (13-2) for linear correlation is also commonly referred to as **Pearson's product moment, r**].

Pearson's product moment

The value of r, the coefficient of linear correlation, for the data in Illustration 13-1 can be found by calculating the two standard deviations and then dividing:

$$s_x = 4.099 \quad \text{and} \quad s_y = 2.098$$

$$r = \frac{0.6}{(4.099)(2.098)} = \textbf{0.07}$$

Finding the correlation coefficient by using formula (13-2) can be a very tedious arithmetic process. The formula can be written in a more workable form, as it was in Chapter 3:

Refer to Chapter 3 for an illustration of the use of this formula.

$$r = \frac{\text{covar}(x, y)}{s_x \cdot s_y} = \frac{\dfrac{\Sigma[(x - \bar{x}) \cdot (y - \bar{y})]}{n - 1}}{s_x \cdot s_y} \tag{13-3}$$

$$= \frac{\text{SS}(xy)}{\sqrt{\text{SS}(x) \cdot \text{SS}(y)}}$$

Formula (13-3) avoids the separate calculations of $\bar{x}$, $\bar{y}$, s_x, and s_y, and more important, the calculations of the deviations from the means. Therefore, formula (13-3) is much easier to use.

EXERCISES

13.1 Explain why $\Sigma(x - \bar{x}) = 0$ and $\Sigma(y - \bar{y}) = 0$.

13.2 Consider a set of paired bivariate data. Describe the relationship of the ordered pairs that will cause $\Sigma[(x - \bar{x}) \cdot (y - \bar{y})]$ to be
  a. positive
  b. negative
  c. near zero

13.3   a. Construct a scatter diagram of the following bivariate data.

	Point									
	A	*B*	*C*	*D*	*E*	*F*	*G*	*H*	*I*	*J*
x	1	1	3	3	5	5	7	7	9	9
y	1	2	2	3	3	4	4	5	5	6

b. Calculate the covariance.
c. Calculate s_x and s_y.
d. Calculate r using formula (13-2).
e. Calculate r using formula (13-3).

13.4   Consider the accompanying bivariate data.

	Point									
	A	*B*	*C*	*D*	*E*	*F*	*G*	*H*	*I*	*J*
x	0	1	1	2	3	4	5	6	6	7
y	6	6	7	4	5	2	3	0	1	1

a. Draw a scatter diagram for the data.
b. Calculate the covariance.
c. Calculate s_x and s_y.
d. Calculate r by formula (13-2).
e. Calculate r by formula (13-3).

13.5   MINITAB was used to complete the preliminary calculations: form the extensions table, calculate the summations Σx, Σy, Σx^2, Σxy, Σy^2, and find the SS(x), SS(y), SS(xy) for the following set of bivariate data. Verify the results.

x	45	52	49	60	67	61
y	22	26	21	28	33	32

With the data listed in C1 and C2, use these MINITAB session commands:

```
NAME C1 'X' C2 'Y' C3 'XSQ' C4 'XY' C5 'YSQ'
LET C3 = C1**2
LET C4 = C1*C2
LET C5 = C2**2
```

(*continued*)

MINITAB (*continued*)
```
SUM C1 C6
SUM C2 C7
SUM C3 C8
SUM C4 C9
SUM C5 C10
NAME K1 'SS(X)' K2 'SS(Y)' K3 'SS(XY)'
LET K1 = SUM(C3)-((SUM(C1)**2)/COUNT(C1))
LET K2 = SUM(C5)-((SUM(C2)**2)/COUNT(C1))
LET K3 = SUM(C4)-((SUM(C1)*SUM(C2))/COUNT(C1))
PRINT C1-C5
PRINT C6-C10
PRINT K1-K3
```

Data Display

Row	X	Y	XSQ	XY	YSQ
1	45	22	2025	990	484
2	52	26	2704	1352	676
3	49	21	2401	1029	441
4	60	28	3600	1680	784
5	67	33	4489	2211	1089
6	61	32	3721	1952	1024

Row	C6	C7	C8	C9	C10
1	334	162	18940	9214	4498

```
SS(X)     347.334
SS(Y)     124.000
SS(XY)    196.000
```

13.6 Use a computer to form the extensions table, calculate the summations Σx, Σy, Σx^2, Σxy, Σy^2, and find the SS(x), SS(y), SS(xy) for the following set of bivariate data. (If you use MINITAB, see Exercise 13.5 for the necessary commands.)

x	11.4	9.4	6.5	7.3	7.9	9.0	9.3	10.6
y	8.1	8.2	5.8	6.4	5.9	6.5	7.1	7.8

13.7 The weight, x, and the waist size, y, were determined for 11 women. The data were as follows:

x	110	143	120	127	143	111	137	154	123	104	128
y	22	29	27	26	27	24	28	28	26	25	27

Verify the correlation coefficient in the following MINITAB output:

See page 154 for information about CORR.

```
MTB > CORRELATION C1, C2

Correlation of C1 and C2 = 0.805
```

13.8 An article in *Physical Therapy* (March 1990) describes two different methods for measuring the differences in leg lengths in a group of ten physical therapy patients. One method (TMM) uses a tape measure, and the other utilizes a radiographic technique. The following data are taken from the paper.

Leg-Length Differences (cm)	
TMM	**Radiographic**
1.30	1.20
−0.10	−0.15
1.50	1.55
0.70	1.15
−2.70	−2.00
2.50	1.15
−0.40	−0.10
−0.70	−0.50
−0.60	−0.05
−1.00	−1.10

  a. Calculate the linear correlation coefficient.
  b. What conclusions might you draw from your answer in (a)?

13.9 a. Calculate the covariance of the set of data (20, 10), (30, 50), (60, 30), (80, 20), (110, 60), and (120, 10).
  b. Calculate the standard deviation of the six x-values and the standard deviation of the six y-values.
  c. Calculate r, the coefficient of linear correlation, for the data in part (a).
  d. Compare these results to those found in the text for Illustration 13-1, page 649.

13.10 A formula that is sometimes given for computing the correlation coefficient is

$$r = \frac{n\left(\sum xy\right) - \left(\sum x\right)\left(\sum y\right)}{\sqrt{n\left(\sum x^2\right) - \left(\sum x\right)^2}\ \sqrt{n\left(\sum y^2\right) - \left(\sum y\right)^2}}$$

Use this expression as well as the formula

$$r = \frac{SS(xy)}{\sqrt{SS(x) \cdot SS(y)}}$$

to compute r for the data in the table.

x	2	4	3	4	0
y	6	7	5	6	3

13.2 | INFERENCES ABOUT THE LINEAR CORRELATION COEFFICIENT

After the linear correlation coefficient, r, has been calculated for the sample data, it seems necessary to ask this question: Does the value of r indicate that there is a linear dependency between the two variables in the population from which the sample was drawn? To answer this question we can perform a hypothesis test. The null hypothesis is "the two variables are linearly unrelated" ($\rho = 0$), where $\boldsymbol{\rho}$ (the lowercase Greek letter rho) is the **linear correlation coefficient for the population**. The alternative hypothesis may be either one-tailed or two-tailed. Most frequently it is two-tailed. However, when we suspect that there is only a positive or only a negative correlation, we should use a one-tailed test. The alternative hypothesis of a one-tailed test is $\rho > 0$ or $\rho < 0$.

ρ (rho)

The critical region for the test is on the right when a positive correlation is expected and on the left when a negative correlation is expected. The test statistic used to test the null hypothesis is the calculated value of r from the sample. Critical values for r are found in Table 10 of Appendix B at the intersection of the column identified by the appropriate value of α and the row identified by the degrees of freedom. The number of degrees of freedom for the r-statistic is 2 less than the sample size, df $= n - 2$.

The rejection of the null hypothesis means that there is evidence of a linear dependency between the two variables in the population. Failure to reject the null hypothesis is interpreted as meaning that linear dependency between the two variables in the population has not been shown.

CAUTION

The sample evidence may say only that the pattern of behavior of the two variables is related in that one can be used effectively to predict the other. *This does not mean that you have established a cause-and-effect relationship.*

Confidence belts

As in other problems, a confidence interval estimate of the population correlation coefficient is sometimes required. It is possible to estimate the value of ρ, the linear correlation coefficient of the population. Usually this is accomplished by using a table showing **confidence belts**. Table 11 in Appendix B gives confidence belts for 95% confidence intervals. This table is a bit tricky to read, so be extra careful when you use it. The next illustration demonstrates the procedure for estimating ρ.

▼ | ILLUSTRATION   13 - 2

A sample of 15 ordered pairs of data have a calculated r-value of 0.35. Find the 95% confidence interval for ρ, the population linear correlation coefficient.

SOLUTION Find $r = 0.35$ at the bottom of Table 11. (See the arrow on Figure 13-5.) Visualize a vertical line through that point. Find the two points where the belts marked for the correct sample size cross the vertical line. The sample size is 15. These two points are circled in Figure 13-5.

FIGURE 13-5
_____

Using Table 11 of
Appendix B,
Confidence Belts
for the Correlation
Coefficient

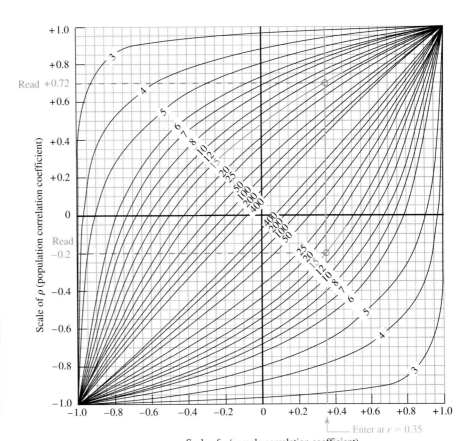

EXERCISE   13.11

Find the 95% inter-
val when a sample
of $n = 25$ results in
$r = 0.35$.

Now look horizontally from the two circled points to the vertical scale on the left and read the confidence interval. The values are 0.72 and -0.20. Thus the 95% confidence interval for ρ, the population coefficient of linear correlation, is **−0.20 to 0.72**.

▲ |

Now let's look at such a hypothesis test.

▼ | ILLUSTRATION   13 - 3

For Illustration 13-1, where $n = 6$, we found $r = 0.07$. Is this significantly different from zero at the 0.02 level of significance?

SOLUTION

STEP 1   Parameter of concern: The linear correlation coefficient for the population, ρ (rho).

STEP 2   Hypotheses:

$$H_0: \rho = 0.$$
$$H_a: \rho \neq 0.$$

STEP 3   Test criteria:
Level of significance: $\alpha = 0.02$.
df $= n - 2 = 6 - 2 = 4$. See the accompanying figure.

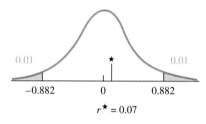

$r^\star = 0.07$

The critical values (-0.882 and 0.882) were obtained from Table 10 of Appendix B.

STEP 4   The calculated value of r, $r^\star$, was found earlier. It is $r^\star = 0.07$.

STEP 5   Results: $r^\star$ is in the noncritical region.

DECISION Fail to reject H_0.

CONCLUSION At the 0.02 level of significance, we have failed to show that x and y are correlated.

▲ |

EXERCISE   13.12

What are the critical values of r for $\alpha = 0.05$ and $n = 20$?

EXERCISES

13.14   Using graphs to illustrate, explain the meaning of a correlation coefficient whose value is (a) -1, (b) 0, (c) $+1$.

13.15   Using graphs to illustrate, explain the meaning of a correlation coefficient whose value is (a) $+0.5$, (b) -0.6.

Case Study

EXERCISE   13.13

The linear relationship between DNA indices appears strong, and the reported correlation coefficient is 0.94. Is that significant? Explain.

DNA Quantitation by Image Cytometry of Touch Preparations from Fresh and Frozen Tissue

The linear relationship between DNA indices from Frozen Touch preps and Fresh Touch preps appears to be very strong.

Quantitation of DNA content by flow cytometry of formalin-fixed, paraffin-embedded tissues can provide satisfactory results but has several disadvantages, including sacrifice of tissue blocks and relatively poor resolution of DNA histograms as a result of cell fragments. Satisfactory results with this method may also depend on optimum fixation time, dehydration, temperature, and embedding. DNA analysis by microspectrophotometry of Feulgen-stained tissue sections taken from paraffin blocks has similar limitations in addition to stereologic problems associated with section thickness and operation of a microspectrophotometer.

DNA quantitation from Feulgen-stained touch preparations correlates well with results of flow cytometry and may be more sensitive in detecting some aneuploid populations. The ability to measure DNA from touch preparations of frozen blocks would facilitate retrospective analysis of tumors for which frozen tissue is available and would provide an alternative method of specimen transport for reference laboratories. . . .

A scatter plot comparing the DNA indices obtained from fresh and frozen tissue samples is shown in Figure 2. In 54 of 59 cases (91.5%) there was excellent agreement between the two methods (Spearman $r = 0.91$; Pearson $r = 0.94$).

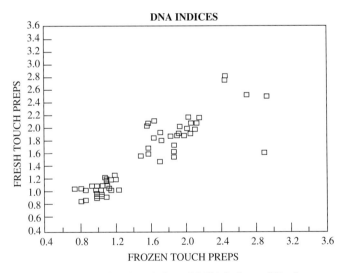

Figure 2   Scatter Plot Showing Correlation of DNA Indices of Fresh and Frozen Touch Preparations

Source: Paula F. Suit, M.D., and Thomas W. Bauer, M.D., Ph.D., *American Journal of Clinical Pathology*, July 1990. Reprinted by permission.

13.16   Use Table 11 of Appendix B to determine a 95% confidence interval for the true population linear correlation coefficient based on the following sample statistics:

   a. $n = 8, r = 0.20$ b. $n = 100, r = -0.40$
   c. $n = 25, r = +0.65$ d. $n = 15, r = -0.23$

13.17   The Test-Retest Method is one way of establishing the reliability of a test. The test is administered and then, at a later date, the same test is readministered to the same individuals. The correlation coefficient is computed between the two sets of scores. The following test scores were obtained in a Test-Retest situation.

First Score	75	87	60	75	98	80	68	84	47	72
Second Score	72	90	52	75	94	78	72	80	53	70

Find r and set a 95% confidence interval for ρ.

13.18   An article in the *Journal for Research in Mathematics Education* (Vol. 19, No. 5, 1988) determined the regression equation connecting the SDMT score, x, and the summative test score, y, for a group of third-graders. The SDMT score is for the Stanford Diagnostic Mathematics Test. The summative test consists of 30 items dealing with problems concerning place value, greatest and smallest numbers, and the use of greater than, less than, and equal to symbols. Suppose, in another study, the data for 10 third-graders were

x	15	20	25	17	20	19	26	25	20	26
y	18	22	25	21	24	18	24	25	21	24

Find the correlation coefficient and find the 95% confidence interval for ρ.

13.19   State the null hypothesis, H_0, and the alternative hypothesis, H_a, that would be used to test the following statements.

   a. The linear correlation coefficient is positive.
   b. There is no linear correlation.
   c. There is evidence of negative correlation.
   d. There is a positive linear relationship.

13.20   Determine the test criteria that would be used in testing each of the following null hypotheses.

   a. $H_0: \rho = 0$ vs. $H_a: \rho \neq 0$, with $n = 18$, $\alpha = 0.05$
   b. $H_0: \rho = 0$ vs. $H_a: \rho > 0$, with $n = 32$, $\alpha = 0.01$
   c. $H_0: \rho = 0$ vs. $H_a: \rho < 0$, with $n = 16$, $\alpha = 0.05$

13.21   A sample of 20 pieces of bivariate data has a linear correlation coefficient of $r = 0.43$. Does this provide sufficient evidence to reject the null hypothesis that $\rho = 0$ in favor of a two-sided alternative? Use $\alpha = 0.10$.

13.22   If a sample of size 18 has a linear correlation coefficient of -0.50, is there significant reason to conclude that the linear correlation coefficient of the population is negative? Use $\alpha = 0.01$.

13.23   A sample of size 10 produced $r = -0.67$. Is this sufficient evidence to conclude that ρ is different from zero at the 0.05 level of significance?

13.24 Is a value of $r = +0.24$ significant in trying to show that ρ is greater than zero for a sample of 62 data at the 0.05 level of significance?

13.25 The *State and Metropolitan Area Data Book—1991* gives the 1988 rate per 100,000 population for physicians and nurses for the West North Central states as follows:

State	Physicians	Nurses
Minnesota	212	787
Iowa	145	805
Missouri	190	745
North Dakota	167	935
South Dakota	138	809
Nebraska	168	725
Kansas	169	676

a. Calculate r.
b. Set a 95% confidence interval of ρ.
c. Describe the meaning of answer (b).
d. Explain the meaning of the width of the interval answer in (b).

13.26 The 1994 *World Almanac and Book of Facts* gives the Nielsen TV ratings for the favorite syndicated programs for 1992–1993. The following table gives these ratings for women, men, teenagers, and children.

Women	Men	Teenagers	Children
11.8	8.0	3.3	3.4
9.9	6.9	3.0	2.3
7.1	9.2	6.4	5.1
6.5	8.2	5.8	4.8
8.4	3.4	2.7	1.7
6.4	5.0	2.1	1.9
4.9	4.7	4.6	5.8
4.1	4.5	6.8	4.7
6.3	4.5	1.7	1.6
5.1	4.2	2.3	1.8
5.6	4.0	1.8	1.6
4.8	4.7	3.1	2.9
4.4	4.1	3.4	3.8
4.2	4.2	4.0	3.4
4.1	3.2	4.6	7.4
4.4	3.1	5.8	4.3
3.7	4.9	3.4	3.2
4.3	2.7	3.4	2.2
3.3	4.6	2.1	2.2
2.5	4.0	5.2	5.6

(continued)

Calculate r and use it to determine a 95% confidence interval on ρ for each of the following cases.

  a. Women and Men
  b. Women and Teenagers
  c. Women and Children
  d. Men and Teenagers
  e. Men and Children
  f. Teenagers and Children
  g. What can be concluded from the above answers? Be specific.

13.27 The 1993 issue of the *Statistical Abstract of the United States* gives the 1992 unemployment rates for males and females for various occupations. The data are given as follows:

Male	Female	Male	Female
3.2	3.0	3.6	4.0
2.8	2.4	5.2	6.2
4.4	3.4	4.8	8.2
6.3	5.6	8.6	7.8
7.0	6.8	4.5	6.1
10.0	7.9	8.9	8.3
5.9	4.9	13.0	10.9
6.1	8.6	11.0	11.0
10.2	11.4	8.1	6.6
15.1	11.7	7.7	10.1

Compute r and test for significant correlation between male and female unemployment rates. Use a level of significance equal to 0.05.

13.28 The 1992 issue of *Crime in the United States* gives the number of rural count arrests for several types of offenses. The following table gives a subset of this table.

Offense Charged	Male	Female
Murder and manslaughter	1314	196
Forcible rape	2917	63
Robbery	2480	224
Aggravated assault	26860	3956
Burglary	29270	2412
Larceny-theft	37343	11149
Motor vehicle theft	7219	976
Arson	1287	207

  a. Calculate the value of r.
  b. Test the null hypothesis that $\rho = 0$ versus $\rho > 0$ at $\alpha = 0.05$.
  c. Explain the meaning of the apparent positive correlation.

13.29   The population (in millions) and the violent crime rate (per 1000) were recorded for ten metropolitan areas. The data are shown in the following table.

Population	10.0	1.3	2.1	7.0	4.4	0.3	0.3	0.2	0.2	0.4
Crime Rate	12.0	9.5	9.2	8.4	8.2	7.3	7.1	7.0	6.9	6.9

Do these data provide evidence to reject the null hypothesis that $\rho = 0$ in favor of $\rho \neq 0$ at $\alpha = 0.05$?

13.30   In a study involving 24 coastal drainage basins in the Mendocino triple junction region of northern California (*Geological Society of America Bulletin*, Nov. 1989), it is reported that the Pearson correlation coefficient between uplift rate and the length of the drainage basin equals 0.16942. Use Table 10 in Appendix B to determine whether these data provide evidence sufficient to reject H_0: $\rho = 0$ in favor of H_a: $\rho \neq 0$ at $\alpha = 0.05$.

13.3 | LINEAR REGRESSION ANALYSIS

Line of best fit

Recall that the **line of best fit** results from an analysis of two or more related variables. (We will restrict our work to two variables. However, on occasion more than two may be mentioned to clarify the analysis.) When two variables are studied jointly, we often would like to control one variable by means of controlling the other. Or we might want to predict the value of a variable based on knowledge about another variable. In both cases we want to find the line of best fit, provided one exists, that will best predict the value of the dependent, or output, variable. Recall that the variable we know or can control is called the *independent*, or *input*, *variable*; the variable resulting from using the equation of the line of best fit is called the *dependent*, or *predicted*, *variable*.

Recall that in Chapter 3 the *method of least squares* was developed. From this concept formulas (3-6) and (3-7) were obtained and are used to calculate b_0 (the y-intercept) and b_1 (the slope of the line of best fit).

$$b_0 = \frac{1}{n}\left(\sum y - b_1 \cdot \sum x\right) \tag{3-6}$$

$$b_1 = \frac{SS(xy)}{SS(x)} \tag{3-7}$$

Then these two coefficients are used to write the equation for the line of best fit in the form

$$\hat{y} = b_0 + b_1 x$$

When the line of best fit is plotted, it does more than just show us a pictorial representation. It tells us two things: (1) whether or not there really is a functional (equational) relationship between the two variables and (2) the quantitative relationship between the two variables. Recall that the line of best fit is of no use when a change in the input variable does not seem to have a definite effect on the output variable. When there is no relationship between the variables, a horizontal line of best fit will result. A horizontal line has a slope of zero, which implies that the value of the input variable has no effect on the output variable. (This idea will be amplified later in this chapter.)

The result of regression analysis is the mathematical equation that is the equation of the line of best fit. We will, as mentioned before, restrict our work to the simple linear case, that is, one input variable and one output variable where the line of best fit is straight. However, you should be aware that not all relationships are of this nature. If the scatter diagram suggests something other than a straight line, we have **Curvilinear regression** **curvilinear regression**. In cases of this type we must introduce terms to higher powers, x^2, x^3, and so on; or other functions, e^x, log x, and so on; or we must introduce other input variables. Maybe two or three input variables would improve the usefulness of our regression equation. These possibilities are examples of curvilinear **Multiple regression** regression and **multiple regression**.

The linear model used to explain the behavior of linear bivariate data *in the population* is

$$y = \beta_0 + \beta_1 x + \epsilon \qquad (13\text{-}4)$$

This equation represents the linear relationship between the two variables in a popula-

Intercept (b_0 or β_0) tion. **β_0 is the y-intercept and β_1 is the slope. ϵ** (lowercase Greek letter epsilon) **is
Slope (b_1 or β_1) the random experimental error** in the observed value of y at a given value of x.
Experimental error The **regression line** from the sample data gives us b_0, which is **our estimate of
(ϵ or e) β_0** and b_1, which is **our estimate of β_1**. The *error ϵ is approximated by $e = y - \hat{y}$,
Regression line the difference between the observed value of y and the predicted value of y, $\hat{y}$, at a
given value of x.

$$e = y - \hat{y} \qquad (13\text{-}5)$$

The random variable e is positive when the observed value of y is larger than the predicted value, $\hat{y}$; e is negative when y is smaller than $\hat{y}$. The sum of the errors for the different values of y for a given value of x is exactly zero. (This is part of the least squares criteria.) Thus the mean value of the experimental error is zero; its variance is σ_ϵ^2. Our next goal is to estimate this variance of the experimental error.

Before we estimate the variance of ϵ, let's try to understand exactly what the error represents. ϵ is the amount of error in our observed value of y. That is, it is the difference between the observed value of y and the mean value of y at that particular value of x. Since we do not know the mean value of y, we will use the regression equation and estimate it with $\hat{y}$, the predicted value of y at this same value of x. Thus the best estimate that we have for ϵ is $e = (y - \hat{y})$, as shown in Figure 13-6.

FIGURE 13-6

The Error e Is
$y - \hat{y}$

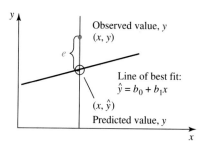

NOTE   *e* is the observed error in measuring *y* at a specified value of *x*.

If we were to observe several values of *y* at a given value of *x*, we could plot a distribution of *y*-values about the line of best fit (about $\hat{y}$, in particular). Figure 13-7 shows a sample of bivariate values for which the value of *x* is the same. Figure 13-8 shows the theoretical distribution of all possible *y*-values at a given *x*-value. A similar distribution occurs at each different value of *x*. The mean of the observed *y*'s at a given value of *x* varies, but it can be estimated by $\hat{y}$.

FIGURE 13-7
_____

Sample of *y*-Values
at a Given *x*

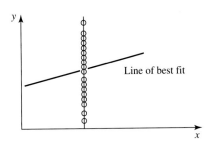

FIGURE 13-8
_____

Theoretical
Distribution of
y-Values for a
Given *x*

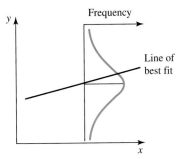

Before we can make any inferences about a regression line, we must assume that the distribution of *y*'s is approximately normal and that the variances of the distributions of *y* at all values of *x* are the same. That is, the standard deviation of the distribution of *y* about $\hat{y}$ is the same for all values of *x*, as shown in Figure 13-9.

FIGURE 13-9
_____

Standard Deviation
of the Distribution
of *y*-Values for
All *x* Is the Same

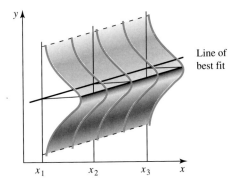

Before looking at the variance of e, let's review the definition of sample variance. The sample variance, s^2, is defined as $\Sigma(x - \bar{x})^2/(n - 1)$, the sum of the squares of each deviation divided by the number of degrees of freedom, $n - 1$, associated with a sample of size n. The variance of y involves an additional complication: There is a different mean for y at each value of x. (Notice the many distributions in Figure 13-9.) However, each of these "means" is actually the predicted value, $\hat{y}$, that corresponds to the x that fixes the distribution. So the **variance of the error e** is estimated by the formula

Variance (s^2 or σ^2)

$$s_e^2 = \frac{\Sigma(y - \hat{y})^2}{n - 2} \quad \boxed{\text{D}}$$

(13-6)

where $n - 2$ is the number of degrees of freedom.

NOTE   The variance of y about the line of best fit is the same as the variance of the error e. Recall that $e = y - \hat{y}$.

Formula (13-6) can be rewritten by substituting $b_0 + b_1x$ for $\hat{y}$. Since $\hat{y} = b_0 + b_1x$, then s_e^2 becomes

$$s_e^2 = \frac{\Sigma(y - b_0 - b_1x)^2}{n - 2}$$

(13-7)

With some algebra and some patience, this formula can be rewritten once again into a more workable form. The form we will use is

$$s_e^2 = \frac{\left(\Sigma y^2\right) - (b_0)\left(\Sigma y\right) - (b_1)\left(\Sigma xy\right)}{n - 2}$$

(13-8)

For ease of discussion let's agree to call the numerator of formulas (13-6), (13-7), and (13-8) the **sum of squares for error (SSE)**.

Sum of squares for error (SSE)

Now let's see how all of this information can be used.

▼| ILLUSTRATION   13 - 4

Suppose that you move into a new city and take a job. You will, of course, be concerned about the problems you will face commuting to and from work. For example, you would like to know how long (in minutes) it will take you to drive to work each morning. Let's use "one-way distance to work" as a measure of where you live. You live x miles away from work and want to know how long it will take you to commute each day. Your new employer, foreseeing this question, has already collected a random sample of data to be used in answering your question. Fifteen of your co-workers were asked to give their one-way travel time and distance to work. The resulting data are shown in Table 13-2. (For convenience the data have been arranged so that the x-values are in numerical order.) Find the line of best fit and the variance of y about the line of best fit, s_e^2.

TABLE 13-2

Data for
Illustration 13-4

Co-worker	Miles (x)	Minutes (y)	x^2	xy	y^2
1	3	7	9	21	49
2	5	20	25	100	400
3	7	20	49	140	400
4	8	15	64	120	225
5	10	25	100	250	625
6	11	17	121	187	289
7	12	20	144	240	400
8	12	35	144	420	1,225
9	13	26	169	338	676
10	15	25	225	375	625
11	15	35	225	525	1,225
12	16	32	256	512	1,024
13	18	44	324	792	1,936
14	19	37	361	703	1,369
15	20	45	400	900	2,025
Total	184	403	2,616	5,623	12,493

SOLUTION The extensions and summations needed for this problem are shown in Table 13-2. The line of best fit, using formulas (2-9), (3-4), (3-7), and (3-6), can now be calculated.

Using formula (2-9),

$$SS(x) = 2616 - \frac{(184)^2}{15} = 358.9333$$

Using formula (3-4),

$$SS(xy) = 5623 - \frac{(184)(403)}{15} = 679.5333$$

Use extra decimal places during these calculations.

Using formula (3-7),

$$b_1 = \frac{679.5333}{358.9333} = 1.893202 = 1.89$$

Using formula (3-6),

$$b_0 = \frac{1}{15}[403 - (1.893202)(184)] = 3.643387$$

$$= \mathbf{3.64}$$

Therefore, the equation for the line of best fit is

$$\hat{y} = 3.64 + 1.89x$$

The variance of y about the regression line is calculated by using formula (13-8).

$$s_e^2 = \frac{\text{SSE}}{n-2} = \frac{\left(\sum y^2\right) - (b_0)\left(\sum y\right) - (b_1)\left(\sum xy\right)}{n-2}$$

$$= \frac{12{,}493 - (3.643387)(403) - (1.893202)(5{,}623)}{15 - 2}$$

$$= \frac{379.2402}{13} = 29.1723$$

$$= \mathbf{29.17}$$

$s_e^2 = 29.17$ is the variance of the 15 e's. In Figure 13-10 the 15 e's are shown as vertical line segments.

NOTE   Extra decimal places are often needed for this type of calculation. Notice that b_1 (1.893202) was multiplied by 5623. If 1.89 had been used instead, that one product would have changed the numerator by approximately 18. That, in turn, would have changed the final answer by almost 1.4, and that is a sizable round-off error.

FIGURE 13-10

The 15 Random Errors as Line Segments

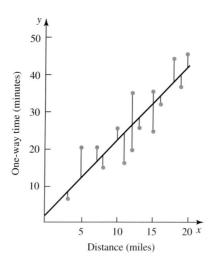

In the sections that follow, the variance of e will be used in much the same way as the variance of x (as calculated in Chapter 2) was used in Chapters 8, 9, and 10 to complete the statistical inferences studied there.

EXERCISES

13.31 Ten salespeople were surveyed and the average number of client contacts per month, x, and the sales volume, y (in thousands), were recorded for each:

x	12	14	16	20	23	46	50	48	50	55
y	15	25	30	30	30	80	90	95	110	130

Refer to the following MINITAB output and verify that the equation of the line of best fit is $\hat{y} = -13.4 + 2.3x$ and that $s_e = 10.17$.

Instructions for REGR are on page 167.

```
MTB > REGRESS Y IN C2 ON 1 PRED IN C1

The regression equation is
C2 = -13.4 + 2.30 C1

Predictor      Coef
Constant     -13.414
C1            2.3028
s = 10.17
```

13.32 A paper titled "Blue Grama Response to Zinc Source and Rates" by E. M. White (*Journal of Range Management*, January 1991) described five different experiments involving herbage yield and zinc application. Some of the data in experiment number two were as follows:

Grams of Zinc per kg Soil	0.0	0.1	0.2	0.4	0.8
Grams of Herbage	3.2	2.8	2.6	2.0	0.1

The paper quotes the correlation coefficient as equal to -0.99 and the line of best fit as $\hat{y} = -3.825x + 3.29$, where $x =$ grams of zinc and $y =$ grams of herbage.
   a. Verify the value for r.
   b. Verify the equation of the line of best fit.

13.33 The computer-science aptitude score, x, and the achievement score, y (measured by a comprehensive final), were measured for 20 students in a beginning computer-science course. The results were as follows. Find the equation of the line of best fit and s_e^2.

x	4	16	20	13	22	21	15	20	19	16	18	17	8	6	5	20	18	11	19	14
y	19	19	24	36	27	26	25	28	17	27	21	24	18	18	14	28	21	22	20	21

13.34   In a regression problem where x = the number of days on a diet and y = the number of pounds lost, we obtain the following:

MINITAB commands for these preliminary calculations are in Exercise 13.5, on page 653.

$$n = 4 \qquad \sum xy = 132$$
$$\sum x = 20 \qquad \sum x^2 = 110$$
$$\sum y = 24 \qquad \sum y^2 = 162$$

a. Find the equation of the line of best fit.
b. How much is the average weight loss in pounds for each additional day on the diet?

13.35   a. Using the ten points shown in the following table, find the equation of the line of best fit, $\hat{y} = b_0 + b_1 x$, and graph it on a scatter diagram.

	Point									
	A	**B**	**C**	**D**	**E**	**F**	**G**	**H**	**I**	**J**
x	1	1	3	3	5	5	7	7	9	9
y	1	2	2	3	3	4	4	5	5	6

b. Find the ordinates $\hat{y}$ for the points on the line of best fit whose abscissas are $x = 1$, 3, 5, 7, and 9.
c. Find the value of e for each of the points in the given data ($e = y - \hat{y}$).
d. Find the variance s_e^2 of those points about the line of best fit by using formula (13-6).
e. Find the variance s_e^2 by using formula (13-8). [Answers to (d) and (e) should be the same.]

13.36   The following data show the number of hours studied for an exam, x, and the grade received on the exam, y (y is measured in 10's; that is, $y = 8$ means that the grade, rounded to the nearest 10 points, is 80).

x	2	3	3	4	4	5	5	6	6	6	7	7	7	8	8
y	5	5	7	5	7	7	8	6	9	8	7	9	10	8	9

a. Draw a scatter diagram of the data.
b. Find the equation of the line of best fit and graph it on the scatter diagram.
c. Find the ordinates $\hat{y}$ that correspond to $x = 2$, 3, 4, 5, 6, 7, and 8.
d. Find the five values of e that are associated with the points where $x = 3$ and $x = 6$.
e. Find the variance s_e^2 of all the points about the line of best fit.

13.4 | INFERENCES CONCERNING THE SLOPE OF THE REGRESSION LINE

Now that the equation of the line of best fit has been determined and the linear model has been verified (by inspection of the scatter diagram), we are ready to determine whether we can use the equation to predict y. We will test the null hypothesis "the equation of the line of best fit is of no value in predicting y given x." That is, the null hypothesis to be tested in "β_1 (the slope of the relationship in the population) is zero." If $\beta_1 = 0$, then the linear equation will be of no real use in predicting y. To test this hypothesis we will use a t test.

Before we look at the hypothesis test, let's discuss the sampling distribution of the slope. If random samples of size n are repeatedly taken from a bivariate population, the calculated slopes, the b_1's, would form a sampling distribution that is approximately normally distributed with a mean of β_1, the population value of the slope, and with a variance of $\sigma_{b_1}^2$, where

$$\sigma_{b_1}^2 = \frac{\sigma_\epsilon^2}{\sum(x - \bar{x})^2} \tag{13-9}$$

provided there is no lack of fit. An appropriate estimator for $\sigma_{b_1}^2$ is obtained by replacing σ_ϵ^2 by s_e^2, the estimate of the variance of the error about the regression line:

$$s_{b_1}^2 = \frac{s_e^2}{\sum(x - \bar{x})^2} \tag{13-10}$$

This formula may be rewritten in the following, more manageable form:

$$s_{b_1}^2 = \frac{s_e^2}{SS(x)} = \frac{s_e^2}{\sum x^2 - \dfrac{(\sum x)^2}{n}} \tag{13-11}$$

Recall we have previously found SS(x), formula (2-9).

NOTE   The "standard error of ___" is the standard deviation of the sampling distribution of ___. Therefore, the standard error of regression (slope) is σ_{b_1} and is estimated by s_{b_1}.

In our illustration of travel times and distances, the variance among the b_1's is estimated by use of formula (13-11):

$$s_{b_1}^2 = \frac{29.1723}{358.9333} = 0.081275 = 0.0813$$

The slope β_1 of the regression line of the population can be estimated by means of a confidence interval. The confidence interval is given by

$$b_1 \pm t\left(n - 2, \frac{\alpha}{2}\right) \cdot s_{b_1} \tag{13-12}$$

The 95% confidence interval for the estimate of the population's slope, β_1, for Illustration 13-4 is

$$1.89 \pm (2.16)(\sqrt{0.0813})$$

$$1.89 \pm 0.6159$$

$$1.89 \pm 0.62$$

1.27   to   **2.51**, the 0.95 confidence interval for β_1

Case Study

13.2

Reexamining the Use of Seriousness Weights in an Index of Crime

Figure 1 is the scatter diagram of a strong linear relationship, $S_1 = -3953.85 + 3.13A_1$. The vertical scale is drawn at $A_1 = 12,600$, and the line of best fit appears to intersect the vertical scale at approximately 35,500.

ABSTRACT   The index of crime has become one of the most important social measurements for political jurisdictions in the United States. To characterize their crime problems, national, state, and local governments rely on a single index of crime, which is invariably constructed from crime statistics reported to the FBI known as Uniform Crime Reports (UCRs). The UCR index, composed of seven general crime categories, is often criticized for failing to account for the relative seriousness of its components. Blumstein (1974) examined whether the national UCR index could be improved by adding crime seriousness weights but found that the weighted index contributed no further information to national crime trends. This study replicated that research using recent Arizona UCRs to address criticisms of Blumstein's study. It also considered the appropriateness of a single index of crime, the UCRs, and how they might best be used. The findings support the conclusions of the original study.

REGRESSION AND CORRELATIONAL MEASURES BETWEEN INDICES

Regression of the Arizona UCR index on the average seriousness index produces the linear relationship depicted in Figure 1. Also shown is the ninety-five percent confidence interval (3.001, 3.262), which is based upon a standard error of .065 on the estimate of the slope. The regression equation for this relationship is

$$S_t = -3953.85 + 3.13A_t.$$

(continued next page)

That is, we can say that the slope of the line of best fit of the population from which the sample was drawn is between 1.27 and 2.51 with 95% confidence.

We are now ready to *test the hypothesis $\beta_1 = 0$*. Let's use the line of best fit determined in Illustration 13-4: $\hat{y} = 3.64 + 1.89x$. That is, we want to determine whether this equation is of any use in predicting travel time y. In this type of hypothesis test, the null hypothesis is always H_0: $\beta_1 = 0$.

STEP 1   Parameter of concern: The slope β_1 of the line of best fit for the population.

STEP 2   Hypothesis: H_0: $\beta_1 = 0$. (This implies that x is of no use in predicting y—that is, that $\hat{y} = \bar{y}$ would be as effective.) The alternative hypothesis can be either one-tailed or two-tailed. If we suspect that the slope is positive, as in Illustration 13-4 (we would expect travel time y to increase as the distance x increased), a one-tailed test is appropriate: H_a: $\beta_1 > 0$.

STEP 3   Level of significance: $\alpha = 0.05$.

EXERCISE 13.37

a. The vertical scale on Figure 1 in Case Study 13-2 is drawn at $A_t = 12,600$, and the line of best fit appears to intersect the vertical scale at approximately 35,500. Verify the coordinates of this point of intersection.

b. The article also gives an interval estimate of (3.001, 3.262). Verify this 95% interval using the information given in the article.

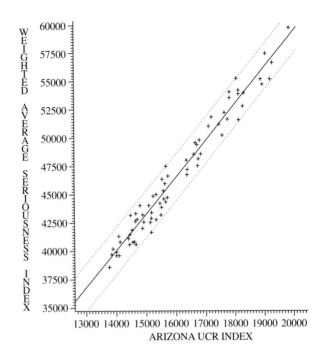

Figure 1 Regression of Average Seriousness and Arizona UCR Indices

Source: Reprinted with permission from the *Journal of Criminal Justice*, Volume 17, Thomas Epperlein and Barbara C. Nienstedt, "Reexamining the Use of Seriousness Weights in an Index of Crime," 1989, Pergamon Press, Inc.

STEP 4   Sample evidence: The test statistic is t with df $= n - 2$ degrees of freedom. df $= n - 2 = 15 - 2 =$ **13**.

The formula used to calculate the value of the *test statistic t* for inferences about the slope is

$$t = \frac{b_1 - \beta_1}{s_{b_1}} \tag{13-13}$$

Using formula (13-13), we find that the observed value of t becomes

$$t = \frac{b_1 - \beta_1}{s_{b_1}} = \frac{1.89 - 0}{\sqrt{0.0813}} = 6.629$$

$$t^\star = \mathbf{6.63}$$

The p-value was calculated using a computer. Instructions are on page 437.

The *p*-value: $p\text{-value} = P(t^\star > 6.63 \mid df = 13) = \mathbf{0.0000076}$

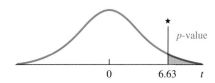

p-value

0   6.63   *t*

STEP 5   Results: *p*-value is less than α.

DECISION Reject H_0

CONCLUSION At the 0.05 level of significance, we conclude that the slope of the line of best fit in the population is greater than zero. The evidence indicates that there is a linear relationship and that the one-way distance (x) is useful in predicting the travel time to work (y).

EXERCISES

13.38   State the null hypothesis, H_0, and the alternative hypothesis, H_a, that would be used to test the following statements.
    a. The slope for the line of best fit is positive.
    b. There is no regression.
    c. There is evidence of negative regression.

13.39   Determine the *p*-value for each of the following situations.
    a. H_a: $\beta_1 > 0$, with $n = 18$, $t^\star = 2.4$
    b. H_a: $\beta_1 \neq 0$, with $n = 15$, $b_1 = 0.16$, $s_{b_1} = 0.08$
    c. H_a: $\beta_1 < 0$, with $n = 24$, $b_1 = -1.29$, $s_{b1} = 0.82$

13.40   Determine the test criteria that would be used in testing each of the following null hypotheses.
    a. H_0: $\beta_1 = 0$ vs. H_a: $\beta_1 \neq 0$, with $n = 18$, $\alpha = 0.05$
    b. H_0: $\beta_1 = 0$ vs. H_a: $\beta_1 > 0$, with $n = 28$, $\alpha = 0.01$
    c. H_0: $\beta_1 = 0$ vs. H_a: $\beta_1 < 0$, with $n = 16$, $\alpha = 0.05$

13.41   Calculate the estimated standard error of regression, s_{b_1}, for the computer-science aptitude score–achievement score relationship in Exercise 13.33.

13.42   Calculate the estimated standard error of regression, s_{b_1}, for the number of hours studied–exam grade relationship in Exercise 13.36.

13.43   The undergraduate grade point average (GPA) and the composite score on the graduate record exam (GRE) were recorded for 15 graduates in a given department at a university. The data are shown in the following table.

GPA (*x*)	2.30	3.65	3.00	2.75	3.10	2.55	2.50	2.30	2.90	3.15	3.25	2.00	2.75	2.65	3.13
GRE (*y*)	925	1300	1150	1400	900	825	950	1050	1200	1200	1100	700	850	990	1000

Verify that the value for testing $\beta_1 = 0$ versus $\beta_1 \neq 0$ is $t^{\star} = 2.70$, as shown on the following MINITAB printout. Complete the test using $\alpha = 0.05$.

Instructions for the REGRess command are on page 167.

```
MTB > REGRESS Y IN C2 ON 1 PRED IN C1

The regression equation is
C2 = 297 + 264 C1

Predictor     Coeff      Stdev     t-ratio       P
Constant      296.9
C1            264.09     97.66      2.70       0.018

s = 158.0
```

13.44 The relationship between the diameter of a spot weld, x, and the shear strength of the weld, y, is very useful. The diameter of the spot weld can be measured after the weld is completed. The shear strength of the weld can be measured only by applying force to the weld until it breaks. Thus it would be very useful to be able to predict the shear strength based only on the diameter. The following data were obtained from several sample welds.

x, Dia. of Weld (.001 in.)	190	215	200	230	209	250	215	265	215	250
y, Shear Strength (lb)	680	1025	800	1100	780	1030	885	1175	975	1300

Complete these questions with the aid of a computer. (If you use MINITAB, see Exercise 13.43 for the necessary commands.)
   a. Draw a scatter diagram.
   b. Find the equation for the line of best fit.
   c. Is the value of b_1 significantly greater than zero at the 0.05 level?
   d. Find the 95% confidence interval for β_1.

13.45 A sample of ten students were asked for the distance and the time required to commute to college yesterday. The data collected are shown in the following table.

Distance (x)	1	3	5	5	7	7	8	10	10	12
Time (y)	5	10	15	20	15	25	20	25	35	35

   a. Draw a scatter diagram of these data.
   b. Find the equation that describes the regression line for these data.
   c. Does the value of b_1 show sufficient strength to conclude that β_1 is greater than zero at the $\alpha = 0.05$ level?
   d. Find the 98% confidence interval for the estimation of β_1. (Retain these answers for use in Exercise 13.49.)

13.46   An article titled "Statistical Approach for the Estimation of Strontium Distribution Coefficient" (November 1993 issue of *Environmental Science & Technology*) reports a linear correlation coefficient of 0.55 between the strontium distribution coefficient (mL/g) and the total aluminum (mmol/100 g-soil) for soils collected from the surface throughout Japan. Consider the following data for ten such samples.

Soil Sample	Strontium Distribution Coefficient	Total Aluminum
1	100	200
2	120	225
3	300	325
4	250	310
5	400	350
6	500	400
7	450	375
8	445	385
9	310	350
10	200	290

Let Y represent the strontium distribution coefficient and X represent the total aluminum.
   a. Find the equation of the line of best fit.
   b. Find a 95% confidence interval for β_1.
   c. Explain the meaning of the interval in (b).

13.47   The September 1994 issue of *Popular Mechanics* gives specifications and dimensions for various jet boats. The following table summarizes some of this information.

Model	Base Price	Engine Horsepower
Baja Blast	8,395	120
Bayliner Jazz	8,495	90
Boston Whaler Rage 15	11,495	115
Dynasty Jet Storm	8,495	90
Four Winds Fling	9,568	115
Regal Rush	9,995	90
Sea-Doo Speedster	11,499	160
Sea Ray Sea Rayder	8,495	90
Seaswirl Squirt	8,495	115
Suga Sand Mirage	8,395	120

   a. Find the equation for the line of best fit. Let X equal the horsepower, and let Y equal the base price.
   b. Find the standard deviation along the line of best fit and the standard error of slope.
   c. Describe the meaning of the two answers in (b).
   d. Is horsepower an effective predictor of the base price? Explain your response using statistical evidence.

13.48   Interest rates are alleged to have an effect on the level of employment. The data in the following table show, by quarters, the bank interest rates on short-term loans and the unemployment rate.

Interest Rate	12.27	12.34	12.31	15.81	15.67	17.75	11.56	15.71	19.91	19.99	21.11
Unemployment Rate	5.9	5.6	5.9	5.9	6.2	7.6	7.5	7.3	7.6	7.2	8.3

a. Calculate the equation of the line of best fit.
b. Does this sample present sufficient evidence at the 0.05 level of significance to reject the null hypothesis (slope is zero) in favor of the alternative hypothesis that the slope is positive? (Retain these answers for use in Exercise 13.50.)

13.5 | CONFIDENCE INTERVAL ESTIMATES FOR REGRESSION

Once the equation for the line of best fit has been obtained and determined usable, we are ready to use the equation to make predictions. There are two different quantities that we can estimate: (1) the mean of the population y-values at a given value of x, written $\mu_{y|x_0}$, and (2) the individual y-value selected at random that will occur at a given value of x, written y_{x_0}. The best **point estimate**, or **prediction, for both** $\mu_{y|x_0}$ **and** y_{x_0} **is** $\hat{y}$. This is the y-value obtained when an x-value is substituted into the equation of the line of best fit. Like other point estimates it is seldom correct. The actual values for both $\mu_{y|x_0}$ and y_{x_0} will vary above and below the calculated value of $\hat{y}$.

Predicted value of y ($\hat{y}$)

Before developing interval estimates for $\mu_{y|x_0}$ and y_{x_0}, recall the development of confidence intervals for the population mean μ in Chapter 8 when the variance was known and in Chapter 9 when the variance was estimated. The sample mean, $\bar{x}$, was the best point estimate of μ. We used the fact that $\bar{x}$ is normally distributed with a standard deviation of $\sigma/\sqrt{n}$ to construct formula (8-1) for the confidence interval for μ. When σ had to be estimated, we used formula (9-1) for the confidence interval.

Confidence interval

The **confidence interval** for $\mu_{y|x_0}$ and the prediction interval for y_{x_0} are constructed in a similar fashion. $\hat{y}$ replaces $\bar{x}$ as our point estimate. If we were to take random samples from the population, construct the line of best fit for each sample, calculate $\hat{y}$ for a given x using each regression line, and plot the various $\hat{y}$-values (they would vary since each sample would yield a slightly different regression line), we would find that the $\hat{y}$-values form a normal distribution. That is, the sampling distribution of $\hat{y}$ is normal, just as the sampling distribution of $\bar{x}$ is normal. What about the appropriate standard deviation of $\hat{y}$? The standard deviation in both cases ($\mu_{y|x_0}$ and y_{x_0}) is calculated by multiplying the square root of the variance of the error by an appropriate correction factor. Recall that the variance of the error, s_e^2, is calculated by means of formula (13-8).

Before looking at the correction factors for the two cases, let's see why they are necessary. Recall that the line of best fit passes through the point $(\bar{x}, \bar{y})$, the centroid. In Section 13.3 we formed a confidence interval for the slope β_1 (see

Illustration 13-4) by using formula (13-12). If we draw lines with slopes equal to the extremes of that confidence interval, 1.27 to 2.51, through the point $(\bar{x}, \bar{y})$ [which is (12.3, 26.9)] on the scatter diagram, we will see that the value for $\hat{y}$ fluctuates considerably for different values of x (Figure 13-11). Therefore, we should suspect a need for a wider confidence interval as we select values of x that are further away from $\bar{x}$. Hence we need a correction factor to adjust for the distance between x_0 and $\bar{x}$. This factor must also adjust for the variation of the y-values about $\hat{y}$.

FIGURE 13-11

Lines Representing
the Confidence
Interval for Slope

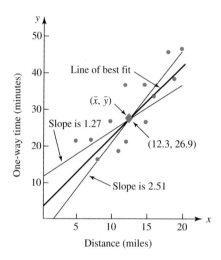

First, let's estimate the *mean value of y* at a given value of x, $\mu_{y|x_0}$. The confidence interval formula is

$$\hat{y} \pm t\left(n - 2, \frac{\alpha}{2}\right) \cdot s_e \cdot \sqrt{\frac{1}{n} + \frac{(x_0 - \bar{x})^2}{\sum(x - \bar{x})^2}} \qquad \textbf{(13-14)}$$

NOTE   The numerator of the second term under the radical sign is the square of the distance of x_0 from $\bar{x}$. The denominator is closely related to the variance of x and has a "standardizing" effect on this term.

Formula (13-14) can be modified to avoid having $\bar{x}$ in the denominator. The new form is

$$\hat{y} \pm t\left(n - 2, \frac{\alpha}{2}\right) \cdot s_e \cdot \sqrt{\frac{1}{n} + \frac{(x_0 - \bar{x})^2}{SS(x)}} \qquad \textbf{(13-15)}$$

Let's compare formula (13-14) with formula (9-1). $\hat{y}$ replaces $\bar{x}$, and

$$s_e \cdot \sqrt{\frac{1}{n} + \frac{(x_0 - \bar{x})^2}{\sum(x - \bar{x})^2}} \qquad \text{(the standard error of } \hat{y})$$

the estimated standard deviation of $\hat{y}$ in estimating $\mu_{y|x_0}$, replaces $s/\sqrt{n}$, the standard deviation of $\bar{x}$. The degrees of freedom are now $n - 2$ instead of $n - 1$ as before. These ideas are explored in the next illustration.

▼ | ILLUSTRATION 13 - 5

Construct a 95% confidence interval for the mean travel time for the co-workers who travel seven miles to work (refer to Illustration 13-4).

SOLUTION

STEP 1 Parameter of concern: $\mu_{y|x = 7}$, the mean travel time for co-workers who travel 7 miles.

STEP 2 Confidence interval criteria: The $1 - \alpha = 0.95$ interval is constructed using formula (13-15).

STEP 3 Sample evidence:
Find s_e: $s_e^2 = 29.17$ (found in Illustration 13-4)

$$s_e = \sqrt{29.17} = \mathbf{5.40}$$

Find $\hat{y}_{x = 7}$: $\hat{y} = 3.64 + 1.89x = 3.64 + 1.89(7) = \mathbf{16.87}$

STEP 4 Confidence interval limits:
Confidence coefficient: $t(13, 0.025) = 2.16$ (from Table 6 in Appendix B)

$$\text{Maximum error: } t\left(n - 2, \frac{\alpha}{2}\right) \cdot s_e \cdot \sqrt{\frac{1}{n} + \frac{(x_0 - \bar{x})^2}{SS(\alpha)}}$$

$$E = (2.16)(5.40)\sqrt{\frac{1}{15} + \frac{(7 - 12.27)^2}{358.9333}}$$

$$E = (2.16)(5.40)\sqrt{0.06667 + 0.07738}$$

$$E = (2.16)(5.40)(0.38)$$

$$E = 4.43$$

Interval limits: $\hat{y} - E$ to $\hat{y} + E$

$16.87 - 4.43$ to $16.87 + 4.43$

12.44 to **21.30**, 95% confidence interval for $\mu_{y|x = 7}$

Confidence belt

This confidence interval is shown in Figure 13-12 by the heavy red vertical line. The **confidence interval belt** showing the upper and lower boundaries of all intervals at 95% confidence is also shown in red. Notice that the boundary lines for x-values far away from $\bar{x}$ become close to the two lines that represent the equations having slopes equal to the extreme values of the 95% confidence interval for the slope (see Figure 13-12).

▲ |

FIGURE 13-12

Confidence Belts
for $\mu_{y|x_0}$

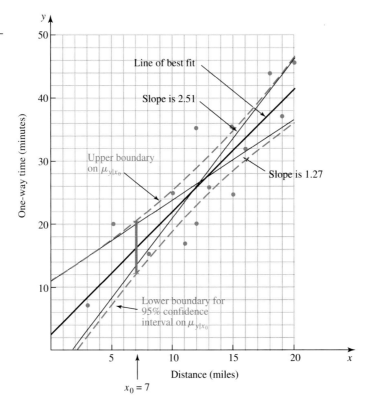

Often when making a prediction, we want to predict the value of an individual y. For example, you live seven miles from your place of business and you are interested in an estimate of how long it will take you to get to work. You are somewhat less interested in the average time for all of those who live seven miles away. The formula for the **prediction interval** of the value of a *single randomly selected y* is

Prediction interval

$$\hat{y} \pm t\left(n - 2, \frac{\alpha}{2}\right) \cdot s_e \cdot \sqrt{1 + \frac{1}{n} + \frac{(x_0 - \bar{x})^2}{SS(x)}} \qquad \textbf{(13-16)}$$

▼ | ILLUSTRATION   13 - 6

What is the 95% prediction interval for the time it will take you to commute to work if you live seven miles away?

SOLUTION

STEP 1   Parameter of concern: $y_{x = 7}$, the travel time for one co-worker who travels 7 miles.

STEP 2   Criteria: $1 - \alpha = 0.95$ and formula (13-16).

STEP 3   Sample evidence: See Illustration 13-5.

$$s_e = 5.40 \qquad \hat{y}_{x=7} = 16.87$$

STEP 4   Prediction interval limits:
Confidence coefficient: $t(13, 0.025) = 2.16$

$$\text{Maximum error: } t\left(n - 2, \frac{\alpha}{2}\right) \cdot s_e \cdot \sqrt{1 + \frac{1}{n} + \frac{(x_0 - \bar{x})^2}{\text{SS}(x)}}$$

$$E = (2.16)\,(5.40)\,\sqrt{1 + 0.06667 + 0.07738}$$
$$E = (2.16)\,(5.40)\,\sqrt{1.14405}$$
$$E = (2.16)\,(5.40)\,(1.0696) = 12.48$$

Interval limits: $\hat{y} - E$ to $\hat{y} + E$

$$16.87 - 12.84 \quad \text{to} \quad 16.87 + 12.84$$

4.39   to   **29.35**, 95% prediction interval for $y_{x=7}$

The prediction interval is shown in Figure 13-13 as the blue vertical line segment at $x_0 = 7$. Notice that it is much longer than the confidence interval for $\mu_{y|x_0 = 7}$. The dashed blue lines represent the upper and lower boundaries of the prediction intervals for individual y-values for all given x-values.

FIGURE 13-13

Prediction Belts
for y_{x_0}

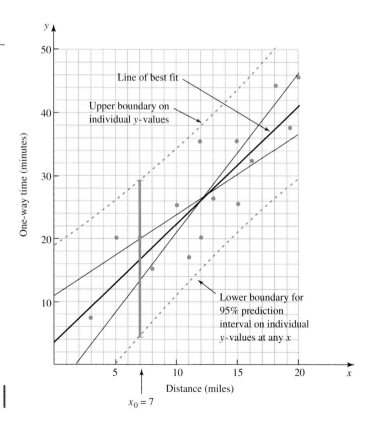

FIGURE 13-14

Confidence Belts
for the Mean Value
of y and Prediction
Belts for Individual
y's.

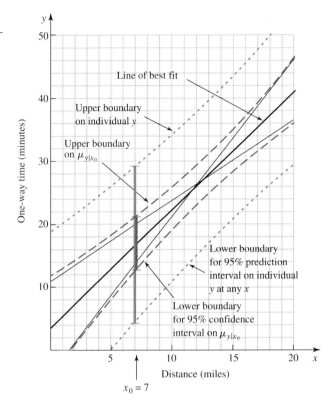

Can you justify the fact that the prediction interval for individual values of y is wider than the confidence interval for the mean values? Think about "individual values" and "mean values" and study Figure 13-14.

The techniques of coding (for large numbers) and the techniques of frequency distributions (for large sets of numbers) are adaptable to the calculations presented in this chapter. Information about these techniques and about models other than linear models may be found in many other textbooks. It is worth noting here that if the data suggest a logarithmic or an exponential functional relationship, a simple change of variable will allow for the use of linear analysis.

There are three basic precautions that you need to be aware of as you work with regression analysis:

1. Remember that the regression equation is meaningful only in the domain of the x variable studied. Estimation outside this domain is extremely dangerous; it requires that we know or assume that the relationship between x and y remains the same outside the domain of the sample data. For example, Joe says that he lives 75 miles from work, and he wants to know how long it will take him to commute. We certainly can use $x = 75$ in all the formulas, but we do not expect the answers to carry the confidence or validity of the

values of x between 3 and 20, which were in the sample. The 75 miles may represent a distance to the heart of a nearby major city. Do you think the estimated times, which were based on local distances of 3 to 20 miles, would be good predictors in this situation? Also, at $x = 0$ the equation has no real meaning. However, although projections outside the interval may be somewhat dangerous, they may be the best predictors available.

2. Don't get caught by the common fallacy of applying the regression results inappropriately. For example, this fallacy would include applying the results of Illustration 13-4 to another company. But suppose that the second company had a city location, whereas the first company had a rural location or vice versa. Do you think the results for a rural location would also be valid for a city location? Basically, the results of one sample should not be used to make inferences about a population other than the one from which the sample was drawn.

3. Don't jump to the conclusion that the results of the regression prove that x causes y to change. (This is perhaps the most common fallacy.) Regressions only measure movement between x and y; they *never prove causation*. A judgment of causation can be made only when it is based on theory or knowledge of the relationship separate from the regression results. The most common difficulty in this regard occurs because of what is called the *missing variable*, or third-variable, *effect*. That is, we observe a relationship between x and y because a third variable, one that is not in the regression, affects both x and y.

MINITAB (Release 10) commands to construct a scatter diagram showing A percent confidence interval belts for estimating the mean value of y (CI), and the prediction belts for estimating individual y-values (PI), from bivariate data where x-values listed in C1 and y-values listed in C2. You may use either subcommand CI or PI or both.

Session commands	*Menu commands*

Enter: %FITLINE C2 C1;
CONFidence A;
CI;
PI.
	Choose: Stat > Regress > Fitted Line Plot
	Enter: Y: C2 X: C1
	          Confidence level: A
	Select: Display options:
	          Confidence bands
	          Prediction bands

COMPUTER SOLUTION MINITAB Printout for parts of Illustrations 13-4, 13-5, and 13-6:

Instructions for PLOT *and* REGR *are on pages 144, 167, and 168.*

With data in columns C1(x) and C2(y):
```
MTB > PLOT C2*C1;
SUBC> TITLe "Commuting To and From Work".
```

A scatter diagram
of the data

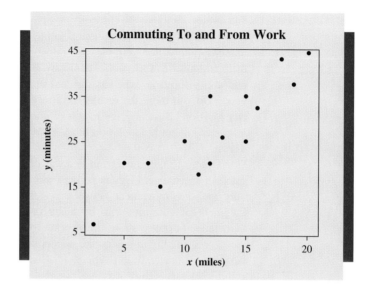

MTB > CORRELATION COEFFICIENT BETWEEN DATA IN C1 AND C2
    Correlation of C1 and C2 = 0.879

Calculated correlation
coefficient,
$r = +0.879$

MTB > REGRESS Y-VARIABLE IN C2 ON 1 X-VARIABLE IN C1
The regression equation is
C2 = 3.64 + 1.89 C1

Equation of line of best
fit, $\hat{y} = 3.64 + 1.89x$;
see page 667, solution
of Illustration 13-4

Calculated values of b_0
and b_1

Calculated value of s_{b_1},
$s_{b_1} =$
$0.285(\sqrt{0.0813})$;
compare to $s_{b_1}^2 =$
0.0813 is found
on page 671.

Predictor	Coef	Stdev	t-ratio	P
Constant	3.643	3.765	0.97	0.351
C1	1.8932	0.2851	6.64	0.000

s = 5.401   R-sq = 77.2%   R-sq(adj) = 75.5%

Analysis of Variance

Calculated $t^\star$ and
p-value for H_0: $\beta_1 = 0$
as found in steps 4 and
5 on pages 673–674.

SOURCE	DF	SS	MS	F	P
Regression	1	1286.5	1286.5	44.10	0.000
Error	13	379.2	29.2		
Total	14	1665.7			

Calculated value of s_e,
$s_e = 5.4011(\sqrt{29.1723})$;
compare to $s_b^2 =$
29.1723 as found in
solution of Illustration
13-4, page 667.

(continued)

MINITAB (*continued*)

Given data ⌐

Obs.	C1	C2	Fit	Stdev.Fit	Residual	St.Resid
1	3.0	7.00	9.32	2.99	−2.32	−0.52
2	5.0	20.00	13.11	2.50	6.89	1.44
3	7.0	20.00	16.90	2.05	3.10	0.62
4	8.0	15.00	18.79	1.85	−3.79	−0.75
5	10.0	25.00	22.58	1.54	2.42	0.47
6	11.0	17.00	24.47	1.44	−7.47	−1.43
7	12.0	20.00	26.36	1.40	−6.36	−1.22
8	12.0	35.00	26.36	1.40	8.64	1.66
9	13.0	26.00	28.26	1.41	−2.26	−0.43
10	15.0	25.00	32.04	1.60	−7.04	−1.36
11	15.0	35.00	32.04	1.60	2.96	0.57
12	16.0	32.00	33.93	1.75	−1.93	−0.38
13	18.0	44.00	37.72	2.15	6.28	1.27
14	19.0	37.00	39.61	2.37	−2.61	−0.54
15	20.0	45.00	41.51	2.61	3.49	0.74

Values of $\hat{y}$ for each given x-value using $\hat{y} = 3.6434 + 1.8932x$ ────────────↑

```
MTB>   %FITLINE C2 C1;
SUBC>  CONFidence 95.0;
SUBC>  CI;
SUBC>  PI;
SUBC>  TITLe 'Commuting To and From Work'.
```

Scatter diagram with confidence interval and prediction interval belts; see Figures 13-12, 13-13, and 13-14.

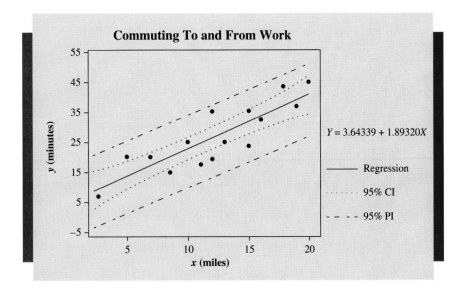

EXERCISES

13.49   Use the data and the answers found in Exercise 13-45 (p. 675) to make the following estimates.
   a. Give a point estimate for the mean time required to commute four miles.
   b. Give a 90% confidence interval for the mean travel time required to commute four miles.
   c. Give a 90% prediction interval for the travel time required for one person to commute the four miles.
   d. Answer (a), (b), and (c) for $x = 9$.

13.50   Use the data and the answers found in Exercise 13.48, (p. 677) to make the following estimates.
   a. Give a point estimate for the mean unemployment rate given that the interest rate is 14.2.
   b. Give a 95% confidence interval for the mean unemployment rate when the interest rate is 14.2.
   c. Give a 95% prediction interval for the unemployment rate when the interest rate is 14.2.

13.51   An experiment was conducted to study the effect of a new drug in lowering the heart rate in adults. The data collected are shown in the following table.

Drug Dose in mg (x)	0.5	0.75	1.00	1.25	1.50	1.75	2.00	2.25	2.50	2.75
Heart-Rate Reduction (y)	10	7	15	12	15	14	20	20	18	21

   a. Find the 95% confidence interval for the mean heart-rate reduction for a dose of 2.00 mg.
   b. Find the 95% prediction interval for the heart-rate reduction expected for an individual receiving a dose of 2.00 mg.

13.52   The relationship between the "strength" and "fineness" of cotton fibers was the subject of a study that produced the following data.

x, Strength	76	69	71	76	83	72	78	74	80	82	90	81	78	80	81	78
y, Fineness	4.4	4.6	4.6	4.1	4.0	4.1	4.9	4.8	4.2	4.4	3.8	4.1	3.8	4.2	3.8	4.2

   a. Draw a scatter diagram.
   b. Find the 99% confidence interval for the mean measurement of fineness for fibers with a strength of 80.
   c. Find the 99% prediction interval for an individual measurement of fineness for fibers with a strength of 75.

13.53 A study in *Physical Therapy* (April 1991) reports on seven different methods to determine crutch length plus two new techniques utilizing linear regression. One of the regression techniques uses the patient's reported height. One hundred and seven individuals were in the study. The mean of the self-reported heights was 68.84 in. The regression equation determined was $y = 0.68x + 4.8$, where y = crutch length and x = self-reported height. The MSE (s_e^2) was reported to be 0.50. In addition, the standard deviation of the self-reported heights was 7.35 in. Use this information to determine a 95% confidence interval estimate for the mean crutch length for individuals who say they are 70 in. tall.

13.54 An article titled "Ailing and Well Babies: 'Gap Is Striking'" appeared in the September 8, 1994, issue of the *Omaha World-Herald*. The article gave weighted median household income in 1989 and percent of households with up-to-date immunizations for children age 2 for Douglas County, which was divided into eight sections. The information is as follows:

Section	Median Household Income	% with Up-to-Date Immunizations
East/Northeast	17,723	43
West/Northeast	27,005	51
North/Central	33,424	62
Northwest	43,337	66
East/Southeast	19,226	46
West/Southeast	29,775	59
South/Central	40,607	65
Southwest	45,496	62

Let Y represent the percent with up-to-date immunizations and X represent the median household income.
    a. Find the equation of the line of best fit.
    b. Find a 95% confidence interval on the mean percent with up-to-date immunizations for families with a median household income equal to $40,000.
    c. Find a 95% prediction interval for the probability that a family will have up-to-date immunizations if their median household income equals $40,000.
    d. Explain the meaning of the intervals found in (b) and (c).

13.55 People not only live longer today but they also are living independently longer. The May/June 1989 issue of *Public Health Reports* included an article titled "A Multistate Analysis of Active Life Expectancy." Two of the variables studied were people's ages at which they became dependent and the number of independent years they had remaining. Suppose the data were as follows:

Age When Became Dependent, x	65	66	67	68	70	72	74	76	78	80	83	85
Independent Years Remaining, y	11.1	10.0	10.4	9.3	8.2	6.8	6.8	4.4	5.4	2.5	2.7	0.9

    a. Draw a scatter diagram.
    b. Calculate the equation for the equation of best fit.
    c. Draw the line of best fit on the scatter diagram.

(continued)

d. For a person who becomes dependent at age 80, how many years of independent living can be expected to remain? Find the answer two different ways: use equation (b) and use the line on the scatter diagram (c).

e. Construct a 99% prediction interval for the number of years of independent living remaining for a person who becomes dependent at age 80.

f. Draw a vertical line segment on the scatter diagram representing the interval found in (e).

13.56   Explain why a 95% confidence interval for the mean value of y at a particular x *is* much narrower than a 95% prediction interval for an individual y-value at the same value of x.

13.57   When $x_0 = \bar{x}$, is the formula for the standard error of $\hat{y}_{x_0}$ what you might have expected it to be, $s \cdot \sqrt{1/n}$?

13.6 | UNDERSTANDING THE RELATIONSHIP BETWEEN CORRELATION AND REGRESSION

Now that we have taken a closer look at both correlation and regression analysis, it is necessary to decide when to use them. Do you see any duplication of work?

The primary use of the linear correlation coefficient is in answering the question "Are these two variables linearly related?" There are other words that may be used to ask this basic question. For example, "Is there a linear correlation between the annual consumption of alcoholic beverages and the salary paid to firemen?"

The linear correlation coefficient can be used to indicate the usefulness of x as a predictor of y in the case where the linear model is appropriate. The test concerning the slope of the regression line (H_0: $\beta_1 = 0$) also tests this same basic concept. Either one of the two is sufficient to determine the answer to this query.

Although the "lack-of-fit" test can be statistically determined, it is beyond the scope of this text. However, we are very likely to carry out this test at a subjective level when we view the scatter diagram. If the scatter diagram is carefully constructed, this subjective decision will determine the mathematical model for the regression line that you believe will fit the data.

The concepts of linear correlation and regression are quite different, because each measures different characteristics. It is possible to have data that yield a strong linear correlation coefficient and have the wrong model. For example, the straight line can be used to approximate almost any curved line if the interval of domain is restricted sufficiently. In such a case the linear correlation coefficient can become quite high, but the curve will still not be a straight line. Figure 13-15 suggests one such interval where r could be significant, but the scatter diagram does not suggest a straight line.

Regression analysis should be used to answer questions about the relationship between two variables. Such questions as "What is the relationship?" "How are two variables related?" and so on, require this regression analysis.

FIGURE 13-15

The Value of r Is
High but the
Relationship Is
Not Linear

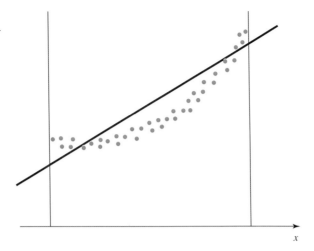

x

IN RETROSPECT

In this chapter we have made a more thorough inspection of the linear relationship between two variables. Although the curvilinear and multiple regression situations were only mentioned in passing, the basic techniques and concepts have been explored. We would only have to modify our mathematical model and our formulas if we wanted to deal with these other relationships.

Although it was not directly emphasized, we have applied many of the topics of earlier chapters in this chapter. The ideas of hypothesis testing and confidence interval were applied to the regression problem. Reference was made to the sampling distribution of the sample slope b_1. This allowed us to make inferences about β_1, the slope of the population from which the sample was drawn. We estimated the mean value of y at a fixed value of x by pooling the variance for the slope with the variance of the y's. This was allowable since they are independent. Recall that in Chapter 10 we presented formulas for combining the variances of independent samples. The idea here is much the same. Finally, we added a measure of variance for individual values of y and made estimates for these individual values of y at fixed values of x.

Case Study 13-2 presents the results of regression analysis on data collected to compare two crime-reporting indices. (Take another look at Case Study 13-2, pp. 672–673.) The scatter diagram very convincingly shows that the two crime indices being compared are related to each other in a very strong and predictable pattern. Thus, as stated, "the weighted index contributed no further information to national trends." Even though the numerical values of the two indices are quite different, the information coming from them is basically the same. Thus, the introduction of the weighted index seems unnecessary since the Uniform Crime Reports index is a recognized standard.

As this chapter ends, you should be aware of the basic concepts of regression analysis and correlation analysis. You should now be able to collect the data for, and do a complete analysis on, any two-variable linear relationship.

CHAPTER EXERCISES

13.58   Answer the following as "sometimes," "always," or "never." Explain each "never" and "sometimes" response.

    a. The correlation coefficient has the same sign as the slope of the least squares line fitted to the same data.

    b. A correlation coefficient of 0.99 indicates a strong causal relationship between the variables under consideration.

    c. An r-value greater than zero indicates that ordered pairs with high x-values will have low y-values.

    d. The two coefficients for the line of best fit have the same sign.

    e. If x and y are independent, then the population correlation coefficient equals zero.

13.59   A study in the *Journal of Range Management* (Sept. 1990) examines the relationships between elements in Russian wild rye. The correlation coefficient between magnesium and calcium was reported to be 0.69 for a sample of size 45. Is there a significant correlation between magnesium and calcium in Russian wild rye (i.e., is $\rho = 0$)?

13.60   A study concerning the plasma concentration of the drug Ranitidine was reported in the *Journal of Pharmaceutical Sciences* (Dec. 1989). The drug was administered (coded I), and the plasma concentration of Ranitidine was followed for twelve hours. The time to the first peak in concentration was called T_{max1}. The same experiment was repeated one week later (coded II). Twelve subjects participated in the study. The correlation coefficient between T_{max1}, I and T_{max1}, II was reported to be 0.818. Use Table 10 in Appendix B to determine bounds on the p-value for the hypothesis test of $H_0: \rho = 0$ versus $H_a: \rho \neq 0$.

13.61   When two dice are rolled simultaneously, it is expected that the result on each roll will be independent of the result on the other. To test this, roll a pair of dice 12 times. Identify one die as x and the other as y. Record the number observed on each of the two dice.

    a. Draw the scatter diagram of x versus y.

    b. If the two do behave independently, what value of r can be expected?

    c. Calculate r.

    d. Test for independence of these dice at $\alpha = 0.10$.

13.62   The coding scheme used for a variable can determine whether a negative or a positive correlation coefficient is obtained between two variables.

    a. Suppose that high school academic performance (HSAP) is coded as: 1 = excellent, 2 = above average, and 3 = average or below. The score made on a final exam in a college algebra course is also recorded. The resulting data are shown in the following table.

HSAP	1	2	3	1	3	2	1	3	3	2
Final Exam Score	79	80	65	89	70	75	90	65	70	80

Compute r for these data.

b. Now suppose that HSAP is coded as: 3 = excellent, 2 = above average, and 1 = average or below. The data would now appear as follows:

HSAP	3	2	1	3	1	2	3	1	1	2
Final Exam Score	79	80	65	89	70	75	90	65	70	80

Compute r for these data.

c. What connection do you see between the two resulting correlation coefficients?

 13.63 One would usually expect that the number of hours studied, x, in preparation for a particular examination would have direct (positive) correlation with the grade, y, attained on that exam. The hours studied and the grades obtained by ten students selected at random from a large class are shown in the following table.

					Student					
	1	**2**	**3**	**4**	**5**	**6**	**7**	**8**	**9**	**10**
Hours Studied (x)	10	6	15	11	7	19	17	3	13	17
Exam Grade (y)	51	36	67	63	44	89	80	26	50	85

a. Draw the scatter diagram and estimate r.
b. Calculate r. (How close was your estimate?)
c. Is there significant positive correlation shown by these data? Test at $\alpha = 0.01$.
d. Find the 95% confidence interval for the true population value of ρ.

 13.64 When buying most items, it is advantageous to buy in as large a quantity as possible. The unit price is usually less for larger quantities. The following data were obtained to test this theory.

Number of Units (x)	1	3	5	10	12	15	24
Cost per Unit (y)	55	52	48	36	32	30	25

a. Calculate r.
b. Does the number of units ordered have an effect on the unit cost? Test at $\alpha = 0.05$.
c. Find the 95% confidence interval estimate for the true value of ρ.

 13.65 The use of electrical stimulation (ES) to increase muscular strength is discussed in the *Journal of Orthopaedic and Sports Physical Therapy* (September 1990). Seventeen healthy volunteers were used in the experiment. Muscular strength, Y, was measured as a torque in foot-pounds, and electrical stimulation, X, was measured in mA (micro-amps). The equation for the line of best fit is given as $Y = 1.8X + 28.7$, and the Pearson correlation coefficient as 0.61.

a. Was the correlation coefficient significantly different from zero? Use $\alpha = 0.05$.
b. Predict the torque for a current equal to 50 mA.

13.66 An article in the *New Zealand Geographer* (Vol. 45, No. 1, 1989) discusses the assumptions and limitations of regression analysis. One of the limitations is that regression models may be used only for predicting values within the range of the combinations of the values for the independent variables. Consider, for example, an experiment in which yield of tomatoes (in pounds per plot), y, and the amount of fertilizer (in ounces), x, are recorded. Suppose the data were as follows:

x	1.1	1.5	1.9	2.1	2.5
y	20.1	22.4	22.0	25.0	30.0

   a. Determine the regression equation and predict the yield for $x = 4.0$.
   b. If the actual yield were 23.5, how might this be explained?

13.67 An article in *Geology* (September 1989) gives the following equation relating pressure, P, and total aluminum content, AL, for 12 Horn blende rims: $P = -3.46(+0.24) + 4.23(+ 0.13)AL$. The quantities shown in parentheses are standard errors for the y-intercept and slope estimates. Find a 95% confidence interval for the slope, β_1.

13.68 The Chapter Case Study (p. 647) contains the Metropolitan Life Insurance Company weight chart for women using three different "frame" categories and using height as the input variable, x.
   a. Construct a scatter diagram showing the intervals for small-frame women.
   b. Construct a scatter diagram showing the intervals for medium-frame women.
   c. Construct a scatter diagram showing the intervals for large-frame women.
   d. Do the variables "height" and "weight" seem to have a linear relationship? Why do the scatter diagrams seem to suggest such a strong relationship?
   e. One inch of height adds how many pounds to the ideal weight? How is the number of pounds per inch related to the line of best fit?

13.69 The following data resulted from an experiment performed for the purpose of regression analysis. The input variable, x, was set at five different levels, and observations were made at each level.

x	**0.5**	**1.0**	**2.0**	**3.0**	**4.0**
y	3.8	3.2	2.9	2.4	2.3
	3.5	3.4	2.6	2.5	2.2
	3.8	3.3	2.7	2.7	2.3
		3.6	3.2	2.3	

   a. Draw a scatter diagram of the data.
   b. Draw the regression line by eye.

c. Place a star, ★, at each level approximately where the mean of the observed y-values is located. Does your regression line look like the line of best fit for these five mean values?

d. Calculate the equation of the regression line.

e. Find the standard deviation of y about the regression line.

f. Construct a 95% confidence interval for the true value of β_1.

g. Construct a 95% confidence interval for the mean value of y at $x = 3.0$. At $x = 3.5$.

h. Construct a 95% prediction interval for an individual value of y at $x = 3.0$. At $x = 3.5$.

 13.70 The following set of 25 scores was randomly selected from a teacher's class list. Let x be the prefinal average and y the final examination score. (The final examination had a maximum of 75 points.)

Student	x	y	Student	x	y
1	75	64	14	73	62
2	86	65	15	78	66
3	68	57	16	71	62
4	83	59	17	86	71
5	57	63	18	71	55
6	66	61	19	96	72
7	55	48	20	96	75
8	84	67	21	59	49
9	61	59	22	81	71
10	68	56	23	58	58
11	64	52	24	90	67
12	76	63	25	92	75
13	71	66			

a. Draw a scatter diagram for these data.

b. Draw the regression line (by eye) and estimate its equation.

c. Estimate the value of the coefficient of linear correlation.

d. Calculate the equation of the line of best fit.

e. Draw the line of best fit on your graph. How does it compare with your estimate?

f. Calculate the linear correlation coefficient. How does it compare with your estimate?

g. Test the significance of r at $\alpha = 0.10$.

h. Find the 95% confidence interval estimate for the true value of ρ.

i. Find the standard deviation of the y-values about the regression line.

j. Calculate a 95% confidence interval estimate for the true value of the slope, β_1.

k. Test the significance of the slope at $\alpha = 0.05$.

l. Estimate the mean final-exam grade that all students with an 85 prefinal average will obtain (95% confidence interval).

m. Using the 95% prediction interval, predict the grade that John Henry will receive on his final, knowing that his prefinal average is 78.

13.71   Twenty-one mature flowers of a particular species were dissected, and the number of stamens and carpels present in each flower were counted. See the following table.

Stamens (x)	Carpels (y)	Stamens (x)	Carpels (y)
52	20	74	29
68	31	38	28
70	28	35	25
38	20	45	27
61	19	72	21
51	29	59	35
56	30	60	27
65	30	73	33
43	19	76	35
37	25	68	34
36	22		

a. Is there sufficient evidence to claim a linear relationship between these two variables at $\alpha = 0.05$?

b. What is the relationship between the number of stamens and carpels in this variety of flower?

c. Is the slope of the regression line significant at $\alpha = 0.05$?

d. Give the 95% confidence interval prediction for the number of carpels that one would expect to find in a mature flower of this variety if the number of stamens were 64.

13.72   It is believed that the amount of nitrogen fertilizer used per acre has a direct effect on the amount of wheat produced. The following data show the amount of nitrogen fertilizer used per test plot and the amount of wheat harvested per test plot.

Pounds of Fertilizer (x)	100 Pounds of Wheat (y)	Pounds of Fertilizer (x)	100 Pounds of Wheat (y)
30	5	70	19
30	9	70	23
30	14	70	31
40	6	80	24
40	14	80	32
40	18	80	35
50	12	90	27
50	14	90	32
50	23	90	38
60	18	100	34
60	24	100	35
60	28	100	39

a. Is there sufficient reason to conclude that the use of more fertilizer results in a higher yield? Use $\alpha = 0.05$.

b. Estimate, with a 98% confidence interval, the mean yield that could be expected if 50 pounds of fertilizer were used per plot.

c. Estimate, with a 98% confidence interval, the mean yield that could be expected if 75 pounds of fertilizer were used per plot.

13.73 The correlation coefficient, r, is related to the slope of best fit, b_1, by the equation

$$r = b_1 \sqrt{\frac{SS(x)}{SS(y)}}$$

Verify this equation using the following data.

x	1	2	3	4	6
y	4	6	7	9	12

13.74 The following equation is known to be true for any set of data: $\Sigma(y - \bar{y})^2 = \Sigma(y - \hat{y})^2 + \Sigma(\hat{y} - \bar{y})^2$. Verify this equation with the following data.

x	0	1	2
y	1	3	2

VOCABULARY LIST

Be able to define each term. Pay special attention to the key terms, which are printed in red. In addition, describe in your own words, and give an example of, each term. Your examples should not be ones given in class or in the textbook.

The bracketed numbers indicate the chapters in which the term previously appeared, but you should define the terms again to show increased understanding of their meaning. Page numbers indicate the first appearance of the term in Chapter 13.

bivariate data [3] (p. 648)
centroid (p. 649)
coefficient of linear correlation [3] (p. 652)
confidence belts (p. 656)
confidence interval [8, 9, 10] (p. 677)
covariance (p. 649)
curvilinear regression (p. 664)
experimental error (ϵ or e) (p. 664)
hypothesis tests [8, 9, 10, 11, 12] (p. 659)
intercept (b_0 or β_0) [3] (p. 664)
line of best fit [3] (p. 663)
linear correlation (p. 648)
linear regression [3] (p. 663)
multiple regression (p. 664)
Pearson's product moment, r (p. 652)

predicted value of μ_y (p. 677)
predicted value of y ($\hat{y}$) (p. 677)
prediction interval (p. 680)
regression line [3] (p. 664)
rho (ρ) (p. 656)
sampling distribution [7, 8, 10] (p. 671)
scatter diagram [3] (p. 648)
slope (b_1 or β_1) [3] (p. 664)
standard error [7, 8, 9] (p. 671)
sum of squares for error (SSE) [1, 2] (p. 666)
variance (s^2 or σ^2) [2, 8, 9, 10, 12] (p. 666)

QUIZ A

Answer "True" if the statement is always true. If the statement is not always true, replace the words shown in bold with words that make the statement always true.

13.1 The error **must be** normally distributed if inferences are to be made.

13.2 Both x and y **must be** normally distributed.

13.3 A high correlation between x and y **proves** that x causes y.

13.4 The values of the input variable **must be** randomly selected to achieve valid results.

13.5 The output variable must be **normally distributed** about the regression line for each value of x.

13.6 **Covariance** measures the strength of the linear relationship and is a standardized measure.

13.7 The **sum of squares for error** is the name given to the numerator portion of the formula for the calculation of the variance of y about the line of regression.

13.8 **Correlation** analysis attempts to find the equation of the line of best fit for two variables.

13.9 There are $n - 3$ degrees of freedom involved with the inferences about the regression line.

13.10 $\hat{y}$ serves as the **point estimate** for both $\mu_{y|x_0}$ and y_{x_0}.

QUIZ B

Answer all questions, showing formulas and work.

It is believed that the amount of nitrogen fertilizer used per acre has a direct effect on the amount of wheat produced. The data show the amount of nitrogen fertilizer used per test plot and the amount of wheat harvested per test plot. All test plots were of the same size.

Pounds of Fertilizer, x	100 Pounds of wheat, y
30	9
30	11
30	14
50	12
50	14
50	23
70	19
70	22
70	31
90	29
90	33
90	35

13.1 Draw a scatter diagram of the data (use graph paper and a straight edge). Be sure to label completely.

13.2   Complete an extensions table.

13.3   Calculate SS(x), SS(xy), SS(y).

13.4   Calculate the linear correlation coefficient, r.

13.5   Determine the 95% confidence interval estimate for the population linear correlation coefficient.

13.6   Calculate the equation for the line of best fit.

13.7   Draw the line of best fit on the scatter diagram (in red ink).

13.8   Calculate the standard deviation of the y-values about the line of best fit.

13.9   Does the value of b_1 show strength significant enough to conclude that the slope is greater than zero at the 0.05 level?

13.10   Determine the 0.95 confidence interval for the mean yield when 85 pounds of fertilizer is used per plot.

13.11   Draw a line on the scatter diagram representing the 95% confidence interval found in (j) (in blue ink).

QUIZ C

13.1   "There is a high correlation between how frequently skiers have their bindings tested and the incidence of lower-leg injuries, according to researchers at the Rochester Institute of Technology. To make sure your bindings release properly when you begin to fall, you should have them serviced by a ski mechanic every 15 to 30 ski days or at least at the start of each ski season." (University of California, Berkeley, "Wellness Letter," Feb. 1991) Explain what two variables are being discussed in this statement and interpret the "high correlation" mentioned.

13.2   Describe why the method used to define the correlation coefficient is referred to as "a product moment."

13.3   If you know the value of r is very close to zero, what value would you anticipate for b_1? Explain why.

13.4   Describe why the method used to find the line of best fit is referred to as "the method of least squares."

13.5   You wish to study the relationship between the amount of sugar contained in a child's breakfast and the child's hyperactivity in school during the four hours after breakfast. You ask 200 mothers of fifth-grade children to keep a careful record of what the child eats and drinks each morning. The parent's report is analyzed and the sugar consumption is determined. During the same time period, data on hyperactivity are collected at school. What statistic will measure the strength and kind of relationship that exists between the amount of sugar and the amount of hyperactivity? Explain why the statistic you selected is appropriate and what value you expect this statistic might have.

13.6   You are interested in studying the relationship between the length of time a person has been supported by welfare and self-esteem. You believe that the longer a person is supported, the lower the self-esteem. What data would you need to collect and what statistics would you calculate if you wish to predict a person's level of self-esteem after having been on welfare for a certain period of time? Explain in detail.

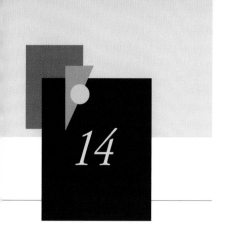

ELEMENTS OF NONPARAMETRIC STATISTICS

STUDY FINDS THAT MONEY CAN'T BUY JOB SATISFACTION

When it comes to getting workers to produce—do their level best—money is less than everything. Feeling appreciated—having a sense of being recognized—is more important.

That is the thesis of Kenneth A. Kovach, of George Mason University. . . .

Kovach reproduces a consequential survey of 30 years ago in which workers and supervisors gave their opinions of "what workers want from their jobs and what management thinks they want." See the accompanying table.

Managements today are more "behavioral" in their approach to managing. Nevertheless, Kovach contends that a wide gap still exists, a gap that is suggested by what workers rank as No. 1 in importance to them ("appreciation") and what supervisors consider No. 1 to workers ("good wages").

	Worker Ranking	Boss Ranking
Full appreciation of work done	1	8
Feeling of being in on things	2	10
Sympathetic help on personal problems	3	9
Job security	4	2
Good wages	5	1
Interesting work	6	5
Promotion and growth in the organization	7	3
Personal loyalty to employees	8	6
Good working conditions	9	4
Tactful disciplining	10	7

Source: Reprinted by permission of the *Philadelphia Inquirer*, 29 December 1976.

It appears that workers and bosses do not agree on what makes workers happy at work. Are these rankings significantly different? (See Exercise 14.64, p. 739.)

● CHAPTER OBJECTIVES

Nonparametric methods of statistics dominate the success story of statistics in recent years. Unlike their parametric counterparts, many of the best-known nonparametric tests, also known as *distribution-free* tests, are founded on a basis of elementary probability theory. The derivation of most of these tests is well within the grasp of the student who is competent in high school algebra and understands binomial probability. Thus, the nonmathematical statistics user is much more at ease with the nonparametric techniques.

This chapter is intended to give you a feeling for basic concepts involved in nonparametric techniques and to show you that nonparametric methods are extremely versatile and easy to use once a table of critical values is developed for a particular application. The selection of the nonparametric methods presented here includes only a few of the common tests and applications. You will learn about the sign test, the Mann–Whitney *U* test, the runs test, and Spearman's rank correlation test. These will be used to make inferences corresponding to both one- and two-sample situations.

●

14.1 | NONPARAMETRIC STATISTICS

Parametric method

Most of the statistical procedures we have studied in this book are known as **parametric methods**. For a statistical procedure to be parametric, either we assume that the parent population is at least approximately normally distributed or we rely on the central limit theorem to give us a normal approximation. This is particularly true of the statistical methods studied in Chapters 8, 9, and 10.

Nonparametric, or distribution-free, methods

The **nonparametric methods**, or **distribution-free methods**, as they are also known, do not depend on the distribution of the population being sampled. The nonparametric statistics are usually subject to much less confining restrictions than are their parametric counterparts. Some, for example, require only that the parent population be continuous.

The recent popularity of nonparametric statistics can be attributed to the following characteristics:

1. Nonparametric methods require few assumptions about the parent population.
2. Nonparametric methods are generally easier to apply than their parametric counterparts.
3. Nonparametric methods are relatively easy to understand.

4. Nonparametric methods can be used in situations where the normality assumptions cannot be made.
5. Nonparametric methods appear to be wasteful of information in that they sacrifice the value of the variable for only a sign or a rank number. However, nonparametric statistics are generally only slightly less efficient than their parametric counterparts.

14.2 | COMPARING STATISTICAL TESTS

Only four nonparametric tests are presented in this chapter. They represent a very small sampling of the many different nonparametric tests that exist. Many of the nonparametric tests can be used in place of certain parametric tests. The question is, then, Which statistical test do we use, the parametric or the nonparametric? Sometimes there is more than one nonparametric test to choose from.

The decision about which test to use must be based on the answer to the question "Which test will do the job best?" First, let's agree that we are dealing with two or more tests that are equally qualified to be used. That is, each test has a set of assumptions that must be satisfied before it can be applied. From this starting point we will attempt to define "best" to mean the test that is best able to control the risks of error and at the same time keep the size of the sample to a number that is reasonable to work with. (Sample size means cost, cost to you or your employer.)

Let's look first at the ability to *control the risk of error*. The risk associated with a type I error is controlled directly by the level of significance α. Recall that P (type I error) $= \alpha$ and P (type II error) $= \beta$. Therefore, it is β that we must control. Statisticians like to talk about power (as do others), and the **power of a statistical test** is defined to be $1 - \beta$. Thus, the power of a test, $1 - \beta$, is the probability that we reject the null hypothesis when we should have rejected it. If two tests with the same α are equal candidates for use, the one with the greater power is the one you would want to choose.

The other factor is the *sample size required* to do a job. Suppose that you set the levels of risk you can tolerate, α and β, and then are able to determine the sample size it would take to meet your specified challenge. The test that required the smaller sample size would then seem to have the edge. Statisticians usually use the term *efficiency* to talk about this concept. **Efficiency** is the ratio of the sample size of the best parametric test to the sample size of the best nonparametric test when compared under a fixed set of risk values. For example, the efficiency rating for the sign test is approximately 0.63. This means that a sample of size 63 with a parametric test will do the same job as a sample of size 100 will do with the sign test.

The power and the efficiency of a test cannot be used alone to determine the choice of test. Sometimes you will be forced to use a certain test because of the data you are given. When there is a decision to be made, the final decision rests in a trade-off of three factors: (1) the power of the test, (2) the efficiency of the test, and (3) the data (and the number of data) available. Table 14-1 shows how the nonparametric tests discussed in this chapter compare with the parametric tests covered in previous chapters.

Power of a statistical test

Efficiency

TABLE 14-1

Comparison of
Parametric and
Nonparametric Tests

Test Situation	Parametric Test	Nonparametric Test	Efficiency of Nonparametric Test
One mean	*t* test (p. 441)	Sign test (p. 702)	0.63
Two independent means	*t* test (p. 516)	*U* test (p. 711)	0.95
Two dependent means	*t* test (p. 503)	Sign test (p. 704)	0.63
Correlation	Pearson's (p. 658)	Spearman test (p. 726)	0.91
Randomness		Runs test (p. 720)	Not meaningful; there is no parametric test for comparison

14.3 | THE SIGN TEST

Sign test

The ordinary **sign test** is a versatile and exceptionally easy-to-apply nonparametric method that *uses only plus and minus signs*. (There are several specific techniques.) The sign test is useful in two situations: (1) a hypothesis test concerning the value of the median for one population and (2) a hypothesis test concerning the median difference (paired difference) for two dependent samples. Both tests are carried out using the same basic procedures and are nonparametric alternatives to the *t* tests used with one mean (Section 9.1) and the difference between two dependent means (Section 10.2).

Single-Sample Hypothesis Test Procedure

The sign test can be used when a random sample is drawn from a population with an unknown **median**, *M*, and the population is assumed to be continuous in the vicinity of *M*. The null hypothesis to be tested concerns the value of the population median *M*. The test may be either one- or two-tailed. This test procedure is presented in the following illustration.

▼| ILLUSTRATION   14 - 1

A random sample of 75 students was selected, and each student was asked to carefully measure the amount of time required to commute from his or her front door to the college parking lot. The data collected were used to test the hypothesis "the median time required for students to commute is 15 minutes" against the alternative that the median is unequal to 15 minutes. The 75 pieces of data were summarized as follows:

    Under 15:   18
    15:         12
    Over 15:   45

Use the sign test to test the null hypothesis against the alternative hypothesis.

SOLUTION The data are converted to (+) and (−) signs. A plus sign will be assigned to each piece of data larger than 15, a minus sign to each piece of data smaller than 15, and a zero to those data equal to 15. The sign test uses only the plus and minus signs; therefore, the zeros are discarded and the usable sample size becomes 63. That is, $n(+) = 45$, $n(−) = 18$, and $n = n(+) + n(−) = 45 + 18 = 63$.

STEP 1   Parameter: Median time to commute.

STEP 2 $H_0: M = 15$
         $H_a: M \neq 15$

STEP 3 $\alpha = 0.05$ for a two-tailed test. The **test statistic** that will be used is the **number of the less frequent sign**, the smaller of $n(+)$ and $n(−)$, which is $n(−)$ for our illustration. We will want to reject the null hypothesis whenever the number of the less frequent sign is extremely small. Table 12 in Appendix B gives the maximum allowable number of the less frequent sign, k, that will allow us to reject the null hypothesis. That is, if the number of the less frequent sign is less than or equal to the critical value in the table, we will reject H_0. If the observed value of the less frequent sign is larger than the table value, we will fail to reject H_0. In the table, n is the total number of signs, not including zeros.

For our illustration, $n = 63$ and the critical value from the table is 23. See the accompanying figure.

Reject H_0	Fail to reject H_0
0 23	24

Number of less frequent sign

STEP 4   The observed value of the test statistic is $x = n(−) = 18$.

STEP 5   Results: x falls in the critical region.

DECISION Reject H_0.

CONCLUSION The sample shows sufficient evidence at the 0.05 level to reject the claim that the median is 15 minutes.

▲

Two-Sample Hypothesis Test Procedure

The sign test may also be applied to a hypothesis test dealing with the *median difference* between paired data that result from *two dependent samples*. A familiar application is the use of before-and-after testing to determine the effectiveness of some activity. In a test of this nature, the signs of the differences are used to carry out the test. Again, zeros are disregarded. The following illustration shows this procedure.

▼ | I L L U S T R A T I O N   14 - 2

A new no-exercise, no-starve weight-reducing plan has been developed and advertised. To test the claim that "you will lose weight within two weeks or . . . ," a local statistician obtained the before-and-after weights of 18 people who had used this plan. Table 14-2 lists the people, their weights, a minus ($-$) for those who lost weight during the two weeks, a 0 for those who remained the same, and a plus ($+$) for those who actually gained weight.

TABLE 14-2

Sample Results for
Illustration 14-2

Person	Weight Before	After	Sign of Difference, Before to After
Mrs. Smith	146	142	$-$
Mr. Brown	175	178	$+$
Mrs. White	150	147	$-$
Mr. Collins	190	187	$-$
Mr. Gray	220	212	$-$
Ms. Collins	157	160	$+$
Mrs. Allen	136	135	$-$
Mrs. Noss	146	138	$-$
Ms. Wagner	128	132	$+$
Mr. Carroll	187	187	0
Mrs. Black	172	171	$-$
Mrs. McDonald	138	135	$-$
Ms. Henry	150	151	$+$
Ms. Greene	124	126	$+$
Mr. Tyler	210	208	$-$
Mrs. Williams	148	148	0
Mrs. Moore	141	138	$-$
Mrs. Sweeney	164	159	$-$

The claim being tested is that people are able to lose weight. The null hypothesis that will be tested is that "there is no weight loss (or the median weight loss is zero)," meaning that only a rejection of the null hypothesis will allow us to conclude in favor of the advertised claim. Actually we will be testing to see whether there are significantly more minus signs than plus signs. If the weight-reducing plan is of absolutely no value, we would expect to find an equal number of plus and minus signs. If it works, there should be significantly more minus signs than plus signs. Thus, the test performed here will be a one-tailed test. (We will want to reject the null hypothesis in favor of the advertised claim if there are "many" minus signs.)

SOLUTION

STEP 1   Parameter: Median weight loss.

STEP 2 H_0: $M = 0$ (no weight loss)

H_a: $M < 0$ (weight loss)

STEP 3   Use $\alpha = 0.05$. $n = 16$ ($n(+) = 5$, $n(-) = 11$)
The critical value from Table 12 shows $k = 4$ as the maximum allowable number. (You must use the $\alpha = 0.10$ column, because the table is set up for two-tailed tests.)

Reject H_0					Fail to reject H_0
0	1	2	3	4	5

Number of less frequent sign

STEP 4 $x = n(+) = 5$

STEP 5   Result: x is in the noncritical region.

DECISION Fail to reject H_0

CONCLUSION The evidence observed is not sufficient to allow us to reject the no-weight-loss null hypothesis at the 0.05 level of significance.

▲

Normal approximation

The sign test may be carried out by means of a **normal approximation** using the standard normal variable z. The normal approximation will be used if Table 12 does not show the particular levels of significance desired or if n is large. z will be calculated by using the formula

$$z^\star = \frac{x' - (n/2)}{\left(\frac{1}{2}\right)\sqrt{n}}$$ (14-1)

(See Note 3 with regard to x'.)

NOTES

1. x may be the number of the less frequent sign or the most frequent sign. You will have to determine this in such a way that the direction is consistent with the interpretation of the situation.
2. x is really a binomial random variable, where $p = \frac{1}{2}$. The sign test statistic satisfies the properties of a binomial experiment (see p. 268). Each sign is the result of an independent trial. There are n trials, and each trial has two

possible outcomes ($+$ or $-$). Since the median is used, the probabilities for each outcome are both $\frac{1}{2}$. Therefore, the mean, μ_x, is equal to

$$\mu_x = n/2 \left[\mu = np = n \cdot \frac{1}{2} = n/2\right]$$

and the standard deviation, σ_x, is equal to

$$\sigma_x = \frac{1}{2}\sqrt{n} \left[\sigma = \sqrt{npq} = \sqrt{n \cdot \frac{1}{2} \cdot \frac{1}{2}} = \frac{1}{2}\sqrt{n}\right]$$

3. x is a discrete variable. But recall that the normal distribution must be used only with continuous variables. However, although the binomial random variable is discrete, it does become approximately normally distributed for large n. Nevertheless, when using the normal distribution for testing, we should make an adjustment in the variable so that the approximation is more accurate. (See Section 6.5, p. 314, on the normal approximation.) This adjustment is illustrated in Figure 14-1 and is called a **continuity correction**. For this discrete variable the area that represents the probability is a rectangular bar. Its width is 1 unit wide, from $\frac{1}{2}$ unit below to $\frac{1}{2}$ unit above the value of interest. Therefore, when z is to be used, we will need to make a $\frac{1}{2}$-unit adjustment before calculating the observed value of z. x' will be the adjusted value for x. If x is larger than $n/2$, $x' = x - \frac{1}{2}$. If x is smaller than $n/2$, $x' = x + \frac{1}{2}$. The test is then completed by the usual procedure.

Continuity correction

FIGURE 14-1

Continuity
Correction

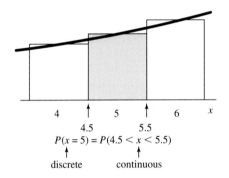

$$P(x = 5) = P(4.5 < x < 5.5)$$

discrete   continuous

▼| ILLUSTRATION   14 - 3

Use the sign test to test the hypothesis that the median number of hours, M, worked by students of a certain college is at least 15 hours per week. A survey of 120 students was taken; a plus sign was recorded if the number of hours the student worked last week was equal to or greater than 15, and a minus sign was recorded if the number of hours was less than 15. Totals showed 80 minus signs and 40 plus signs.

SOLUTION

STEP 1   Parameter: Median number of hours.

STEP 2 H_0: $M = 15$ ($\geq$)   (at least as many plus signs as minus signs)
H_a: $M < 15$ (fewer plus signs than minus signs)

STEP 3 $\alpha = 0.05$

x is the number of plus signs. The critical value is shown in the accompanying figure.

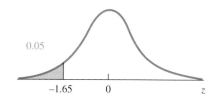

STEP 4

$$z^\star = \frac{x' - (n/2)}{\left(\frac{1}{2}\right)\sqrt{n}} = \frac{40.5 - 60}{\left(\frac{1}{2}\right)\sqrt{120}} = \frac{-19.5}{\left(\frac{1}{2}\right)\sqrt{10.95}}$$

$$= \frac{-19.5}{5.475} = -3.562$$

$$z^\star = -3.56$$

STEP 5 Result: $z^\star$ is in the critical region.

DECISION Reject H_0.

CONCLUSION At the 0.05 level, there are significantly more minus signs than plus signs, thereby implying that the median is less than the claimed 15 hours.

Confidence Interval Procedure

The sign test techniques can be applied to obtain a confidence interval estimate for the unknown population median, M. To accomplish this we will need to arrange the data in ascending order (smallest to largest). The data are identified as x_1 (smallest), $x_2, x_3, \ldots, x_n$ (largest). The critical value, k, for the maximum allowable number of signs, obtained from Table 12 in Appendix B, tells us the number of positions to be dropped from each end of the ordered data. The remaining extreme values become the bounds of the $1 - \alpha$ confidence interval. That is, the lower boundary for the confidence interval is x_{k+1}, the $(k + 1)$th piece of data; the upper boundary is x_{n-k}, the $(n - k)$th piece of data. The following illustration will clarify this procedure.

▼ ILLUSTRATION 14 - 4

Suppose that we have 12 pieces of data in ascending order ($x_1, x_2, x_3, \ldots, x_{12}$) and we wish to form a 95% confidence interval for the population median. Table 12 shows a critical value of 2 ($k = 2$) for $n = 12$ and $\alpha = 0.05$ for a hypothesis test. This means that the critical region for a two-tailed hypothesis test would contain the last two values on each end (x_1 and x_2 on the left; x_{11} and x_{12} on the right). The noncritical region is then from x_3 to x_{10}, inclusive. The 95% confidence interval is then x_3 to x_{10}, expressed as

$$x_3 \quad \text{to} \quad x_{10} \qquad \text{95\% confidence interval for } M$$

In general, the two pieces of data that bound the confidence interval will occupy positions $k + 1$ and $n - k$, where k is the critical value read from Table 12. Thus

$$x_{k+1} \quad \text{to} \quad x_{n-k} \qquad 1 - \alpha \text{ confidence interval for } M$$

If the normal approximation is to be used (including the continuity correction), the position numbers become

$$\left(\tfrac{1}{2}\right)n \pm \left(\tfrac{1}{2} + \tfrac{1}{2} \cdot z(\alpha/2) \cdot \sqrt{n}\right) \tag{14-2}$$

The interval is

$$x_L \quad \text{to} \quad x_U \qquad 1 - \alpha \text{ confidence interval for } M$$

where

$$L = \frac{n}{2} - \frac{1}{2} - \frac{z(\alpha/2)}{2} \cdot \sqrt{n}$$

$$U = \frac{n}{2} + \frac{1}{2} + \frac{z(\alpha/2)}{2} \cdot \sqrt{n}$$

(L should be rounded down and U should be rounded up to be sure that the level of confidence is at least $1 - \alpha$.)

▼ | ILLUSTRATION   14 - 5

Estimate the population median with a 95% confidence interval for a given set of 60 pieces of data: $x_1, x_2, x_3, \ldots, x_{59}, x_{60}$.

SOLUTION When we use formula (14-2), the position numbers L and U are

$$\left(\tfrac{1}{2}\right)(60) \pm \left[\tfrac{1}{2} + \tfrac{1}{2}(1.96)\sqrt{60}\right]$$
$$30 \pm [0.50 + 7.59]$$
$$30 \pm 8.09$$

Thus,

$$L = 30 - 8.09 = 21.91 \quad \text{(21st piece of data)}$$
$$U = 30 + 8.09 = 38.09 \quad \text{(39th piece of data)}$$

Therefore

▲ | $$x_{21} \quad \text{to} \quad x_{39} \qquad 95\% \text{ confidence interval for } M$$

EXERCISES

14.1 State the null hypothesis, H_0, and the alternative hypothesis, H_a, that would be used to test the following statements.

  a. The median value is at least 32.

  b. People prefer the taste of the bread made with the new recipe.

  c. There is no change in weight from weigh-in until after two weeks of the diet.

14.2 Determine the test criteria that would be used to test the null hypothesis for the following multinomial experiments.

a. H_0: $P(+) = 0.5$ vs. H_a: $P(+) \neq 0.5$, with $n = 18$, $\alpha = 0.05$
b. H_0: $P(+) = 0.5$ vs. H_a: $P(+) > 0.5$, with $n = 78$, $\alpha = 0.05$
c. H_0: $P(+) = 0.5$ vs. H_a: $P(+) < 0.5$, with $n = 38$, $\alpha = 0.05$
d. H_0: $P(+) = 0.5$ vs. H_a: $P(+) \neq 0.5$, with $n = 148$, $\alpha = 0.05$

14.3 Use the sign test to test the hypothesis that the median daily high temperature in the city of Rochester, New York, during the month of December is 48 degrees. The following daily highs were recorded on 20 randomly selected days during December.

47	46	40	40	46	35	34	59	54	33
65	39	48	47	46	46	42	36	45	38

a. State the null hypothesis for the test.
b. State specifically what it is that you are actually testing when using the sign test.
c. Complete the test for $\alpha = 0.05$, and carefully state your findings.

14.4 *USA Today* reported on May 8, 1991, "Teacher's pay increases 5.4%." The following sample of average teacher salary increases was taken from the annual report of the National Education Association.

1.6	11.4	4.2	8.7	4.5	6.4	5.9	3.7	4.5
4.6	4.4	4.1	7.6	5.5	10.6	2.8	6.8	3.0
3.5	7.9	8.0	4.3	2.4	5.0			

Do the sample data suggest that the claim "median pay increase is 5.4%" be rejected at the 0.05 level of significance?

14.5 In order to test the null hypothesis that there is no difference in the ages of husbands and wives, the following data were collected.

Husbands	28	45	40	37	25	42	21	22	54	47	35	62	29	44	45	38	59
Wives	26	36	40	42	28	40	20	24	50	54	37	60	25	40	34	42	49

Does the sign test show that there is a significant difference in the ages of husbands and wives at $\alpha = 0.05$?

14.6 Sixteen students were given an elementary statistics test on a hot day. Eight randomly selected students took the test in a room that was not air-conditioned. Then, after a short break, they completed a similar test in an air-conditioned room. The procedure was reversed for the other eight students.

Student	1	2	3	4	5	6	7	8	9	10	11	12	13	14	15	16
Not Air-Conditioned	52	90	63	74	87	77	92	72	77	94	67	86	78	84	57	55
Air-Conditioned	49	94	60	78	93	77	93	74	78	93	78	89	92	83	49	68

Does the sample provide sufficient reason to conclude that the use of air conditioning on a hot day has an effect on test grades? Use $\alpha = 0.05$.

14.7   An article titled "Venocclusive Disease of the Liver: Development of a Model for Predicting Fatal Outcome After Marrow Transplantation" (*Journal of Clinical Oncology*, September 1993) gives the median age of 355 patients who underwent marrow transplantation at the Fred Hutchinson Cancer Research Center as 30 years. A sample of 100 marrow transplantation patients were recently selected for a study, and it was found that 40 of the patients were over 30 and 60 were under 30 years of age. Test the null hypothesis that the median age of the population from which the 100 patients were selected equals 30 years versus the alternative that the median does not equal 30 years. Use $\alpha = 0.05$.

14.8   An article titled "Naturally Occurring Anticoagulants and Bone Marrow Transplantation: Plasma Protein C Predicts the Development of Venocclusive Disease of the Liver" (*Blood*, June 1993) compared baseline values for antithrobin III with antithrobin III values 7 days after a bone marrow transplant for 45 patients. The differences were found to be nonsignificant. Suppose 17 of the differences were positive and 28 were negative. The null hypothesis is that the median difference is zero, and the alternative hypothesis is that the median difference is not zero. Use the 0.05 level of significance. Give the critical value for the test, the observed value of the statistic, and your conclusion.

14.9   A blind taste test was used to determine people's preference for the taste of the old cola and new cola. The results were

    645 preferred the new
    583 preferred the old
    272 had no preference

Is the preference for the taste of the new cola significantly greater than one-half? Use $\alpha = 0.01$.

14.10 A taste test was conducted with a regular beef pizza. Each of 133 individuals was given two pieces of pizza, one with a whole-wheat crust and the other with a white crust. Each person was then asked whether she or he preferred whole-wheat or white crust. The results were

    65 preferred whole-wheat to white crust
    53 preferred white to whole-wheat crust
    15 had no preference

Is there sufficient evidence to verify the hypothesis that whole-wheat crust is preferred to white crust at the $\alpha = 0.05$ level of significance?

14.11 Determine the 95% confidence interval for the medium daily high temperature in Rochester during December based on the sample given in Exercise 14.3.

14.12 Find the 75% confidence interval for the median swim time for a swimmer whose recorded times are

24.7	24.7	24.6	25.5	25.7	25.8	26.5	24.5	25.3
26.2	25.5	26.3	24.2	25.3	24.3	24.2	24.2	

14.13 Use the data in Exercise 14.6 to find a 95% confidence interval for the median difference in scores attained in the air-conditioned room and in the not-air-conditioned room.

14.14 A sample of the daily rental-car rates for a compact car was collected in order to estimate the average daily cost of renting a compact car.

33.93	35.00	36.99	32.99	36.93	29.00	34.95	23.99
43.93	44.95	28.95	22.99	37.93	37.00	35.99	36.99
30.93	28.95	29.99	25.99	39.93	40.50	28.90	23.80
26.93	23.70	26.99	21.94	47.93	40.00	29.94	28.99
23.93	22.70	26.99	25.48	31.93	31.90	31.92	29.99

Find the 99% confidence interval for the median daily rental cost.

14.15 According to an article in a *Newsweek* special issue (Fall/Winter 1990), 51.1% of 17-year-olds answered the following question correctly:

$$\text{If}\quad 7X + 4 = 5X + 8, \quad \text{then } X =$$

☐ 1   ☐ 2   ☐ 4   ☐ 6

Suppose we wished to test the null hypothesis that one-half of all 17-year-olds could solve the problem above against the alternative hypothesis, "the proportion who can solve differs from one-half." Furthermore, suppose we asked 75 randomly selected 17-year-olds to solve the problem. Let + represent a correct solution and − represent an incorrect solution. If we obtain 25 + signs and 50 − signs, do we have sufficient evidence to show the proportion who can solve is different than one-half? Use $\alpha = 0.05$.

14.16 According to an article in *USA Today* (6-7-91), "only 46% of high school seniors can solve problems involving fractions, decimals and percentages." Suppose we wish to test the null hypothesis "one-half of all seniors can solve problems involving fractions, decimals and percentages" against an alternative that the proportion who can solve differs from one-half. Let + represent passed and − represent failed the test on fractions, decimals, and percentages. If a random sample of 1500 students is tested, what value of x, the number of the least frequent sign, will be the critical value at the 0.05 level of significance?

14.4 | THE MANN–WHITNEY U TEST

Mann–Whitney U test

The **Mann–Whitney U test** is a nonparametric alternative for the t test for the **difference between two independent means**. It can be applied when we have two independent random samples (independent within each sample as well as between samples) in which the random variable is continuous. This test is often applied in situations in which the two samples are drawn from the same population but different "treatments" are used on each set. We will demonstrate the procedure in the following illustration.

▼ | I L L U S T R A T I O N   14 - 6

In a large lecture class, when the instructor gives a one-hour exam, she gives two "equivalent" examinations. Students in even-numbered seats take exam A, and those in the odd-numbered seats take exam B. It is reasonable to ask, "Are these two different exam forms equivalent?" Assuming that the odd- or even-numbered seats had no effect, we would want to test the hypothesis "the test forms yielded scores that had identical distributions."

To test this hypothesis, the following two random samples were taken.

A	52	78	56	90	65	86	64	90	49	78
B	72	62	91	88	90	74	98	80	81	71

The size of the individual samples will be called n_a and n_b; actually, it makes no difference which way these are assigned. In our illustration they both have the value 10.

The first thing that must be done with the entire sample (all $n_a + n_b$ pieces of data) is to order it into one sample, smallest to largest:

49	52	56	62	64	65	71	72	74	78
78	80	81	86	88	90	90	90	91	98

Rank number

Each piece of data is then assigned a **rank number**. The smallest (49) is assigned rank 1, the next smallest (52) is assigned rank 2, and so on, up to the largest, which is assigned rank $n_a + n_b$ (20). Ties are handled by assigning to each of the tied observations the mean rank of those rank positions that they occupy. For example, in our illustration there are two 78s; they are the 10th and 11th pieces of data. The mean rank for each is then $(10 + 11)/2 = 10.5$. In the case of the three 90s, the 16th, 17th, and 18th pieces of data, each is assigned 17, since $(16 + 17 + 18)/3 = 17$. The rankings are shown in Table 14-3.

TABLE 14-3

Ranked Data for
Illustration 14-6

Ranked Data	Rank	Source	Ranked Data	Rank	Source
49	1	A	78	10.5	A
52	2	A	80	12	B
56	3	A	81	13	B
62	4	B	86	14	A
64	5	A	88	15	B
65	6	A	90	17	A
71	7	B	90	17	A
72	8	B	90	17	B
74	9	B	91	19	B
78	10.5	A	98	20	B

Figure 14-2 shows the relationship between the two sets of data, first by using the data values and second by comparing the rank numbers for each piece of data.

FIGURE 14-2

Comparing the Data
of Two Samples

Do you see a differ-
ent relationship be-
tween the two sets
of data? Explain.

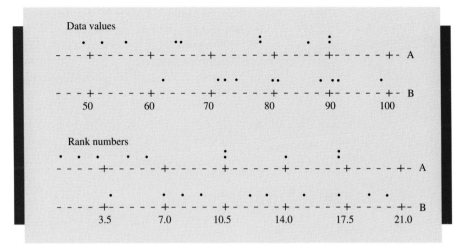

Test statistic

The calculation of the test statistic U is a two-step procedure. We first determine the sum of the ranks for each of the two samples. Then, using the two sums of ranks, we calculate two **U scores** for each sample. The **smaller U score is the test statistic.**
The sum of ranks for R_a for sample A is computed as

$$R_a = 1 + 2 + 3 + 5 + 6 + 10.5 + 10.5 + 14 + 17 + 17 = 86$$

The sum of ranks R_b for sample B is

$$R_b = 4 + 7 + 8 + 9 + 12 + 13 + 15 + 17 + 19 + 20 = 124$$

The U score for each sample is obtained by using the following pair of formulas.

$$U_a = n_a \cdot n_b + \frac{(n_b)(n_b + 1)}{2} - R_b \tag{14-3}$$

$$U_b = n_a \cdot n_b + \frac{(n_a)(n_a + 1)}{2} - R_a \tag{14-4}$$

$U^\star$, the test statistic, will be the smaller of U_a and U_b.
For our illustration, we obtain

$$U_a = (10)(10) + \frac{(10)(10 + 1)}{2} - 124 = 31$$

$$U_b = (10)(10) + \frac{(10)(10 + 1)}{2} - 86 = 69$$

Therefore,

$$U^\star = 31$$

Before we carry out the test for this illustration, let's try to understand some of the underlying possibilities. Recall that the null hypothesis is that the distributions are the same and that we will most likely want to conclude from this that the means are approximately equal. Suppose for a moment that they are indeed quite different—say, all of one sample comes before the smallest piece of data in the second sample when they are ranked together. This would certainly mean that we want to reject the null hypothesis. What kind of a value can we expect for U in this case? Suppose, in Illustration 14-6, that the ten A values had ranks 1 through 10 and the ten B values had ranks 11 through 20. Then we would obtain

$$R_a = 55 \qquad R_b = 155$$

$$U_a = (10)(10) + \frac{(10)(10 + 1)}{2} - 155 = 0$$

$$U_b = (10)(10) + \frac{(10)(10 + 1)}{2} - 55 = 100$$

Therefore,

$$U^\star = 0$$

Suppose, on the other hand, that both samples were perfectly matched, that is, a score in each set identical to one in the other. Now what would happen?

54	54	62	62	71	71	72	72	· · ·
A	B	A	B	A	B	A	B	· · ·
1.5	1.5	3.5	3.5	5.5	5.5	7.5	7.5	· · ·

$$R_a = R_b = 105$$

$$U_a = U_b = (10)(10) + \frac{(10)(10 + 1)}{2} - 105 = 50$$

Therefore,

$$U^\star = 50$$

If this were the case, we certainly would want to reach the decision "fail to reject the null hypothesis."

Note that the sum of the two U's $(U_a + U_b)$ will always be equal to the product of the two sample sizes $(n_a \cdot n_b)$. For this reason we need to concern ourselves with only one of them, the smaller one.

Now let's return to the solution of Illustration 14-6. In order to complete our hypothesis test, we need to be able to determine a critical value for U. Table 13 in Appendix B gives us the critical value for some of the more common testing situations as long as both samples are reasonably small. Table 13 shows only the critical region in the left-hand tail, and the null hypothesis will be rejected if the observed value for U is less than or equal to the value read from the table. For our example, $n_a = n_b = 10$; at $\alpha = 0.05$ in a two-tailed test, the critical value is 23. We observed a value of $U^\star = 31$, and therefore we make the decision "fail to reject the null hypothesis." This means that we do not have sufficient evidence to reject the "equivalent" hypothesis.

▲ |

If the *samples are larger than size 20*, we may make the test decision with the aid of the standard normal variable, z. This is possible since the distribution of U is approximately normal with a mean

$$\mu_U = \frac{n_a \cdot n_b}{2} \tag{14-5}$$

and a standard deviation

$$\sigma_U = \sqrt{\frac{n_a n_b (n_a + n_b + 1)}{12}} \tag{14-6}$$

The null hypothesis is then tested by using the test statistic $z^\star$,

$$z^\star = \frac{U - \mu_U}{\sigma_U} \tag{14-7}$$

in the typical fashion. The test statistic z may be used whenever n_a and n_b are both greater than 10.

▼| ILLUSTRATION   14 - 7

A dog-obedience trainer is training 27 dogs to obey a certain command. The trainer is using two different training techniques, the reward-and-encouragement method (I) and the no-reward method (II). The following table shows the number of obedience sessions that were necessary before the dogs would obey the command. Does the trainer have sufficient evidence to claim that the reward method will, on the average, require less training time ($\alpha = 0.05$)?

I	29	27	32	25	27	28	23	31	37	28	22	24	28	31	34
II	40	44	33	26	31	29	34	31	38	33	42	35			

SOLUTION

STEP 1   Parameters: Average training time for each technique.

STEP 2 H_0: The average amount of training time required is the same for both methods.

H_a: The reward method requires less time on the average.

STEP 3 $\alpha = 0.05$.
Test criteria: The critical value is shown in the accompanying figure.

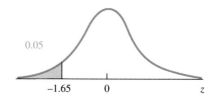

Case Study   Health Beliefs and Practices of Runners Versus Nonrunners

Valerie Walsh used the Mann–Whitney U *statistic to conclude that runners place a greater value on personal health than do nonrunners.*

Returns of mailed questionnaires from 77 runner and 63 nonrunner respondents showed that runners placed a statistically higher value on health and performed greater numbers of health-related behaviors. Major differences were found in nutrition, exercise, and medical awareness and self-care. No major differences were found in addictive substance use, stress management, or safety practices. A number of concerns regarding runners' health practices were identified, including running while ill or in pain, incidence of injuries, negative feelings when unable to run, neglect of a conscious cool-down period, low weight levels, and a tendency to increase workouts following perceived dietary indiscretions. . . .

The purpose of this investigation was to explore the differences between runners and nonrunners in terms of specific health beliefs and behaviors. It was hypothesized that there is a difference between runners and nonrunners in the relative value placed on personal health. . . .

EXERCISE 14.18

Calculate the value of $z^\star$ and the *p*-value that corresponds to 1876.5 (Case Study 14-1) for a two-tailed test using the Mann–Whitney *U*.

RESULTS

The first hypothesis, that there is a difference between runners and nonrunners in the relative value placed on personal health, was tested using the Mann–Whitney *U*, with alpha set at .05, and was accepted, $p < .019$. The value of *U* was found to be 1876.5. Greater value was placed on personal health by the runners than by the nonrunners. . . .

DISCUSSION

Runners placed a higher value on health and performed more health-related behaviors than did nonrunners. Although these results might have been anticipated in light of findings from other studies (Dawber, 1980; Paffenbarger et al., 1977), the information derived from them regarding health behaviors of runners was limited and inferential at best. The findings from this investigation, however, are congruent with those of Blair et al. (1981), in that runners exert tighter control over the types of nutrients they consume than do nonrunners.

Source: Valerie R. Walsh, *Nursing Research*, November/December 1985, Vol. 34, no. 6.

STEP 4   The two sets of data are ranked jointly and ranks are assigned, as shown in Table 14-4, page 717. Then the sums are

$$R_{\text{I}} = 1 + 2 + 3 + 4 + 6.5 + \cdots + 20.5 + 23 = 151.0$$
$$R_{\text{II}} = 5 + 11.5 + \cdots + 26 + 27 = 227.0$$

TABLE 14-4

Rankings for
Training Methods

Number of Sessions	Group	Rank		Number of Sessions	Group	Rank	
22	I	1		31	II	15	14.5
23	I	2		31	II	16	14.5
24	I	3		32	I	17	
25	I	4		33	II	18	18.5
26	II	5		33	II	19	18.5
27	I	6	6.5	34	I	20	20.5
27	I	7	6.5	34	II	21	20.5
28	I	8	9	35	II	22	
28	I	9	9	37	I	23	
28	I	10	9	38	II	24	
29	I	11	11.5	40	II	25	
29	II	12	11.5	42	II	26	
31	I	13	14.5	44	II	27	
31	I	14	14.5				

Now by using formulas (14-3) and (14-4), we can obtain the value of the U scores:

$$U_{\mathrm{I}} = (15)(12) + \frac{(12)(12 + 1)}{2} - 227 = 180 + 78 - 227 = 31$$

$$U_{\mathrm{II}} = (15)(12) + \frac{(15)(15 + 1)}{2} - 151 = 180 + 120 - 151 = 149$$

Therefore,

$$U^{\star} = 31$$

Now we use formulas (14-5), (14-6), and (14-7) to determine the z statistic.

$$\mu_U = \frac{12 \cdot 15}{2} = 90$$

$$\sigma_U = \sqrt{\frac{(12)(15)(12 + 15 + 1)}{12}} = \sqrt{\frac{(180)(28)}{12}} = \sqrt{420} = 20.49$$

$$z^{\star} = \frac{31 - 90}{20.49} = \frac{-59}{20.49} = -2.879$$

$$z^{\star} = -2.88$$

STEP 5   Result: $z^{\star}$ is in the critical region.

DECISION Reject H_0.

CONCLUSION At the 0.05 level of significance, the data show sufficient evidence to conclude that the reward method does, on the average, require less training time.

MINITAB (Release 10) commands to complete the Mann-Whitney test for data listed in C1 and C2. The sum of ranks for the first listed sample (W) and the p-value are reported.

Session commands

Enter: **MANN**-Whitney C1 C2

Menu commands

Choose:
        Stat > Nonpara > Mann-Whitney
      Enter: First Sample: C1
             Second Sample: C2

EXERCISES

14.19 State the null hypothesis, H_0, and the alternative hypothesis, H_a, that would be used to test the following statements.
    a. There is a difference in the value of the variable between the two groups of subjects.
    b. The average value is not the same for both groups.
    c. The blood pressure for group A is higher than for group B.

14.20 Determine the test criteria that would be used to test the following hypotheses for experiments involving two independent samples:
    a. H_0: Average(A) = Average(B)
       H_a: Average (A) > Average(B)
       with $n_A = 18$, $n_B = 15$, and $\alpha = 0.05$.
    b. H_0: The average score is the same for both groups.
       H_a: Group I average scores are less than those for group II,
       with $n_I = 78$, $n_{II} = 45$, and $\alpha = 0.05$.

14.21 An article in the *International Journal of Sports Medicine* (July 1994) discusses the use of the Mann–Whitney U test to compare the total cholesterol (mg/dL) of 35 adipose boys with that of 27 adipose girls. No significant difference was found between the two groups with respect to total cholesterol. A similar study involving 6 adipose boys and 8 adipose girls gave the following total-cholesterol values.

Adipose Boys	175	185	160	200	170	150		
Adipose Girls	160	190	175	190	185	150	140	195

Use the Mann–Whitney U test to test the research hypothesis that the total-cholesterol values differ for the two groups, using the 0.05 level of significance.

14.22 The July 4, 1994, issue of *Newsweek* gives quotes by several tobacco executives. Cigarette makers point out that some brands have more nicotine and tar than others do. Consider a study designed to compare the nicotine content of two different brands of cigarettes. The nicotine content was determined for 25 cigarettes of brand A and 25 cigarettes of brand B. The sum of ranks for brand A equals 688, and the sum of ranks for brand B equals 587. Use the Mann–Whitney U statistic to test the null hypothesis that the average nicotine content is the same for the two brands versus the alternative that the average nicotine content differs. Use $\alpha = 0.01$.

14.23   Pulse rates were recorded for 16 men and 13 women. The results are shown in the following table.

Males	61	73	58	64	70	64	72	60
	65	80	55	72	56	56	74	65

Females	83	58	70	56	76	64	80
	68	78	108	76	70	97	

These data were used to test the hypothesis that the distribution of pulse rates differs for men and women. The following MINITAB output printed out the sum of ranks for males (W = 192.0) and the *p*-value of 0.0373. Verify these two values.

```
With the data in C1 and C2;
MTB > MANN-WHITNEY ON DATA IN C1, C2

Mann-Whitney Confidence Interval and Test

C1     N = 16    MEDIAN =    64.50
C2     N = 13    MEDIAN =    76.00

W = 192.0
TEST OF ETA1 = ETA2 VS. ETA1 N.E. ETA2 IS SIGNIFICANT AT 0.0373
```

14.24   The following set of data represents the ages of drivers involved in automobile accidents. Do these data present sufficient evidence to conclude that there is a difference in the average age of men and women drivers involved in accidents? Use a two-tailed test at $\alpha = 0.05$.

Men	70	60	77	39	36	28	19	40
	23	23	63	31	36	55	24	76

Women	62	46	43	28	21	22	27	42
	21	46	33	29	44	29	56	70

a. State the null hypothesis that is being tested.
b. Complete the test using a computer.

14.25   A study titled "Textbook Pictures and First-Grade Children's Perception of Mathematical Relationships" by Patricia Campbell (*Journal for Research in Mathematics Education*, November 1978) investigated the influence of artistic style and the number of pictures on first-grade children's perception of mathematics textbook pictures. Analysis with the Mann–Whitney *U* test indicated that students who initially viewed and described sequences of pictures had

(continued)

significantly higher story-response scores than those students who viewed only single pictures. Consider the following data from two such groups. Group 1 is the group who viewed sequences of pictures, and group 2 is the group who viewed only single pictures.

Group 1	30	35	40	42	40	45	36
Group 2	25	32	27	39	30		

Using the Mann–Whitney U test, determine if the group 1 scores are significantly higher than the group 2 scores. Use $\alpha = 0.05$.

 14.26   To determine whether there is a difference in the breaking strength of lightweight monofilament fishing line produced by two companies, a researcher obtained the following test data (pounds of force required to break the line). Use the Mann–Whitney U test and $\alpha = 0.05$ to determine whether there is a difference.

A	12.4	11.9	11.8	13.5	11.6	12.0	12.9	11.3	13.8	12.3		
B	12.7	10.7	13.2	11.8	11.8	12.1	12.5	11.9	12.8	11.2	11.5	11.3

14.5 | THE RUNS TEST

The **runs test** is most frequently used to test the **randomness of data** (or lack of randomness). A **run** is a sequence of data that possesses a common property. One run ends and another starts when a piece of data does not display the property in question. The test statistic in this test is V, the number of runs observed.

▼| ILLUSTRATION   14 - 8

To illustrate the idea of runs, let's draw a sample of ten single-digit numbers from the telephone book, using the next-to-last digit from each of the selected telephone numbers.

Sample: 2, 3, 1, 1, 4, 2, 6, 6, 6, 7

Let's consider the property of "odd" (o) or "even" (e). The sample, as it was drawn, becomes $e, o, o, o, e, e, e, e, e, o$, which displays four runs.

$$e \quad o \quad o \quad o \quad e \quad e \quad e \quad e \quad e \quad o$$

▲| Thus, $V = 4$.

In Illustration 14-8, if the sample contained no randomness, there would be only two runs—all the evens, then all the odds, or the other way around. We would also not expect to see them alternate—odd, even, odd, even. The maximum number of possible runs would be $n_1 + n_2$, or less (provided n_1 and n_2 are not equal), where n_1 and n_2 are the number of data that have each of the two properties being identified.

We will often want to interpret the maximum number of runs as a rejection of a null hypothesis of randomness, since we often want to test randomness of the data in reference to how they were obtained. For example, if the data alternated all the way down the line, we might suspect that the data had been tampered with. There are many aspects to the concept of randomness. The occurrence of odd and even as discussed in Illustration 14-8 is one aspect. Another aspect of randomness that we might wish to check is the ordering of fluctuations of the data above (a) or below (b) the mean or median of the sample.

▼| ILLUSTRATION   14 - 9

Consider the sequence that results from determining whether each of the data points in the sample of Illustration 14-8 is above or below the median value. Test the null hypothesis that this sequence is random. Use $\alpha = 0.05$.

SOLUTION

STEP 1   Parameter of concern: Randomness of the values above or below the median.

STEP 2 H_0: The numbers in the sample form a random sequence with respect to the two properties "above" and "below" the median value.

H_a: The sequence is not random.

Sample: 2,   3,   1,   1,   4,   2,   6,   6,   6,   7

First we must rank the data and find the median. The ranked data are 1, 1, 2, 2, 3, 4, 6, 6, 6, 7. Since there are ten pieces of data, the median is at the $i = 5.5$ position. Thus, $\tilde{x} = (3 + 4)/2 = 3.5$. By comparing each number in the original sample to the value of the median, we obtain the following sequence of a's (above) and b's (below).

b   b   b   b   a   b   a   a   a   a

We observe $n_a = 5$, $n_b = 5$, and 4 runs. So $V = 4$.

If n_1 and n_2 are both less than or equal to 20 and a two-tailed test at $\alpha = 0.05$ is desired, then Table 14 in Appendix B will give us the two critical values for the test. For our illustration with $n_a = 5$ and $n_b = 5$, Table 14 shows critical values of 2 and 10. This means that if 2 or fewer, or 10 or more, runs are observed, the null hypothesis will be rejected. If between 3 and 9 runs are observed, we will fail to reject the null hypothesis.

Step 3 $\alpha = 0.05$.

Test criteria: A two-tailed test is used. The critical values for V are found in Table 14 (see the accompanying figure). The critical values are 2 and 10.

Reject H_0	Fail to reject H_0		Reject H_0
2	3	9	10

V, number of runs

Step 4   Four runs were observed; $V^\star = 4$.

Step 5   Result: $V^\star$ is in the noncritical region.

Decision Fail to reject H_0.

▲   Conclusion We are unable to reject the hypothesis of randomness at the 0.05 level of significance.

To complete the hypothesis test about randomness when n_1 and n_2 are larger than 20 or when α is other than 0.05, we will use z, the standard normal random variable. V is approximately normally distributed with a mean of μ_V and a standard deviation of σ_V. The formulas are as follows:

$$\mu_V = \frac{2n_1 \cdot n_2}{n_1 + n_2} + 1 \qquad (14\text{-}8)$$

$$\sigma_V = \sqrt{\frac{(2n_1 \cdot n_2)(2n_1 \cdot n_2 - n_1 - n_2)}{(n_1 + n_2)^2(n_1 + n_2 - 1)}} \qquad (14\text{-}9)$$

$$z^\star = \frac{V^\star - \mu_V}{\sigma_V} \qquad (14\text{-}10)$$

▼   Illustration   14 - 10

Test the null hypothesis that the sequence that results from classifying the sample data in Illustration 14-8 as "odd" or "even" is a random sequence. Use $\alpha = 0.10$.

Solution

Step 1   Randomness of the even and odd numbers.

Step 2 H_0: The sequence of odd and even occurrences is a random sequence.
H_a: The sequence is not random.

STEP 3 $\alpha = 0.10$.

A two-tailed test is to be used. The test criteria are shown in the following figure.

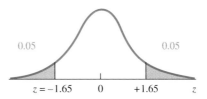

$$z = -1.65 \qquad 0 \qquad +1.65 \qquad z$$

0.05   0.05

STEP 4   The sample and the sequence of odd and even properties are shown in Illustration 14-8. $n_1 = n(\text{even}) = 6$ and $n_2 = n(\text{odd}) = 4$. There are 4 runs, so $V^\star = 4$.

$$\mu_V = \frac{2n_1 n_2}{n_1 + n_2} + 1 = \frac{2 \cdot 6 \cdot 4}{6 + 4} + 1 = \frac{48}{10} + 1 = 5.8$$

$$\sigma_V = \sqrt{\frac{(2n_1 n_2)(2n_1 n_2 - n_1 - n_2)}{(n_1 + n_2)^2 (n_1 + n_2 - 1)}}$$

$$= \sqrt{\frac{(2 \cdot 6 \cdot 4)(2 \cdot 6 \cdot 4 - 6 - 4)}{(6 + 4)^2 (6 + 4 - 1)}}$$

$$= \sqrt{\frac{(48)(38)}{(10)^2(9)}} = \sqrt{\frac{1824}{900}} = \sqrt{2.027} = 1.42$$

$$z^\star = \frac{V^\star - \mu_V}{\sigma_V} = \frac{4.0 - 5.8}{1.42} = \frac{-1.8}{1.42} = -1.268$$

$$z^\star = -1.27$$

STEP 5   Result: z is in the noncritical region.

DECISION Fail to reject H_0.

CONCLUSION We are unable to reject the hypothesis of randomness at the 0.10 level of significance.

▲|

MINITAB (Release 10) commands to complete the runs test for data above and below sample mean or above and below sample median K with data listed in C1. The number of runs and the p-value are reported.

Session commands

Enter: RUNS C1
    or
    RUNS K C1

Menu commands

Choose: Stat > Nonparametrics > Runs
Enter: Variable: C1
Select: Above and below mean or
        Above and below: K

EXERCISES

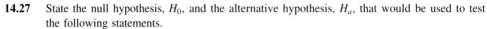

14.27   State the null hypothesis, H_0, and the alternative hypothesis, H_a, that would be used to test the following statements.

    a. The data did not occur in a random order about the median.

    b. The sequence of odd and even is not random.

    c. The gender of customers entering a grocery store was recorded; the entry is not random in order.

14.28   Determine the test criteria that would be used to test the null hypothesis for the following multinomial experiments.

    a. H_0: The results collected occurred in random order above and below the median.

      H_a: The results were not random, with $n(A) = 14$, $n(B) = 15$, and $\alpha = 0.05$.

    b. H_0: The two properties alternated randomly.

      H_a: The two properties didn't occur in random fashion, with $n(I) = 78$, $n(II) = 45$, and $\alpha = 0.05$.

14.29   A manufacturing firm hires both men and women. The following shows the sex of the last 20 individuals hired (M = male, F = female).

M	M	F	M	F	F	M	M	M	M
M	M	F	M	M	F	M	M	M	M

At the $\alpha = 0.05$ level of significance, are we correct in concluding that this sequence is not random?

14.30   A student was asked to perform an experiment that involved tossing a coin 25 times. After each toss, the student recorded the results. The following data were reported (H = heads, T = tails).

H	T	H	T	H	T	H	T	H	H	T	T	H
H	T	T	H	T	H	T	H	T	H	T	H	

Use the runs test at a 5% level of significance to test the student's claim that the results reported are random.

14.31   The following are 24 consecutive downtimes (in minutes) of a particular machine.

20	33	33	35	36	36	22	22	25	27	30	30
30	31	31	32	32	36	40	40	50	45	45	40

The null hypothesis of randomness is to be tested against the alternative that there is a trend. A MINITAB analysis of the number of runs above and below the median follows. Confirm that the number of runs is 4, and compute the value of $z^\star$ and the p-value. Would you reject the hypothesis of randomness?

```
With data in C1;
MTB > RUNS ABOVE OR BELOW 32.5 FOR DATA IN C1

   K = 32.5000

   THE OBSERVED NO. OF RUNS = 4
   THE EXPECTED NO. OF RUNS = 13.0000
   12 OBSERVATIONS ABOVE K      12 BELOW
                    THE TEST IS SIGNIFICANT AT 0.0002
```

14.32 The article "Water Boosts Hemoglobin's Love for Oxygen" (*Science News*, March 30, 1991) discusses the ability of the iron-rich protein pigment hemoglobin to carry oxygen throughout the body. The article states that the protein's conversion to an oxygen-loving state involves between 60 and 80 water molecules. Suppose 20 different determinations of the number of water molecules needed resulted in the following data.

79	75	69	70	70	65	75	75	65	70
60	62	63	63	67	65	70	60	65	62

  a. Determine the median and the number of runs above and below the median.
  b. Use the runs test to test these data for randomness about the median.
  c. State your conclusion.

14.33 The following data were collected in an attempt to show that the number of minutes the city bus is late is steadily growing larger. The data are in order of occurrence.

6	1	3	9	10	10	2	5	5	6
12	3	7	8	9	4	5	8	11	14

At $\alpha = 0.05$, do these data show sufficient lack of randomness to support the claim?

14.34 Mrs. Brown attended a special reading course that claimed to improve reading speed and comprehension. She took a pretest (on which she scored 106) and a test at the end of each of 15 sessions. Her test scores are shown in the following table. Do these data support the claim of improvement at the $\alpha = 0.025$ level?

Session	1	2	3	4	5	6	7	8	9	10	11	12	13	14	15
Score	109	108	110	112	108	111	112	113	110	112	115	114	116	116	118

14.35 A USA Snapshot® (*USA Today*, 10-14-94) titled "School buildings aging" gives the average age for schools in 5 cities as follows: Washington, D.C., 75 years, St. Louis 74 years, San Diego 30 years, Baltimore 30 years, and Fresno, California, 30 years. The following ages of school buildings were collected in the sequence given for Spokane, Washington: 5, 13, 25, 45, 15, 17, 22, 35, 16, 23, 36, 22, 35, and 35.
  a. Determine the median and the number of runs above and below the median.
  b. Use the runs test to test these data for randomness about the median. Use $\alpha = 0.05$.

14.36 An article titled "Clintonomics Hurt Middle Class, Poor" (*USA Today*, 10-17-94) states that the median income for 1993 equals $36,959. A random sample of 250 incomes has a median value different from any of the 250 incomes in the sample. The data contains 105 runs above and below the median. Use the above information to test the null hypothesis that the incomes in the sample form a random sequence with respect to the two properties above and below the median value versus the alternative that the sequence is not random at $\alpha = 0.05$.

14.37   The number of absences recorded at a lecture that met at 8 AM on Mondays and Thursdays last semester were (in order of occurrence)

| 5 | 16 | 6 | 9 | 18 | 11 | 16 | 21 | 14 | 17 | 12 | 14 | 10 |
| 6 | 8 | 12 | 13 | 4 | 5 | 5 | 6 | 1 | 7 | 18 | 26 | 6 |

Do these data show a randomness about the median value at $\alpha = 0.05$? Complete this test by using (a) critical values from Table 14 in Appendix B and (b) the standard normal distribution.

14.38   In an attempt to answer the question "Does the husband (h) or wife (w) do the family banking?" the results of a sample of 28 married customers doing the family banking show the following sequence of arrivals at the bank.

| w | w | w | w | h | w | h | h | h | h | w | w | w | w |
| w | h | h | w | w | w | h | h | h | h | w | h | h | w |

Do these data show lack of randomness with regard to whether the husband or wife does the family banking? Use $\alpha = 0.05$.

14.6 | RANK CORRELATION

The rank correlation coefficient was developed by C. Spearman in the early 1900s. It is a nonparametric alternative to the linear correlation coefficient that was discussed in Chapters 3 and 13. Only rankings are used in the calculation of this coefficient. If the data are quantitative, each of the two variables must be ranked separately.

Spearman rank correlation coefficient

The **Spearman rank correlation coefficient, r_s,** is found by using the formula

$$r_s = 1 - \frac{6\sum(d_i)^2}{n(n^2 - 1)} \qquad \textbf{(14-11)}$$

where d_i is the difference in the rankings and n is the number of pairs of data. The value of r_s will range from -1 to $+1$ and will be used in much the same manner as Pearson's linear correlation coefficient was used previously.

The null hypothesis that we will be testing is "there is no correlation between the two rankings." The alternative hypothesis may be either two-tailed, "there is correlation," or one-tailed, if we anticipate the existence of either positive or negative correlation. The critical region will be on the side(s) corresponding to the specific alternative that is expected. For example, if we suspect negative correlation, the critical region will be in the left-hand tail.

▼| ILLUSTRATION   14 - 11

Let's consider a hypothetical situation in which four judges rank five contestants in a contest. Let's identify the judges as A, B, C, and D and the contestants as a, b, c, d, and e. Table 14-5 lists the awarded rankings.

TABLE 14-5

Rankings for Five Contestants

Contestant	Judge A	B	C	D
a	1	5	1	5
b	2	4	2	2
c	3	3	3	1
d	4	2	4	4
e	5	1	5	3

When we compare judges A and B, we see that they ranked the contestants in exactly the opposite order—perfect disagreement (see Table 14-6). From our previous work with correlation, we expect the calculated value for r_s to be exactly -1 for these data.

TABLE 14-6

Rankings of A and B

Contestant	A	B	$d_i = A - B$	$(d)^2$
a	1	5	-4	16
b	2	4	-2	4
c	3	3	0	0
d	4	2	2	4
e	5	1	4	16
				40

$$r_s = 1 - \frac{6\left[\sum(d_i)^2\right]}{n(n^2 - 1)} = 1 - \frac{(6)(40)}{(5)(5^2 - 1)} = 1 - \frac{240}{120} = -1$$

When judges A and C are compared, we see that their rankings of the contestants are identical (see Table 14-7). We would expect to find a calculated correlation coefficient of $+1$ for these data.

TABLE 14-7

Rankings of A and C

Contestant	A	C	$d_i = A - C$	$(d)^2$
a	1	1	0	0
b	2	2	0	0
c	3	3	0	0
d	4	4	0	0
e	5	5	0	0
				0

$$r_s = 1 - \frac{(6)(0)}{(5)(5^2 - 1)} = 1 - 0 = 1$$

By comparing the rankings of judge A with those of judge B and then with those of judge C, we have seen the extremes: total agreement and total disagreement. Now let's compare the rankings of judge A with those of judge D (see Table 14-8). There seems to be no real agreement or disagreement here. Let's compute r_s:

TABLE 14-8

Rankings of A and D

Contestant	A	D	$d_i = A - D$	$(d)^2$
a	1	5	−4	16
b	2	2	0	0
c	3	1	2	4
d	4	4	0	0
e	5	3	2	4
				24

$$r_s = 1 - \frac{(6)(24)}{(5)(5^2 - 1)}$$

$$= 1 - \frac{144}{120}$$

$$= 1 - 1.2 = -\mathbf{0.2}$$

This is fairly close to zero, which is what we should have suspected since there was no real agreement or disagreement.

The test of significance will result in a failure to reject the null hypothesis when r_s is close to zero and will result in a rejection of the null hypothesis in cases where r_s is found to be close to +1 or −1. The critical values found in Table 15 of Appendix B are positive critical values only. Since the null hypothesis is "the population correlation coefficient is zero" (that is, $\rho_s = 0$), we have a symmetric test statistic. Hence we need only add a plus or minus sign to the value found in the table, as appropriate. This will be determined by the specific alternative that we have in mind.

In our illustration, the critical values for a two-tailed test at $\alpha = 0.10$ are ±0.900. (Remember that n represents the number of pairs.) If the calculated value for r_s is between 0.9 and 1.0 or between −0.9 and −1.0, we will reject the null hypothesis in favor of the alternative "there is a correlation."

▲

Use MINITAB's CORR with two columns of rank numbers to calculate r_s.

The Spearman rank coefficient is defined by formula (3-1) [Pearson's product moment, r], with rankings used in place of the quantitative x- and y-values. If there are no ties in the rankings, formula (14-11) is equivalent to formula (3-1). Formula (14-11) provides us with a much easier procedure to use for calculating the r_s statistic. When there are only a few ties, it is common practice to use formula (14-11). The resulting value of r_s is not exactly the same; however, it is generally considered to be an acceptable estimate. Illustration 14-12 shows the procedure for handling ties and uses formula (14-11) for the calculation of r_s. For comparison, Exercise 14-47 (p. 734) asks you to calculate r_s using formula (3-1).

If ties occur in either set of the ordered pairs of rankings, assign each tied observation the mean of the ranks that would have been assigned had there been no ties, as was done in the Mann–Whitney U test.

▼| ILLUSTRATION   14 - 12

Students who finish exams more quickly than the rest of the class are often thought to be smarter. The following set of data shows the score and order of finish for 12 students on a recent one-hour exam. At the 0.05 level, do these data support the alternative hypothesis that the first students to complete an exam have higher grades?

Order of Finish	1	2	3	4	5	6	7	8	9	10	11	12
Exam Score	90	74	76	60	68	86	92	60	78	70	78	64

SOLUTION

STEP 1   Parameter of concern: Correlation coefficient between score and order of finish.

STEP 2 H_0: Order of finish has no relationship to exam score.
H_a: First to finish tend to have higher grades.

STEP 3 $\alpha = 0.05$.
Test criteria: $n = 12$; the critical region is shown in the following diagram.

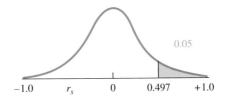

STEP 4   Rank the scores from higher to lowest, assigning the highest score the rank number 1, as shown.

92	90	86	78	78	76	74	70	68	64	60	60
1	2	3	4	5	6	7	8	9	10	11	12
			4.5	4.5						11.5	11.5

Case Study

14.2

Effects of Therapeutic Touch on Tension Headache Pain

Elizabeth Keller and Virginia Bzdek made use of the Spearman rank correlation coefficient to determine the influence of the subjects' education, age, and several other attributes on the results of therapeutic touch. They reported several observed values and the associated p-values.

Therapeutic touch (TT) is a modern derivative of the laying on of hands that involves touching with the intent to help or heal. This study investigated the effects of TT on tension headache pain in comparison with a placebo simulation of TT. Sixty volunteer subjects with tension headaches were randomly divided into treatment and placebo groups. The McGill–Melzack Pain Questionnaire was used to measure headache pain levels before each intervention, immediately afterward, and 4 hours later. . . .

EXERCISE 14.39

"The highest correlation was between the PRI and the NWC, $r = .79$, $p < .0001$. The lowest correlation was between NWC and the PPI, $r = .30$, $p < .02$." These statements are from Case Study 14-2. Explain the meaning of this quote.

HYPOTHESES

I. Tension headache pain will be reduced following therapeutic touch, and the initial reduction will be maintained for a 4-hour period.

II. Subjects who receive therapeutic touch will experience greater tension headache pain reduction than subjects receiving a placebo simulation of therapeutic touch.

III. Subjects who receive therapeutic touch will maintain greater tension headache pain reduction than subjects receiving a placebo simulation of therapeutic touch 4 hours following the intervention. . . .

Findings

Spearman correlation coefficients were calculated to determine the influence of the subject's education, age, sex, practice of meditation, religion, and level of initial skepticism toward TT. The differences in the MMPQ pain scores were not significantly correlated with any belief system or demographic variable in either group. The one exception was the posttest difference on the PPI scale in the placebo group, which was inversely correlated with years of education, $r = -.53$, $p = .002$.

The MMPQ was internally consistent in this study to a statistically significant degree between all three subscales on the pretest, posttest, and delayed posttest. The highest correlation was between the PRI and the NWC, $r = .79$, $p < .0001$. The lowest correlation was between NWC and the PPI, $r = .30$, $p < .02$. When one subscale reported a significant difference between the groups, the other two scales also reported significant differences in the same direction. Based on these observations of internal consistency, the MMPQ was considered to be a reliable instrument in this study.

Source: Elizabeth Keller and Virginia M. Bzdek, Nursing Research, March/April 1986, Vol. 35, no. 2.

The rankings and the preliminary calculations are shown in Table 14-9.

TABLE 14-9

Rankings for Test Scores and Differences

Test Score Rank	Order of Finish	Difference (d_i)	$(d_i)^2$
1	7	−6	36.00
2	1	1	1.00
3	6	−3	9.00
4.5	9	−4.5	20.25
4.5	11	−6.5	42.25
6	3	3	9.00
7	2	5	25.00
8	10	−2	4.00
9	5	4	16.00
10	12	−2	4.00
11.5	4	7.5	56.25
11.5	8	3.5	12.25
			235.00

Using formula (14-11), we obtain

$$r_s = 1 - \frac{(6)(235.0)}{(12)(143)} = 1 - \frac{1410}{1716} = 1 - 0.822$$

$$= \mathbf{0.178}$$

STEP 5   Result: r_s falls in the noncritical region.

DECISION Fail to reject H_0.

CONCLUSION There is not sufficient evidence presented by these sample data to enable us to conclude that the first students to finish have the higher grades at the 0.05 level of significance.

EXERCISES

14.40 State the null hypothesis, H_0, and the alternative hypothesis, H_a, that would be used to test the following statements.
 a. There is no relationship between the two rankings.
 b. The two variables are unrelated.
 c. There is a positive correlation between the two variables.
 d. Age has a decreasing effect on monetary value.

14.41 Determine the test criteria that would be used to test the null hypothesis for the following multinomial experiments.
   a. H_0: No relationship between the two variables.
      H_a: There is a relationship,
      with $n = 14$ and $\alpha = 0.05$.
   b. H_0: No correlation.
      H_a: Positively correlated,
      with $n = 27$ and $\alpha = 0.05$.
   c. H_0: Variable A has no effect on variable B.
      H_a: Variable B decreases as A increases,
      with $n = 18$ and $\alpha = 0.01$.

14.42 The following data involving molecular weight and percent absorption in humans for 13 compounds are extracted from an article about nasal absorption of drugs (*Journal of Pharmaceutical Sciences*, July 1987):

Molecular Wt.	362	1007	1056	1069	1069	1182	1182	1182	1209	1297	1337	4800	5000
% Absorption	40	10	10	15	40	1	2	2.5	15	2.5	5	2	1

The article states that the Spearman rank correlation coefficient equals -0.67. Verify this result.

14.43 The following data represent the ages of 12 subjects and the mineral concentration (in parts per million) in their tissue samples.

Age (x)	82	83	64	53	47	50	70	62	34	27	75	28
Mineral Concentration (y)	170	40	64	5	15	5	48	34	3	7	50	10

Refer to the following MINITAB output and verify that the Spearman rank correlation coefficient equals 0.753.

```
With data in C1 and C2, find ranks and use Pearson's
correlation coefficient.

MTB > RANK C1, PUT RANKS IN C3
MTB > RANK C2, PUT RANKS IN C4
MTN > CORRELATION C3,C4

Correlation of C3 and C4 = 0.753
```

14.44 The 1994 issue of the *World Almanac and Book of Facts* gives the Nielsen rankings for America's favorite prime-time television programs, 1992–1993. The top ten programs are listed below along with the rankings assigned by a panel of educators.

Program	Nielsen Ranking	Panel Ranking
60 Minutes	1	1
Roseanne	2	9.5
Home Improvement	3	2
Murphy Brown	4	6
Murder, She Wrote	5	7
Coach	6	3
Monday Night Football	7	8
CBS Sunday Movie	8.5	5
Cheers	8.5	9.5
Full House	10	4

Compute the Spearman rank coefficient. Test the null hypothesis that there is no relationship between the Nielsen rankings and the panel rankings versus the alternative that there is a relationship between them. Use $\alpha = 0.05$.

14.45 An article titled "The Graduate Record Examination as an Admission Requirement for the Graduate Nursing Program" (*Journal of Professional Nursing*, October 1994) reported a significant correlation between undergraduate GPA and GPA at graduation from a graduate nursing program. The following data were collected on ten nursing students who graduated from a graduate nursing program.

Undergraduate GPA	GPA at Graduation
3.5	3.4
3.1	3.2
2.7	3.0
3.7	3.6
2.5	3.1
3.3	3.4
3.0	3.0
2.9	3.4
3.8	3.7
3.2	3.8

Compute the Spearman rank coefficient and test the null hypothesis of no relationship versus a positive relationship. Use a level of significance equal to 0.05.

14.46 An article in *Self* magazine (February 1991) discusses the relationship between the pace of life and the coronary-heart-disease death rate. New York City, for example, was ranked as only the third-fastest-paced city but number one for deadly heart attacks. Suppose the data from another such study involving eight cities were as follows:

City	Rank for Pace-of-Life	Rank for Heart-Disease Death Rate
Salt Lake City	4	7
Buffalo	2	2
Columbus	5	8
Worcester	6	4
Boston	1	6
Providence	8	3
New York	3	1
Paterson	7	5

Find the Spearman rank correlation coefficient.

14.47 Using formula (3-2), calculate the Spearman rank correlation coefficient for the data in Illustration 14-12, page 729. Recall that formula (3-2) is equivalent to the definition formula (3-1) and that rank numbers must be used with this formula in order for the resulting statistic to be Spearman's r_s.

14.48 Refer to the bivariate data shown in the following table.

x	-2	-1	1	2
y	4	1	1	4

a. Construct a scatter diagram.
b. Calculate Spearman's correlation coefficient, r_s [formula (14-11)].
c. Calculate Pearson's correlation coefficient, r [formula (3-2)].
d. Compare the two results from (b) and (c). Do the two measures of correlation measure the same thing?

IN RETROSPECT

In this chapter you have become acquainted with some of the basic concepts of nonparametric statistics. While learning about the use of nonparametric methods and specific nonparametric tests of significance, you should have also come to realize and understand some of the basic assumptions that are needed when the parametric techniques of the earlier chapters are encountered. You now have seen a variety of tests, many of which somewhat duplicate the job done by others. What you must keep in mind is that you should use the best test for your particular needs. The power of the test and the cost of sampling, as related to size and availability of the desired response variable, will play important roles in determining the specific test to be used.

The article at the beginning of this chapter illustrates only one of many situations in which a nonparametric test can be used. Spearman's rank correlation test may be used to statistically compare the two rankings. They appear to be quite different, but are they significantly different? The answer to this is left for you to determine in Exercise 14.64.

CHAPTER EXERCISES

14.49 Is the absentee rate in the 8 AM statistics class the same as in the 11 AM statistics class? The following sample of the daily number of absences was taken from the attendance records of the two classes.

| | Day | | | | | | | | | | | |
Class	1	2	3	4	5	6	7	8	9	10	11	12
8 AM	0	1	3	1	0	2	4	1	3	5	3	2
11 AM	1	0	1	0	1	2	3	0	1	3	2	1

Is there sufficient reason to conclude that there is more absence in the 8 AM class? Use $\alpha = 0.05$.

14.50 Track coaches, runners, and fans talk a lot about the "speed of the track." The surface of the track is believed to have a direct effect on the amount of time that it takes a runner to cover the required distance. To test this effect, ten runners were asked to run a 220-yard sprint on each of two tracks. Track A is a cinder track, and track B is made of a new synthetic material. The running times are given in the following table. Test the claim that the surface on track B is conducive to faster running times.

| | Runner | | | | | | | | | |
Track	1	2	3	4	5	6	7	8	9	10
A	27.7	26.8	27.0	25.5	26.6	27.4	27.2	27.4	25.8	25.1
B	27.0	26.7	25.3	26.0	26.1	25.3	26.7	27.1	24.8	27.1

a. State the null and alternative hypotheses being tested. Complete the test using $\alpha = 0.05$.

b. State your conclusions.

14.51 A test that measures computer anxiety was administered to students in a statistics course that uses statistical packages on the computer. The test was given at the beginning and end of the course. The hypothesis being studied was that computer anxiety would be reduced as the students used computers in the course. (A high score on the test indicates high computer anxiety.) Can we conclude that computer anxiety was reduced during this course? Use $\alpha = 0.05$.

Beginning	68	79	49	34	60	87	68	70	60	67	39	59	42	40	62	80	62	63	102	42	68	70	44	63	55
End	77	78	32	40	60	68	87	88	62	68	40	60	31	39	62	63	58	59	110	49	54	67	43	54	41

14.52 A candy company has developed two new chocolate-covered candy bars. Six randomly selected people all preferred candy bar I. Is this statistical evidence, at $\alpha = 0.05$, that the general public will prefer candy bar I?

14.53 While trying to decide on the best time to harvest his crop, a commercial apple farmer recorded the day on which the first apple on the top half and the first apple on the bottom half of 20 randomly selected trees were ripe. The variable x was assigned a value of 1 on the first day that the first ripe apple appeared on 1 of the 20 trees. The days were then numbered sequentially. The observed data are shown in the following table. Do these data provide convincing evidence that the apples on the top of the trees start to ripen before the apples on the bottom half? Use $\alpha = 0.05$.

										Tree										
Position	1	2	3	4	5	6	7	8	9	10	11	12	13	14	15	16	17	18	29	20
Top	5	6	1	4	5	3	6	7	8	5	8	6	4	7	8	10	3	2	9	7
Bottom	6	5	5	7	3	6	6	8	9	4	10	7	5	11	6	11	5	6	9	8

14.54 A sample of 32 students received the following grades on an exam.

41	42	48	46	50	54	51	42
51	50	45	42	32	45	43	56
55	47	45	51	60	44	57	57
47	28	41	42	54	48	47	32

a. Does this sample show that the median score for the exam differs from 50? Use $\alpha = 0.05$.

b. Does this sample show that the median score for the exam is less than 50? Use $\alpha = 0.05$.

14.55 An article in the journal *Sedimentary Geology* (Vol. 57, 1988) compares a measure called the *roughness coefficient* for translucent and opaque quartz sand grains. If you measured the roughness coefficient for 20 sand grains of each type (translucent and opaque), for what values of the Mann–Whitney U statistic would you reject the null hypothesis in a two-tailed test with alpha equal to 0.05?

14.56 Twenty students were randomly divided into two equal groups. Group 1 was taught an anatomy course using a standard lecture approach. Group 2 was taught using a computer-assisted approach. The test scores on a comprehensive final exam were as follows:

Group 1	75	83	60	89	77	92	88	90	55	70
Group 2	77	92	90	85	72	59	65	92	90	79

Test to test the claim that a computer-assisted approach produces higher achievement (as measured by final exam scores) in anatomy courses than does a lecture approach. Use $\alpha = 0.05$.

14.57 The use of nuclear magnetic resonance (NMR) spectroscopy for detection of malignancy is discussed in the journal *Clinical Chemistry* (Vol. 34, No. 3, 1988). The line width at the half height of peaks in the NMR spectra is measured. The spectra is produced from assaying plasma from an individual. Suppose the following line widths were obtained from a normal group and a group known to have malignancies. Would you reject a two-tailed research hypothesis at the 0.05 level of significance?

Normal Group	35.1	32.9	30.6	30.5	30.9
Malignancy Group	28.5	29.5	30.7	27.5	28.0

14.58 A firm is currently testing two different procedures for adjusting the cutting machines used in the production of greeting cards. The results of two samples show the following recorded adjustment times.

Method 1	17	15	14	18	16	15	17	18	15	14	14	16	15		
Method 2	14	14	13	13	15	12	16	14	16	13	14	13	12	15	17	13

Is there sufficient reason to conclude that method 2 requires less time (on the average) than method 1 at the 0.05 level of significance?

14.59   The fixed interest rates charged on auto loans vary between banks and other lending institutions. Some of the rates charged by several banks and credit unions that were listed in the *Democrat and Chronicle* (5-6-91) were as follows:

Banks	12.50	12.50	11.10	12.00	12.25
	11.25	12.25	10.99	11.25	11.25
	12.90	11.40	12.25	11.25	11.98
Credit Unions	9.75	10.25	11.25	9.90	10.40
	11.00	10.25	9.95	10.80	8.90
	10.90	11.75			

Do these data present sufficient evidence to conclude that the credit unions charge a lower fixed auto-loan rate than the banks charge? Use $\alpha = 0.05$.

14.60   Two table-tennis-ball manufacturers have agreed that the quality of their products can be measured by the height to which the balls rebound. A test is arranged, the balls are dropped from a constant height, and the rebound heights are measured. The results (in inches) are shown in the following table. Manufacturer A claims, "The results show my product to be superior." Manufacturer B replies, "I know of no statistical test that supports this claim." Can you find a test that supports A's claim?

A	14.0	12.5	11.5	12.2	12.4	12.3	11.8	11.9	13.7	13.2
B	12.0	12.5	11.6	13.3	13.0	13.0	12.1	12.8	12.2	12.6

a. Does the appropriate parametric test show that A's product is superior? [What parametric test (or tests) is appropriate, and what exactly does it show?]

b. Does the appropriate nonparametric test show that A's product is superior?

14.61   Consider the following sequence of defective parts (d) and nondefective parts (n) produced by a machine.

n	n	n	d	n	n	n	n	n	d	n	n
n	n	n	n	n	d	n	d	n	n	n	n

Can we reject the hypothesis of randomness at $\alpha = 0.05$?

14.62   A patient was given two different types of vitamin pills, one containing iron and one iron-free. The patient was instructed to take the pills on alternate days. To free himself from remembering which pill he needed to take, he mixed all of the pills together in a large bottle. Each morning he took the first pill that came out of the bottle. To see whether this was a random process, for 25 days he recorded an "I" each morning that he took a vitamin with iron and an "N" for no iron.

Day	1	2	3	4	5	6	7	8	9	10	11	12	13
Type	I	I	N	I	I	N	N	I	N	N	N	N	N

Day	14	15	16	17	18	19	20	21	22	23	24	25
Type	I	I	I	N	I	I	I	I	N	I	I	N

Is there sufficient reason to reject the null hypothesis that the vitamins were taken in random order at the 0.05 level of significance?

 14.63 Can today's high temperature be effectively predicted using yesterday's high? Pairs of yesterday's and today's high temperatures were randomly selected. The results are shown in the following table. Do the data present sufficient evidence to justify the statement "Today's high temperature tends to correlate with yesterday's high temperature"? Use $\alpha = 0.05$.

Pairs of Days																		
Reading	1	2	3	4	5	6	7	8	9	10	11	12	13	14	15	16	17	18
Yesterday's	40	58	46	33	40	51	55	81	85	83	89	64	73	63	46	58	28	69
Today's	40	56	34	59	46	51	74	77	83	84	85	68	65	60	54	62	34	66

 14.64 The article in the Chapter Case Study (p. 699) shows the rankings assigned to ten components of job satisfaction. Do the rankings assigned by the workers and the boss show a significant difference in what each thinks is important? Test by using $\alpha = 0.05$.

 14.65 In a study to see whether spouses are consistent in their preferences for television programs, a market research firm asked several married couples to rank a list of 12 programs (1 represents the highest score; 12 represents the lowest). The average ranks for the programs, rounded to the nearest integer, were as follows:

Program												
Rank	1	2	3	4	5	6	7	8	9	10	11	12
Husbands	12	2	6	10	3	11	7	1	9	5	8	4
Wives	5	4	1	9	3	12	2	8	6	10	7	11

Is there significant evidence of negative correlation at the 0.01 level of significance?

14.66 Nonparametric tests are also called *distribution-free* tests. However, the normal distributions are used in the inference-making procedures.

a. To what does the *distribution-free* term apply? (The population? The sample? The sampling distribution?) Explain.

b. What is it that has the normal distribution? Explain.

VOCABULARY LIST

Be able to define each term. Pay special attention to the key terms, which are printed in red. In addition, describe in your own words, and give an example of, each term. Your examples should not be ones given in class or in the textbook.

The bracketed numbers indicate the chapters in which the term previously appeared, but you should define the terms again to show increased understanding of their meaning. Page numbers indicate the first appearance of the term in Chapter 14.

binomial random variable [5, 9] (p. 705)
continuity correction [6] (p. 706)
correlation [3, 13] (p. 726)
dependent sample [10] (p. 704)
distribution-free test (p. 700)
efficiency (p. 701)
independent sample [10] (p. 711)
Mann–Whitney U test (p. 711)
median, M [2] (p. 702)
nonparametric test (p. 700)
normal approximation [6] (p. 705)

paired data [3, 10, 13] (p. 726)
parametric test (p. 700)
power (p. 701)
randomness [2, 7] (p. 720)
rank (p. 712)
run (p. 720)
runs test (p. 720)
sign test (p. 702)
Spearman rank correlation
  coefficient (p. 726)
test statistic [8, 9, . . .] (p. 703)

QUIZ A

Answer "True" if the statement is always true. If the statement is not always true, replace the words shown in bold with words that make the statement always true.

14.1 One of the advantages that the nonparametric tests have is the necessity for **less restrictive** assumptions.

14.2 The sign test is a possible replacement for the **F test**.

14.3 The **sign test** can be used to test the randomness of a set of data.

14.4 If a tie occurs in a set of ranked data, the data that form the tie are **removed from the set**.

14.5 Two dependent **means** can be compared nonparametrically by using the sign test.

14.6 The sign test is a possible alternative to the Student's t test for **one mean value**.

14.7 The **run test** is a nonparametric alternative to the difference between two independent means.

14.8 The **confidence level** of a statistical hypothesis test is measured by $1 - \beta$.

14.9 Spearman's rank correlation coefficient is an alternative to using the **linear correlation coefficient**.

14.10 The **efficiency** of a nonparametric test is the probability that a false null hypothesis is rejected.

QUIZ B

14.1 The weights of nine people before they stopped smoking and five weeks after they stopped smoking are as follows:

Person	1	2	3	4	5	6	7	8	9
Before	148	176	153	116	128	129	120	132	154
After	155	178	151	120	130	136	126	128	158

Find the 95% confidence interval estimate for the average weight change.

14.2 The following data show the weight gains for 20 laboratory mice, half of which were fed one diet and half a different diet. Test to determine if the difference in weight gain is significant at $\alpha = 0.05$.

Diet A	41	40	36	43	36	43	39	36	24	41
Diet B	35	34	27	39	31	41	37	34	42	38

14.3 A large textbook publishing company hired nine new sales representatives three years ago. At the time of hire, the nine were rated according to their potential. Now three years later the company president wants to know how well their potential rates correlate with their sales totals for the three years.

Sales Rep.	a	b	c	d	e	f	g	h	i
Potential	2	5	6	1	4	3	9	8	7
Sales Tot.	450	410	350	345	330	400	250	310	270

Is there significant correlation at the 0.05 level?

14.4 The new school principal thought there might be a pattern to the order in which discipline problems arrived at his office. He had his secretary record the grade level of the students as they arrived.

9	10	11	9	12	11	9	10	10	11	10	11	10	10	11
12	12	9	9	11	12	10	9	12	10	11	12	11	10	10

At the 0.05 level, is there significant evidence of randomness?

QUIZ C

14.1   What advantages do nonparametric statistics have over parametric methods?

14.2   Explain how the sign test is based on the binomial distribution and is often approximated by the normal distribution.

14.3   Why does the sign test use a null hypothesis about the median instead of the mean like a t test uses?

14.4   Explain why a nonparametric test is not as sensitive to an extreme datum as a parametric test might be.

14.5   A restaurant has collected data on which of two seating arrangements its customers prefer. In a sign test to determine if one seating arrangement is significantly preferred, the null hypothesis would be

      a. $\mu = 0$
      b. $\mu = 0.5$
      c. $p = 0$
      d. $p = 0.5$

Explain your choice.

Sample mean:

$$\bar{x} = \frac{\Sigma x}{n} \quad \textbf{(2-1)} \quad \text{or} \quad \frac{\Sigma xf}{\Sigma f} \quad \textbf{(2-11)}$$

Depth of sample median: $\quad d(\tilde{x}) = (n+1)/2 \quad \textbf{(2-2)}$

Range: $\quad H - L \quad \textbf{(2-4)}$

Sample variance:

$$s^2 = \frac{\Sigma(x - \bar{x})^2}{n - 1} \quad \textbf{(2-6)}$$

$$\text{or} \quad s^2 = \frac{\Sigma x^2 - \dfrac{(\Sigma x)^2}{n}}{n - 1} \quad \textbf{(2-10)}$$

$$\text{or} \quad s^2 = \frac{\Sigma x^2 f - \dfrac{(\Sigma xf)^2}{\Sigma f}}{\Sigma f - 1} \quad \textbf{(2-12)}$$

Sample standard deviation: $\quad s = \sqrt{s^2} \quad \textbf{(2-7)}$

Chebyshev's Theorem: $\quad$ at least $1 - (1/k^2) \quad \textbf{(p. 104)}$

Sum of squares of x:

$SS(x) = \Sigma x^2 - ((\Sigma x)^2/n) \quad \textbf{(2-9)}$

Sum of squares of y:

$SS(y) = \Sigma y^2 - ((\Sigma y)^2/n) \quad \textbf{(3-3)}$

Sum of squares of xy:

$SS(xy) = \Sigma xy - ((\Sigma x \cdot \Sigma y)/n) \quad \textbf{(3-4)}$

Pearson's Correlation Coefficient:

$r = SS(xy)/\sqrt{SS(x) \cdot SS(y)} \quad \textbf{(3-2)}$

Equation for line of best fit: $\quad \hat{y} = b_0 + b_1 x \quad \textbf{(p. 160)}$

Slope for line of best fit: $\quad b_1 = SS(xy)/SS(x) \quad \textbf{(3-7)}$

y-intercept for line of best fit:

$b_0 = [\Sigma y - b_1 \cdot \Sigma x]/n \quad \textbf{(3-6)}$

Empirical (observed) probability:

$P'(A) = n(A)/n \quad \textbf{(4-1)}$

Theoretical probability for equally likely sample space:

$P(A) = n(A)/n(S) \quad \textbf{(4-2)}$

Complement Rule:

$P(\text{not A}) = P(\overline{A}) = 1 - P(A) \quad \textbf{(4-3)}$

General Addition Rule:

$P(A \text{ or } B) = P(A) + P(B) - P(A \text{ and } B) \quad \textbf{(4-4a)}$

Special Addition Rule for mutually exclusive events:

$P(A \text{ or } B \text{ or } ... \text{ or } D) = P(A) + P(B) + \cdots + P(D) \quad \textbf{(4-4c)}$

General Multiplication Rule:

$P(A \text{ and } B) = P(A) \cdot P(B \mid A) \quad \textbf{(4-7a)}$

Special Multiplication Rule for independent events:

$P(A \text{ and } B \text{ and } ... \text{ and } D) = P(A) \cdot P(B) \cdots P(D) \quad \textbf{(4-8b)}$

Conditional Probability:

$P(A \mid B) = P(A \text{ and } B)/P(B) \quad \textbf{(4-6b)}$

Bayes's Rule:

$P(A_i \mid B) = [P(A_i) \cdot P(B \mid A_i)]/\Sigma[P(A_i) \cdot P(B \mid A_i)] \quad \textbf{(4-9)}$

Mean of discrete random variable:

$\mu = \Sigma[xP(x)] \quad \textbf{(5-1)}$

Variance of discrete random variable:

$\sigma^2 = \Sigma[x^2 P(x)] - \{\Sigma[xP(x)]\}^2 \quad \textbf{(5-3a)}$

Standard deviation of discrete random variable:

$\sigma = \sqrt{\sigma^2} \quad \textbf{(5-4)}$

Factorial: $\quad n! = (n)(n-1)(n-2) \cdots 2 \cdot 1 \quad \textbf{(p. 270)}$

Binomial coefficient:

$$\binom{n}{x} = \frac{n!}{x! \cdot (n-x)!} \quad \textbf{(5-6)}$$

Binomial probability function:

$P(x) = \binom{n}{x} \cdot p^x \cdot q^{n-x}, \quad x = 0, \ldots, n \quad \textbf{(5-5)}$

Mean of binomial random variable: $\quad \mu = np \quad \textbf{(5-7)}$

Standard deviation, binomial random variable:

$\sigma = \sqrt{npq} \quad \textbf{(5-8)}$

Standard score: $\quad z = (x - \mu)/\sigma \quad \textbf{(6-3)}$

Standard score for $\bar{x}$: $\quad z = \dfrac{\bar{x} - \mu}{\sigma/\sqrt{n}} \quad \textbf{(7-2)}$

Confidence interval estimate for mean, μ (σ known):

$\bar{x} \pm z(\alpha/2) \cdot (\sigma/\sqrt{n}) \quad \textbf{(8-1)}$

Sample size for $1 - \alpha$ confidence estimate for μ:

$n = [z(\alpha/2) \cdot \sigma/E]^2 \quad \textbf{(8-3) (p. 375)}$

Calculated test statistic for H_0: $\mu = \mu_0$ (σ known):

$z^\star = (\bar{x} - \mu_0)/\sigma/\sqrt{n} \quad \textbf{(8-4) (p. 391)}$

Confidence interval estimate for mean, μ (σ unknown):

$\bar{x} \pm t(\text{df}, \alpha/2) \cdot (s/\sqrt{n}) \quad$ with df $= n - 1 \quad \textbf{(9-1)}$

Calculated test statistic for H_0: $\mu = \mu_0$ (σ unknown):

$t^\star = \dfrac{\bar{x} - \mu_0}{s/\sqrt{n}} \quad$ with df $= n - 1 \quad \textbf{(9-2)}$

Confidence interval estimate for proportion, p:

$p' \pm z(\alpha/2) \cdot \sqrt{(p'q')/n}, \quad p' = x/n \quad \textbf{(9-6)}$

Calculated test statistic for H_0: $p = p_0$:

$z^\star = (p' - p_0)/\sqrt{(p_0 q_0/n)}, \quad p' = x/n \quad \textbf{(9-9)}$

Calculated test statistic for H_0: $\sigma^2 = \sigma_0^2$ or $\sigma = \sigma_0$:

$\chi^{2\star} = (n-1)s^2/\sigma_0^2, \quad \text{df} = n - 1 \quad \textbf{(9-10)}$

Mean difference between two dependent samples:

$\quad$ **Paired difference:** $\quad d = x_1 - x_2 \quad \textbf{(10-1)}$

$\quad$ **Sample mean of paired differences:**

$\bar{d} = \Sigma d/n \quad \textbf{(10-2)}$

$\quad$ **Sample standard deviation of paired differences:**

$$s_d = \sqrt{\frac{\Sigma d^2 - \dfrac{(\Sigma d)^2/n}{}}{n - 1}} \quad \textbf{(10-3)}$$

Confidence interval estimate for mean, μ_d:

$\bar{d} \pm t(\text{df}, \alpha/2) \cdot s_d/\sqrt{n}$ **(10-4)**

Calculated test statistic for H_0: $\mu_d = \mu_0$:

$t^\star = (\bar{d} - \mu_d)/(s_d/\sqrt{n})$, $\text{df} = n - 1$ **(10-5)**

Difference between means of two independent samples:

Degrees of freedom:

$\text{df} = \text{smaller of } (n_1 - 1) \text{ or } (n_2 - 1)$ **(p. 513)**

Confidence interval estimate for $\mu_1 - \mu_2$:

$(\bar{x}_1 - \bar{x}_2) \pm t(\text{df}, \alpha/2) \cdot \sqrt{(s_1^2/n_1) + (s_2^2/n_2)}$ **(10-8)**

Calculated test statistic for H_0: $\mu_1 - \mu_2 = (\mu_1 - \mu_2)_0$:

$t^\star = [(\bar{x}_1 - \bar{x}_2) - (\mu_1 - \mu_2)_0]/\sqrt{(s_1^2/n_1) + (s_2^2/n_2)}$

(10-9)

Ratio of variances between two independent samples:

Calculated test statistic for H_0: $\sigma_1^2/\sigma_2^2 = 1$:

$F^\star = s_1^2/s_2^2$ **(10-10)**

Difference between proportions of two independent samples:

Confidence interval estimate for $p_1 - p_2$:

$(p_1' - p_2') \pm z\left(\dfrac{\alpha}{2}\right) \cdot \sqrt{\dfrac{p_1'q_1'}{n_1} + \dfrac{p_2'q_2'}{n_2}}$ **(10-12)**

Pooled observed probability:

$p_p' = (x_1 + x_2)/(n_1 + n_2)$ **(10-14)**

$q_p' = 1 - p_p'$ **(10-15)**

Calculated test statistic for H_0: $p_1 - p_2 = 0$:

$z^\star = \dfrac{p_1' - p_2'}{\sqrt{(p_p')(q_p')\left[\left(\dfrac{1}{n_1}\right) + \left(\dfrac{1}{n_2}\right)\right]}}$ **(10-16)**

Calculated test statistic for enumerative data:

$\chi^2 \star = \Sigma[(O - E)^2/E]$ **(11-1)**

Multinomial experiment:

Degrees of freedom: $\text{df} = k - 1$ **(11-2)**

Expected frequency: $E = n \cdot p$ **(11-3)**

Test for independence or Test of homogeneity:

Degrees of freedom:

$\text{df} = (r - 1) \cdot (c - 1)$ **(11-4)**

Expected value: $E = (R \cdot C)/n$ **(11-5)**

Mathematical model:

$x_{c,k} = \mu + F_c + \epsilon_{k(c)}$ **(12-13)**

Total sum of squares:

$\text{SS(total)} = \Sigma(x^2) - \dfrac{(\Sigma x)^2}{n}$ **(12-2)**

Sum of squares due to factor:

$\text{SS(factor)} = \left[\left(\dfrac{C_1^2}{k_1}\right) + \left(\dfrac{C_2^2}{k_2}\right) + \left(\dfrac{C_3^2}{k_3}\right) + \cdots\right] - \left[\dfrac{(\Sigma x)^2}{n}\right]$ **(12-3)**

Sum of squares due to error:

$\text{SS(error)} =$
$\Sigma(x^2) - [(C_1^2/k_1) + (C_2^2/k_2) + (C_3^2/k_3) + \cdots]$ **(12-4)**

Degrees of freedom for total:

$\text{df(total)} = n - 1$ **(12-6)**

Degrees of freedom for factor:

$\text{df(factor)} = c - 1$ **(12-5)**

Degrees of freedom for error:

$\text{df(error)} = n - c$ **(12-7)**

Mean square for factor:

$\text{MS(factor)} = \text{SS(factor)}/\text{df(factor)}$ **(12-10)**

Mean square for error:

$\text{MS(error)} = \text{SS(error)}/\text{df(error)}$ **(12-11)**

Calculated test statistic for H_0: Mean value is same at all levels: $F^\star = \text{MS (factor)}/\text{MS(error)}$ **(12-12)**

Covariance of x and y:

$\text{covar}(x, y) = \Sigma[(x - \bar{x})(y - \bar{y})]/(n - 1)$ **(13-1)**

Pearson's Correlation Coefficient:

$r = \text{covar}(x, y)/(s_x \cdot s_y)$ **(13-2)**

or $r = \text{SS}(xy)/\sqrt{\text{SS}(x) \cdot \text{SS}(y)}$ **(3-2)** or **(13-3)**

Experimental error: $e = y - \hat{y}$ **(13-5)**

Variance of error ϵ: $s_e^2 = \Sigma(y - \hat{y})^2/(n - 2)$

(13-6)

or $s_e^2 = \dfrac{(\Sigma y^2) - (b_0)(\Sigma y) - (b_1)(\Sigma xy)}{n - 2}$ **(13-8)**

Standard deviation about the line of best fit:

$s_e = \sqrt{s_e^2}$ **(p. 666)**

Square of standard error of regression:

$s_{b_1}^2 = \dfrac{s_e^2}{\text{SS}(x)} = \dfrac{s_e^2}{\Sigma x^2 - [(\Sigma x)^2/n]}$ **(13-11)**

Confidence interval estimate for β_1:

$b_1 \pm t(\text{df}, \alpha/2) \cdot s_{b_1}$ **(13-12)**

Calculated test statistic for H_0: $\beta_1 = \beta_1$:

$t^\star = (b_1 - \beta_1)/s_{b_1}$ with $\text{df} = n - 2$ **(13-13)**

Confidence interval estimate for mean value of y at x_0:

$\hat{y} \pm t(n - 2, \alpha/2) \cdot s_e \cdot \sqrt{\dfrac{1}{n} + \dfrac{(x_0 - \bar{x})^2}{\text{SS}(x)}}$ **(13-15)**

Prediction interval estimate for y at x_0:

$\hat{y} \pm t\left(n - 2, \dfrac{\alpha}{2}\right) \cdot s_e \cdot \sqrt{1 + \dfrac{1}{n} + \dfrac{(x_0 - \bar{x})^2}{\text{SS}(x)}}$ **(13-16)**

Mann-Whitney U test:

$U_a = n_a \cdot n_b + [(n_b) \cdot (n_b + 1)/2] - R_b$ **(14-3)**

$U_b = n_a \cdot n_b + [(n_a) \cdot (n_a + 1)/2] - R_a$ **(14-4)**

Spearman's rank correlation coefficient:

$r_s = 1 - \left[\dfrac{6\Sigma d^2}{n(n^2 - 1)}\right]$ **(14-11)**

Working with Your Own Data

Many variables in everyday life can be treated as bivariate. Often two such variables have a mathematical relationship that can be approximated by means of a straight line. The following demonstrates such a situation.

A | The Age and Value of Peggy's Car

Peggy would like to sell her 1990 Corvette, and she needs to know what price to ask for it in order to write a newspaper advertisement. The car is in average condition, and Peggy expects to get an average price for it ("average for a Corvette!"). She must answer the question "What is an average asking price for a 1990 Corvette?"

Inspection of many classified sections of newspapers turned up only three advertisements for 1990 Corvettes. The prices listed varied a great deal. Peggy finally decided that, in order to determine an accurate selling price, she would define two variables and collect several pairs of values based on the following definitions.

POPULATION   Used Chevrolet Corvettes advertised for sale by individual owners, dealers not included.

INDEPENDENT VARIABLE, x The age of the car as measured in years and defined by

$$x = \text{(present calendar year)} - \text{(year of manufacture)} + 1$$

EXAMPLE   During 1995 Peggy's 1990 Corvette is considered to be 6 years old.

$$x = (1995 - 1990) + 1 = 5 + 1 = 6$$

DEPENDENT VARIABLE, y The advertised asking price.

The table below lists the data collected in April 1995.

Year of Manufacture:	1979	1995	1991	1992	1987	1982	1994	1985
Asking Price:	$6,500	$34,900	$22,900	$27,500	$19,500	$13,500	$28,900	$15,750
Year:	1987	1988	1983	1986	1987	1991	1993	1989
Price:	$15,500	$16,700	10,900	14,500	13,900	27,500	24,900	23,500
Year:	1990	1989	1991	1982	1992	1988	1991	1989
Price:	23,900	18,500	21,900	10,500	24,900	17,900	26,250	19,950
Year:	1990	1986	1990	1988	1985	1990	1988	1992
Price:	25,900	16,900	20,500	20,500	12,000	24,500	21,500	29,250
Year:	1984	1989	1993	1995	1994	1993	1985	1994
Price	14,000	21,500	26,900	30,500	31,000	31,900	19,500	32,995

1. Construct and label a scatter diagram of Peggy's data.
2. Determine the equation for the line of best fit.
3. Draw the line of best fit on the scatter diagram.
4. Test the equation of the line of best fit to see whether the linear model is appropriate for the data. Use $\alpha = 0.05$.
5. Construct a 95% confidence interval for the mean advertised price for 1990 Corvettes.
6. Draw a line segment on the scatter diagram that represents the interval estimate found for question 5.
7. What does the value of the slope, b_1, represent? Explain.
8. What does the value of the y-intercept, b_0, represent? Explain.

B | YOUR OWN INVESTIGATION

Identify a situation of interest to you that can be investigated statistically using bivariate data. (Consult your instructor for specific guidance.)

1. Define the population, the independent variable, the dependent variable, and the purpose for studying these two variables as a regression analysis.
2. Collect 15 to 20 ordered pairs of data.
3. Construct and label a scatter diagram of your data.
4. Determine the equation for the line of best fit.
5. Draw the line of best fit on the scatter diagram.
6. Test the equation of the line of best fit to see whether the linear model is appropriate for the data. Use $\alpha = 0.05$.
7. Construct a 95% confidence interval for the mean value of the dependent variable at the following value of x: Let x be equal to one-third the sum of the lowest value of x in your sample and twice the largest value. That is,

$$x = \frac{L + 2H}{3}$$

8. Draw a line segment on the scatter diagram that represents the interval estimate found for question 7.
9. What does the value of the slope, b_1, represent? Explain.
10. What does the value of the y-intercept, b_0, represent? Explain.

A

BASIC PRINCIPLES OF COUNTING

In order to find the probability of many events, it is necessary to determine the number of possible outcomes for the experiment involved. This requires us to enumerate (obtain a "count" of) the possibilities. This "count" can be obtained by using one of two methods: (1) list all the possibilities and then proceed to count them (1, 2, 3, . . .); or (2) since it is often not necessary to delineate (obtain a representation of) all possibilities, the count can be determined by calculating its numerical value. In this section, we are going to learn three commonly used methods for obtaining the count by calculation: the fundamental technique and two specific techniques.

▼ ILLUSTRATION A-1

An automobile dealer offers one of its small sporty models with two transmission options (standard or automatic) and in one of three colors (black, red, or white). How many different choices of transmission and color combinations are there for the customer?

SOLUTION The number of choices available can easily be found by listing and counting them. There are six.

Standard, black	Automatic, black
Standard, red	Automatic, red
Standard, white	Automatic, white

The possible choices can also be demonstrated by use of a tree diagram.

Trans. Opt.	Color Opt.	Possible Choices
standard	black	standard, black
	red	standard, red
	white	standard, white
automatic	black	automatic, black
	red	automatic, red
	white	automatic, white

NOTE   More information and additional illustrations of tree diagrams can be found in Chapter 4 and in the *Statistical Tutor*.

Each of the two transmission choices can be paired with any one of three colors; thus there are 2×3 or six different possible choices. This suggests the following rule:

FUNDAMENTAL COUNTING RULE

If an experiment is composed of two trials, where one of the trials (single action or choice) has m possible outcomes (results) and the other trial has n possible outcomes, then when the two trials are performed together, there are

$$m \times n \qquad \text{(A-1)}$$

possible outcomes for the experiment.

In Illustration A-1, $m = 2$ (the number of transmission choices) and $n = 3$ (the number of color choices). Using the Fundamental Counting Rule (formula A-1), the number of possible choices available to a customer is

$$m \times n = 2 \times 3 = 6$$

This fundamental counting rule may be extended to include experiments that have more than two trials.

GENERAL COUNTING RULE

If an experiment is composed of k trials performed in a definite order, where the first trial has n_1 possible outcomes, the second trial has n_2 possible outcomes, the third trial has n_3 outcomes, and so on, then the number of possible outcomes for the experiment is

$$n_1 \times n_2 \times n_3 \times \cdots \times n_k. \qquad \text{(A-2)}$$

▼ Illustration   A - 2

In many states, automobile license plates use three letters followed by three numerals to make up the "license plate number." (There are other combinations of letters and numerals used; however, let's focus only on this six-character "number" for now.) If we assume that any one of the 26 letters may be used for each of the first three characters and that any one of the 10 numerals 0 through 9 can be used for each of the last three characters, how many different license plate numbers are possible?

SOLUTION There are 26 possible choices for the first letter ($n_1 = 26$), 26 possible choices for the second letter ($n_2 = 26$), and 26 possible choices for the third letter ($n_3 = 26$). In similar fashion, there are 10 choices for the numeral to be used for each of the fourth ($n_4 = 10$), fifth ($n_5 = 10$), and sixth ($n_6 = 10$) characters. Therefore, using the General Counting Rule (formula A-2), we find there are

$$26 \times 26 \times 26 \times 10 \times 10 \times 10 = \textbf{17,576,000}$$

▲ different "license plate numbers" using this six-character scheme.

▼ Illustration   A - 3

How many different "license plate numbers" are possible if the non-zero numerals are used for the three leading characters, letters are used for the three trailing characters, and the letters are not allowed to repeat?

SOLUTION There are 9 possible choices for each of the first three characters (since only 1 through 9 may be used). Thus, $n_1 = 9$, $n_2 = 9$, and $n_3 = 9$. The fourth character may be chosen from any one of the 26 letters ($n_4 = 26$). However, the fifth character must be chosen from any one of the 25 letters not previously used ($n_5 = 25$), and the sixth character must be chosen from the 24 letters not previously used ($n_6 = 24$). Applying the General Counting Rule (A-2), we find there are

$$9 \times 9 \times 9 \times 26 \times 25 \times 24 = \textbf{11,372,400}$$

▲ different "license plate numbers" using this second six-character scheme.

We are now ready to investigate two additional concepts commonly encountered when enumerating possibilities: *permutations* and *combinations*. A permutation is a collection of distinct objects arranged in a specific order, while a combination is a collection of distinct objects without any specific order.

▼ Illustration   A - 4

There are four flags of different colors (one each of red, white, blue, and green) in a box, and you are asked to select any three. If you select {red, white, green} you have the same combination of colors as {green, red, white}. This question does not require or distinguish between different "orders" or arrangements; thus each set of ▲ flags is one combination.

I L L U S T R A T I O N   A - 5

There are four flags of different colors (one each of red, white, blue, and green) in a box, and you are asked to select any three of them and make a "signal" by hanging the three different flags, one above the other, on a flagpole. Since red over green over white is different from green over white over red, order is important and each

▲| possible signal is one permutation.

PERMUTATIONS

▼| I L L U S T R A T I O N   A - 6

Select four different letters from the English alphabet and arrange them in any specific order. As a result of following these instructions, Barbara created the "four-letter word" BSJT. Rob created the word EOST. Steve selected KOCM. How many different "four-letter words" can be created?

Each of these "words" is a *permutation* of four letters selected from the set of 26 different letters forming the alphabet.

PERMUTATION

An ordered arrangement of a set of distinct objects. That is, there is a first object, a second object, a third object, and so on; and each object is distinctly different from the others.

The number of permutations that can be formed is calculated using an adaptation of the General Counting Rule.

PERMUTATION FORMULA

The number of permutations that can be formed using r different objects selected from a set of n distinct objects (symbolized by $_nP_r$ and read "the number of permutations of n objects selected r at a time") is

$$_nP_r = n \times (n - 1) \times (n - 2) \times \cdots \times (n - r + 1) \qquad \text{(A-3)}$$

or, in factorial notation,

$$_nP_r = \frac{n!}{(n - r)!} \qquad \text{(A-4)}$$

NOTE   More information and additional illustrations of factorial notation can be found in Chapter 5 and in the *Statistical Tutor*. Remember: $0! = 1$.

Let's continue with the solution of Illustration A-6. Since the 4 letters were selected from the 26 letters of the alphabet, the value of $r = 4$ (the number of selections) and $n = 26$ (the number of objects available for selection). Using formula A-3,

$$_nP_r = n \times (n - 1) \times \cdots \times (n - r + 1):$$
$$_{26}P_4 = 26 \times 25 \times \cdots \times (26 - 4 + 1)$$
$$= 26 \times 25 \times 24 \times 23 = \mathbf{358{,}800}$$

Or using formula A-4,

$$_nP_r = \frac{n!}{(n - r)!}:$$

$$_{26}P_4 = \frac{26!}{(26 - 4)!} = \frac{26!}{22!}$$

$$= \frac{26 \times 25 \times 24 \times 23 \times 22 \times 21 \times \cdots \times 1}{22 \times 21 \times \cdots \times 1}$$

$$= \frac{26 \times 25 \times 24 \times 23 \times (22 \times 21 \times \cdots \times 1)}{(22 \times 21 \times \cdots \times 1)}$$

$$= 26 \times 25 \times 24 \times 23 = \mathbf{358{,}800}$$

▲ |

▼ | ILLUSTRATION   A - 7

A group of eight finalists in a ceramic art competition are to be awarded five prizes—first, second, and so on. How many different ways are there to award these five prizes?

SOLUTION Since the prizes are ordered, this is a permutation of $n = 8$ different people, taken 5 at a time (only five prizes, and each prize is distinctly different from the others). Using formula (A-3), we find there are

$$_nP_r = n \times (n - 1) \times \cdots \times (n - r + 1):$$
$$_8P_5 = 8 \times 7 \times \cdots \times (8 - 5 + 1)$$
$$= 8 \times 7 \times 6 \times 5 \times 4 = \mathbf{6{,}720}$$

▲ | different possible ways of awarding these five prizes.

COMBINATIONS

▼ | ILLUSTRATION   A - 8

Select a set of four different letters from the English alphabet. As a result of following this instruction, Kevin selected A, E, R, and T. Karen selected D, E, N, and Q. Sue selected R, E, A, and T. Notice that Kevin and Sue selected the same set of letters, even though they selected them in different orders. These three people have selected two different sets of four letters. How many different sets of four letters can be selected?

SOLUTION Each of these "sets" of four letters represents a *combination* of $r = 4$ objects having been selected from a set of $n = 26$ distinct objects.

COMBINATION

A set of distinct objects without regard to an arrangement or an order. That is, the membership of the set is all that matters.

The number of combinations that can be selected is related to the number of permutations. In Illustration A-6, we found that there were 358,800 permutations of four letters possible. Many permutations were "words" formed from the same set of four letters. For example, the set of four letters A, B, C, and D can be used to form many permutations ("words"):

ABCD	ABDC	ACBD	ACDB	ADBC	ADCB
BACD	BADC	BCAD	BCDA	BDAC	BDCA
CBAD	CBDA	CABD	CADB	CDBA	CDAB
DBCA	DBAC	DCBA	DCAB	DABC	DACB

There are 4! ($4 \times 3 \times 2 \times 1$) or 24 different permutations for this set of four letters. Every other set of four letters can also be used to form 24 permutations. Therefore, if we divide the number of permutations possible (358,800) by the number of permutations each set has (24), the quotient will be the number of different sets (combinations) possible. That is, there are 14,950 (358,800/24) combinations of four letters possible. This concept is generalized in the following formula:

COMBINATION FORMULA

The number of combinations of r objects that can be selected from a set of n distinct objects (symbolized by $_nC_r$, and read "the number of combinations of n things taken r at a time") is

$$_nC_r = \frac{n(n-1)(n-2)\cdots(n-r+1)}{r!} \qquad \text{(A-5)}$$

or, in factorial notation,

$$_nC_r = \frac{n!}{(n-r)! \times r!} \qquad \text{(A-6)}$$

Let's continue with the solution of Illustration A-8 using these new formulas. First using formula (A-5),

$$_nC_r = \frac{n(n-1)(n-2) \cdots (n-r+1)}{r!}:$$

$$_{26}C_4 = \frac{26(25)(24) \cdots (26-4+1)}{4!} = \frac{26 \times 25 \times 24 \times 23}{4 \times 3 \times 2 \times 1}$$

$$= \frac{358,800}{24} = \mathbf{14,950}$$

Using formula (A-6),

$$_nC_r = \frac{n!}{(n-r)! \times r!}:$$

$$_{26}C_4 = \frac{26!}{(26-4)! \times 4!} = \frac{26!}{22! \times 4!}$$

$$= \frac{26 \times 25 \times 24 \times 23 \times 22 \times 21 \times \cdots \times 2 \times 1}{(22 \times 21 \times \cdots \times 2 \times 1)(4 \times 3 \times 2 \times 1)}$$

$$= \frac{26 \times 25 \times 24 \times 23 \times (22 \times 21 \times \cdots \times 2 \times 1)}{(22 \times 21 \times \cdots \times 2 \times 1)(4 \times 3 \times 2 \times 1)}$$

$$= \frac{358,800}{24} = \mathbf{14,950}$$

▲ |

▼ | I L L U S T R A T I O N   A - 9

A department has 30 members and a committee of 5 people is needed to carry out a task. How many different possible committees are there?

SOLUTION As stated, there is no specific assignment or order to the members of the committee; therefore, each possible committee is a combination and $n = 30$ (the number of people eligible to be selected), and $r = 5$ (the number to be selected).

$$_nC_r = \frac{n!}{(n-r)! \times r!}:$$

$$_{30}C_5 = \frac{30!}{(30-5)! \times 5!} = \frac{30!}{25! \times 5!}$$

$$= \frac{30 \times 29 \times 28 \times 27 \times 26 \times 25 \times 24 \times \cdots \times 2 \times 1}{(25 \times 24 \times \cdots \times 2 \times 1)(5 \times 4 \times 3 \times 2 \times 1)}$$

$$= \frac{30 \times 29 \times 28 \times 27 \times 26 \times (25 \times 24 \times \cdots \times 2 \times 1)}{(25 \times 24 \times \cdots \times 2 \times 1)(5 \times 4 \times 3 \times 2 \times 1)}$$

$$= \frac{17,100,720}{120} = \mathbf{142,506}$$

▲ | It is possible to select 142,506 different committees of 5 people from this department of 30 people.

NOTE   The number of combinations $_nC_r$ and the binomial coefficient $\binom{n}{r}$ or $\binom{n}{x}$ are numerically equivalent.

The three "counting" formulas described above in this section (formulas A-2, A-4, and A-6) can be and often are used together to solve problems.

▼| ILLUSTRATION   A - 10

A department has 30 members and a committee is needed to carry out a task. The committee is to be composed of a chairperson and four members. How many different possible committees are there?

SOLUTION   This problem is solved by treating it in two parts: consider the chairperson position and the committee members are two separate parts to be combined using the Fundamental Counting Rule ($m \times n$). Let m be the number of possible choices for the chairperson. Since any one of the 30 department members could serve as the chair, $m = 30$. Let n be the number of four-person committees that can be selected from the remaining 29 department members. Since these four have no specific assignment, the number of possibilities, n, is the number of combinations of 29 things taken 4 at a time.

$$n(\text{committees}) = m \times n$$
$$= 30 \times {}_{29}C_4$$
$$= 30 \times \frac{29 \times 28 \times 27 \times 26 \times (25 \times 24 \times \cdots \times 2 \times 1)}{(25 \times 24 \times \cdots \times 2 \times 1) \times 4 \times 3 \times 2 \times 1}$$
$$= 30 \times 23{,}751 = \mathbf{712{,}530}$$

▲| 712,530 different committees of size 5 with an assigned chair are possible.

▼| ILLUSTRATION   A - 11

A department has 30 members and a committee is needed to carry out a task. The committee is to be composed of two co-chairpersons and three members. How many different possible committees are there?

SOLUTION   This problem is solved by treating it in two parts: consider the selecting of two co-chairpersons and then the remaining committee members as two separate parts to be combined using the Fundamental Counting Rule ($m \times n$). Let m be the number of possible choices for the co-chairpersons. This is like a committee of two, since there is no further distinction between them; therefore $m = {}_{30}C_2$, since any two of the 30 department members could serve as the co-chairs. Let n be the number of three-person committees that can be selected from the remaining 28 department members. Since these three have no specific assignment, the number of possibilities, n, is the number of combinations of 28 things taken 3 at a time.

$$n(\text{committees}) = m \times n$$

$$= {}_{30}C_2 \times {}_{28}C_3$$

$$= \frac{30!}{28! \times 2!} \times \frac{28!}{25! \times 3!}$$

$$= \frac{(30 \times 29 \times 28 \times 27 \times 26 \times 25 \times \cdots \times 1) \times (28 \times 27 \times \cdots \times 1)}{(28 \times 27 \times \cdots \times 1) \times (2 \times 1) \times (25 \times \cdots \times 1) \times (3 \times 2 \times 1)}$$

$$= 15 \times 29 \times 28 \times 9 \times 13$$

$$= \mathbf{1{,}425{,}060}$$

▲ | 1,425,060 different committees of size 5 with assigned co-chairpersons are possible.

▼

EXERCISES

A.1   A long weekend of three days is being planned. The three days are to be spent taking scenic drives through the countryside and ending each day at a motel where reservations have been previously made for Friday and Saturday nights (will be home Sunday night). There are three scenic routes that may be traveled on Friday, two choices for Saturday, and three scenic route choices for Sunday's return trip. How many different trips are possible if
   a. all the scenic options are considered?
   b. one of the Friday routes has been previously driven, and is not a choice?
   c. on Sunday, it is decided to drive straight home and not take a scenic route?

A.2   a. Show that formulas (A-3) and (A-4) are equivalent.
   b. Show that formulas (A-5) and (A-6) are equivalent.

A.3   Explain why each of the following pairs of "counts" are equal:
   a. ${}_{n}P_{n}$ and ${}_{n}P_{n-1}$ b. ${}_{n}P_{1}$ and ${}_{n}C_{1}$
   c. ${}_{n}C_{r}$ and ${}_{n}C_{n-r}$ d. ${}_{n}C_{r}$ and the binomial coefficient $\binom{n}{r}$.

A.4   A department of 30 people is to select a committee of 5 persons. How many different committees are possible if the committee is composed of
   a. a chairperson, a secretary, and three others?
   b. two co-chairs and three others?
   c. two co-chairs, a secretary, and two others?

A.5   License plates are to be "numbered" using a combination of letters and numerals. How many different "numbers" are possible if each of the following sets of restrictions is used?
   a. Six characters using any combination or arrangement of the 26 letters and 10 single-digit numerals.
   b. Six characters using any combination or arrangement of letters and single-digit numerals, except that "zero" and "one" are not to be used because they are hard to distinguish from "o" and "i."

    c. Six characters using letters for the two leading characters and the 10 single-digit numerals for the four trailing characters.

    d. Six characters using the 10 single-digit numerals for the four leading characters and letters for the two trailing characters.

    e. Six characters using the 10 single-digit numerals for the four leading characters and letters for the two trailing characters, except that "zero" cannot be the leading character.

A.6 Mathew has six shirts, four pairs of pants, and five pairs of socks clean and ready to wear. How many different "outfits" can he assemble if

    a. he wears one item from each category?

    b. he wears one specific shirt and one item from the other two categories?

    c. he only wears two of the shirts with one specific pair of pants and no socks, but the rest are worn in any complete combination?

A.7 Five cards are to be randomly selected from a standard bridge deck of 52 cards.

    a. How many different "hands" of five cards are possible?

    b. How many different "hands" of five cards are possible if the first drawn is an ace?

    c. How many different "hands" of five cards are possible if the first card drawn is an ace and the remaining four are not aces?

    d. How many different "hands" of five cards are possible if the first card drawn is a club and the remaining four are not clubs?

    e. How many different "hands" of five cards are possible if the first card drawn is an ace and the remaining four are not clubs?

APPENDIX B

TABLES

TABLE 1   Random Numbers

10	09	73	25	33	76	52	01	35	86	34	67	35	48	76	80	95	90	91	17	39	29	27	49	45
37	54	20	48	05	64	89	47	42	96	24	80	52	40	37	20	63	61	04	02	00	82	29	16	65
08	42	26	89	53	19	64	50	93	03	23	20	90	25	60	15	95	33	43	64	35	08	03	36	06
99	01	90	25	29	09	37	67	07	15	38	31	13	11	65	88	67	67	43	97	04	43	62	76	59
12	80	79	99	70	80	15	73	61	47	64	03	23	66	53	98	95	11	68	77	12	17	17	68	33
66	06	57	47	17	34	07	27	68	50	36	69	73	61	70	65	81	33	98	85	11	19	92	91	70
31	06	01	08	05	45	57	18	24	06	35	30	34	26	14	86	79	90	74	39	23	40	30	97	32
85	26	97	76	02	02	05	16	56	92	68	66	57	48	18	73	05	38	52	47	18	62	38	85	79
63	57	33	21	35	05	32	54	70	48	90	55	35	75	48	28	46	82	87	09	83	49	12	56	24
73	79	64	57	53	03	52	96	47	78	35	80	83	42	82	60	93	52	03	44	35	27	38	84	35
98	52	01	77	67	14	90	56	86	07	22	10	94	05	58	60	97	09	34	33	50	50	07	39	98
11	80	50	54	31	39	80	82	77	32	50	72	56	82	48	29	40	52	42	01	52	77	56	78	51
83	45	29	96	34	06	28	89	80	83	13	74	67	00	78	18	47	54	06	10	68	71	17	78	17
88	68	54	02	00	86	50	75	84	01	36	76	66	79	51	90	36	47	64	93	29	60	91	10	62
99	59	46	73	48	87	51	76	49	69	91	82	60	89	28	93	78	56	13	68	23	47	83	41	13
65	48	11	76	74	17	46	85	09	50	58	04	77	69	74	73	03	95	71	86	40	21	81	65	44
80	12	43	56	35	17	72	70	80	15	45	31	82	23	74	21	11	57	82	53	14	38	55	37	63
74	35	09	98	17	77	40	27	72	14	43	23	60	02	10	45	52	16	42	37	96	28	60	26	55
69	91	62	68	03	66	25	22	91	48	36	93	68	72	03	76	62	11	39	90	94	40	05	64	18
09	89	32	05	05	14	22	56	85	14	46	42	75	67	88	96	29	77	88	22	54	38	21	45	98
91	49	91	45	23	68	47	92	76	86	46	16	28	35	54	94	75	08	99	23	37	08	92	00	48
80	33	69	45	98	26	94	03	68	58	70	29	73	41	35	54	14	03	33	40	42	05	08	23	41
44	10	48	19	49	85	15	74	79	54	32	97	92	65	75	57	60	04	08	81	22	22	20	64	13
12	55	07	37	42	11	10	00	20	40	12	86	07	46	97	96	64	48	94	39	28	70	72	58	15
63	60	64	93	29	16	50	53	44	84	40	21	95	25	63	43	65	17	70	82	07	20	73	17	90
61	19	69	04	46	26	45	74	77	74	51	92	43	37	29	65	39	45	95	93	42	58	26	05	27
15	47	44	52	66	95	27	07	99	53	59	36	78	38	48	82	39	61	01	18	33	21	15	94	66
94	55	72	85	73	67	89	75	43	87	54	62	24	44	31	91	19	04	25	92	92	92	74	59	73
42	48	11	62	13	97	34	40	87	21	16	86	84	87	67	03	07	11	20	59	25	70	14	66	70
23	52	37	83	17	73	20	88	98	37	68	93	59	14	16	26	25	22	96	63	05	52	28	25	62
04	49	35	24	94	75	24	63	38	24	45	86	25	10	25	61	96	27	93	35	65	33	71	24	72
00	54	99	76	54	64	05	18	81	59	96	11	96	38	96	54	69	28	23	91	23	28	72	95	29
35	96	31	53	07	26	89	80	93	54	33	35	13	54	62	77	97	45	00	24	90	10	33	93	33
59	80	80	83	91	45	42	72	68	42	83	60	94	97	00	13	02	12	48	92	78	56	52	01	06
46	05	88	52	36	01	39	09	22	86	77	28	14	40	77	93	91	08	36	47	70	61	74	29	41
32	17	90	05	97	87	37	92	52	41	05	56	70	70	07	86	74	31	71	57	85	39	41	18	38
69	23	46	14	06	20	11	74	52	04	15	95	66	00	00	18	74	39	24	23	97	11	89	63	38
19	56	54	14	30	01	75	87	53	79	40	41	92	15	85	66	67	43	68	06	84	96	28	52	07
45	15	51	49	38	19	47	60	72	46	43	66	79	45	43	59	04	79	00	33	20	82	66	95	41
94	86	43	19	94	36	16	81	08	51	34	88	88	15	53	01	54	03	54	56	05	01	45	11	76

For specific details on the use of this table, see the *Statistical Tutor*.

TABLE 1   (Continued)

98	08	62	48	26	45	24	02	84	04	44	99	90	88	96	39	09	47	34	07	35	44	13	18	80
33	18	51	62	32	41	94	15	09	49	89	43	54	85	81	88	69	54	19	94	37	54	87	30	43
80	95	10	04	06	96	38	27	07	74	20	15	12	33	87	25	01	62	52	98	94	62	46	11	71
79	75	24	91	40	71	96	12	82	96	69	86	10	25	91	74	85	22	05	39	00	38	75	95	79
18	63	33	25	37	98	14	50	65	71	31	01	02	46	74	05	45	56	14	27	77	93	89	19	36
74	02	94	39	02	77	55	73	22	70	97	79	01	71	19	52	52	75	80	21	80	81	45	17	48
54	17	84	56	11	80	99	33	71	43	05	33	51	29	69	56	12	71	92	55	36	04	09	03	24
11	66	44	98	83	52	07	98	48	27	59	38	17	15	39	09	97	33	34	40	88	46	12	33	56
48	32	47	79	28	31	24	96	47	10	02	29	53	68	70	32	30	75	75	46	15	02	00	99	94
69	07	49	41	38	87	63	79	19	76	35	58	40	44	01	10	51	82	16	15	01	84	87	69	38
09	18	82	00	97	32	82	53	95	27	04	22	08	63	04	83	38	98	73	74	64	27	85	80	44
90	04	58	54	97	51	98	15	06	54	94	93	88	19	97	91	87	07	61	50	68	47	66	46	59
73	18	95	02	07	47	67	72	62	69	62	29	06	44	64	27	12	46	70	18	41	36	18	27	60
75	76	87	64	90	20	97	18	17	49	90	42	91	22	72	95	37	50	58	71	93	82	34	31	78
54	01	64	40	56	66	28	13	10	03	00	68	22	73	98	20	71	45	32	95	07	70	61	78	13
08	35	86	99	10	78	54	24	27	85	13	66	15	88	73	04	61	89	75	53	31	22	30	84	20
28	30	60	32	64	81	33	31	05	91	40	51	00	78	93	32	60	46	04	75	94	11	90	18	40
53	84	08	62	33	81	59	41	36	28	51	21	59	02	90	28	46	66	87	95	77	76	22	07	91
91	75	75	37	41	61	61	36	22	69	50	26	39	02	12	55	78	17	65	14	83	48	34	70	55
89	41	59	26	94	00	39	75	83	91	12	60	71	76	46	48	94	97	23	06	94	54	13	74	08
77	51	30	38	20	86	83	42	99	01	68	41	48	27	74	51	90	81	39	80	72	89	35	55	07
19	50	23	71	74	69	97	92	02	88	55	21	02	97	73	74	28	77	52	51	65	34	46	74	15
21	81	85	93	13	93	27	88	17	57	05	68	67	31	56	07	08	28	50	46	31	85	33	84	52
51	47	46	64	99	68	10	72	36	21	94	04	99	13	45	42	83	60	91	91	08	00	74	54	49
99	55	96	83	31	62	53	52	41	70	69	77	71	28	30	74	81	97	81	42	43	86	07	28	34
33	71	34	80	07	93	58	47	28	69	51	92	66	47	21	58	30	32	98	22	93	17	49	39	72
85	27	48	68	93	11	30	32	92	70	28	83	43	41	37	73	51	59	04	00	71	14	84	36	43
84	13	38	96	40	44	03	55	21	66	73	85	27	00	91	61	22	26	05	61	62	32	71	84	23
56	73	21	62	34	17	39	59	61	31	10	12	39	16	22	85	49	65	75	60	81	60	41	88	80
65	13	85	68	06	87	60	88	52	61	34	31	36	58	61	45	87	52	10	69	85	64	44	72	77
38	00	10	21	76	81	71	91	17	11	71	60	29	29	37	74	21	96	40	49	65	58	44	96	98
37	40	29	63	97	01	30	47	75	86	56	27	11	00	86	47	32	46	26	05	40	03	03	74	38
97	12	54	03	48	87	08	33	14	17	21	81	53	92	50	75	23	76	20	47	15	50	12	95	78
21	82	64	11	34	47	14	33	40	72	64	63	88	59	02	49	13	90	64	41	03	85	65	45	52
73	13	54	27	42	95	71	90	90	35	85	79	47	42	96	08	78	98	81	56	64	69	11	92	02
07	63	87	79	29	03	06	11	80	72	96	20	74	41	56	23	82	19	95	38	04	71	36	69	94
60	52	88	34	41	07	95	41	98	14	59	17	52	06	95	05	53	35	21	39	61	21	20	64	55
83	59	63	56	55	06	95	89	29	83	05	12	80	97	19	77	43	35	37	83	92	30	15	04	98
10	85	06	27	46	99	59	91	05	07	13	49	90	63	19	53	07	57	18	39	06	41	01	93	62
39	82	09	89	52	43	62	26	31	47	64	42	18	08	14	43	80	00	93	51	31	02	47	31	67
59	58	00	64	78	75	56	97	88	00	88	83	55	44	86	23	76	80	61	56	04	11	10	84	08
38	50	80	73	41	23	79	34	87	63	90	82	29	70	22	17	71	90	42	07	95	95	44	99	53
30	69	27	06	68	94	68	81	61	27	56	19	68	00	91	82	06	76	34	00	05	46	26	92	00
65	44	39	56	59	18	28	82	74	37	49	63	22	40	41	08	33	76	56	76	96	29	99	08	36
27	26	75	02	64	13	19	27	22	94	07	47	74	46	06	17	98	54	89	11	97	34	13	03	58
91	30	70	69	91	19	07	22	42	10	36	69	95	37	28	28	82	53	57	93	28	97	66	62	52
68	43	49	46	88	84	47	31	36	22	62	12	69	84	08	12	84	38	25	90	09	81	59	31	46
48	90	81	58	77	54	74	52	45	91	35	70	00	47	54	83	82	45	26	92	54	13	05	51	60
06	91	34	51	97	42	67	27	86	01	11	88	30	95	28	63	01	19	89	01	14	97	44	03	44
10	45	51	60	19	14	21	03	37	12	91	34	23	78	21	88	32	58	08	51	43	66	77	08	83
12	88	39	73	43	65	02	76	11	84	04	28	50	13	92	17	97	41	50	77	90	71	22	67	69
21	77	83	09	76	38	80	73	69	61	31	64	94	20	96	63	28	10	20	23	08	81	64	74	49
19	52	35	95	15	65	12	25	96	59	86	28	36	82	58	69	57	21	37	98	16	43	59	15	29
67	24	55	26	70	35	58	31	65	63	79	24	68	66	86	76	46	33	42	22	26	65	59	08	02
60	58	44	73	77	07	50	03	79	92	45	13	42	65	29	26	76	08	36	37	41	32	64	43	44
53	85	34	13	77	36	06	69	48	50	58	83	87	38	59	49	36	47	33	31	96	24	04	36	42
24	63	73	87	36	74	38	48	93	42	52	62	30	79	92	12	36	91	86	01	03	74	28	38	73
83	08	01	24	51	38	99	22	28	15	07	75	95	17	77	97	37	72	75	85	51	97	23	78	67
16	44	42	43	34	36	15	19	90	73	27	49	37	09	39	85	13	03	25	52	54	84	65	47	59
60	79	01	81	57	57	17	86	57	62	11	16	17	85	76	45	81	95	29	79	65	13	00	48	60

From tables of the RAND Corporation. Reprinted from Wilfred J. Dixon and Frank J. Massey, Jr., *Introduction to Statistical Analysis*. 3rd ed. (New York: McGraw-Hill, 1969), pp. 446–447. Reprinted by permission of the RAND Corporation.

TABLE 2 Binomial Probabilities $\left[\binom{n}{x} \cdot p^x q^{n-x}\right]$

n	x	0.01	0.05	0.10	0.20	0.30	0.40	0.50	0.60	0.70	0.80	0.90	0.95	0.99	x
2	0	.980	.902	.810	.640	.490	.360	.250	.160	.090	.040	.010	.002	0+	0
	1	.020	.095	.180	.320	.420	.480	.500	.480	.420	.320	.180	.095	.020	1
	2	0+	.002	.010	.040	.090	.160	.250	.360	.490	.640	.810	.902	.980	2
3	0	.970	.857	.729	.512	.343	.216	.125	.064	.027	.008	.001	0+	0+	0
	1	.029	.135	.243	.384	.441	.432	.375	.288	.189	.096	.027	.007	0+	1
	2	0+	.007	.027	.096	.189	.288	.375	.432	.441	.384	.243	.135	.029	2
	3	0+	0+	.001	.008	.027	.064	.125	.216	.343	.512	.729	.857	.970	3
4	0	.961	.815	.656	.410	.240	.130	.062	.026	.008	.002	0+	0+	0+	0
	1	.039	.171	.292	.410	.412	.346	.250	.154	.076	.026	.004	0+	0+	1
	2	.001	.014	.049	.154	.265	.346	.375	.346	.265	.154	.049	.014	.001	2
	3	0+	0+	.004	.026	.076	.154	.250	.346	.412	.410	.292	.171	.039	3
	4	0+	0+	0+	.002	.008	.026	.062	.130	.240	.410	.656	.815	.961	4
5	0	.951	.774	.590	.328	.168	.078	.031	.010	.002	0+	0+	0+	0+	0
	1	.048	.204	.328	.410	.360	.259	.156	.077	.028	.006	0+	0+	0+	1
	2	.001	.021	.073	.205	.309	.346	.312	.230	.132	.051	.008	.001	0+	2
	3	0+	.001	.008	.051	.132	.230	.312	.346	.309	.205	.073	.021	.001	3
	4	0+	0+	0+	.006	.028	.077	.156	.259	.360	.410	.328	.204	.048	4
	5	0+	0+	0+	0+	.002	.010	.031	.078	.168	.328	.590	.774	.951	5
6	0	.941	.735	.531	.262	.118	.047	.016	.004	.001	0+	0+	0+	0+	0
	1	.057	.232	.354	.393	.303	.187	.094	.037	.010	.002	0+	0+	0+	1
	2	.001	.031	.098	.246	.324	.311	.234	.138	.060	.015	.001	0+	0+	2
	3	0+	.002	.015	.082	.185	.276	.312	.276	.185	.082	.015	.002	0+	3
	4	0+	0+	.001	.015	.060	.138	.234	.311	.324	.246	.098	.031	.001	4
	5	0+	0+	0+	.002	.010	.037	.094	.187	.303	.393	.354	.232	.057	5
	6	0+	0+	0+	0+	.001	.004	.016	.047	.118	.262	.531	.735	.941	6
7	0	.932	.698	.478	.210	.082	.028	.008	.002	0+	0+	0+	0+	0+	0
	1	.066	.257	.372	.367	.247	.131	.055	.017	.004	0+	0+	0+	0+	1
	2	.002	.041	.124	.275	.318	.261	.164	.077	.025	.004	0+	0+	0+	2
	3	0+	.004	.023	.115	.227	.290	.273	.194	.097	.029	.003	0+	0+	3
	4	0+	0+	.003	.029	.097	.194	.273	.290	.227	.115	.023	.004	0+	4
	5	0+	0+	0+	.004	.025	.077	.164	.261	.318	.275	.124	.041	.002	5
	6	0+	0+	0+	0+	.004	.017	.055	.131	.247	.367	.372	.257	.066	6
	7	0+	0+	0+	0+	0+	.002	.008	.028	.082	.210	.478	.698	.932	7
8	0	.923	.663	.430	.168	.058	0.17	.004	.001	0+	0+	0+	0+	0+	0
	1	.075	.279	.383	.336	.198	.090	.031	.008	.001	0+	0+	0+	0+	1
	2	.003	.051	.149	.294	.296	.209	.109	.041	.010	.001	0+	0+	0+	2
	3	0+	.005	.033	.147	.254	.279	.219	.124	.047	.009	0+	0+	0+	3
	4	0+	0+	.005	.046	.136	.232	.273	.232	.136	.046	.005	0+	0+	4
	5	0+	0+	0+	.009	.047	.124	.219	.279	.254	.147	.033	.005	0+	5
	6	0+	0+	0+	.001	.010	.041	.109	.209	.296	.294	.149	.051	.003	6
	7	0+	0+	0+	0+	.001	.008	.031	.090	.198	.336	.383	.279	.075	7
	8	0+	0+	0+	0+	0+	.001	.004	.017	.058	.168	.430	.663	.923	8

For specific details on the use of this table, see p. 272.

TABLE 2 (Continued)

n	x	0.01	0.05	0.10	0.20	0.30	0.40	p 0.50	0.60	0.70	0.80	0.90	0.95	0.99	x
9	0	.914	.630	.387	.134	.040	.010	.002	0+	0+	0+	0+	0+	0+	0
	1	.083	.299	.387	.302	.156	.060	.018	.004	0+	0+	0+	0+	0+	1
	2	.003	.063	.172	.302	.267	.161	.070	.021	.004	0+	0+	0+	0+	2
	3	0+	.008	.045	.176	.267	.251	.164	.074	.021	.003	0+	0+	0+	3
	4	0+	.001	.007	.066	.172	.251	.246	.167	.074	.017	.001	0+	0+	4
	5	0+	0+	.001	.017	.074	.167	.246	.251	.172	.066	.007	.001	0+	5
	6	0+	0+	0+	.003	.021	.074	.164	.251	.267	.176	.045	.008	0+	6
	7	0+	0+	0+	0+	.004	.021	.070	.161	.267	.302	.172	.063	.003	7
	8	0+	0+	0+	0+	0+	.004	.018	.060	.156	.302	.387	.299	.083	8
	9	0+	0+	0+	0+	0+	0+	.002	.010	.040	.134	.387	.630	.914	9
10	0	.904	.599	.349	.107	.028	.006	.001	0+	0+	0+	0+	0+	0+	0
	1	.091	.315	.387	.268	.121	.040	.010	.002	0+	0+	0+	0+	0+	1
	2	.004	.075	.194	.302	.233	.121	.044	.011	.001	0+	0+	0+	0+	2
	3	0+	.010	.057	.201	.267	.215	.117	.042	.009	.001	0+	0+	0+	3
	4	0+	.001	.011	.088	.200	.251	.205	.111	.037	.006	0+	0+	0+	4
	5	0+	0+	.001	.026	.103	.201	.246	.201	.103	.026	.001	0+	0+	5
	6	0+	0+	0+	.006	.037	.111	.205	.251	.200	.088	.011	.001	0+	6
	7	0+	0+	0+	.001	.009	.042	.117	.215	.267	.201	.057	.010	0+	7
	8	0+	0+	0+	0+	.001	.011	.044	.121	.233	.302	.194	.075	.004	8
	9	0+	0+	0+	0+	0+	.002	.010	.040	.121	.268	.387	.315	.091	9
	10	0+	0+	0+	0+	0+	0+	.001	.006	.028	.107	.349	.599	.904	10
11	0	.895	.569	.314	.086	.020	.004	0+	0+	0+	0+	0+	0+	0+	0
	1	.099	.329	.384	.236	.093	.027	.005	.001	0+	0+	0+	0+	0+	1
	2	.005	.087	.213	.295	.200	.089	.027	.005	.001	0+	0+	0+	0+	2
	3	0+	.014	.071	.221	.257	.177	.081	.023	.004	0+	0+	0+	0+	3
	4	0+	.001	.016	.111	.220	.236	.161	.070	.017	.002	0+	0+	0+	4
	5	0+	0+	.002	.039	.132	.221	.226	.147	.057	.010	0+	0+	0+	5
	6	0+	0+	0+	.010	.057	.147	.226	.221	.132	.039	.002	0+	0+	6
	7	0+	0+	0+	.002	.017	.070	.161	.236	.220	.111	.016	.001	0+	7
	8	0+	0+	0+	0+	.004	.023	.081	.177	.257	.221	.071	.014	0+	8
	9	0+	0+	0+	0+	.001	.005	.027	.089	.200	.295	.213	.087	.005	9
	10	0+	0+	0+	0+	0+	.001	.005	.027	.093	.236	.384	.329	.099	10
	11	0+	0+	0+	0+	0+	0+	0+	.004	.020	.086	.314	.569	.895	11
12	0	.886	.540	.282	.069	.014	.002	0+	0+	0+	0+	0+	0+	0+	0
	1	.107	.341	.377	.206	.071	.017	.003	0+	0+	0+	0+	0+	0+	1
	2	.006	.099	.230	.283	.168	.064	.016	.002	0+	0+	0+	0+	0+	2
	3	0+	.017	.085	.236	.240	.142	.054	.012	.001	0+	0+	0+	0+	3
	4	0+	.002	.021	.133	.231	.213	.121	.042	.008	.001	0+	0+	0+	4
	5	0+	0+	.004	.053	.158	.227	.193	.101	.029	.003	0+	0+	0+	5
	6	0+	0+	0+	.016	.079	.177	.226	.177	.079	.016	0+	0+	0+	6
	7	0+	0+	0+	.003	.029	.101	.193	.227	.158	.053	.004	0+	0+	7
	8	0+	0+	0+	.001	.008	.042	.121	.213	.231	.133	.021	.002	0+	8
	9	0+	0+	0+	0+	.001	.012	.054	.142	.240	.236	.085	.017	0+	9
	10	0+	0+	0+	0+	0+	.002	.016	.064	.168	.283	.230	.099	.006	10
	11	0+	0+	0+	0+	0+	0+	.003	.017	.071	.206	.377	.341	.107	11
	12	0+	0+	0+	0+	0+	0+	0+	.002	.014	.069	.282	.540	.886	12

TABLE 2 (Continued)

n	x	0.01	0.05	0.10	0.20	0.30	0.40	0.50	0.60	0.70	0.80	0.90	0.95	0.99	x
13	0	.878	.513	.254	.055	.010	.001	0+	0+	0+	0+	0+	0+	0+	0
	1	.115	.351	.367	.179	.054	.011	.002	0+	0+	0+	0+	0+	0+	1
	2	.007	.111	.245	.268	.139	.045	.010	.001	0+	0+	0+	0+	0+	2
	3	0+	.021	.100	.246	.218	.111	.035	.006	.001	0+	0+	0+	0+	3
	4	0+	.003	.028	.154	.234	.184	.087	.024	.003	0+	0+	0+	0+	4
	5	0+	0+	.006	.069	.180	.221	.157	.066	.014	.001	0+	0+	0+	5
	6	0+	0+	.001	.023	.103	.197	.209	.131	.044	.006	0+	0+	0+	6
	7	0+	0+	0+	.006	.044	.131	.209	.197	.103	.023	.001	0+	0+	7
	8	0+	0+	0+	.001	.014	.066	.157	.221	.180	.069	.006	0+	0+	8
	9	0+	0+	0+	0+	.003	.024	.087	.184	.234	.154	.028	.003	0+	9
	10	0+	0+	0+	0+	.001	.006	.035	.111	.218	.246	.100	.021	0+	10
	11	0+	0+	0+	0+	0+	.001	.010	.045	.139	.268	.245	.111	.007	11
	12	0+	0+	0+	0+	0+	0+	.002	.011	.054	.179	.367	.351	.115	12
	13	0+	0+	0+	0+	0+	0+	0+	.001	.010	.055	.254	.513	.878	13
14	0	.869	.488	.229	.044	.007	.001	0+	0+	0+	0+	0+	0+	0+	0
	1	.123	.359	.356	.154	.041	.007	.001	0+	0+	0+	0+	0+	0+	1
	2	.008	.123	.257	.250	.113	.032	.006	.001	0+	0+	0+	0+	0+	2
	3	0+	.026	.114	.250	.194	.085	.022	.003	0+	0+	0+	0+	0+	3
	4	0+	.004	.035	.172	.229	.155	.061	.014	.001	0+	0+	0+	0+	4
	5	0+	0+	.008	.086	.196	.207	.122	.041	.007	0+	0+	0+	0+	5
	6	0+	0+	.001	.032	.126	.207	.183	.092	.023	.002	0+	0+	0+	6
	7	0+	0+	0+	.009	.062	.157	.209	.157	.062	.009	0+	0+	0+	7
	8	0+	0+	0+	.002	.023	.092	.183	.207	.126	.032	.001	0+	0+	8
	9	0+	0+	0+	0+	.007	.041	.122	.207	.196	.086	.008	0+	0+	9
	10	0+	0+	0+	0+	.001	.014	.061	.155	.229	.172	.035	.004	0+	10
	11	0+	0+	0+	0+	0+	.003	.022	.085	.194	.250	.114	.026	0+	11
	12	0+	0+	0+	0+	0+	.001	.006	.032	.113	.250	.257	.123	.008	12
	13	0+	0+	0+	0+	0+	0+	.001	.007	.041	.154	.356	.359	.123	13
	14	0+	0+	0+	0+	0+	0+	0+	.001	.007	.044	.229	.488	.869	14
15	0	.860	.463	.206	.035	.005	0+	0+	0+	0+	0+	0+	0+	0+	0
	1	.130	.366	.343	.132	.031	.005	0+	0+	0+	0+	0+	0+	0+	1
	2	.009	.135	.267	.231	.092	.022	.003	0+	0+	0+	0+	0+	0+	2
	3	0+	.031	.129	.250	.170	.063	.014	.002	0+	0+	0+	0+	0+	3
	4	0+	.005	.043	.188	.219	.127	.042	.007	.001	0+	0+	0+	0+	4
	5	0+	.001	.010	.103	.206	.186	.092	.024	.003	0+	0+	0+	0+	5
	6	0+	0+	.002	.043	.147	.207	.153	.061	.012	.001	0+	0+	0+	6
	7	0+	0+	0+	.014	.081	.177	.196	.118	.035	.003	0+	0+	0+	7
	8	0+	0+	0+	.003	.035	.118	.196	.177	.081	.014	0+	0+	0+	8
	9	0+	0+	0+	.001	.012	.061	.153	.207	.147	.043	.002	0+	0+	9
	10	0+	0+	0+	0+	.003	.024	.092	.186	.206	.103	.010	.001	0+	10
	11	0+	0+	0+	0+	.001	.007	.042	.127	.219	.188	.043	.005	0+	11
	12	0+	0+	0+	0+	0+	.002	.014	.063	.170	.250	.129	.031	0+	12
	13	0+	0+	0+	0+	0+	0+	.003	.022	.092	.231	.267	.135	.009	13
	14	0+	0+	0+	0+	0+	0+	0+	.005	.031	.132	.343	.366	.130	14
	15	0+	0+	0+	0+	0+	0+	0+	0+	.005	.035	.206	.463	.860	15

TABLE 3   Areas of the Standard Normal Distribution

The entries in this table are the probabilities that a random variable with a standard normal distribution assumes a value between 0 and z; the probability is represented by the shaded area under the curve in the accompanying figure. Areas for negative values of z are obtained by symmetry. See page 292 for specific instructions for use of this table.

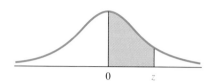

z	0.00	0.01	0.02	0.03	Second Decimal Place in z 0.04	0.05	0.06	0.07	0.08	0.09
0.0	0.0000	0.0040	0.0080	0.0120	0.0160	0.0199	0.0239	0.0279	0.0319	0.0359
0.1	0.0398	0.0438	0.0478	0.0517	0.0557	0.0596	0.0636	0.0675	0.0714	0.0753
0.2	0.0793	0.0832	0.0871	0.0910	0.0948	0.0987	0.1026	0.1064	0.1103	0.1141
0.3	0.1179	0.1217	0.1255	0.1293	0.1331	0.1368	0.1406	0.1443	0.1480	0.1517
0.4	0.1554	0.1591	0.1628	0.1664	0.1700	0.1736	0.1772	0.1808	0.1844	0.1879
0.5	0.1915	0.1950	0.1985	0.2019	0.2054	0.2088	0.2123	0.2157	0.2190	0.2224
0.6	0.2257	0.2291	0.2324	0.2357	0.2389	0.2422	0.2454	0.2486	0.2517	0.2549
0.7	0.2580	0.2611	0.2642	0.2673	0.2704	0.2734	0.2764	0.2794	0.2823	0.2852
0.8	0.2881	0.2910	0.2939	0.2967	0.2995	0.3023	0.3051	0.3078	0.3106	0.3133
0.9	0.3159	0.3186	0.3212	0.3238	0.3264	0.3289	0.3315	0.3340	0.3365	0.3389
1.0	0.3413	0.3438	0.3461	0.3485	0.3508	0.3531	0.3554	0.3577	0.3599	0.3621
1.1	0.3643	0.3665	0.3686	0.3708	0.3729	0.3749	0.3770	0.3790	0.3810	0.3830
1.2	0.3849	0.3869	0.3888	0.3907	0.3925	0.3944	0.3962	0.3980	0.3997	0.4015
1.3	0.4032	0.4049	0.4066	0.4082	0.4099	0.4115	0.4131	0.4147	0.4162	0.4177
1.4	0.4192	0.4207	0.4222	0.4236	0.4251	0.4265	0.4279	0.4292	0.4306	0.4319
1.5	0.4332	0.4345	0.4357	0.4370	0.4382	0.4394	0.4406	0.4418	0.4429	0.4441
1.6	0.4452	0.4463	0.4474	0.4484	0.4495	0.4505	0.4515	0.4525	0.4535	0.4545
1.7	0.4554	0.4564	0.4573	0.4582	0.4591	0.4599	0.4608	0.4616	0.4625	0.4633
1.8	0.4641	0.4649	0.4656	0.4664	0.4671	0.4678	0.4686	0.4693	0.4699	0.4706
1.9	0.4713	0.4719	0.4726	0.4732	0.4738	0.4744	0.4750	0.4756	0.4761	0.4767
2.0	0.4772	0.4778	0.4783	0.4788	0.4793	0.4798	0.4803	0.4808	0.4812	0.4817
2.1	0.4821	0.4826	0.4830	0.4834	0.4838	0.4842	0.4846	0.4850	0.4854	0.4857
2.2	0.4861	0.4864	0.4868	0.4871	0.4875	0.4878	0.4881	0.4884	0.4887	0.4890
2.3	0.4893	0.4896	0.4898	0.4901	0.4904	0.4906	0.4909	0.4911	0.4913	0.4916
2.4	0.4918	0.4920	0.4922	0.4925	0.4927	0.4929	0.4931	0.4932	0.4934	0.4936
2.5	0.4938	0.4940	0.4941	0.4943	0.4945	0.4946	0.4948	0.4949	0.4951	0.4952
2.6	0.4953	0.4955	0.4956	0.4957	0.4959	0.4960	0.4961	0.4962	0.4963	0.4964
2.7	0.4965	0.4966	0.4967	0.4968	0.4969	0.4970	0.4971	0.4972	0.4973	0.4974
2.8	0.4974	0.4975	0.4976	0.4977	0.4977	0.4978	0.4979	0.4979	0.4980	0.4981
2.9	0.4981	0.4982	0.4982	0.4983	0.4984	0.4984	0.4985	0.4985	0.4986	0.4986
3.0	0.4987	0.4987	0.4987	0.4988	0.4988	0.4989	0.4989	0.4989	0.4990	0.4990
3.1	0.4990	0.4991	0.4991	0.4991	0.4992	0.4992	0.4992	0.4992	0.4993	0.4993
3.2	0.4993	0.4993	0.4994	0.4994	0.4994	0.4994	0.4994	0.4995	0.4995	0.4995
3.3	0.4995	0.4995	0.4995	0.4996	0.4996	0.4996	0.4996	0.4996	0.4996	0.4997
3.4	0.4997	0.4997	0.4997	0.4997	0.4997	0.4997	0.4997	0.4997	0.4997	0.4998
3.5	0.4998	0.4998	0.4998	0.4998	0.4998	0.4998	0.4998	0.4998	0.4998	0.4998
3.6	0.4998	0.4998	0.4999	0.4999	0.4999	0.4999	0.4999	0.4999	0.4999	0.4999
3.7	0.4999									
4.0	0.49997									
4.5	0.499997									
5.0	0.4999997									

TABLE 4   Critical Values of Standard Normal Distribution

A | ONE-TAILED SITUATIONS

The entries in this table are the critical values for z for which the area under the curve representing α is in the right-hand tail. Critical values for the left-hand tail are found by symmetry.

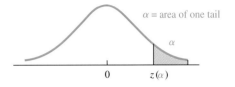

α = area of one tail

Amount of α in one tail

α	0.25	0.10	0.05	0.025	0.02	0.01	0.005
$z(\alpha)$	0.67	1.28	1.65	1.96	2.05	2.33	2.58

One-tailed example:
$\alpha = 0.05$
$z(\alpha) = z(0.05) = 1.65$

B | TWO-TAILED SITUATIONS

The entries in this table are the critical values for z for which the area under the curve representing α is split equally between the two tails.

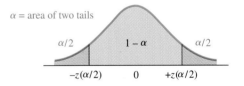

α = area of two tails

Amount of α in two tails

α	0.25	0.20	0.10	0.05	0.02	0.01
$z(\alpha/2)$	1.15	1.28	1.65	1.96	2.33	2.58
$1 - \alpha$	0.75	0.80	0.90	0.95	0.98	0.99

Area in the "center"

Two-tailed example:
$\alpha = 0.05$ or $1 - \alpha = 0.95$
$\alpha/2 = 0.025$
$z(\alpha/2) = z(0.025) = 1.96$

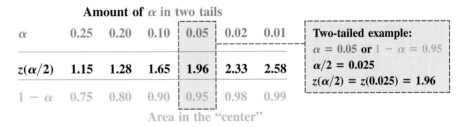

For specific details on the use of this table, see page 366.

TABLE 5   p-Values for Standard Normal Distribution

The entries in this table are the p-values related to the right-hand tail for the calculated z^* for the standard normal distribution. For specific instructions on using this table see page 393.

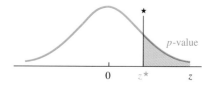

z^*	p-value	z^*	p-value	z^*	p-value	z^*	p-value
0.00	0.5000	1.00	0.1587	2.00	0.0228	3.00	0.0013
0.05	0.4801	1.05	0.1469	2.05	0.0202	3.05	0.0011
0.10	0.4602	1.10	0.1357	2.10	0.0179	3.10	0.0010
0.15	0.4404	1.15	0.1251	2.15	0.0158	3.15	0.0008
0.20	0.4207	1.20	0.1151	2.20	0.0139	3.20	0.0007
0.25	0.4013	1.25	0.1056	2.25	0.0122	3.25	0.0006
0.30	0.3821	1.30	0.0968	2.30	0.0107	3.30	0.0005
0.35	0.3632	1.35	0.0885	2.35	0.0094	3.35	0.0004
0.40	0.3446	1.40	0.0808	2.40	0.0082	3.40	0.0003
0.45	0.3264	1.45	0.0735	2.45	0.0071	3.45	0.0003
0.50	0.3085	1.50	0.0668	2.50	0.0062	3.50	0.0002
0.55	0.2912	1.55	0.0606	2.55	0.0054	3.55	0.0002
0.60	0.2743	1.60	0.0548	2.60	0.0047	3.60	0.0002
0.65	0.2578	1.65	0.0495	2.65	0.0040	3.65	0.0001
0.70	0.2420	1.70	0.0446	2.70	0.0035	3.70	0.0001
0.75	0.2266	1.75	0.0401	2.75	0.0030	3.75	0.0001
0.80	0.2119	1.80	0.0359	2.80	0.0026	3.80	0.0001
0.85	0.1977	1.85	0.0322	2.85	0.0022	3.85	0.0001
0.90	0.1841	1.90	0.0287	2.90	0.0019	3.90	0+
0.95	0.1711	1.95	0.0256	2.95	0.0016	3.95	0+

TABLE 6   Critical Values of Student's t-Distribution

The entries in this table, $t(\text{df}, \alpha)$, are the critical values for Student's t-distribution for which the area under the curve in the right-hand tail is α. Critical values for the left-hand tail are found by symmetry. For specific details on the use of this table, see page 435.

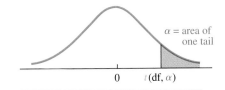

α = area of one tail

$0 \qquad t(\text{df}, \alpha)$

	Amount of α in One Tail					
	0.25	0.10	0.05	0.025	0.01	0.005
	Amount of α in Two Tails					
df	0.50	0.20	0.10	0.05	0.02	0.01
3	0.765	1.64	2.35	3.18	4.54	5.84
4	0.741	1.53	2.13	2.78	3.75	4.60
5	0.729	1.48	2.02	2.57	3.37	4.03
6	0.718	1.44	1.94	2.45	3.14	3.71
7	0.711	1.42	1.89	2.36	3.00	3.50
8	0.706	1.40	1.86	2.31	2.90	3.36
9	0.703	1.38	1.83	2.26	2.82	3.25
10	0.700	1.37	1.81	2.23	2.76	3.17
11	0.697	1.36	1.80	2.20	2.72	3.11
12	0.696	1.36	1.78	2.18	2.68	3.05
13	0.694	1.35	1.77	2.16	2.65	3.01
14	0.692	1.35	1.76	2.14	2.62	2.98
15	0.691	1.34	1.75	2.13	2.60	2.95
16	0.690	1.34	1.75	2.12	2.58	2.92
17	0.689	1.33	1.74	2.11	2.57	2.90
18	0.688	1.33	1.73	2.10	2.55	2.88
19	0.688	1.33	1.73	2.09	2.54	2.86
20	0.687	1.33	1.72	2.09	2.53	2.85
21	0.686	1.32	1.72	2.08	2.52	2.83
22	0.686	1.32	1.72	2.07	2.51	2.82
23	0.685	1.32	1.71	2.07	2.50	2.81
24	0.685	1.32	1.71	2.06	2.49	2.80
25	0.684	1.32	1.71	2.06	2.49	2.79
26	0.684	1.32	1.71	2.06	2.48	2.78
27	0.684	1.31	1.70	2.05	2.47	2.77
28	0.683	1.31	1.70	2.05	2.47	2.76
29	0.683	1.31	1.70	2.05	2.46	2.76
30	0.683	1.31	1.70	2.04	2.46	2.75
35	0.682	1.31	1.69	2.03	2.44	2.73
40	0.681	1.30	1.68	2.02	2.42	2.70
50	0.679	1.30	1.68	2.01	2.40	2.68
70	0.678	1.29	1.67	1.99	2.38	2.65
100	0.677	1.29	1.66	1.98	2.36	2.63
df > 100	0.675	1.28	1.65	1.96	2.33	2.58

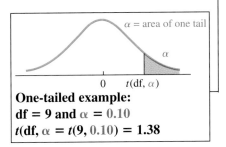

α = area of one tail

α

$0 \qquad t(\text{df}, \alpha)$

One-tailed example:
df = 9 and $\alpha = 0.10$
$t(\text{df}, \alpha = t(9, 0.10) = 1.38$

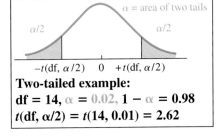

α = area of two tails

$\alpha/2 \qquad\qquad \alpha/2$

$-t(\text{df}, \alpha/2) \quad 0 \quad +t(\text{df}, \alpha/2)$

Two-tailed example:
df = 14, $\alpha = 0.02$, $1 - \alpha = 0.98$
$t(\text{df}, \alpha/2) = t(14, 0.01) = 2.62$

TABLE 7

Probability-Values
for Student's
t-distribution

The entries in this table are the p-values
related to the right-hand tail for the calculated t^* value for the t-distribution of df
degrees of freedom.

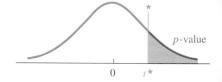

p-value

0 t^*

Degrees of Freedom

t^*	3	4	5	6	7	8	10	12	15	18	21	25	29	35	df≥45
0.0	0.500	0.500	0.500	0.500	0.500	0.500	0.500	0.500	0.500	0.500	0.500	0.500	0.500	0.500	0.500
0.1	0.463	0.463	0.462	0.462	0.462	0.461	0.461	0.461	0.461	0.461	0.461	0.461	0.461	0.460	0.460
0.2	0.427	0.426	0.425	0.424	0.424	0.423	0.423	0.422	0.422	0.422	0.422	0.422	0.421	0.421	0.421
0.3	0.392	0.390	0.388	0.387	0.386	0.386	0.385	0.385	0.384	0.384	0.384	0.383	0.383	0.383	0.383
0.4	0.358	0.355	0.353	0.352	0.351	0.350	0.349	0.348	0.347	0.347	0.347	0.346	0.346	0.346	0.346
0.5	0.326	0.322	0.319	0.317	0.316	0.315	0.314	0.313	0.312	0.312	0.311	0.311	0.310	0.310	0.310
0.6	0.295	0.290	0.287	0.285	0.284	0.283	0.281	0.280	0.279	0.278	0.277	0.277	0.277	0.276	0.276
0.7	0.267	0.261	0.258	0.255	0.253	0.252	0.250	0.249	0.247	0.246	0.246	0.245	0.245	0.244	0.244
0.8	0.241	0.234	0.230	0.227	0.225	0.223	0.221	0.220	0.218	0.217	0.216	0.216	0.215	0.215	0.214
0.9	0.217	0.210	0.205	0.201	0.199	0.197	0.195	0.193	0.191	0.190	0.189	0.188	0.188	0.187	0.186
1.0	0.196	0.187	0.182	0.178	0.175	0.173	0.170	0.169	0.167	0.165	0.164	0.163	0.163	0.162	0.161
1.1	0.176	0.167	0.161	0.157	0.154	0.152	0.149	0.146	0.144	0.143	0.142	0.141	0.140	0.139	0.139
1.2	0.158	0.148	0.142	0.138	0.135	0.132	0.129	0.127	0.124	0.123	0.122	0.121	0.120	0.119	0.118
1.3	0.142	0.132	0.125	0.121	0.117	0.115	0.111	0.109	0.107	0.105	0.104	0.103	0.102	0.101	0.100
1.4	0.128	0.117	0.110	0.106	0.102	0.100	0.096	0.093	0.091	0.089	0.088	0.087	0.086	0.085	0.084
1.5	0.115	0.104	0.097	0.092	0.089	0.086	0.082	0.080	0.077	0.075	0.074	0.073	0.072	0.071	0.070
1.6	0.104	0.092	0.085	0.080	0.077	0.074	0.070	0.068	0.065	0.064	0.062	0.061	0.060	0.059	0.058
1.7	0.094	0.082	0.075	0.070	0.066	0.064	0.060	0.057	0.055	0.053	0.052	0.051	0.050	0.049	0.048
1.8	0.085	0.073	0.066	0.061	0.057	0.055	0.051	0.049	0.046	0.044	0.043	0.042	0.041	0.040	0.039
1.9	0.077	0.065	0.058	0.053	0.050	0.047	0.043	0.041	0.038	0.037	0.036	0.035	0.034	0.033	0.032
2.0	0.070	0.058	0.051	0.046	0.043	0.040	0.037	0.034	0.032	0.030	0.029	0.028	0.027	0.027	0.026
2.1	0.063	0.052	0.045	0.040	0.037	0.034	0.031	0.029	0.027	0.025	0.024	0.023	0.022	0.022	0.021
2.2	0.058	0.046	0.040	0.035	0.032	0.029	0.026	0.024	0.022	0.021	0.020	0.019	0.018	0.017	0.016
2.3	0.052	0.041	0.035	0.031	0.027	0.025	0.022	0.020	0.018	0.017	0.016	0.015	0.014	0.014	0.013
2.4	0.048	0.037	0.031	0.027	0.024	0.022	0.019	0.017	0.015	0.014	0.013	0.012	0.012	0.011	0.010
2.5	0.044	0.033	0.027	0.023	0.020	0.018	0.016	0.014	0.012	0.011	0.010	0.010	0.009	0.009	0.008
2.6	0.040	0.030	0.024	0.020	0.018	0.016	0.013	0.012	0.010	0.009	0.008	0.008	0.007	0.007	0.006
2.7	0.037	0.027	0.021	0.018	0.015	0.014	0.011	0.010	0.008	0.007	0.007	0.006	0.006	0.005	0.005
2.8	0.034	0.024	0.019	0.016	0.013	0.012	0.009	0.008	0.007	0.006	0.005	0.005	0.005	0.004	0.004
2.9	0.031	0.022	0.017	0.014	0.011	0.010	0.008	0.007	0.005	0.005	0.004	0.004	0.004	0.003	0.003
3.0	0.029	0.020	0.015	0.012	0.010	0.009	0.007	0.006	0.004	0.004	0.003	0.003	0.003	0.002	0.002
3.1	0.027	0.018	0.013	0.011	0.009	0.007	0.006	0.005	0.004	0.003	0.003	0.002	0.002	0.002	0.002
3.2	0.025	0.016	0.012	0.009	0.008	0.006	0.005	0.004	0.003	0.002	0.002	0.002	0.002	0.001	0.001
3.3	0.023	0.015	0.011	0.008	0.007	0.005	0.004	0.003	0.002	0.002	0.002	0.001	0.001	0.001	0.001
3.4	0.021	0.014	0.010	0.007	0.006	0.005	0.003	0.003	0.002	0.002	0.001	0.001	0.001	0.001	0.001
3.5	0.020	0.012	0.009	0.006	0.005	0.004	0.003	0.002	0.002	0.001	0.001	0.001	0.001	0.001	0.001
3.6	0.018	0.011	0.008	0.006	0.004	0.004	0.002	0.002	0.001	0.001	0.001	0.001	0.001	0+	0+
3.7	0.017	0.010	0.007	0.005	0.004	0.003	0.002	0.002	0.001	0.001	0.001	0.001	0+	0+	0+
3.8	0.016	0.010	0.006	0.004	0.003	0.003	0.002	0.001	0.001	0.001	0.001	0+	0+	0+	0+
3.9	0.015	0.009	0.006	0.004	0.003	0.002	0.001	0.001	0.001	0.001	0+	0+	0+	0+	0+
4.0	0.014	0.008	0.005	0.004	0.003	0.002	0.001	0.001	0.001	0+	0+	0+	0+	0+	0+

For specific instructions on using this table see page 442.

TABLE 8

Critical Values of χ^2 ("Chi-Square") Distribution

The entries in this table, $\chi^2(df, \alpha)$, are the critical values for the χ^2 distribution for which the area under the curve to the right is α. For specific details on use of this table, see page 472.

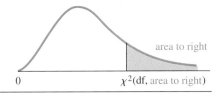

area to right

$\chi^2(df, \text{area to right})$

	Area to the Right												
	0.995	0.99	0.975	0.95	0.90	0.75	0.50	0.25	0.10	0.05	0.025	0.01	0.005
	Area in Left-hand Tail						Median	Area in Right-hand Tail					
df	0.005	0.01	0.025	0.05	0.10	0.25	0.50	0.25	0.10	0.05	0.025	0.01	0.005
1	0.0000393	0.000157	0.000982	0.00393	0.0158	0.101	0.455	1.32	2.71	3.84	5.02	6.63	7.88
2	0.0100	0.0201	0.0506	0.103	0.211	0.575	1.39	2.77	4.61	5.99	7.38	9.21	10.6
3	0.0717	0.115	0.216	0.352	0.584	1.21	2.37	4.11	6.25	7.82	9.35	11.3	12.8
4	0.207	0.297	0.484	0.711	1.06	1.92	3.36	5.39	7.78	9.49	11.1	13.3	14.9
5	0.412	0.554	0.831	1.15	1.61	2.67	4.35	6.63	9.24	11.1	12.8	15.1	16.8
6	0.676	0.872	1.24	1.64	2.20	3.45	5.35	7.84	10.6	12.6	14.5	16.8	18.6
7	0.990	1.24	1.69	2.17	2.83	4.25	6.35	9.04	12.0	14.1	16.0	18.5	20.3
8	1.34	1.65	2.18	2.73	3.49	5.07	7.34	10.2	13.4	15.5	17.5	20.1	22.0
9	1.73	2.09	2.70	3.33	4.17	5.90	8.34	11.4	14.7	16.9	19.0	21.7	23.6
10	2.16	2.56	3.25	3.94	4.87	6.74	9.34	12.5	16.0	18.3	20.5	23.2	25.2
11	2.60	3.05	3.82	4.57	5.58	7.58	10.34	13.7	17.3	19.7	21.9	24.7	26.8
12	3.07	3.57	4.40	5.23	6.30	8.44	11.34	14.8	18.5	21.0	23.3	26.2	28.3
13	3.57	4.11	5.01	5.89	7.04	9.30	12.34	16.0	19.8	22.4	24.7	27.7	29.8
14	4.07	4.66	5.63	6.57	7.79	10.2	13.34	17.1	21.1	23.7	26.1	29.1	31.3
15	4.60	5.23	6.26	7.26	8.55	11.0	14.34	18.2	22.3	25.0	27.5	30.6	32.8
16	5.14	5.81	6.91	7.96	9.31	11.9	15.34	19.4	23.5	26.3	28.8	32.0	34.3
17	5.70	6.41	7.56	8.67	10.1	12.8	16.34	20.5	24.8	27.6	30.2	33.4	35.7
18	6.26	7.01	8.23	9.39	10.9	13.7	17.34	21.6	26.0	28.9	31.5	34.8	37.2
19	6.84	7.63	8.91	10.1	11.7	14.6	18.34	22.7	27.2	30.1	32.9	36.2	38.6
20	7.43	8.26	9.59	10.9	12.4	15.5	19.34	23.8	28.4	31.4	34.2	37.6	40.0
21	8.03	8.90	10.3	11.6	13.2	16.3	20.34	24.9	29.6	32.7	35.5	38.9	41.4
22	8.64	9.54	11.0	12.3	14.0	17.2	21.34	26.0	30.8	33.9	36.8	40.3	42.8
23	9.26	10.2	11.7	13.1	14.8	18.1	22.34	27.1	32.0	35.2	38.1	41.6	44.2
24	9.89	10.9	12.4	13.8	15.7	19.0	23.34	28.2	33.2	36.4	39.4	43.0	45.6
25	10.5	11.5	13.1	14.6	16.5	19.9	24.34	29.3	34.4	37.7	40.6	44.3	46.9
26	11.2	12.2	13.8	15.4	17.3	20.8	25.34	30.4	35.6	38.9	41.9	45.6	48.3
27	11.8	12.9	14.6	16.2	18.1	21.7	26.34	31.5	36.7	40.1	43.2	47.0	49.6
28	12.5	13.6	15.3	16.9	18.9	22.7	27.34	32.6	37.9	41.3	44.5	48.3	51.0
29	13.1	14.3	16.0	17.7	19.8	23.6	28.34	33.7	39.1	42.6	45.7	49.6	52.3
30	13.8	15.0	16.8	18.5	20.6	24.5	29.34	34.8	40.3	43.8	47.0	50.9	53.7
40	20.7	22.2	24.4	26.5	29.1	33.7	39.34	45.6	51.8	55.8	59.3	63.7	66.8
50	28.0	29.7	32.4	34.8	37.7	42.9	49.33	56.3	63.2	67.5	71.4	76.2	79.5
60	35.5	37.5	40.5	43.2	46.5	52.3	59.33	67.0	74.4	79.1	83.3	88.4	92.0
70	43.3	45.4	48.8	51.7	55.3	61.7	69.33	77.6	85.5	90.5	95.0	100.0	104.0
80	51.2	53.5	57.2	60.4	64.3	71.1	79.33	88.1	96.6	102.0	107.0	112.0	116.0
90	59.2	61.8	65.6	69.1	73.3	80.6	89.33	98.6	108.0	113.0	118.0	124.0	128.0
100	67.3	70.1	74.2	77.9	82.4	90.1	99.33	109.0	118.0	124.0	130.0	136.0	140.0

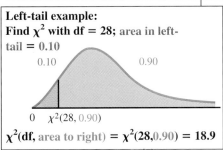

Left-tail example:
Find χ^2 with df = 28; area in left-tail = 0.10

0.10   0.90

0   $\chi^2(28, 0.90)$

$\chi^2(df, \text{area to right}) = \chi^2(28, 0.90) = 18.9$

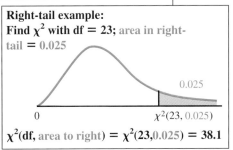

Right-tail example:
Find χ^2 with df = 23; area in right-tail = 0.025

0.025

0        $\chi^2(23, 0.025)$

$\chi^2(df, \text{area to right}) = \chi^2(23, 0.025) = 38.1$

TABLE 9a

Critical Values of
the *F* Distribution
($\alpha = 0.05$)

The entries in this table are critical values of *F* for which the area under the curve to the right is equal to 0.05.

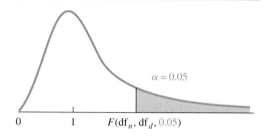

$\alpha = 0.05$

$0 \qquad 1 \qquad F(\mathrm{df}_n, \mathrm{df}_d, 0.05)$

Degrees of Freedom for Numerator

	1	2	3	4	5	6	7	8	9	10
1	161	200	216	225	230	234	237	239	241	242
2	18.5	19.0	19.2	19.2	19.3	19.3	19.4	19.4	19.4	19.4
3	10.1	9.55	9.28	9.12	9.01	8.94	8.89	8.85	8.81	8.79
4	7.71	6.94	6.59	6.39	6.26	6.16	6.09	6.04	6.00	5.96
5	6.61	5.79	5.41	5.19	5.05	4.95	4.88	4.82	4.77	4.74
6	5.99	5.14	4.76	4.53	4.39	4.28	4.21	4.15	4.10	4.06
7	5.59	4.74	4.35	4.12	3.97	3.87	3.79	3.73	3.68	3.64
8	5.32	4.46	4.07	3.84	3.69	3.58	3.50	3.44	3.39	3.35
9	5.12	4.26	3.86	3.63	3.48	3.37	3.29	3.23	3.18	3.14
10	4.96	4.10	3.71	3.48	3.33	3.22	3.14	3.07	3.02	2.98
11	4.84	3.98	3.59	3.36	3.20	3.09	3.01	2.95	2.90	2.85
12	4.75	3.89	3.49	3.26	3.11	3.00	2.91	2.85	2.80	2.75
13	4.67	3.81	3.41	3.18	3.03	2.92	2.83	2.77	2.71	2.67
14	4.60	3.74	3.34	3.11	2.96	2.85	2.76	2.70	2.65	2.60
15	4.54	3.68	3.29	3.06	2.90	2.79	2.71	2.64	2.59	2.54
16	4.49	3.63	3.24	3.01	2.85	2.74	2.66	2.59	2.54	2.49
17	4.45	3.59	3.20	2.96	2.81	2.70	2.61	2.55	2.49	2.45
18	4.41	3.55	3.16	2.93	2.77	2.66	2.58	2.51	2.46	2.41
19	4.38	3.52	3.13	2.90	2.74	2.63	2.54	2.48	2.42	2.38
20	4.35	3.49	3.10	2.87	2.71	2.60	2.51	2.45	2.39	2.35
21	4.32	3.47	3.07	2.84	2.68	2.57	2.49	2.42	2.37	2.32
22	4.30	3.44	3.05	2.82	2.66	2.55	2.46	2.40	2.34	2.30
23	4.28	3.42	3.03	2.80	2.64	2.53	2.44	2.37	2.32	2.27
24	4.26	3.40	3.01	2.78	2.62	2.51	2.42	2.36	2.30	2.25
25	4.24	3.39	2.99	2.76	2.60	2.49	2.40	2.34	2.28	2.24
30	4.17	3.32	2.92	2.69	2.53	2.42	2.33	2.27	2.21	2.16
40	4.08	3.23	2.84	2.61	2.45	2.34	2.25	2.18	2.12	2.08
60	4.00	3.15	2.76	2.53	2.37	2.25	2.17	2.10	2.04	1.99
120	3.92	3.07	2.68	2.45	2.29	2.18	2.09	2.02	1.96	1.91
∞	3.84	3.00	2.60	2.37	2.21	2.10	2.01	1.94	1.88	1.83

Degrees of Freedom for Denominator (left axis label)

For specific details on the use of this table, see p. 533.

TABLE 9a

(Continued)

		Degrees of Freedom for Numerator								
		12	15	20	24	30	40	60	120	∞
	1	244	246	248	249	250	251	252	253	254
	2	19.4	19.4	19.4	19.5	19.5	19.5	19.5	19.5	19.5
	3	8.74	8.70	8.66	8.64	8.62	8.59	8.57	8.55	8.53
	4	5.91	5.86	5.80	5.77	5.75	5.72	5.69	5.66	5.63
	5	4.68	4.62	4.56	4.53	4.50	4.46	4.43	4.40	4.37
	6	4.00	3.94	3.87	3.84	3.81	3.77	3.74	3.70	3.67
	7	3.57	3.51	3.44	3.41	3.38	3.34	3.30	3.27	3.23
	8	3.28	3.22	3.15	3.12	3.08	3.04	3.01	2.97	2.93
	9	3.07	3.01	2.94	2.90	2.86	2.83	2.79	2.75	2.71
Degrees of Freedom for Denominator	10	2.91	2.85	2.77	2.74	2.70	2.66	2.62	2.58	2.54
	11	2.79	2.72	2.65	2.61	2.57	2.53	2.49	2.45	2.40
	12	2.69	2.62	2.54	2.51	2.47	2.43	2.38	2.34	2.30
	13	2.60	2.53	2.46	2.42	2.38	2.34	2.30	2.25	2.21
	14	2.53	2.46	2.39	2.35	2.31	2.27	2.22	2.18	2.13
	15	2.48	2.40	2.33	2.29	2.25	2.20	2.16	2.11	2.07
	16	2.42	2.35	2.28	2.24	2.19	2.15	2.11	2.06	2.01
	17	2.38	2.31	2.23	2.19	2.15	2.10	2.06	2.01	1.96
	18	2.34	2.27	2.19	2.15	2.11	2.06	2.02	1.97	1.92
	19	2.31	2.23	2.16	2.11	2.07	2.03	1.98	1.93	1.88
	20	2.28	2.20	2.12	2.08	2.04	1.99	1.95	1.90	1.84
	21	2.25	2.18	2.10	2.05	2.01	1.96	1.92	1.87	1.81
	22	2.23	2.15	2.07	2.03	1.98	1.94	1.89	1.84	1.78
	23	2.20	2.13	2.05	2.01	1.96	1.91	1.86	1.81	1.76
	24	2.18	2.11	2.03	1.98	1.94	1.89	1.84	1.79	1.73
	25	2.16	2.09	2.01	1.96	1.92	1.87	1.82	1.77	1.71
	30	2.09	2.01	1.93	1.89	1.84	1.79	1.74	1.68	1.62
	40	2.00	1.92	1.84	1.79	1.74	1.69	1.64	1.58	1.51
	60	1.92	1.84	1.75	1.70	1.65	1.59	1.53	1.47	1.39
	120	1.83	1.75	1.66	1.61	1.55	1.50	1.43	1.35	1.25
	∞	1.75	1.67	1.57	1.52	1.46	1.39	1.32	1.22	1.00

TABLE 9b

Critical Values of
the F Distribution
($\alpha = 0.025$)

The entries in this table are critical
values of F for which the area
under the curve to the right is
equal to 0.025.

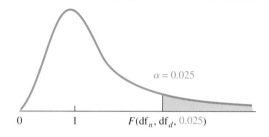

$\alpha = 0.025$

$0 \qquad 1 \qquad F(df_n, df_d, 0.025)$

Degrees of Freedom for Numerator

		1	2	3	4	5	6	7	8	9	10
Degrees of Freedom for Denominator	1	648	800	864	900	922	937	948	957	963	969
	2	38.5	39.0	39.2	39.2	39.3	39.3	39.4	39.4	39.4	39.4
	3	17.4	16.0	15.4	15.1	14.9	14.7	14.6	14.5	14.5	14.4
	4	12.2	10.6	9.98	9.60	9.36	9.20	9.07	8.98	8.90	8.84
	5	10.0	8.43	7.76	7.39	7.15	6.98	6.85	6.76	6.68	6.62
	6	8.81	7.26	6.60	6.23	5.99	5.82	5.70	5.60	5.52	5.46
	7	8.07	6.54	5.89	5.52	5.29	5.12	4.99	4.90	4.82	4.76
	8	7.57	6.06	5.42	5.05	4.82	4.65	4.53	4.43	4.36	4.30
	9	7.21	5.71	5.08	4.72	4.48	4.32	4.20	4.10	4.03	3.96
	10	6.94	5.46	4.83	4.47	4.24	4.07	3.95	3.85	3.78	3.72
	11	6.72	5.26	4.63	4.28	4.04	3.88	3.76	3.66	3.59	3.53
	12	6.55	5.10	4.47	4.12	3.89	3.73	3.61	3.51	3.44	3.37
	13	6.41	4.97	4.35	4.00	3.77	3.60	3.48	3.39	3.31	3.25
	14	6.30	4.86	4.24	3.89	3.66	3.50	3.38	3.28	3.21	3.15
	15	6.20	4.77	4.15	3.80	3.58	3.41	3.29	3.20	3.12	3.06
	16	6.12	4.69	4.08	3.73	3.50	3.34	3.22	3.12	3.05	2.99
	17	6.04	4.62	4.01	3.66	3.44	3.28	3.16	3.06	2.98	2.92
	18	5.98	4.56	3.95	3.61	3.38	3.22	3.10	3.01	2.93	2.87
	19	5.92	4.51	3.90	3.56	3.33	3.17	3.05	2.96	2.88	2.82
	20	5.87	4.46	3.86	3.51	3.29	3.13	3.01	2.91	2.84	2.77
	21	5.83	4.42	3.82	3.48	3.25	3.09	2.97	2.87	2.80	2.73
	22	5.79	4.38	3.78	3.44	3.22	3.05	2.93	2.84	2.76	2.70
	23	5.75	4.35	3.75	3.41	3.18	3.02	2.90	2.81	2.73	2.67
	24	5.72	4.32	3.72	3.38	3.15	2.99	2.87	2.78	2.70	2.64
	25	5.69	4.29	3.69	3.35	3.13	2.97	2.85	2.75	2.68	2.61
	30	5.57	4.18	3.59	3.25	3.03	2.87	2.75	2.65	2.57	2.51
	40	5.42	4.05	3.46	3.13	2.90	2.74	2.62	2.53	2.45	2.39
	60	5.29	3.93	3.34	3.01	2.79	2.63	2.51	2.41	2.33	2.27
	120	5.15	3.80	3.23	2.89	2.67	2.52	2.39	2.30	2.22	2.16
	∞	5.02	3.69	3.12	2.79	2.57	2.41	2.29	2.19	2.11	2.05

For specific details on the use of this table, see page 533.

TABLE 9b

(Continued)

		Degrees of Freedom for Numerator							
	12	15	20	24	30	40	60	120	∞
1	977	985	993	997	1,001	1,006	1,010	1,014	1,018
2	39.4	39.4	39.4	39.5	39.5	39.5	39.5	39.5	39.5
3	14.3	14.3	14.2	14.1	14.1	14.0	14.0	13.9	13.9
4	8.75	8.66	8.56	8.51	8.46	8.41	8.36	8.31	8.26
5	6.52	6.43	6.33	6.28	6.23	6.18	6.12	6.07	6.02
6	5.37	5.27	5.17	5.12	5.07	5.01	4.96	4.90	4.85
7	4.67	4.57	4.47	4.42	4.36	4.31	4.25	4.20	4.14
8	4.20	4.10	4.00	3.95	3.89	3.84	3.78	3.73	3.67
9	3.87	3.77	3.67	3.61	3.56	3.51	3.45	3.39	3.33
10	3.62	3.52	3.42	3.37	3.31	3.26	3.20	3.14	3.08
11	3.43	3.33	3.23	3.17	3.12	3.06	3.00	2.94	2.88
12	3.28	3.18	3.07	3.02	2.96	2.91	2.85	2.79	2.72
13	3.15	3.05	2.95	2.89	2.84	2.78	2.72	2.66	2.60
14	3.05	2.95	2.84	2.79	2.73	2.67	2.61	2.55	2.49
15	2.96	2.86	2.76	2.70	2.64	2.59	2.52	2.46	2.40
16	2.89	2.79	2.68	2.63	2.57	2.51	2.45	2.38	2.32
17	2.82	2.72	2.62	2.56	2.50	2.44	2.38	2.32	2.25
18	2.77	2.67	2.56	2.50	2.44	2.38	2.32	2.26	2.19
19	2.72	2.62	2.51	2.45	2.39	2.33	2.27	2.20	2.13
20	2.68	2.57	2.46	2.41	2.35	2.29	2.22	2.16	2.09
21	2.64	2.53	2.42	2.37	2.31	2.25	2.18	2.11	2.04
22	2.60	2.50	2.39	2.33	2.27	2.21	2.14	2.08	2.00
23	2.57	2.47	2.36	2.30	2.24	2.18	2.11	2.04	1.97
24	2.54	2.44	2.33	2.27	2.21	2.15	2.08	2.01	1.94
25	2.51	2.41	2.30	2.24	2.18	2.12	2.05	1.98	1.91
30	2.41	2.31	2.20	2.14	2.07	2.01	1.94	1.87	1.79
40	2.29	2.18	2.07	2.01	1.94	1.88	1.80	1.72	1.64
60	2.17	2.06	1.94	1.88	1.82	1.74	1.67	1.58	1.48
120	2.05	1.95	1.82	1.76	1.69	1.61	1.53	1.43	1.31
∞	1.94	1.83	1.71	1.64	1.57	1.48	1.39	1.27	1.00

Degrees of Freedom for Denominator

From E. S. Pearson and H. O. Hartley, *Biometrika Tables for Statisticians*, vol. I (1958), pp. 159–163. Reprinted by permission of the Biometrika Trustees.

TABLE 9c

Critical Values of
the *F* Distribution
($\alpha = 0.01$)

The entries in the table are critical
values of *F* for which the area
under the curve to the right is equal
to 0.01.

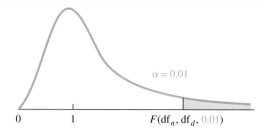

$\alpha = 0.01$

0   1 $F(df_n, df_d, 0.01)$

Degrees of Freedom for Numerator

	1	2	3	4	5	6	7	8	9	10
1	4,052	5,000	5,403	5,625	5,764	5,859	5,928	5,982	6,024	6,056
2	98.5	99.0	99.2	99.2	99.3	99.3	99.4	99.4	99.4	99.4
3	34.1	30.8	29.5	28.7	28.2	27.9	27.7	27.5	27.3	27.2
4	21.2	18.0	16.7	16.0	15.5	15.2	15.0	14.8	14.7	14.5
5	16.3	13.3	12.1	11.4	11.0	10.7	10.5	10.3	10.2	10.1
6	13.7	10.9	9.78	9.15	8.75	8.47	8.26	8.10	7.98	7.87
7	12.2	9.55	8.45	7.85	7.46	7.19	6.99	6.84	6.72	6.62
8	11.3	8.65	7.59	7.01	6.63	6.37	6.18	6.03	5.91	5.81
9	10.6	8.02	6.99	6.42	6.06	5.80	5.61	5.47	5.35	5.26
10	10.0	7.56	6.55	5.99	5.64	5.39	5.20	5.06	4.94	4.85
11	9.65	7.21	6.22	5.67	5.32	5.07	4.89	4.74	4.63	4.54
12	9.33	6.93	5.95	5.41	5.06	4.82	4.64	4.50	4.39	4.30
13	9.07	6.70	5.74	5.21	4.86	4.62	4.44	4.30	4.19	4.10
14	8.86	6.51	5.56	5.04	4.70	4.46	4.28	4.14	4.03	3.94
15	8.68	6.36	5.42	4.89	4.56	4.32	4.14	4.00	3.89	3.80
16	8.53	6.23	5.29	4.77	4.44	4.20	4.03	3.89	3.78	3.69
17	8.40	6.11	5.19	4.67	4.34	4.10	3.93	3.79	3.68	3.59
18	8.29	6.01	5.09	4.58	4.25	4.01	3.84	3.71	3.60	3.51
19	8.19	5.93	5.01	4.50	4.17	3.94	3.77	3.63	3.52	3.43
20	8.10	5.85	4.94	4.43	4.10	3.87	3.70	3.56	3.46	3.37
21	8.02	5.78	4.87	4.37	4.04	3.81	3.64	3.51	3.40	3.31
22	7.95	5.72	4.82	4.31	3.99	3.76	3.59	3.45	3.35	3.26
23	7.88	5.66	4.76	4.26	3.94	3.71	3.54	3.41	3.30	3.21
24	7.82	5.61	4.72	4.22	3.90	3.67	3.50	3.36	3.26	3.17
25	7.77	5.57	4.68	4.18	3.86	3.63	3.46	3.32	3.22	3.13
30	7.56	5.39	4.51	4.02	3.70	3.47	3.30	3.17	3.07	2.98
40	7.31	5.18	4.31	3.83	3.51	3.29	3.12	2.99	2.89	2.80
60	7.08	4.98	4.13	3.65	3.34	3.12	2.95	2.82	2.72	2.63
120	6.85	4.79	3.95	3.48	3.17	2.96	2.79	2.66	2.56	2.47
∞	6.63	4.61	3.78	3.32	3.02	2.80	2.64	2.51	2.41	2.32

Degrees of Freedom for Denominator

For specific details on the use of this table, see page 533.

TABLE 9c

(Continued)

		\multicolumn{9}{c}{**Degrees of Freedom for Numerator**}
		12	**15**	**20**	**24**	**30**	**40**	**60**	**120**	**∞**
\multirow{25}{*}{**Degrees of Freedom for Denominator**}	1	6,106	6,157	6,209	6,235	6,261	6,287	6,313	6,339	6,366
	2	99.4	99.4	99.4	99.5	99.5	99.5	99.5	99.5	99.5
	3	27.1	26.9	26.7	26.6	26.5	26.4	26.3	26.2	26.1
	4	14.4	14.2	14.0	13.9	13.8	13.7	13.7	13.6	13.5
	5	9.89	9.72	9.55	9.47	9.38	9.29	9.20	9.11	9.02
	6	7.72	7.56	7.40	7.31	7.23	7.14	7.06	6.97	6.88
	7	6.47	6.31	6.16	6.07	5.99	5.91	5.82	5.74	5.65
	8	5.67	5.52	5.36	5.28	5.20	5.12	5.03	4.95	4.86
	9	5.11	4.96	4.81	4.73	4.65	4.57	4.48	4.40	4.31
	10	4.71	4.56	4.41	4.33	4.25	4.17	4.08	4.00	3.91
	11	4.40	4.25	4.10	4.02	3.94	3.86	3.78	3.69	3.60
	12	4.16	4.01	3.86	3.78	3.70	3.62	3.54	3.45	3.36
	13	3.96	3.82	3.66	3.59	3.51	3.43	3.34	3.25	3.17
	14	3.80	3.66	3.51	3.43	3.35	3.27	3.18	3.09	3.00
	15	3.67	3.52	3.37	3.29	3.21	3.13	3.05	2.96	2.87
	16	3.55	3.41	3.26	3.18	3.10	3.02	2.93	2.84	2.75
	17	3.46	3.31	3.16	3.08	3.00	2.92	2.83	2.75	2.65
	18	3.37	3.23	3.08	3.00	2.92	2.84	2.75	2.66	2.57
	19	3.30	3.15	3.00	2.92	2.84	2.76	2.67	2.58	2.49
	20	3.23	3.09	2.94	2.86	2.78	2.69	2.61	2.52	2.42
	21	3.17	3.03	2.88	2.80	2.72	2.64	2.55	2.46	2.36
	22	3.12	2.98	2.83	2.75	2.67	2.58	2.50	2.40	2.31
	23	3.07	2.93	2.78	2.70	2.62	2.54	2.45	2.35	2.26
	24	3.03	2.89	2.74	2.66	2.58	2.49	2.40	2.31	2.21
	25	2.99	2.85	2.70	2.62	2.53	2.45	2.36	2.27	2.17
	30	2.84	2.70	2.55	2.47	2.39	2.30	2.21	2.11	2.01
	40	2.66	2.52	2.37	2.29	2.20	2.11	2.02	1.92	1.80
	60	2.50	2.35	2.20	2.12	2.03	1.94	1.84	1.73	1.60
	120	2.34	2.19	2.03	1.95	1.86	1.76	1.66	1.53	1.38
	∞	2.18	2.04	1.88	1.79	1.70	1.59	1.47	1.32	1.00

TABLE 10

Critical Values of r
When $\rho = 0$

The entries in this table are the critical values of r for a two-tailed test at α. For simple correlation, df $= n - 2$, where n is the number of pairs of data in the sample. For a one-tailed test, the value of α shown at the top of the table is double the value of α being used in the hypothesis test.

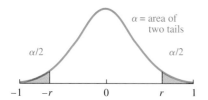

df	α 0.10	0.05	0.02	0.01
1	0.988	0.997	1.000	1.000
2	0.900	0.950	0.980	0.990
3	0.805	0.878	0.934	0.959
4	0.729	0.811	0.882	0.917
5	0.669	0.754	0.833	0.874
6	0.621	0.707	0.789	0.834
7	0.582	0.666	0.750	0.798
8	0.549	0.632	0.716	0.765
9	0.521	0.602	0.685	0.735
10	0.497	0.576	0.658	0.708
11	0.476	0.553	0.634	0.684
12	0.458	0.532	0.612	0.661
13	0.441	0.514	0.592	0.641
14	0.426	0.497	0.574	0.623
15	0.412	0.482	0.558	0.606
16	0.400	0.468	0.542	0.590
17	0.389	0.456	0.528	0.575
18	0.378	0.444	0.516	0.561
19	0.369	0.433	0.503	0.549
20	0.360	0.423	0.492	0.537
25	0.323	0.381	0.445	0.487
30	0.296	0.349	0.409	0.449
35	0.275	0.325	0.381	0.418
40	0.257	0.304	0.358	0.393
45	0.243	0.288	0.338	0.372
50	0.231	0.273	0.322	0.354
60	0.211	0.250	0.295	0.325
70	0.195	0.232	0.274	0.302
80	0.183	0.217	0.256	0.283
90	0.173	0.205	0.242	0.267
100	0.164	0.195	0.230	0.254

For specific details on the use of this table, see p. 656.

TABLE 11   Confidence Belts for the Correlation Coefficient $(1 - \alpha) = 0.95$

The numbers on the curves are sample sizes.

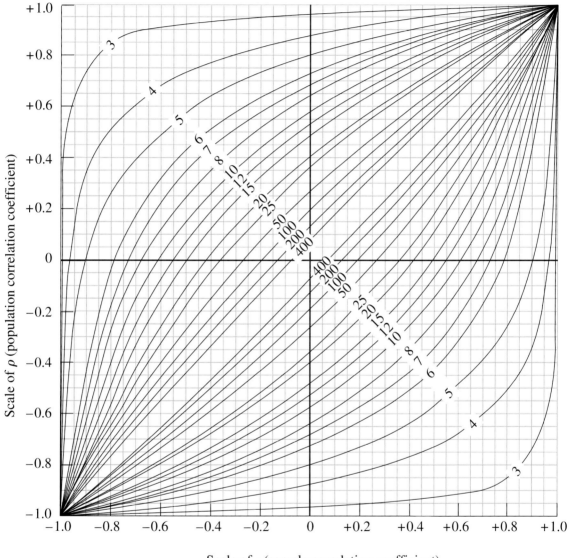

Scale of ρ (population correlation coefficient)

Scale of r (sample correlation coefficient)

For specific details on the use of this table see p. 657.

TABLE 12

Critical Values of the Sign Test

The entries in this table are the critical values for the number of the least frequent sign for a two-tailed test at α for the binomial $p = 0.5$. For a one-tailed test, the value of α shown at the top of the table is double the value of α being used in the hypothesis test.

n	0.01	0.05	0.10	0.25	n	0.01	0.05	0.10	0.25
1					51	15	18	19	20
2					52	16	18	19	21
3				0	53	16	18	20	21
4				0	54	17	19	20	22
5			0	0	55	17	19	20	22
6		0	0	1	56	17	20	21	23
7		0	0	1	57	18	20	21	23
8	0	0	1	1	58	18	21	22	24
9	0	1	1	2	59	19	21	22	24
10	0	1	1	2	60	19	21	23	25
11	0	1	2	3	61	20	22	23	25
12	1	2	2	3	62	20	22	24	25
13	1	2	3	3	63	20	23	24	26
14	1	2	3	4	64	21	23	24	26
15	2	3	3	4	65	21	24	25	27
16	2	3	4	5	66	22	24	25	27
17	2	4	4	5	67	22	25	26	28
18	3	4	5	6	68	22	25	26	28
19	3	4	5	6	69	23	25	27	29
20	3	5	5	6	70	23	26	27	29
21	4	5	6	7	71	24	26	28	30
22	4	5	6	7	72	24	27	28	30
23	4	6	7	8	73	25	27	28	31
24	5	6	7	8	74	25	28	29	31
25	5	7	7	9	75	25	28	29	32
26	6	7	8	9	76	26	28	30	32
27	6	7	8	10	77	26	29	30	32
28	6	8	9	10	78	27	29	31	33
29	7	8	9	10	79	27	30	31	33
30	7	9	10	11	80	28	30	32	34
31	7	9	10	11	81	28	31	32	34
32	8	9	10	12	82	28	31	33	35
33	8	10	11	12	83	29	32	33	35
34	9	10	11	13	84	29	32	33	36
35	9	11	12	13	85	30	32	34	36
36	9	11	12	14	86	30	33	34	37
37	10	12	13	14	87	31	33	35	37
38	10	12	13	14	88	31	34	35	38
39	11	12	13	15	89	31	34	36	38
40	11	13	14	15	90	32	35	36	39
41	11	13	14	16	91	32	35	37	39
42	12	14	15	16	92	33	36	37	39
43	12	14	15	17	93	33	36	38	40
44	13	15	16	17	94	34	37	38	40
45	13	15	16	18	95	34	37	38	41
46	13	15	16	18	96	34	37	39	41
47	14	16	17	19	97	35	38	39	42
48	14	16	17	19	98	35	38	40	42
49	15	17	18	19	99	36	39	40	43
50	15	17	18	20	100	36	39	41	43

*For specific details on the use of this table, see pp. 703 and 707. From Wilfred J. Dixon and Frank J. Massey, Jr., *Introduction to Statistical Analysis*, 3d ed. (New York: McGraw-Hill, 1969), p. 509. Reprinted by permission.

TABLE 13 Critical Values of U in the Mann-Whitney Test

A. The entries are the critical values of U for a one-tailed test at 0.025 or for a two-tailed test at 0.05.

n_2 \ n_1	1	2	3	4	5	6	7	8	9	10	11	12	13	14	15	16	17	18	19	20
1																				
2								0	0	0	0	1	1	1	1	1	2	2	2	2
3					0	1	1	2	2	3	3	4	4	5	5	6	6	7	7	8
4				0	1	2	3	4	4	5	6	7	8	9	10	11	11	12	13	13
5			0	1	2	3	5	6	7	8	9	11	12	13	14	15	17	18	19	20
6			1	2	3	5	6	8	10	11	13	14	16	17	19	21	22	24	25	27
7			1	3	5	6	8	10	12	14	16	18	20	22	24	26	28	30	32	34
8		0	2	4	6	8	10	13	15	17	19	22	24	26	29	31	34	36	38	41
9		0	2	4	7	10	12	15	17	20	23	26	28	31	34	37	39	42	45	48
10		0	3	5	8	11	14	17	20	23	26	29	33	36	39	42	45	48	52	55
11		0	3	6	9	13	16	19	23	26	30	33	37	40	44	47	51	55	58	62
12		1	4	7	11	14	18	22	26	29	33	37	41	45	49	53	57	61	65	69
13		1	4	8	12	16	20	24	28	33	37	41	45	50	54	59	63	67	72	76
14		1	5	9	13	17	22	26	31	36	40	45	50	55	59	64	67	74	78	83
15		1	5	10	14	19	24	29	34	39	44	49	54	59	64	70	75	80	85	90
16		1	6	11	15	21	26	31	37	42	47	53	59	64	70	75	81	86	92	98
17		2	6	11	17	22	28	34	39	45	51	57	63	67	75	81	87	93	99	105
18		2	7	12	18	24	30	36	42	48	55	61	67	74	80	86	93	99	106	112
19		2	7	13	19	25	32	38	45	52	58	65	72	78	85	92	99	106	113	119
20		2	8	13	20	27	34	41	48	55	62	69	76	83	90	98	105	112	119	127

B. The entries are the critical values of U for a one-tailed test at 0.05 or for a two-tailed test at 0.10.

n_2 \ n_1	1	2	3	4	5	6	7	8	9	10	11	12	13	14	15	16	17	18	19	20
1																			0	0
2					0	0	0	1	1	1	1	2	2	2	3	3	3	4	4	4
3			0	0	1	2	2	3	3	4	5	5	6	7	7	8	9	9	10	11
4			0	1	2	3	4	5	6	7	8	9	10	11	12	14	15	16	17	18
5		0	1	2	4	5	6	8	9	11	12	13	15	16	18	19	20	22	23	25
6		0	2	3	5	7	8	10	12	14	16	17	19	21	23	25	26	28	30	32
7		0	2	4	6	8	11	13	15	17	19	21	24	26	28	30	33	35	37	39
8		1	3	5	8	10	13	15	18	20	23	26	28	31	33	36	39	41	44	47
9		1	3	6	9	12	15	18	21	24	27	30	33	36	39	42	45	48	51	54
10		1	4	7	11	14	17	20	24	27	31	34	37	41	44	48	51	55	58	62
11		1	5	8	12	16	19	23	27	31	34	38	42	46	50	54	57	61	65	69
12		2	5	9	13	17	21	26	30	34	38	42	47	51	55	60	64	68	72	77
13		2	6	10	15	19	24	28	33	37	42	47	51	56	61	65	70	75	80	84
14		2	7	11	16	21	26	31	36	41	46	51	56	61	66	71	77	82	87	92
15		3	7	12	18	23	28	33	39	44	50	55	61	66	72	77	83	88	94	100
16		3	8	14	19	25	30	36	42	48	54	60	65	71	77	83	89	95	101	107
17		3	9	15	20	26	33	39	45	51	57	64	70	77	83	89	96	102	109	115
18		4	9	16	22	28	35	41	48	55	61	68	75	82	88	95	102	109	116	123
19	0	4	10	17	23	30	37	44	51	58	65	72	80	87	94	101	109	116	123	130
20	0	4	11	18	25	32	39	47	54	62	69	77	84	92	100	107	115	123	130	138

For specific details on the use of this table, see p. 714.

Reproduced from the *Bulletin of the Institute of Educational Research at Indiana University*, vol. 1, no. 2; with the permission of the author and the publisher.

TABLE 14   Critical Values for Total Number of Runs (V)

The entries in this table are the critical values for a two-tailed test using $\alpha = 0.05$. For a one-tailed test, $\alpha = 0.025$ and use only one of the critical values: the smaller critical value for a left-hand critical region, the larger for a right-hand critical region.

The smaller of n_1 and n_2 / The larger of n_1 and n_2

smaller \ larger	5	6	7	8	9	10	11	12	13	14	15	16	17	18	19	20
2							2/6	2/6	2/6	2/6	2/6	2/6	2/6	2/6	2/6	2/6
3		2/8	2/8	2/8	2/8	2/8	2/8	2/8	2/8	3/8	3/8	3/8	3/8	3/8	3/8	3/8
4	2/9	2/9	2/10	3/10	3/10	3/10	3/10	3/10	3/10	3/10	3/10	4/10	4/10	4/10	4/10	4/10
5	2/10	3/10	3/11	3/11	3/12	3/12	4/12	4/12	4/12	4/12	4/12	4/12	4/12	5/12	5/12	5/12
6		3/11	3/12	4/13	4/13	4/13	4/13	5/14	5/14	5/14	5/14	5/14	5/14	5/14	6/14	6/14
7			3/13	4/13	4/14	5/14	5/14	5/14	5/15	5/15	6/15	6/16	6/16	6/16	6/16	6/16
8				4/14	5/14	5/15	5/15	6/16	6/16	6/16	6/16	6/17	7/17	7/17	7/17	7/17
9					5/15	5/16	6/16	6/16	6/17	7/17	7/18	7/18	7/18	8/18	8/18	8/18
10						6/16	6/17	7/17	7/18	7/18	7/18	8/19	8/19	8/19	8/20	9/20
11							7/17	7/18	7/19	8/19	8/19	8/20	9/20	9/20	9/21	9/21
12								7/19	8/19	8/20	8/20	9/21	9/21	9/21	10/22	10/22
13									8/20	9/20	9/21	9/21	10/22	10/22	10/23	10/23
14										9/21	9/22	10/22	10/23	10/23	11/23	11/24
15											10/22	10/23	11/23	11/24	11/24	12/25
16												11/23	11/24	11/25	12/25	12/25
17													11/25	12/25	12/26	13/26
18														12/26	13/26	13/27
19															13/27	13/27
20																14/28

*See p.721 in regard to critical values.

From C. Eisenhart and F. Swed, "Tables for testing randomness of grouping in a sequence of alternatives," The Annals of Statistics, vol. 14 (1943): 66–87. Reprinted by permission.

TABLE 15 Critical Values of Spearman's Rank Correlation Coefficient

The entries in this table are the critical values of r_s for a two-tailed test at α. For a one-tailed test, the value of α shown at the top of the table is double the value of α being used in the hypothesis test.

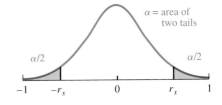

n	$\alpha = 0.10$	$\alpha = 0.05$	$\alpha = 0.2$	$\alpha = 0.01$
5	0.900	—	—	—
6	0.829	0.886	0.943	—
7	0.714	0.786	0.893	—
8	0.643	0.738	0.833	0.881
9	0.600	0.700	0.783	0.833
10	0.564	0.648	0.745	0.794
11	0.536	0.618	0.736	0.818
12	0.497	0.591	0.703	0.780
13	0.475	0.566	0.673	0.745
14	0.457	0.545	0.646	0.716
15	0.441	0.525	0.623	0.689
16	0.425	0.507	0.601	0.666
17	0.412	0.490	0.582	0.645
18	0.399	0.476	0.564	0.625
19	0.388	0.462	0.549	0.608
20	0.377	0.450	0.534	0.591
21	0.368	0.438	0.521	0.576
22	0.359	0.428	0.508	0.562
23	0.351	0.418	0.496	0.549
24	0.343	0.409	0.485	0.537
25	0.336	0.400	0.475	0.526
26	0.329	0.392	0.465	0.515
27	0.323	0.385	0.456	0.505
28	0.317	0.377	0.448	0.496
29	0.311	0.370	0.440	0.487
30	0.305	0.364	0.432	0.478

From E. G. Olds, "Distribution of sums of squares of rank differences for small numbers of individuals," *Annals of Statistics,* vol. 9 (1938), pp. 138–148, and amended, vol. 20 (1949), pp. 117–118. Reprinted by permission.
For specific details on the use of this table, see p. 728.

Answers to Selected Exercises

Chapter 1

1.1 a. King-Size Company customers   b. 10,000
c. 99% of the 10,000 customers considered airline seating cramped.
d. More than one answer was allowed.

1.2 a. American men
b. The samples described represent only part of the population, leaving large proportions of the population with no representation.

1.3 The percentage of people indicating negative responses increased considerably.

1.4 a. to determine the value that teachers, parents, and students put on various subjects
b. Teachers value writing more so than parents and students do.
c. 35% of the teachers picked writing as the most important skill, whereas only 20% of the parents and 21% of the students picked writing.

1.7 a. inferential   b. descriptive

1.9 a. Insurance companies have a major problem.
b. 48.7%   c. 21%   d. 48.7% vs. 21%
e. No, percentages alone do not indicate size of the numbers.

1.11 Measure the seat dimensions on the various airplanes and compare them to the dimensions that passengers do find comfortable.

1.13 marital status, ZIP code, gender, highest level of education

1.14 a. (1) Did most of your family eat dinner together last night?
(2) importance of eating dinner with your family
(3) number of evenings your family ate dinner together in the past 7 days
(4) length of dinner time
b. attribute, attribute, numerical, numerical

1.15 annual income, age, distance to store, amount spent

1.16 a. Height, build, arm length, club length
b. 4.087 yd/in., 3.827 yd/in.

1.17 a. all individuals who have hypertension and use prescription drugs to control it
b. the 5000 people in the study
c. the proportion of the population for which the drug is effective
d. the proportion of the sample for which the drug is effective, 80%
e. no

1.19 a. all assembled parts from the assembly line
b. infinite   c. the parts checked
d. attribute, attribute, quantitative

1.21 a. attribute
b. the percentage of all American households familiar with the food pyramid
c. the percentage of 372 households familiar with the food pyramid, 63%

1.23 a. numerical   b. attribute   c. numerical
d. attribute   e. numerical   f. numerical

1.25   a. The population contains all objects of interest; the sample contains only those actually studied.
      b. convenience, availability, practicality

1.27   the football players, because their weights cover a wider range of values

1.29   A lack of variability would indicate all students attained very similar scores.

1.31   a. all adults who deal with stockbrokers
      b. gender, answers to questions dealing with risk tolerance, information about explanations regarding investments and bonds, number of interruptions

1.32   a. American voters
      b. In 1992, 69% of Americans felt that it was the responsibility of government to take care of people who cannot take care of themselves.
      c. Only people with very strong opinions on the subjects will phone in.

1.33   Magazine/published *mail-in* surveys and radio and television broadcast surveys where the listener *calls-in* by telephone.

1.35   a. There are three choices for the first pick and then still three choices for the second pick; therefore, 3 times 3 gives 9 different samples.
      b. (1, 1), (1, 2), (1, 3), (1, 4), (2, 1), (2, 2), (2, 3), (2, 4), (3, 1), (3, 2), (3, 3), (3, 4), (4, 1), (4, 2), (4, 3), (4, 4)
      c. (1, 1, 1), (1, 1, 2), (1, 1, 3), (1, 2, 1), (1, 2, 2), (1, 2, 3), (1, 3, 1), (1, 3, 2), (1, 3, 3), (2, 1, 1), (2, 1, 2), (2, 1, 3), (2, 2, 1), (2, 2, 2), (2, 2, 3), (2, 3, 1), (2, 3, 2), (2, 3, 3), (3, 1, 1), (3, 1, 2), (3, 1, 3), (3, 2, 1), (3, 2, 2), (3, 2, 3), (3, 3, 1), (3, 3, 2), (3, 3, 3)

1.37   when the population is extremely large or spread out

1.39   a. judgment sample
      b. No, statistical inference requires probability samples.

1.41   Only people with telephones and listed phone numbers will be considered.

1.43   a. probability   b. statistics

1.45   a. statistics   b. statistics
      c. probability   d. statistics

1.47   Draw graphs, print charts, calculate statistics.

1.49   Each student's answers will be different. A few possibilities:

      a. color of hair, major, gender, marital status
      b. number of courses taken, number of credit hours, height, weight, distance from hometown to college, cost of textbooks

1.51   a. $T = 3$ is a piece of data.
      b. What is the average number of times per week the people in the sample went shopping?
      c. What is the average number of times per week that people (all people) go shopping?

1.53   a. all the Alzheimer's patients in the U.S.
      b. the cost in medical expenses and lost productivity per patient per year
      c. the total cost per year for all Alzheimer's patients in the U.S.
      d. the total cost per year for the Alzheimer's patients used as a sample

1.55   a. young men who do not go to college and take restaurant jobs
      b. the 2000 individuals in the study
      c. Most likely, the sample was a judgment sample.

1.57   a. business travelers
      b. No, percentages were given based on the sample, not total counts.
      c. Women might be more likely to: not want to dine alone, feel that room service is a special treat, feel safer dining in their room.
      d. Women business travelers are more likely to use room service.

1.59   Each will have different examples.

1.61   Each will have different examples.

Chapter 2

2.1

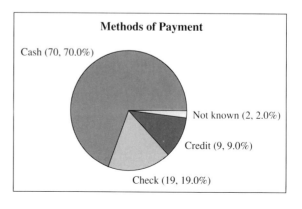

Methods of Payment

Cash (70, 70.0%)
Not known (2, 2.0%)
Credit (9, 9.0%)
Check (19, 19.0%)

2.2

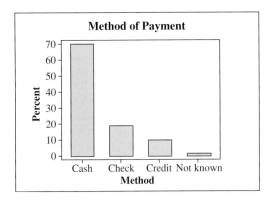

2.3 The circle graph makes it easier to visually compare the relative sizes of the parts to each other and the size of each part to the whole.

2.4

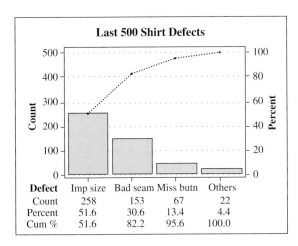

2.5

Points Scored per Game by Basketball Team

2.6 96: leaf of 6 is placed on the 9 stem
66: leaf of 6 is placed on the 6 stem

2.7 Points scored per game

3	6
4	6
5	6 4 5 4 2 1
6	1 1 8 0 6 1 4
7	1

2.8 Each leaf value represents the ones or units position.

2.9 a.

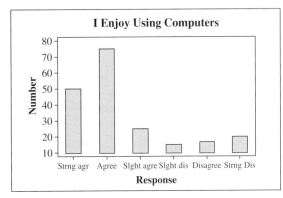

b. 25%, 37.5%, 12.5%, 7.5%, 7.5%, 10%

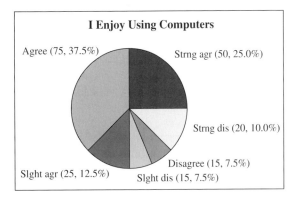

c. The circle graph makes it easy to visually compare the relative sizes of the parts to each other and the size of each part to the whole.

2.11 a. The frequencies are 54, 138, 120, 276, and 12.

  b.

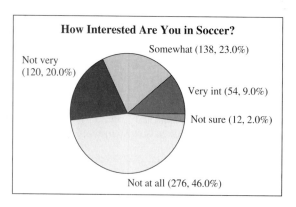

  c.

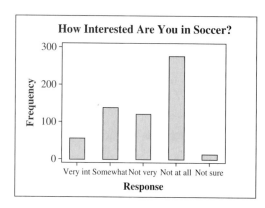

2.13

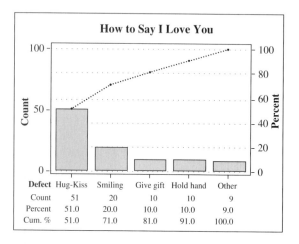

2.15 a.

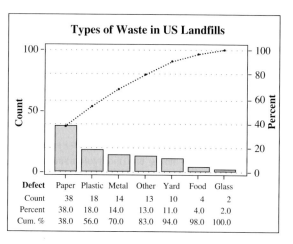

  b. The Pareto diagram puts the categories in order, largest first. A very large group classified as "other" in the middle is not consistent with the Pareto format.

2.17

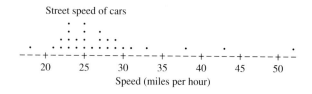

2.19 a. 15

  b. 11.2, 11.2, 11.3, 11.4, 11.7

  c. 15.6   d. 13.7, 3

2.21 One-way travel time

0	5 5
1	5 5 5 0
2	0 0 5 0 5 5 5 5 5 0 5 0 0 0 0 5 0 0 0
3	0 5 0
4	0 5

2.23 a. Width of stems is 0.1.

  b. Each leaf represents the hundredths place.

  c. 16   d. 5.97, 6.01, 6.04, 6.08

2.25

x	f
0	2
1	5
2	3
3	0
4	2
	12

2.26
a. f is frequency; values of 70 or more but less than 80 occurred 8 times.
b. $\Sigma f = 19$; $\Sigma f =$ sum of all the frequencies, or number of data.
c. 19

2.27
a. $5 \le x < 8$

b.

Class Interval	Frequency
$0 \le x < 2$	30
$2 \le x < 5$	35
$5 \le x < 8$	75
$8 \le x < 11$	60
$11 \le x$	300
	500

2.28
a. class #4; $65 \le x < 75$
b. all values greater than or equal to 65 and also less than 75 (does not include 75)
c. difference between upper- and lower-class boundaries
   i. subtracting the lower-class boundary from the upper-class boundary for any one class
   ii. subtracting a lower-class boundary from the next consecutive lower-class boundary
   iii. subtracting an upper-class boundary from the next consecutive upper-class boundary
   iv. subtracting a class mark from the next consecutive class mark

2.29

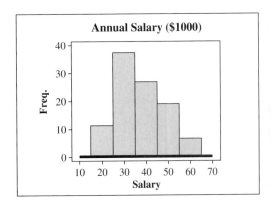

2.30 same information; differ in use of class labels and frequency scale

2.31 symmetric: weight of dry cereal per box, breaking strength of something
uniform: number resulting from tossing a die
skewed right: salaries
skewed left: hour exam scores
bimodal: heights, weights for groups containing both males and females
J-shaped: amount of television watched per day

2.32

Class Boundaries	Cumulative Frequency
$15 \le x < 25$	12
$25 \le x < 35$	49
$35 \le x < 45$	75
$45 \le x < 55$	94
$55 \le x \le 65$	100

2.33

Class Boundaries	Cum. Rel. Frequency
$15 \le x < 25$	0.12
$25 \le x < 35$	0.49
$35 \le x < 45$	0.75
$45 \le x < 55$	0.94
$55 \le x \le 65$	1.00

2.34

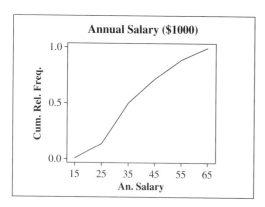

2.35

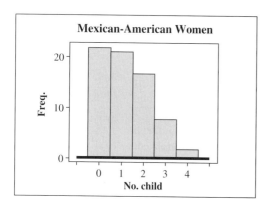

2.37

a. Age	Freq.	b. Age	Rel. Freq.	c. Age	Cum. Rel. Freq.
17	1	17	0.02	17	0.02
18	3	18	0.06	18	0.08
19	16	19	0.32	19	0.40
20	10	20	0.20	20	0.60
21	12	21	0.24	21	0.84
22	5	22	0.10	22	0.94
23	1	23	0.02	23	0.96
24	2	24	0.04	24	1.00
	50		1.00		

d.

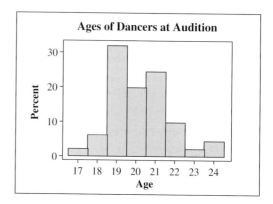

e.

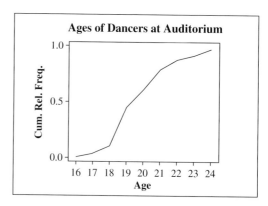

2.39   a. 12–16   b. 2, 6, 10, 14, 18, 22, 26   c. 4.0
d. 0.08, 0.16, 0.16, 0.40, 0.12, 0.06, 0.02
e.

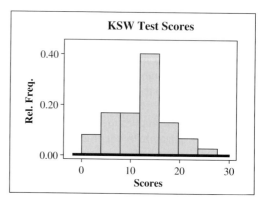

2.41 a.

Class Limits	Frequency
12–18	1
18–24	14
24–30	22
30–36	8
36–42	5
42–48	3
48–54	2

b. 6   c. 27, 24, 30

d.

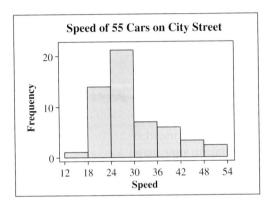

2.43 a.

Third-Graders at Roth Elementary School

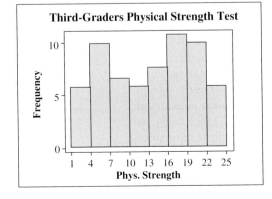

b.

c.

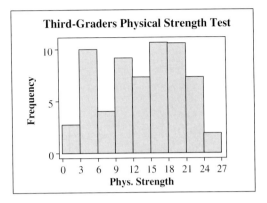

d.

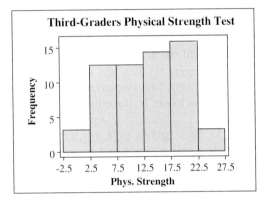

e.

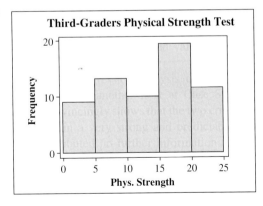

f. There seems to be an indication of a bimodal distribution.

(continued)

g. The class boundaries used determine where data values fall, causing different groupings of the data, which in turn results in different distributions.

2.45 a.

Class Limits	Frequency
2.00–3.00	4
3.00–4.00	13
4.00–5.00	12
5.00–6.00	8
6.00–7.00	2
7.00–8.00	1
	40

b.

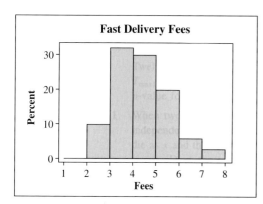

2.47 a.

Number	Cum. Rel. Freq.
0	0.030
1	0.060
2	0.121
3	0.273
4	0.379
5	0.515
6	0.636
7	0.803
8	0.909
9	0.954
10	0.969
11	0.999 *round-off error

b.

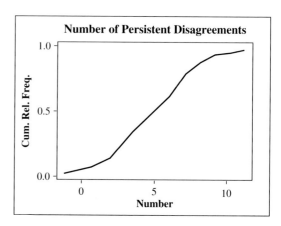

2.49   2.25

2.50   73

2.51   4.55

2.52   2

2.53 $5.38

2.54   8.2, 8.5, 9, 8.0

2.55   a. 6.0   b. 7   c. no mode   d. 5.5

2.57   a. 6.9   b. 7   c. 7   d. 7.5

2.59   5.35%

2.61   a. 30.05, 30, 30, 29.5

b.

Police Recruits

$ = mean, $\bar{x} = 30.05$
★ = midrange = 29.5
= median = mode = 30

Exercise Capacity (minutes)

c. All locate the "center." All are close in value; there are no mavericks, and the data are somewhat evenly distributed about the mean.

2.63   a. mean; total number of cars $\Sigma x = n \cdot \bar{x}$

b. Since the data are whole numbers, "1.9" cannot be: the median or midrange because they would be a whole number or a "0.5" number; the mode because it would be a whole number.

c. 438 spaces

2.65   a.

Third-Graders at Roth Elementary School

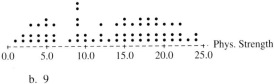

b. 9

c.

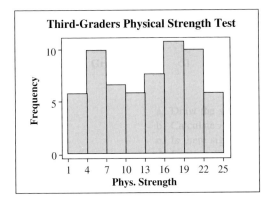

d. appears bimodal

e. Dotplot shows the mode to be 9, which is in the 7–10 class, while the histogram shows the two modal classes to be 4–7 and 16–19. The mode is not in either modal class.

f. No, there is only one numerical value per class.

g. The mode is the single data value that occurs most often, whereas a modal class results from data forming a cluster of data values, not necessarily all of one value.

2.67 $19.28; range measures the spread.

2.68   a. $x = 45$ is 12 units above the mean.
b. $x = 84$ is 20 units below the mean.

2.69   The mean is the "balance point" or "center of gravity" of the data values.

2.70   11.5

2.71   11.5

2.73   a. 5   b. 3.2   c. 1.8

2.75   a. 3.1   b. 3.1   c. 1.8

2.77   a. 22,153.6   b. 22,153.6

2.79   a. 9   b. 7.8   c. 2.8
d.

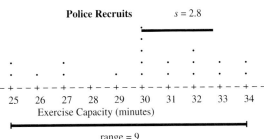

e. Except for values $x = 28$ and 30, the distribution looks rectangular. Range is a little more than 3 standard deviations.

2.81

	Σx	$\Sigma(x - \bar{x})$	$\Sigma\lvert x - \bar{x}\rvert$	$\Sigma(x - \bar{x})^2$	**Range**
Set 1:	250	0	14	54	9
Set 2:	250	0	46	668	35

The values of $\Sigma\lvert x - \bar{x}\rvert$, $SS(x)$ and range reflect the fact that there is more variability in data set 2.

2.83   There are many different possible answers.
a. 100, 100, 100, 100, [100]; $s = 0.0$
b. 99, 99.5, 100.5, 100, [100]; $s = 0.57$
c. 107, 95, 94, 108, [100]; $s = 6.53$
d. 75, 78, 123, 124, [100]; $s = 23.53$

2.85   a. Command #1—k1 is the mean found by summing the data and dividing by the number of data.
#2—the mean is subtracted from each piece of data and the differences are placed in column 2.
#3—each of the differences is squared and the squares are placed in column 3.
#4—k2 is the variance, the sum of all the squared differences divided by ($n - 1$).

(continued)

#5—k3 is the standard deviation, the square root of the variance.
#6—results.

Row	C1	C2	C3
1	7	-0.2	0.04
2	6	-1.2	1.44
3	10	2.8	7.84
4	7	-0.2	0.04
5	5	-2.2	4.84
6	9	1.8	3.24
7	3	-4.2	17.64
8	7	-0.2	0.04
9	5	-2.2	4.84
10	13	5.8	33.64

```
#7—results
k1 7.20000
k2 8.17778
k3 2.85968

#8—results
Column standard deviation
Standard deviation of C1 = 2.8597.
```

b.

```
Let k1 = sum(c1)/count(c1)—steps 1
and 2.
Let c2 = c1 — k1—step 3.
Let c3 = c2**2—step 4.
Let k2 = sum(c3)/(count(c1) — 1)—
step 5.
```

c. The procedure goes through the data list repeatedly; it counts, it sums, it finds deviations, it squares each deviation, and so on.

2.87 a.

x	f	xf	x^2f
0	1	0	0
1	3	3	3
2	8	16	32
3	5	15	45
4	3	12	48
Σ	20	46	128

b. 20, 46, 128
c. 4 is a possible data value; 8 is the number of times a "2" occurred; Σf = sum of the frequencies = sample size; Σxf = sum of the xf products, the sum of the data.

2.88 a. Sum of the x-column has no meaning unless each value occurred only once.
b. Each data value is multiplied by how many times it occurred. Summing these products will give the sum of all the data.

2.89 2.3

2.90 1.2

2.91 1.1

2.92 12.9, 21.7, 4.7

2.93 1.8, 1.8, 1.4

2.95 a. \$3,387,312   b. \$27,539   c. academic
d. \$2111

2.97 12.2, 3.5

2.99 21.00, 9.95

2.101 29.2, 8.2

2.103 a. 63.9, 65, 65, 55   b. 238.25, 15.4

2.105 a. 40.7, 6.43   b. 41.2, 6.2
c. $\bar{x}$: % diff. = 1.2%, and s: % diff. = 3.1%
d. $\Sigma x = 286$, $\Sigma x^2 = 11694$; $\Sigma xf = 294$, $\Sigma x^2 f = 12348$
e. Grouping considers the middle value only for each class.
f. All statistics are estimates of their corresponding population parameters.
g. Grouped will be easier if doing the work by hand. There is no advantage to using grouped technique when using a computer.

2.107 91 is in the 44th position from the low value of 39.
91 is in the 7th position from the high value of 98.

2.108 64, 70

2.109 88.5, 95

2.110 The distribution needs to be symmetric about the mean.

2.111

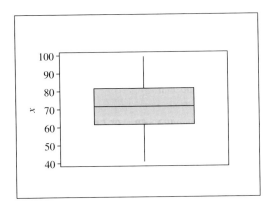

2.112 1.67, −0.75

2.113 a. 3.8, 5.6 b. 4.7 c. 3.5, 4.0, 6.9

2.115 a. 33.0 b. 34.8 c. 33.65

d.

30.1	31.3	33.0	36.0	39.5
L	Q_1	$\bar{x}$	Q_3	H

e.

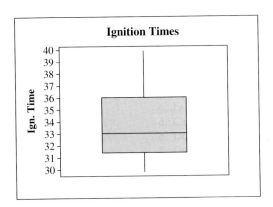

2.117 a. 1.0 b. 0.0 c. 2.25 d. −1.25

2.119 680

2.121 a. 152 is 1.5 standard deviations above the mean.

b. The score is 2.1 standard deviations below the mean.

c. the number of standard deviations from the mean

2.123 A

2.125 from 175 through 225 words inclusive

2.126 at least 93.75%

2.127 a. 50% b. 68% c. 84%

2.128 94

2.129 a. Range is approximately 6 times the standard deviation for a normal distribution.

b. Approximate the standard deviation by dividing the range by 6.

2.131 a. at most 11% b. at most 6.25%

2.133 a. at least 75% b. approximately 95%

2.135 a.

x	f
1	7
2	10
3	22
4	8
5	7
6	2
7	3
8	0
9	1
Σ	60

b. 3.4, 1.7 c. 1.7, 5.1 d. 47, 78%
e. 0.0, 6.8 f. 56, 93% g. −1.7, 8.5
h. 59, 98.3%
i. 93% is at least 75% and 98.3% is at least 89%; both agree with Chebyshev's theorem.
j. 78%, 93%, and 98.3% are not approximately equal to the 68%, 95%, and 99.7% of the empirical rule.

2.137   a. Results will vary, but expect similar results.

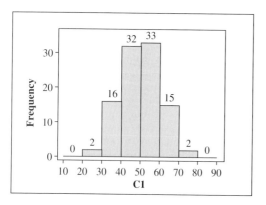

65% is within one standard deviation of mean, 96% is within two, 100% within three; extremely close to the empirical rule.

2.139   yes

2.140   32% is more than half.

2.141   Figure 1 emphasizes the variation between the numbers; Figure 2 shows the relative size of the numbers.

2.142   The graph is a bar graph; each bar shows the relative frequency of fatal injuries for each age class.

2.143   a. The class width is not uniform
             b. See Case Study 2-6, page 113.

2.145   a. increased   b. unchanged   c. unchanged
             d. increased   e. increased   f. increased
             g. increased

2.147   a. 4.56   b. 1.34   c. close to 4%

2.149   a. 44.6   b. 10.9

2.151   a.

x	3	4	5	6	7	8	9	10	11	12	13	14	Σ
f	1	2	3	4	5	5	6	7	8	9	6	2	58

b. 58 litters
c. f is the number of litters; Σf is the total number of litters.

2.153   79.9, 12.4

2.155   a. set of all residents living in the U.S., ownership of home.

b. They are 4 of several hundred statistics for the several hundred sample cities.
c. cannot determine, not enough information
d. no

2.157   a. 15.4   b. 16   c. 16   d. 11.5

2.159   a.

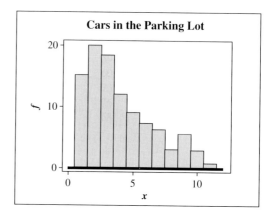

b. mean = 4.0, median = 3, mode = 2, midrange = 6, midquartile = 3.75
c. 2, 5.5   d. 1, 1   e. 10, 6.7, 2.6

2.161   a.

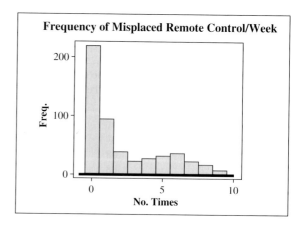

b. 1.988, 1, 0, 4.5   c. 6.4608, 2.5418
d. 0, 4, 6   e. 2   f. 0, 0, 1, 4, 9

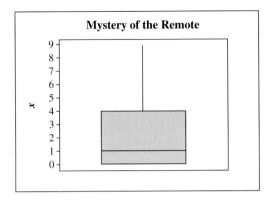

Mystery of the Remote

b. Histogram; the height of a bar is a more visible demonstration of density than a collection of dots.
c. $\bar{x} = 41.44$, $\tilde{x} = 35.50$, midrange $= 57.5$
d. Median; it is not affected by the skewness of the extreme values.
e. range $= 109$, $s^2 = 823.4$, $s = 28.7$
f. standard deviation   g. 17, 58
h. "Middle class" versus "lower" and "upper" quarters. Accordingly, "middle class" are families earning between \$17,000 and \$58,000.

2.167 \$142, \$175

2.163 a.

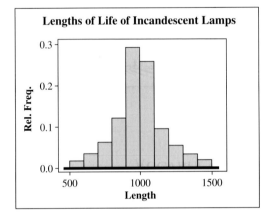

Lengths of Life of Incandescent Lamps

2.169 a.

Class Limits	f
220.5–221.5	7
221.5–222.5	4
222.5–223.5	10
223.5–224.5	7
224.5–225.5	0
225.5–226.5	2
226.5–227.5	1
227.5–228.5	2
Σ	33

b.

b. 995.9   c. 169.2

2.165 a. 1994 Annual Family Incomes

Fam Incom

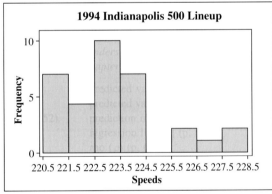

1994 Indianapolis 500 Lineup

c. 223.27, 1.94
2.171 a. 98th   b. 16th
2.173 58, 0, 9.8, 6.9, 181
2.175 a. at least 75%   b. at least 89%
2.177 99.7% of a normal distribution is between z-scores of −3 and +3.
2.179 a. 3978.1   b. 203.9   c. 3570.3 to 4385.9

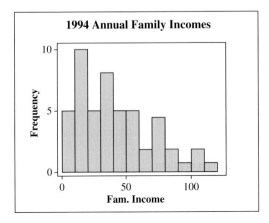

1994 Annual Family Incomes

2.181   a. 1993 Export Business

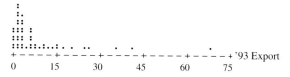

    b. 8.34, 3.25   c. 12.53
    d. Mean; the total of the data values is used to calculate the mean.
    e. Median; the mean is increased disproportionally by the few very large values.

2.183   a. Including A and B, the distribution seems to be skewed to the right.
    b. Excluding A and B, the distribution seems to be approximately normal—that is, mounded and approximately symmetrical about the middle.
    c. If the distribution in (b) is normal, then A is not a typical occurrence.

2.185   a. 70, 77.5, 77.5, 77.5, 85 yields $s = 5.30$.
    b. 70, 76, 85, 89, 95 yields $s = 10.02$.
    c. 70, 85, 90, 99, 110 yields $s = 15.02$.
    d. In order to increase the standard deviation, the data had to become more dispersed.

2.187   a. The Sagarin College Basketball Ratings

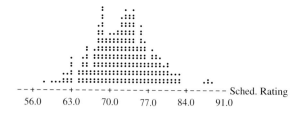

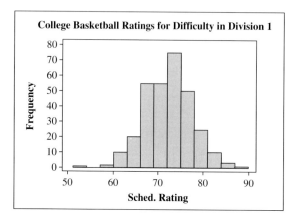

b., c., d. Samples will vary; however, expect the results to resemble the population.

2.189   As the sample size increases, the closer the distribution is to a normal distribution.

Chapter 3

3.1

	Very Good	Good	Other	Marginal Total
Husband	23.0%	17.3%	9.7%	50%
Wife	24.0%	17.7%	8.3%	50%
Marginal total	47.0%	35.0%	18.0%	100%

3.2

	Very Good	Good	Other	Marginal Total
Husband	46.0%	34.6%	19.4%	100%
Wife	48.0%	35.4%	16.6%	100%
Marginal total	47.0%	35.0%	18.0%	100%

The table shows the distribution of ratings for husband and wife separately.

3.3

	Very Good	Good	Other	Marginal Total
Husband	48.9%	49.4%	53.9%	50%
Wife	51.1%	50.6%	46.1%	50%
Marginal total	100%	100%	100%	100%

The table shows the distribution of husbands and wives for each of the ratings.

3.4   January Unemployment Rates

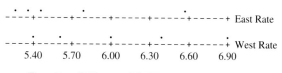

East: $\bar{x} = 5.72$, $\tilde{x} = 5.5$; West: $\bar{x} = 6.06$, $\tilde{x} = 6.0$

3.5 The largest frequency numbers appear at the lower levels for 14 months word comprehension and the lower levels for 20 months word comprehension.

3.6 The input variable most likely would be height.

3.7

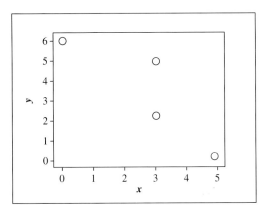

3.8 a.

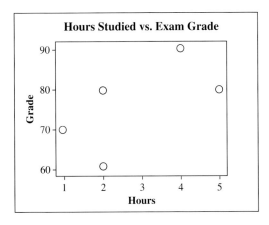

b. As hours studied increased, the exam grades tended to also increase.

3.9 a.

		Gender		
		Male	Female	Marginal Total
Made a purchase	yes	10	14	24
	no	17	19	36
Marginal Total		27	33	60

b.

		Gender		
		Male	Female	Marginal Total
Made a purchase	yes	16.7%	23.3%	40%
	no	28.3%	31.7%	60%
Marginal Total		45.0%	55.0%	100%

c.

		Gender		
		Male	Female	Marginal Total
Made a purchase	yes	37%	42%	40%
	no	63%	58%	60%
Marginal Total		100%	100%	100%

d.

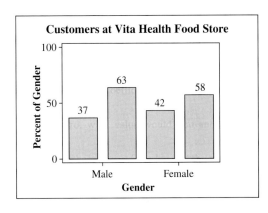

3.11 a. 3000
b. Two variables, political affiliation and news information preferred, are collected from each person. Both are qualitative.
c. 950   d. 50%   e. 25%

3.13 a.

```
Min. Deposit
5  .                        .              . . : :.. :
   -+-----+-----+-----+-----+-----+-----Yield
Min. Deposit
10                              . :.. ::  .
   _+-----+-----+-----+-----+-----+-----Yield
Min. Deposit
25                     .    . :..:. . . . .
   _+-----+-----+-----+-----+-----+-----Yield
     3.90    4.20    4.50    4.80    5.10    5.40
```

b. Rates for minimum deposits of:

	$500	$1000	$2500
High	5.26	5.45	5.50
Q_3	5.19	5.32	5.24
$\bar{x}$	5.09	5.15	5.10
Q_1	4.955	5.09	5.00
Low	4.00	5.00	4.75

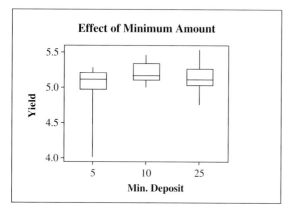

Effect of Minimum Amount

c. The middle 50% of the three sets are fairly similar. The $500 set has two extremely low values, making for a wider skewed distribution. Overall, minimum deposit appears to have little effect on interest rate.

3.15 a.

Heights of World Cup Soccer Players

Weights of World Cup Soccer Players

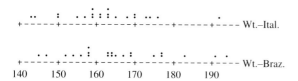

Ages of World Cup Soccer Players

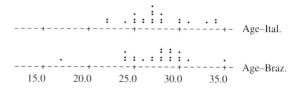

b. No; height, weight, and age had basically the same spread of data with peaks in the same locations.

c. The data cannot be paired between the two teams.

3.17

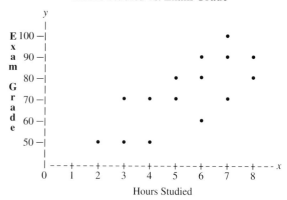

Hours Studied vs. Exam Grade

3.19

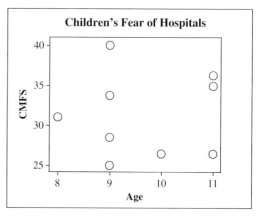

Children's Fear of Hospitals

3.21 a.

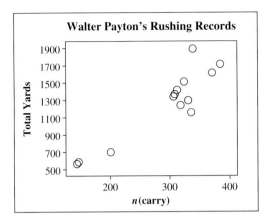

Walter Payton's Rushing Records

b. As number of carries increased so did the number of total yards. There are also two separate groups of data points.

c. Two of the "low" group come from the first and last year of his career.

3.23 a.

x	y	x^2	xy	y^2
2	80	4	160	6400
5	80	25	400	6400
1	70	1	70	4900
4	90	16	360	8100
2	60	4	120	3600
14	380	50	1110	29400

$SS(x) = 10.8$, $SS(y) = 520$, $SS(xy) = 46$

b. 0.61

3.24 a.

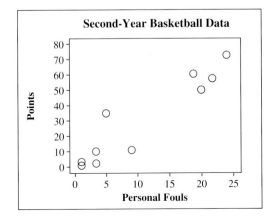

Second-Year Basketball Data

b. 0.94

c. No, a player who scores many points is most likely in the game more, therefore more likely to have more fouls. A strong correlation coefficient indicates a strong linear relationship, not necessarily a cause-and-effect situation.

3.25 −0.75, 0.00, +0.75

3.27 little or no linear correlation

3.29 a. 10.4   b. 238.9   c. 6.6   d. 0.13

3.31 a. 2/3 or 0.7   b. 0.74

3.33 a. 7.449   b. 912.9   c. −79.09   d. −0.96

3.35 a   −0.707

b. As work satisfaction decreased, the inclination to leave increased.

3.37 a. 13,717, 1396.9; 15,298, 858.0; 14,257, 919.0

b. Summations, Σ's, are sums of data; sums of squares, SS( ), are parts of formulas.

3.38 a. 28.1, 47.9

b. Yes, the line of best fit is made up of all points that satisfy its equation.

3.39 a. $\hat{y} = 64.1 + 4.26x$

b. Points (1, 68.4) and (3, 76.9) can be used to locate the line.

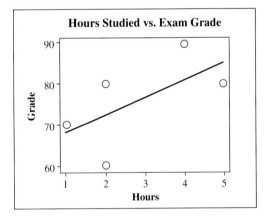

Hours Studied vs. Exam Grade

c. Yes, as the hours studied increased, the exam grades appeared to increase, also.

3.40 a. 41   b. no

c. 41 is the average number of sit-ups expected for students who do 40 push-ups.

3.41   a.

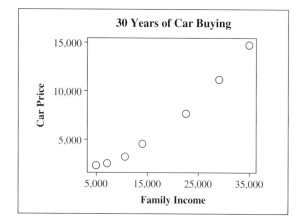

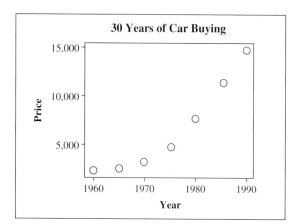

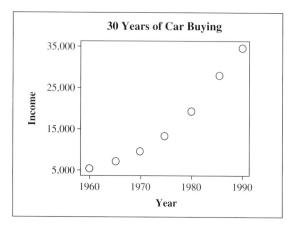

b. The patterns are definitely elongated; however, they do not appear to be linear.

c. yes

3.42   The vertical scale is located at $x = 58$; not the y-axis.

3.43   5.83, -260.61

3.45   a. the total monthly telephone cost when x is equal to zero

b. the rate at which the total phone bill increases for each additional long-distance call

3.47   6.81 or $68,100

3.49   1.1583

3.51 $\hat{y} = 0.34 + 0.22x$

3.53   a.

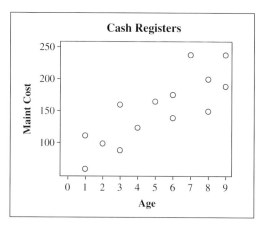

b. $\hat{y} = 78.95 + 14.69x$ c. $196.50

d. the expected average cost of maintenance for all 8-year-old cash registers

3.55   a. 0.427   b. $\hat{y} = 19.1 + 0.488x$

3.57   a.

	Rural	Non	Total
Parents	75	30	105
Relatives	46	32	78
Friends	24	44	68
Combination	37	57	94
Other	2	7	9
Total	184	170	354

b.

	Rural	Non	Total
Parents	21.2%	8.5%	29.7%
Relatives	13.0%	9.0%	22.0%
Friends	6.8%	12.4%	19.2%
Combination	10.5%	16.1%	26.6%
Other	0.6%	2.0%	2.6%
Total	52.1%	48.0%	100.1%*

*round-off error

c.

	Rural	Non	Total
Parents	71.4%	28.6%	100%
Relatives	59.0%	41.0%	100%
Friends	35.3%	64.7%	100%
Combination	39.4%	60.6%	100%
Other	22.2%	77.8%	100%
Total	52.0%	48.0%	100%

d.

	Rural	Non	Total
Parents	40.8%	17.6%	29.7%
Relatives	25.0%	18.8%	22.0%
Friends	13.0%	25.9%	19.2%
Combination	20.1%	33.5%	26.6%
Other	1.1%	4.1%	2.5%
Total	100.0%	100.0%	100.0%

e.

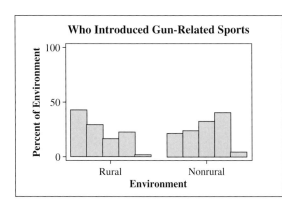

3.59 a.

	Less Than 6 Mo	6 Mo–1 Yr	More Than 1 Yr	Total
Under 28	413	192	295	900
28–40	574	208	218	1000
Over 40	653	288	259	1200
Total	1640	688	772	3100

b.

	Less Than 6 Mo	6 Mo–1 Yr	More Than 1 Yr	Total
Under 28	13.3%	6.2%	9.5%	29.0%
28–40	18.5%	6.7%	7.0%	32.2%
Over 40	21.1%	9.3%	8.4%	38.8%
Total	52.9%	22.2%	24.9%	100%

c.

	Less Than 6 Mo	6 Mo–1 Yr	More Than 1 Yr	Total
Under 28	45.9%	21.3%	32.8%	100%
28–40	57.4%	20.8%	21.8%	100%
Over 40	54.4%	24.0%	21.6%	100%
Total	52.9%	22.2%	24.9%	100%

d.

	Less Than 6 Mo	6 Mo–1 Yr	More Than 1 Yr	Total
Under 28	25.2%	27.9%	38.2%	29.0%
28–40	35.0%	30.2%	28.2%	32.3%
Over 40	39.8%	41.9%	33.6%	38.7%
Total	100%	100%	100%	100%

(continued)

e.

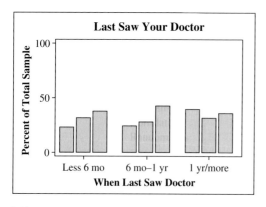

3.61   a. correlation   b. regression   c. correlation
d. regression   e. correlation

3.63   1.00, $\hat{y} = 1.0 + 2.0x$

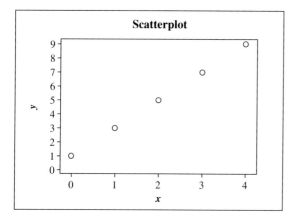

3.65   Many answers are possible.

3.67   a. $r = 0.9993$, $\hat{y} = -160.4 + 4.79x$

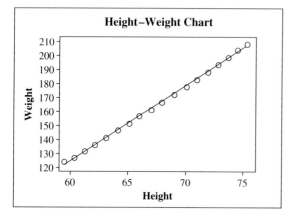

b. There is a strong positive linear relationship since r is so close to a $+1$.
c. an average increase of 4.79 pounds per inch height
d. The chart has an upper limit of 174 lb for weight. Substituting 70 inches into $\hat{y} = -160.4 + 4.79x$ yields a predicted upper limit of 174.9 lb.

3.69   a. 0.427   b. $\hat{y} = 3.5 + 0.19x$

3.71   a.

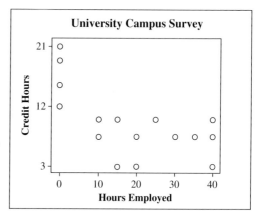

b. -0.68
c. r suggests a negative correlation. However, the scatter diagram appears to show two subpopulations: those employed for $x = 0$ hours and those employed for $x > 0$ hours, suggesting that correlation analysis is inappropriate.

3.73   a. Numerator of formula (3-1):
$$\text{Numerator} = \Sigma(x - \bar{x})(y - \bar{y})$$
$$= \Sigma[xy - \bar{x}y - x\bar{y} + \bar{x}\bar{y}]$$
$$= \Sigma xy - \bar{x} \cdot \Sigma y - \bar{y} \cdot \Sigma x + n\bar{x}\bar{y}$$
$$= \Sigma xy - [(\Sigma x/n) \cdot \Sigma y] - [(\Sigma y/n) \cdot \Sigma x]$$
$$+ [n \cdot (\Sigma x/n)(\Sigma y/n)]$$
$$= \Sigma xy - [(\Sigma x \cdot \Sigma y/n) - (\Sigma x \cdot \Sigma y/n)$$
$$+ (\Sigma x \cdot \Sigma y/n)]$$
$$= \Sigma xy - [(\Sigma x \cdot \Sigma y)/n]$$
$$= SS(xy)$$
Denominator of formula (3-1):
$$\text{Denominator} = (n - 1)s_x s_y$$
$$= (n - 1) \cdot \sqrt{SS(x)/(n - 1)}$$
$$\cdot \sqrt{SS(y)/(n - 1)}$$
$$= \sqrt{SS(x) \cdot SS(y)}$$
Therefore, formula (3-1) is equivalent to formula (3-2).

b. The numerators of formulas (3-5) and (3-7) were shown to be equal in part (a). The denominators are equal by definition (formula 2-9).

Chapter 4

4.1 Each student will get different results; answers will be fractions with denominators of 10.

4.3 Each student will get different results; answers will be fractions with denominators of 25.

4.5 Each student will get different results.

4.7 0.225

4.8 All three are calculated by dividing the experimental count by the sample size.

4.9 You can expect a 1 to occur approximately 1/6th of the time when you roll a die repeatedly.

4.11 a. 0.18 b. 0.000095

4.13 Each student will get different results.

4.15 Each student will get different results.

4.17 {0, 1, 2, 3, 4, 5, 6, 7, 8, 9}

4.18

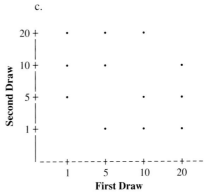

4.19

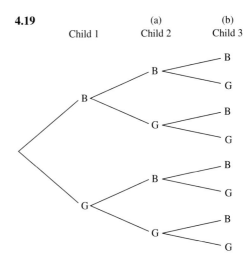

4.20 a. $S = \{\$1, \$5, \$10, \$20\}$

b.

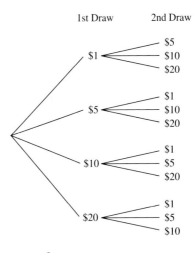

c.

4.21 {JH, JC, JD, JS, QH, QC, QD, QS, KH, KC, KD, KS}

4.23 {(MM,MM), (MM,BC), (MM,SC), (MM,O), (BC,MM), (BC,BC), (BC,SC), (BC,O), (SC,MM), (SC,BC), (SC,SC), (SC,O), (O,MM), (O,BC), (O,SC), (O,O)}

4.25 {H1, H2, H3, H4, H5, H6, T1, T2, T3, T4, T5, T6}

4.27 a. {HH, HT, T1, T2, T3, T4, T5, T6}   b. 0.5

4.29 Everyone's results will be different.

4.31 4/36, 5/36, 6/36, 5/36, 4/36, 3/36, 2/36, 1/36

4.32 4/5

4.33 a. 2/9   b. 7 : 2

4.34 a. 0.0004267   b. 12 : 7   c. 37 : 1

4.35 40/52

4.37 1/4, 1/4, 1/2

4.39 a. 1/6   b. 3/6   c. 4/6   d. 3/6

4.41 1/7, 2/7, 4/7

4.43 The three success ratings appear to be mutually exclusive and all-inclusive events. If this is true, none of the three sets of probabilities are appropriate. The total must be exactly 1.0, and each probability must be between 0.0 and 1.0.

4.45 a. 0.55   b. 0.40

4.47 a. 1 : 4   b. 1/750,001   c. 20/21

4.49 0.89

4.51 a. Yes, a student cannot be both male and female.
b. No, a student can be both male and registered for statistics.
c. No, a student can be both female and registered for statistics.
d. Yes, together they form the entire sample space.
e. No, not all students are included.
f. Yes, no common elements are shared by the two events.
g. No, two mutually exclusive events do not necessarily make up the whole sample space.

4.52 a. A & C and A & E are mutually exclusive because they cannot occur at the same time.
b. 12/36, 11/36, 10/36

4.53 a. not mutually exclusive
b. not mutually exclusive
c. not mutually exclusive
d. mutually exclusive

4.55 If two events are mutually exclusive, then there is no intersection.

4.57 a. 0.7   b. 0.6   c. 0.7   d. 0.0

4.59 No. *Female* students can be *working* students.

4.61 a. 0.000000143   b. 0.999999857   c. 0.000002143

4.63 4%

4.65 a. 0.45   b. 0.40   c. 0.55
d. No, $P(S) \neq P(S|F) \neq P(S|M)$

4.66 0.28

4.67 0.15

4.68 a. 0.40   b. yes

4.71 a. independent   b. not independent
c. independent   d. independent
e. not independent   f. not independent

4.73 a. 0.12   b. 0.4   c. 0.3

4.75 a. 0.5   b. 0.667   c. no

4.77 a. independent   b. independent   c. dependent

4.79 a. 0.5041   b. 0.0841   c. 0.357911

4.81 a. 0.5375   b. 0.175   c. 0.6125

4.83 a. 0.36   b. 0.16   c. 0.48

4.85 a. 3/5   b. 0.16, 0.48, 0.36

4.87 0.3276

4.89 Illustrations 4-13 and 4-14 use a stepwise thought process to calculate the probabilities. Illustration 4-15 uses a formula that takes care of all the steps.

4.90 a. Switch strategy:

$P(\text{win}|\text{when switch}) = 2/3$
b. Do not switch strategy:

$P(\text{win}|\text{when do not switch}) = 1/3$
c. $P(\text{win}|\text{switch}) = 2/3$ (a) and
$P(\text{win}|\text{not switch}) = 1/3$.

d. First, there are no rules that require Monty to open a door revealing a goat or that require him to give the contestant a chance to switch. Further, Monty knows where the car is, he has a pocket full of money, and he plans to make his show "fun to watch." The assumptions made above are not true. Further, probabilities are not appropriate when the player can be manipulated into doing almost anything by the way Monty "plays" the contestant.

4.91 0.7

4.93 a. 0.15   b. 0.65   c. 0.7   d. 0.5   e. 0.7
f. No. Independent events intersect.

4.95 a. 20/56   b. 30/56   c. 6/56

4.97 a. 1/2, 1/4, 1/8   b. 9/16, 9/32, 9/64

4.99 8/30

4.101 a. 0.9   b. 0.075   c. 0.025

4.103 a. 0.625   b. 0.25   c. not independent

4.105 a. 0.53   b. 0.38

4.107 a. True. Law of large numbers.
b. True. Law of large numbers.
c. False. The irregularity will be the order of occurrence of each individual outcome, not the overall summary of all 100 million outcomes. The law of large numbers says to expect about one-half of the 100 million to be H's and the other half to be T's.
d. False. The probability of a head occurring on any individual toss is the same as on any other toss, provided the tossing is done in a fair way.

4.109 a. 0.30   b. 0.40
c. Not independent; $P(A \text{ and } B) = 0.233$, which is not equal to $P(A) \cdot P(B) = 0.12$.
d. Blue eyes and brown eyes are mutually exclusive events. They are not complementary since not everyone was classified as having brown or blue eyes. Since they are mutually exclusive, they cannot be independent events.

4.111 a. False   b. True   c. False   d. False

4.113 a. {GGG, GGR, GRG, GRR, RGG, RGR, RRG, RRR}
b. 3/8   c. 7/8

4.115 a. 0.3168   b. 0.4659
c. No. $P(A)$ does not equal $P(A|B)$.
d. No. $P(A \text{ and } B)$ does not equal 0.0.

e. The two events could not both happen at the same time.

4.117 0.080

4.119 0.28

4.121 a. 0.429   b. 0.476   c. 0.905

4.123 0.300

4.125 a. 0.30   b. 0.60   c. 0.10   d. 0.60
e. 0.333   f. 0.25

4.127 a. 0.531   b. 0.262   c. 0.047

4.129 0.592

4.131 0.061

4.133 a. 0.988   b. 0.500

4.135 a. 0.60   b. 0.648   c. 0.710
d. (a) 0.70, (b) 0.784, (c) 0.874
e. (a) 0.90, (b) 0.972, (c) 0.997
f. The larger the number of games in the series, the greater the chance that the "best" team will win. The greater the difference between the two teams' individual chances, the more likely the "best" team wins.

Chapter 5

5.1 The number of courses per student. Possible values: $x = 1, 2, 3 \ldots n$.

5.2 The combined weight for books and supplies being carried to class. Possible values: 0 to 30 pounds for most students.

5.3 discrete, continuous

5.4 a. "Score" is an integer.
b. "Number of minutes to commute to work" is a measurement with fractional values.

5.5 Number of children per family; discrete; $x = 0, 1, 2, 3, \ldots, n$.

5.7 Distance from center to arrow; continuous; possible values: $x = 0$ to n, $n =$ radius of target.

5.9

x	0	1
$P(x)$	1/2	1/2

5.10

x	1	2	3	4	5	6
$P(x)$	1/6	1/6	1/6	1/6	1/6	1/6

5.11   a.

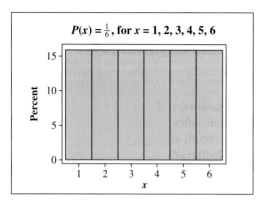

b. Uniform

5.12   a.

x	$P(x)$
0	0.05
1	0.12
2	0.15
3	0.25
4	0.21
5	0.10
6	0.05
7	0.04
8	0.02
9	0.01

b. $5\% = 0.05 \approx 1/21$; 21 days = 3 weeks

5.13

x	0	1	2	3
$P(x)$	0.20	0.30	0.40	0.10

Each $P(x)$ is between 0.0 and 1.0, and the sum of all $P(x)$'s is 1.0.

5.15   a.

x	$P(x)$
1	0.12
2	0.18
3	0.28
4	0.42
Σ	1.00

$P(x)$ is a probability function:
1. Each $P(x)$ is between 0 and 1.
2. The sum of the $P(x)$'s is 1.

b.

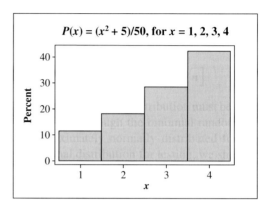

5.17   No, the percentages are for each age group.

5.19   a. Everyone's generated values will be different.
     c.

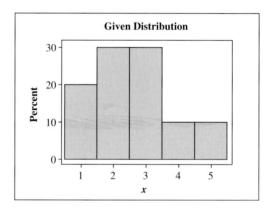

d. The sample distribution is expected to be similar to the given distribution.

5.21
$$\sigma^2 = \Sigma[(x - \mu)^2 \cdot P(x)]$$
$$= \Sigma[x^2 - 2x\mu + \mu^2) \cdot P(x)]$$
$$= \Sigma[x^2 \cdot P(x) - 2x\mu \cdot P(x) + \mu^2 \cdot P(x)]$$
$$= \Sigma[x^2 \cdot P(x)] - 2\mu \cdot \Sigma[x \cdot P(x)]$$
$$+ \mu^2 \cdot [\Sigma P(x)]$$
$$= \Sigma[x^2 \cdot P(x)] - 2\mu \cdot \{\mu\} + \mu^2 \cdot [1]$$
$$= \Sigma[x^2 \cdot P(x)] - 2\mu^2 + \mu^2$$
$$= \Sigma[x^2 \cdot P(x)] - \mu^2$$
$$\text{or} \quad \Sigma[x^2 \cdot P(x)] - \{\Sigma[x \cdot P(x)]\}^2$$

5.22

x	$P(x)$	$xP(x)$	$x^2P(x)$
1	1/6	1/6	1/6
2	2/6	4/6	8/6
3	3/6	9/6	27/6
Σ	6/6 = 1.0 ck	14/6 = 2.33	36/6 = 6.0

5.23 2.33

5.24 0.55556

5.25 0.745

5.26 These sums would represent each number occurring one time; not meaningful.

5.27 2.0, 1.0

5.29 a.

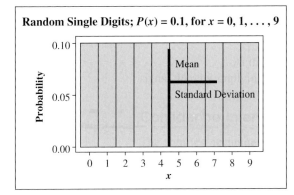

Random Single Digits; $P(x) = 0.1$, **for** $x = 0, 1, \ldots, 9$

b. 4.5, 2.87
c. See graph in part (a).   d. 100%

5.31 a. 2.0, 1.4   b. 1, 2, 3, and 4   c. 0.9

5.33 a. 2.44, 0.73
b. The mean and standard deviation will be smaller than the true values.

5.35 Each question is in itself a separate trial with its own outcome having no effect on the outcomes of the other questions.

5.36 There are four different ways that one correct and three wrong answers can be obtained in four questions, each with the same probability.

5.37 The 1/3 is the probability of success on one question; 4 is the number of independent trials; on the average, one should be able to guess 1/3 of all questions correctly.

5.38 Answers will vary.

5.39 Property 1: One trial is the flip of one coin, $n = 50$; trials are independent because the outcome on any one toss has no effect on the probabilities for the other tosses.
Property 2: Two outcomes on each trial: success = H, failure = T.
Property 3: $p = P(\text{heads}) = 1/2$ and $q = P(\text{tails}) = 1/2 [p + q = 1]$
Property 4: x = the number of heads for the experiment, 0 to 4.

5.40 a. 24   b. 4

5.41 0.125, 0.375, 0.125

5.42 a. 0.0146, 0.00098
b. $\Sigma P(x) = 0.99998 \approx 1$ and $0 \leq$ each $P(x) \leq 1$

5.43 0.007

5.44 0.240

5.45 a. 24   b. 5,040   c. 1   d. 360   e. 10
f. 15   g. 0.0081   h. 35   i. 10   j. 1
k. 0.4096   l. 0.16807

5.47 $n = 100$ trials (shirts), two outcomes (first quality or irregular), $p = P(\text{irregular})$, $x = n(\text{irregular})$; any integer value from 0 to 100.

5.49 a. The trials are not independent.
b. $n = 4$, the number of trials; two outcomes, ace and not ace; $p = P(\text{ace}) = 4/52$, $q = 48/52$ (the trials are independent); $x = n(\text{aces})$, 0, 1, 2, 3, or 4.

5.51

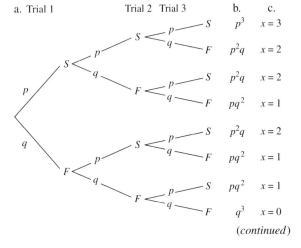

a. Trial 1   Trial 2 Trial 3   b.   c.
p^3 $x = 3$
p^2q $x = 2$
p^2q $x = 2$
pq^2 $x = 1$
p^2q $x = 2$
pq^2 $x = 1$
pq^2 $x = 1$
q^3 $x = 0$

(continued)

e. $P(x) = \binom{3}{x}p^x q^{3-x}$, for $x = 0, 1, 2, 3$

5.53   a. 0.4116   b. 0.384   c. 0.5625
d. 0.329218   e. 0.375   f. 0.0046296

5.55   By inspecting the function, we see the binomial properties: $n = 5$; $p = 1/2$, $q = 1/2$; the two exponents x and $5 - x$ add up to $n = 5$; x can be any integer from 0 to 5.
By inspecting the probability distribution:

x	$T(x)$
0	1/32
1	5/32
2	10/32
3	10/32
4	5/32
5	1/32
Σ	32/32 = 1.0

It is a probability distribution.

1. Each $T(x)$ is between 0 and 1.
2. $\Sigma T(x) = 1.0$.

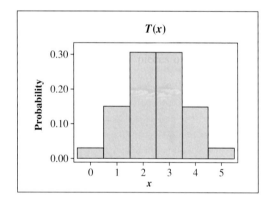

5.57   0.930

5.59   0.668

5.61   a. 0.590   b. 0.918

5.63   0.0011

5.65   a. 0.463   b. 0.537

5.67   0.984

5.69   0.1035

5.71   a. 0.2757   b. 0.1865   c. 0.5425   d. 0.6435

5.73

x	$P(x)$
1	0.0000
2	0.0003
3	0.0015
4	0.0056
5	0.0157
6	0.0353
7	0.0652
8	0.1009
9	0.1328
10	0.1502
11	0.1471
12	0.1254
13	0.0935
14	0.0611
15	0.0351
16	0.0177
17	0.0079
18	0.0031
19	0.0010
20	0.0003
21	0.0001
22	0.0000

5.75   0.665261

5.77   The number of defective items should be fairly small and therefore easier to count.

5.79   18, 2.7

5.80   a. 0.55, 0.72

b.

x	$P(x)$
0	0.569
1	0.329
2	0.087
3	0.014
4	0.001
5	0+
Σ	1.0

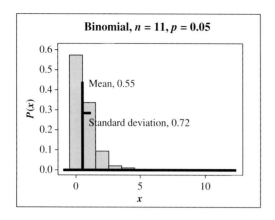

Binomial, $n = 11, p = 0.05$

Mean, 0.55

Standard deviation, 0.72

5.81 $\mu = 0.55, \sigma = 0.72$

5.83   a. 7.7, 2.7   b. 24.0, 4.7   c. 44.0, 2.3

5.85   400, 0.5

5.87   a. Assume x is approximately binomial:

$$P(x) = \binom{5}{x}(0.75)^x(0.25)^{5-x} \text{ for}$$

$$x = 0, 1, \ldots, 5$$

x	$P(x)$
0	0.00098
1	0.01465
2	0.08789
3	0.26367
4	0.39551
5	0.23730
Σ	1.00000

b.

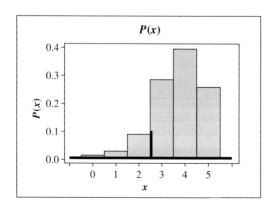

$P(x)$

c. 3.75, 0.97

5.89   150, 10.2

5.91   1. Each probability, $P(x)$, is a value between 0 and 1.
2. The sum of all the $P(x)$ is exactly 1.

5.93 $f(x)$ is a probability function.

5.95   a. 0.1   b. 0.4   c. 0.6

5.97   a. 0.27   b. cannot be answered
c. 0.42   d. No
e. yes; $\Sigma P(x) = 1$ and $0 \leq$ each $P(x) \leq 1$
f. 2.54, 1.37

5.99   a. 0.930   b. 0.264

5.101   a. 0.116   b. 0.0541

5.103   The four-engine plane has a higher probability of a successful flight.

5.105   0.063

5.107   On first selection, $P(\text{defective}) = 3/10$; on next selection, $P(\text{defective})$ is either 3/9 or 2/9 depending on the first selection; the trials are not independent.

5.109   2600

5.111   a. $p^3 + 3p^2q$ b. 0.028   c. 0.896
d. when p is greater than 0.5   e. 0, 0.5, 1.0

5.113   The Tool Shop has a slightly higher mean profit.

5.115

$$\mu = \Sigma[x \cdot P(x)]$$
$$= (1) \cdot (1/n) + (2) \cdot (1/n) + \cdots + (n) \cdot (1/n)$$
$$= (1/n) \cdot [1 + 2 + 3 + \cdots + n]$$
$$= (1/n) \cdot [(n)(n + 1)/2]$$
$$= (n + 1)/2$$

Chapter 6

6.1   0.4147

6.2   0.0212

6.3   0.9582

6.4   0.4177

6.5   0.0630

6.6   0.8571

6.7   0.2144

6.8   0.84

6.9 −1.15 and +1.15

6.11 a. 0.4032 b. 0.3997 c. 0.4993 d. 0.4761

6.13 a. 0.4394 b. 0.0606 c. 0.9394 d. 0.8788

6.15 a. 0.7737 b. 0.8981 c. 0.8983 d. 0.3630

6.17 a. 0.5000 b. 0.1469 c. 0.9893 d. 0.9452
e. 0.0548

6.19 a. 0.4906 b. 0.9725 c. 0.4483 d. 0.9306

6.21 a. 1.14 b. 0.47 c. 1.66 d. 0.86
e. 1.74 f. 2.23

6.23 a. 1.65 b. 1.96 c. 2.33

6.25 −1.28 or +1.28

6.27 −0.67 and +0.67

6.29 1.28, 1.65, 2.33

6.31 2.88

6.32 0.3944

6.33 0.8944

6.34 86

6.35 a. −2.00 b. 1.33

6.36 89

6.37 $29,008

6.38 Everyone's results will be different. Commands needed are:

```
RANDom 100 C1;
NORMal 50 12.
PRINt C1
```

6.39 Everyone's results will be different. Commands needed are:

```
PDF C1 C2;
NORMal 50 12.
PRINt C1 C2
```

6.40 Everyone's results will be different. Commands needed are:

```
PLOT C2*C1;
CONNect.
```

6.41 0.2316

6.43 a. 0.0038 = 0.38% b. 0.00003 = 0.003%

6.45 a. 0.6826 b. 0.9544 c. 0.9974
d. 0.6826 ≈ 68%; 0.9544 ≈ 95%;
0.9974 ≈ 99.74%

6.47 a. 0.2857 b. 0.0002

6.49 a. 0.1131 b. 0.0505 c. 4.64 minutes

6.51 a. 89.6 b. 79.2 c. 57.3

6.53 20.26

6.55 a. 0.0401 ≈ 4% b. 0.0179 ≈ 1.8%

6.57 a. 0.056241 b. 0.505544 c. 0.438215
d. 0.0559, 0.5077, 0.4364
e. Round-off errors; specifically in (d) when z is calculated.

6.59 Everyone's generated values will be different, but should have a mean and standard deviation close to 100 and 16, respectively, and should be approximately normally distributed.

6.61

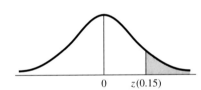

6.62

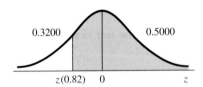

6.63 1.04

6.64 −0.92

6.65 a. $z(0.03)$ b. $z(0.14)$ c. $z(0.75)$
d. $z(0.13)$ e. $z(0.91)$ f. $z(0.82)$

6.67 a. 1.96 b. 1.65 c. 2.33

6.69 a. 1.28, 1.65, 1.96, 2.05, 2.33, 2.58
b. −2.58, −2.33, −2.05, −1.96, −1.65, −1.28

6.71 a. 0.4602 b. 1.28 c. 0.5199 d. −1.65

6.73 no; $np = 2$ is less than 5.

6.74 0.0149

6.75 a. Rule of thumb: $np = 3$ and $nq = 7$; therefore, the approximation is not appropriate, since $np < 5$. The distribution is skewed.

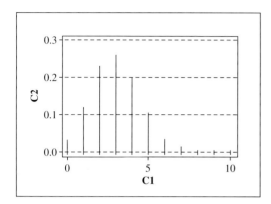

b. $np = 0.5$ and $nq = 99.5$; therefore, the approximation is not appropriate, since $np < 5$. Distribution is J-shaped.

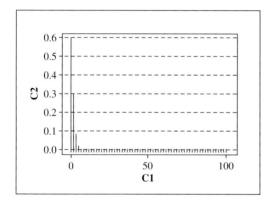

c. $np = 50$ and $nq = 450$; therefore, the approximation is appropriate, since both $np > 5$ and $nq > 5$. The distribution appears to be approximately normal.

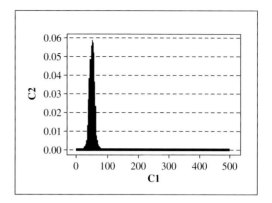

d. $np = 10$ and $nq = 40$; therefore, the approximation is appropriate, since both $np > 5$ and $nq > 5$. The distribution appears to be approximately normal.

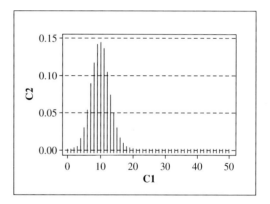

6.77 0.1822

6.79 0.9429

6.81 0.8078

6.83 0.0049

6.85 0.7053

6.87 at least 0.75, 0.9544

6.89 a. 1.26 b. 2.16 c. 1.13

6.91 a. 0.0930 b. 0.9684

6.93 a. 1.175 or 1.18 b. 0.58
c. −1.04 d. −2.33

6.95 a. Yes, it is calculated by dividing the mental age by the chronological age.

b. ≈ 68% of the data within 1 standard deviation of mean; mounded distribution; approximately symmetrical distribution; middle 50% is bounded by $Q1$ and $Q3$ and their z-scores are $+0.625$ and -0.625, fairly close to ± 0.67 of the normal distribution.

c. IQ score between 118 and 130; approx. 10%

d. 25%   e. 0.0301

6.97 a. 0.0158   b. 1.446 years

6.99 10.033

6.101 a. Both $np = 7.5$ and $nq = 17.5$ are greater than 5.

b. 7.5, 2.29

6.103 a. Commands needed:

```
SET C1
0:50
END
PDF C1 C2;
BINOmial 50 0.1.
PRINt C1 C2
```

where C1 contains the numbers between 0 and 50.

b. 0.77023   c. 0.751779

6.105 a. $P(0) + P(1) + P(2) + \cdots + P(75)$

b. 0.9856   c. 0.9873

6.107 a. 5.534

b. Everyone's generated values will vary. Commands needed:

```
RANDom 40 C1;
NORMal 5.534 0.2.
HIST C1;
CUTPoint 5:6.2/0.05.
```

c. Answers will vary, but most of them will have 0 or 1 of the 40 overflow.

d. Yes, the setting seems to work.

6.109 0.0322

6.111 a. 0.0386   b. 0.1762   c. 0.9868

6.113 0.0015

Chapter 7

7.1 $P(0) = 0.04$, since only 1 of the 25 samples has an $\bar{x} = 0$; $1/25 = 0.04$. $P(2) = 0.12$, since there are 3 of the 25 samples with $\bar{x} = 2$; $3/25 = 0.12$.

7.2 $\mu = 3.0$ and $\sigma = 1.41$

7.3

Class	Freq.
1.8–2.2	3
2.2–2.6	5
2.6–3.0	6
3.0–3.4	6
3.4–3.8	5
3.8–4.2	4
4.2–4.6	1
Σ	30

7.4 a.

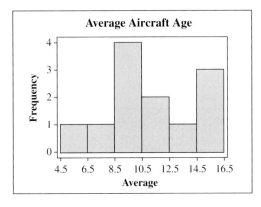

b. Samples are not all the same size; the airlines have different size fleets.

7.5 2.98, 0.638

7.7 a.

11	31	51	71	91
13	33	53	73	93
15	35	55	75	95
17	37	57	77	97
19	39	59	79	99

b.

$\bar{x}$	1	2	3	4	5	6	7	8	9
$P(\bar{x})$	0.04	0.08	0.12	0.16	0.20	0.16	0.12	0.08	0.04

c.

R	0	2	4	6	8
P(R)	0.20	0.32	0.24	0.16	0.08

7.9 Every student will have different results.
    c. normal: mounded, symmetric, centered around the 4.5
    d. same as part (c)

7.11 Every student will have different results.
    d. approximately normal: mounded, symmetric, and centered near 4.5

7.13 6.25, 4.167, 2.50

7.14 a. 1
    b. As n increases, the value of a fraction decreases.

7.15 a. 500   b. 5   c. approximately normal

7.17 a. approximately normal   b. \$31.65   c. \$1.00

7.19 Every student will have different results.
    d. $\bar{\bar{x}}$ should be approximately 20, $s_{\bar{x}}$ should be approximately 1.84, and the histogram should be approximately normal.

7.21 Both 90 and 110 are 2 standard errors away from the mean and $P(0 < z < 2) = 0.4772$.

7.22 0.9699

7.23 38.73 inches

7.25 a. approximately normal, $\mu = 69$, $\sigma = 4$
    b. 0.4013
    c. approximately normal, mean of 69, standard error of 1
    d. 69, 1.0   e. 0.1587   f. 0.0228

7.27 a. 0.3830   b. 0.9938   c. 0.3085   d. 0.0031

7.29 a. 0.2743   b. 0.0359
    c. Yes, but not exact; wind speeds have a mounded but skewed distribution.
    d. Allowed the use of the normal probability distribution to estimate the probabilities.

7.31 a. 0.6390
    b. Yes, the sample size is $n = 250$.
    c. It meant that the mean is approximately equal to the median.
    d. Probably not since salaries usually form a skewed distribution.
    e. It probably increased the probability since \$381 is within the interval.

7.33 Everybody will get different answers.
    a. 0.6826

7.35 a. 0.0643   b. 0.6103   c. 0.9974
    d. In parts (a) and (b), x is given to be normal; in (c) the sampling distribution is normal.
    e. Parts (a) and (b) are distributions of individual x-values, while (c) is a sampling distribution of $\bar{x}$-values.

7.37 0.9956

7.39 a. 0.1498   b. 0.0089

7.41 0.9544

7.43 0.0228

7.45 0.0023

7.47 a. approximately 1.000   b. 0.9772

7.49 a. 0.0571   b. 134.78 seconds

7.51 a. 8.0, 2
    b.

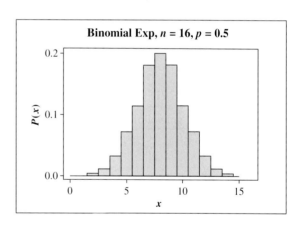

Binomial Exp, $n = 16$, $p = 0.5$

c., d. Every student will have different results.
    e. The empirical distribution should look very much like the distribution in part (b). The mean of the sample means is expected to be approximately equal to 8, and the standard deviation of the sample means is expected to be approximately equal to 0.4.

7.53 a. 99.7% of the sample means will be within 3 standard deviations of the mean: 40.5 to 49.5.
    b. 1.3   c. 4.5
    d. It is 3 multiples of the standard error.

Chapter 8

8.1 difficulty in obtaining a very large sample; cost of sampling; destruction of product in cases like the rivets illustration

8.2 $\sigma_{\bar{x}} = \sigma/\sqrt{n} = 18/\sqrt{36} = 18/6 = 3$

8.4 a. 33
b. 95% of the time the parameter being estimated will be a value within the resulting interval.
c. 52 is the point estimate for μ; ± 5 is one-half of the interval width; 52 ± 5 represents the interval with bounds 47 and 57.

8.5 a. $24 =$ sample size $= n$;
$4'11'' =$ sample mean $= \bar{x}$
b. $16 =$ population standard deviation $= \sigma$
c. $190 =$ sample variance $= s^2$
d. $69 =$ population mean $= \mu$

8.7 a. II has the lower variability; both are unbiased.
b. II is unbiased; they are equally variable.
c. Neither; I is negatively biased with less variability, while II is only slightly positively biased with a larger variability.

8.9 $\mu_{\bar{x}} = \mu$; $\sigma_{\bar{x}} = \sigma/\sqrt{n}$ decreases as n increases

8.11 $160,750

8.13 See the four definitions.

8.14 a. 75.92   b. 0.368   c. 75.552 to 76.288

8.15 The numbers are calculated for samples of snow for some of the months; it would be impossible to gather this information from every inch of snow.

8.16 Each student will have different results; however, 4.5 should be in the interval about 90% of the time.

8.17 56

8.19 Either the sampled population is normally distributed or the random sample is sufficiently large for the central limit theorem to hold.

8.21 a. 25.76 to 31.64
b. Yes; the sampled population is normally distributed.

8.23 a. 125.58 to 131.42
b. Yes; the sample size is sufficiently large to satisfy the CLT.

8.25 a. the mean amount spent for textbooks per student
b. 151.63 to 164.97

8.27 a. 25.3   b. 24.29 to 26.31
c. 23.97 to 26.63

8.29 a. 55.20   b. 8.546   c. 46.654 to 63.746

8.31 a. $5173.97 to $6026.03
b. The resulting interval is likely to have a level of confidence that is less than 90%.

8.33 27

8.35 106

8.37 H_0: The system is reliable vs. H_a: The system is not reliable

8.38 H_a: Teaching techniques have a significant effect on students' exam scores.

8.39
Type A correct decision:
Truth of situation: the party will be a dud.
Conclusion: the party will be a dud.
Action: did not go [avoided dud party].

Type B correct decision:
Truth of situation: the party will be a great time.
Conclusion: the party will be a great time.
Action: did go [party was great time].

Type I error:
Truth of situation: the party will be a dud.
Conclusion: the party will be a great time.
Action: did go [party was a dud].

Type II error:
Truth of situation: the party will be a great time.
Conclusion: the party will be a dud.
Action: did not go [missed great party].

8.40 You missed a great time.

8.41 α is the probability of rejecting a TRUE null hypothesis; $1 - \beta$ is the probability of rejecting a FALSE null hypothesis; they are two distinctly different acts that both result in rejecting the null hypothesis.

8.42 The smaller the probability of an event is, the less often it occurs.

8.43 a. H_0: Special delivery mail does not take too much time.
H_a: Special delivery mail takes too much time.
b. H_0: The new design is not more comfortable.
H_a: The new design is more comfortable.

c. H_0: Cigarette smoke has no effect on the quality of a person's life.
H_a: Cigarette smoke has an effect on the quality of a person's life.
d. H_0: The hair conditioner is not effective on "split ends."
H_a: The hair conditioner is effective on "split ends."

8.45 a. H_0: The victim is alive.
H_a: The victim is not alive.
b. Type A correct decision: The victim is alive and is treated as though alive.
Type I error: The victim is alive but is treated as though dead.
Type II error: The victim is dead but is treated as if alive.
Type B correct decision: The victim is dead and treated as dead.
c. The type I error is very serious; the victim may very well die without the attention that is not being received. The type II error is not as serious; the victim is receiving attention that is of no value. This would be serious only if there were other victims that needed this attention.

8.47 a. A type I error occurs when it is determined that the majority of Americans do not favor laws against assault weapons when in fact the majority do favor such laws. A type II error occurs when it is determined that the majority of Americans do favor laws against assault weapons when in fact they do not favor such laws.
b. A type I error occurs when it is determined that the fast food is low salt when in fact it is not low salt. A type II error occurs when it is determined that the fast food is not low salt when in fact it is low salt.
c. A type I error occurs when it is determined that the building must be demolished when in fact it should not be demolished. A type II error occurs when it is determined that the building must not be demolished when in fact it should be demolished.
d. A type I error occurs when it is determined that there is waste in government spending when in fact there is no waste. A type II error occurs when it is determined that there is no waste in government spending when in fact there is waste.

8.49 a. type I b. type II c. type I d. type II
8.51 a. The type I error is very serious; willing to allow it only 1 chance in 1000.
b. The type I error is somewhat serious; willing to allow it 1 chance in 20.
c. The type I error is not at all serious; willing to allow it 1 chance in 10.
8.53 a. α b. β
8.55 a. The experimenter is thoroughly convinced the alternative hypothesis can be shown to be true; thus, when the decision *reject H_0* is attained, the experimenter will want to proclaim victory. Thus, the conclusion is a fairly strong statement; "the evidence shows beyond a shadow of a doubt (is significant) that the alternative hypothesis is correct."
b. The experimenter is thoroughly convinced the alternative hypothesis can be shown to be true; thus, when the decision *fail to reject H_0* is attained, the experimenter is disappointed and will want to say something like "okay, so this evidence was not significant, I'll try again tomorrow." Thus, the statement of the conclusion is a fairly mild statement like "the evidence was not sufficient to show the alternative hypothesis to be correct."
8.57 a. 0.1151 b. 0.2119
8.59 H_0: The mean shearing strength is at least 925 lb.
H_a: The mean shearing strength is less than 925 lb.
8.60 H_0: $\mu = 54.4$ vs. H_a: $\mu \neq 54.4$
8.61 a. H_0: $\mu = 1.25$ ($\leq$) vs. H_a: $\mu > 1.25$
b. H_0: $\mu = 335$ ($\geq$) vs. H_a: $\mu < 335$
c. H_0: $\mu = 230,000$ vs. H_a: $\mu \neq 230,000$
8.62 Type A correct decision: The mean shearing strength is at least 925 lb, and it is decided that it is at least 925 lb.
Type I error: The mean shearing strength is at least 925 lb, and it is decided that it is less than 925 lb.
Type II error: The mean shearing strength is less than 925 lb, and it is decided that it is greater than or equal to 925 lb.
Type B correct decision: The mean shearing strength is less than 925 lb, and it is decided that it is less than 925 lb.
Type II error: you buy and use weak rivets.

8.63 −1.46

8.64 a. 0.0107   b. 0.0359

8.65 a. Fail to reject H_0 b. Reject H_0

8.66 a. 2.0, 0.0228   b. −1.33, 0.0918
c. −1.6, 0.1096

8.67 0.2714

8.68 The p-value measures the likeliness of the sample results based on a true null hypothesis.

8.69 n is the number of data values; MEAN calculated using formula $\Sigma x/n$; STDEV calculated using formula $\sqrt{\Sigma(x - \bar{x})^2/(n - 1)}$; SEMEAN calculated using formula $\sigma/\sqrt{n}$; z calculated using formula $(\bar{x} - \mu)/(\sigma/\sqrt{n})$; p-value calculated using formula $P(z < -1.50)$

8.70 Results will vary; expect results similar to those shown in Table 8-7.

8.71 a. H_0: $\mu = 26$ yr ($\leq$) vs. H_a: $\mu > 26$
b. H_0: $\mu = 36.7$ lb ($\geq$) vs. H_a: $\mu < 36.7$
c. H_0: $\mu = 1600$ hr ($\geq$) vs. H_a: $\mu < 1600$
d. H_0: $\mu = 210$ lb ($\leq$) vs. H_a: $\mu > 210$
e. H_0: $\mu = 570$ lb/unit vs. H_a: $\mu \neq 570$

8.73 A type I error: decision of reject H_0 was reached, interpreted as "mean hourly charge is less than \$50 per hour" when in fact the mean hourly charge is at least \$50 per hour.
A type II error: decision of fail to reject H_0 was reached, interpreted as "mean hourly charge is at least \$50 per hour" when in fact the mean hourly charge is less than \$50 per hour.

8.75 a. 1.26   b. 1.35   c. 2.33   d. −0.74

8.77 a. reject H_0 or fail to reject H_0
b. The calculated p-value is smaller than α; the calculated p-value is larger than α.

8.79 a. Fail to reject H_0 b. Reject H_0

8.81 a. 0.0694   b. 0.1977   c. 0.2420
d. 0.0174   e. 0.3524

8.83 a. 1.57   b. −2.13   c. −2.87 or 2.87

8.85 a. H_0: $\mu = 525$ vs. H_a: $\mu < 525$
b. Fail to reject H_0; the sample mean is not significantly less than 525.

8.87 a. the mean test score for all elementary education majors
b. H_0: $\mu = 35.70$ vs. H_a: $\mu < 35.70$
c. −5.86, +0.0000
d. Reject H_0; the mean test score is less than 35.70.

8.89 $z^\star = 10.51$; $\mathbf{P} = 0.0000+$; reject H_0; the mean number is more than 12.6.

8.91 H_0: The mean shearing strength is at least 925 lb
H_a: The mean shearing strength is less than 925 lb

8.92 H_0: $\mu = 9$ ($\leq$) vs. H_a: $\mu > 9$

8.93 a. H_0: $\mu = 1.25$ ($\geq$) vs. H_a: $\mu < 1.25$
b. H_0: $\mu = 335$ vs. H_a: $\mu \neq 335$
c. H_0: $\mu = 230,000$ ($\leq$) vs. H_a: $\mu > 230,000$

8.94 See 8.62.

8.95 −1.10

8.96 a. Reject H_0 b. Fail to reject H_0

8.97 $z \leq -2.33$

8.98 $z \geq 2.05$

8.99 Approx. 120 lb

8.100 Results will vary; expect results to be similar to those in Table 8-10.

8.101 a. H_0: $\mu = 16$ yr ($\geq$) vs. H_a: $\mu < 16$
b. H_0: $\mu = 6$ ft 6 in. ($\leq$) vs. H_a: $\mu > 6$ ft 6 in.
c. H_0: $\mu = 285$ ft ($\geq$) vs. H_a: $\mu < 285$
d. H_0: $\mu = 0.375$ yr ($\leq$) vs. H_a: $\mu > 0.375$
e. H_0: $\mu = 200$ units vs. H_a: $\mu \neq 200$

8.103 a. Decide that mean minimum plumber's call is more than \$85 when in fact is it not more than \$85.
b. Decide that mean minimum plumber's call is at most \$85 when in fact it is greater than \$85.

8.105 a. The set of all values of the test statistic that will cause us to reject H_0.
b. The value(s) of the test statistic that forms the boundary to the critical region.

8.107 a. $z \leq -1.65$ or $z \geq 1.65$ b. $z \geq 2.33$
c. $z \leq -1.65$ d. $z \leq -2.58$ or $z \geq 2.58$

8.109 21,004.133

8.111 a. 3.0   b. yes

8.113 a. Reject H_0 or Fail to reject H_0
b. Calculated test statistic falls in the critical region; or calculated test statistic falls in the noncritical region.

8.115 a. H_0: $\mu = 15.0$ cm vs. H_a: $\mu \neq 15.0$
b. Reject H_0; there is sufficient evidence to conclude that the mean is different than 15.0.

8.117 H_0: $\mu = 36.7$ vs. H_a: $\mu < 36.7$; $z \leq -2.33$; $z^\star = -2.59$; reject H_0; the population mean is significantly less than the claimed mean of 36.7.

8.119 H_0: $\mu = 7.0$ vs. H_a: $\mu < 7.0$; $z \leq -1.65$; $z^\star = -3.79$; reject H_0; the population mean is significantly less than the claimed national mean of 7.0 days.

8.121 a. 32.0 b. 2.4 c. 64 d. 0.90 e. 1.65 f. 0.3 g. 0.495 h. 32.495 i. 31.505

8.123 a. $\mu = 100$ b. $\mu \neq 100$ c. 0.01 d. 100 e. 96 f. 12 g. 1.70 h. -2.35 i. 0.0188 j. Fail to reject H_0 k. $\alpha = 0.01$

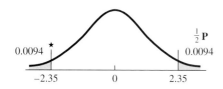

8.125 Alpha determines the size of the critical region.

8.127 a. 40.5 to 42.5
b. $z^\star = 3.00$; $\mathbf{P} = 0.0026$; reject H_0; there is sufficient evidence to support the contention that the mean is not equal to 40.
c. $\pm z(0.025) = \pm 1.96$; $z^\star = 3.00$; reject H_0; there is sufficient evidence to support the contention that the mean is not equal to 40.
d.

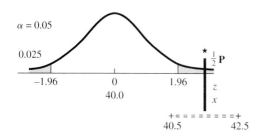

$z^\star = 3.00$ is in the critical region and $\mathbf{P}$ is less than α; and $\mu = 40$ is not within the interval estimate of 40.5 to 42.5.

8.129 a. 9.75 to 9.99 b. 9.71 to 10.03
c. The increased confidence level widened the interval.

8.131 a. 69.89 to 75.31
b. Yes. 75.0 falls within the interval.

8.133 3.87 to 4.13

8.135 489.2 to 540.8

8.137 60

8.139 Probably not. Individuals who can afford to buy homes in an area having a mean value of $150,000 can also afford homes in an area that has a mean value which is only a few thousand dollars more in cost.

8.141 a. 1.65
b. H_a: $\mu > 32$; $z \geq 1.65$; $z^\star = 1.00$; fail to reject H_0

8.143 a. H_0: $\mu = 0.50$ vs. H_a: $\mu \neq 0.50$
b. 0.2112 c. $z = \pm 2.33$

8.145 H_a: $\mu > 9$; $z^\star = 3.14$; $\mathbf{P} = 0.0008$; reject H_0

8.147 H_a: $\mu > 0.3125$; $z^\star = 2.79$; $\mathbf{P} = 0.0026$; reject H_0

8.149 a. H_a: $r > a$. Failure to reject H_0 will result in the drug being marketed. Because of the high current mortality rate, burden of proof is on the old, ineffective drug.
b. H_a: $r < a$. Failure to reject H_0 will result in the new drug not being marketed. Because of the low current mortality rate, burden of proof is on the new drug.

8.151 125.10 to 132.94

8.153 H_a: $\mu \neq 18$; fail to reject H_0. The sample mean is not significantly different than the population mean of 18.

8.155 Every student will have different results, but they should be similar to the following.
a. Commands needed:

```
RANDom 50 C1-C28;
NORMal 19 4.
RMEAn C1-C28 C29
LET C30=((C29-18)/(4/SQRT(28)))
```

b. $34/50 = 68\%$; an empirical p-value.
c. Critical points $= \pm 2.58$; $45/50 = 90\%$ fall in the noncritical region; an empirical β, probability of type II error.

Chapter 9

9.1 Pick any three numbers; the fourth must be the negative of the sum of the first three.

9.2 The bottom row of Table 6 is identical to the $z(\alpha)$ values in Table 4A.

9.3 a. 2.68   b. 2.07

9.4 a. -1.33 b. -2.82

9.5 ± 2.18

9.6 15.60 to 17.8

9.7 1.20, 0.124, fail to reject H_0

9.8 -1.30, 0.220, fail to reject H_0

9.9 Using Table 6: $0.10 < P < 0.25$. Using Table 7: $0.143 < P < 0.167$.

9.10 $t \leq -2.82$, -1.24, fail to reject H_0

9.11 $t \leq -2.40$ or $t \geq 2.40$

9.12 c. probably 0.05

9.13 a. 1.71   b. 1.37   c. 2.60   d. 2.08
e. -1.72 f. -2.06 g. -2.47 h. 2.01

9.15 7

9.17 a. -2.49 b. 1.71   c. -0.685

9.19 a. Symmetric about mean; mean is 0.
b. Standard deviation of t-distribution is greater than 1; t-distribution is different for each different sample size while there is only one z-distribution.

9.21 51.84 to 56.16

9.23 39.3 to 40.7

9.25 a. 13.42   b. 6.36   c. 11.75 to 15.09

9.29 a. H_0: $\mu = 11$ ($\geq$) vs. H_a: $\mu < 11$
b. H_0: $\mu = 54$ ($\leq$) vs. H_a: $\mu > 54$
c. H_0: $\mu = 75$ vs. H_a: $\mu \neq 75$

9.31 a. $t \leq -2.14$ or $t \geq 2.14$
b. $t \geq 2.49$ c. $t \leq -1.74$
d. $t \geq 2.42$

9.33 H_0: $\mu = 25$ (at least) vs. H_a: $\mu < 25$ (less than); $t^\star = -2.74$; reject H_0
a. $0.005 < P < 0.007$ b. $t \leq -2.52$

9.35 a. The midrange is close in value to the mean. H_0: $\mu = 38.4\%$ vs. H_a: $\mu \neq 38.4\%$; $t^\star = 1.08$; fail to reject H_0
b. $0.288 < P < 0.338$
c. $t \leq -2.14$ or $t \geq 2.14$

9.37 H_0: $\mu = 35$ (reasonable) vs. H_a: $\mu \neq 35$ (not reasonable), $t^\star = 1.02$, fail to reject H_0
a. $0.322 < P < 0.364$
b. $t \leq -2.57$ or $t \geq 2.57$

9.41 H_0: $\mu = 18.810$ vs. H_a: $\mu \neq 18.810$, $t^\star = 0.84$, $P = 0.41$, fail to reject H_0

9.43 a. law of large numbers
b. The mean of the p' distribution is p, the parameter being estimated.

9.44 $\sqrt{npq}/n = \sqrt{npq}/\sqrt{n^2} = \sqrt{npq/n^2} = \sqrt{pq/n}$

9.45 0.189 to 0.271

9.46 2401

9.47 1277

9.48 $z^\star = -1.52$, $P = 0.0643$, fail to reject H_0

9.49 $z \geq 1.65$, $z^\star = 1.78$, reject H_0

9.50 The confidence interval, 0.052 to 0.118, does not contain the hypothesized 0.15.

9.51 a. 0.018   b. 0.9464

9.52 a. 0.094
b. If there is no probability of risk of lung cancer for passive smokers, then $p = 0$.
c. $p > 0$, in hopes of showing there is an effect due to passive smoking.
d. 2.02   e. 0.0217
f. 0.006 to 0.374, -0.029 to 0.409
g. Often in mathematics the negative sign is used to indicate a decrease or reduction in a quantity.

9.53 a. $x =$ number of successes, $n =$ sample size $=$ number of independent trials.
b. 0.30

9.55 a. $p + q = 1$
b. $p + q = 1$ is equivalent to $q = 1 - p$
c. 0.4   d. 0.727

9.57 a. 0.21   b. 0.043

9.59 0.206 to 0.528

9.61 0.37 to 0.43

9.63 0.015

9.65 a. 0.162, 0.438   b. 0.562, 0.838
c. 0.239, 0.761   d. 0.418, 0.582
e. 0.474, 0.526
f. The two answers are symmetric about 0.5.
g. As sample size increased, interval became narrower.

9.67 a. 0.5005   b. 0.003227   c. 0.4942 to 0.5068
d.–f. Each student will have different results.

9.69 a. 915   b. 229   c. 1825
d. Increasing the level of confidence increases the sample size.
e. Increasing the maximum error decreases the sample size.

9.71 1068

9.73 a. H_0: $p = P(\text{work}) = 0.60 \, (\leq)$ vs.
H_a: $p > 0.60$
b. H_0: $p = P(\text{win tonight}) = 0.50 \, (\geq)$ vs.
H_a: $p < 0.50$
c. H_0: $p = P(\text{interested in quitting}) = 1/3 \, (\leq)$
vs. H_a: $p > 1/3$
d. H_0: $p = P(\text{believe in spanking}) = 0.50 \, (\geq)$
vs. H_a: $p < 0.50$
e. H_0: $p = P(\text{vote for}) = 0.50 \, (\leq)$ vs.
H_a: $p > 0.50$
f. H_0: $p = P(\text{seriously damaged}) = 3/4 \, (\geq)$ vs.
H_a: $p < 3/4$
g. H_0: $p = P(\text{H}\,|\,\text{tossed fairly}) = 0.50$ vs.
H_a: $p \neq 0.50$
h. H_0: $p = P(\text{odd}\,|\,\text{random}) = 0.50$ vs.
H_a: $p \neq 0.50$

9.75 a. 0.1388   b. 0.0238   c. 0.1635   d. 0.0559

9.77 a. 0.017   b. 0.085   c. 0.101   d. 0.004

9.79 a. correctly fail to reject H_0 b. 0.036
c. commit a type II error   d. 0.128

9.81 H_0: $p = 0.90$, H_a: $p < 0.90$, $z^\star = -4.82$,
reject H_0
a. **P** = 0.000003   b. $z \leq -1.65$

9.83 H_0: $p = 0.60$ [will receive 60% of vote],
H_a: $p < 0.60$ [will receive less than 60%],
$z^\star = -2.04$, reject H_0
a. **P** = 0.0207   b. $z \leq -1.65$

9.85 H_0: $p = 0.27$, H_a: $p < 0.27$, $z^\star = -4.47$,
P = 0.000003, reject H_0

9.87 a. 23.2   b. 23.3

9.88 a. 3.94   b. 8.64

9.89 H_0: $\sigma^2 = 532$ vs. H_a: $\sigma^2 > 532$, $\chi^{2\star} = 25.08$,
$0.05 < \mathbf{P} < 0.10$, fail to reject H_0

9.90 $0.02 < \mathbf{P} < 0.05$

9.91 H_0: $\sigma^2 = 0.25$ vs. H_a: $\sigma^2 < 0.25$, $\chi^2 \leq 3.33$,
$\chi^{2\star} = 3.46$, fail to reject H_0

9.92 a. 1.72   b. 3.58
c. increased standard deviation
d. The standard deviation doubled in size.

9.93 K1 is population variance, K2 is the degrees of freedom, df = $n - 1$, K3 is the standard deviation of the data in C1, C2 is the chi square value, $\chi^2 = (n - 1)s^2/\sigma^2$, C3 is the cumulative probability, C4 is the probability to the right of the C2 value.

9.95 a. 30.1   b. 13.3   c. 7.56   d. 43.2

9.97 a. 11.1   b. 11.1   c. 9.24

9.99 0.94

9.101 a. H_0: $\sigma = 24 \, (\leq)$ vs. H_a: $\sigma > 24$
b. H_0: $\sigma = 0.5 \, (\leq)$ vs. H_a: $\sigma > 0.5$
c. H_0: $\sigma = 10$ vs. H_a: $\sigma \neq 10$
d. H_0: $\sigma^2 = 18 \, (\geq)$ vs. H_a: $\sigma^2 < 18$
e. H_0: $\sigma^2 = 0.025$ vs. H_a: $\sigma^2 \neq 0.025$
f. H_0: $\sigma^2 = 34.5 \, (\leq)$ vs. H_a: $\sigma^2 > 34.5$

9.103 a. $0.02 < \mathbf{P} < 0.05$ b. 0.01
c. $0.05 < \mathbf{P} < 0.10$ d. $0.025 < \mathbf{P} < 0.05$

9.105 H_0: $\sigma = 8$, H_a: $\sigma \neq 8$, $\chi^{2\star} = 29.3$,
$0.01 < \mathbf{P} < 0.02$, reject H_0

9.107 a. Allows the use of the chi-square distribution to calculate probabilities.
b. Graph the rates for given days.
c. H_0: $\sigma = 0.5\%$, H_a: $\sigma \neq 0.5\%$, $\chi^{2\star} = 67.76$,
reject H_0
d. **P** = 0+   d. $\chi^2 \leq 5.63$ or $\chi^2 \geq 26.1$

9.109 $s^2 = 17.4595$; H_0: $\sigma = 3.50$ [not different],
H_a: $\sigma \neq 3.50$ [differs], $\chi^{2\star} = 19.95$, fail to reject H_0
a. $0.20 < \mathbf{P} < 0.50$
b. $\chi^2 \leq 5.63$ or $\chi^2 \geq 26.1$

9.111 a. Results will vary for each student.
b. They are approximately the same.
c. The mean and median are approximately equal to the degrees of freedom. The mode gets closer to the number of degrees of freedom as the sample size increases.
d. The distributions are no longer skewed to the right. They appear to be symmetric with the mean $\approx$ median $\approx$ mode $\approx$ df.

9.113 1480 to 1620

9.115 $t^\star = -2.75$, $0.004 < \mathbf{P} < 0.005$

9.117 $\bar{x} = 3.44$, $s = 0.653$; H_0: $\mu = 3.8$, H_a: $\mu < 3.8$
[lower], $t^\star = -1.91$, $0.034 < \mathbf{P} < 0.043$,
reject H_0

9.119 H_0: $\mu = 80$ [same as state], H_a: $\mu \neq 80$ [different], $t^\star = -4.47$, reject H_0
a. **P** = 0+   b. $t \leq -2.09$ or $t \geq 2.09$

9.121 a. $n = 800$; trial = one person surveyed; success = doctor should not be prosecuted; $p = P$(not prosecuted); $x = n$(not prosecuted) = 0, 1, 2, . . . , 800

b. 2/3 is the point estimate and was obtained by dividing the number who said "doctor should not be prosecuted" by the number of people polled.

c. 0.033

d. Rounded to the nearest half of 1%, it is the same.

9.123 0.088 to 0.232

9.125 0.09, 0.16, 0.21, 0.24, 0.25, 0.24, 0.21, 0.16, 0.09

9.127 a. 0.027

b. To the nearest percent, they are equal.

c. 2213

9.129 0.0261

9.131 13

9.133 H_0: $\sigma = 12$, H_a: $\sigma \neq 12$ [different], $\chi^{2\star} = 42.0$, $P = 0.6844$, fail to reject H_0

9.135 a. H_0: $\sigma = 0.5$ [in control], H_a: $\sigma > 0.5$ [out of control], $\chi^{2\star} = 87.8$, $\mathbf{P} < 0.005$, reject H_0

b. H_0: $\mu = 27.5$, H_a: $\mu \neq 27.5$ [out of control], $t^\star = 0.82$, $0.376 < \mathbf{P} < 0.430$, fail to reject H_0

9.137 67,756 to 82,244

Chapter 10

10.1 Divide the class into two groups, males and females. Randomly select from each group.

10.2 Randomly select a set of students, obtaining the two heights from each.

10.3 Information obtained from one would not be independent from the other twin.

10.5 Dependent samples. Each person provided one piece of data for each sample.

10.7 Independent samples. Each set of 10 specimens forms a separate sample.

10.9 a. The two sets are selected so that there is no relationship between the two resulting sets.

b. If they were husband and wife, brother and sister, or related in some way.

10.11 1, 1, 0, 2, −1

10.12 0.6, 1.14

10.13 2.13; number of standard errors

10.14 a. 4.24 to 8.36

b. Larger n resulted in a narrower confidence interval.

10.15 Confidence Intervals

Variable	N	Mean	StDev	SEMean	95% C.I.
C3	6	3.0	3.35	1.37	(−0.51, 6.51)

10.16 H_a: $\mu_d > 0$, $t^\star = 2.45$, $0.033 < \mathbf{P} < 0.037$, reject H_0

10.17 T-Test of the Mean

Test of mu = 0.00 vs. mu < 0.00

Variable	N	Mean	StDev	SEMean	T	P-value
C3	6	−2.33	4.41	1.80	−1.30	0.125

10.18 H_a: $\mu_d > 0$, $t^\star = 1.35$, fail to reject H_0

10.19 a. 1.0   b. −1.53 to 3.53

10.21 −1.03 to 8.53

10.23 −0.143 to 1.743

10.25 a. H_0: $\mu_d = 10$, H_a: $\mu_d > 10$, d = posttest−pretest

b. H_0: $\mu_d = 10$, H_a: $\mu_d < 10$, d = after−before

c. H_0: $\mu_d = 12$, H_a: $\mu_d < 12$, d = before−after

d. H_0: $\mu_d = 0$, H_a: $\mu_d \neq 0$, d = after−before

10.27 H_a: $\mu_d > 0$, $t^\star = 3.067$, $\mathbf{P} \approx 0.002$, reject H_0

10.31 H_a: $\mu_d > 0$, $t^\star = 4.08$, $\mathbf{P} < 0.005$, reject H_0

10.33 4.92

10.34 Case I: df will be between 17 and 40. Case II: df = 17.

10.35 −6.3 to 16.3

10.36 1.24

10.37 0.210

10.38 With the smaller degrees of freedom, df = 9, a higher calculated value is needed, making it more difficult to reject H_0.

10.39 a. The samples are taken from two different sets of schools.

b. The probability that there is no difference between the two types of schools is about 0.

c. $0.8 < t^\star < 0.9$

d. A five-point scale was used, and no indication of normality is given.

10.41 0.54 to 0.66

10.45 0.22 to 3.78

10.47 −0.75 to 8.75

10.49 a. $H_0: \mu_1 - \mu_2 = 0$ vs. $H_a: \mu_1 - \mu_2 \neq 0$

b. $H_0: \mu_1 - \mu_2 = 0$ vs. $H_a: \mu_1 - \mu_2 > 0$

c. $H_0: \mu_1 - \mu_2 = 20$ vs. $H_a: \mu_1 - \mu_2 > 20$

d. $H_0: \mu_A - \mu_B = 50$ vs. $H_a: \mu_A - \mu_B < 50$

10.51 a. 0.125   b. 0.012   c. 0.092   d. 0.084

10.53 $H_a: \mu_2 - \mu_1 > 0$, $t^\star = 2.84$, $0.008 < P < 0.012$, reject H_0

10.55 $\mu_E - \mu_C > 0$, $t^\star = 9.97$, $P = 0+$, reject H_0

10.57 b. $0.554 < P < 0.624$ c. $0.560 < P < 0.626$

10.59 $H_a: \mu_B - \mu_A > 0$, $t^\star = 1.98$, $0.037 < P < 0.047$, reject H_0

10.61 a. 15.53, 1.98, 1.41   b. 12.53, 1.98, 1.41

c. $H_a: \mu_A - \mu_B \neq 0$, $t^\star = 5.84$, $P < 0.002$, reject H_0

10.63 Everybody will get different results, but expect them to be similar to the following:

a. $N(100, 20)$

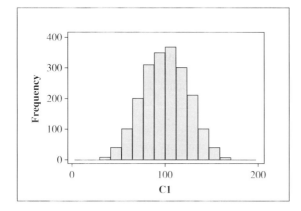

$N(120, 20)$

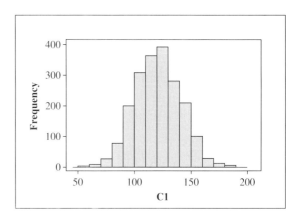

b. normal in shape, mean of 20, standard error of 10

d. 100 values for $\bar{x}_1 - \bar{x}_2$

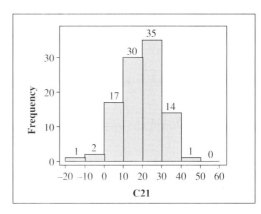

e. For the above empirical sampling distribution, the mean is 19.51 and the standard error is 10.71. There is 65%, 96%, and 99% of the values within one, two, and three standard errors of the expected mean of 20.

f. Expected similar results on repeated trials.

10.65 Everybody will get different results, but they will look very similar to the results found in Exercise 10.63. It turns out that the $t^\star$-statistic works quite well even when the assumptions are not met.

10.67 Divide both sides by σ_p^2.

10.68   3.03, 3.78

10.69 $H_0: \sigma_m/\sigma_p = 1$ vs. $H_a: \sigma_m/\sigma_p > 1$

10.70   1.52

10.71   3.37

10.72   a. $H_0: \sigma_N^2 = \sigma_A^2$ vs. $H_a: \sigma_N^2 > \sigma_A^2$
b. 0.5% chance of rejecting H_0 when it is in fact true
c. 0.0058

10.73   0.495

10.74   Multiply it by 2.

10.75   a. $H_0: \sigma_A^2 = \sigma_B^2$ vs. $H_a: \sigma_A^2 \neq \sigma_B^2$
b. $H_0: \sigma_I = \sigma_{II}$ vs. $H_a: \sigma_I > \sigma_{II}$
c. $H_0: \sigma_A^2/\sigma_B^2 = 1$ vs. $H_a: \sigma_A^2/\sigma_B^2 \neq 1$
d. $H_0: \sigma_D^2/\sigma_C^2 = 1$ vs. $H_a: \sigma_D^2/\sigma_C^2 > 1$

10.77   a. 2.51   b. 2.20   c. 2.91   d. 4.10
e. 2.67   f. 3.77   g. 1.79   h. 2.99

10.79 $H_a: \sigma_k^2 \neq \sigma_m^2$, $F^\star = 1.33$, $\mathbf{P} > 0.10$, fail to reject H_0

10.81   a. $H_a: \sigma_y > \sigma_o$, $F^\star = 12.35$, $\mathbf{P} < 0.01$, reject H_0
b. 0+

10.83   MINITAB verify

10.85   Everybody will get different results, but they all can be expected to look very similar to the following.
a. $N(100, 20)$

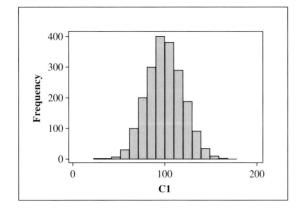

$N(120, 20)$

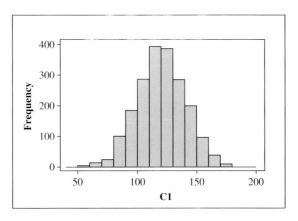

These two very large samples strongly suggest that we are sampling two normal populations.

Chapter 11

11.1 $9 + 3 + 3 + 1 = 16$ parts total; thus: 9/16, 3/16, 3/16, and 1/16

11.2 $E = 556(9/16)$; $O - E = 315 - 312.75$; $(O - E)^2/E = (2.25)^2/312.75$

11.3   Trial: each adult surveyed; variable: reason why we rearrange furniture. Multiple answers: bored, moving, new furniture, and so on.

11.4   Multiple answers are possible.

11.5   a. $H_0: P(1) = P(2) = P(3) = P(4) = P(5) = 0.2$
H_a: The numbers are not equally likely.
b. $H_0: P(1) = 2/8, P(2) = 3/8, P(3) = 2/8, P(4) = 1/8$
H_a: Probabilities are distributed differently than listed in the null hypothesis.
c. $H_0: P(E) = 0.16, P(G) = 0.38, P(F) = 0.41, P(P) = 0.05$
H_a: Percentages are different than specified in H_0.

11.7   a. $H_0: P(A) = P(B) = P(C) = P(D) = P(E) = 0.2$
b. $\chi^2 = \Sigma[(O - E)^2/E]$
c. H_0: Equal preference; $\chi^{2\star} = 4.40$;
(1) $0.25 < \mathbf{P} < 0.50$ or (2) $\chi^2 \geq 7.78$; fail to reject H_0.

11.11 H_0: Preferences are the same; $\chi^{2\star} = 12.01$; $\mathbf{P} = 0.017$ or $\chi^2 \geq 9.49$; reject H_0.

11.13 H_0: $P(0) = P(1) = \ldots = P(9)$; $\chi^{2\star} = 6.00$; $0.50 < \mathbf{P} < 0.75$ or $\chi^2 \geq 16.9$; fail to reject H_0.

11.15
a. near zero, fail to reject null hypothesis
b. "He did not roll the die!"; zero; "the data and the theory expressed by the null hypothesis are too much alike for chance to have had any role." Either the experimenter produced data that agreed or the maker of the hypothesis saw the data and then made up the theory.
c. Hopefully the experimenter will use random data.

11.17 10

11.18 a. 113 b. 122 c. 300 d. 68.23

11.19
a. A distribution for each region of New York State is compared.
b. Four distributions of yes, no, and not sure answers are being compared.

11.20 a.

Reason	Men	Women
No time	0.36	0.37
In good health	0.39	0.21
Poor health	0.11	0.15
Dislike exercise	0.04	0.08
Too lazy	0.01	0.06
Other	0.09	0.13

b.

Reason	Men	Women
No time	360	370
In good health	390	210
Poor health	110	150
Dislike exercise	40	80
Too lazy	10	60
Other	90	130

c. H_0: The reasons are distributed the same; $\chi^{2\star} = 116.6$; $\mathbf{P} = 0+$ or $\chi^2 \geq 11.1$; reject H_0.

11.21
a. H_0: Voters' preference and party affiliation are independent.
H_a: Voters' preference and party affiliation are not independent.
b. H_0: The distribution is the same for all three.
H_a: The distribution is not the same for all three.

c. H_0: The proportion of yeses is the same in all categories sampled.
H_a: The proportion of yeses is not the same in all categories.

11.23 H_0: The number of defective items is independent of the day of the week; $\chi^{2\star} = 8.548$; $\mathbf{P} = 0.074$ or $\chi^2 \geq 9.49$; fail to reject H_0.

11.27 H_0: The distribution of reactions is the same for both groups; $\chi^{2\star} = 22.56$;
a. $\mathbf{P} = 0+$ or
b. $\chi^2 \geq 4.61$; reject H_0.

11.29 H_0: The distributions are the same for all regions; $\chi^{2\star} = 21.695$; $\mathbf{P} = 0.001$ or $\chi^2 \geq 12.6$; reject H_0.

11.31 Information to be collected from doctors: (1) Does the doctor deliver babies? (2) How old is the doctor? (3) How many babies has the doctor delivered by Cesarean? (4) How many by natural childbirth?
Use a contingency table; classes of doctor's age form the columns, and the method of birth form the rows. The entries would be the number of babies in each category. Use chi-square to analyze the resulting counts.

11.33 H_0: $1 : 3 : 4$ proportions; $\chi^{2\star} = 10.33$
a. $0.005 < \mathbf{P} < 0.01$ or
b. $\chi^2 \geq 5.99$; reject H_0.

11.35 H_0: The distribution of deaths is the same; $\chi^{2\star} = 62.85$; $\mathbf{P} = 0+$ or $\chi^2 \geq 11.1$; reject H_0.

11.37 H_0: The proportions are the same for all age groups; $\chi^{2\star} = 3.904$; p-value $= 0.273$ or $\chi^2 \geq 7.82$; fail to reject H_0.

11.39 H_0: Proportion popped is the same for all brands; $\chi^{2\star} = 2.839$
a. $0.25 < \mathbf{p} < 0.50$ or
b. $\chi^2 \geq 7.82$; fail to reject H_0.

11.41
a. 30.07, 30.93; 15.28, 15.72; 15.77, 16.23; 8.87, 9.13
b. 9.46
c. H_0: The number of children is distributed the same for both groups of women.
H_a: The number of children is distributed differently for the two groups.
d. $\chi^2 \geq 7.82$
e. Reject H_0. The distributions are significantly different.

11.43 a. H_0: The distribution is the same for all regions; $\chi^{2\star} = 7.788$; $0.25 < \mathbf{P} < 0.50$ or $\chi^2 \geq 12.8$; fail to reject H_0.
  b. 4.296. The calculated chi-square is smaller; this should be expected since as n increases a fixed difference in proportion becomes more significant.

11.45 H_0: The response and the school's location are independent; $\chi^{2\star} = 3.651$; $\mathbf{P} = 0.456$ or $\chi^2 \geq 9.49$; fail to reject H_0.

11.47 H_0: Rate of absenteeism is the same for all groups; $\chi^{2\star} = 27.24$
  a. $\mathbf{P} = 0+$ or
  b. $\chi^2 \geq 11.3$; reject H_0.

11.49 Two of many possible answers: 3 values each of 86 and 114, or 1 value of 135 with 5 values of 93.

Chapter 12

12.1 units produced per hour at each temperature level

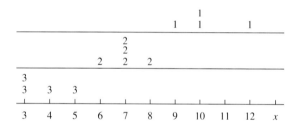

There appears to be a difference between the three sets.

12.2 H_0: $\mu_1 = \mu_2 = \mu_3$. H_a: Means are not all the same.

12.3 0.0052

12.4 a. From looking at the graphic in Case Study 12-1, it would seem that type of felony does have an effect on the average amount of time required to move the charge through trial court.
  b. The variability of the times is needed.

12.5 The "amount of money spent per week" was collected along with the teenagers' gender and age. The money amounts were separated into subsamples according to the teenagers' age category for presentation and analyzing. ANOVA "subdivides" the data into categories of the factor.

12.6 a. 0   b. 2   c. 4   d. 31   e. 393

12.7 a. H_0: $\mu_1 = \mu_2 = \mu_3 = \mu_4 = \mu_5$ vs. H_a: Means not all equal
  b. H_0: $\mu_1 = \mu_2 = \mu_3 = \mu_4$ vs. H_a: Means not all equal
  c. H_0: $\mu_1 = \mu_2 = \mu_3 = \mu_4$ vs. H_a: Means not all equal
  d. H_0: $\mu_1 = \mu_2 = \mu_3$ vs. H_a: Means not all equal

12.9 a. $F \geq 3.34$ b. $F \geq 5.99$ c. $F \geq 3.44$

12.11 a. 0.04 of the probability distribution associated with F and a true null hypothesis is more extreme than $F^\star$.
  b. Reject the null hypothesis.
  c. Fail to reject the null hypothesis.

12.13 a. The test factor has no effect on the mean at the tested levels.
  b. The test factor does have an effect on the mean at the tested levels.
  c. $F^\star$ must fall in the critical region; that is, the variance between levels is much larger than the variance within the levels.
  d. The tested factor has a significant effect on the variable.
  e. $F^\star$ must fall in the noncritical region; that is, the variance between levels is not larger than the variance within the levels.
  f. The tested factor does not have a significant effect on the variable.

12.15 H_0: The mean values for workers are all equal; H_a: Mean values are not all equal, $\mathbf{P} = 0.041$ or $F(2, 12, 0.05) = 3.89$, $F^\star = 4.22$, reject H_0.

12.17

Source	DF	SS	MS	F	p
Factor	2	94.5	47.3	4.62	0.033
Error	12	122.8	10.2		
Total	14	217.3			

$F(2, 12, 0.05) = 3.89$, reject H_0.

12.21 a. 120   b. 3
  d. Yes. The p-value is very small.
  e. No. The p-value is too large.

12.23 a. H_0: The mean amount of salt is the same in all tested brands of peanut butter.
H_a: The mean amount is not the same.
b. $F(2, 15, 0.05) \geq 3.68$
c. Fail to reject H_0. There is no significant difference in the mean amounts of salt in the tested brands.
d. Since it is quite large (much larger than 0.05), it tells us to "fail to reject H_0."

12.25 H_0: The mean amount of relief time is the same for all four drugs, $F^\star = 12.50$, **P** = 0.001 or $F \geq 3.49$, reject H_0.

12.27 H_0: The mean amount dispensed by the machines are all equal, $F^\star = 31.6$, **P** = 0+ or $F \geq 5.21$, reject H_0.

12.29 a. x = tillage plot yield

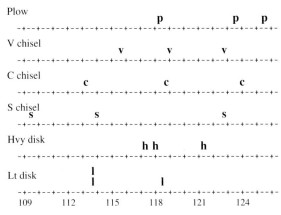

b. H_0: The mean yields for tillage methods are all equal [tillage tool used has no effect on yield].
H_a: The mean yields for tillage methods are not all equal [tillage tool used does effect the yield].
c. $F^\star = 1.23$, **P** = 0.353 or $F(5, 12, 0.05) = 3.11$, fail to reject H_0.

12.31 H_0: The mean typing speeds are equal on the two types of machines, $F^\star = 5.08$, **P** = 0.048 or $F \geq 4.96$, reject H_0.

12.33 H_0: The mean family income is the same for all three counties, $F^\star = 13.83$, **P** = 0+ or $F \geq 3.40$, reject H_0.

12.35 H_0: The mean percentage change is the same for the three cities, $F^\star = 2.91$, **P** = 0.085 or $F \geq 6.36$, fail to reject H_0.

12.37 SS(error) = 29.3333

12.39 82.4752

12.41 H_0: The six different brands of golf balls withstood the durability test equally well, as measured by the mean number of hits before failure, $F^\star = 5.30$, **P** = 0.001 or $F \geq 2.48$, reject H_0.

Chapter 13

13.1 The summation of the deviations about the mean is zero.

13.3 a.

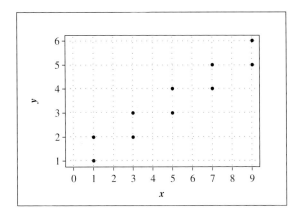

b. 4.44   c. 2.981, 1.581
d. 0.943   e. 0.943

13.9 a. 60   b. 40.99, 20.98   c. 0.07
d. the same

13.11 −0.05 to 0.65

13.12 ±0.444

13.13 Yes, the critical value is approximately 0.231.

13.15   a.

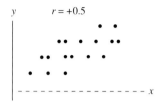

b.

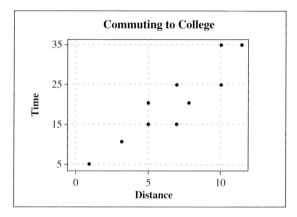

13.17   0.955; 0.78 to 0.98

13.19   a. $H_0: \rho = 0$ vs. $H_a: \rho > 0$
b. $H_0: \rho = 0$ vs. $H_a: \rho \neq 0$
c. $H_0: \rho = 0$ vs. $H_a: \rho < 0$
d. $H_0: \rho = 0$ vs. $H_a: \rho > 0$

13.21 $H_a: \rho \neq 0.0$, critical values ± 0.378, $r^\star = 0.43$, reject H_0.

13.23 $H_a: \rho \neq 0.0$, critical values ± 0.632, $r^\star = -0.67$, reject H_0.

13.25   a. $r = -0.177$ b. -0.78 to 0.62
c. The population correlation coefficient is between -0.78 and 0.62.
d. The interval is very wide, largely due to the small sample size.

13.27 $r = 0.859$, critical values are ± 0.444, unemployment rates are significantly correlated.

13.29 $H_a: \rho \neq 0.0$, critical values ± 0.632, $r^\star = 0.798$, reject H_0.

13.33 $\hat{y} = 16.34 + 0.4245x$, 21.088

13.35   a. $\hat{y} = 1.0 + 0.5x$ b. 1.5, 2.5, 3.5, 4.5, 5.5
c. $-.5, .5, -.5, .5, -.5, .5, -.5, .5, -.5, .5$
d. 0.3125   e. 0.3125

13.37   a. $S_1 = -3953.85 + 3.13(12,600) = 35484.15 \approx 35,500$
b. $n = 72$, 3.130 ± 0.129

13.39   a. 0.0145   b. 0.0668   c. 0.0653

13.41   0.1894

13.45   a.

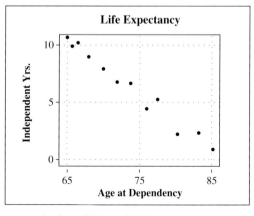

b. $\hat{y} = 2.38 + 2.664x$
c. $H_a: \beta_1 > 0$, critical values 1.86, $t^\star = 6.55$, reject H_0.
d. 1.48 to 3.84

13.47   a. $\hat{y} = 5937 + 30.73x$ b. 1135, 17.16
c. 1135 measures the variability of the ordered pairs about the line of best fit; 17.16 is the estimate for the standard deviation of the sampling distribution of the possible values of slope for all possible samples of size 10.
d. $H_a: \beta_1 > 0$, $t^\star = 1.79$, **P** = 0.0556, fail to reject H_0.

13.49   a. 13.04   b. 9.81 to 16.27   c. 4.69 to 21.39
d. 26.36; 23.41 to 29.31; 18.11 to 34.61

13.51   a. 15.4 to 19.1   b. 11.6 to 22.8

13.53   52.3 to 52.5

13.55   a.

b. $\hat{y} = 42.76 - 0.492x$

c.

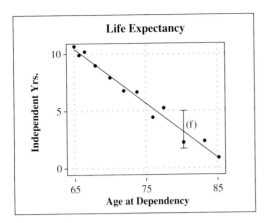

Life Expectancy

(f)

d. 3.4   e. 1.1 to 5.7

13.57 Yes; $s\sqrt{\dfrac{1}{n}} = s/\sqrt{n}$

13.59 Yes; H_a: $\rho > 0$, $\alpha = 0.05$, critical value ≈ 0.25, $r^\star = 0.69$, reject H_0.

13.61 a. Results will vary; however, most of the time the data will indicate no relationship (independence).
b. near zero
c., d. The absolute value of r should be expected to be larger than the table value 10% of the time.

13.63 a.

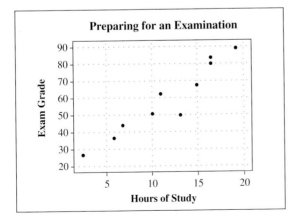

Preparing for an Examination

b. 0.961
c. H_a: $\rho > 0.0$, critical value $= 0.716$, $r^\star = 0.961$, reject H_0.
d. 0.85 to 0.99

13.65 a. Yes; H_a: $\rho \neq 0$, critical values ± 0.482, $r^\star = 0.61$, reject H_0.
b. 118.7

13.67 3.94 to 4.52

13.69 a.

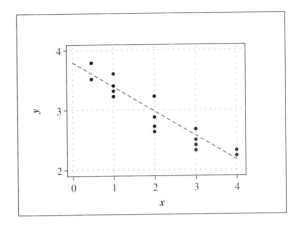

c.

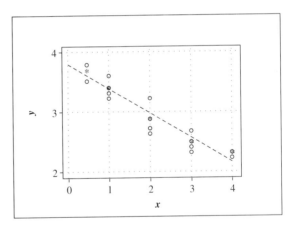

d. $\hat{y} = 3.79 - 0.415x$ e. 0.1942
f. -0.496 to -0.334
g. 2.44 to 2.68; 2.21 to 2.51
h. 2.13 to 2.99; 1.92 to 2.80

13.71 a. H_a: $\rho \neq 0.0$, critical value ± 0.433, $r^\star = 0.5133$, reject H_0.
b. $\hat{y} = 16.40 + 0.189x$
c. H_a: $\beta_1 > 0$, $\mathbf{P} < 0.01$, critical value $= 1.73$, $t^\star = 2.61$, reject H_0.
d. 18.53 to 38.47

13.73 $b_1 = 1.5811$, $r = 0.9973$

Chapter 14

14.1 a. H_a: Median < 32
b. H_a: P(prefer) > 0.50
c. H_a: P(+) $\neq 0.5$; (+) $=$ gain

14.3 a. H_0: Median $= 48$
b. H_0: P(+) $= 0.5$; (+) $=$ above 48
c. H_a: Median $\neq 48$; $x = 3$; reject H_0.

14.5 H_a: There is a difference in their ages;
(+) $=$ husband older; $x = 6$; crit. reg. $x \leq 3$;
fail to reject H_0.

14.7 H_0: Median $= 30$; crit. reg., ≤ 39, $x = 40$,
fail to reject H_0

14.9 H_a: P(prefer) > 0.5; crit. reg. $z(0.01) \geq 2.33$;
(+) $=$ prefer new; $x = 645$; $z^\star = 1.74$; fail to
reject H_0

14.11 39 to 47

14.13 -1 to 6; $d =$ air $-$ not air

14.15 H_a: P(+) $\neq 0.5$; (+) $=$ correct solution;
$x = 25$; crit. reg.: $x \leq 28$; reject H_0

14.17 The two sets of data seem to be similar, with
the exception that sample B seems to be "slid"
to the right by about ten points.

14.19 a. H_a: The average value is different for the
groups.
b. H_a: The average value is not the same for the
groups.
c. H_a: The average blood pressure for group A
is higher than that for group B.

14.21 H_0: Average value is same for both groups;
crit. val.: 8; $U^\star = 23$; fail to reject H_0

14.23 MINITAB verify—answers given in exercise.

14.25 H_a: Group 1 scores are higher; crit. reg.: $U \leq 6$;
$U^\star = 4.5$; reject H_0

14.27 a. H_a: The data did not occur in a random
order.
b. H_a: Not in random order.
c. H_a: The order of entry was not random.

14.29 H_a: The hiring sequence is not of random order;
crit. reg.: $V \leq 4$ or $V \geq 12$; $V^\star = 9$; fail to
reject H_0

14.31 Minitab verify—answers given in exercise. Yes,
reject the null hypothesis.

14.33 H_a: Lack of randomness (a trend, an increase
in wait time); crit. reg.: $V \leq 3$ or $V \geq 10$;
$V^\star = 8$; fail to reject H_0

14.35 a. Median $= 22.5$; $V = 8$
b. Crit. val. are 3 and 13; fail to reject H_0

14.37 H_a: Lack of randomness
a. Crit. reg.: $V \leq 8$ or $V \geq 20$; $V^\star = 9$; fail to
reject H_0
b. $z^\star = -2.00$; reject H_0

14.39 The p-values indicate that these two correlations
are significant, as long as α is greater than 0.02.
Thus the implication is that all the correlations
completed were significant.

14.41 a. Crit. reg.: $r_s \leq -0.545$ or $r_s \geq 0.545$
b. Crit. reg.: $r_s \geq 0.323$
c. Crit. reg.: $r_s \leq -0.564$

14.45 Crit. val.: 0.564; $r^\star = 0.732$; reject H_0

14.47 $r_s = 0.175$

14.49 H_a: There is a higher absentee rate at 8 AM;
(+) $=$ more absent at 8 AM class; $x = 2$;
crit. reg.: $x \leq 2$; reject H_0

14.51 H_a: Computer anxiety was reduced;
(+) $=$ begin–end > 0; $x = 10$;
crit. reg.: $x \leq 7$; fail to reject H_0

14.53 (+) $=$ top ripened first; H_a: top ripens first;
$x = 4$; crit. reg.: $x \leq 5$; reject H_0

14.55 Reject for $U \leq 127$

14.57 a. $U^\star = 2$ b. Crit. reg.: $U \leq 2$; yes

14.59 H_a: Credit unions charge lower fixed rates than
banks; crit. reg.: $U \leq 55$; $U^\star = 12$; reject H_0

14.61 H_a: Lack of randomness; crit. val.: $V = 4$ and
10; $V^\star = 9$; fail to reject H_0

14.63 H_a: $\rho_s > 0$ (positive correlation);
crit. reg.: $r_s \geq 0.399$; $r_s^\star = 0.880$; reject H_0

14.65 H_a: $\rho_s < 0$ (negative correlation); $r_s \leq -0.703$;
$r_s^\star = 0.168$; fail to reject H_0

Appendix A

A.1 a. 18   b. 12   c. 6

A.3 a. $_nP_n = n!$ and $_nP_{n-1} = n \times n - 1 \times n - 2 \times \cdots \times 2$. $_nP_{n-1}$ is the same product as $_nP_n$, only the last factor of 1 is missing.

b. Both have value n.

c. $$_nC_r = \frac{n!}{r!(n-r)!} \quad \text{and}$$

$$_nC_{n-r} = \frac{n!}{(n-r)!(n-(n-r))!} = \frac{n!}{(n-r)!r!}$$

d. They are two different names and applications for the same techniques.

A.5 a. 2,176,782,336 assuming zero and 0 can be used as first characters.

b. 1,544,804,416, assuming "0" can be the first character.

c. 6,760,000   d. 6,760,000   e. 6,084,000

A.7 a. $_{52}C_5 = 2,598,960$

b. $_4C_1 \times _{51}C_4 = 4 \times 249,900 = 999,600$

c. $_4C_1 \times _{48}C_4 = 4 \times 194,580 = 778,320$

d. $_{13}C_1 \times _{39}C_4 = 13 \times 82,251 = 1,069,263$

e. $_1C_1 \times _{39}C_4 + _3C_1 \times _{38}C_4 = 303,696$

ANSWERS TO CHAPTER QUIZZES

Quiz A Answers

Only the replacement for the word(s) in boldface type is given. (If the statement is true, no answer is shown. If the statement is false, a replacement is given.)

Chapter 1 Quiz A, Page 34

1.1 descriptive

1.2 inferential

1.4 sample

1.5 population

1.6 attribute or qualitative

1.7 quantitative

1.9 random

Chapter 2 Quiz A, Page 128

2.1 median

2.2 dispersion

2.3 never

2.5 zero

2.6 higher than

Chapter 3 Quiz A, Page 180

3.1 regression

3.2 strength of the

3.3 +1 or −1

3.5 positive

3.7 positive

3.8 −1 and +1

3.9 output or predicted value

Chapter 4 Quiz A, Page 246

4.1 any number value between 0 and 1, inclusive

4.4 simple

4.5 seldom

4.6 sum to 1.0

4.7 dependent

4.8 complementary

4.9 mutually exclusive or dependent

4.10 multiplication rule

Chapter 5 Quiz A, Page 286

5.1 continuous

5.3 one

5.5 exactly two

5.6 binomial

5.7 one success occurring on 1 trial

5.8 population

5.9 population parameters

Chapter 6 Quiz A, Page 324

6.1 its mean
6.4 one standard deviation
6.6 right
6.7 zero, 1
6.8 some (many)
6.9 mutually exclusive events
6.10 normal

Chapter 7 Quiz A, Page 351

7.1 is not
7.2 some (many)
7.3 population
7.4 divided by $\sqrt{n}$
7.5 decreases
7.6 approximately normal
7.7 sampling
7.8 means
7.9 random

Chapter 8 Quiz A, Page 427

8.1 alpha
8.2 alpha
8.3 sample distribution of the mean
8.7 type II error
8.8 beta
8.9 correct decision
8.10 critical (rejection) region

Chapter 9 Quiz A, Page 491

9.2 Student's t
9.3 chi-square
9.4 to be rejected
9.6 t score
9.7 $n - 1$
9.9 $\sqrt{pq/n}$
9.10 z(normal)

Chapter 10 Quiz A, Page 563

10.1 two independent means
10.3 F distribution
10.4 Student's t-distribution
10.5 Student's t
10.7 nonsymmetric (or skewed)
10.9 decreases

Chapter 11 Quiz A, Page 608

11.1 one less than
11.3 expected
11.4 contingency table
11.6 test of homogeneity
11.8 approximated by chi-square

Chapter 12 Quiz A, Page 642

12.2 mean square
12.3 SS(factor) or MS(factor)
12.5 reject H_0
12.7 the number of factor levels less one
12.8 mean
12.9 need to
12.10 does not indicate

Chapter 13 Quiz A, Page 696

13.2 need not be
13.3 does not prove
13.4 need not be
13.6 the linear correlation coefficient
13.8 regression
13.9 $n - 2$

Chapter 14 Quiz A, Page 740

14.2 t-test
14.3 runs test
14.4 assigned equal ranks
14.7 Mann-Whitney U test
14.8 power
14.10 power

Quiz B Answers

Chapter 1 Quiz B, Page 35

1. a. A   b. A   c. B   d. B   e. C

2. c, g, h, b, e, a, d, f

Chapter 2 Quiz B, Page 129

1. a. 30   b. 46   c. 91   d. 15   e. 1   f. 61
    g. 75   h. 76   i. 91   j. 106 or 114

2. a. two items purchased
    b. Nine people purchased 3 items each.
    c. 40   d. 120   e. 5   f. 2   g. 3   h. 3
    i. 3.0   j. 1.795   k. 1.34

3. a. 6.7   b. 7   c. 8   d. 6.5   e. 5   f. 6
    g. 3.0   h. 1.7   i. 5

4. a. -1.5 b. 153

Chapter 3 Quiz B, Page 180

1. a. B, D, A, C   b. 12   c. 10   d. 175   e. N
    f. (125, 13)   g. N   h. P

2. Someone made a mistake in arithmetic. r must be
    between -1 and $+1$.

3. a. 12   b. 10   c. 8   d. 0.73   e. 0.67
    f. 4.33   g. $\hat{y} = 4.33 + 0.67x$

Chapter 4 Quiz B, Page 246

1. a. $\frac{4}{8}$ b. $\frac{4}{8}$ c. $\frac{2}{8}$ d. $\frac{6}{8}$ e. $\frac{2}{8}$ f. $\frac{6}{8}$ g. 0
    h. $\frac{6}{8}$ i. $\frac{1}{8}$ j. $\frac{5}{8}$ k. $\frac{2}{4}$ l. 0   m. $\frac{1}{2}$ n. no (e)
    o. yes (g)   p. no(i)   q. yes (a, k)   r. no (b, l)
    s. yes (a, m)

2. a. 0   b. 0.7   c. 0   d. no (c)

3. a. 0.14   b. 0.76   c. 0.2   d. no (a)

4. a. 0.7   b. 0.5   c. no, $P(E \text{ and } F) = 0.2$
    d. yes, $P(E) = P(E|F)$

5. a. 0.4   b. 0.5   c. no, $P(G \text{ and } H) = 0.1$
    d. no, $P(G)$ not equal to $P(G|H)$

6. 0.51

Chapter 5 Quiz B, Page 286

1. a. Each $P(x)$ is between zero and 1, and the sum of
    all $P(x)$ is exactly one.

b. 0.2   c. 0   d. 0.8   e. 3.2   f. 1.25

2. a. 0.230   b. 0.085   c. 1.2   d. 1.04

Chapter 6 Quiz B, Page 324

1. a. 0.4922   b. 0.9162   c. 0.1020   d. 0.9082

2. a. 0.63   b. -0.95 c. 1.75

3. a. $z(0.8100)$ b. $z(0.2830)$

4. 0.7910

5. 28.03

6. a. 0.0569   b. 0.9890   c. 537   d. 417   e. 605

Chapter 7 Quiz B, Page 352

1. a. 0.4364   b. 0.2643

2. a. 0.0918   b. 0.9525

3. 0.6247

Chapter 8 Quiz B, Page 427

1. 4.72 to 5.88

2. a. $H_0: \mu = 245$, $H_a: \mu > 245$
    b. $H_0: \mu = 4.5$, $H_a: \mu < 4.5$
    c. $H_0: \mu = 35$, $H_a: \mu \neq 35$

3. a. 0.05, z, $z \leq -1.65$
    b. 0.05, z, $z \geq +1.65$
    c. 0.05, z, $z \leq -1.96$ or $z \geq +1.96$

4. a. 1.65   b. 2.33   c. 1.18   d. -1.65
    e. -2.05 f. -0.67

5. a. $z^\star = 2.50$ b. 0.0062

6. $H_0: \mu = 1520$ vs. $H_a: \mu < 1520$,
    crit. reg. $z \leq -2.33$, $z^\star = -1.61$, fail to reject H_0

Chapter 9 Quiz B, Page 492

1. a. 2.05   b. -1.73 c. 14.6

2. a. 28.6   b. 1.44   c. 27.16 to 30.04

3. 0.528 to 0.752

4. a. $H_0: \mu = 225$, $H_a: \mu > 225$
    b. $H_0: \sigma = 3.7$, $H_a: \sigma < 3.7$
    c. $H_0: p = 0.40$, $H_a: p \neq 0.40$

5. a. 0.05, z, $z \leq -1.65$
    b. 0.05, z, $z \geq +1.65$
    c. 0.05, t, $t \leq -2.08$ or $t \geq +2.08$
    d. 0.05, χ^2, $\chi^2 \leq 14.6$ or $\chi^2 \geq 43.2$

6. $H_0: \mu = 26$ vs. $H_a: \mu < 26$, crit. reg. $t \le -1.71$, $t^\star = -1.86$, reject H_0

7. $H_0: \sigma = 0.1$ vs. $H_a: \sigma > 0.1$, crit. reg. $\chi^2 \ge 21.1$, $\chi^{2\star} = 23.66$, reject H_0

8. $H_0: p = 0.50$ vs. $H_a: p > 0.50$, crit. reg. $z \ge 2.05$, $z^\star = 1.29$, fail to reject H_0

Chapter 10 Quiz B, Page 564

1. a. $H_0: \mu_N - \mu_A = 0$, $H_a: \mu_N - \mu_A \ne 0$
   b. $H_0: \sigma_o/\sigma_m = 1.0$, $H_a: \sigma_o/\sigma_m > 1.0$
   c. $H_0: p_m - p_f = 0$, $H_a: p_m - p_f \ne 0$

2. a. $t \ge 1.69$ b. $F, F \ge 2.11$
   c. $z, z \le -1.96$ or $z \ge +1.96$
   d. $t, t \le -2.05$ or $t \ge +2.05$
   e. $t, df = 7$ and $t \ge 1.89$

3. a. 1.75   b. 2.13   c. 2.42   d. 4.50

4. $H_0: \mu_L - \mu_P = 0$ vs. $H_a: \mu_L - \mu_P > 0$, crit. reg. $t \ge +1.83$, $t^\star = 0.979$, fail to reject H_0

5. $H_0: \mu_d = 0$ vs. $H_a: \mu_d > 0$, crit. reg. $t \ge 1.89$, $t^\star = 1.88$, fail to reject H_0

6. 0.072 to 0.188

Chapter 11 Quiz B, Page 608

1. a. H_0: Digits generated occur with equal probability.
   H_a: Digits do not occur with equal probability.
   b. H_0: Votes were cast independently of party affiliation.
   H_a: Votes were not cast independently of party affiliation.
   c. H_0: The crimes distributions are the same for all four cities.
   H_a: The crimes distributions are not all the same.

2. a. 4.40   b. 35.7

3. $H_0: P(1) = P(2) = P(3) = \frac{1}{3}$
   H_a: preferences not all equal, $\chi^{2\star} = 3.78$;
   $0.10 < \mathbf{P} < 0.25$ or crit. reg. $\chi^2 \ge 5.99$;
   fail to reject H_0

4. a. H_0: The distribution is the same for all types of soil.
   H_a: The distributions are not all the same.
   b. 25.622   c. 13.746
   d. $0.005 < \mathbf{P} < 0.01$ e. $\chi^2 \ge 9.49$
   f. Reject H_0. There is sufficient evidence to show that the growth distribution is different for at least one of the three soil types.

Chapter 12 Quiz B, Page 642

1. a. T   b. T   c. F   d. T   e. T   f. T   g. F
   h. F   i. F   j. F   k. T   l. F   m. F
   n. F   o. T

2. a. 72   b. 72   c. 22   d. 4   e. 4.5

Chapter 13 Quiz B, Page 696

1.

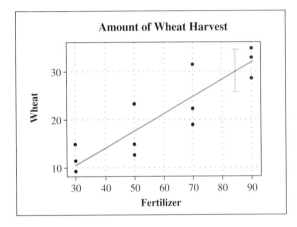

Amount of Wheat Harvest

Wheat / Fertilizer

2. $\Sigma x = 720$, $\Sigma y = 252$, $\Sigma x^2 = 49,200$, $\Sigma xy = 17,240$, $\Sigma y^2 = 6,228$

3. $SS(x) = 6000$, $SS(y) = 936$, $SS(xy) = 2120$

4. 0.895

5. 0.65 to 0.97

6. $\hat{y} = -0.20 + 0.353x$

7. See red line in figure above.

8. 4.324

9. Yes; $H_0: \beta_1 = 0$ vs $H_a: \beta_1 > 0$, $t^\star = 6.33$, reject H_0

10. 25.63 to 33.98

11. See blue vertical segment in figure above.

Chapter 14 Quiz B, Page 741

1. -2 to $+7$

2. H_0: No difference in weight gain.
   H_a: There is a difference in weight gain, crit. val.: 23, $U^\star = 32.5$, fail to reject H_0

3. H_0: no correlation
    H_a: correlated, crit. val.: ± 0.683, $r_s^* = -0.70$,
    reject H_0. Yes, there is significant correlation.

4. $(+)$ = higher grade level than previous problem
    $(-)$ = lower grade level than previous problem
    H_0: $P(+) = 0.5$
    H_a: $P(+) = 0.5$, crit. val.: 7, $x = 11$, fail to reject
    H_0. This sample does not show a significant pattern.

Quiz C Answers

Chapter 1 Quiz C, Page 35

1. See definitions; examples will vary. Note: *population* is set of ALL possible, while *sample* is the actual set of subjects studied.

2. See definitions; examples will vary. Note: *variable* is the idea of interest, while *data* are the actual values obtained.

3. See definitions; examples will vary. Note: *data* is the value describing one source, the *statistic* is a value (usually calculated) describing all the data in the sample, the *parameter* is a value describing the entire population (usually unknown).

4. Every element of the population has an equal chance of being selected.

Chapter 2 Quiz C, Page 130

1. a. 98   b. 50   c. 121   d. 100
2. a. $32,000, $26,500, $20,000, $50,000
    b.

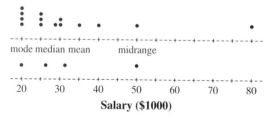

    Salary ($1000)

    c. Mr. VanCott—midrange; business manager—mean; foreman—median; new worker—mode
    d. The distribution is *J*-shaped.

3. There is more than one possible answer for these.
    a. 12, 12, 12   b. 15, 20, 25
    c. 12, 15, 15, 18   d. 12, 15, 16, 25, 25
    e. 12, 12, 15, 16, 17   f. 20, 25, 30, 32, 32, 80

4. A is right; B is wrong; standard deviation will not change.

5. B is correct. For example, if standard deviation is $5, then the variance, (standard deviation)^2, is "25 dollars squared." Who knows what "dollars squared" are?

Chapter 3 Quiz C, Page 182

1. Young children have small feet and probably tend to have less mathematics ability, while adults have larger feet and would tend to have more ability.

2. Student B is correct. -1.78 can occur only as a result of faulty arithmetic.

3. These answers will vary, but should somehow include the basic thought:
    a. strong negative   b. strong positive
    c. no correlation   d. no correlation
    e. impossible value, bad arithmetic

4. There is more than one possible answer for these.
    a. (1,1), (2,1), (3,1)   b. (1,1), (3,3), (5,5)
    c. (1,5), (3,3), (5,1)   d. (1,1), (5,1), (1,5), (5,5)

Chapter 4 Quiz C, Page 247

1. Check the weather reports for a long period of time and determine the relative frequency with which each occurs.

2. Student B is right. *Mutually exclusive* means no intersection, while *independence* means one event does not affect the probability of the other.

3. These answers will vary, but should somehow include the basic thought:
    a. no common occurrence
    b. either event has no effect on the probability of the other
    c. the relative frequency with which the event occurs
    d. probability that an event will occur even though the conditional event has previously occurred

Chapter 5 Quiz C, Page 287

1. n independent repeated trials of two outcomes; the two outcomes are "success" and "failure"; $p = P(\text{success})$ and $q = P(\text{failure})$ and $p + q = 1$; $x = n(\text{success}) = 0, 1, 2, \ldots, n$.

2. Student B is correct. The sample mean and standard deviation are statistics found using formulas studied in Chapter 2. The probability distributions studied in Chapter 5 are theoretical populations and their means and standard deviations are parameters.

3. Student B is correct. There are no restrictions on the values of the variable x.

Chapter 6 Quiz C, Page 325

1. This answer will vary but should somehow include the basic properties: bell-shaped, mean of 0, standard deviation of 1.

2. This answer will vary but should somehow include the basic ideas: it is a z-score, α represents the area under the curve and to the right of z.

3. All normal distributions have the same shape and probabilities relative to the z-score.

Chapter 7 Quiz C, Page 352

1. In this case each head produced one piece of data, the estimated length of the line. The CLT assures us that the mean value of a sample is far less variable than individual values of the variable x.

2. All samples must be of one fixed size.

3. Student A is correct. A population distribution is a distribution formed by all x values that make up the entire population.

4. Student A is correct. The standard error is found by dividing the standard deviation by the square root of the *sample size*.

Chapter 8 Quiz C, Page 428

1. a. H_0 – (a), H_a – (b)   b. 4   c. 2
   d. $P(\text{type I error})$ is alpha, decreases; $P(\text{type II error})$ increases

Chapter 9 Quiz C, Page 493

1. If the distribution is normal, six standard deviations is approximately equal to the range.

2. B

3. They are both correct.

4. When the sample size, n, is large, the critical value of t is estimated by using the critical value from the standard normal distribution of z.

5. Student A

6. Student B is right. It is significant at the 0.01 level of significance.

7. Student A is correct.

8. It depends on what it means to improve the confidence interval. For most purposes, an increased sample size would be the best improvement.

Chapter 10 Quiz C, Page 565

1. independent

2. One possibility: Test all students before the course starts, then randomly select 20 of those who finish the course and test them afterwards. Use the before scores for these 20 as the before sample.

3. For starters, if the two independent samples are of different sizes, the techniques for dependent samples could not be completed. They are testing very different concepts, the "mean of the differences of paired data" and the "difference between two mean values."

4. It is only significant if the calculated t-score is in the critical region. The variation among the data and their relative size will play a role.

5. The 80 scores actually are two independent samples of size 40. A test to compare the mean scores of the two groups could be completed.

6. A fairly large sample of both Catholic and non-Catholic families would need to be taken, and the number of each whose children attended private schools would need to be obtained. The difference between two proportions could then be estimated.

Chapter 11 Quiz C, Page 609

1. Similar in that there are n repeated independent trials. Different in that the binomial has two possible outcomes, while the multinomial has several. Each possible outcome has a probability and these probabilities sum to 1 for each different experiment, both for binomial and multinomial.

2. The test of homogeneity compares several distributions in a side-by-side comparison, while the test for independence tests the independence of the two factors that create the rows and columns of the contingency table.

3. Student A is right in that the calculations are completed in the same manner. Student B is correct in that the test of independence starts with one large sample and homogeneity has several samples.

4. a. If a chi-square test is to be used, the results of the four questions would be pooled to estimate the expected probability.
   b. Use a chi-square test for homogeneity.

Chapter 12 Quiz C, Page 643

1. This answer will vary but should somehow include the basic ideas: It is the comparison of several mean values that result from testing some statistical population by measuring a variable repeatedly at each of the several levels for which the factor is being tested.

2. a. $x_{r,k} = \mu + F$ scrubber $+ \epsilon_{k(r)}$
   b. H_0: The mean amount of emissions is the same for all three scrubbers tested.
   H_a: The mean amounts are not all equal.

c.

Source	df	SS	MS
Scrubber	2	12.80	6.40
Error	13	33.63	2.59
Total	15	46.44	

d. $F(2, 13, 0.05) = 3.81$, $F^\star = 2.47$, fail to reject H_0. The difference in the mean values for the scrubbers is not significant.

e.

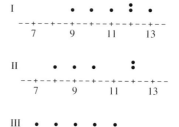

Chapter 13 Quiz C, Page 697

1. Variable 1: The frequency of skiers having their bindings tested
   Variable 2: The incidence of lower-leg injury
   The statement implies that as the frequency with which the bindings are tested increases, the frequency of lower-leg injury decreases; thus the strong correlation must be negative for these variables.

2. A "moment" is the distance from the mean, and the product of both the horizontal moment and the vertical moment is summed in calculating the correlation coefficient.

3. A value close to zero, also. The formulas used to calculate both values have the same numerator, namely SS(xy).

4. The vertical distance from a potential line of best fit to the data point is measured by $(y - \hat{y})$. The line of best fit is defined to be the line that results in the smallest possible total when the squared values of $(y - \hat{y})$ are totaled. Thus "the method of least squares."

5. The strength of the linear relationship could be measured with the correlation coefficient.

6. A random sample will be needed from the population of interest. The data collected need to be for the variables length of time on welfare and the measure of current level of self-esteem.

Chapter 14 Quiz C, Page 742

1. The nonparametric statistics do not require assumptions about the distribution of the variable.

2. The sign test is a binomial experiment of n trials (the n data observations) with two outcomes for each data [$(+)$ or $(-)$], and $p = (+) = 0.5$. The variable x is the number of the least frequent sign.

3. The median is the middle value such that 50% of the distribution is larger in value and 50% is smaller in value.

4. The extreme value in a set of data can have a sizeable effect on the mean and standard deviation in the parametric methods. The nonparametric methods typically use rank numbers. The extreme value with ranks is either 1 or n, and neither change if the value is more extreme.

5. d; $p = P(+) = P$(prefer seating arrangement A) $= 0.5$, no preference

Index of MINITAB Commands

INDEX

GLOSSARY OF SYMBOLS

$\overline{A}$	Complement of set A
ANOVA	Analysis of variance
α(alpha)	Probability of a type I error
β(beta)	Probability of a type II error
$1 - \beta$	Power of a statistical test
β_0	y-intercept of the true linear relationship
β_1	Slope of the true linear relationship
b_0	y-intercept for the line of best fit for the sample data
b_1	Slope for the line of best fit for the sample data
$_nC_r$	Number of combinations of n things r at a time
C_j	Column total
c	Column number or class width
d	Difference in value between two paired pieces of data
$\overline{d}$	Mean value of observed differences d
$d(\ )$	Depth of
df or df$(\ )$	Number of degrees of freedom
d_i	Difference in the rankings of the ith element
E	Expected frequency or maximum error of estimate
e	Error (observed)
ϵ(epsilon)	Experimental error
ϵ_{ij}	Amount of experimental error in the value of the jth piece of data in the ith row
F	F distribution statistic
$F(\mathrm{df}_n, \mathrm{df}_d, \alpha)$	Critical value for the F distribution
f	Frequency
H	Value of the largest-valued piece of data in a sample
H_a	Alternative hypothesis
H_0	Null hypothesis
i	Index number when used with Σ notation
i	Position number for a particular data
k	Identifier for the kth percentile
k	Number of cells or variables
L	Value of the smallest-valued piece of data in a sample
m	Number of classes
MS$(\ )$	Mean square
μ(mu)	Population mean
μ_d	Mean value of the paired differences
$\mu_{\overline{x}}$	Mean of the distribution of all possible $\overline{x}$'s
$\mu_{y|x_0}$	Mean of all y values at the fixed value of x, x_0
μ_v	Mean number of runs for the sampling distribution of number of runs
M	Population median
n	(Sample size) number of pieces of data in one sample
$n(\ )$	Cardinal number of
$\binom{n}{r}$	Binomial coefficient or number of r successes in n trials
O	Observed frequency
$\mathbf{P}$	Probability value or p-value
$P(A|B)$	Conditional probability, the probability of A given B
$P(a < x < b)$	Probability that x has a value between a and b
P_k	kth percentile
$_nP_r$	Number of permutations of n things r at a time

p or $P(\ )$	Theoretical probability of an event or proportion of time that a particular event occurs	s_d	Standard deviation of the observed differences d
p' or $P'(\ )$	Empirical (experimental) probability or a probability estimate from observed data	$s_{b_1}^2$	Square of the standard error for repeated observed values of the slope for the line of best fit
p_p	Pooled estimate for the proportion	T	Grand total
Q_1	First quartile	t	Student's t-distribution statistic
Q_3	Third quartile	$t(\text{df}, \alpha/2)$	Critical value for Student's t-distribution
q	($q = 1 - p$) probability that an event does not occur	U	Mann-Whitney U statistic
q'	($q' = 1 - p'$) observed proportion of time that an event did not occur	V	Number of runs
		χ^2	Chi-square statistic
		$\chi^2(\text{df}, \alpha)$	Critical value of chi-square distribution
R	Range of the data	x	Value of a single piece of data or class mark
R_i	Row total		
R^2	Coefficient of determination	$\bar{x}$	Sample mean
ρ(rho)	Population linear correlation coefficient	$\tilde{x}$	Sample median
		x_{ij}	Value of the jth piece of data in the ith row
r	Linear correlation coefficient for the sample data or row number	x_0	A given value of the variable x
		$\hat{y}$	Predicted value of y for a given x
r_s	Spearman's rank correlation coefficient	y_{x_0}	Individual value of y at the x value of x_0
Σ(capital sigma)	Summation notation		
SS( )	Sum of squares	z	Standard score
s^2	Sample variance	$z(\alpha/2)$	Critical value of z
σ (lowercase sigma)	Population standard deviation	$\star$(star)	Identifies the calculated value of any test statistic
$\sigma_{\bar{x}}$	Standard error for means, the standard deviation of the distribution of all possible $\bar{x}$'s	$|\ |$	Absolute value of a number
		$=$	Equal to
		$\neq$	Not equal to
$\sigma_{p'}$	Standard error for proportions	$<$	Less than
$\sigma_{\mu r}^2$	Variance among the means of the r rows (ANOVA)	$\leq$	Less than or equal to
		$>$	Greater than
σ_v	Standard error for the number of runs for the sampling distribution of number of runs	$\geq$	Greater than or equal to
		$\approx$	Approximately equal to
s	Sample standard deviation	$\sqrt{\ }$	Square root